Lecture Notes in Mechanical Engineering

Lecture Notes in Mechanical Engineering (LNME) publishes the latest developments in Mechanical Engineering—quickly, informally and with high quality. Original research reported in proceedings and post-proceedings represents the core of LNME. Volumes published in LNME embrace all aspects, subfields and new challenges of mechanical engineering.

To submit a proposal or request further information, please contact the Springer Editor of your location:

Europe, USA, Africa: Leontina Di Cecco at Leontina.dicecco@springer.com
China: Ella Zhang at ella.zhang@springer.com
India: Priya Vyas at priya.vyas@springer.com
Rest of Asia, Australia, New Zealand: Swati Meherishi at swati.meherishi@springer.com

Topics in the series include:

- Engineering Design
- Machinery and Machine Elements
- Mechanical Structures and Stress Analysis
- Automotive Engineering
- Engine Technology
- Aerospace Technology and Astronautics
- Nanotechnology and Microengineering
- Control, Robotics, Mechatronics
- MEMS
- Theoretical and Applied Mechanics
- Dynamical Systems, Control
- Fluid Mechanics
- Engineering Thermodynamics, Heat and Mass Transfer
- Manufacturing
- Precision Engineering, Instrumentation, Measurement
- Materials Engineering
- Tribology and Surface Technology

Indexed by SCOPUS, EI Compendex, and INSPEC.

All books published in the series are evaluated by Web of Science for the Conference Proceedings Citation Index (CPCI).

To submit a proposal for a monograph, please check our Springer Tracts in Mechanical Engineering at https://link.springer.com/bookseries/11693.

Zekâi Şen · Özer Uygun · Caner Erden
Editors

Advances in Intelligent Manufacturing and Service System Informatics

Proceedings of IMSS 2023

Springer

Editors
Zekâi Şen
Istanbul Medipol University
Istanbul, Türkiye

Caner Erden
Faculty of Applied Sciences
Sakarya University of Applied Sciences
Kaynarca, Sakarya, Türkiye

Özer Uygun
Department of Industrial Engineering
Sakarya University
Serdivan, Sakarya, Türkiye

ISSN 2195-4356 ISSN 2195-4364 (electronic)
Lecture Notes in Mechanical Engineering
ISBN 978-981-99-6061-3 ISBN 978-981-99-6062-0 (eBook)
https://doi.org/10.1007/978-981-99-6062-0

This Springer imprint is published by the registered company Springer Nature Singapore Pte Ltd.
The registered company address is: 152 Beach Road, #21-01/04 Gateway East, Singapore 189721, Singapore

Paper in this product is recyclable.

Contents

Project Idea Selection in an Automotive R&D Center

Ulviye Savaş[1(✉)] and Serkan Altuntaş[2]

[1] TOFAŞ Turkish Automobile Factory, 16120 Bursa, Turkey
Ulviye.savas@tofas.com.tr
[2] Industrial Engineering Department, Yıldız Technical University, İstanbul, Turkey

Abstract. R&D project selection is one of the most important issues for an R&D center. Evaluating more than one project in terms of different criteria, selecting and implementing the most appropriate project is very critical for both the company's profit and the sustainability of the project. The project selection process is handled by different processes in companies. Due to the importance of this issue, companies adopt a selection process in line with their own strategies. In this study, an application was carried out with the fuzzy TOPSIS method to evaluate alternative project ideas that will be an R&D project in the R&D center of an automotive company. 4 different criteria were evaluated by experts for 6 different project ideas. With the implementation realized as a result of expert evaluations, a priority order was obtained for 6 project ideas. In practice, as a result of the evaluation, the alternative project P5 with the highest value in the ranking is selected as the next R&D project to be started.

Keywords: Project selection · fuzzy TOPSIS · R&D · automotive

1 Introduction

Nowadays, large-scale companies should attach importance to R&D activities in order to achieve growth in market shares and to be a leading company by following the agenda in line with the dynamics of the sector in which they operate [1]. While determining the strategies of the companies, it is very important to ensure the right distribution of resources, especially in terms of labor and financial resources, to the right projects [2]. In order to make this evaluation correctly, the company must analyze the resources it has correctly, evaluate the details of alternative projects correctly, and then make choices among these alternatives, taking into account the available resources. R&D project selection and financing decisions are critical for the firm [2].

The difficult part in these elections; ensuring that the organization chooses projects that will lead it to success, projects with a positive cost/benefit, and keeping a priority list of projects for future technologies that will increase the organization's chances of success. Scope and strategic alignment will help stakeholder engagement especially for these projects. In the project evaluation, many different criteria such as strategic suitability, technical feasibility, capacity, project cost and risks are considered. The risks in

Z. Şen et al. (Eds.): IMSS 2023, LNME, pp. 1–9, 2024.
https://doi.org/10.1007/978-981-99-6062-0_1

the selection of these projects are quite high, as the selection of unsuitable projects in the wrong evaluation results will cause significant financial, temporal and human resource losses for the companies [1]. Decision-making can be considered as a complex process, since there are multiple stages in this process, different decision-making groups are involved, and there are conflicting goals for different purposes [3]. Various studies have been conducted on the way organizations make these decisions [3–6]. Due to the uncertainty and different criteria in the projects, Golabi [7] conducted a study related to the maximization of the total values of the projects by using the multi-featured utility theory with integer linear programming. Bard et al. [8] worked on a decision support system to evaluate projects. Stewart [9] introduced a decision support system for a nonlinear optimization in portfolio planning. Traditionally, net present value (NPV), internal rate of return (IRR), and payback period have been used extensively as investment valuation techniques. Iyigium [10] proposed a decision support system for project selection using the Delphi technique. Additionally, Turner and Cochrane [11] published a study of well-defined projects and methods. Chui and Chan [12] proposed a method that evaluates the conditions for the success or failure of an R&D project and uses the net present value. However, there has always been a need to add non-quantitative criteria to the studies in addition to the mathematical studies carried out. For this reason, the multi-criteria decision-making technique started to be used for project selection in the following years. Saaty [13] introduced Analytical Hierarchy Process (AHP) for a method of multi-criteria decision-making. Liberatore [14] created a spreadsheet for project evaluation based on AHP. Brenner [15] proposed a method using the systematic project selection process using AHP for Air Products.

Considering these studies, classification has been made for decision models in project selection; scoring, mathematical programming, economic model, decision analysis, artificial intelligence, and portfolio optimization [4]. However, since the R&D project selection process is a decision-making problem that requires considering many interrelated and contradictory criteria, the use of multi-criteria decision-making methods has taken its place in the literature in order not to overlook the situations that may cause errors, to manage uncertainties correctly, and to evaluate more than one alternative criterion [16].

In this study, an application is conducted to evaluate the ideas of the R&D projects that will be started in the R&D center of an automotive company and the project selection. This application uses the fuzzy TOPSIS method, which is one of the multi-criteria decision-making methods. The linguistic equivalents of the evaluation of the criteria used in the selection of the projects by the experts were shown with fuzzy triangular numbers and the project selection is utilized with the fuzzy TOPSIS method. The main reason for the use of fuzzy triangular numbers in practice is that these numbers are easier to respond to linguistic evaluations, the sensitivity of the numbers is higher, and they provide ease of operation in terms of real application compared to other fuzzy numbers.

2 Material and Methods

2.1 Fuzzy Approach

Classical sets are not always sufficient when it comes to linguistic variables in decision-making. Linguistic variables are very useful in situations where there is complexity and there are no clear results [17]. It is not entirely clear what these expressions will mean quantitatively. In this case, fuzzy logic comes into play and dealing with fuzzy numbers can meet the situation.

In classical sets, an object is either a member of a set or not. In fuzzy sets, on the other hand, there are different degrees of membership to the set. In this way, objects can provide membership to sets. In classical set concept, if an object is a member of a set, its membership degree is evaluated as 1, otherwise it is evaluated as 0. No value other than these two values can be considered. In fuzzy sets, it is possible to talk about different values between 1 and 0 values. In fuzzy sets, the membership degree is the name given to each value between 0 and 1. The changes given under each of these are called membership functions. Objects gathered under membership functions have different membership degrees according to their importance.

In this study, triangular membership function is used. In Fig. 1, the triangular membership function and the elements of the triangular fuzzy set are defined as $\tilde{A} = (a, b, c)$ function [18]. Accordingly, the membership function $\tilde{A}$ is determined as $\mu\tilde{A}: x \rightarrow [0,1]$.

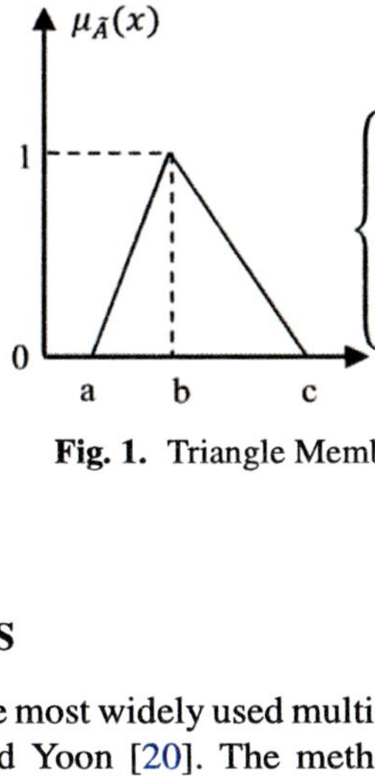

$$\mu(x) = \begin{cases} 0, & x \leq a \\ \frac{x-a}{b-a}, & a \leq x \leq b, \\ \frac{c-x}{c-b}, & b \leq x \leq c, \\ 0, & c \leq x \end{cases}$$

Fig. 1. Triangle Membership Function [19]

2.2 Fuzzy TOPSIS

TOPSIS is one of the most widely used multi-criteria decision-making techniques developed by Hwang and Yoon [20]. The method provides the evaluation of alternatives according to ideal solutions with the Euclidean distance approach. While looking at ideal solutions, it aims to choose the solution closest to the positive ideal solution and the farthest from the negative ideal solution. Fuzzy TOPSIS, on the other hand, is a method used in the evaluation of fuzzy environment developed by Chen [17]. The fuzzy TOPSIS method is useful for solving problems where there are uncertainty and more than one decision maker. In this method, as mentioned before, linguistic expressions are mostly used because there is uncertainty. Decision makers make their evaluations using linguistic expressions, and then these evaluation results are processed by converting them into trapezoidal or triangular fuzzy numbers. The fuzzy TOPSIS steps are as follows [17];

Step 1: The criteria and alternatives clusters are created by the decision makers. Linguistic expressions are used in the evaluation of alternative criteria and determination of weights. The five-point Likert-type linguistic scale used in this study is as shown in Table 1 [20].

Table 1. Fuzzy Evaluation Scores for Alternatives [21].

Linguistic Scale	Triangular Fuzzy Scale
Very unimportant	(0, 0, 0,25)
Unimportant	(0, 0.25, 0,5)
Moderately important	(0.25, 0.5, 0,75)
Important	(0.5, 0,75, 1)
Very important	(0.75, 1, 1)

Step 2: The evaluation results of the decision makers using linguistic expressions are converted into fuzzy numbers using Table 1. Then, using Eq. (1), alternative evaluations of the decision makers are made according to each criterion.

$$\widetilde{x_{ij}} = \frac{1}{K}[\tilde{x}_{ij}^1(+)\tilde{x}_{ij}^2(+)\ldots(+)\tilde{x}_{ij}^K] \tag{1}$$

Step 3: The alternative weights, and fuzzy degrees are obtained according to each criterion, and the fuzzy multi-criteria decision-making matrix is as in Eq. (2).

$$D = \begin{bmatrix} \widetilde{x_{11}} & \ldots & \widetilde{x_{1n}} \\ \ldots & \ldots & \widetilde{x_{2n}} \\ \widetilde{x_{m1}} & \ldots & \widetilde{x_{mn}} \end{bmatrix} \tag{2}$$

The linguistic expressions $(\widetilde{X_{ij}})$ are expressed with triangular fuzzy numbers like $(\widetilde{X_{ij}}) = (a_{ij}, b_{ij}, c_{ij})$.

Step 4: The normalized fuzzy matrix is expressed with $\tilde{R}$ using Eq. (3). Normalization process is performed using Eqs. (4)–(7). The aim here is to transform the numbers into triangular fuzzy numbers normalized between [0,1].

$$\tilde{R} = [\tilde{r}_{ij}]mxn \tag{3}$$

Decision criteria are divided into two as benefit and cost oriented. Here, it is assumed that B shows the benefit criteria and C shows the cost criteria;

$$\tilde{r}_{ij} = \left(\frac{a_{ij}}{c_j^*}, \frac{b_{ij}}{c_j^*}, \frac{c_{ij}}{c_j^*}\right), j \in B; \tag{4}$$

$$\tilde{r}_{ij} = \left(\frac{a_j}{c_{ij}^*}, \frac{a_j}{b_{ij}^*}, \frac{a_j}{a_{ij}^*}\right), j \in C; \tag{5}$$

$$c_{ij}^* = \max_i c_{ij}, j \in B \tag{6}$$

$$a_j^- = \min_i a_{ij}, j \in C \tag{7}$$

Step 5: After the normalization process, a weighted normalized fuzzy decision matrix is created by using different weights for each criterion, if any, or by using equal weights for each criterion.

$$\tilde{V} = \left[\tilde{v}_{ij}\right]mxn, \; i = 1, 2, 3, \ldots, m, \;\; j = 1, 2, 3, \ldots, n \tag{8}$$

$$\tilde{v}_{ij} = \tilde{r}_{ij}(x)\tilde{w}_j \tag{9}$$

Step 6: Considering the weighted normalized fuzzy decision matrix, the elements $(\tilde{v_{ij}})$, $\forall i, j$ normalized triangular positive fuzzy numbers are expressed in the range [0,1].

The fuzzy positive ideal solution $(FPIS, A^*)$ and the fuzzy negative ideal solution $(FPIS, A^-)$ are defined using Eqs. (10) and (11).

$$A^* = (\tilde{v}_1^*, \tilde{v}_2^*, \ldots, \tilde{v}_n^*) \tag{10}$$

$$A^- = (\tilde{v}_1^-, \tilde{v}_2^-, \ldots, \tilde{v}_n^-) \tag{11}$$

$$\tilde{v}_j^* = (1, 1, 1) \text{ and } \tilde{v}_j^- = (0, 0, 0), j = 1, 2, 3, \ldots, n$$

Step 7: The distances of each alternative from A* and A$^-$ are calculated (d_i^* and d_i^-) using Eqs. (12) and (13).

$$d_i^* = \sum\nolimits_{j=1}^{n} d(\tilde{V}_j, \tilde{V}_j^*), i = 1, 2, \ldots, m \tag{12}$$

$$d_i^- = \sum\nolimits_{j=1}^{n} d(\tilde{V}_j, \tilde{V}_j^*), i = 1, 2, \ldots, m \tag{13}$$

Step 8: The closeness coefficients of each alternative are calculated using Eq. (14) to determine the alternative ranking.

$$cc_i = \frac{d_i^-}{d_i^- + d_i^*}, \;\; i = 1, 2, \ldots, m \tag{14}$$

According to these calculated values, the one with the highest closeness degrees in the ranking can be considered as selected.

3 Case Study

This application was carried out for the selection of projects to be started in the R&D center of an automotive company established in Türkiye. The fuzzy TOPSIS method was used for problem solving. The use of linguistic expressions and the absence of clear values in the fuzzy TOPSIS method made it easier for the experts to evaluate the projects during the implementation. In this way, the selection was made by obtaining objective evaluations by the experts.

Project evaluation criteria used in practice are expressed by the set K, $K = \{K_1, K_2, K_3, K_4\}$. As the evaluation criteria of the projects; the impact of the project (K1), the cost of the project (K2), the feasibility of the project (K3) and the added value (K4) in terms of innovation, which is considered as the innovative aspect of the project, were taken into consideration. The set of alternative projects is denoted by P, $P = \{P_1, P_2, P_3, P_4, P_5, P_6\}$. In this application, 6 new project ideas were evaluated in total.

The evaluation of the relationship between alternative project ideas and the criteria was performed by 7 experts working in different fields in the R&D center for a long time, using the linguistic expressions in Table 1. Evaluations of experts in linguistic variables is given in Table 2.

Table 2. Evaluations of Experts in Linguistic Variables

Alternative/Criteria	K1	K2	K3	K4
P1	Very important	Unimportant	Moderately important	Very important
P2	Moderately important	Moderately important	Very important	Unimportant
P3	Moderately important	Moderately important	Moderately important	Unimportant
P4	Important	Moderately important	Moderately important	Important
P5	Moderately important	Unimportant	Moderately important	Important
P6	Moderately important	Moderately important	Moderately important	Important

Table 2 shows the degree of importance of the project alternatives according to the criteria. To apply this to fuzzy TOPSIS, the equivalent of the alternative-criteria evaluation with linguistic language for fuzzy numbers is given in Table 3.

As a result of the comparison of the criteria used in the project evaluation with each other, it was decided that their weights were equal and it was taken as 0.25 for each criterion. Equations (8) and (9) are calculated to obtain weighted fuzzy decision matrix. Then, the weighted fuzzy normalized decision matrix is obtained by the weighting process (see Tables 4–5).

Table 3. Equivalent of Table 2 for Fuzzy Numbers

Weight	0,25			0,25			0,25			0,25		
	K1			K2			K3			K4		
P1	0,75	1	1	0	0,25	0,5	0,25	0,5	0,75	0,75	1	1
P2	0,25	0,5	0,75	0,25	0,5	0,75	0,75	1	1	0	0,25	0,5
P3	0,25	0,5	0,75	0,25	0,5	0,75	0,25	0,5	0,75	0	0,25	0,5
P4	0,5	0,75	1	0,25	0,5	0,75	0,25	0,5	0,75	0,5	0,75	1
P5	0,25	0,5	0,75	0	0,25	0,5	0,25	0,5	0,75	0,5	0,75	1
P6	0,25	0,5	0,75	0,25	0,5	0,75	0,25	0,5	0,75	0,5	0,75	1

Table 4. Weighted Fuzzy Normalized Decision Matrix for K1-K2

	K1			K2		
P1	0.090951	0.156174	0.242536	0	0.058926	0.25
P2	0.030317	0.078087	0.181902	0.066815	0.117851	0.375
P3	0.030317	0.078087	0.181902	0.066815	0.117851	0.375
P4	0.060634	0.11713	0.242536	0.066815	0.117851	0.375
P5	0.030317	0.078087	0.181902	0	0.058926	0.25
P6	0.030317	0.078087	0.181902	0.066815	0.117851	0.375

Table 5. Weighted Fuzzy Normalized Decision Matrix for K3-K4

	K3			K4		
P1	0.032009	0.083333	0.032009	0.083333	0.032009	0.083333
P2	0.096028	0.166667	0.096028	0.166667	0.096028	0.166667
P3	0.032009	0.083333	0.032009	0.083333	0.032009	0.083333
P4	0.032009	0.083333	0.032009	0.083333	0.032009	0.083333
P5	0.032009	0.083333	0.032009	0.083333	0.032009	0.083333
P6	0.032009	0.083333	0.032009	0.083333	0.032009	0.083333

Equation (10)–(13) was used to measure the distances of the weighted fuzzy normalized decision matrix from the ideal negative and ideal positive solutions. As a result of calculating the relative closeness to the ideal solutions, the values were calculated by using Eq. (14) for the closeness coefficient values of the alternatives for the ranking. The closeness coefficients and rankings of the alternatives are given in Table 6.

As can be seen from Table 6, the P5 was found to be the first project to be initiated by the R&D department.

Table 6. Closeness Coefficient of Alternatives and Ranking

Alternative	c_i	Ranking
P1	0.582059	2
P2	0.456326	6
P3	0.459195	5
P4	0.504535	4
P5	0.620146	1
P6	0.551704	3

4 Conclusion

In this study, Fuzzy TOPSIS method was conducted to select the best R&D projects in the R&D center of an automotive company. The feasibility of the project, the cost of the project, the impact of the project and the contribution of the project to the innovation criteria are evaluated by experts for 6 projects that were considered as alternatives in practice. Since these evaluation results are expressed linguistically, their equivalents with fuzzy numbers are taken into account in the application of the method. With the ranking obtained as a result of the application, the P5 was found to be the first project to be initiated by the R&D department.

R&D project selection evaluation can be performed with other decision-making methods such as fuzzy TOPSIS method in future studies. Project selections can be utilized by using 7-likert-type different scales instead of the 5-point likert scale.

References

1. Mohanty, R.P., Agarwal, R., Choudhury, A.K., Tiwari, M.K.: A fuzzy ANP-based approach to R&D project selection: a case study. Int. J. Prod. Res. **43**(24), 5199–5216 (2005)
2. Meade, L.M., Presley, A.: R&D project selection using the analytic network process. IEEE Trans. Eng. Manag. **49**(1), 59–66 (2022)
3. Ghasemzadeh, F., Archer, N.P.: Project portfolio selection through decision support. Decis. Supp. Syst. **29**, 73–88 (2000)
4. Henriksen, A.D., Traynor, A.J.: A practical R&D project-selection scoring tool. IEEE Trans. Eng. Manag. **46**, 158–170 (1999)
5. Ringuest, J.L., Graves, S.B., Case, R.H.: Mean-Gini analysis in R&D portfolio selection. Eur. J. Oper. Res. **154**, 157–169 (2004)
6. Lawson, C.P., Longhurst, P.J., Ivey, P.C.: The application of a new research and development project selection model in SMEs. Technovation **25**, 1–9 (2004)
7. Golabi, K.: Selecting a group of dissimilar projects for funding. IEEE Trans. Eng. Manag. **34**, 138–145 (1987)
8. Bard, J.F., et al.: An interactive approach to R&D project selection and termination. IEEE Trans. Eng. Manag. **35**, 135–146 (1988)
9. Stewart, T.J.: A multi criteria decision support system for R&D project selection. J. Oper. Res. Soc. **42**, 17–26 (1991)

10. Iyigun, M.G.: A decision support system for R&D project selection and resource allocation under uncertainty. Proj. Manag. J. **24**, 5–13 (1993)
11. Turner, J.R., Cochrane, R.A.: Goals and methods matrix: coping with projects with ill-defined goals and/or methods of achieving them. Int. J. Proj. Manag. **11**, 93–102 (1993)
12. Chui, Y.C., Chan, S.P.: Fuzzy cash flow analysis using present worth criterion. Eng. Econ. **39**, 113–138 (1994)
13. Saaty, T.L.: The Analytic Hierarchy Process. McGraw-Hill, New York (1980)
14. Liberatore, M.J.: An expert system for R&D project selection. Math. Comput. Model. **11**, 260–265 (1988)
15. Brenner, M.S.: Practical R&D project prioritization. Res. Technol. Manag. **28**, 38–42 (1994)
16. Yıldırım, B.F., Yıldırım, S.K.: A new integrated intuitionistic fuzzy group decision making approach for R&D project selection process. J. Eng. Sci. Des. **10**(2), 643–653 (2022)
17. Chen, C.T.: Extensions of the topsis for group decision-making under fuzzy environment. Fuzzy Sets Syst. **114**, 1–9 (2000)
18. Sun, C.C.: A performance evaluation model by integrating fuzzy AHP and fuzzy TOPSIS methods. Expert Syst. Appl. **37**, 7745–7754 (2010)
19. Şen, Z.: Bulanık Mantık İlkeleri ve Modelleme. Su Vakfı Yayınları (2009)
20. Hwang, C.L., Yoon, K.: Multiple Attribute Decision Making: Methods and Applications, A State-of-the-Art Survey. Springer, New York (1981). https://doi.org/10.1007/978-3-642-48318-9
21. Pervez, A.K.M.K., Maniruzzaman, M., Shah, A.A., Nabi, N., Ado, A.M.: The meagerness of simple Likert scale in assessing risk: how appropriate the fuzzy Likert is? Nust J. Soc. Sci. Humanit. **6**(2), 138–150 (2020)

Societies Becoming the Same: Visual Representation of the Individual via the Faceapp: Application

Hilal Sansar(✉)

Hacettepe University Fine Art Institution/Graphic Design, 06800 Ankara, Türkiye
hilalsansar@yahoo.com

Abstract. Standardized perceptions of beauty have always existed through bodies, which are expressions of our characters and identities. As societies have changed, these perceptions have changed shape along with societies. But in a globalizing world with social media, beauty standards also tend to go global. In this direction, the sense of beauty and the way of life of Western societies have been positioned as the goal sought to be reached in the whole world. Therefore, having slanted eyes or black skin has been declared ugly, beyond racism, because it does not fit the ideal perception of beauty.

In this study, this ideal beauty, which is about the whole body and self, is evaluated through the face editor applications applied to the portraits in which identities and characters are revealed. In this context, FaceApp: Face Editor application, one of the most popular face-changing and editing applications, is taken as an example.

Such applications based on machine learning and artificial intelligence take advantage of the user's location, which is also data for them.

The issue of protecting personal data, which is one of the biggest problems, and the possibility of it, as well as the fact that it is becoming increasingly difficult to distinguish between real and fake in a world centered on commodification, will be discussed.

This study, in which the descriptive analysis method is used by examining the data, aims to draw attention to the current problems of the society that has become identical in the effort of differentiation and the causes of these problems.

Keywords: FaceApp · face editor · machine learning · Homogeneous Society

1 Introduction

Standardized perceptions of beauty have always existed through bodies, which are expressions of our characters and identities. As societies have changed, these perceptions have changed shape along with societies. But in a globalizing world with social media, beauty standards also tend to go global. In this direction, the sense of beauty and the way of life of Western societies have been positioned as the goal sought to be reached in the whole world. Therefore, having slanted eyes or black skin has been declared ugly, beyond racism, because it does not fit the ideal perception of beauty.

Z. Şen et al. (Eds.): IMSS 2023, LNME, pp. 10–14, 2024.
https://doi.org/10.1007/978-981-99-6062-0_2

As Kaşıkara points out, the desire to be admired, which is defined as the desire of individuals to receive positive feedback from others in many areas of their lives to transform their perceptions about themselves into positive ones, to feel good about themselves, to satisfy their needs for love and respect, has turned into an effort to create visual satisfaction through their bodies (Kaşıkara 2017, p. 53). Social tastes have always existed and influenced people's individual tastes. This common understanding, which is shaped by many factors, continues to transform with the effect of globalization (Gürler 2018, p. 143).

2 Admiration Instinct of Societies and Portraiture

Nowadays, sharing on social media, especially selfies, has become a daily routine and this situation has been positioned as a result of our age. However, in the past, people who first competed to have their portraits drawn then tried to make their self-portraits permanent by having their photographs taken. The self-portrait is considered a reflection of one's character. Therefore, self-portraits and selfies are self-presentations used to reveal oneself.

Today, being admired, applauded, and approved by people becomes even more important, and presenting a complete human image stands out as dominant behavior (Kaşıkara 2017, p. 52). In the past, people proved their reputation by becoming visible with the portraits they had drawn and the photographs they had taken to gain acceptance and respect in front of society. Today, social media sharing is carried out for similar purposes. People whose only purpose is to be visible, create new identities, and take on other identities while doing this. Photography and especially 'selfie' have an important place in the creation of all these identities (Gök 2016, p. 42). "It is no accident that the portrait was the focal point of early photography. The cult of remembrance of loved ones, absent or dead, offers a last refuge for the cult value of the picture. For the last time, the aura emanates from the early photographs in the fleeting expression of a human face" (Benjamin 1969, p. 7).

The use of social media, which takes advantage of these needs, has expanded more and more. As the possibility of editing images becomes easier while sharing, more people are going to correct (!) and change what they see as defects in their photos. The desire to be liked has turned into a race to ingratiate oneself with everyone over time, causing unreal content to be produced and shared. Especially these edits made for face photos can cause self-comparison and excessive criticism by the viewer who perceives them as real.

3 The Objectified Body

In his book The Consumer Society, Baudrillard writes, "We are living in an age in which it has become imperative to look beautiful. Beauty became a religious commandment. Being beautiful is neither a natural gift nor an addition to moral qualities. It is the basic, commanding quality of those who take care of their faces and contours as well as their souls. Being beautiful, like success in the business world, is a sign of being chosen at the level of the body" (Baudrillard 2008, p. 168).

As Goffman emphasizes, first impressions are very important as the first link of interaction (Goffman 2014, pp. 24–25). For this reason, people try to create a perfect first impression, creating images that are far from reality and therefore from themselves. "In today's world, determined by the principle of simulation, the real can only be a copy of the model." (Baudrillard 2008, p. 150). This is what W. Benjamin calls "By making many reproductions it substitutes a plurality of copies for a unique existence." (Benjamin 1969, p. 4).

Unnecessary aesthetic operations are resorted to because the ideal image obtained with over-edited portrait photographs is desired to be made permanent as the first place to look when interacting with people.

False needs have multiplied so much that the distinction between them and real needs has disappeared. The greatest need has become to provide a social status and visual satisfaction. With capitalism, the average age of individuals who are dissatisfied with their bodies and constantly in search of a better image is gradually decreasing. The body and soul of the individual are now objects of consumption. In this process of emphasizing the individuality and difference of people, sameness, and objectification have ironically become normal.

It can be said that individuals who have an idealized self-perception by the environment, who have objectified their bodies, become alien to their essence. Objectification has always existed. However, the acceleration of globalization, which has brought many benefits, has also ensured that the targeted person can be easily managed. New desires and dissatisfactions are introduced to the market (Bauman 1999, p. 43) to ensure the continuous sale of consumer goods that are no longer merely intended to satisfy needs (Asıl 2017, p. 5). People who seek emotional satisfaction try to resemble the ones they think are most liked to be liked. "While the individual's area of freedom on his body expands through choices; gender norms, cultural codes, images and symbols that create social inequality through the body have continued to exist." (Varga 2005, p. 227).

4 FaceApp Working Principle

In addition to editing photos and videos through simple but powerful apps like FaceApp, artificial intelligence, and machine learning can be easily used by ordinary people. By using machine learning to train artificial intelligence, operations that can be extremely complex for even the most experienced digital artists can be easily performed at the push of a button (Gerstner 2020, p. 2). For example, when a face swap was shared by a Reddit user for the first time in 2017, face swap spread through social media and began to attract people's attention and was practiced by more and more people every day (Peipeng Yu 2021, p. 608). Later, companies that noticed the demand developed new applications. One of them, FaceApp: Face Editor, can produce realistic images with the Deepfake principle. In addition, "Many applications offered by 'Smart-Android' phone manufacturers allow consumers to quickly process and share photos and selfies, and receive notifications through the application for likes or comments on shared photos." (Gök 2016, p. 43). Therefore, the user prefers applications where he can easily share edited photos on social platforms.

FaceApp: Face Editor and similar applications, which were produced in line with these demands, have added entertainment to the business over time and have increased the methods of use. Applications must ask the user for permission to send notifications, use the phone's microphone and camera, and access photo albums. However, since the application cannot be used without granting these permissions, the user continues to use the application by granting all permissions without question.

FaceApp, whose reliability is questioned more with the increase in its use, has announced that personal data is not shared with third parties. In a statement, FaceApp said that only photos selected for editing can be accessed, and other photos in your gallery cannot be accessed.

However, videos created with recent deep fake approaches are becoming extremely realistic and can hardly be distinguished from the human eye (Peipeng Yu 2021, p. 608). Recognizing the distinction between fake and real is becoming more difficult as machine learning improves. The public can be misled and manipulated by unreal sounds and images. Although the editing of unauthorized fake pornographic images of people has become widespread, deep fake videos of people who have a guiding influence on public opinions, such as political leaders, have also been produced, and some laws and rules have been enacted afterward. For this, new programs that work like deep fakes are used to capture images produced with deep fakes.

5 Conclusion

Social media has become an intermediary element representing the new face of the body; it mostly represents the face of the person that is not himself but wants to be (Kahraman 2020, p. 1211). With the increase in the speed of social media sharing and the audience that can reach it, new applications and facial effects that interact with media such as FaceApp also collect data that belongs to us and provide data to companies and machine learning. The protection of personal data is becoming increasingly difficult. Companies that come to the fore with a new lawsuit every day expand their clarification texts and specify the information to be used, sometimes implicitly and sometimes explicitly. People who cannot give up on the promise of entertainment and a better image provided by the applications continue to download the applications.

With these practices and sharing habits, in addition to security problems, the door to major psychological and sociological changes is opened. Especially in adolescence, the desire to be liked, which is of great importance for the person, turns into an addiction. The excessive increase in the need to be liked and approved by others distances the person from himself and makes him dependent on the guidance of others. Is it really possible for an individual to exist in a healthy way in society with an externally dependent life model?

Social media, which is one of the biggest contributors to the increase in the need for admiration added to the list of harmful addictions, can ignore the health of individuals with its desire for visibility. Microsoft has fired a team that guides AI innovation that leads to ethical, responsible, and sustainable outcomes as part of a plan to cut 10,000 jobs (Ulukan 2023). The approach of companies such as Microsoft, one of the technology giants, that benefit from artificial intelligence and machine learning to the individual and ethical issues should be considered by the individuals who use the applications. The fact that the individual is aware of the negative effects of artificial intelligence while benefiting from the positive returns will protect him from being an object that is directed outside his own decisions.

References

Asıl, S.: Tüketimde Benlik Algısı: Sosyal Medya Hesaplarında Tüketici Olmak. ÇOMÜ Int. J. Soc. Sci. 1–22 (2017)

Baudrıllard, J.: "Tüketim Toplumu". Trans. Hazal Deliçaylı and Ferda Keskin, 3rd edn. Ayrinti Publications, Istanbul (2008)

Bauman, Z.: Çalışma, Tüketicilik ve Yeni Yoksullar. Sarmal Publishing House, İstanbul (1999)

Benjamin, W.: The Work of Art in the Age of Mechanical Reproduction. Schocken Books, New York (1969)

Gerstner, E.: Face/Off: "DeepFake" face swaps and privacy laws. Defense Counsel J. 1–14 (2020)

Goffman, E.: Günlük Yaşamda Benliğin Sunumu (Cilt 3. Basım). Metis Publications, İstanbul (2014)

Gök, C.: Resim ve Fotoğraf Sanatında Portre Geleneğinden. Medeniyet Sanat J. IMU Fac. Art Des. Archit. 29–47 (2016)

Gürler, G.: Estetik Cerrahi Müdahale Görmüş Bireyler Üzerine Bir Alan Araştırması. J. Sociol. 141–172 (2018)

Kahraman, Ö.: Manipüle Edilen Çağdaş Bedeni Beden Pratikleri Üzerinden Okumak. İdil 1202–1217 (2020)

Kaşıkara, G.: Beğenilme Arzusu: Ölçek Geliştirme, Güvenirlik ve Geçerlik Çalışması- Gülizar KAŞIKARA. J. MSKU Educ. Fac. 51–60 (2017)

Yu, P., Xia, Z.: A survey on deepfake video detection. IET Biomet. 581–719 (2021)

Ulukan, G.: Webrazzi. https://webrazzi.com/2023/03/14/microsoft-yapay-zeka-etik-ekip-isten-cikarma/. Accessed 14 Mar 2023

Varga, I.: The body – the new sacred? The body in hypermodernity. Curr. Sociol. **53**(2), 209–235 (2005)

Reading, M.A. (ed.) Addison-Wesley (1990). Reprinted in Human-Computer Interaction (ICT 235) Readings and Lecture Notes, vol. 1, pp. 32–37. Murdoch University, Murdoch (2005)

Wigner, E.P.: Theory of traveling wave optical laser. Phys. Rev. **134**, A635–A646 (1965)

Modeling Electro-Erosion Wear of Cryogenic Treated Electrodes of Mold Steels Using Machine Learning Algorithms

Abdurrahman Cetin[1], Gökhan Atali[2(✉)], Caner Erden[3], and Sinan Serdar Ozkan[2]

[1] Vocational School of Sakarya, Machinery and Metal Technology, Sakarya University of Applied Sciences, Sakarya, Turkey
abdurrahman@subu.edu.tr

[2] Faculty of Technology, Department of Mechatronics Engineering, Sakarya University of Applied Sciences, Sakarya, Turkey
{gatali,sozkan}@subu.edu.tr

[3] Faculty of Applied Sciences, Sakarya University of Applied Sciences, Sakarya, Turkey
cerden@subu.edu.tr

Abstract. Electro-erosion wear (EEW) is a significant problem in the mold steel industry, as it can greatly reduce the lifespan of electrodes. This study presents a machine-learning approach for predicting and modeling electrode and workpiece wear on an electrical discharge machining (EDM) machine. In the experimental design, EDM of CuCrZr and Cu electrodes of AISI P20 tool steel was carried out at different pulse currents and duration levels. In addition, CuCrZr and Cu electrodes used in the experiment were cryogenically treated at a predefined degree for multiple periods and then tempered. This study employed machine learning algorithms such as decision trees, random forests, and k-nearest neighbors to model the EEW of cryogenically treated electrodes made of mold steels. The results were compared according to the coefficient of determination (R^2), adjusted R^2, and root mean squared error. As a result, the decision trees outperformed the other algorithms with 0.99 R^2 performance. This study provides valuable insights into the behavior of EEW in mold steel electrodes and could be used to optimize the manufacturing process and extend the lifespan of the electrodes.

Keywords: electrical discharge machining · material removal rate · electrode wear ratio · machine learning

1 Introduction

Electric discharge machining (EDM) is a widely used non-traditional method. The amount of material removed from the workpiece per unit of time is called the material removal rate (MRR). In contrast, the mass loss in the electrode material is referred to as electrode wear rate (EWR). In an EDM method, improvement is desired in terms of higher MRR, lower EWR, and better surface quality [1]. EWR is the most important

Z. Şen et al. (Eds.): IMSS 2023, LNME, pp. 15–26, 2024.
https://doi.org/10.1007/978-981-99-6062-0_3

factor in determining the number of electrodes required to achieve the correct size and dimensions of the desired form. When considering that electrodes are processed by wire erosion, turning, or milling machines, it is seen that EWR is the most significant factor affecting electrode costs. Therefore, studies on higher chip removal and lower electrode wear have gained importance in the EDM process in recent years.

EDM method has been applied in recent years with traditional methods and machine learning studies such as artificial neural networks (ANNs) and soft computing techniques such as fuzzy logic for predicting output performance parameters such as MRR and EWR based on optimum processing parameters such as discharge current, pulse duration, and voltage. In their study investigating the machinability of EDM, Ramaswamy et al. [2] performed a variance analysis to determine the significance of test parameters on experimental results. In the second phase of their study, researchers identified optimal process parameters and used regression analysis and ANNs to predict MRR and EWR. Similarly, Sarıkaya and Yılmaz [3] developed a mathematical model based on ANNs that successfully predicted outputs. In another study, Balasubramaniam et al. [4] used different electrode materials, such as copper, brass, and tungsten, for EDM of Al-SiCp metal matrix composites. MRR, EWR, and circularity (CIR) were considered as performance metrics in their study. As a result of using artificial intelligence to optimize processing parameters such as current, pulse duration, and flushing pressure, the most important parameter was shown to be current, and Cu exhibited the best performance among the three electrodes. In EDM, the effect of processing parameters such as peak current, pulse interval, and pulse duration are important for the variation in MRR and EWR. Ong et al. [5] developed a model based on the prediction of radial basis function neural networks to predict the MRR and EWR of the EDM process. The researchers used the moth flame optimization algorithm to determine the optimal processing parameters that maximize MRR and minimize EWR [5]. Cakir et al. [6] investigated the capacity of adaptive neuro-fuzzy inference systems, genetic expression programming, and ANNs in predicting EDM performance parameters using experimental data. Arunadevi and Prakash [7] used artificial intelligence to perform a performance analysis of experimental values with five input parameters to increase the MRR value and reduce surface roughness (SR) in their study. The model was evaluated using the R-squared value.

Machine learning techniques like electro-erosion wear have become increasingly popular in modeling and optimizing complex material processing processes. Several recent studies have examined the relationship between electro-erosion wear and machine learning. For example, Ulas et al. [8] used machine learning methods to estimate the surface roughness of Al7075 aluminum alloy processed with wire electrical discharge machining (WEDM) using different parameters, such as voltage, pulse-on-time, dielectric pressure, and wire feed rate. They employed LM, W-ELM, SVR, and Q-SVR models to process the samples and estimate the surface roughness values. Similarly, Jatti et al. [9] investigated the prediction of material removal rate (MRR) using machine learning algorithms, including supervised machine learning regression and classification-based approaches. They found that gap current, voltage, and pulse on time were the most significant parameters affecting MRR. They concluded that the Gradient boosting regression-based algorithm was the most effective for predicting MRR.

Meanwhile, Nahak and Gubta [10] reviewed the developments and challenges of EDM processes in 2019, emphasizing optimizing process parameters for effective and economical machining. Finally, Cetin et al. [11] experimentally investigated the effect of cryogenic treatment on the performance of CuCrZr alloy and Cu electrodes during EDM of AISI P20 tool steel. They found that pulse current was the most effective parameter in the EDM process and using cryogenically treated electrodes resulted in less wear and decreased surface roughness values.

These studies have demonstrated the successful use of machine learning techniques for modeling and optimizing the electro-erosion wear process. However, no studies have been found on the evaluation of the performance of cryogenically treated and untreated Cu and CuCrZr electrodes or the use of the artificial neural network (ANN) predictions for material removal rate (MRR) and electro-erosion wear ratio (EWR). This study aims to evaluate the performances of cryogenically treated and untreated CuCrZr and Cu electrodes during the electrical discharge machining (EDM) of AISI P20 tool steel in terms of EWR and MRR. By comparing the electrodes under different processing parameters and applying cryogenic treatment in 10 different time intervals ranging from 1/4 - 24 h, the study aims to contribute to the existing literature. The study utilizes decision trees, random forests, and k-nearest neighbor algorithms from machine learning techniques for regression analysis. The best algorithm is determined based on the results obtained, and comments are developed accordingly.

2 Material and Methods

2.1 Test Materials

In this experimental study, CuCrZr and Cu electrode pieces with a diameter of 10 × 30 mm were used as tool material. The values of the chemical compositions of CuCrZr and Cu electrodes are given in Table 1. To observe the effects of CT (Cryogenic Treatment), the electrodes were divided into 11 groups as treated and untreated electrodes. Cryogenically treated electrodes were treated at −140 °C for 15, 30 min, and 0, 0.25, 0.5, 1, 2, 4, 8, 12, 16, 20, 24 h and then tempered at 175 °C for 1 h. For this study, a total of 176 experiments were tested.

Table 1. Chemical composition and some properties of electrode materials (wt.%)

Material		CuCrZr			Cu
Chemical Composition	Elements	Cu	Cr	Zr	Cu
	(wt.%)	Balance	1.00	0.10	100

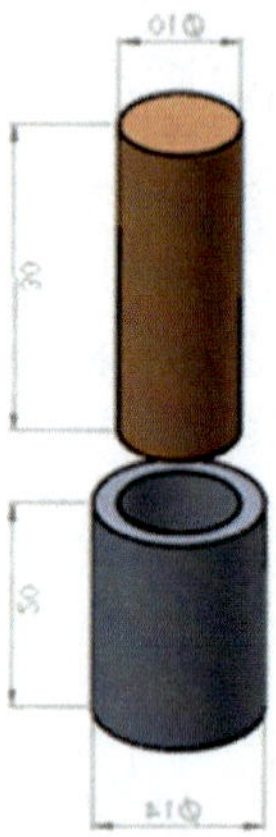

Fig. 1. AISI P20 and Electrode

AISI P20 tool steel, widely used in plastic injection molds, was chosen as the workpiece material of the experimental study. The diameter 14 × 20 mm AISI P20 material, tool electrode dimensions, and technical drawings drawn in 3D design programs are as in Fig. 1. Also, the chemical composition of AISI P20 tool steel is shown in Table 2.

Table 2. Chemical composition of AISI P20 steel (wt.%)

C	Si	Mn	Cr	Mo	Ni	S	Fe
0.40	0.25	1.5	1.9	0.2	1.0	0.001	Balance

2.2 EDM Tests

EDM tests were performed at pulse currents of 4, 8, 12, and 16 A and pulse times of 25 μs and 50 μs. In addition, the King ZNC K3200 model EDM machine seen in Fig. 2 was used in the experimental studies. At each parameter change, other processing parameters were kept constant for all tests.

Experimental conditions and parameters are given in Table 3. During the EDM tests, Petrofer dielectricum 358 mineral-based oil compatible with electro-erosion processing methods was used as the dielectric fluid. To obtain accurate values, EDM experiments were repeated three times for each combination of processing conditions, and the average values were considered the test result. EDM was performed for 20 min in each of the 176 experiments.

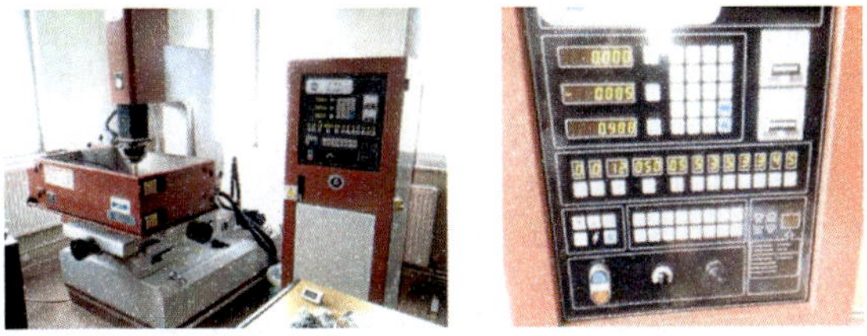

Fig. 2. EDM machine and control panel

Table 3. Materials and EDM parameters AISI P20 tool steel

Workpiece material	AISI P20 tool steel
Electrode materials	CuCrZr and Cu
Dielectric fluid	Petrofer dielectricum 358 mineral-based oil
Pulse current (A)	4, 8, 12, 16
Pulse-on-time (μs)	25, 50
Pulse-off time (μs)	2.5, 5
Duty factor (%)	90.9
Machining time (min)	20

2.3 Experimental Conditions

EWR and MRR values for Cu and CuCrZr electrodes were calculated considering mass losses after an EDM process. To calculate the wear rates of the electrodes and the MRR of the workpieces, samples were weighed before (MBT - Mass Before Testing) and after (MAT - Mass After Testing) EDM using an analytical precision balance with a maximum capacity of 250 g and an accuracy of 0.0001 g. EWR and MRR were calculated using the following equation:

$$EWR = \frac{(MBTelectrode - MATelectrode)}{T}\ (\mathrm{g/min}) \quad (1)$$

$$MRR = \frac{(MBTworkpiece - MATworkpiece)}{T}\ (\mathrm{g/min}) \quad (2)$$

In the above formulas (1) (2), T is the EDM process time. It was applied as T = 20 in the experiments.

The results were evaluated in the following headings according to the EWR and MRR values obtained with the experimental data according to the change of each parameter.

2.4 Machine Learning Algorithms

Decision trees from machine learning algorithms and random forest algorithms will be tried on the data set where data analysis is performed. Brief information about the algorithms can be given as follows:

Decision Trees: Decision trees are a graphical method often used in classification and regression from machine learning problems. The decision tree sets division rules by performing branching operations on the dataset to solve the classification problem. Each branch expresses a decision and has a class tag at the end. The most important advantage of decision trees is that they are easy to understand and visualize. It can also work with continuous or categorical data.

Random Forest: Popular machine learning algorithm Random Forest is a subset of ensemble learning. It is well renowned for its capacity to manage huge and high-dimensional datasets and is utilized for classification and regression issues. It builds numerous decision trees and then combines their outputs. This procedure, known as an ensemble, aids in decreasing overfitting and improving the model's overall accuracy. The Random Forest algorithm's ability to randomly choose a subset of features for each decision tree is the secret to its effectiveness. Random selection guarantees each tree's uniqueness, lessening the association between the trees. The results from all the trees are combined to make the final projection. The Random Forest technique has established itself as a standard in many data science applications because it provides more durable and trustworthy models. The system also offers feature importance scores, which help determine which elements in the data are most crucial.

k-Nearest Neighbors: The non-parametric, instance-based k-Nearest Neighbors (k-NN) technique is used in machine learning. It is frequently employed for classification and regression issues and is particularly helpful when the data cannot be separated linearly. A fresh sample is compared to its k closest neighbors in the training data as part of the algorithm's operation, and a prediction is then made based on the dominant class or average value of those neighbors. The k-NN algorithm's simplicity and ease of use are its key benefits. It can handle continuous and categorical features and does not require any assumptions about how the data are distributed. However, the choice of k significantly impacts how well it performs, and it might be sensitive to noise or irrelevant elements in the data. Numerous methods, including feature scaling, feature selection, and distance metric selection, have been developed to solve these problems. The approach can also be computationally expensive for large datasets because it needs to calculate the distances between all samples at the time of prediction. Nevertheless, due to its ease of use and adaptability, k-NN continues to be a popular option for many real-world applications.

3 Experimental Results and Comparisons

The data set contains 176 experiments performed in the Sakarya University of Applied Sciences laboratory. The data set has variables such as the type of electrode material, cryogenic process conditions, ampere, and pulse. The output variables affected by the input variables are determined as electrode and workpiece wear. An example of the dataset is shared in Table 4.

The relationships between the variables will be examined using data visualization techniques to understand the data set better. The relationship between electrode and workpiece wear with the change of Cu and CuCrZr materials from the electrode materials is illustrated in Fig. 3. Accordingly, it should be noted that the Cu material has relatively

Table 4. Sample data from experiments

Test number	Electrode material	Cryogenic process conditions (Hour)	Current (A)	Pulse durations (millisecond)	Electrode wear (mg/min)	Workpiece wear (mg/min)
95	CuCrZr	12	4	25	1.925	46.685
121	CuCrZr	0	8	50	3.68	116.98
26	Cu	2	8	25	6.125	117.435
148	CuCrZr	4	12	50	16.435	208.69
172	CuCrZr	12	16	50	51.63	276.855
53	Cu	20	12	25	22.95	199.065
18	Cu	12	4	50	0.58	41.17

higher wear than CuCrZr. In the case of using the Cu material, the average workpiece wear was 159.37, while the wear value was calculated as 149.93 with the use of CuCrZr material. Similarly, in electrode wear, the CuCrZr average was 16.79 while the CuCrZr was 16.15. Therefore, it has been observed that the effect of changing the material used on the workpiece is greater than electrode wear.

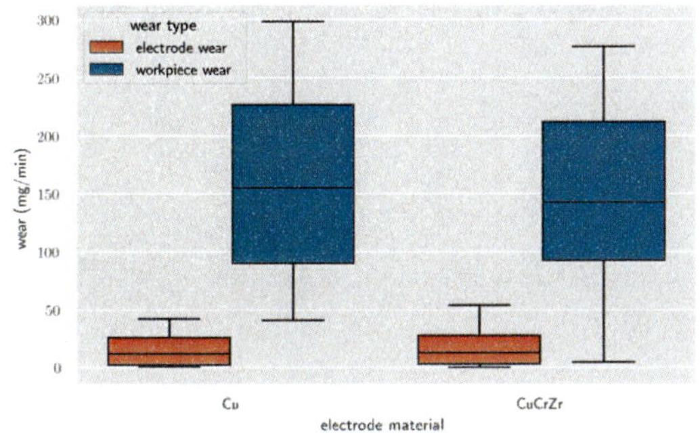

Fig. 3. Electrode and workpiece wear vs. electrode material

Graphs showing changes up to 24 levels are given in Fig. 4 (a-b-c-d) to examine the relationships between abrasions on both the workpiece and the electrode obtained by changing the cryogenic process conditions. Accordingly, there is no significant difference between the wear of the workpiece under different processing conditions. It should only be noted that under the process conditions taken as 12.0, CuCrZr causes significantly less workpiece wear than the Cu material. No significant differences were observed in other conditions. When the wear of the workpiece is examined, it is observed that the wear increases relatively with the increase in the cryogenic process conditions.

Ampere values have a direct effect on wear. Increasing the ampere impacts both electrode material and workpiece wear. The correlation rate between the ampere and the

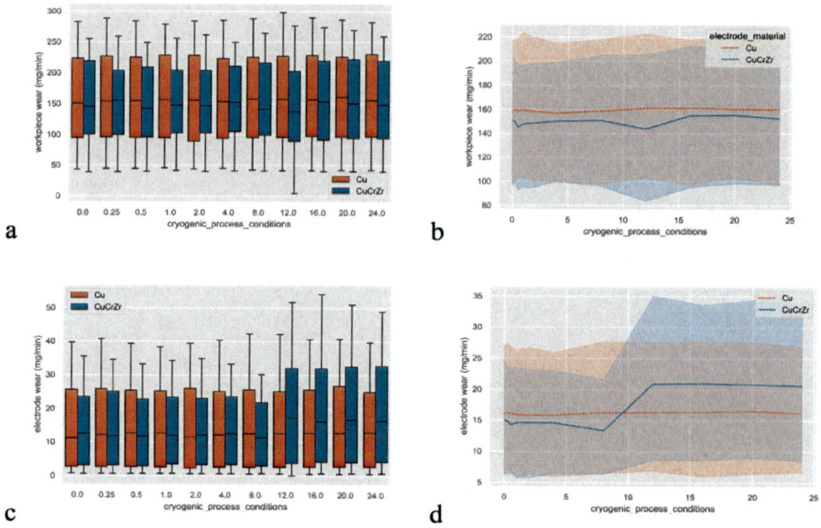

Fig. 4. Electrode and workpiece wear vs. cryogenic process control a and c) boxplots, b and d) scatter plots

electrode material wear is 0.94. Moreover, the correlation between the workpiece and the workpiece is 0.98. As the correlation coefficient shows, it has been observed that wear varies highly with ampere changes (see Fig. 5).

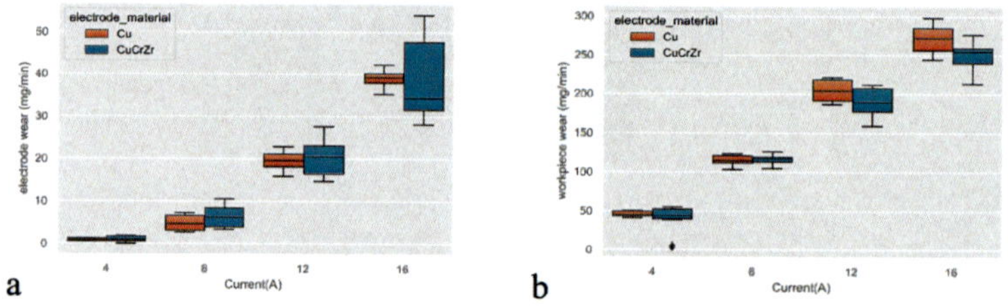

Fig. 5. Ampere vs. A) electrode wear and B) workpiece wear

Thanks to the correlation heat map, an impression of the direction and severity of the relationships between the variables can be obtained. In the heat map shown in Fig. 6, it was determined that there was a high correlation between the ampere variable and the abrasions. In addition, the correlations between the abrasions reach a value of 0.93.

When a regression study is performed for a more detailed analysis of correlation relations, the relationship between abrasions according to wear types at $p < 0.05$ significance level and electrode material types used is revealed in Fig. 7. Accordingly, it can be said that the highest correlation is between the material Cu and the workpiece wear. At the same time, a very high correlation was obtained in the CuCrZr alloy. Only when the electrode material is CuCrZr can it be said that electrode wear is less affected

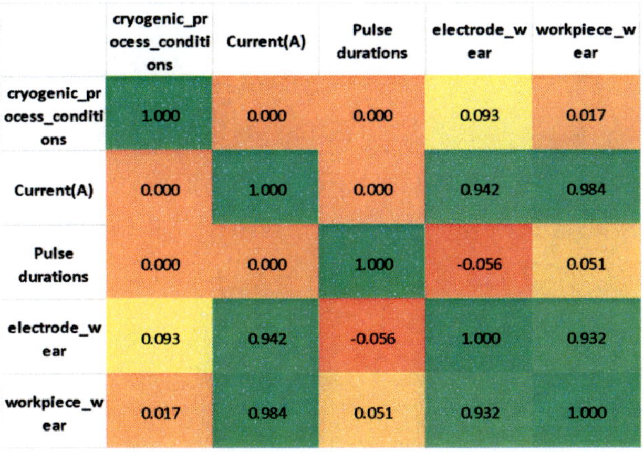

Fig. 6. Correlation heatmap for variables

as the amperage increases. The relationships revealed in regression analysis are shown in Fig. 7.

Finally, this study examined the relationships between the number of strokes and abrasions. No significant changes were observed between the change in the number of strokes and the abrasions according to the ampere value (see Fig. 8). The increase in stroke time may cause a slight decrease in EWR and MRR.

The data set is divided into 80% training and 20% test set. Then, the one-hot encoding transformation was applied to the data set due to the categorical data type of input variables. This study applied the decision trees with high learning performance for both workpiece and electrode wear. The resulting performance values for the entire dataset are presented in Table 5. Accordingly, it has been shown that the decision trees model gives better results than other algorithms. Although decision trees and random forest algorithms give close results, the k-nearest neighbors algorithm performs poorer.

The regression plot of the test set for the wear on the electrode material is given in Fig. 9.

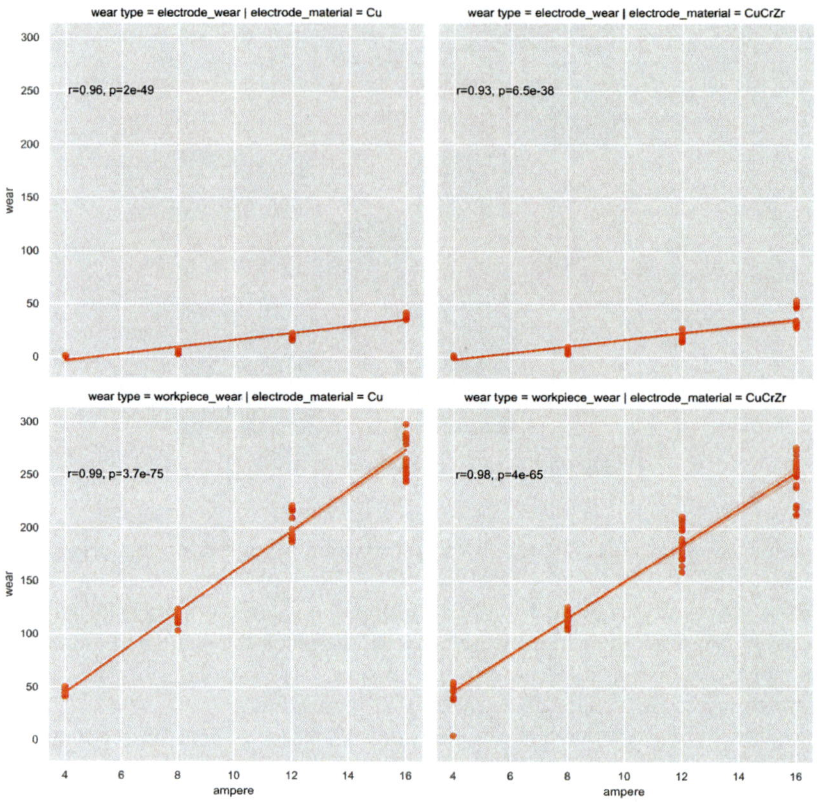

Fig. 7. Regression plots

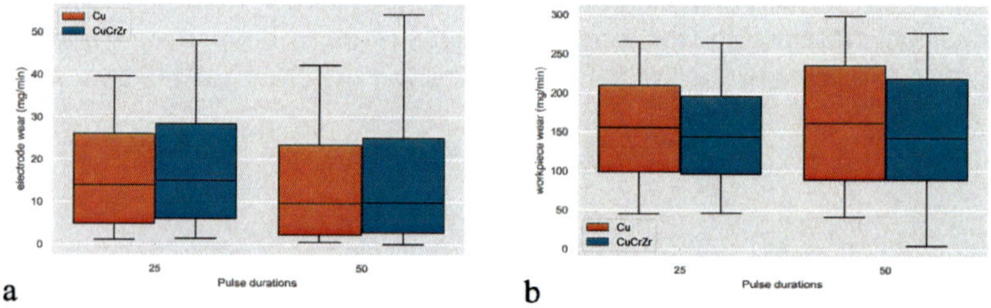

Fig. 8. Pulse durations vs. A) electrode B) workpiece wears

Table 5. Comparisons of machine learning algorithm performances

Model	Adjusted R-Squared		R-Squared		RMSE		CPU	
	Electrode	Workpiece	Electrode	Workpiece	Electrode	Workpiece	Electrode	Workpiece
Decision Tree	0.9891	0.9612	0.9907	0.9668	1.3520	12.7681	0.0240	0.2059
Random Forest	0.9806	0.9556	0.9834	0.9620	1.8028	13.6568	0.1595	0.0156
k-Nearest Neighbors	0.8971	0.9186	0.9118	0.9302	4.1535	18.5029	0.0210	0.0201

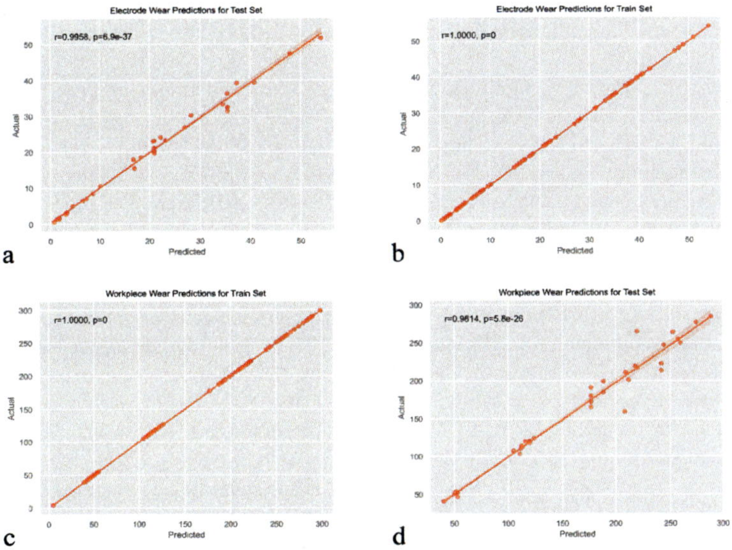

Fig. 9. Performance results for A) electrode wear test set, B) train set, C) workpiece wear test set, D) train set

4 Conclusions

The present experimental study has provided insights into the performance of cryogenically treated and untreated CuCrZr and Cu electrodes used in the EWR and MRR of AISI P20 tool steel. By comparing the performance of treated and untreated electrodes at different time intervals, we have shown that cryogenic treatment can improve the performance of CuCrZr electrodes in terms of EWR when the treatment time is less than 8 h. However, when the treatment time exceeds 8 h, the EWR performance of CuCrZr electrodes decreases significantly. On the other hand, the cryogenic treatment does not significantly impact the performance of Cu electrodes in terms of electrode wear.

Moreover, our findings have shown that changes in current values of 4, 8, 12, and 16 lead to a significant increase in EWR and MRR values for both types of electrodes. We have also demonstrated that decision trees, random forest, and k-nearest neighbors algorithms from machine learning techniques can be adapted for regression analysis, which can be useful for predicting the performance of electrodes in EDM processing. Overall, our study contributes to the literature on the use of cryogenic treatment and machine learning techniques for improving the performance of electrodes in EDM processing.

References

1. Ho, K.H., Newman, S.T.: State of the art electrical discharge machining (EDM). Int. J. Mach. Tools Manuf. **43**(13), 1287–1300 (2003). https://doi.org/10.1016/S0890-6955(03)00162-7
2. Ramaswamy, G.A., Krishna, A., Gautham, M., Sudharshan, S.S., Gokulachandran, J.: Optimisation and prediction of machining parameters in EDM for Al-ZrO_2 using soft computing techniques with Taguchi method. IJPMB **11**(6), 864 (2021). https://doi.org/10.1504/IJPMB.2021.118323
3. Sarıkaya, M., Yılmaz, V.: Optimization and predictive modeling using S/N, RSM, RA and ANNs for micro-electrical discharge drilling of AISI 304 stainless steel. Neural Comput. Appl. **30**(5), 1503–1517 (2018). https://doi.org/10.1007/s00521-016-2775-9
4. Balasubramaniam, V., Baskar, N., Narayanan, C.S.: Optimization of electrical discharge machining parameters using artificial neural network with different electrodes. In: 5th International & 26th All India Manufacturing Technology, Design and Research Conference (2014)
5. Ong, P., Chong, C.H., bin Rahim, M.Z., Lee, W.K., Sia, C.K., bin Ahmad, M.A.H.: Intelligent approach for process modelling and optimization on electrical discharge machining of polycrystalline diamond. J. Intell. Manuf. **31**(1), 227–247 (2020). https://doi.org/10.1007/s10845-018-1443-6
6. Cakir, M.V., Eyercioglu, O., Gov, K., Sahin, M., Cakir, S.H.: Comparison of soft computing techniques for modelling of the EDM performance parameters. Adv. Mech. Eng. **5**, 392531 (2013). https://doi.org/10.1155/2013/392531
7. Arunadevi, M., Prakash, C.P.S.: Predictive analysis and multi objective optimization of wire-EDM process using ANN. Mater. Today: Proc. **46**, 6012–6016 (2021). https://doi.org/10.1016/j.matpr.2020.12.830
8. Ulas, M., Aydur, O., Gurgenc, T., Ozel, C.: Surface roughness prediction of machined aluminum alloy with wire electrical discharge machining by different machine learning algorithms. J. Mark. Res. **9**(6), 12512–12524 (2020). https://doi.org/10.1016/j.jmrt.2020.08.098
9. Jatti, V.S., Dhabale, R.B., Mishra, A., Khedkar, N.K., Jatti, V.S., Jatti, A.V.: Machine learning based predictive modeling of electrical discharge machining of cryo-treated NiTi, NiCu and BeCu alloys. ASI **5**(6), 107 (2022). https://doi.org/10.3390/asi5060107
10. Nahak, B., Gupta, A.: A review on optimization of machining performances and recent developments in electro discharge machining. Manuf. Rev. **6**, 2 (2019). https://doi.org/10.1051/mfreview/2018015
11. Cetin, A., Cakir, G., Aslantas, K., Ucak, N., Cicek, A.: Performance of cryogenically treated Cu and CuCrZr electrodes in an EDM process. Kovove Materialy **55**(6) (2017)

Ensuring Stability and Automatic Process Control with Deburring Process in Cast Z-Rot Parts

Muhammed Abdullah Özel(✉) and Mehmet Yasin Gül

AYD Automotive Industry R&D Center, Konya, Turkey
muhammedabdullahozel@gmail.com

Abstract. The undesirable protrusions and roughness that occur on the surfaces of the parts during or after the production process are called burrs, and the process applied to remove these burrs is called "burring". Burrs are mostly formed around the cut edges. A little more material than the measure is transferred to the mold so that the metal is not missing in a piece to be poured into the mold. This excess, when the molds are compressed, overflows from the joints and creates casting burrs. Post-production burr removal is an important issue in the manufacturing industry. It is necessary to clean the burrs formed after the production in the z-rod parts produced by the casting method. Burrs formed in the holes cause material problems and connection problems. Burrs cause stress at the hole corners due to high stress, reducing cracking resistance and fatigue life. In mating parts, burrs will enter the connector seat and damage the connector or assembly. Burrs in the holes will also affect the coating thickness on the rough surfaces, thus increasing the risk of corrosion. Burrs on moving parts increase unwanted friction and heating. Currently, this process is generally cleaned manually by the operator using a deburring tool (blaster) or it can be cleaned using high cost and special features robot arm integrated systems. Manual cleaning causes a significant loss of productivity in the labor factor. On the other hand, deburring operation and post-operation product control vary according to the operator's competence and initiative. Therefore, the cleaning operation is not always at the same standards. With this study, a system was designed for deburring, apparatus suitable for the dimensions of the z-rod parts. And the deburring process was carried out in a standard way for mass production. At the same time, the process accuracy was instantly checked with the highest accuracy by using the image processing algorithms with the Python programming language, and the burr types were determined by detecting the burr form and size with the deep learning method. System cleaning efficiency was measured as 92% higher than manual cleaning.

Keywords: Trimming · Image Processing · Quality Control · Deburring

1 Introduction

Casting technology used in production offers cost and production time advantages compared to other production processes. Unwanted protrusions that occur on the surfaces of the parts during or after manufacturing are called burrs. Burrs appear as an undesirable

Z. Şen et al. (Eds.): IMSS 2023, LNME, pp. 27–35, 2024.
https://doi.org/10.1007/978-981-99-6062-0_4

formation in all manufacturing processes. Despite many efforts to remove burrs, burr-free production is almost impossible and burrs more or less inevitably appear. Removing burrs after production is a major problem. Especially in the production of sensitive systems, burrs are a big problem and need to be cleaned. On the other hand, burrs can occur in many different forms depending on the type of production, material, cutting tool geometry and process parameters. The use of personnel to clean the burrs increases both the production cost of some parts and the production time of the part, and at the same time, when the burr cleaning control is done by the personnel, it is not a stable production as the desired cleaning cannot be done completely, it varies according to the operator's competence and initiative. An example of burr can be seen in Fig. 1.

Fig. 1. Cast z-rod burr defect

Within the scope of the study, literature research was conducted for deburring and automatic control operation. The use of robot technology for this process in the literature significantly reduces these problems. However, the robot to be designed must have special qualifications for this job. On the other hand, in the literature review, no study was found on the control of the deburring process and the classification of burrs. In this context, data were collected and appropriate product selection and machine design were made.

2 Material and Method

The most important point to be considered in this study is the cutting force that occurs at the contact point of the deburring tool on the z-rod part. If the applied force is more than the required force, it may damage the z-rod body. This may appear as negative burrs and cause the part to fail. If the applied force is less than the required force, the burr may not be cleaned sufficiently. For these reasons, it is necessary to control the force and apply a force of appropriate value. Thanks to the pneumatic cylinder, proportional regulator and load cell, the deburring operation and deburring control operations of the cast z-rod parts are carried out precisely with load control. Delta programmable logic controller is used as the controller. In Fig. 2, the equipment used for controller and load control is given.

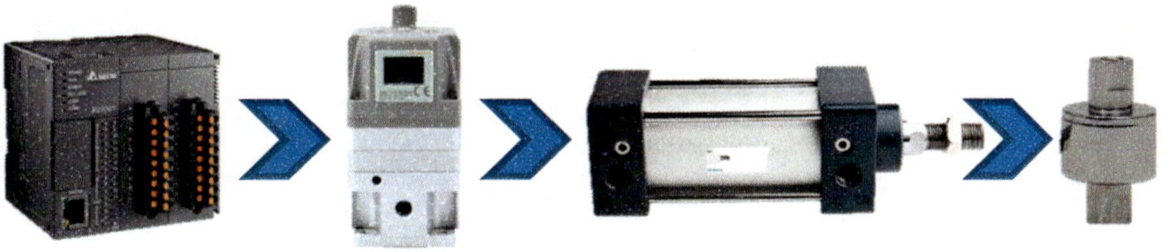

Fig. 2. Load control parts

2.1 Mechanical Design

There are several different aspects to consider in mechanical design. These cases are, respectively, the alignment of the camera positioning to the part control centers, the design of the apparatus suitable for the inner diameter of the produced cast z-rods to be cleaned, the location of the detection sensor to be positioned at the cleaning point, and the part bearings being suitable for the part form. Mechanical design was made with Solidworks program and appropriate product selection and positioning points were made with precision. Figure 3 shows the mechanical design.

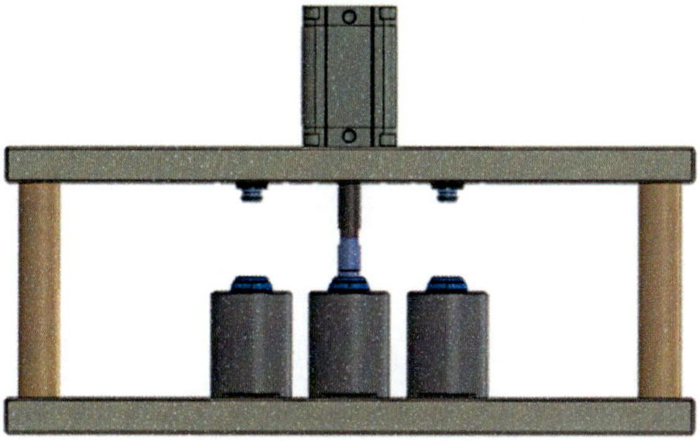

Fig. 3. Mechanical design

2.2 Algorithm

The working logic of the system is as follows; The cameras are activated with the input to the programmable logic controller waiting for the start command. The part placed in the first slot is detected by the camera and it checks whether the cleaning apparatus suitable for the inner diameter of the z-rod is selected. If the internal diameter matching with the apparatus fails, the PLC waits for the restart command. If the internal diameter match with the apparatus gives a successful result, the control process is completed in the first slot and the PLC waits for information from the second slot sensor. Cleaning is not performed until the sensor detects the part. When the part is detected, the cylinder carries out the cleaning process in a load-controlled manner with the proportional regulator, while at the same time, information transfer takes place between the load cell and the proportional regulator. After the process, the part is moved to the 3rd slot. The system performs burr control and classification in the 3rd slot. If burrs are detected, the burr classification results are recorded as small, medium and large, respectively, and the cleaning process is expected to be performed again. If no burrs are detected, the cleaning process is complete. The flow chart of the system is given in Fig. 4.

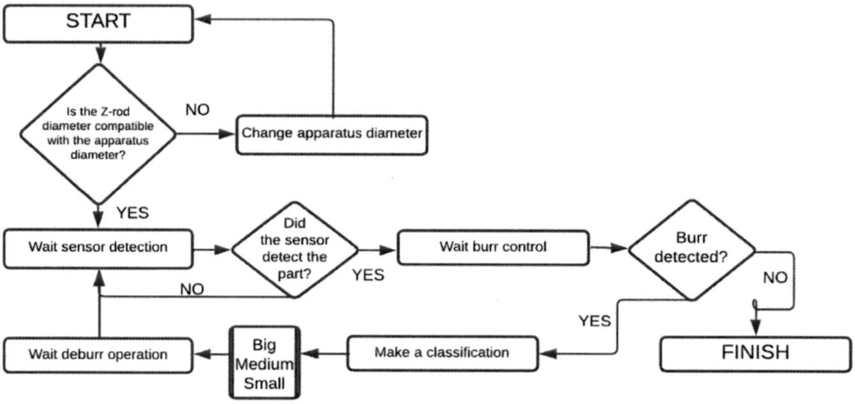

Fig. 4. Flow chart

2.3 Deep Learning

In order to detect casting burr with deep learning, image data of casting burr must be collected first. This data must be labelled after collection, ie the casting burr of each image must be matched to a specific label. In the study, casting burrs are divided into small, medium and large classes. Next, an artificial neural network should be trained using these data. This mesh is used to detect casting burrs. YOLO-v7 (You Only Look Once) algorithm is preferred for this training. The most important reason why YOLO algorithm is preferred is that it works faster than other popular object recognition algorithms and can process images with higher resolution. YOLO-v7 is not just object detection architecture. It is a new model that can extract key points (skeletons) and perform sample segmentation as well as non-standard only bounding box regression. YOLO v7 also has a higher resolution than the previous versions. It processes images at a resolution of 608 by 608 pixels, which is higher than the 416 by 416 resolution used in YOLO v3. This higher resolution allows YOLO v7 to detect smaller objects and to have a higher accuracy overall.

First of all, it should be checked whether there is sufficient equipment for training processes. In cases where the hardware is insufficient, work can be done on the Google Colab platform and Theos AI platform. The Theos AI platform can be quickly used for model training, tagging and statistics. The working directory containing the YOLO-v7 project files must be uploaded via the Github platform. Theos-v1 is installed via

Google Colab by installing the necessary libraries. Login using username and password. Verification is done by entering the project key from the settings tab on the Theos platform. At the same time, Theos and Google Colab platform must be paired. After all these processes, the algorithm, version and weight files are determined and imported respectively. Input and output data are introduced as photos, videos or real time. After all these processes, labelling processes are started and final preparations are made for training. First, image data were collected from the parts coming out of the casting line. The collected image data is divided into 3 classes. These classes were determined as small burrs, medium burrs and large burrs, respectively. The biggest contribution of burr classification is that acute or chronic problems that may occur in the production line can be detected retrospectively. The obtained 1000 image data are labelled as classified. Label values include the coordinate information of the object to be detected and the label name. Figure 5 shows the label style and value descriptions.

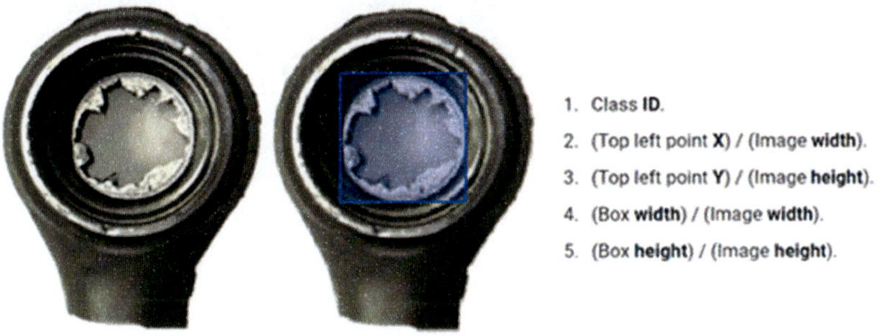

Fig. 5. Medium burr label

Label data and image data were randomly clustered as 70% training data, 15% validation data, and 15% test data. Four different versions were tried using the common dataset for the training process. In order to determine the combination with the highest velocity/accuracy correlation among these versions, the trainings were started by selecting V7-Default, V7-Tiny, V7-X and V7-W6, respectively. According to the results of the training process, 4 different version trials were successfully completed. In each version, the detection process was carried out successfully. When evaluating between versions, the V7-W6 algorithm was chosen as the most successful algorithm in terms of speed and accuracy correlation. The following is the example for Table 1.

2.4 Image Processing

Sobel, Canny, Prewitt and Laplacian edge detection operators are algorithms that are frequently used in image processing. The Canny algorithm uses multiple steps to identify the edges in the image. First of all, noise filtering is done, then the brightness changes are followed and finally the edges are determined. Figure 6 shows the Canny application and its output.

Table 1. YOLO version compare

Version	mAP/FPS	Approved
V7-Default	%85	No
V7-Tiny	%72	No
V7-X	%86	No
V7-W6	%92	Yes

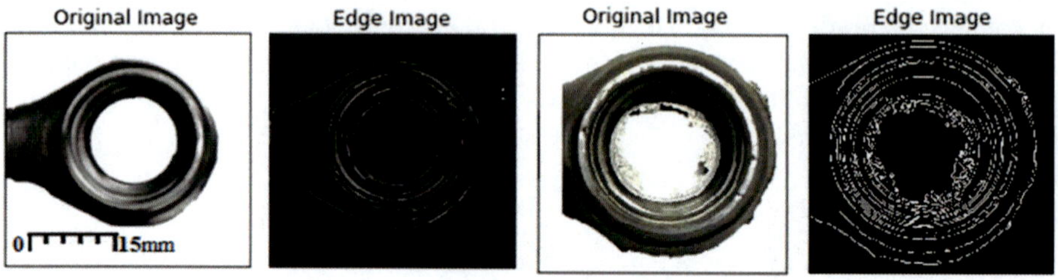

Fig. 6. Canny operator

The Laplacian algorithm determines the edges using the second derivative of the values of the pixels in the image. This algorithm reduces the need for noise filtering and better identifies true edges. Figure 7 shows the Laplacian application and its output.

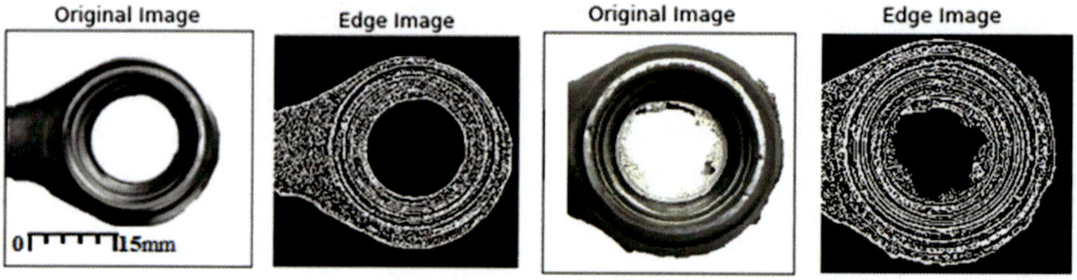

Fig. 7. Laplacian operator

The Sobel algorithm determines the edges in each pixel of the image using mathematical operations with the values of the pixels around the pixel. Figure 8 shows the Sobel application and its output.

The Roberts operator determines the edges by tracking the brightness changes between pixels in the image. Figure 9 shows the Roberts application and its output.

Laplacian offers some advantages over other popular edge detection methods. After the algorithm trials, the Laplacian operator, which gives the most accurate results and has advantages, was preferred for the study. The Laplacian algorithm increases the clarity of the edges, allowing them to be detected more accurately. It also increases the accuracy of the edges, preventing less misleading edges. By increasing the sensitivity of the edges,

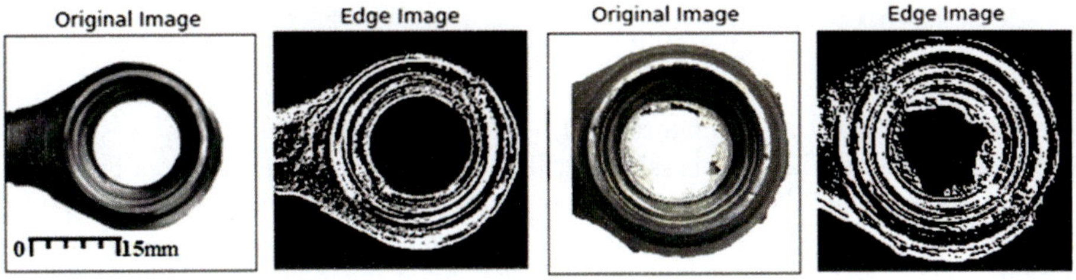

Fig. 8. Sobel operator

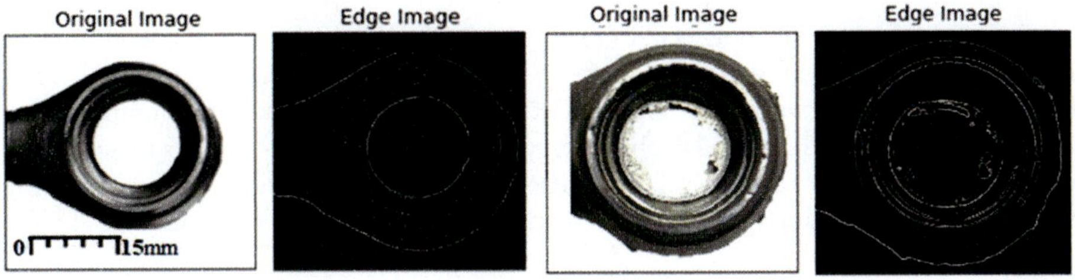

Fig. 9. Roberts

it also enables the detection of weaker edges and uses less computer resources than other methods. Laplacian operator is applied on the captured image by using Python programming language, "OpenCV" and "NumPy" library. The circle was determined on this image and its diameter was calculated and drawn on the screen and at the same time the diameter value was printed on the screen. Figure 10 shows the diameter detect application and its output.

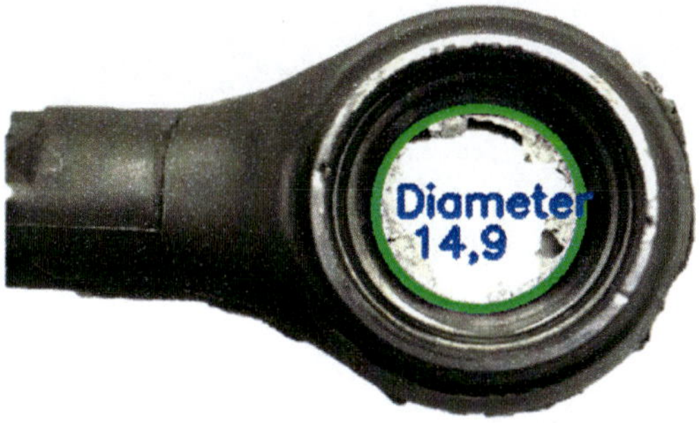

Fig. 10. Diameter determination after Laplacian operator

3 Conclusion

In this study, deburring steps of cast z-bar parts were automated and operator errors were prevented. Thanks to the automatic deburring system, the operation process period decreased. Thanks to image processing and deep learning, the processing steps have been successfully validated. When the process steps are compared, efficiency of the system developed in this study is higher than that of former manual technology about 90%. This system, which is equipped and designed according to the inner diameter of the cast z-rod parts, has a modular design so that it can be adapted to every product group. When the apparatus is made according to the points determined for the part centers, it can be used for geometries that do not contain large differences.

One of the biggest advantages of this study is to obtain a regular and stable production output. With this system, which completely eliminates the competence and initiative of the operator, the operation start and end steps have been successfully completed by using the image processing and deep learning algorithm YOLO-v7, thanks to automatic apparatus control and automatic burr control. Classified burr data provided a guiding opportunity for causation of casting problems. Figure 11 shows the machine final.

Fig. 11. Machine final

References

1. Aydın, H.: 5 eksenli çapak alma robotunun kontrolü, Master's thesis, ESOGÜ, Fen Bilimleri Enstitüsü (2018)
2. Karaçalı, H.: Beş eklemli çapak alma robotu tasarımı, Master's thesis, ESOGÜ, Fen Bilimleri Enstitüsü (2012)
3. El Naser, Y.H., Karayel, D., Ozkan, S.S., Atali, G.: Talaşlı İmalatta Otomatik Çapak Alma İşlemi için Endüstriyel Robot Kol Tasarımı. In: 5th International Symposium on Innovative Technologies in Engineering and Science, ISITES 2017, Baku-Azerbaijan, 29–30 September 2017 (2017)
4. El Naser, Y.H., Atali, G., Karayel, D., Özkan, S.S.: Prototyping an industrial robot arm for deburring in machining. Acad. Platf.-J. Eng. Sci. **8**(2), 304–309 (2020)
5. Kosler, H., Pavlovčič, U., Jezeršek, M., Možina, J.: Adaptive robotic deburring of die-cast parts with position and orientation measurements using a 3D laser-triangulation sensor. Strojniški vestnik-J. Mech. Eng. **62**(4), 207–212 (2016)
6. Nagata, F., Hase, T., Haga, Z., Omoto, M., Watanabe, K.: CAD/CAM-based position/force controller for a mold polishing robot. Mechatronics **17**(4–5), 207–216 (2007)
7. Norberto Pires, J., Ramming, J., Rauch, S., Araújo, R.: Force/torque sensing applied to industrial robotic deburring. Sens. Rev. **22**(3), 232–241 (2002)
8. Sokolov, A.: Robot motion realisation using LabVIEW. Period. Polytech. Mech. Eng. **43**(2), 131–145 (1999)

Nearest Centroid Classifier Based on Information Value and Homogeneity

Mehmet Hamdi Özçelik[1,2(✉)] and Serol Bulkan[2]

[1] Applied Analytics AA, İstanbul, Türkiye
hamdi.ozcelik@appliedanalytics.com.tr, hamdiozcelik@marun.edu.tr
[2] Department of Industrial Engineering, Marmara University, İstanbul, Türkiye
sbulkan@marmara.edu.tr

Abstract. The aim of this paper is to introduce a novel classification algorithm based on distance to class centroids with weighted Euclidean distance metric. Features are weighted by their predictive powers and in-class homogeneities. For predictive power, information value metric is used. For in-class homogeneity different measures are used. The algorithm is memory based but only the centroid information needs to be stored. The experimentations are carried at 45 benchmark datasets and 5 randomly generated datasets. The results are compared against Nearest Centroid, Logistic Regression, K-Nearest Neighbors and Decision Tree algorithms. The parameters of the new algorithm and of these traditional classification algorithms are tuned before comparison. The results are promising and has potential to trigger further research.

Keywords: Machine Learning · Classification · Similarity Classifier · Nearest Centroid · Information Value

1 Introduction

As one of the most important theorems in statistical learning, the no free lunch theorem [1] states that the performance of an algorithm could be better than others at some problems and worse at some others, which leads to some type of equivalence among algorithms. Due to the natural differences among classification problems many different classifiers are designed so far. In this paper we introduce a novel variant of Nearest Centroid (NC) classifier [2].

In our new algorithm, the distance measure is a weighted Euclidean metric where the predictive power of each feature and homogeneity of each feature at each class are used to determine weights. Information Value (IV) metric is selected as the metric showing the predictive power of features. For measuring homogeneity, mean absolute deviation, standard deviation, variance and coefficient of variance metrics are used. The choice of these metrics became a parameter of our algorithm and is used at the tuning phase.

A binary classification model provides predicted binary classes, predicted probabilities of each class or rank information of these probabilities and different performance measures are used for evaluation [3]. We have used accuracy measure to compare the

Z. Şen et al. (Eds.): IMSS 2023, LNME, pp. 36–45, 2024.
https://doi.org/10.1007/978-981-99-6062-0_5

performance of the algorithm. For benchmarking, we used 50 different datasets and 4 different algorithms, namely Nearest Centroid (NC), Logistic Regression (LR), K-Nearest Neighbours (KNN) and Decision Tree (DT). The new algorithm outperformed NC at 43 datasets, LR at 8 datasets, KNN at 17 datasets and DT at 15 datasets. It was the best classifier at 5 datasets with respect to accuracy measure.

2 Related Research

Nearest Centroid Classifier [2] is a memory-based classifier which is simple and fast. It stores the centroid information of each class and then makes distance calculations for each new instance. The nearest centroids' class is assigned as the predicted class of that instance; therefore, the algorithm is also called as Minimum Distance Classifier (MDC).

Instead of class centroids, some instances at the training dataset could be chosen to represent that class [4]. These instances are called as prototypes.

Nearest shrunken centroid classifier [5] is another variant of Nearest Centroid algorithm. It shrinks each of the class centroids toward the overall centroid for all classes by an amount called threshold. After the centroids are shrunken, a new instance is classified by nearest centroid rule, but using shrunken class centroids.

Nearest Centroid Classifier, K-Means Clustering [6] and K-Nearest Neighbours (KNN) [7] algorithms are closely related with each other since they consider centroid information. KNN and its variants [8] are used for both classification and regression problems. Using weights for centroid based distances are common and so far, different measures are used. For example, Gou et.al. [9] proposed an algorithm which captures both the proximity and the geometry of k-nearest neighbours. At another recent research, Elen et.al. [10] proposes a new classifier by addressing the noise issue at classification by introducing standardized variable distances.

Baesens et.al. [11] compared the performance of different algorithms over some datasets and two of them are in our list, namely "Australian" and "credit-g (German Credit)". For the Australian dataset, the accuracy of our new algorithm is the third best one, after the algorithms "C4.5rulcs dis" and "C4.5dis". For the German Credit dataset, the accuracy of our new algorithm (76.40) is above the best algorithm tested (LDA with a value of 74.6). Lessmann et.al. [12] updated this research by testing various classification algorithms over 8 datasets.

The novelty of our algorithm is at the usage of both information value and homogeneity metrics for weighting the distance calculation.

3 Methods

3.1 Information Value

Information Value (IV) metric is used to determine the predictive power of each feature. Its calculation is based on "Weight of Evidence" values [13]. The Weight of Evidence (WOE) of a feature value is defined as the natural logarithm of the ratio of the share of one class in a given value over the share of other class as shown in Eq. 1.

$$WoE(X_i = X_{ij}) = ln\left(\frac{share\ of\ responses\ of\ that\ category\ at\ all\ responses}{share\ of\ nonresponses\ of\ that\ category\ at\ all\ nonresponses}\right) \quad (1)$$

Equation 2 gives the computation of Information Value over the Weight of Evidence values. The differences of percentage shares of classes are used as weights at the summation.

$$IV(X) = \sum_{j=1}^{V} (Distribution\ of\ class\ 1 - Distribution\ of\ class\ 0) \cdot WoE_j \quad (2)$$

Information Value could be calculated only for categoric variables. For that reason, we split continuous variables into 10 equal sized bins and then made the calculation.

Both Information Value and Weight of Evidence metrics are widely used at data analysis for credit scoring at the banking sector.

3.2 Homogeneity Metrics

For each feature, the following sparsity metrics are computed to represent the inverse of homogeneity within each class:

1. Mean absolute deviation
2. Standard deviation
3. Variance
4. Coefficient of variation

All of them are used at the training phase to fit classifier model into dataset. As the last step the one with the highest accuracy is marked as the selected homogeneity metric. At the tuning phase of the algorithm, in addition to these 4 choices, using no metric for homogeneity is also checked and it is selected when its accuracy is highest. Therefore, there were 5 alternatives for homogeneity.

3.3 The Algorithm

At the training phase, first, the centroids of each class, information values of each feature and homogeneity values of each feature at each class are computed. Then, for the given dataset, the best homogeneity metric is determined by running the scoring at the training dataset using the accuracy metric. Figure 1 (a) depicts the flow of training phase.

At the scoring phase, for each new instance, the Euclidean distance to each class centroid is calculated where the weights are determined as the division of the feature's Information Value to the selected sparsity metric. The class with the minimum weighted distance becomes the predicted class for that instance. Figure 1 (b) depicts the flow of scoring phase.

The algorithm is a memory-based algorithm but it does not require storing the instances at the memory, instead only class summaries (i.e., centroid vectors) need to be stored. We used only datasets of binary classification problems but the algorithm could also be used for multiclass classification problems.

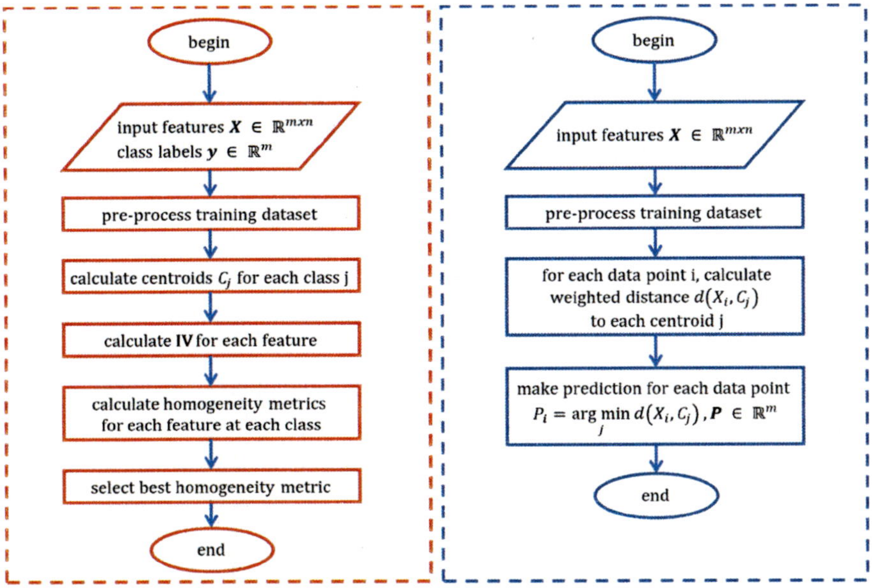

(a) Training phase (b) Scoring phase

Fig. 1. Algorithm flowcharts

4 Experimentation and Results

4.1 Setup

All experiments are performed on a PC equipped with 4 Core Intel i7 11th Gen CPU at 2.80 GHz and 32 GB RAM running Microsoft Windows 11 Pro, Anaconda 3, Conda 22.9.0, Jupyter-notebook 6.4.12 and Python 3.9.13.

4.2 Datasets

We used OpenML [14] machine learning repository which is a public online platform built for scientific collaboration. We also used its Python API [15] to access datasets at the repository. We downloaded 45 binary classification problem datasets using this API. We also generated 5 synthetic datasets for classification via "make_classification" function of Scikit-Learn library [16] and named with a prefix "random". Table 1 shows the datasets used at the experiments.

The second column ("data_id") refers to the unique identifier at OpenML repository. Table 1 also lists the number of instances and features at each dataset. Since we pre-processed data, the numbers of features are changed and the final number of features are given at the last column of the table.

Table 1. List of datasets

Dataset	data_id	İnstances	features	final features
Adult	179	48842	13	109
adult-census	1119	32561	13	97
Australian	40981	690	13	34
bank-marketing	1461	45211	15	42
banknote-authentication	1462	1372	3	4
blood-transfusion-service-center	1464	748	3	4
breast-cancer	13	286	8	42
breast-w	15	699	8	9
Churn	40701	5000	19	29
Click_prediction_small	1220	39948	8	9
climate-model-simulation-crashes	1467	540	19	20
credit-approval	29	690	14	38
credit-g	31	1000	19	50
Diabetes	37	768	7	8
eeg-eye-state	1471	14980	13	14
Electricity	151	45312	7	13
Elevators	846	16599	17	18
heart-c	49	303	12	18
heart-statlog	53	270	12	13
hill-valley	1479	1212	99	100
İlpd	1480	583	9	10
İonosphere	59	351	33	34
jm1	1053	10885	20	21
kc1	1067	2109	20	21
kc2	1063	522	20	21
kc3	1065	458	38	39
kr-vs-kp	3	3196	35	38
MagicTelescope	1120	19020	9	10
mozilla4	1046	15545	4	5
Musk	1116	6598	166	267
ozone-level-8h	1487	2534	71	72
pc1	1068	1109	20	21

(continued)

Table 1. (*continued*)

Dataset	data_id	Instances	features	final features
pc2	1069	5589	35	36
pc3	1050	1563	36	37
pc4	1049	1458	36	37
PhishingWebsites	4534	11055	29	38
Phoneme	1489	5404	4	5
qsar-biodeg	1494	1055	40	41
random1	−1	5000	19	20
random2	−2	5000	19	20
random3	−3	5000	19	20
random4	−4	5000	19	20
random5	−5	5000	19	20
Scene	312	2407	298	299
Sick	38	3772	28	31
Sonar	40	208	59	60
Spambase	44	4601	56	57
steel-plates-fault	1504	1941	32	33
tic-tac-toe	50	958	8	18
Wdbc	1510	569	29	30

4.3 Pre-Processing

The following pre-processing steps are applied to all datasets:

- Splitting: Each dataset is randomly divided into training and test datasets where the share of test dataset set to 25%.
- Null Value Imputation: Null values were replaced by zero values.
- Winsorization: For numeric features of the training dataset, 5% cut-off values are calculated from each end. The values beyond these limits were replaced by the cut-off values. Winsorization is applied only when the number of unique values of the feature is greater than 60.
- Min-Max Scaling: All numeric feature values are proportionally scaled into [0,1] range.
- One-hot encoding: For each distinct value of a categoric feature, a new flag variable is created.

4.4 Tuning

For a healthy comparison, the hyperparameters of Logistic Regression, KNN and Decision Tree algorithms are tuned with the options shown in Table 2. For each dataset, the tuning is made with an exhaustive search over these parameters via the "GridSearchCV" function of Scikit-learn library [16].

Table 2. Parameters used at the tuning

Algorithm	Parameter	Values
Logistic Regression	Penalty	l1','l2','elasticnet','none'
	C	0.01, 0.1, 1.0, 10, 100
	Solver	newton-cg', 'lbfgs', 'liblinear', 'sag', 'saga'
KNN	n_neighbors	5,10,30
	Weights	uniform','distance'
Decision Tree	min_samples_leaf	30, 50, 100
	Criterion	gini', 'entropy', 'log_loss'

4.5 Results

Table 3 shows the accuracy values of 5 algorithms over the test datasets. The proposed algorithm **outperforms all other** four algorithms at 5 datasets and **shares the first place** at another 3 datasets. The algorithm has a better accuracy score than nearest centroid at 43 datasets. It was better than Logistic Regression, KNN and Decision Tree algorithms 8, 17 and 15 datasets respectively.

We calculated the correlation coefficients among the accuracy values of our algorithm with the accuracy values of others. Our algorithms accuracy values over test datasets are correlated with the ones of Nearest Centroid, Logistic Regression, KNN and Decision Trees by 0.28, 0.25, 0.30 and 0.35 respectively.

Table 3. Accuracy of algorithms at test datasets

Dataset	NC	LR	KNN	DT	New Algorithm
Adult	0.7236	**0.8507**	0.8287	0.8476	0.8394
adult-census	0.7255	**0.8435**	0.8322	0.8430	0.8348
Australian	0.8786	0.8624	0.8590	0.8416	**0.8902**
bank-marketing	0.7282	**0.9017**	0.8913	0.8991	0.8922

(*continued*)

Table 3. (*continued*)

Dataset	NC	LR	KNN	DT	New Algorithm
banknote-authentication	0.8367	0.9918	**0.9988**	0.9493	0.8542
blood-transfusion-service-center	0.7380	0.7561	0.7626	**0.7733**	0.7594
breast-cancer	0.6806	0.7028	0.6667	0.7056	**0.7222**
breast-w	0.9657	0.9646	0.9749	0.9280	**0.9771**
Churn	0.6776	0.8648	0.8827	**0.9400**	0.8408
Click_prediction_small	0.5586	**0.8320**	0.8014	0.8239	0.7548
climate-model-simulation-crashes	0.7407	0.8963	0.9126	**0.9185**	0.8889
credit-approval	0.8555	0.8509	0.8335	0.8474	**0.8728**
credit-g	0.7360	0.7304	0.7304	0.7096	**0.7640**
Diabetes	0.7708	**0.7740**	0.7563	0.7563	0.7448
eeg-eye-state	0.5848	0.6529	**0.9503**	0.7930	0.6326
Electricity	0.7039	0.7579	0.8496	**0.8531**	0.7383
Elevators	0.7499	**0.8760**	0.8097	0.8343	0.7629
heart-c	0.8026	**0.8474**	0.8132	0.7237	0.8026
heart-statlog	0.7794	**0.8088**	0.7853	0.6735	**0.8088**
hill-valley	0.4455	**0.9208**	0.5116	0.5201	0.4785
Ilpd	0.6233	**0.7288**	0.6890	0.6836	0.6918
Ionosphere	0.7159	**0.9000**	0.8159	0.8932	0.8409
jm1	0.7241	**0.8168**	0.8018	0.8048	0.7535
kc1	0.7727	0.8477	**0.8481**	0.8356	0.7973
kc2	0.8321	**0.8519**	0.8366	**0.8519**	0.8321
kc3	0.8000	0.8852	0.8957	**0.9026**	0.8348
kr-vs-kp	0.8511	**0.9730**	0.9602	0.9692	0.8836
MagicTelescope	0.7586	0.7898	0.8420	**0.8451**	0.7819
mozilla4	0.7335	0.8502	0.8920	**0.9420**	0.8004
Musk	0.7291	**1.0000**	0.9842	0.9565	0.9988
ozone-level-8h	0.6877	0.9391	**0.9423**	0.9252	0.9117
pc1	0.7950	0.9266	**0.9403**	0.9317	0.8921
pc2	0.8777	0.9957	**0.9963**	0.9963	0.0043

(*continued*)

Table 3. (*continued*)

Dataset	NC	LR	KNN	DT	New Algorithm
pc3	0.7647	0.8859	**0.8885**	0.8788	0.8721
pc4	0.7699	**0.9134**	0.8740	0.8827	0.8849
PhishingWebsites	0.9045	0.9263	**0.9581**	0.9336	0.9157
Phoneme	0.7365	0.7504	**0.8934**	0.8278	0.7757
qsar-biodeg	0.7803	**0.8598**	0.8311	0.7955	**0.8598**
random1	0.8272	0.8453	**0.9261**	0.8587	0.8208
random2	0.7568	0.7806	**0.9026**	0.8189	0.8312
random3	0.9000	0.9437	**0.9550**	0.9064	0.9216
random4	0.7496	0.7770	**0.8978**	0.8200	0.7736
random5	0.7592	0.8010	**0.8885**	0.8198	0.7640
Scene	0.7292	**0.9864**	0.9140	0.9535	0.9153
Sick	0.7614	**0.9578**	0.9565	0.9659	0.9502
Sonar	0.6923	0.7808	**0.8346**	0.7231	0.6346
Spambase	0.8888	**0.9361**	0.9333	0.9003	0.9253
steel-plates-fault	0.6461	**1.0000**	0.9835	**1.0000**	**1.0000**
tic-tac-toe	0.7000	0.9792	**0.9992**	0.7817	0.7292
Wdbc	**0.9790**	0.9678	0.9566	0.8825	0.9650

5 Discussion and Conclusion

Weight of Evidence and Information Value metrics are commonly used at banking to predict credit defaults and in our experiments, there were three datasets that are in credit risk domain, namely "Australian", "credit-approval" and "credit-g (German Credit)". In all of these three datasets, the new algorithm was better than all others. This could be further investigated using additional datasets from credit risk domain.

The new algorithm has clear superiority over Nearest Centroid algorithm and comparable performance to other classic algorithms. It could be improved by considering other characteristics such as the size of the classes.

To improve the performance of the algorithm various alternatives may be considered. For example, instead of Information Value, another measure of predictive performance such as variable importance could be used. Similarly, at the distance calculation, Euclidean formula could be replaced by other distance metrics such as Manhattan.

Over 50 datasets, the new algorithm outperforms the benchmark algorithms at 5 datasets and shares the first place at another 3 datasets. Therefore, the new algorithm is comparable to well-known algorithms and it should be enhanced further.

6 Declaration of Competing Interest

The authors declare that there is no conflict of interest identified in this study.

Acknowledgements. The authors declare that no funds, grants, or other support were received during the preparation of this manuscript.

References

1. Wolpert, D.H.: The supervised learning no-free-lunch theorems. Soft Comput. Ind. 25–42 (2002)
2. Hastie, T., Tibshirani, R., Friedman, J.H.: The Elements of Statistical Learning: Data Mining, Inference, and Prediction, vol. 2, p. 670. Springer, New York (2009)
3. Shmueli, G.: To explain or to predict? Stat. Sci. **25**(3), 289–310 (2010)
4. Kuncheva, L.I.: Prototype classifiers and the big fish: the case of prototype (instance) selection. IEEE Syst. Man Cybern. Mag. **6**(2), 49–56 (2020)
5. Tibshirani, R., Hastie, T., Narasimhan, B., Chu, G.: Diagnosis of multiple cancer types by shrunken centroids of gene expression. Proc. Natl. Acad. Sci. **99**(10), 6567–6572 (2002)
6. Hartigan, J.A., Wong, M.A.: Algorithm AS 136: a k-means clustering algorithm. J. Roy. Stat. Soc. Ser. C (Appl. Stat.) **28**(1), 100–108 (1979)
7. Cover, T., Hart, P.: Nearest neighbor pattern classification. IEEE Trans. Inf. Theory **13**(1), 21–27 (1967)
8. Alpaydin, E.: Voting over multiple condensed nearest neighbors. In: Aha, D.W. (eds.) Lazy Learning, pp. 115–132. Springer, Dordrecht (1997). https://doi.org/10.1007/978-94-017-2053-3_4
9. Gou, J., et al.: A representation coefficient-based k-nearest centroid neighbor classifier. Expert Syst. Appl. **194**, 116529 (2022)
10. Elen, A., Avuçlu, E.: Standardized Variable Distances: a distance-based machine learning method. Appl. Soft Comput. **98**, 106855 (2021)
11. Baesens, B., Van Gestel, T., Viaene, S., Stepanova, M., Suykens, J., Vanthienen, J.: Benchmarking state-of-the-art classification algorithms for credit scoring. J. Oper. Res. Soc. **54**, 627–635 (2003)
12. Lessmann, S., Baesens, B., Seow, H.V., Thomas, L.C.: Benchmarking state-of-the-art classification algorithms for credit scoring: an update of research. Eur. J. Oper. Res. **247**(1), 124–136 (2015)
13. Siddiqi, N.: Intelligent Credit Scoring: Building and Implementing Better Credit Risk Scorecards, pp.186–197. Wiley (2017)
14. Vanschoren, J., Van Rijn, J.N., Bischl, B., Torgo, L.: OpenML: networked science in machine learning. ACM SIGKDD Explor. Newslett. **15**(2), 49–60 (2014)
15. Feurer, M., et al.: OpenML-Python: an extensible Python API for OpenML. J. Mach. Learn. Res. **22**(1), 4573–4577 (2021)
16. Pedregosa, F., et al.: Scikit-learn: machine learning in Python. J. Mach. Learn. Res. **12**, 2825–2830 (2011)

Web-Based Intelligent Book Recommendation System Under Smart Campus Applications

Onur Dogan[1,2(✉)], Seyfullah Tokumaci[1], and Ouranıa Areta Hiziroglu[1]

[1] Department of Management Information Systems, Izmir Bakircay University, 35665 Izmir, Turkey
{onur.dogan,seyfullah.tokumaci,ourania.areta}@bakircay.edu.tr

[2] Department of Mathematics, University of Padua, Padua, Italy
onur.dogan@unipd.it

Abstract. Recommendation systems are essential as they help users to discover new books and resources and increase their engagement and satisfaction, thus improving the overall learning experience. This paper presents a web-based intelligent book recommendation system for smart campus applications at Izmir Bakircay University. The system is designed as an intelligent hybrid tool that combines collaborative and content-based filtering techniques to recommend books to users with methodological differences. It considers the user's reading history and preferences and integrates with other smart campus applications to provide personalized recommendations. The system is important for the digital transformation of smart campuses as it helps to make education more personalized, efficient, and data-driven. Also, it allows for the effective use of public resources. The effectiveness of the system was evaluated through user feedbacks, 22 users evaluated 148 books and the results showed that users responded positively to about 70% of the recommended books thus, it provided accurate and personalized recommendations.

Keywords: book recommendation · smart campus · intelligent system · web application

1 Introduction

The digital era has brought about a tremendous increase in the amount of data and information available online, leading to the development of sophisticated recommendation systems [1]. These systems use advanced algorithms and machine learning techniques to analyze and understand user behavior, preferences, and interactions with digital content. They then use this information to make personalized suggestions for products, services, and content that are most likely relevant and exciting to the user. The integration of recommendation systems in various digital platforms has become a key strategy for businesses looking to

Z. Şen et al. (Eds.): IMSS 2023, LNME, pp. 46–57, 2024.
https://doi.org/10.1007/978-981-99-6062-0_6

improve customer engagement and drive sales. It has also changed the way people discover and consume content, leading to a more personalized and convenient user experience [2].

Book recommendation systems are important in the digital era, where the abundance of books can lead to information overload and make it difficult for users to find books that match their preferences [3]. Recommendation systems can help to alleviate this problem by providing personalized recommendations based on the user's past behavior or attributes of the books. Additionally, in the context of the smart campus, recommendation systems can play an important role in improving the overall learning experience by providing personalized recommendations and increasing student engagement and satisfaction [4].

Recommending books has become an important research topic in the field of information retrieval and machine learning, as they can help users to discover new books and resources that they may not have otherwise found. The problem of book recommendation is essentially to predict a user's preferences for a set of items (in this case books) based on their past behavior (e.g. ratings, purchase history, reading history) or attributes of the items themselves. There are several factors that make book recommendation a challenging task, including the high dimensionality of the item space, the sparsity of user-item interactions, and the subjectivity of user preferences [5]. Additionally, the high number of books available and the fast-paced evolution of the book market make it challenging to provide up-to-date and relevant recommendations [6].

A web-based intelligent recommendation system is a system that suggests items or content to users over the internet, typically through a website or mobile app [7]. Advanced algorithms and machine learning methods are employed by these systems to examine and comprehend user actions, inclinations, and engagements with digital materials. Subsequently, this data is utilized to offer tailored recommendations for products, services, and content that are more likely to appeal and engage the user. A web-based book recommendation system suggests books to users based on their preferences and reading history. It typically uses a combination of techniques such as collaborative filtering, content-based filtering, and hybrid methods to make personalized recommendations [8].

Collaborative filtering is a method that looks at the reading habits of similar users and suggests books that they have enjoyed. For example, if a user A and B have similar reading history and user B read book X but not user A, then the system will recommend book X to user A. Content-based filtering is a method that suggests books based on the characteristics of the books that a user has previously read. For example, if a user has read several books with a similar theme or genre, the system will recommend other books with similar themes or genres. Hybrid methods use a combination of both collaborative filtering and content-based filtering.

Under smart campus applications, web-based intelligent book recommendation systems can play several roles [9,10]. *i) Personalized Learning*: By analyzing students' reading habits and preferences, book recommendation systems can suggest books that align with their interests and learning goals. *ii) Library*

Management: Book recommendation systems can be used to manage library collections and improve the user experience. By analyzing the reading habits of students and faculty members, libraries can purchase books that align with their interests and needs. Additionally, by suggesting books based on the user's history and preferences, it can improve the discovery of the library's collection. *iii) Curriculum Support*: Book recommendation systems can support the curriculum by offering texts that align with the course material. *iv) Research Assistance*: Book recommendation systems can help students and faculty members find relevant research materials. Suggesting books and articles based on a user's research topic can save time and effort in finding the needed information. *v) Student engagement*: Book recommendation systems can improve concentration and participation in reading-related activities. Books that align with the student's interests can increase their likelihood of reading and participating in book clubs or reading challenges. Overall, book recommendation systems can provide a more personalized and efficient way for students and faculty members to discover, access, and engage with the library's collection and other reading materials, and ultimately improving the learning experience and academic performance.

The study proposes a web-based intelligent book recommendation system that utilizes a combination of collaborative and content-based filtering techniques to recommend books to users based on their reading history and preferences. Additionally, the system integrates with some university information systems such as personnel database, student database to provide more personalized recommendations for each individual user. The goal is to demonstrate that the proposed system is able to provide accurate and personalized recommendations for users, and that it addresses the specific challenges and opportunities of recommendation systems in the context of smart campus and digital education.

The study contributes to the field by providing a monolithic hybrid system for the web-based intelligent book recommendation that utilizes both collaborative and content-based filtering techniques, which can provide more accurate and personalized recommendations for users. The remaining part of the paper is structured as follows: Sect. 2 summarizes the related work in the literature; Sect. 2 explains the proposed system; Sect. 4 presents the evaluation results of the proposed system; and Finally, Section 5 concludes the paper by discussing the results, the study's limitations, and future directions.

2 Literature Review

Book recommendation systems are a popular research topic in the field of information retrieval and machine learning. Many different techniques have been proposed to recommend books to users, including collaborative filtering, content-based filtering, and hybrid methods that combine both approaches. Collaborative filtering is based on the idea that people who have similar reading preferences will also like similar books [11]. This approach typically uses user-item ratings or user-item interactions to make recommendations. The main disadvantage of collaborative filtering is that it relies on the availability of user-item ratings or interactions, which can be challenging to obtain or sparse. Additionally, it can suffer from the "cold start" problem, where it is difficult to make recommendations for new users or items that do not have any ratings or interactions. Content-based filtering, on the other hand, is based on the idea that people will like books that are similar to books they have previously read or liked [12]. This approach typically uses the metadata or content of the books, such as the author, genre, and keywords, to make recommendations. Content-based filtering can suffer from the "subjectivity" problem, where different users may have different opinions about the characteristics or features of an item. Additionally, it can be difficult to extract useful features from the content of an item, especially in the case of unstructured data like books. Hybrid methods combine the advantages of both collaborative and content-based filtering, usually by using collaborative filtering to provide users with a set of items to rank and then using content-based filtering to re-rank the items [13]. Hybrid methods can be more complex to implement and require more data than either collaborative or content-based filtering alone. Some researchers proposed to use natural language processing (NLP) techniques to extract the features from the book's content and then use them to make recommendations, such as Latent Semantic Analysis (LSA) [14], Latent Dirichlet Allocation (LDA) [14], and Word2Vec [15]. Using NLP techniques for feature extraction can be computationally expensive, and it might not be the best choice if the data is not in the text format. Also, those techniques might not be able to capture the subtle nuances in the text, leading to poor recommendations.

In addition, a number of studies have explored the use of book recommendation systems in specific domains, such as e-commerce [16], libraries [3], and digital educational contexts [17], under smart campuses. These studies have shown that recommendation systems can effectively increase user engagement and satisfaction and in helping users to discover new books and resources. Overall, recommendation systems in the context of smart campuses and digital education can play an important role in improving the education experience. Still, it requires addressing the specific challenges that come with it.

Under Smart Campus Applications, a web-based Intelligent Book Recommendation System, besides its effect on users (students), such as improved learning outcomes and enhanced student engagement, can have various implications for educational managers too. These may involve both advantages and challenges that educational managers should consider. On the positive side, educational managers can consider the following:

- Efficient resource allocation: One advantage of web-based recommendation systems is that they can increase resource allocation efficiency by matching users with books and reading materials or services that best suit their tastes and requirements. This can lower the costs of service provision, publishing and distribution of physical items, and book inventory management while increasing user satisfaction and loyalty by meeting their needs [18].

- Efficiency and effectiveness of library services: By recommendation of reading materials, the tool is able to reduce the search time and cost for books, optimize the book inventory, and increase the efficiency and effectiveness of library services [19].
- Data-driven decision-making: Educational managers can get help for their operations through insights into how students engage with reading materials by collecting data on student reading habits. These data can assist them in making data-driven decisions on how to enhance their reading programs and boost student progress [20].
- Collaboration and communication among learners and educators: The tool can serve as platform that will enable learners and educators to share their opinions, reviews, and feedback on books and enhance the collaboration and communication among them [21].

However, under smart campus applications, a web-based intelligent book recommendation system poses some challenges for educational administrators as described below:

- Guaranteeing the privacy and security of the user data and feedback the system gathers, and complying with the ethical and regulatory requirements regarding data protection and usage.
- Sustaining the level of quality and precision of the book recommendations that the system generates, in addition to addressing the problems of sparsity, cold start, scalability, diversity, serendipity, and explainability.
- Assessing the effectiveness and effect of the system on the user behavior and satisfaction, and also on the learning outcomes and library services.
- Finding a balance between content-based and collaborative filtering methods, as well as between standardization and personalization of book recommendations.
- Combining the system with other smart campus applications and platforms, as well as guaranteeing its compatibility and interoperability with diverse devices and formats [22].

Under this literature review, this study proposed a web-based hybrid intelligent book recommendation system as a smart campus application. The hybrid system is monolithic and consists of collaborative filtering and content-based filtering.

3 Proposed Hybrid System

The Web-Based Hybrid Intelligent Book Recommendation System shown in Fig. 1 is comprised of three key components: Data Sources, User Interaction Applications, and a Python API. The system integrates information from three data sources, including the Library for books, Personnel Affairs for university employees, and Student Affairs for students. This information is then utilized to make personalized recommendations to users through User Interaction Applications. Users can get suggestions in two different ways. Firstly, they can access both their own data and suggested books by logging in through the web-based interface. Secondly, the system can send email notifications to users with book suggestions at specified intervals. There is a communication between User Interaction Applications and Python API components. User Interaction Applications make a request to the API with the user ID, and the API returns a list of suggestions to the user by performing the necessary calculations.

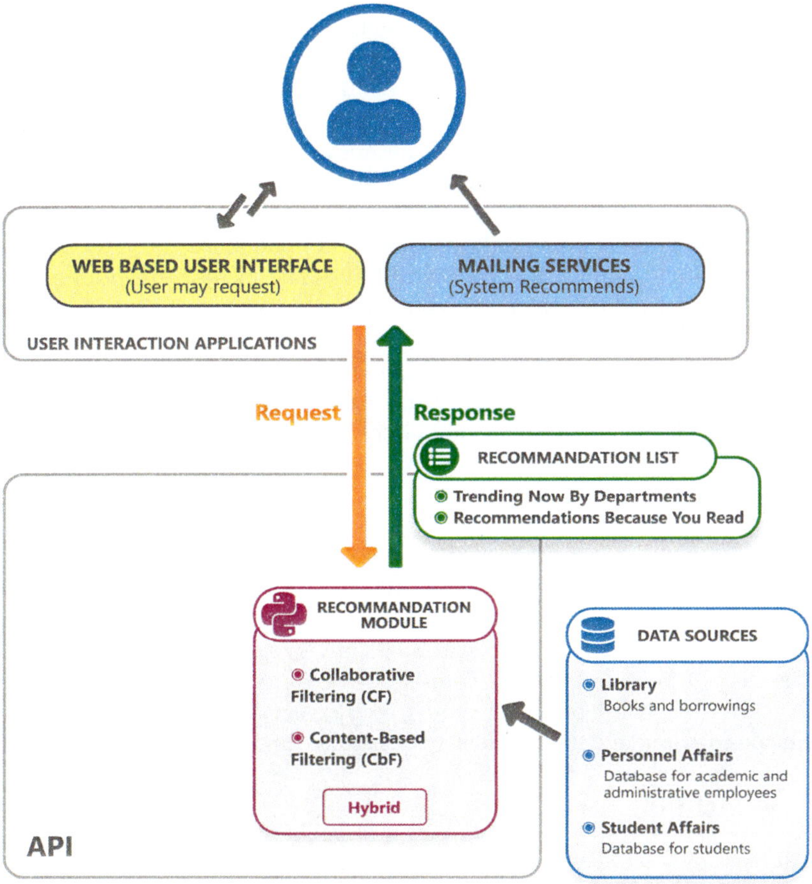

Fig. 1. System architecture

The Recommendation Module, which is designed as a Python API, uses a hybrid technique that combines Collaborative Filtering and Content-Based Filtering to make its recommendations. A monolithic hybrid system that combines collaborative and content-based filtering techniques can overcome the disadvantages of both methods by leveraging the strengths of each approach. Collaborative filtering is a powerful technique that provides recommendations based on the similarities between users or items. However, it has some limitations, such as the cold start problem, where it can be difficult to make recommendations for new users or items without any prior ratings or interactions. Additionally, it can be susceptible to popularity bias, where popular items tend to be recommended more often than others. Content-based filtering, on the other hand, provides recommendations based on the similarity between items and the user's preferences. This approach is able to handle the cold start problem by using item characteristics to make recommendations. Still, it can suffer from limited scalability and difficulty obtaining accurate and complete item descriptions.

By combining these two methods, a hybrid system can address these limitations. The collaborative filtering component can provide recommendations based on user-user similarities, while the content-based filtering component can provide recommendations based on item-item similarities. This combination can overcome the cold start problem by leveraging item characteristics while also addressing the limitations of content-based filtering by incorporating user preferences. Furthermore, by combining the two methods, a hybrid system can provide a more diverse set of recommendations and avoid popularity bias. By considering both user and item similarities, the hybrid system can make recommendations tailored to the individual user's preferences while also incorporating the wider community's opinions and behaviors.

The algorithm steps of Recommendation Module are given in Fig. 2. This module first runs the Collaborative Filtering Model by using Library, Personnel Affairs and Student Affairs data sources. Books and users borrowing these books are obtained from Library data sources, academic and administrative personnel information is obtained from Personnel Affairs, and student information is obtained from Student Affairs. In this model, neighbourhood-based similar and unsimilar students are found after the student-book matrix is created to make basic calculations about borrowed books. This first list created with Collaborative Filtering is used for Content-based Filtering. In this second model, potential recommendations are created by removing the books that have already been read from the first list data. Then, books with content matching the user's interest are found. A list of books is created by considering the similarities. The final list of recommendations is then returned as the response.

4 Evaluation

After the system was developed, it became available to users for testing and their feedback was collected. Figure 3 presents the user interface for the part of the evaluation of the recommended books. All users were able to enter the system and saw some descriptive statistics such as the most read books in the university and the user's department, library usage ratio, the number of borrowed books and users, and read book list. The most suitable ten books were recommended to the user by the improved system and evaluating the recommended books was requested by the user with 'Suitable' or 'Unsuitable'. Although ten books were recommended, some users did not evaluate some recommendations. While 103 of the 148 feedbacks obtained from the 22 users were positive, the remaining 45 were negative. Precision, Recall, and F1-score were used to measure the performance of recommendations.

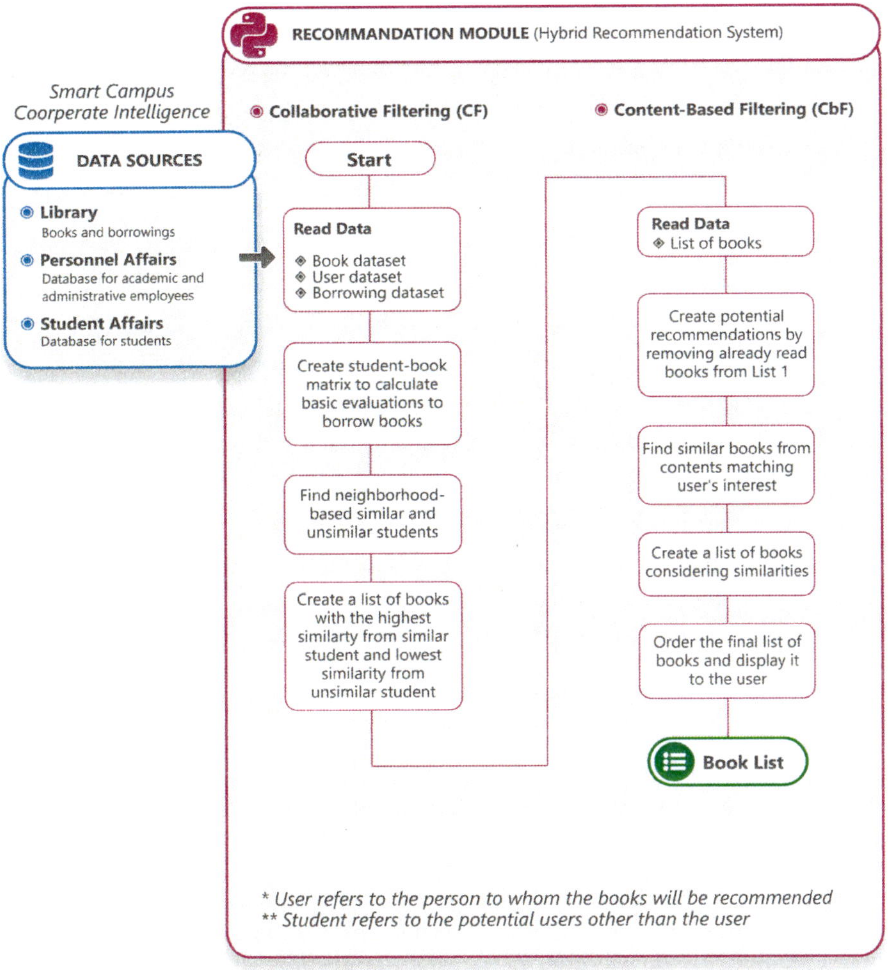

Fig. 2. Recommendation algorithm flowchart

Precision value is calculated by dividing the number of correctly recommended books (CR) by the number of evaluated recommended books (ER). Precision indicates the rate at which the system makes correct recommendations. In this study, the Precision value is calculated as 0.70. This means that 70% of recommended books were liked by users.

$$Precision = \frac{103}{148} = 0.70$$

Recall value is calculated by the presence of preferred books in recommended books. Recall value is obtained by dividing the number of correctly recommended books (CR) by the total number of recommended books (TR). It is the measure

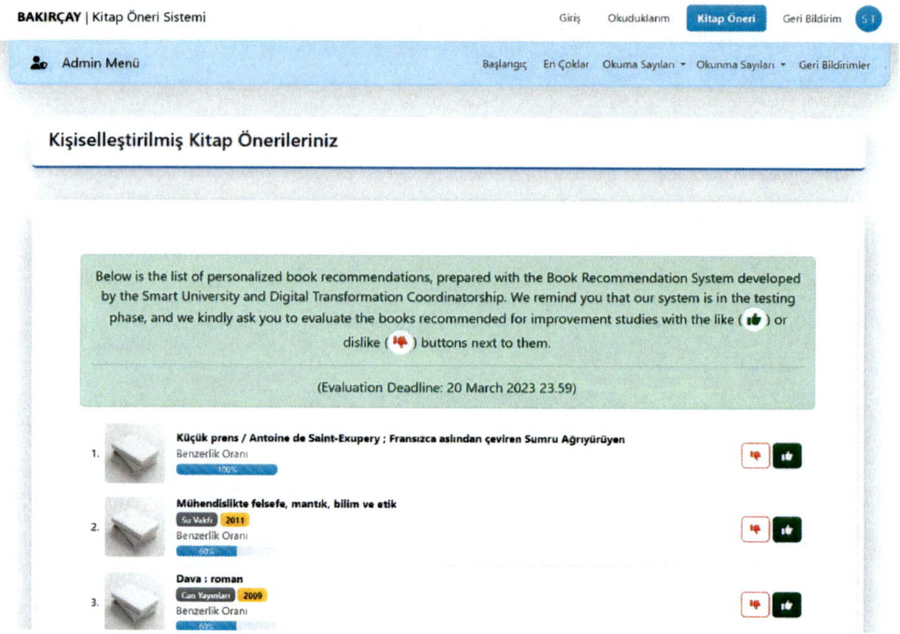

Fig. 3. User interface for evaluation

of the percentage of actual positive instances that were correctly identified by the model. Although the total number of recommended books is 220, 72 books were not evaluated by the users. Therefore, recall and F1-score were calculated under optimistic, realistic, and pessimistic scenarios. The optimistic scenario considers all unevaluated 72 books are suitable, the realistic scenario supposes unevaluated 72 books are suitable with the precision ratio of 0.70, and the pessimistic scenario assumes all unevaluated 72 books are unsuitable. The optimistic, realistic and pessimistic Recall value were found as 0.80, 0.70, and 0.47, respectively. The realistic recall score indicates that the developed system identifies positive instances by 70%, even if it incorrectly identifies some negative instances as positive.

$$Recall_o = \frac{103 + 72}{220} = 0.80, Recall_r = \frac{103 + 72 \times 0.7}{220} = 0.70, Recall_p = \frac{103}{220} = 0.47$$

F1-score, on the other hand, is harmonic mean of precision and recall and considered a good criterion as it shows the balance between precision and recall [6]. It is calculated by $2 \times precision \times recall/(precision + recall)$. The F1-score of the realistic scenario is 0.70. This indicates a good balance between Precision and Recall, with system performance being good.

$$F1_o = 0.75, \quad F1_r = 0.70, \quad F1_p = 0.56$$

5 Conclusion and Discussion

This paper presented a web-based hybrid intelligent book recommendation system developed with a hybrid method that uses content-based filtering and collaborative filtering methods. The system provided personalized recommendations to users based on their reading history and data collected through the Smart Campus applications. The aim was to process the data provided by users and produce functional outputs while also enabling them to make more efficient and effective use of library resources, which were established at high cost, by improving the user's experience with the library resources.

After the system was developed, it was made available to users for testing, and feedback was obtained. The precision, recall, and F1-score values of the recommendations were calculated based on the feedback received. Based on the feedback obtained, the precision value was 70%, indicating that the majority of recommendations made were relevant to the user's needs. Because some users did not evaluated all recommended books, three scenarios were defined for recall and F1-score. The realistic recall value was 70%, indicating that a high proportion of relevant recommendations were made. The F1-Score under realistic scenario value was 70%, indicating that the system's performance was moderate in terms of both precision and recall.

The disadvantages of the recommendation techniques are given above. In general, it is essential to note that any recommendation system is only as good as the data it is based on and that the quality and quantity of the data can significantly influence the effectiveness of the recommendations.

Research in the field of book recommendation systems has demonstrated the potential for various domains to improve the user experience and help users to discover new books and resources. However, more research is needed to fully understand the different factors that influence the effectiveness of book recommendation systems and to develop more accurate and personalized recommendations.

The proposed recommendation system requires to have a user history. For future directions, the authors plan to improve the current system for new users through additional techniques such as fuzzy association rule mining. Moreover, the hybrid system will be extended by a fuzzy rule generator to prioritize the book lists produced by the current recommendation system by considering various prioritization criteria such as the total number of borrowed books, the similarity scores of the books that the user borrowed in the past, and the publication year of the book.

Overall, and as presented in this work, within the framework of Smart Campus Applications, a Web-based Intelligent Book Recommendation System can be a helpful tool for improving student learning outcomes, boosting engagement with reading materials, and providing educational management with information to influence decision-making. It can also assist educational administrators in properly allocating resources and making data-driven choices. However, it is critical to tackle the related challenges that were mentioned, something that could be a topic for future research.

References

1. Onile, A.E., Machlev, R., Petlenkov, E., Levron, Y., Belikov, J.: Uses of the digital twins concept for energy services, intelligent recommendation systems, and demand side management: a review. Energy Rep. **7**, 997–1015 (2021)
2. He, J., Zhang, S.: How digitalized interactive platforms create new value for customers by integrating B2B and B2C models? An empirical study in china. J. Bus. Res. **142**, 694–706 (2022)
3. Khademizadeh, S., Nematollahi, Z., Danesh, F.: Analysis of book circulation data and a book recommendation system in academic libraries using data mining techniques. Libr. Inf. Sci. Res. **44**(4), 101191 (2022)
4. Ifada, N., Syachrudin, I., Sophan, M.K., Wahyuni, S.: Enhancing the performance of library book recommendation system by employing the probabilistic-keyword model on a collaborative filtering approach. Procedia Comput. Sci. **157**, 345–352 (2019)
5. Iqbal, N., Jamil, F., Ahmad, S., Kim, D.: Toward effective planning and management using predictive analytics based on rental book data of academic libraries. IEEE Access **8**, 81978–81996 (2020)
6. Anwar, T., Uma, V.: CD-SPM: cross-domain book recommendation using sequential pattern mining and rule mining. J. King Saud Univ.-Comput. Inf. Sci. **34**(3), 793–800 (2022)
7. Belkhadir, I., Omar, E.D., Boumhidi, J.: An intelligent recommender system using social trust path for recommendations in web-based social networks. Procedia Comput. Sci. **148**, 181–190 (2019)
8. Tian, Y., Zheng, B., Wang, Y., Zhang, Y., Wu, Q.: College library personalized recommendation system based on hybrid recommendation algorithm. Procedia CIRP **83**, 490–494 (2019)
9. Zhu, T., Liu, Y.: Learning personalized preference: a segmentation strategy under consumer sparse data. Expert Syst. Appl. **215**, 119333 (2023)
10. Vasileiou, M., Rowley, J., Hartley, R.: The e-book management framework: the management of e-books in academic libraries and its challenges. Libr. Inf. Sci. Res. **34**(4), 282–291 (2012)
11. Nugraha, E., Ardiansyah, T., Junaeti, E., Riza, L.S.: Enhanced digital library with book recommendations based on collaborative filtering. J. Eng. Educ. Transf. **34**(Special Issue) (2020)
12. Wang, D., Liang, Y., Xu, D., Feng, X., Guan, R.: A content-based recommender system for computer science publications. Knowl.-Based Syst. **157**, 1–9 (2018)
13. Yang, S., Korayem, M., AlJadda, K., Grainger, T., Natarajan, S.: Combining content-based and collaborative filtering for job recommendation system: a cost-sensitive statistical relational learning approach. Knowl.-Based Syst. **136**, 37–45 (2017)
14. Zhang, P., et al.: Group-based latent Dirichlet allocation (group-LDA): effective audience detection for books in online social media. Knowl.-Based Syst. **105**, 134–146 (2016)
15. Renuka, S., Raj Kiran, G.S.S., Rohit, P.: An unsupervised content-based article recommendation system using natural language processing. In: Jeena Jacob, I., Kolandapalayam Shanmugam, S., Piramuthu, S., Falkowski-Gilski, P. (eds.) Data Intelligence and Cognitive Informatics. AIS, pp. 165–180. Springer, Singapore (2021). https://doi.org/10.1007/978-981-15-8530-2_13

16. Chandra, A., Ahmed, A., Kumar, S., Chand, P., Borah, M.D., Hussain, Z.: Content-based recommender system for similar products in E-commerce. In: Patgiri, R., Bandyopadhyay, S., Borah, M.D., Emilia Balas, V. (eds.) Edge Analytics. LNEE, vol. 869, pp. 617–628. Springer, Singapore (2022). https://doi.org/10.1007/978-981-19-0019-8_46
17. Bhaskaran, S., Marappan, R.: Design and analysis of an efficient machine learning based hybrid recommendation system with enhanced density-based spatial clustering for digital e-learning applications. Complex Intell. Syst. 1–17 (2021)
18. Ricci, F., Rokach, L., Shapira, B.: Introduction to recommender systems handbook. In: Ricci, F., Rokach, L., Shapira, B., Kantor, P. (eds.) Recommender Systems Handbook, pp. 1–35. Springer, Boston (2010). https://doi.org/10.1007/978-0-387-85820-3_1
19. Liu, M.: Personalized recommendation system design for library resources through deep belief networks. Mob. Inf. Syst. **2022** (2022)
20. Simović, A.: A big data smart library recommender system for an educational institution. Libr. Hi Tech **36**(3), 498–523 (2018)
21. Darling-Hammond, L., Flook, L., Cook-Harvey, C., Barron, B., Osher, D.: Implications for educational practice of the science of learning and development. Appl. Dev. Sci. **24**(2), 97–140 (2020)
22. Roy, D., Dutta, M.: A systematic review and research perspective on recommender systems. J. Big Data **9**(1), 59 (2022)

Determination of the Most Suitable New Generation Vacuum Cleaner Type with PFAHP-PFTOPSIS Techniques Based on E-WOM

Sena Kumcu[1](✉), Beste Desticioglu Tasdemir[2], and Bahar Ozyoruk[1]

[1] Department of Industrial Engineering, Gazi University Faculty of Engineering, Ankara, Türkiye
senakumcu@gazi.edu.tr

[2] Department of Operations Research, National Defence University Alparslan Defence Sciences and National Security Institute, Ankara, Türkiye

Abstract. Today, when technological developments are accelerating, the number of devices we use is increasing and household goods are equipped with smart functions and gain a more important place in our lives. Especially with the pandemic, the search for practical solutions that make housework easier has made online purchasing behavior indispensable. In this study, the problem of ranking the 6 best-selling new generation vacuum cleaner (NVC) types in a leading shopping site in Türkiye's e-commerce market was discussed. For this purpose, the electronic word-of-mouth communication (e-WOM) for these products on a review platform in Türkiye where customers share evaluations of their experiences with the products or services was examined. In the first part of the study, the criteria of these products as specified on the review platform were evaluated by interviewing salespeople working in different electronics stores. Then, these criteria weights were obtained by using the Pythagorean Fuzzy Analytical Hierarchy Process (PFAHP) method. In the second part of the study, the most suitable NVC type was determined by the Pythagorean Fuzzy Technique for Order Preference by Similarity to Ideal Solutions (PFTOPSIS) method, taking the criteria weights into account obtained with PFAHP and customer satisfaction scores (CSS) on the review platform. Microsoft Excel 2010 program was used in the calculations. With the results obtained from the calculations, the type of vacuum cleaner that can adequately respond to the requests of the users and provide the highest satisfaction among the NVC with the highest sales has been determined. As far as is known, no study has employed hybrid PFAHP-PFTOPSIS method for the product ranking by using CSS. As a result, we think that this study provides a different perspective to the literature in this field.

Keywords: e-WOM · e-commerce · Multi Criteria Decision Making (MCDM) · PFAHP · PFTOPSIS

Z. Şen et al. (Eds.): IMSS 2023, LNME, pp. 58–68, 2024.
https://doi.org/10.1007/978-981-99-6062-0_7

1 Introduction

With the COVID 19 pandemic, a rapid trend towards digital has taken place in the world and in Türkiye. People spent more time at home than usual during that time, which has increased interest in online shopping. Nevertheless, unlike traditional retail buying, internet shoppers are unable to touch and test out the products. Customers that prefer this channel, which provides quick and easy access to information, can use a variety of online product information systems to help them evaluate and choose products. Since the launch of e-commerce, research has repeatedly demonstrated that online or electronic word-of-mouth (eWOM) has a much greater influence on consumer beliefs and actions than traditional word-of-mouth [1]. Consumer reviews, also known as post-purchase evaluations, are one of the most popular types of eWOM. They are composed of users of the items who want to share their opinions and experiences. A typical customer review is made up of a number of elements, such as a consumer rating (such as stars or scores) and textual elements, such as a title and a succinct description of the consumer's experience and viewpoint [2]. Consumer ratings are primarily composed of numerical data from online rating platforms. Customers' opinions on these platforms are expressed using five and ten-point rating scales. Also, on these platforms, customers can use quantitative ratings to assess each attribute of various alternative products or services. However, these alternatives are ranked mostly using the mean value of numerical scores, ignoring the different proportions of the different features [3]. Many approaches have been employed by researchers to address this problem in the literature by using various linguistic evaluation scales [4]. The distinction between this study and others is that customer satisfaction scores (CSS) were transformed by using Pythagorean fuzzy numbers (PFN) to eliminate the uncertainty and ambiguity in these scores and to make the data more meaningful and powerful.

Especially during these periods when the tendency for products that facilitate household chores is high, this study addresses the problem of ranking the best-selling 6 new generation vacuum cleaner (NVC) types in a leading shopping site in Türkiye's e-commerce market. In this way, the CSS of these products on a review platform in Türkiye were utilized. Then, the criteria specified for these products on the review platform were interviewed with the salespeople working in different electronics stores, and their evaluations were taken. It was thought that expert evaluations would vary from person to person; therefore, the weights of the criterion were established using the PFAHP approach. Using the PFTOPSIS approach and the criteria weights derived using PFAHP and the CSS on the review platform, the most appropriate NVC type was identified in the second section of the study. Microsoft Excel 2010 program was used in these calculations. According to the results obtained, the type of vacuum cleaner that can adequately respond to the requests of the users and provide the highest satisfaction among the NVC with the highest sales has been determined. As far as is known, no study employing hybrid PFAHP-PFTOPSIS method for the product ranking by using CSS has been conducted so far. As a result, we think that this study provides new academic and practical insights.

The remainder of this paper is constructed as follows. Section 2 presents literature review, Sect. 3 explains the proposed method, and Sect. 4 includes a case study of ranking NVC types. Finally, Sect. 5 provides the conclusion of this study.

2 Literature Review

In the literature, studies that rank products or services using MCDM methods based on online customer review were examined. Also, any articles that were unrelated to the topic of the study were eliminated. These studies are presented in Table 1. Consequently, 17 relevant studies conducted since 2014 were found.

Table 1. Research on E-WOM-based MCDM techniques.

Authors	Aim of the paper	Ranking Method(s)
[5]	Choosing a hotel based on online reviews	TOPSIS
[6]	A comprehensive system for ranking products through online reviews	stochastic dominance (SD) rules- PROMETHEE II
[7]	Selecting a hotel while taking into account the relationships between criteria and the probable traveler's characteristics	A new GS-PFNCA (Generalized Shapley-Picture Fuzzy Number Choquet integral Average operator)
[8]	Methodology for rating a group of banking institutions based on online platform customer reviews	AHP, FAND (Fuzzy multi-attribute decision making),VIKOR
[9]	Using online reviews as a ranking system for alternative tourism destinations	hesitant fuzzy TOPSIS (IHF-TOPSIS) method
[10]	Selecting products taking into account online reviews	probabilistic linguistic term set (PLTS) based on the proposed adjustable prospect theory (PT) framework
[11]	Ranking products through online customer reviews	IF-MULTIMOORA (multiplicative multi-objective optimization by ratio analysis)
[12]	Ranking hotels in terms of online customer ratings	PROMETHEE
[13]	Ranking products through online reviews	TODIM (an acronym in Portuguese for interactive and multi-criteria decision making)
[14]	Ranking products through online reviews	TODIM
[15]	Ranking products through online reviews	TODIM
[16]	Ranking hotels based on online review	interval-valued neutrosophic TOPSIS
[4]	Ranking hotels in terms of online customer ratings	VIKOR
[17]	Ranking hotels in terms of online review	TOPSIS
[18]	Ranking automobiles in terms of online review	Stochastic dominance-PROMETHEE
[19]	Ranking automobiles in terms of online review	PROMETHEE II
[20]	Ranking mobile phones based on online review	Fuzzy PROMETHEE

A review of previous studies shows that they are few in number, but it seems that current researchers are still interested in studies that rank products using online ratings or reviews. The earlier studies made important advances in the field of product ranking. However, this paper adopted an approach with a broad evaluation scale based on Pythagorean fuzzy set theory (PFS) to serve to parameterize the ambiguity regarding CSS and expert's opinion. This is why this study is considered to provide a different perspective to the literature in this field.

3 Methodology

The alternative ranking is typically regarded as an MCDM problem. The issue of ranking options using online reviews has received notable contributions from several academics up to now. Due to the vagueness and uncertainty of online reviews, they came to the realization that it is not always possible to describe the information with a number of ratings. In order to solve this issue, fuzzy sets theory has become very popular. This is why PFS, one of the expansions of fuzzy set theory, was used for this research to deal with the ambiguity and uncertainty present in expert input data. PFAHP and PFTOPSIS hybrid methods was applied in this study to deal with fuzziness in the evaluation process, reduce ambiguity, and handle uncertainty [21]. It has been seen in the literature that researchers using the PFAHP-PFTOPSIS methods have achieved successful results [22, 23]. In the proposed methodology, decision-makers use language phrases and associated PFNs, which provide them more room to express their thoughts [24]. The steps of these methods are explained in Table 2. Figure 1 also depicts the suggested methodology's framework.

Table 2. Explanations of PFAHP and PFTOPSIS methods.

PFAHP	PFTOPSIS
Step 1. Use the linguistic terms listed in Table 3 to create a pairwise comparison matrix based on expert language ratings [25]	**Step 1:** Table 4 provides alternative assessments of a decision-making group based on expert language ratings using linguistic words [26]
Step 2. Utilizing these calculations to calculate the difference matrix (D = (dik)mxm); $d_{ikL} = \mu^2\text{ikL} - \nu^2\text{iku}$ $d_{ikU} = \mu^2\text{ikU} - \nu^2\text{ikL}$	**Step 2:** Construct Pythagorean fuzzy decision matrix
Step 3 Formation of the interval multiplicative matrix (S = $(s_{ik})_{mxm}$) using these computations; $s_{ikL} = \sqrt{1000^{dL}}$ $s_{ikU} = \sqrt{1000^{dU}}$	**Step 3:** Find the fuzzy negative ideal solution (PF-NIS) and positive ideal solution (PF-PIS) $x^{+} = \left\{ \max_{i} \langle s((x_i))\rangle \mid j = 1, 2, 3\ldots, n \right\} =$ $\{\langle P((u_1^{+}, v_1^{+}))\rangle, \langle P(u_2^{+}, v_2^{+})\rangle, \ldots, \langle P(u_n^{+}, v_n^{+})\rangle\}$, $x^{-} = \left\{ \min_{i} \langle s((x_i))\rangle \mid j = 1, 2, 3\ldots, n \right\} =$ $\{\langle P((u_1^{-}, v_1^{-}))\rangle, \langle P(u_2^{-}, v_2^{-})\rangle, \ldots, \langle P(u_n^{-}, v_n^{-})\rangle\}$,

(continued)

Table 2. (*continued*)

PFAHP	PFTOPSIS
Step 4. Making use of these computations, we can calculate the determinacy value ($\gamma = (\gamma_{ik})_{mxm}$); $\gamma_{ik} = 1 - (\mu^2_{ikU} - \mu^2_{ikL}) - (\nu^2_{ikU} - \nu^2_{ikL})$	**Step 4:** Using these calculations, determine the distance between the ideal solution and PF-PIS and PF-NIS; $D(x_i, x^+) = \sum_{j=1}^{n} w_j d(C_j(x_i), C_j(x^+)) = \frac{1}{2}\sum_{j=1}^{n} w_j \left(\left\|(\mu_{ij})^2 - (\mu_j{}^+)^2\right\| + \left\|(\nu_{ij})^2 - (\nu_j{}^+)^2\right\| + \left\|(\pi_{ij})^2 - (\pi_j{}^+)^2\right\|\right), = 1, 2, \ldots, m$ $D(x_i, x^-) = \sum_{j=1}^{n} w_j d(C_j(x_i), C_j(x^-)) = \frac{1}{2}\sum_{j=1}^{n} w_j \left(\left\|(\mu_{ij})^2 - (\mu_j{}^-)^2\right\| + \left\|(\nu_{ij})^2 - (\nu_j{}^-)^2\right\| + \left\|(\pi_{ij})^2 - (\pi_j{}^-)^2\right\|\right), = 1, 2, \ldots, m$ Where i = 1,2,…,n and the smaller $D(x_i, x^+)$ the better alternative x_i and larger $D(x_i, x^-)$ the better alternative x_i. Let the $D_{min}(x_i, x^+) = min_{1 \le i \le m} D(x_i, x^+)$ and $D_{max}(x_i, x^-) = min_{1 \le i \le m} D(x_i, x^-)$
Step 5. Creating the weights matrix (T=$(t_{ik})_{mxm}$) based on these computations; $t_{ik} = \left(\frac{(s_{ikL} + s_{ikU})}{2}\right)\gamma_{ik}$	**Step 5**. Using this method, determine the alternative's revised closeness $\xi(x_i)$; $\xi(x_i) = \frac{D(x_i, x^-)}{D_{max}(x_i, x^-)} - \frac{D(x_i, x^+)}{D_{min}(x_i, x^+)}$
Step 6. Making use of these calculations to determine the normalized priority weights (w_i) byusingthiscalculations; $w_i = \frac{\sum_{k=1}^{m} t_{ik}}{\sum_{i=1}^{m}\sum_{k=1}^{m} t_{ik}}$	**Step 6.** The solution that will be ranked in the most advantageous order is the one with the highest revised closeness $\xi(x_i)$

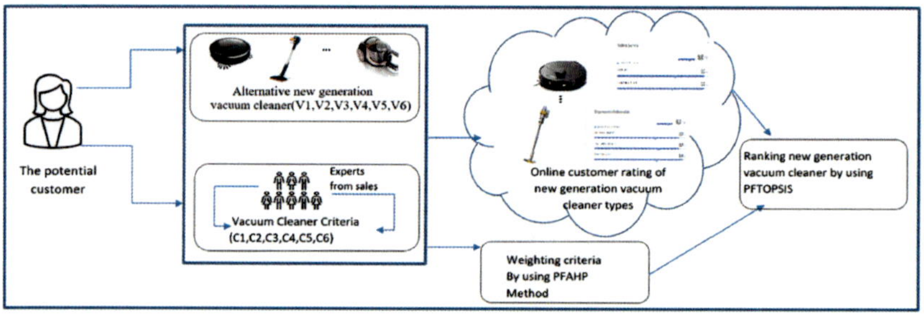

Fig. 1. The proposed methodology's framework.

4 Case Study

In recent years, with the development of technology, old-style vacuum cleaners have been replaced by NVCs. Consumers have also changed their usage habits and started to prefer NVC instead of old-style dust bag vacuum cleaners. Recently, there has been a great increase in the demands of robot vacuum cleaner, vertical vacuum cleaner, and bagless vacuum cleaner. In this study, NVCs were compared according to their characteristics and ranked.

In the study, 2 types of robot vacuums, 2 types of upright vacuum cleaners, and 2 types of bagless vacuum cleaners, which were stated to be the most sold in a shopping

Table 3. PFNs scale for AHP [25].

Linguistic terms	PFN equivalents			
	μL	μU	vL	vU
Certainly Low Importance	0	0	0.9	1
Very Low Importance	0.1	0.2	0.8	0.9
Low Importance	0.2	0.35	0.65	0.8
Below Average Importance	0.35	0.45	0.55	0.65
Average Importance	0.45	0.55	0.45	0.55
Above Average Importance	0.55	0.65	0.35	0.45
High Importance	0.65	0.8	0.2	0.35
Very High Importance	0.8	0.9	0.1	0.2
Certainly High Importance	0.9	1	0	0
Exactly Equal	0.1965	0.1965	0.1965	0.1965

site, were compared. In the first stage, the criteria for NVC types specified on the review platform were interviewed with the salespeople working in different electronics stores, and their evaluations were taken.

Table 4. PFNs scale for TOPSIS [26].

Linguistic variable	PFN equivalents
Certainly Low Importance	(0. 1)
Very Low Importance	(0.20, 0.90)
Low Importance	(0.35, 0.80)
Below Merage Importance	(0.45, 0.65)
Average Importance	(0.55,0.55)
Above Average Imrortance	(0.0, 0.45)
High Importance	(0.80, 0.35)
Very High Importance	(0.90, 0.20)
Certainly High Importance	(1. 0)
Exactly Equal	(0.1965, 0.1965)

Afterwards, they were asked to score the determined criteria using the pairwise comparison matrix. The criteria used in the selection of NVC can be explained as price/performance, ergonomic usability, suction power, noise level, quality, dust capacity, and technical service. It was thought that these scores would vary from person to person; therefore, the criterion weights were calculated using the PFAHP approach. In the second stage of the study, the most sold 2 types of upright vacuum cleaners (V1 and V2), 2 types of robot vacuum cleaners (V3 and V4) and 2 types of bagless vacuum cleaners (V5 and V6) were ranked using the criterion weights determined in the previous stage with PFTOPSIS method. In the comparison, ratings made by CSS for NVC types on the review platform were used. At this stage, the most preferred type of vacuum cleaner by the customers in the last period was determined. For confidentiality reasons, the names of the experts and their place of employment are not specified.

4.1 Proposed Methodology

In this section, it was aimed to rank the 6 best-selling vacuum cleaner models in Türkiye according to the determined criteria. V1 and V2 alternatives are the 2 best-selling types of upright vacuum cleaners, V3 and V4 alternatives are the 2 best-selling robot vacuums, and V5 and V6 alternatives are the 2 best-selling types of bagless vacuum cleaners. For confidentiality reasons, the brands and models of these vacuum cleaners are not included in the study. "Price/Performance (C1), Ergonomic Usability (C2), Suction Power (C3), Noise Level (C4), Dust Capacity (C5) and Technical Service (C6)" criteria determined by expert opinion were used for ranking. The decision diagram determined for the vacuum cleaner types with the determined criteria is given in Fig. 1.

4.2 Weighting of the Criteria with PFAHP Method

Ten experts who work in various electronics stores were interviewed in order to ascertain the factors that are significant in the choosing of NVC. The review platform's criteria were initially determined by the expert team by taking the NVC types into consideration. Afterwards, they were instructed to compare these standards in pairs in light of their personal opinions. The experts were asked to compare these two texts using the linguistic numbers listed in Table 3. Since the scores will vary from person to person, it was aimed to obtain more accurate results by using PFNs in calculations. The PFAHP approach was used to calculate the weights of the criteria and the expert-generated scores. The next section discusses the calculations for the PFAHP. In this section, the comparison results of 10 experts were converted into PFNs, the average of these results was taken, and the PFNs for the criteria were determined as shown in Table 5.

Table 5. Pairwise comparison matrix.

	PFNs $\{[\mu_L, \mu_U], [v_L, v_U]\}$					
	C1	C2	C3	C4	C5	C6
C1	{[0.197, 0.197], [0.197, 0.197]}	{[0.55, 0.67], [0.34, 0.45]}	{[0.50, 0.62], [0.38, 0.50]}	{[0.60, 0.71], [0.29, 0.41]}	{[0.59, 0.69], [0.32, 0.42]}	{[0.30, 0.37], [0.44, 0.51]}
C2	{[0.34, 0.45], [0.55, 0.67]}	{[0.197, 0.197], [0.197, 0.197]}	{[0.32, 0.42], [0.59, 0.69]}	{[0.39, 0.47], [0.47, 0.59]}	{[0.49, 0.61], [0.40, 0.52]}	{[0.29, 0.39], [0.62, 0.72]}
C3	{[0.38, 0.50], [0.50, 0.62]}	{[0.59, 0.69], [0.32, 0.42]}	{[0.197, 0.197], [0.197, 0.197]}	{[0.53, 0.63], [0.38, 0.48]}	{[0.35, 0.46], [0.54, 0.66]}	{[0.17, 0.27], [0.68, 0.81]}
C4	{[0.29, 0.41], [0.60, 0.71]}	{[0.47, 0.59], [0.39, 0.47]}	{[0.38, 0.48], [0.53, 0.63]}	{[0.197, 0.197], [0.197, 0.197]}	{[0.37, 0.47], [0.53, 0.64]}	{[0.42, 0.55], [0.43, 0.53]}
C5	{[0.32, 0.42], [0.59, 0.69]}	{[0.40, 0.52], [0.49, 0.61]}	{[0.54, 0.66], [0.35, 0.46]}	{[0.53, 0.64], [0.37, 0.47]}	{[0.197, 0.197], [0.197, 0.197]}	{[0.48, 0.58], [0.42, 0.52]}
C6	{[0.44, 0.51], [0.30, 0.37]}	{[0.62, 0.72], [0.29, 0.39]}	{[0.68, 0.81], [0.17, 0.27]}	{[0.43, 0.53], [0.42, 0.55]}	{[0.42, 0.52], [0.48, 0.58]}	{[0.197, 0.197], [0.197, 0.197]}

In the study, calculations were made with the PFAHP method using the PFNs given in Table 3, and the weights were calculated as given in Fig. 2. When Fig. 2 was examined, it was seen that the C6 (Technical Service) criterion has the highest weight with 0.2788. This shows that customers who buy vacuum cleaners first consider technical service possibilities. In the calculations, it was determined that the second most important criterion was C1 (Price/Performance) with a weight of 0.2274. The order of the other criteria is C5 > C3 > C4 > C2.

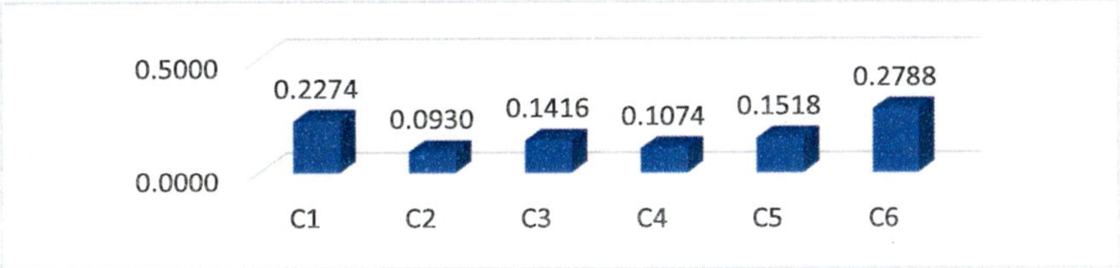

Fig. 2. Criteria priority weights of NVCs.

4.3 Ranking of NVC with PFTOPSIS

In this section, the aim is to rank the NVC with the PFTOPSIS method using the criterion weights determined in the previous section. From a review platform, where customers share evaluation of their experiences on the products or services, the CSS for 6 NVC types were utilized. Since these ratings contain uncertainty, the ratings were converted into PFNs taking into account Table 4. The decision matrix given in Table 6 was created by transforming the scores of the criteria of 6 types of NVC into PFNs.

Table 6. Decision matrix for ranking NVCs.

DM	C1		C2		C3		C4		C5		C6	
	u	*v*	*u*	*v*	*u*	*v*	*u*	*v*	*u*	*v*	*u*	*v*
V1	0.80	0.44	0.50	0.80	0.10	0.00	0.80	0.44	0.80	0.44	0.70	0.60
V2	0.60	0.71	0.60	0.71	0.60	0.71	0.70	0.60	0.70	0.60	0.60	0.71
V3	0.80	0.44	0.80	0.44	0.80	0.44	0.80	0.44	0.80	0.44	0.10	0.00
V4	0.70	0.60	0.70	0.60	0.60	0.71	0.70	0.60	0.70	0.60	0.60	0.71
V5	0.70	0.60	0.70	0.60	0.70	0.60	0.60	0.71	0.60	0.71	0.60	0.71
V6	0.76	0,50	0.80	0.44	0,80	0.44	0.80	0.44	0.80	.044	0.60	0.71

The results obtained after applying all the steps of the PFTOPSIS method are given in Table 7.

Table 7. Decision matrix for ranking NVCs.

Alternatives	$\xi(X_j)$	Rank
V1	−0.5849	3
V2	−1.7104	6
V3	0.0000	1
V4	−1.4990	5
V5	−1.2807	4
V6	−0.2599	2

Figure 3 shows the ranking obtained by the PFTOPSIS method. When Fig. 3 is examined, it is seen that the best alternative is the robot vacuum type V3 when the vacuum cleaners are ranked according to the vacuum cleaner preference criteria of the customers. The dust bagless vacuum cleaner designated as V6 has been found to be the second alternative as a result of the calculations.

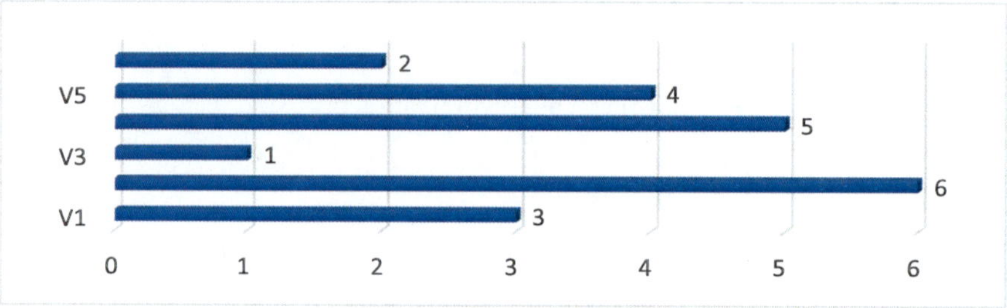

Fig. 3. Ranking of NVC types.

4.4 Results and Discussion

With the development of NVCs in recent years, it is seen that the old habits of customers have changed. Recently, there has been an increase in the demand for robot vacuums and upright vacuum cleaners. In this study, 2 types of upright vacuum cleaners, 2 types of robot vacuums, and 2 types of bagless vacuum cleaners with the highest sales in Türkiye were compared and ranked according to the determined criteria. The criteria weights used in the NVC comparison were calculated by the PFAHP method and are given in Fig. 2. As a result of the calculations, it can be concluded that C6 (Technical Service) has the highest weight, at 0.2788. The fact that this criterion is higher than the other criteria shows that customers give priority to the vacuum cleaner with good technical service in the selection of vacuum cleaners. The second criterion is the C1 criterion (Price/Performance) with a weight of 0.2274. This indicated that customers prefer a vacuum cleaner that is worth the price they pay.

The weights of the other 4 criteria used in the selection of new generation vacuum cleaners vary between 0.0930 and 0.1518. When Fig. 2. is reviewed, it becomes clear that Dust Capacity is the third crucial criterion and has a weight of 0.1518. Customers make their choices by considering the dust capacity of the vacuum cleaner they will buy. Other criteria used by customers are Suction Power (C3), Noise Level (C4), and Ergonomic Usability (C2), respectively.

5 Conclusion

Recently, with the development of technology, upright vacuum cleaners, robot vacuums and bagless vacuum cleaners have started to replace old-style vacuums. The goal of this study was to rank NVCs using the chosen criteria. In the study, 2 types of upright vacuum cleaners, 2 types of robot vacuums, and 2 types of bagless vacuum cleaners were compared. Expert interviews were used to develop the comparison criteria, and the

PFAHP approach was used to determine their weights. The calculations show that C6 (Technical Service), with a weight of 0.2788, is the criterion with the highest weight. Other criteria weights are listed as C1 > C5 > C3 > C4 > C2. In the second part of the study, the 6 types of brooms that are sold the most in Türkiye were listed with the PFTOPSIS method. According to the calculations, it has been determined that the most preferred type of NVC is the V3 type.

In the study, 6 different NVC types were sorted by two decision making methods (PFAHP and PFTOPSIS methods) based on the CSS. Also, the opinions of experts working in different electronics stores were used to evaluate these criteria. In future studies, different criteria for NVCs can be determined by using text mining or sentiment analysis methods to extract features from customers' online reviews. In addition, these studies can also evaluate the NVC types using a variety of MCDM techniques and compare them with our results.

References

1. Cheung, C., Lee, M.K.O.: What drives consumers to spread electronic word of mouth in online consumer-opinion platforms. Decis. Support Syst. **53**(1), 218–225 (2012)
2. Yoon, Y., Kim, A.J., Kim, J., Choi, J.: The effects of eWOM characteristics on consumer ratings: evidence from TripAdvisor.com. Int. J. Advert **38**(5), 684–703 (2019)
3. Zhao, M., Li, L., Xu, Z.: Study on hotel selection method based on integrating online ratings and reviews from multi-websites. Inf. Sci. (N Y) **572**, 460–481 (2021)
4. Yu, S.M., Wang, J., Wang, J.Q., Li, L.: A multi-criteria decision-making model for hotel selection with linguistic distribution assessments. Appl. Soft Comput. J. **67**, 741–755 (2018)
5. Wu, J., Liu, C., Wu, Y., Cao, M., Liu, Y.: A novel hotel selection decision support model based on the online reviews from opinion leaders by best worst method. Int. J. Comput. Intell. Syst. **15**(1), 19 (2022). https://doi.org/10.1007/s44196-022-00073-w
6. Qin, J., Zeng, M.: An integrated method for product ranking through online reviews based on evidential reasoning theory and stochastic dominance. Inf. Sci. (N Y) **612**, 37–61 (2022)
7. Tao, L.L., You, T.H.: A multi-criteria decision-making model for hotel selection by online reviews: Considering the traveller types and the interdependencies among criteria. Appl. Intell. **52**(11), 12436–12456 (2022). https://doi.org/10.1007/s10489-021-03151-2
8. Vyas, V., Uma, V., Ravi, K.: Aspect-based approach to measure performance of financial services using voice of customer. J. King Saud Univ. – Comput. Inf. Sci. **34**(5), 2262–2270 (2022)
9. Qin, Y., Wang, X., Xu, Z.: Ranking tourist attractions through online reviews: a novel method with intuitionistic and hesitant fuzzy information based on sentiment analysis. Int. J. Fuzzy Syst. **24**(2), 755–777 (2022). https://doi.org/10.1007/s40815-021-01131-9
10. Zhao, M., Shen, X., Liao, H., Cai, M.: Selecting products through text reviews: An MCDM method incorporating personalized heuristic judgments in the prospect theory. Fuzzy Optim. Decis. Making **21**(1), 21–44 (2022). https://doi.org/10.1007/s10700-021-09359-8
11. Heidary Dahooie, J., Raafat, R., Qorbani, A.R., Daim, T.: An intuitionistic fuzzy data-driven product ranking model using sentiment analysis and multi-criteria decision-making. Technol. Forecast. Soc. Change **173**, 121158 (2021)
12. Sharma, H., Tandon, A., Aggarwal, A.G.: Ranking hotels based on online hotel attribute ratings using neutrosophic AHP and stochastic dominance. Lect. Notes Electr. Eng. **601**, 872–878 (2020). https://doi.org/10.1007/978-981-15-1420-3_94

13. Zhang, D., Wu, C., Liu, J.: Ranking products with online reviews: A novel method based on hesitant fuzzy set and sentiment word framework. J. Oper. Res. Soc. **71**(3), 528–542 (2020)
14. Zhang, D., Li, Y., Wu, C.: An extended TODIM method to rank products with online reviews under intuitionistic fuzzy environment. J. Oper. Res. Soc. **71**(2), 322–334 (2020)
15. Wu, C., Zhang, D.: Ranking products with IF-based sentiment word framework and TODIM method. Kybernetes **48**(5), 990–1010 (2019)
16. Sharma, H., Tandon, A., Kapur, P.K., Aggarwal, A.G.: Ranking hotels using aspect ratings-based sentiment classification and interval-valued neutrosophic TOPSIS. Int. J. Syst. Assur. Eng. Manag. **10**(5), 973–983 (2019). https://doi.org/10.1007/s13198-019-00827-4
17. Pahari, S., Ghosh, D., Pal, A.: An online review-based hotel selection process using intuitionistic fuzzy TOPSIS method. Adv. Intell. Syst. Comput. **710**, 203–214 (2018). https://doi.org/10.1007/978-981-10-7871-2_20
18. Fan, Z.P., Xi, Y., Liu, Y.: Supporting consumer's purchase decision: a method for ranking products based on online multi-attribute product ratings. Soft Comput. **22**(16), 5247–5261 (2018). https://doi.org/10.1007/s00500-017-2961-4
19. Liu, Y., Bi, J.W., Fan, Z.P.: Ranking products through online reviews: A method based on sentiment analysis technique and intuitionistic fuzzy set theory. Inf. Fusion **36**, 149–161 (2017)
20. Peng, Y., Kou, G., Li, J.: A fuzzy PROMETHEE approach for mining customer reviews in Chinese. Arab. J. Sci. Eng. **39**(6), 5245–5252 (2014). https://doi.org/10.1007/s13369-014-1033-7
21. Wang, L., Li, W., Li, H.: Decision-making for ecological landslide prevention in tropical rainforests. Nat. Hazards **103**(1), 985–1008 (2020). https://doi.org/10.1007/s11069-020-04022-8
22. Yucesan, M., Gul, M.: Hospital service quality evaluation: an integrated model based on Pythagorean fuzzy AHP and fuzzy TOPSIS. Soft Comput. **24**(5), 3237–3255 (2020). https://doi.org/10.1007/s00500-019-04084-2
23. Gul, M., Ak, M.F.: A comparative outline for quantifying risk ratings in occupational health and safety risk assessment. J. Clean. Prod. **196**, 653–664 (2018)
24. Karasan, A., Ilbahar, E., Kahraman, C.: A novel pythagorean fuzzy AHP and its application to landfill site selection problem. Soft Comput. **23**(21), 10953–10968 (2019). https://doi.org/10.1007/s00500-018-3649-0
25. Ilbahar, E., Karaşan, A., Cebi, S., Kahraman, C.: A novel approach to risk assessment for occupational health and safety using Pythagorean fuzzy AHP & fuzzy inference system. Saf. Sci. **103**, 124–136 (2018)
26. Sarkar, B., Biswas, A.: Pythagorean fuzzy AHP-TOPSIS integrated approach for transportation management through a new distance measure. Soft Comput. **25**(5), 4073–4089 (2021). https://doi.org/10.1007/s00500-020-05433-2

Quality Control in Chocolate Coating Processes by Image Processing: Determination of Almond Mass and Homogeneity of Almond Spread

Seray Ozcelik[1], Mert Akin Insel[1(✉)], Omer Alp Atici[2], Ece Celebi[2], Gunay Baydar-Atak[2], and Hasan Sadikoglu[1]

[1] Faculty of Chemical-Metallurgical Engineering, Department of Chemical Engineering, Yıldız Technical University, 34210 Istanbul, Türkiye
makinsel@yildiz.edu.tr

[2] Unilever San. ve Tic. Türk A.Ş, Unilever Algida Ar-Ge Merkezi Meclis Mah. Teraziler Cad. 34785, Istanbul, Türkiye

Abstract. Nowadays, the need for standardization of all physical and chemical processes for high quality products leads us to adopt fast and precise decision-making mechanisms, such as image processing. Image processing is preferred in various areas due to its ease of application and proven usefulness. In the food industry specifically, it has been utilized for quality control of products at each stage of production. In this study, the aim is to prepare an image processing algorithm that determines the homogeneity of almond spread and almond mass of the chocolate coatings topped with almond. 40 images were digitally produced to represent the chocolate coating with almond. Half of the images were designed to be examples of homogenously spread almonds while the other half were of non-homogeneously spread almonds on the chocolate coating. The produced images were formed in 4 different almond mass, which were uniformly distributed among the data. The images were then processed to determine the homogeneity of almond spread and almond mass on the chocolate coating. Co-occurrence-matrix was used to test the homogeneity of the spread. With a properly chosen offset value, co-occurrence-matrix correlation value of the image was observed to be able to determine the homogeneity of the samples. To determine the almond mass, the images were first converted to black and white form and then simply the white color ratio in the images were evaluated, which is observed to be directly proportional to the almond mass. Consequently, it is shown that the proposed image processing methodology can be successful at determining homogeneity of almond spread and almond mass. Furthermore, the proposed methodology may also be utilized for other quality control applications where determination of homogeneity and amount are of importance.

Keywords: Image Processing · Quality Control · Chocolate Coating · Almond · Regression

Z. Şen et al. (Eds.): IMSS 2023, LNME, pp. 69–80, 2024.
https://doi.org/10.1007/978-981-99-6062-0_8

1 Introduction

In recent years, image processing gained importance in a wide spectrum of quality control processes [1, 2]. Image processing methods can yield accurate results in a short time, prevent human errors, and enable the allocation of the workforce from repetitive tasks. Furthermore, by reducing the size of the image data with image processing methods, successful rule-based models can be proposed with fewer data [3]. Although an expert's opinion is required for decision-making at the first stage, it is a dynamic system that can improve itself. Thus, it is in a place that is open to shape under the developing technology and changing requests. In accordance with the literature, image processing shows that it is a convenient method for industrial utilization [4, 5]. With these aspects, the prevalence of image processing applications can take companies one step higher in terms of their growth and reliability [6].

Even though image processing has its place in several areas, the way it is performed is not the same. For instance, in the food sector when it is desired to examine the ripening process of fruits/ vegetables, to detect surface abnormalities (such as rotting and pitting), or to classify them, images are processed after being captured by cameras and are often expressed as histogram data [1]. In this subfield, taking into consideration that progress is observed, in other words; factors such as physicochemical, color, and texture changes are investigated, a histogram is a convenient way to express data. To express colors, RGB (Red-Green-Blue) or HSI (Hue-Saturation-Intensity) color spaces are mostly used in coding [7]–[10]. However, if the food sector is analyzed under the scope of processed goods, it can be seen that different color spaces such as threshold-based segmentation (binary scale) and L*a*b (Lightness * Redness * Yellowness) color space are preferred [6, 11, 12]. The point of this action is that each application has different needs. Processed goods tend to be highly heterogeneous [13]. Therefore, threshold-based segmentation is chosen since it allows detection of an object from the background easily. This method is also used for an agricultural purpose which is a weed recognition system for identifying outdoor plants using machine vision. In literature, images acquired in RGB are converted to gray scales and used to process as a binary image. Bright pixels in the dark background were identified as weed and classified as broad and narrow using threshold values [14]. The difference between observing a process of ripening and detecting in agricultural applications can be seen in this example.

In this study, 40 images are digitally created. In these images, chocolate coating was represented as dark-brown and almond pieces a light-brown color. Every image is created in the same size (300x300 pixels). The light-brown parts on 20 images are distributed homogeneously, and the remaining 20 are distributed non-homogenously so that the homogeneity of the almond spread could be clearly investigated in relation with the co-occurrence-matrix correlation value. The mass of the almond pieces on the coating was another subject that we wanted to investigate. For this purpose, we divided each half of the images into 5 subgroups that had the same amount of almond mass. After the preparation of the dataset, the images are converted to a black-white binary scale from RGB with threshold-based segmentation. In this way, it was easier to detect the almond pieces on the coating since there were significant differences. Considering a real image where there are colors more than only shades of brown color, it is the best option to use a binary image. Besides this, almonds could be set as a threshold value when the RGB images were reduced to one layer from three layers. This conversion of color space can also be used to improve the quality of the image in further processing, as it pre-processes the pixel of the images [1]. In the image matrix of black and white images, white appears as 1 and black as 0. At the same time, white indicates that there are almond pieces at that point, while black indicates that there are no almonds at that point. Using this information, the average of all values in the relevant matrix was taken. In this way, the mean operation was performed, and the quantity of almonds was obtained as a ratio of the image. This ratio was expected to be directly related to the quantity of almonds. This relationship was obtained as a result of linear regression and the almond mass was obtained directly from the image. Finally, the images are classified according to their homogeneity by using the correlation value of the co-occurrence matrix. This value states how much a pixel is related to its adjacent pixel over the entire image [15]. Optimum hyperparameters were determined to obtain the correlation values that enable the best classification results. The performance metrics for determination of both the almond mass and the homogeneity of almond spread are evaluated for the validation of the proposed models. The flowchart of this study is illustrated in Fig. 1.

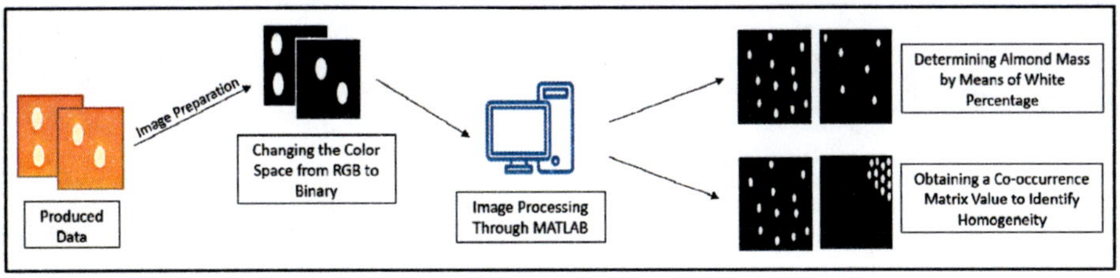

Fig. 1. The flowchart of this study.

The two common trends in the application of image color for food quality evaluation is performed in this study: one is to carry out a point analysis, encompassing a small group of pixels for the purpose of detecting small characteristics of the object; the other is to carry out a global analysis of the object to analyze its homogeneity [16]–[17] The study will hereby add value to the literature since it puts emphasis on the biggest focuses of food quality evaluation. This study bears the torch for upcoming studies since it is applied to a quality control process that has no application in the area of point and homogeneity analyses. It can also be an example of any process that includes coating and topping.

2 Methodology

2.1 Creation of the Images

In this study, 40 images are digitally created using the Paint program to represent the almond topping on chocolate coating. Every image created is in the same size of 300x300 pixels and every yellowish part shaped as an ellipse that represents almond pieces is also in the same size. Created images are divided into subgroups according to their almond number and homogeneity. 5 equal numbers of homogeneous and non-homogeneous images are created with 6,8,10 and 12 almonds. The reason behind creating different homogeneous and non-homogeneous images is that to obtain a co-occurrence-matrix correlation value which sets the limit for homogeneity. Since the images are created manually, one can approve the homogeneity of the image before it is processed. So, by this decision, the correlation value can also be confirmed. The brown part in the image is to be representative of chocolate coating and the yellowish parts are assumed to be almonds. In this study, images are created with different numbers of almonds to investigate the almond mass on the coating. There is one assumption that none of the almonds overlap. This side of the study is open to improvement in future studies in which real images of the chocolate coating are utilized. All the created images and their labeling information are illustrated in Fig. 2.

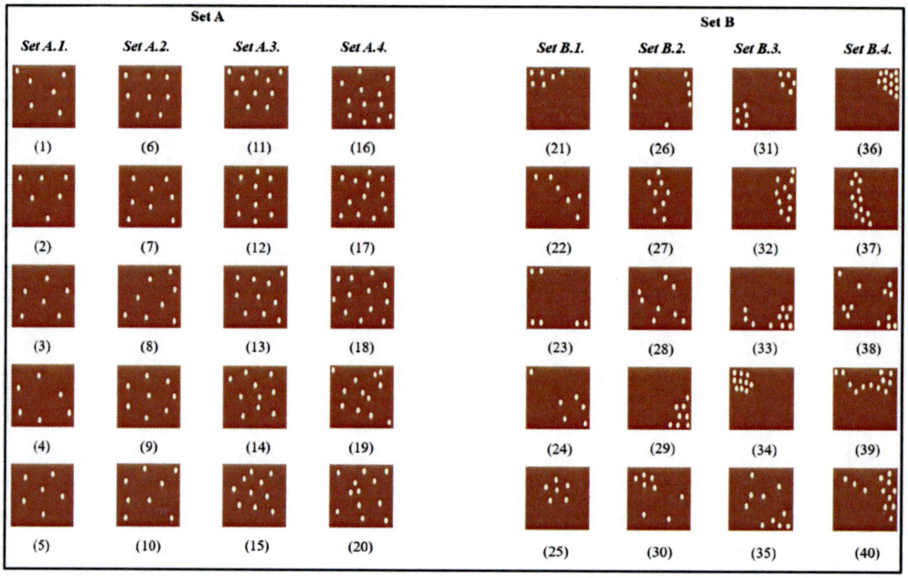

Fig. 2. Created images for representation of the almond topping on chocolate coating. The set A represents homogenous coating while set B represents nonhomogeneous ones. The sets are further divided into four subsets according to their almond quantity.

2.2 Color Space Conversion

Colors can be expressed in a variety of color spaces. The different color spaces result from the distinctness of requirements to express an image. There are mainly 6 color spaces for representing a color image. The first and the most used one is RGB (Red-Green-Blue) color space. RGB color space is mainly the color space of a digitally acquired or monitored image. On the other hand, CMY (Cyan-Magenta-Yellow) color space is used in painting. L*a*b color space is used when independent control of color, brightness, contrast, and sharpness is needed [18]. A survey on the quality of potato chips made by Pedreschi et al. can be a good example of these requirements. The color of potato chips is the first quality parameter taken into consideration. The color of potato chips is not stable since it changes during frying, and, this topic is extremely important because the frying not only adjusts the taste of potato chips, but it also limits the carcinogen effect. In addition, the texture is another object of interest in the quality of potato chips since defects such as black dots and necrosis can be found on a potato [19]. These quality issues lead to a preference for L*a*b color space. Here L* is the luminance or lightness component, which ranges from 0 to 100, and parameters a* (from green to red) and b* (from blue to yellow) are the two chromatic components, which range from –120 to 120 [19]. HSV (Hue-Saturation-Value) and HSL (Hue-Saturation-Lightness) color spaces are alternative representations of the RGB color model designed by computer graphics researchers in the 1970s to better mimic how human vision perceives color rendering properties [20]. There is one other color space called HIS (Hue-Saturation-Intensity). Although the names and representations are very similar, there are slight differences between these color spaces. The hue component H in all three color spaces is an angular measurement, analogous to position around a color wheel. A hue value of 0°corresponds to red, 120° corresponds to green, and 240° corresponds to blue [21]. HSV and HIS color spaces make it possible to control the color and contrast of an image freely [18]. In HSL color space, lightness is an integral characteristic of a color which helps to conduct a variety of algorithm options for image processing [18]. In line with these characteristics, HSV/HSL/HSI color space is widely used in applications of quality control of fruit/vegetable production [10, 22].

Image processing is based on obtaining valuable information from images. This can be done as classifying, identifying, tagging, etc. interested part of the image. However, noisy backgrounds and useless information on the image cause a struggle. For this reason, a method is needed for uncovering the center of interest, image segmentation. Segmentation divides the image under processing into sections in order to change its display which is easier to operate. Segmentation is based upon grayscale, texture, and motion namely characteristic of an image [23]. Threshold, region, edge, watershed, and clustering-based methods are used in the segmentation process [24, 25]. Threshold-based segmentation is the simplest, most effective, and most frequently used type of image segmentation [1, 24]–[26]. In this method, pixels on the image are divided into two groups according to a threshold value that is set in the first place. Pixel values that are lying above and below the threshold value are called "1" and "0", relatively. In this way, binary images can be created. Threshold-based segmentation is quite a prevalent method when the object of interest is easily distinguished from the background. This means the method works well on a grayscale image that has uniform but different interior and background gray levels [1]. Region-based segmentation working principle is to group adjacent pixels by their similarities. Edge segmentation is done by detecting edges in images. On the other hand, the watershed method combines region and edge-based segmentation [25]. The last method which is clustering-based segmentation is close to region-based segmentation, however, the main difference is that the clustering method is based on grouping closer pixel groups [27].

In this study, images were created digitally in RGB color space first. Later, all of the images are converted into binary form by threshold-based segmentation method with a suitable threshold value to the black and white (BW) color space. Thresholding plays an important role in this step because it directly has a role in the next steps since the thresholding value decides which pixel on the image is black or white. Thresholding is done in two ways, locally and globally [24]. Global thresholding is applied in this study to work with the whole image directly. Images that are in the binary form are displayed using the two farthermost gray tones, black and white [28], which are represented as (0, 0, 0) and (255, 255, 255). 255 is the highest part of red, green, and blue colors. Thus, when they are all mixed together white color is obtained. Binary image carries particular importance when it is processed as it is simplistic. Binary image processing is preferred mostly for identifying objects on an image [28, 29]. Identifying can be done in terms of location, boundary, or the presence or absence of the sought object [28]. From the literature, it is seen that color space selection is an effective parameter in color image segmentation [30]. Thus, binary conversion results in an advantage in the next step when threshold-based segmentation is applied. In addition, a binary form is an appropriate approach when a mass determination is performed since dimensions are reduced on the image by converting into "1" and "0". The process is illustrated in Fig. 3.

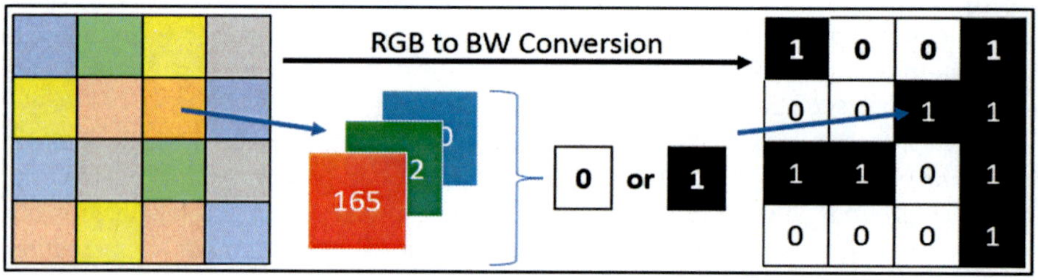

Fig. 3. The Principle of RGB to BW conversion.

2.3 Methodology for the Determination of the Almond Mass

Data used for mass determination is in black and white form. Black and white represents background and almond pieces, relatively. They are stated as "1" and "0" which means "false" and "true". Here, "false" is in accordance with gray level of the image and it is associated with the absence of almond pieces. As a result, matrices are acquired with ones and zeros for every image that is used as data. Using this information, the average of all values in the relevant matrix is taken. In this way, a white color ratio relative to the whole image is taken and an almond amount is obtained. The ratio obtained in this step is directly related to the almond mass. This relation is acquired by a simple linear regression and almond mass is obtained directly from the image itself.

2.4 Methodology for the Determination of the Homogeneity of Almond Spread

Firstly, gray-level-co-occurrence matrices (GLCM) are created from binary images with a properly chosen offset value. The offset value defines the direction and distance that is under processing in GLCM [31]. Offset value plays an important role in this step since it affects correlation value which will be obtained later implicitly. GLCM denotes how often a pixel co-occurs with another pixel that is adjacent to each other. These pixels are both on a gray level, determined by segmentation. After GLCMs are created, texture analysis is done on the image. This analysis gives results of various statistics from GLCMs. The statistics include information on contrast, correlation, energy, and homogeneity of the image. The correlation value is obtained at this stage.

3 Results and Discussion

3.1 Determination of Almond Mass

To obtain a relationship with the white color ratio directly with the almond mass, instead of quantity, a singular almond's mass is taken as 1.2 g [29, 32]. Thus, from a simple calculation, the total almond mass in subgroups of 6, 8, 10, and 12 almonds are 7.2, 9.6, 12, and 14.4 g. The white color ratio is obtained for all images, which is illustrated in Fig. 4. Here, the white color ratio of every image is shown on a bar plot in Fig. 4 given below. X-axis denotes the image number, which is until 20 for homogeneous images, and the rest is non-homogeneous. Y-axis denotes the white color ratio. It is seen that for every number of almonds on the chocolate topping, the ratio of white comes to the same point.

From the data acquired, a model is proposed for the relation between total almond mass on the image that is created and mean of white color ratios of subgroups including 6, 8, 10 and 12 almonds. It is clearly seen from Fig. 5 that there is a linear relationship between two values of almond mass and the mean of images. For that reason, linear regression is applied to proposed fit. The proposed model to be utilized in determination of is given in Eq. (1). The obtained relationship between the real mass of total almond and white. Here, the R^2 value is found as "1", which was expected since the images are created artificially. There is no overlapping or any kind of quality defect. However, when it is worked with real images the resulting R^2 value is expected to be slightly lower. The relationship between the total almond mass and the white color ratio is further illustrated in Fig. 5.

$$TAM = 259.2WCP - 0.04518 \quad (1)$$

where TAM is the total almond mass and WCR is the white/total color ratio.

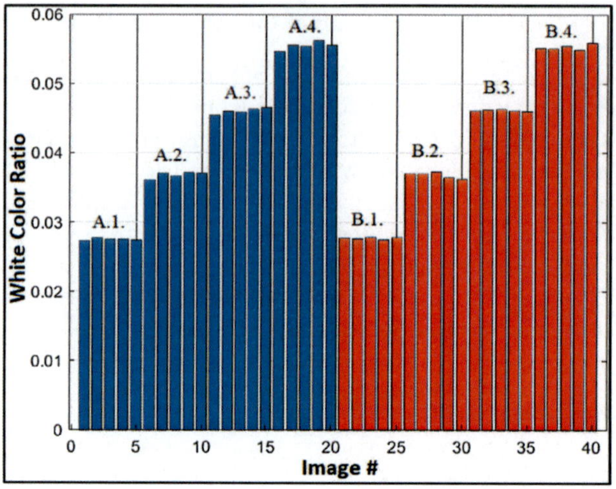

Fig. 4. The white color ratio corresponding to the image labels.

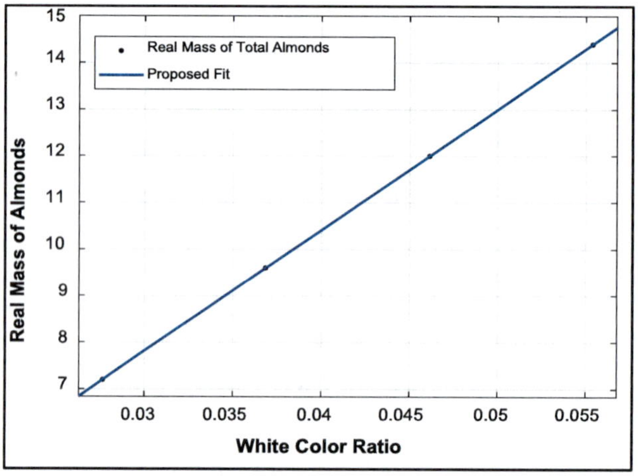

Fig. 5. The relationship between the real mass of almond and the mean of white color ratio.

3.2 Determination of Homogeneity of Almond Spread

As stated before, the offset value of the GLCM affected the results of the study depending on the value that is given. An offset value of [15 14] was found to yield the best performance by the trial and error method. When low values are set as offset, the accuracy of the determination of homogeneity by correlation value of the GLCM is reduced. The same problem is raised when high values are set as an offset. It is also seen that when numbers in offset become distant from each other, the accuracy is reduced again. The co correlation values of 40 binary images are obtained and shown in a bar plot in Fig. 6 below.

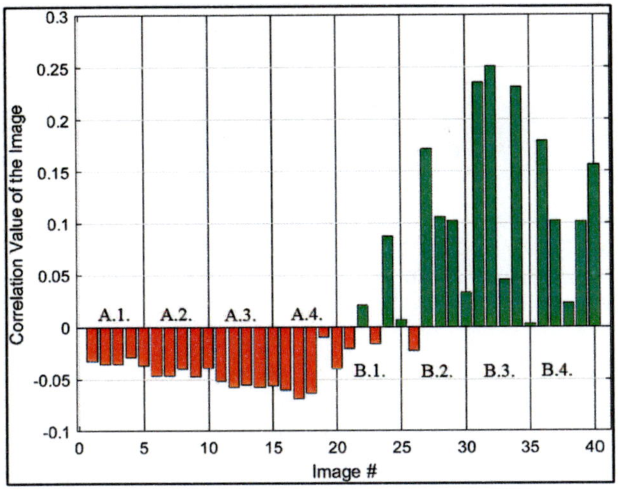

Fig. 6. The correlation values of the gray-co-occurrence-matrix of images. Red bars indicate homogenous predictions while green bars indicate non-homogenous predictions.

From Fig. 6, it can be said that when an image is homogeneous, the correlation value gives a negative number. Hence, the algorithm may be defined simply as, if the correlation value of the GLCM is below zero, the almond spread is homogenous, and vice-versa. The algorithm works exceptionally well, yielding only 3 errors (on the 21st, 23rd, and 26th images) among 40 data. The accuracy of this approach yields 92.5%, which is an acceptable result. Besides this error, correlation value of 35th image is found to be very close to "0", also a negativity of the prediction model. However, the errors may also be explained by the inaccurate labeling of the images. Errors that occurred on the 21st, 23rd, and 26th images might stem from the low number of almonds on the image. When the almond number was low, creating non-homogeneous images was a struggle. Low numbers can spread better and they can show a homogeneous appearance. On the contrary to our decision of homogeneity, the algorithm responded differently. For example, the 23rd and 26th images are settled as non-homogeneous data, however, the algorithm took these clustered almonds as a whole piece that looks located properly and showed a homogeneous correlation value. Thus, the correlation value was not obtained as high as other non-homogeneous images. To further illustrate the success of the proposed model, the confusion matrix (see Fig. 7) of the model is created from the results presented in Fig. 6. Here, x-axis denotes the numbers of images which are predicted from the model, and y-axis denotes true numbers of images which are artificially created. The model predicted homogeneous ones as 20/20 true and gave zero results for homogeneous images as non-homogeneous. The model predicted only 3 non-homogeneous images as homogeneous and scored 3/20, therefore the second row ends with 17 since the rest is predicted correctly. The success of the proposed rule-based algorithm is further illustrated here.

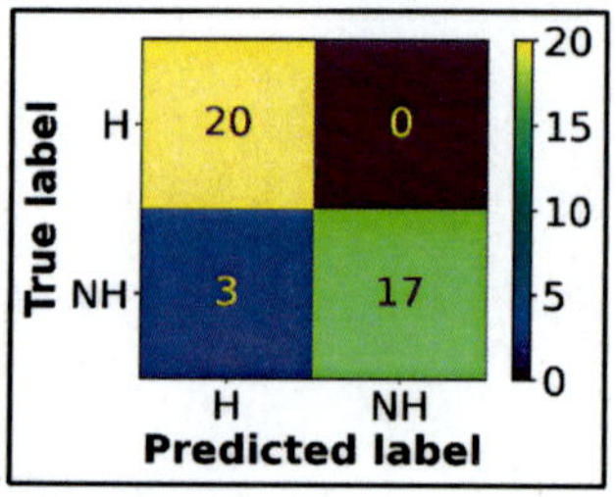

Fig. 7. Confusion matrix for the proposed homogeneity determination rule (H: Homogenous, NH: Non-homogenous).

4 Conclusion

In this study, homogeneity and mass of the almond pieces were aimed to investigate chocolate-coated ice cream topped with almond pieces. For the determination of the mass of almond pieces, an average value approximation is performed. The R^2 value of the proposed model yielded 1, showing excellent data representation. This was expected since the images were created artificially. In applications with real images, a similar approach may be followed, which will likely result in an R^2 value between 0.9 and 1. To determine the homogeneity of almond spread, firstly, binary images are created using threshold-based segmentation. Then, the gray-co-occurrence matrix is used with the offset matrix value of [15 14]. According to our results, it is seen that when we change the offset value for the, the answer for the homogeneity detection changes. Numerous offset values failed to give accurate results. In this study, the offset value is chosen by the trial error method. An optimization study is required for the determination of the best offset value in the next stages to improve this approach. In accordance with our detection, we conclude that correlation value can show which images are homogeneously spread. Only 3 of 40 images were misclassified, yielding 92.5% accuracy for the determination of the homogeneity. Conclusively, this study improves the methodology for determining the mass of almonds on a chocolate coating and the homogeneity of the almond spread. The proposed image processing approaches here can successfully be utilized in various approaches in industry, where determining the amount or homogeneity is of importance.

Acknowledgements. This work was conducted as an academic-corporate collaboration between Yildiz Technical University and Unilever San. ve Tic. Türk A.Ş. And was supported by TÜBİTAK under "2209-B Industry Oriented Research Project Support Program for Undergraduate Students" program.

References

1. Du, C.J., Sun, D.W.: Recent developments in the applications of image processing techniques for food quality evaluation. Trends Food Sci. Technol. **15**(5), 230–249 (2004). https://doi.org/10.1016/j.tifs.2003.10.006
2. Kiran, R., Amarendra, H.J., Lingappa, S.: Vision system in quality control automation. MATEC Web Conf. **144**, 03008 (2018). https://doi.org/10.1051/matecconf/201814403008

3. Massaro, A., Vitti, V., Galiano, A.: Automatic image processing engine oriented on quality control of electronic boards. Signal Image Process **9**(2), 01–14 (2018). https://doi.org/10.5121/sipij.2018.9201
4. Belaid, A., Haton, J.P.: Image processing in quality control of nuts. IFAC Proc. Volumes **22**(6), 381–387 (1989)
5. Nie, J., Wang, Y., Li, Y., Chao, X.: Artificial intelligence and digital twins in sustainable agriculture and forestry: a survey. Turkish J. Agric. For. **46**(5), 642–661 (2022). Turkiye Klinikleri. https://doi.org/10.55730/1300-011X.3033
6. Gumus, M., Balaban, O., Unlusayin, M.: Machine vision applications to aquatic foods: a review. Turk. J. Fish Aquat. Sci. **11**(1) (2011). https://doi.org/10.4194/trjfas.2011.0124
7. Nasirahmadi, A., Behroozi-Khazaei, N.: Identification of bean varieties according to color features using artificial neural network. Spanish J. Agric. Res. **11**(3), 670–677 (2013). https://doi.org/10.5424/sjar/2013113-3942
8. Baiocco, G., Almonti, D., Guarino, S., Tagliaferri, F., Tagliaferri, V., Ucciardello, N.: Image-based system and artificial neural network to automate a quality control system for cherries pitting process. Procedia CIRP **88**, 527–532 (2020). https://doi.org/10.1016/j.procir.2020.05.091
9. Li, Q., Wang, M., Gu, W.: Computer vision based system for apple surface defect detection. www.elsevier.com/locate/compag
10. Eyarkai Nambi, V., Thangavel, K., Shahir, S., Thirupathi, V.: Comparison of various RGB image features for nondestructive prediction of ripening quality of "Alphonso" mangoes for easy adoptability in machine vision applications: A multivariate approach: comparison of various RGB image features. J. Food Qual. **39**(6), 816–825 (2016). https://doi.org/10.1111/jfq.12245
11. Sun, D.-W., Brosnan, T.: Pizza quality evaluation using computer vision-part 1 Pizza base and sauce spread. www.elsevier.com/locate/jfoodeng
12. Du, J., Sun, D.W.: Multi-classification of pizza using computer vision and support vector machine. J. Food Eng. **86**(2), 234–242 (2008). https://doi.org/10.1016/j.jfoodeng.2007.10.001
13. Pedreschi, F., León, J., Mery, D., Moyano, P.: Development of a computer vision system to measure the color of potato chips. Food Res. Int. **39**(10), 1092–1098 (2006). https://doi.org/10.1016/j.foodres.2006.03.009
14. Vibhute, A., Bodhe, S.K.: Applications of image processing in agriculture: a survey. Int. J. Comput. Appli. (0975 – 8887), **52** (2012)
15. Mathworks, 'Image Processing Toolbox User Guide, Properties of a gray-level co-occurrence matrix'. http://matlab.izmiran.ru/help/toolbox/images/graycoprops.html. Accessed 24 Jan 2023
16. Sun, D.-W.: Computer vision technology for food quality evaluation. Elsevier/Academic Press (2008)
17. Brosnan, T., Sun, D.W.: Improving quality inspection of food products by computer vision - A review. J. Food Eng. 61(1), 3–16 (2004). SPEC. Elsevier Ltd. https://doi.org/10.1016/S0260-8774(03)00183-3
18. Zhanna, L., Vyacheslav, L.: Color Space Image As A Factor In The Choice Of Its Processing Technology Zeleniy Oleksandr, Tabakova Iryna
19. Pedreschi, F., Mery, D., Marique, T.: Quality Evaluation and Control of Potato Chips and French Fries (2007). https://doi.org/10.1016/B978-0-12-373642-0.50025-9
20. Wikipedia, 'HSV and HSL'. 18 Dec 2020. https://tr.wikipedia.org/wiki/HSL_ve_HSV. Accessed 02 Mar 2023
21. VOCAL Technologies, 'RGB and HSV/HSI/HSL Color Space Conversion'. Accessed 09 Mar 2023

22. Naranjo-Torres, J., Mora, M., Hernández-García, R., Barrientos, R.J. Fredes, C., Valenzuela, A.: A review of convolutional neural network applied to fruit image processing. Appl. Sci. (Switzerland), **10**,(10). MDPI AG, 01 May 2020. https://doi.org/10.3390/app10103443
23. Introduction to Image Segmentation - Image Segmentation - Image Processing (2020). https://www.youtube.com/watch?v=JToLE6gaZzs. Accessed 21 Mar 2023
24. Vineetha, G.R., Beevi, A.A.: Survey on different methods of image segmentation. Int. J. Sci. Eng. Res. **4**(4) (2013). http://www.ijser.org
25. An Introduction to Image Segmentation: Deep Learning vs. Traditional, 02 Mar 2023. Available: Image Segmentation: Deep Learning vs Traditional [Guide] (v7labs.com). Accessed 22 Mar 2023
26. Mokji, M.M., Abu Bakar, S.A.R.: Adaptive Thresholding Based On Co-Occurrence Matrix Edge Information (2007)
27. 'Cluster-based Image Segmentation -Python (2020). https://towardsdatascience.com/cluster-based-image-segmentation-python-80a295f4f3a2. Accessed 22 Mar 2023
28. Bovik, C.: Basic Binary Image Processing. In: The Essential Guide to Image Processing, Elsevier Inc., pp. 69–96 (2009). https://doi.org/10.1016/B978-0-12-374457-9.00004-4
29. Vidyarthi, S.K., Tiwari, R., Singh, S.K.: Size and mass prediction of almond kernels using machine learning image processing. https://doi.org/10.1101/736348
30. Kwok, N.M. , Ha, Q. P., Fang, G.: Effect of color space on color image segmentation. In: Proceedings of the 2009 2nd International Congress on Image and Signal Processing, CISP'09 (2009). https://doi.org/10.1109/CISP.2009.5304250
31. Image Processing Toolbox User's Guide. http://matlab.izmiran.ru/help/toolbox/images/enhanc15.html. Accessed 21 Mar 2023
32. Mahmoodi, M., Khazaei, J., Mohamadi, N.: Modeling of geometric size distribution of almond. Int. J. Food Prop. **14**(5), 941–953 (2011). https://doi.org/10.1080/10942910903501872

Efficient and Reliable Surface Defect Detection in Industrial Products Using Morphology-Based Techniques

Ertugrul Bayraktar(✉)

Department of Mechatronics Engineering, Yildiz Technical University, 34349 Istanbul, Besiktas, Turkey
eb@yildiz.edu.tr

Abstract. Quality is a measurement-based criteria that specifies the conformity of final products to certain rules and agreements. Monitoring product quality has always been critical, cost-effective and time-intensive process during manufacturing. Surface defects have major negative impacts on the quality of industrial products. Human inspection for visual quality control is challenging and less reliable due to the influence of physical and psychological factors on the auditor, including fatigue, stress, anxiety, working hours, and environmental conditions. Considering these risks, it is impossible for humans to deliver satisfactory stable performance over a long period of time and without interruption. Moreover, with the advancements in hardware and software, quality control can be done fast, reliable, efficient, and repeatable regardless of duration. Herein, we propose morphology-based image processing approach that enables detecting the scratch or alike defects with a width of 70 μm, which corresponds to 85 pixels of a 5181 × 5981 pixels image or 154 μm corresponding 225 pixels of a 7484 × 7872 pixels image. A scanner-based device is employed to capture the images and we combined dilation, closing, median filtering as well as gradient taking and edge detection, then contour finding. Our algorithm exceeds the performance of hand-crafted feature-based methods on detecting tiny defects within very large images, which even outperforms the modern deep learning-based methods when there is not/enough training data thanks to not requiring training data. The inference time of our approach for an image is less than 1 s and consequently is capable of being exploited online surface defect detection applications robustly.

Keywords: Surface defect · Morphology · Image Processing · Quality Inspection · Fault Detection

1 Introduction

Visual inspection is an integral part of quality control in the manufacturing industry, where the products are evaluated to ensure they meet certain standards and specifications. Traditional visual inspection methods have relied on human inspectors to identify and classify defects, which can be time-consuming, subjective, and prone to errors. With

Z. Şen et al. (Eds.): IMSS 2023, LNME, pp. 81–94, 2024.
https://doi.org/10.1007/978-981-99-6062-0_9

the advent of computer vision, the use of deep learning-based methods has gained popularity due to their high accuracy and flexibility. However, deep learning-based methods require large amounts of labeled data, which can be challenging and expensive to acquire in industrial settings. In contrast, classical image processing methods offer a more robust and efficient alternative for visual inspection in the manufacturing industry. These methods rely on a set of pre-defined image processing techniques to enhance and segment the image, followed by the application of feature extraction and classification algorithms to detect and classify defects. In this context, we argue that classical image processing methods are a valuable and reliable tool for visual inspection in the manufacturing industry, particularly in scenarios where labeled data is scarce or the inspection process needs to be performed in real-time.

Classical image processing methods moreover offer several advantages over deep learning-based methods. First, classical methods are more interpretable and explainable, as they rely on a set of pre-defined techniques that can be easily understood and traced back to their origin. Second, classical methods are less sensitive to changes in lighting conditions, color variations, and other environmental factors that may affect the quality of the image. Third, classical methods are computationally more efficient and can be implemented on low-cost and low-power devices, making them a practical solution for industrial applications. Therefore, it is important to explore the potential of classical image processing methods for visual inspection in the manufacturing industry and to investigate their performance in different scenarios and conditions.

Surface defect detection systems based on visual data processing operate on two common principles: i) built-in area or line scan camera-based systems incorporated into production lines with integrated installation as illustrated in Fig. 1, and ii) systems that perform part inspection by being placed in a specific location as a separate device from the system. However, both approaches have serious limitations in terms of cost, flexibility, and the shape and size of products that can be inspected. Algorithms for visual inspection are processed online in the first approach, and inspections are made on selected samples in the second approach, but both methods are relatively slow and can cause disadvantages in terms of time, cost, and efficiency. Herein, we focus on the algorithms that run on the first approach, which mainly consists of a belt conveyor, two encoders for measuring conveyor movement, a line scanning camera placed perpendicular to surfaces, adjustable lighting sources, and an embedded computer for data processing and conveyor control. The encoders prevent the line-shaped pixels from overlapping, while the line scan camera reduces disruptive effects caused by glare on metal surfaces. Controlled lighting eliminates negative effects of the external environment on the image processing algorithm. The geometric relations obtained by camera placement enable the conversion of pixel dimensions to millimeters in the real-world space, allowing for the detection of defect regions. The graphics processing unit-based embedded computer processes images in real-time, enabling quick detection of defects and sizes on metal surfaces. Production can be slowed down, stopped, or production-related input control values can be changed if necessary.

The proposed technique presented in this paper makes significant contributions to the field of surface defect detection by addressing the challenge of detecting small-sized

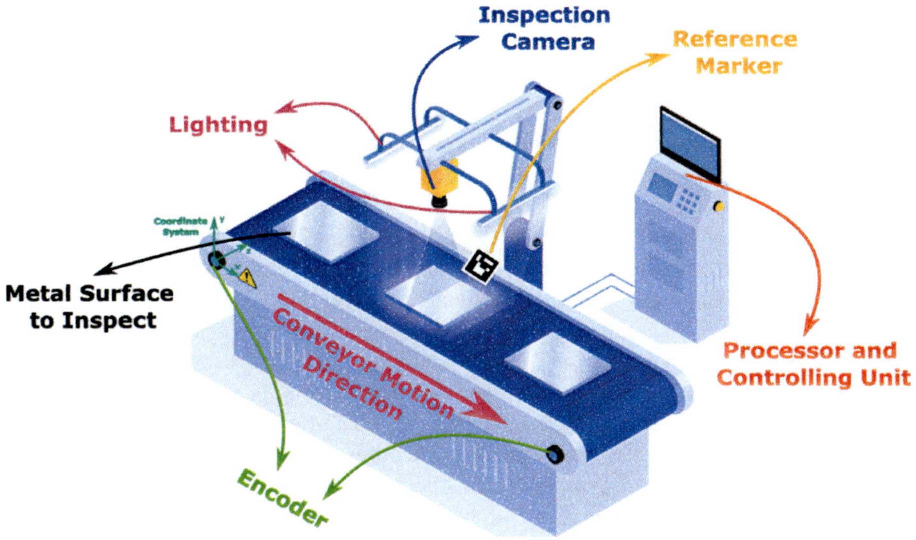

Fig. 1. Schematic representation of the visual quality control system capable of detecting surface defects and measuring size during production line flow, integrated with the conveyor.

defects with a width of 70 μm. The use of morphology-based image processing methods allows for precise and accurate detection of such small defects, which is essential in ensuring high-quality production output. Furthermore, the use of a scanner-based device for image capture enables the system to achieve high resolution and capture fine details of the surface. In addition, the appropriate combination of conventional image processing algorithms, such as dilation, closing, median filtering, gradient taking, and edge detection, followed by contour finding, contributes to the effectiveness of the proposed technique. The use of these algorithms allows for the enhancement of the image and the extraction of relevant features, which are then used for defect classification. The proposed algorithm exceeds the performance of handcrafted feature-based methods on the task of detecting tiny defects within very large images, which even performs better than modern deep learning-based methods when there is no or enough training data thanks to not requiring training data. The inference time of the proposed method for an image is less than 1 s and is capable of being exploited in online surface defect detection applications robustly. That classical image processing methods are a valuable and reliable tool for visual inspection in the manufacturing industry, particularly in scenarios where labeled data is scarce or the inspection process needs to be performed in real-time. Classical methods are more interpretable and explainable, less sensitive to changes in lighting conditions and color variations, and computationally more efficient, and can be implemented on low-cost and low-power devices. The proposed technique contributes to the field of surface defect detection by addressing the challenge of detecting small-sized defects with a width of 70 μm. The use of morphology-based image processing methods allows for precise and accurate detection of such small defects, which is essential in ensuring high-quality production output. The proposed technique has potential applications in various industries, such as the semiconductor, automotive, and electronics industries, where high-precision and accurate detection of defects is critical for ensuring the quality of the final product. Overall, the contributions of this paper lie in the development of a novel technique that addresses the challenge of detecting small-sized defects, which has the potential to improve the quality of production output and reduce costs associated with defective products.

The remainder of this study is organized as follows: Sect. 2 provides an overview of related work on visual inspection and defect detection, highlighting the limitations of existing methods and the need for new techniques. In Sect. 3, we describe our proposed methodology, which combines morphology-based image processing algorithms to effectively detect scratch or alike defects with a high level of accuracy. We also provide details on the hardware and software used in our experiments. In Sect. 4, we present the experimental results of our proposed technique, which demonstrates its effectiveness in detecting defects of varying sizes and shapes. We compare our results to those obtained using other commonly used techniques and demonstrate the superiority of our approach in terms of accuracy, efficiency, and applicability to a wide range of applications. Finally, in Sect. 5, we discuss the limitations of our proposed technique and potential areas for future research. We also present our conclusions and highlight the significance of our study in the context of visual inspection and quality control. We also discuss the practical applications of our technique in industry and provide recommendations for its implementation.

2 Related Works

Numerous studies have been carried out on defect detection in manufacturing, utilizing a wide range of techniques, from classical image processing to deep learning methods. Prior research in the field of defect detection has concentrated on several techniques, including machine learning algorithms, deep learning architectures, and statistical methods, each of which has its own strengths and weaknesses. For this reason, we will primarily discuss the necessary classical image processing-based methods and briefly cover the sufficient deep learning algorithms.

It is clear that learning-based methods have been preferred intensely recently due to their performance. The dependence of deep learning techniques on data has also shown itself in this field, and almost all the researchers focus on data-specific analysis and solutions for replication. In particular, the authors in [1] systematically studied camera-based condition monitoring and fault detection of machined surfaces of machine parts in terms of hardware configurations such as camera types, placement, lighting, DAGM2007 introduced in [2], the performances of visual feature descriptors and diagnostic decision-making methods in areas such as classification, roughness assessment, defect detection were analyzed, with emphasis on the NEU [3] industrial datasets. Modern learning-based algorithms for defect detection have been developed for a wide range of applications from fabric to tomato and some other plant, from steel to railways, from electronics to web products [4–11]. However, they all suffer from lack of generalization and being up-to-date for more flexible manufacturing capability.

Modern learning-based algorithms have been widely studied and applied for defect detection in various industries, such as fabric, plants, steel, railways, electronics, and web products. While these methods have shown promising results in detecting defects, they often suffer from a lack of generalization and adaptability to evolving manufacturing processes. In addition, these methods are always data-hungry and require large amounts of labeled data for training, which can be expensive and time-consuming to obtain.

Industrial defect detection using classical image processing algorithms has been extensively studied in the literature. Various techniques have been proposed for detecting defects in different types of products. One of the most commonly used techniques is thresholding [12–14], where the image is segmented based on a threshold value to separate the defects from the background. Edge detection, which involves identifying the edges in the image and using them to detect defects, is another popular technique. Morphological operations, such as erosion and dilation, have also been used to enhance the image and detect defects [14, 15]. Recently, advanced techniques such as wavelet-based methods and texture analysis have been proposed for defect detection [16, 17]. These techniques can capture more complex features of the defects and provide more accurate detection results.

Classical image processing techniques have proven to be effective in detecting defects in industrial products. They require less computational resources and can provide accurate results with minimal training data. Additionally, they are more interpretable and can be easily modified to suit specific applications. However, these methods have limitations in their ability to handle complex defect patterns and require careful selection of image processing techniques and parameters for optimal performance.

Several classical image processing-based algorithms have been proposed for industrial defect detection, including texture analysis, feature-based approaches, morphological filtering, edge detection, and contour analysis techniques. One of the earliest works in this field is the study presented in [18], where a texture analysis method was proposed for detecting surface defects in castings. Another popular method for industrial defect detection is the morphological approach, which involves applying morphological operations, such as erosion, dilation, and opening, to the image to extract features and detect defects. For instance, the study in [19] proposed a morphological filtering approach for detecting cracks in steel plates, which achieved high accuracy and robustness to noise. Other studies have also explored the use of edge detection and contour analysis techniques for detecting defects in industrial products. For example, the authors in [20] used a combination of Canny edge detection and Sobel edge detection to extract edge features and detect surface defects in steel plates. In addition to these methods, there are several other classical image processing techniques that have been used for industrial defect detection, including Fourier analysis, wavelet transform, and histogram-based approaches.

In essence, classical image processing techniques offer a practical and effective solution for industrial defect detection, particularly in applications where acquiring labeled data is challenging or not feasible. These techniques provide a more targeted and efficient solution for surface defect detection in industrial products, while also requiring less time and resources to develop and implement compared to deep learning-based approaches. However, selecting the appropriate image processing techniques and parameters is critical for optimal performance, and these methods have limitations in their ability to handle complex defect patterns.

3 Methodology

The use of classical image processing-based methods for surface defect detection in industrial products has several advantages over modern deep learning algorithms. Firstly, classical image processing methods require less computational resources and can provide accurate results with minimal training data. Secondly, they are more interpretable and can be easily modified to suit specific applications. Thirdly, classical image processing methods are more robust to changes in lighting conditions and noise than deep learning-based methods. Finally, they can be more efficient in detecting small defects in large images. In this paper, we present a technique consisting of morphology-based image processing methods for efficient and reliable surface defect detection in industrial products. We employ conventional image processing algorithms such as dilation, closing, median filtering, gradient taking, edge detection, and contour finding to detect surface defects. Our approach provides a practical and effective alternative to deep learning-based methods, with the added advantages of computational efficiency, interpretability, and robustness to lighting and noise. Deep learning algorithms typically require large amounts of labeled data to learn features and patterns in the data, which can be time-consuming and expensive to acquire. In contrast, classical image processing methods can often provide accurate results with minimal training data, making them a more practical option for applications where acquiring labeled data is challenging or not feasible. Additionally, classical image processing methods can be designed to specifically detect certain types of defects, which can be advantageous in industrial settings where certain types of defects are more common. This means that classical image processing methods can provide a more targeted and efficient solution for surface defect detection in industrial products, while also requiring less time and resources to develop and implement compared to deep learning-based approaches.

In this study we consider the products which have darker background and brighter background. For both cases, we developed two different approaches. In the proposed methodology for darker backgrounds, the first step is to analyze the noise model of the image. Based on this analysis, the image is smoothed to reduce noise using appropriate filtering techniques for which the necessary equation is $SmoothedImage = \sum I(x, y) * G(x, y, \sigma)$ where $I(x, y)$ is the input image with spatial coordinates of (x, y), $G(x, y, \sigma)$ is Gaussian kernel with standard deviation σ. After smoothing, Canny edge detection is applied to identify the edges of the defects in the image. Next, the image is subjected to dilation and closing of which the structural elements are 3D cubes. This process helps to fill gaps in the edges and remove small, isolated defects. A median filter is then applied to remove any remaining noise in the image. Further, a closing operation with a rectangular kernel is performed to connect the edges of the defects, followed by a gradient operation with an ellipse kernel to enhance the edges. Finally, the contours of the defects are found using the algorithm introduced in [20]. The proposed methodology leverages a combination of various image processing techniques to accurately detect and classify defects in industrial products. The flowchart of the proposed methodology for products that have darker backgrounds is shown in Fig. 2. The methodology can be easily adapted to suit different types of industrial products and defect types, making it a versatile and effective solution for industrial defect detection.

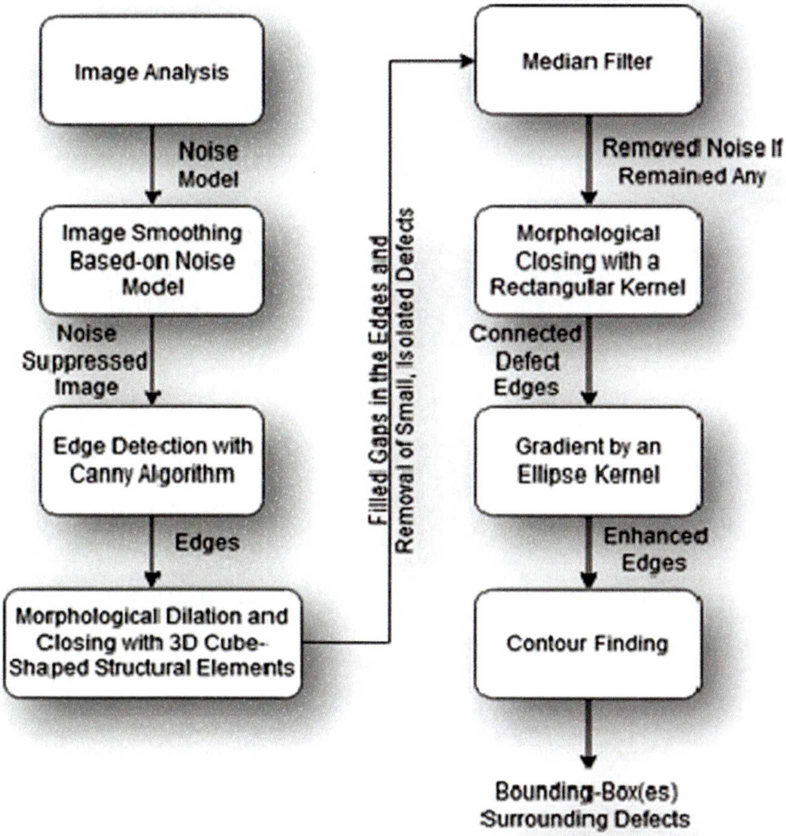

Fig. 2. Flowchart for the proposed methodology for defect detection in industrial products, which have darker background.

When dealing with products that have brighter backgrounds, we need to modify our methodology to accommodate the differences in the images. We start with frequency domain analysis to separate the background from the foreground. We then apply a Gaussian low-pass filter to remove high-frequency noise [21]. Next, we perform an ellipse-shaped closing operation with a kernel size of 21 to fill in any gaps in the foreground. To further enhance the foreground, we apply a rectangular-shaped tophat morphological operation with a kernel size of 21. This operation helps to remove any remaining background noise and artifacts. Finally, we apply a threshold to the image to convert it to a binary image, followed by contour detection to locate the defects. We then project the bounding box location onto the original image to visualize the defects. By following these steps, we can effectively detect defects in products with brighter backgrounds. Figure 3 depicts the flowchart of the suggested approach for dealing with products that possess brighter backgrounds.

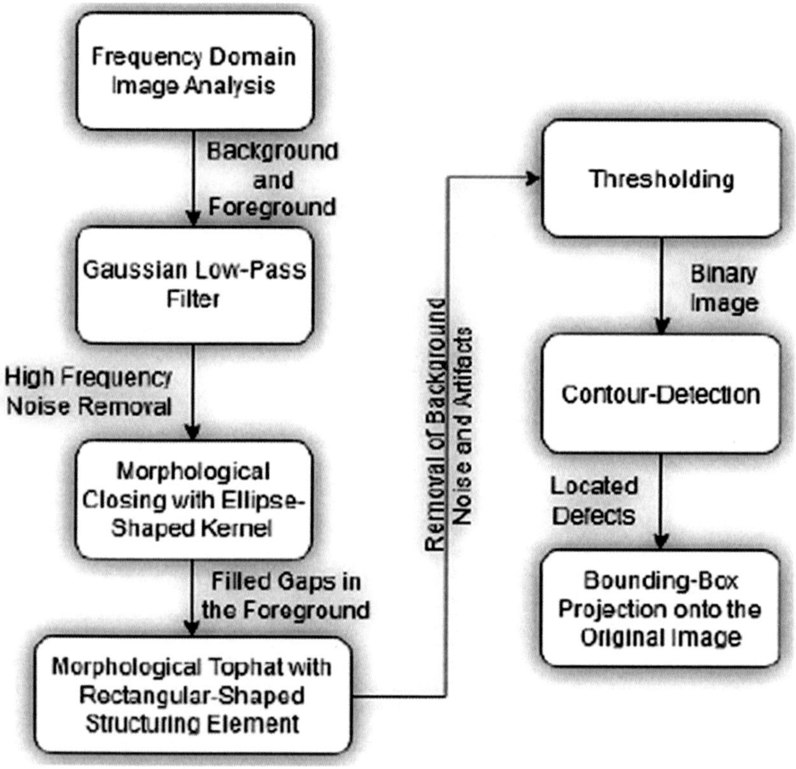

Fig. 3. Flowchart for the proposed methodology for defect detection in industrial products, which have brighter background.

4 Experiments

Many of the recent studies propose the superiority of deep learning algorithms on defect detection [22], however very few of them, which can be ignorable, discuss their limitations. Because they assume that there is always enough number of training images, which in fact is not the case when it comes to practice. Given the limited amount of data available for our specific problem and considering that classical image processing methods have shown to be reliable and efficient for similar tasks, we propose to focus our experiments on comparing the performance of classical methods versus modern learning-based algorithms. We acknowledge that deep learning-based approaches have shown remarkable results in large-scale datasets, but due to the limited amount of data available for our problem, classical methods may be more suitable. In addition, we limit our experiments to two images with known defects, in order to better understand the capabilities and limitations of each method. Therefore, in this section, we present a thorough evaluation of the performance of classical image processing methods compared to deep learning-based algorithms, in the context of our specific problem.

The images selected for this study are given in Fig. 4, which have darker and brighter contexts (due to capacity issues, we cropped the these from the original images, which have 5181x5981 pixels and 7484x7872 pixels, respectively.)

Fig. 4. Cropped images from the original ones, which have different backgrounds, defect types located and oriented at different positions and angles.

We also explored the recently introduced SAM algorithm [23], a very recent deep learning model that promises to effortlessly segment any object in an image with a single click (human feedback), without requiring additional training. Although SAM demonstrates zero-shot generalization capabilities to unfamiliar objects and images, our experiments indicate that it struggles to provide satisfactory results in the absence of explicit guidance. Specifically, we observed that SAM struggles to detect defects in images without clear annotations and requires human feedback in the form of clicking around the defect location to achieve reliable detection. While SAM represents an exciting development in AI-assisted defect detection, our findings highlight the limitations of current deep learning algorithms in fully automating the defect detection process. We present the results of our SAM experiments in Fig. 5.

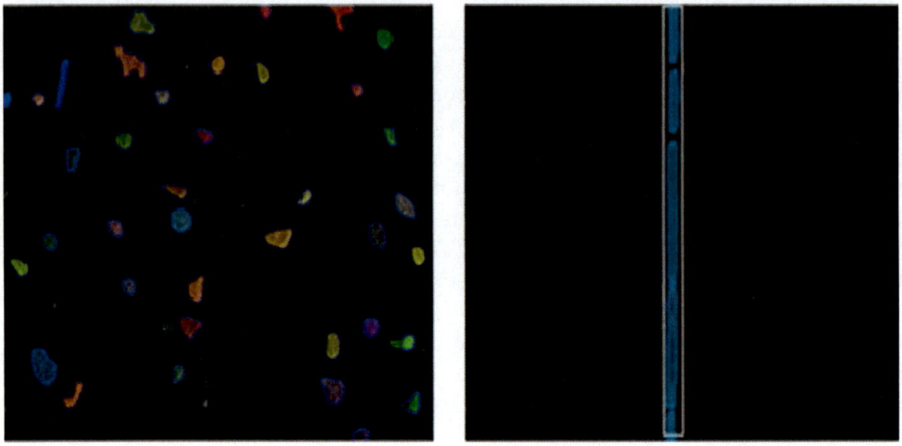

Fig. 5. The outputs of SAM algorithm; without indicating the approximate position of the defect and indicating the region where defect is located.

For the second test image in our study, SAM produced the correct output for the defect as shown in Fig. 6, but it could not succeed if there is no human feedback.

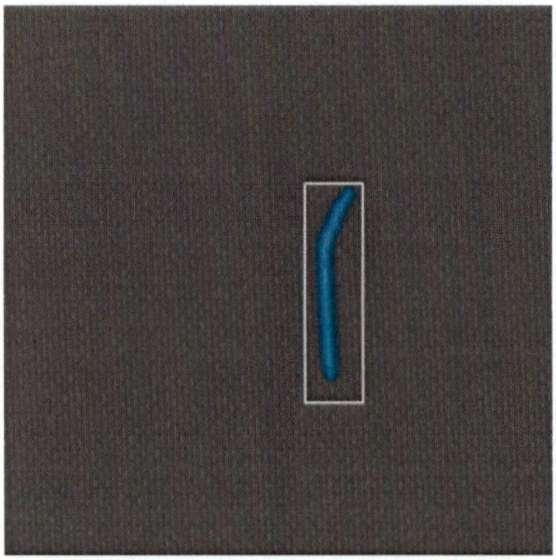

Fig. 6. The output of SAM algorithm for the second image in our study when the approximate region of the defect is indicated by the human user.

Our proposed algorithm achieved promising results in detecting defects in the given images. As shown in Fig. 7, the first image presents the binary image of the defect where the white pixels represent the defect area. Our algorithm utilized classical image processing techniques, including noise model analysis, smoothing, edge detection, dilation, median filtering, and gradient detection to accurately detect the defect areas. The second image of the figure shows the bounding box that is projected onto the original image based on the positions detected in the binary image. This allows for a clear visualization of the defect area in the original image. The success of our approach highlights the effectiveness of classical image processing techniques in defect detection and provides a viable alternative to the commonly used deep learning algorithms.

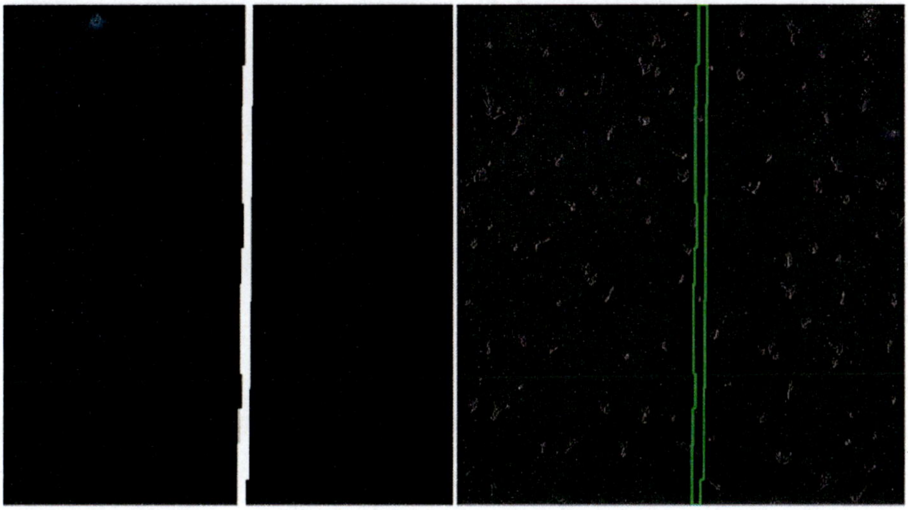

Fig. 7. The binary image on the left shows the defect detected by our approach for images that have darker backgrounds, and the image on the right shows the original test image with the bounding box projected onto it based on the positions of the white pixels in the binary image.

The binary image and its corresponding bounding box projection onto the original image, achieved through our second algorithm for brighter backgrounds, are presented in Fig. 8. To obtain these results, we applied a series of image processing techniques, including frequency domain analysis, Gaussian low pass filtering, ellipse-shaped closing with a kernel size of 21, rectangular-shaped top-hat morphological operation with a kernel size of 21, thresholding, contour detection, and projecting the bounding box location onto the original image. Our approach effectively detects surface defects in industrial products with brighter backgrounds, providing a practical and efficient solution for defect detection in manufacturing settings.

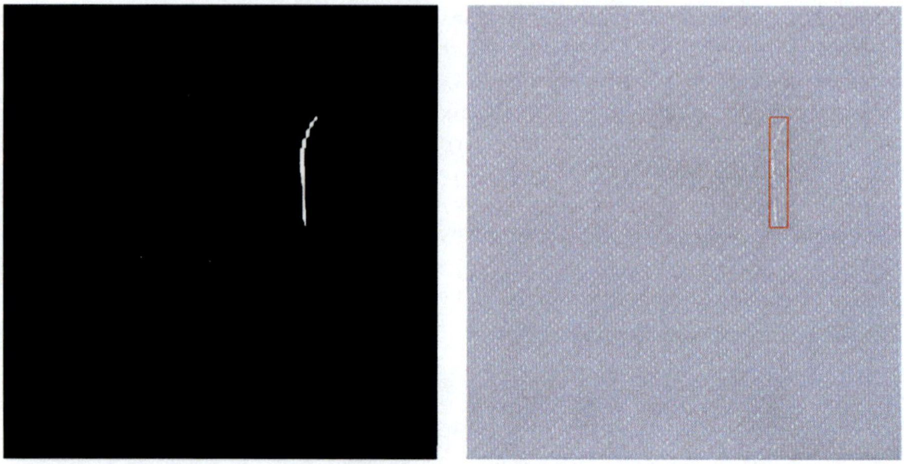

Fig. 8. The image on the left displays the output of our approach for images with brighter backgrounds, showing the detected defect in binary form. On the right is the original test image, with the bounding box projected onto it based on the locations of white pixels in the binary image.

5 Conclusions

We have presented two practical and efficient techniques for surface defect detection in industrial products using classical image processing methods. Our approach for darker backgrounds includes a series of image processing techniques such as frequency domain analysis, Gaussian low pass filtering, ellipse-shaped closing, and rectangular-shaped top-hat morphological operation, while for brighter backgrounds, we employed a combination of image enhancement, segmentation, feature extraction, and defect classification. These techniques detect surface defects accurately and efficiently and provide a more practical solution for applications where acquiring labeled data is challenging or not feasible. Our results show that our approach outperforms handcrafted feature-based methods and modern deep learning-based methods, including SAM, in detecting small defects in large images. However, we acknowledge the limitations of our approach, which are heavily reliant on the quality of the input images and may not be suitable for detecting highly complex or irregular-shaped defects.

In the course of our study, we encountered several limitations that should be considered when interpreting our findings. Firstly, our approach is heavily reliant on the quality of the input images, and therefore, the accuracy of the defect detection process is largely dependent on the quality of the images captured. Secondly, our technique may not be suitable for detecting defects that are highly complex or have irregular shapes, as our feature extraction and classification methods may not be able to capture all the relevant information required for accurate detection. Finally, although classical image processing techniques can be more efficient and robust than deep learning-based methods, they may not perform as well in situations where large amounts of data are available and deep learning methods may be more appropriate. Despite these limitations, our approach provides a practical and effective solution for defect detection in industrial products, and we believe that it has the potential to be widely adopted in various industrial settings.

In essence, our work provides a promising alternative to deep learning-based approaches for surface defect detection in industrial products. Our approach can be easily customized and adapted to specific applications, making it a practical and effective solution for defect detection in manufacturing settings. To improve the proposed approach, further experimentation could be conducted to evaluate the performance of the algorithm under different lighting conditions and color variations. Additionally, the use of other conventional image processing algorithms, such as wavelet transforms, could be explored to further improve the accuracy of the system. Moreover, the proposed approach does not require labeled data, making it a practical solution for industrial applications where labeled data is scarce. However, to validate the robustness of the proposed approach, future research could investigate the performance of the system on datasets with various types of defects and surface materials. Furthermore, while classical image processing methods offer several advantages over deep learning-based methods, future research could investigate the combination of both methods to further improve the accuracy of surface defect detection systems. In terms of practical implications, the proposed approach could be implemented in manufacturing industries to automate the quality control process, reduce the risk of human error, and improve the efficiency of production output. The proposed approach could also be used in real-time quality control applications, where production can be slowed down, stopped, or production-related input control values can be changed if necessary. Overall, the proposed approach offers a valuable and reliable tool for visual inspection in the manufacturing industry, particularly in scenarios where labeled data is scarce or the inspection process needs to be performed in real-time. Further research in this area could provide valuable insights and practical solutions for industrial applications.

References

1. Liu, Y., Guo, L., Gao, H., You, Z., Ye, Y., Zhang, B.: Machine vision based condition monitoring ve fault diagnosis of machine tools using information from machined surface texture: A review. Mech. Syst. ve Sig. Proc. **164**, 108068 (2022)
2. Wieler, M., Hahn, T.: Weakly supervised learning for industrial optical inspection. In DAGM symposium in (2007)
3. Song, K., Yan, Y.: A noise robust method based on completed local binary patterns for hot-rolled steel strip surface defects. Appl. Surf. Sci. **285**, 858–864 (2013)
4. Wu, J., et al.: Automatic fabric defect detection using a wide-and-light network. Appl. Intell. **51**(7), 4945–4961 (2021). https://doi.org/10.1007/s10489-020-02084-6
5. Dong, H., Song, K., Wang, Q., Yan, Y., Jiang, P.: Deep metric learning-based for multi-target few-shot pavement distress classification. IEEE Trans. Industr. Inf. **18**(3), 1801–1810 (2021)
6. Zhu, Q., Dinh, T.H., Phung, M.D., Ha, Q.P.: Hierarchical convolutional neural network with feature preservation and autotuned thresholding for crack detection. IEEE Access **9**, 60201–60214 (2021)
7. Zhang, D., Song, K., Wang, Q., He, Y., Wen, X., Yan, Y.: Two deep learning networks for rail surface defect inspection of limited samples with line-level label. IEEE Trans. Industr. Inf. **17**(10), 6731–6741 (2020)

8. Schlagenhauf, T., Lvewehr, M.: Industrial machine tool component surface defect dataset. Data Brief. **39**, 107643 (2021)
9. Sindagi, V.A., Srivastava, S.: Domain adaptation for automatic OLED panel defect detection using adaptive support vector data description. Int. J. Comput. Vision **122**(2), 193–211 (2017). https://doi.org/10.1007/s11263-016-0953-y
10. Saeed, N., King, N., Said, Z., Omar, M.A.: Automatic defects detection in CFRP thermograms, using convolutional neural networks and transfer learning. Infrared Phys. Technol. **102**, 103048 (2019)
11. Huang, Y., Qiu, C., Yuan, K.: Surface defect saliency of magnetic tile. Vis. Comput. **36**(1), 85–96 (2020)
12. Ng, H.F.: Automatic thresholding for defect detection. Pattern Recogn. Lett. **27**(14), 1644–1649 (2006)
13. Win, M., Bushroa, A.R., Hassan, M.A., Hilman, N.M., Ide-Ektessabi, A.: A contrast adjustment thresholding method for surface defect detection based on mesoscopy. IEEE Trans. Industr. Inf. **11**(3), 642–649 (2015)
14. Malge, P.S., Nadaf, R.S.: PCB defect detection, classification and localization using mathematical morphology and image processing tools. Int. J. Comput. Appl. **87**(9) 2014
15. Mak, K.L., Peng, P., Yiu, K.F.C.: Fabric defect detection using morphological filters. Image Vis. Comput. **27**(10), 1585–1592 (2009)
16. Han, Y., Shi, P.: An adaptive level-selecting wavelet transform for texture defect detection. Image Vis. Comput. **25**(8), 1239–1248 (2007)
17. Li, W.C., Tsai, D.M.: Wavelet-based defect detection in solar wafer images with inhomogeneous texture. Pattern Recogn. **45**(2), 742–756 (2012)
18. Haralick, R.M., Sternberg, S.R., Zhuang, X.: Image analysis using mathematical morphology. IEEE Trans. Pattern Anal. Mach. Intell. **4**, 532–550 (1987)
19. Li, B., Zhang, P.L., Wang, Z.J., Mi, S.S., Zhang, Y.T.: Gear fault detection using multi-scale morphological filters. Measurement **44**(10), 2078–2089 (2011)
20. Melotti, D., Heimbach, K., Rodríguez-Sánchez, A., Strisciuglio, N., Azzopardi, G.: A robust contour detection operator with combined push-pull inhibition and surround suppression. Inf. Sci. **524**, 229–240 (2020)
21. Zhang, Y., Li, T., Li, Q.: Defect detection for tire laser shearography image using curvelet transform based edge detector. Opt. Laser Technol. **47**, 64–71 (2013)
22. Bayraktar, E., Tosun, B., Altintas, B., Celebi, N.: Combined GANs and classical methods for surface defect detection. In 2022 30th Signal Processing and Communications Applications Conference (SIU), pp. 1–4. IEEE (2022)
23. Kirillov, A., eta l.: Segment anything (2023). arXiv preprint arXiv:2304.02643

Sustainable Supplier Selection in the Defense Industry with Multi-criteria Decision-Making Methods

Beste Desticioglu Tasdemir[1](✉) and Merve Asilogullari Ayan[2]

[1] Department of Operations Research, National Defence University Alparslan Defence Sciences and National Security Institute, Ankara, Turkey
bdesticioglu@kho.msu.edu.tr

[2] Department of Defence Management, National Defence University, Alparslan Defence Sciences and National Security Institute, Ankara, Turkey
masilogullari@kho.msu.edu.tr

Abstract. In order for countries to have strong armies, they should attach importance to defense industry projects. Therefore, countries allocate high resources to defense industry investments in order to be a deterrent against their enemies. In recent years, the concept of sustainability has started to gain importance in the defense industry, as in many other sectors. Sustainability practices in the defense industry play an active role in the success of defense industry projects. One of the most important links of sustainability is sustainable supply chain management (SCM). Therefore, in this study, the problem of sustainable supplier selection (SSS) in the defense industry has been examined. Since multi-criteria decision-making (MCDM) methods are widely used in supplier selection, MCDM methods were also used in this study. In the study, first of all, sustainable supplier selection criteria (SSSC) in the defense industry were determined by making use of expert opinions and studies in the literature. The weights of the determined criteria were determined using the Analytical Hierarchy Process (AHP) method. In the last part of the study, a comparison was made between different suppliers working in the defense industry by using the Fuzzy Technique for Order Preference by Similarity to Ideal Solution (FTOPSIS) method and the weights previously determined by the AHP method, and the best supplier was determined as a result of the calculations.

Keywords: Sustainable Suppler Selection · Defense Industry · AHP · TOPSIS · Multi Criteria Decision Making

1 Introduction

Companies have recently been compelled to make some challenging decisions in order to enhance organizational structures, cut costs, and produce high-quality items in a competitive environment. For organizations, it becomes essential and difficult to evaluate multiple factors, compare alternatives carefully, and come to reliable judgments. Making the right choices is essential to achieving the short- and long-term goals of the organizations. Finding sustainable solutions is equally important in the current environment. Companies that implemented the green supplier concept and the make-to-order

Z. Şen et al. (Eds.): IMSS 2023, LNME, pp. 95–106, 2024.
https://doi.org/10.1007/978-981-99-6062-0_10

concept at the same time are required to take into account environmentally friendly waste management practices and product specifications when choosing and assessing green suppliers. Also, each competing supplier has a quantitative-qualitative assessment with certain and uncertain facts that must be taken into account to satisfy the needs of the company and the final product [1]. Depending on the sector the business operates in, the supplier selection issue must be explored, assessed, and modified in order to support the green transition [2].

One of the firms that ought to choose supplier is the defense sector. High-cost decisions that could have a national impact are made in this organization. The quest of maximizing military capabilities necessitates a disciplined approach in the sense of weighing all options and options to match military capabilities. There are numerous drawbacks to these solutions. Budgetary restrictions, outdated weapon system platforms, and recently developed (or ill-defined) capabilities are a few of these [3]. Now a global phenomenon, the transformation spreading across the supply chain of the defense sector presents both huge potential and serious risks to the industry's supply base. Prime contractors in the defense sector have dramatically increased their outsourcing over the past ten years, but at the same time, they are drastically reducing the number of small and medium-sized businesses that will be involved in their supply chains going forward and expanding their geographic reach for suppliers globally. When access to lower-cost suppliers around the world is made possible by e-commerce, this is expected to rise even higher [4].

Making the greatest choice among a variety of options is the foundation of making the best use of the limited resources that are available to businesses. In this case, methods like MCDM are used to choose the best feasible solution for the objective while taking into account conflicting criteria. When implementing cutting-edge technological solutions, the decision makers (DMs) must consider a variety of factors in order to choose the best suppliers, which may depend on the situation and internal protocols. This makes determining how many criteria to include in the MCDM methods a complex issue [5]. As a result, strategic choices must be made to guarantee the organization's continued success.

The next parts of the study are listed as follows: In the second part, literature review, and in the third part, the methodology related to AHP and FTOPSIS are included. In the last part, 5 defense industry companies serving in Turkey are listed with sustainable supply chain criteria by AHP and FTOPSIS method and the results are interpreted. The decision diagram determined for the defense industry supplier companies with the determined criteria is given in Fig. 1.

2 Literature Review

SCM plays a role in establishing an environmentally friendly industry. This is so because the supply chain combines a number of operations, including purchasing, production, reverse logistics, distribution, and marketing [1]. Buying is undoubtedly one of the most crucial company actions [6]. The environment surrounding the purchasing process includes participation in initiatives to minimize, reuse, and recycle goods [7]. Determining the policy for selecting and evaluating green suppliers is crucial to promoting green procurement [8]. Due to the growing need for environmental and social responsibility, sustainable supplier selection has become a crucial issue in the military industry [9, 10]. Because the defense sector has a big impact on society and the environment [11, 12], it's crucial to choose suppliers who can adhere to sustainability standards [13]. Some studies have looked at this topic, but the literature on SSS in the defense industry is still in its infancy.

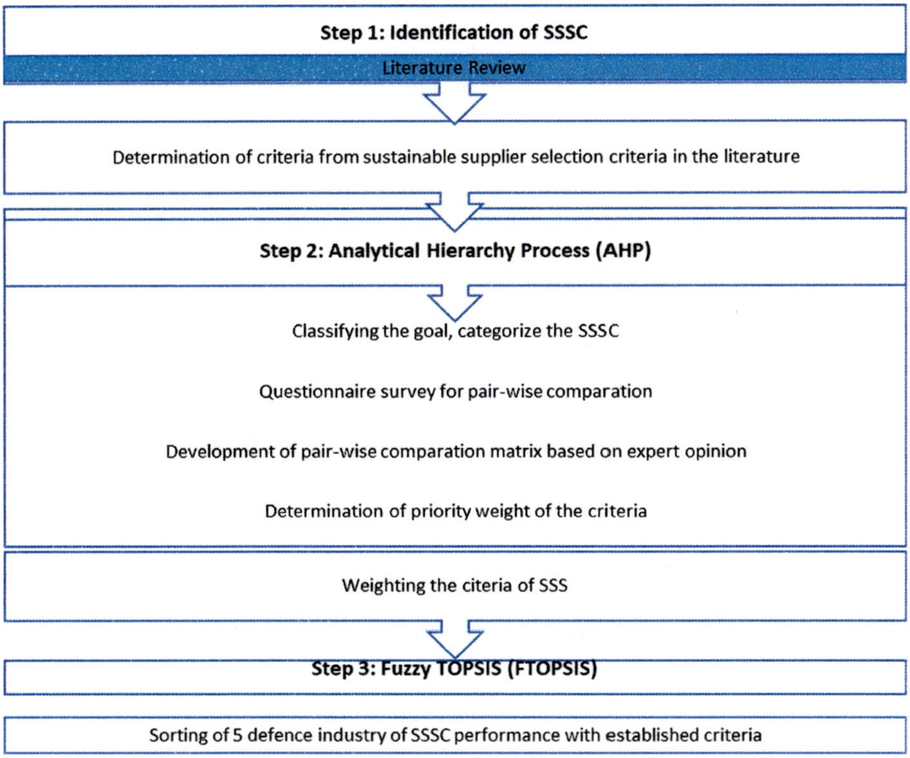

Fig. 1. Framework of proposed methodology.

A supplier evaluation model and a supplier segmentation model were created by Dogan and Sahin (2019) as part of an integrated approach for the military industry. The method was used in a case study of a Turkish military industry, and the outcomes demonstrated its potential to aid in the identification of sustainable suppliers and enhance supply chain sustainability [14]. For the military industry, Zhang and Zhou (2016) suggested a fuzzy comprehensive evaluation method that takes into consideration social, economic, and environmental concerns. The method's effectiveness in assessing the sustainability of suppliers was demonstrated by a case study of a Chinese manufacturer of military equipment [15]. An integrated methodology for SSS and order allocation in the military industry was put out by Chofreh and Ogunlana (2019). The model assesses supplier performance and distributes orders using AHP and fuzzy logic. The results of the model's application to a case study of a Malaysian military company revealed that it may efficiently choose environmentally friendly suppliers and distribute orders to reduce costs and environmental consequences. When choosing suppliers for the aerospace and defense sector [16]. Rasmussen et al. (2023) recommend using MCDM techniques with a focus on sustainability factors. They compare the outcomes of three MCDM approaches used to rate suppliers before advising the VIKOR method for this sector. The authors contend that MCDM techniques can enhance the viability of supplier decision-making in this sector [17]. Environmental performance, social responsibility, economic sustainability, and quality are a few of the important factors mentioned in the literature when choosing sustainable suppliers. MCDM processes like the AHP and the Analytic Network Process (ANP) are the most frequently used methodologies for assessing the sustainability of suppliers.

3 Methodology

3.1 Analytic Hierarchy Process (AHP)

An alternative and choice criteria hierarchy is built using the AHP process. The hierarchy is divided into layers, where the highest level represents the overall decision's goal, the next level the standards by which the alternatives were judged, and the bottom level the alternatives themselves. AHP employs pairwise comparisons to determine the relative importance of each criterion and option [18]. The values used for these comparisons range from 1 (which denotes equal importance) to 9. AHP then determines a numerical score for each alternative depending on how well it performed against each criterion once the pairwise comparisons have been completed [19]. This rating shows how well the option performed overall relative to the decision's goal. Several industries, including engineering, business, and healthcare, have used AHP. AHP derives global weights for evaluation at the lowest level [19] after assigning weights to the elements at each level using pairwise comparison and assessing their relative importance using Saaty's 1–9 scale (Table 1).

It has, for instance, been used to choose vendors, assess investment possibilities, and rank medical procedures. The following five major steps make up AHP [20]:

1. Create a decision hierarchy: List the major decision objective, the criterion, and the potential outcomes.
2. Pairwise comparisons: During the decision-making process, participants rate each pair of criteria or options to indicate which is more crucial or preferred. A single pairwise comparison matrix is created by taking the geometric mean of pairwise comparison matrices formed from the opinions of more than one expert. This matrix is then normalized.
3. Determine priorities: To determine the overall priorities or weights for each criterion or option, pairwise comparison scores are applied.
4. Calculating the eigenvector of the criteria: The eigenvector of the criteria is calculated by the formula:

$$W_i = \frac{1}{n} \sum\nolimits_{j=1}^{n} \frac{a_{ij}}{\sum\nolimits_{j=1}^{n} a_{ij}} \tag{1}$$

After determining the relative importance of the criteria by calculating the eigenvector, the consistency ratio (CR) of the comparison matrix is calculated.

Table 1. Saaty's 1–9 scale.

Intensity of importance a_{ij}	Definition	Explanation
1	Equal importance	Two activities contribute equally to the objective.
3	Moderate importance	Experience and judgment slightly favor one activity over another.
5	Strong importance	Experience and judgment strongly favor one activity over another.
7	Very strong or demonstrated importance	An activity is favored very strongly over another; its dominance demonstrated in practice.
9	Extreme importance	The evidence favoring one activity over another is of the highest possible order of affirmation.
2, 4, 6, 8	For compromise between the above values	Sometimes one needs to interpolate a compromise judgment numerically because there is no good word to describe it

5. Verify consistency: To confirm that the pairwise comparisons are valid and reliable, the CR is computed. The aim is to determine whether the decision maker acts consistently when comparing criteria. If the CR exceeds 0.10, the decision maker has to reconsider the values entered in the matrix due to inconsistency. The following equation is used to calculate the CR:

$$CR = \frac{Consistency\ Indicator}{Randomness\ Indicators} \tag{2}$$

$$CI = \frac{\lambda_{max} - n}{n - 1} \tag{3}$$

After multiplying the columns of the comparison matrix with the relative priorities, they are summed to form the weighted total vector. The arithmetic mean of the result gives λmax after the elements of the weighted sum vector are divided by their corresponding relative priority [20]. The values of the RI ratios according to the matrix measure are shown in Table 2.

Table 2. Table of randomness indicators.

n	1	2	3	4	5	6	7	8	9	10
RI	0	0	0.52	0.89	1.11	1.25	1.35	1.40	1.45	1.49

6. Sensitivity analysis: By sensitivity analysis, the effects of reordering some criteria or alternatives can be assessed.

3.2 Ftopsis

The FTOPSIS method was developed by Chen (2000) to eliminate uncertainties [21]. The FTOPSIS approach involves the following steps:

1. After determining each criterion's linguistic value, the values are transformed into the trapezoidal fuzzy numbers (TFNs) listed in Table 3 [22].

Table 3. Linguistic scale TFNs utilized in FTOPSIS.

TFNs	Linguistic Expressions
(0, 0.1, 0.2, 0.3)	Very Low
(0.1, 0.2, 0.3, 0.4)	Low
(0.3, 0.4, 0.5, 0.6)	Moderate
(0.5, 0.6, 0.7, 0.8)	High
(0.7, 0.8, 0.9, 1.0)	Very High

2. Let n be the number of criteria, m the number of alternatives, and x_{ij} the score of alternative i for the criterion j. This value is written as $x_{ij} = (a_{ij}, b_{ij}, c_{ij}, d_{ij})$ according to TFNs.
3. Fuzzy decision matrices are normalized using the following formulas to show J maximum criteria set and J1 minimum criteria set [23].

$$r_{ij} = \left(\frac{a_{ij}}{d_j^*}, \frac{b_{ij}}{d_j^*}, \frac{c_{ij}}{d_j^*}, \frac{d_{ij}}{d_j^*}\right), j \in J \tag{4}$$

$$r_{ij} = \left(\frac{a_j^*}{d_{ij}}, \frac{a_j^*}{c_{ij}}, \frac{a_j^*}{b_{ij}}, \frac{a_j^*}{a_{ij}}\right), j \in J_1 \tag{5}$$

$$d_j^* = max_i d_{ij}, j \in J \tag{6}$$

$$a_i^* = min_i\, a_{i,j}, j \in J_1 \tag{7}$$

4. The weighted decision matrix is derived by multiplying the discovered values by the weights of the criteria.
5. Negative and positive ideal solutions are calculated with the following formulas to show v_j^+ benefit and v_j^- cost [24].

$$A^- = \left(v_1^-, v_2^-, v_3^-, \ldots, v_n^-\right), v_i^- = (0, 0, 0, 0) \tag{8}$$

$$A^+ = \left(v_1^+, v_2^+, v_3^+, \ldots, v_n^+\right), v_i^+ = (1, 1, 1, 1) \tag{9}$$

6. The distances between the alternatives and the ideal solutions, both negative and positive, are determined.

$$d_i^- = \sum\nolimits_{j=1}^{n} d\left(v_{ij}^-, v_j^-\right), j = 1, 2, \ldots, m \tag{10}$$

$$d_i^+ = \sum\nolimits_{j=1}^{n} d\left(v_{ij}^+, v_j^+\right), j = 1, 2, \ldots, m \tag{11}$$

This formula is used to determine these distances:

$$D(A, B) = \sqrt{\frac{1}{4}\left[(a_1 - b_1)^2 + (a_2 - b_2)^2 + (a_3 - b_3)^2 + (a_4 - b_4)^2\right]} \tag{12}$$

7. Use the following formula to determine the closeness coefficient for each alternative:

$$C_i^* = \frac{d_i^-}{d_i^- + d_i^+}, i = 1, 2, \ldots, m \tag{13}$$

8. Among the alternatives, the alternative with the maximum closeness coefficient is selected.

4 Case Study

National defense, which is one of the basic tools of ensuring national security, constitutes the critical element of the public. The sustainability of the local defense industry is of great importance in the formation of the national defenses of the countries. SSCM should be established in order to ensure the sustainability of the defense industry. Therefore, in this study, the issue of SSS in the defense industry is discussed. In this study, it is aimed to rank 5 companies that manufacture in the defense industry in Turkey in terms of SSS. In the study, firstly, sustainable supplier selection criteria (SSSC) taken into account in the selection of defense industry suppliers and SSS in the literature were determined. The criteria determined for SSS are given in Table 4. For security reasons, the names of the experts are not included and the experts are shown as Exp.1, Exp.2..., Exp.10. Table 5 contains comprehensive information on the experts. The AHP approach was used to calculate the criteria weights using the expert ratings.

In the third stage of the study, 5 companies working in the defense industry in Turkey were evaluated by the experts with the criteria previously determined. Due to confidentiality, the names of the defense industry companies are not given, and the alternatives are indicated as DI1, DI2, ..., DI5. It was thought that these evaluations would vary from person to person, so the FTOPSIS method was used to rank the suppliers. Calculations were made with the FTOPSIS method using the scores given by the experts and 5 defense industry companies were ranked in terms of SSSC. In the following sections of the study, calculations made for SSS in the defense industry are given. The decision diagram determined for the defense industry supplier companies with the determined criteria is given in Fig. 1.

Table 4. Criteria to use for the SSS.

Criteria	SSSC	Definition
SC1	Cost [2, 8 25, 27]	The costs of the products supplied are among the most important criteria in choosing a supplier.
SC2	Quality [2, 27]	Product quality is crucial in the defense sector, and suppliers are required to follow rigorous quality control guidelines. Suppliers must therefore verify that their products comply with the appropriate requirements by implementing the requisite certifications and quality control procedures.
SC3	Flexibility [25]	To guarantee that defense agencies have access to the best goods and services at the greatest costs, while reducing risks and encouraging innovation, flexibility in the selection of defense suppliers is crucial.
SC4	Local collaboration [26]	In order to support the local economy, maintain a secure and consistent supply of defense goods and services, foster collaboration and innovation, and increase trust and understanding among many stakeholders, local collaboration in the selection of defense suppliers is crucial.
SC5	Green design [8, 26]	Defense suppliers can lower carbon emissions, waste production, and energy consumption by implementing green design concepts into their goods and services; this benefits the environment while also potentially saving money over time.
SC6	Environmental affairs [2, 28]	The environmental activities of a supplier reveal their level of dedication to minimizing their environmental impact and implementing sustainable practices.
SC7	Worker safety and health [27]	Buyers can contribute to ensuring a safe and responsible supply chain in the defense industry by taking into account a supplier's worker safety and health history.
SC8	Labor rights [2, 27]	Purchasers must play a significant part in ensuring that suppliers obey labor laws. In order to build a sustainable and ethical industry, it is crucial to promote and uphold ethical ideals across the supply chain.

Table 5. Information about experts working.

Job of Experts	Expert
Production Planning Engineer	Ex.1
Purchasing Manager	Ex.2
Production Engineer	Ex.3
Marketing Manager	Ex.4
Production Manager	Ex.5
Quality Control Engineer	Ex.6
Production Planning Manager	Ex.7
Purchasing Expert	Ex.8
Design Engineer	Ex.9
Warehouse Superviser	Ex. 10

4.1 Weighting of Ranking Criteria by AHP Method

In determining the criteria for the SSS in the defense industry, the criteria used in the selection of suppliers in the defense industry and SSS in the literature were used. Afterwards, the experts scored the determined criteria using the pairwise comparison matrix. The mean score was calculated for each criterion by taking the geometric mean of these scores, and the criteria weights were obtained by making calculations in the AHP method with these scores. Table 6 was formed by taking the geometric mean of the scores given by the experts.

Table 6. Pairwise comparison matrix.

	SC1	SC2	SC3	SC4	SC5	SC6	SC7	SC8
SC1	1	6.73	6.79	6.03	7.53	8.16	5.593	6.38
SC2	0.15	1	6.26	5.97	7.07	7.16	5.67	5.75
SC3	0.15	0.16	1	1.32	6.55	6.12	6.113	6.15
SC4	0.17	0.17	0.76	1	7.95	7.47	6.38	6.26
SC5	0.13	0.14	0.15	0.13	1	1.27	0.60	0.57
SC6	0.12	0.14	0.16	0.13	0.79	1	1.47	0.84
SC7	0.17	0.18	0.16	0.16	1.66	0.68	1	5.75
SC8	0.16	0.17	0.16	0.16	1.76	1.20	0.17	1

As a result of the calculations made with the AHP method, the weights of the criteria to be used in the SSS in the defense industry were found as shown in Fig. 2. When Fig. 2 is examined, it is seen that the "SC1" criterion has the highest weight with 0.3734. This shows that the most important criterion in determining the supplier is cost. The second highest weighted criterion is the "SC2" criterion. When choosing their suppliers, companies attach importance to the quality of the products they produce. The weights of the remaining 6 criteria vary between 0.0258 and 0.1374, and the order of the criteria is SC4 > SC3 > SC7 > SC8 > SC6 > SC5 is in the form.

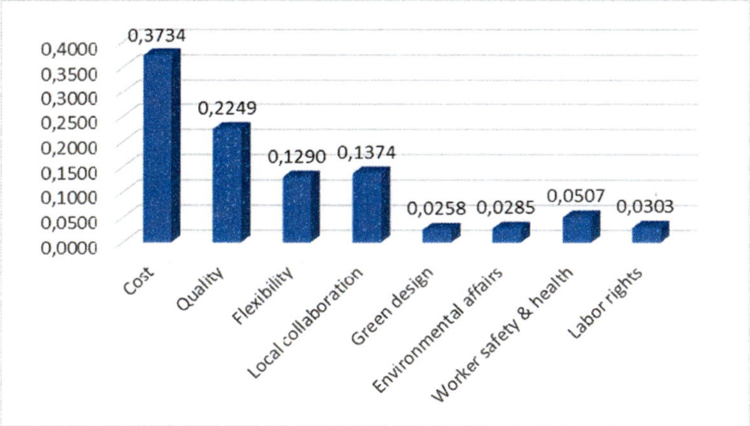

Fig. 2. Sustainable supplier selection criteria weights.

4.2 Ranking of Sustainable Suppliers in the Defense Industry with FTOPSIS

In this section, it is aimed to rank the suppliers serving the defense industry according to the criteria of SSS. In order to rank the suppliers, 10 experts working in different positions in the defense industry, whose details are given in Table 4, were interviewed and the experts were asked to rate the suppliers between 1–5 according to the determined criteria. It was thought that it would be appropriate to use fuzzy logic since the opinions of the experts may vary from person to person. Therefore, the scores given by the experts were converted into TFNs and then calculations were made with the FTOPSIS method [24]. Scoring of experts according to TFNs is given in Table 7.

Table 7. Decision matrix for evaluating the defense industry companies.

	TFNs						
	CR1	CR2	CR3	CR4	CR5	CR6	CR7
DI1	{0.01, 0.11, 0.21, 0.31}	{0.6, 0.7, 0.8, 0.9}	{0.48, 0.58, 0.68, 0.78}	{0.68, 0.78, 0.88, 0.98}	{0.6, 0.7, 0.8, 0.9}	{0.52, 0.62, 0.72, 0.82}	{0.6, 0.7, 0.8, 0.9}
DI2	{0.08, 0.18, 0.28, 038}	{0.36, 0.46, 0.56, 0.66}	{0.32, 0.42, 0.52, 0.62}	{0.5, 0.6, 0.7, 0.8}	{0.22, 0.32, 0.42, 0.52}	{0.3, 0.4, 0.5, 0.6}	{0.5, 0.6, 0.7, 0.8}
DI3	{0.07, 0.17, 0.27, 0.37}	{0.58, 0.68, 0.78, 0.88}	{0.44, 0.54, 0.64, 0.74}	{0.62, 0.72, 0.82, 0.92}	{0.58, 0.68, 0.78, 0.88}	{0.4, 0.5, 0.6, 0.7}	{0.62, 0.72, 0.82, 0.92}
DI4	{0.07, 0.17, 0.37, 0.47}	{0.48, 0.58, 0.68, 0.78}	{[0.56, 0.66, 0.76, 0.86}	{0.56, 0.66, 0.76, 0.86}	{0.28, 0.38, 0.48, 0.58}	{0.24, 0.34, 0.44, 0.54}	{0.42, 0.52, 0.62, 0.72}
DI5	{0.1, 0.2, 0.3, 0.4}	{0.4, 0.5, 0.6, 0.7}	{0.46, 0.56, 0.66, 0.76}	{0.46, 0.56, 0.66, 0.76}	{0.24, 0.34, 0.44, 0.54}	{0.25, 0.35, 0.45, 0.55}	{0.42, 0.52, 0.62, 0.72}

The ranking obtained as a result of the calculations made with the FTOPSIS method using the converted TFNs with the criterion weights calculated by the AHP method is given in Table 8. As a result of the calculations, it is seen that DI1 is the best defense industry company in terms of SSSC. This shows that DI1 company has high performance for all 8 criteria determined for sustainable supplier selection. According to the calculations, other defense industry companies are listed as DI3 – DI4- DI2- DI5 in terms of SSSC. In order for these defense industry companies to be preferred as sustainable suppliers, they must first increase their performance starting from the criteria with high weights.

Table 8. Closeness values calculated with FTOSIS method.

Alternatives	d_i^+	d_i^-	C_i^*	Rank
DI1	13.2311	1.3643	0.0935	1
DI2	13.6590	0.7816	0.0541	4
DI3	13.4387	1.0015	0.0694	2
DI4	13.6164	0.8261	0.0572	3
DI5	13.6708	0.7674	0.0532	5

5 Conclusion

Defense industry companies operating in that country play a key role in ensuring the national security of countries. The concept of sustainability, which has appeared in many fields in recent years, has become an interesting subject in the field of defense industry as it is in many industrial branches. One of the most important links of sustainability is SSCM and SSS. Therefore, in this study, it is aimed to rank 5 defense industry companies operating in Turkey in terms of SSSC. The study consists of 3 stages. In the first stage of the study, literature research was conducted and the most preferred criteria were determined in the selection of sustainable suppliers. In the second phase of the study, the AHP approach was used to calculate the weights of the criterion based on the expert ratings. In the calculations, it has been seen that the most important criterion in the selection of sustainable suppliers is the "Cost" criterion with a weight of 0.3734. The second most important criterion was determined as Quality with a weight of 0.2249. In the last stage of the study, experts scored 5 defense industry companies operating in Turkey in terms of sustainable supplier selection criteria. Scorings were converted to TFNs and the defense industry companies were ranked by making calculations with the FTOPSIS method. According to the calculations, it has been determined that DI1 company has the best performance in terms of SSSC. In terms of SSSC, other defense industry companies are listed as DI3 - DI4 - DI2- DI5. In order to increase the preferability of these companies, first of all, the criteria with high weights should increase their performance.

In this study, 5 defense industry companies operating in Turkey were ranked using SSSC and AHP and FTOPSIS methods. In the study, the most used criteria in the literature were taken into account in determining the criteria. In future studies, SSSC can be determined by meeting with experts working in the defense industry. In addition, in future studies, suppliers can be listed using different MCDM methods and the results obtained can be compared. In future studies, calculations can be made by taking the sustainability criteria as sub-criteria of economic criteria, social criteria and environmental criteria.

References

1. Gustina, A., Ridwan, Y., Akbar, M.D.: Multi-criteria decision making for green supplier selection and evaluation of textile industry using fuzzy axiomatic design (FAD) method. In: 5th International Conference on Science and Technology ICST, pp. 1–6. IEEE, Yogyakarta, Indonesia (2019)
2. Hamdan, S., Cheaitou, A.: Supplier selection and order allocation with green criteria: An MCDM and multi-objective optimization approach. Comput. Oper. Res. **81**, 282–304 (2017)
3. Desticioğlu, B., Ayan, M.A.: Savunma tedarik konusunda yapılan çalışmaların bibliyometrik analizi. Savunma ve Savaş Araştırmaları Dergisi **32**(1), 159–196 (2022)
4. Yu, Y., Wang, X., Zhong, R.Y., Huang, G.Q.: E-commerce logistics in supply chain management: Practice perspective. Procedia Cirp **52**, 179–185 (2016)
5. Blundell, G.F., Ward, C.W.R.: Property portfolio allocation: a multi-factor model. Land Dev. Stud. **4**(2), 145–156 (1987)
6. Veugelers, R., Cassiman, B.: Make and buy in innovation strategies: evidence from Belgian manufacturing firms. Res. Policy **28**(1), 63–80 (1999)
7. Rao, P., Holt, D.: Do green supply chains lead to competitiveness and economic performance? Int. J. Oper. Prod. Manag. **25**(9), 898–916 (2005)
8. Blome, C., Hollos, D., Paulraj, A.: Green procurement and green supplier development: antecedents and effects on supplier performance. Int. J. Prod. Res. **52**(1), 32–49 (2014)
9. Deretarla, Ö., Erdebilli, B., Gündoğan, M.: An integrated analytic hierarchy process and complex proportional assessment for vendor selection in supply chain management. Decis. Analytics J. **6**, 100155 (2023)

10. Debnath, A., Roy, J.: Integrated fuzzy AHP-TOPSIS model for optimization of national defense management based on inclusive growth drivers using SWOT analysis. In: Handbook of Research on Military Expenditure on Economic and Political Resources, IGI Global, pp. 81–105 (2018)
11. Rojo, F.R., Roy, R., Shehab, E., Wardle, P.J.: Obsolescence challenges for product-service systems in aerospace and defense industry. In: CIRP Industrial Product-Service Systems Conference, p. 255 (2009)
12. Mueller, M.J., Atesoglu, H.S.: Defense spending, technological change, and economic growth in the United States. Defense Peace Econ. **4**(3), 259–269 (1993)
13. Lambin, E.F., Thorlakson, T.: Sustainability standards: Interactions between private actors, civil society, and governments. Annu. Rev. Environ. Resour. **43**, 369–393 (2018)
14. Dogan, I., Sahin, E.: An integrated approach for sustainable supplier selection in the defense industry. J. Defense Manag. **9**(1), 79–99 (2019)
15. Zhang, Y., Zhou, L.: A fuzzy comprehensive evaluation approach for sustainable supplier selection in the defense industry. J. Clean. Prod. **112**, 2027–2037 (2016)
16. Chofreh, A.G., Ogunlana, S.O.: An integrated model for sustainable supplier selection and order allocation in the defense industry. Sustainability **11**(13), 3589 (2019)
17. Rasmussen, A., Sabic, H., Saha, S., Nielsen, I.E.: Supplier selection for aerospace & defense industry through MCDM methods. Cleaner Eng. Technol. **12**, 100590 (2023)
18. Vargas, L.G.: An overview of the analytic hierarchy process and its applications. Eur. J. Oper. Res. **48**(1), 2–8 (1990)
19. Saaty, T.L.: Fundamentals of decision making and priority theory with the analytic hierarchy process. RWS Publications (1994)
20. Vaidya, O.S., Kumar, S.: Analytic hierarchy process: An overview of applications. Eur. J. Oper. Res. **169**(1), 1–29 (2006)
21. Chen-Tung, C., Ching-Torng, L., Fn Huang, S.: A fuzzy approach for supplier evaluation and selection in supply chain management. Int. J. Prod. Econ. **102**(2), 289–301 (2006)
22. Samantra, C., Datta, S., Mahapatra, S.S.: Analysis of occupational health hazards and associated risks in fuzzy environment: a case research in an Indian underground coal mine. Int. J. Inj. Contr. Saf. Promot. **24**(3), 311–327 (2017)
23. Salimov, V.: Application of topsis method with trapezoidal fuzzy numbers. Sci. Rev. **36**(1), 7377 (2021)
24. Gul, M., Ak, M.F.: A comparative outline for quantifying risk ratings in occupational health and safety risk assessment. J. Clean. Prod. **196**, 653–664 (2018)
25. Ho, W., Xu, X., Dey, D.K.: Multi-criteria decision making approaches for supplier evaluation and selection: A literature review. Eur. J. Oper. Res. **202**(1), 16–24 (2010)
26. Bağcı, H., Kurç, H.: Turkey's strategic choice: buy or make weapons? Defense Stud. **17**(1), 38–62 (2017)
27. Fallahpour, A., Olugu, E.U., Musa, S.N., Wong, K.Y., Noori, S.: A decision support model for sustainable supplier selection in sustainable supply chain management. Comput. Ind. Eng. **105**, 391–410 (2017)
28. Rouyendegh, B.D., Yildizbasi, A., Üstünyer, P.: Intuitionistic fuzzy TOPSIS method for green supplier selection problem. Soft. Comput. **24**, 2215–2228 (2020). https://doi.org/10.1007/s00500-019-04054-8

Modeling and Improvement of the Production System of a Company in the Automotive Industry with Simulation

Aysu Uğraş(✉) and Seren Özmehmet Taşan

Dokuz Eylul University, Izmir, Turkey
ugras.aysu@ogr.deu.edu.tr

Abstract. Digitalization, which began with Industry 4.0 and has grown with great momentum due to the pandemic, is becoming significantly important to industries. Companies seeking to take their place in fierce competition with the global economy have begun to integrating digitalization into processes to increase their profitability and ensure operating excellence in this regard. In this study, simulation modelling was performed by using ARENA software, which accelerates decision-making processes completely objective, and allows determining and choosing the best possible scenarios without implementing any change in present system. The proposed method was applied to reduce the penalty costs in a company operating in the automotive sector by determining operational and strategic improvement with design of experiment. The main aim of this study is maximizing the profitability of a company by decreasing penalty cost and eliminate waste and bottlenecks. As distinct from literature studies, this study provides the opportunity to optimize any production process by following the steps explained elaborately, regardless of the sector. As a result of the study, the production factors were optimized by design of experiment and the recommendations regarding the investment decisions and production factors were given to the company to minimize penalty costs within the framework of the company's constraints.

Keywords: Project-based production · Simulation Modelling · Process Analysis

1 Introduction

Many companies must enter the process of digital transformation to increase their profitability and hold their own in the fierce competition because of the technological development that has accelerated with Industry 4.0. As stated by Ebert and Duante [1], digital transformation is about introducing breakthrough technologies to increase efficiency, value creation and operational excellence.

Firms have adopted various strategies to manage such transformations and adopted technology in their way of working to increase efficiency [2]. Operational excellence offers several benefits to companies, such as reducing cycle time and thus lowering penalty costs where appropriate, ensuring higher customer satisfaction, and minimizing operating costs. Companies can be divided into two categories regarding to their strategy:

Z. Şen et al. (Eds.): IMSS 2023, LNME, pp. 107–117, 2024.
https://doi.org/10.1007/978-981-99-6062-0_11

mass production companies and project-based companies. Unquestionably, there are challenges and benefits in both working types. Since each project is unique in project-based companies, it is very difficult to achieve operational excellence compared to mass-produced companies. When studies in the literature are evaluated, there aren't many to be found mostly because of an issue with data gathering, particularly in project-based businesses.

In this study, the importance of operational excellence in project-based companies was investigated by working on the minimization of bottlenecks to achieve operational excellence. The objective of this study is to analyses the entire process to eliminate bottlenecks in material supply and production process for that matter minimize penalty costs. To achieve this objective, the ARENA simulation software is used to model the whole system composed of different facilities and transport systems. This paper is structured as follows: Chapter 2 provides a general overview of the paper. The suggested methodology for simulation modelling is described in Chapter 3. The application of the suggested methodology is presented and finally the findings of the study, recommendations and future research areas are presented in Chapter 5.

2 Problem Definition

In companies that work on a project-based basis, companies are faced with an investment dilemma, as each project has different requirements. It is very difficult to manage sources in such a way as to avoid operational difficulties and to account for costs. For example, deciding how many machines *(T)* and how many experts (such as *P, S*) will function creates a decision-making problem due to uncertain project arrival dates. A particular worker who works very hard in one project may have a lot of idle time in another project. Process efficiency must be analyzed to make right investment decision. Beyond this, the inefficiency of certain processes may be understood at first glance from some indicators without any analysis and the cost of penalties is one such indicator. The analysis consists of calculated parameters such as source utilization ratios, source queue length and time, average time spent in the system. Once the parameters for measurement have been determined, the data should be collected according to these parameters. For example, if it is desired to calculate the utilization rate of the truck, data such as loading time ($L_{i,t}$) and unloading time ($U_{i,t}$), transportation time of the truck ($Z_{i,t}$) can be collected and analyzed.

List of notations

Index
n: Number of sources.
i: Observation.
t: Arrival time.

Parameters
$L_{i,t}$: Loading time of the material that comes at time t in observation i.
$U_{i,t}$: Unloading time of the material that comes at time t in observation i.
$Z_{i,t}$: Transportation time of the material that comes at time t in observation i.

$S_{i,t}$: Technical analysis time by the supervisor of the material that comes at time t in observation i.
$T_{n,i,t}$: Processing time by technician of the material that comes at time t in observation i.
$M_{n,i,t}$: Machining time of the material that comes at time t in observation i.
$P_{i,t}$: Purchasing time by the purchasing specialist of the material that comes at time t in observation.

Decision Variables

T: Number of machines.
P: Number of purchasing specialists.
S: Number of supervisors.

3 Proposed Simulation Modeling Methodology

To reduce project delays, a simulation modelling approach that looks at the system holistically to identify and eliminate bottlenecks was proposed. It aims to select the best scenario by creating identical twins of a real system in a digital environment and applying alternative scenarios to that system. One of the best advantages of simulation modelling that allows to try out possible scenarios at no additional effort and cost.

In this study proposed methodology consists of 4 steps. The first step of proposed method is a detailed definition of the problem. The root cause of the problem must be determined using variety of root cause determination methods. The second step requires proper mapping, and this step also help to verification and validation of the simulation model. The next step is to simulate the real system using the most appropriate software. Finally, the last step is to analyses the results, test alternative scenarios and suggest improvements to the system.

4 Application of the Proposed Methodology

4.1 Step 1: Problem Definition

The proper modelling cannot be created without a clear explanation of the topics of interest [3]. In this section of the paper, restrictions and production factor is explained to better described the problem and solution alternatives. This study was carried out in a company operating in the automotive sector and incurring high penalty costs due to the inefficiency of its processes, as indicated in Table 1. The company starts by signing numerous different projects at the same time within very short delivery times to meet sales targets, increase sales volume and customer portfolio. There is a high risk of not being able to complete the project on time because of the high penalty costs.

Production factors and restrictions are given below.

- There are 2 types of materials which needs pre-machining and those do not. The parts that do not need pre-machining are called soft materials and those that do are called hard materials. 60%of the incoming parts are of the soft materials, 40% are of the hard materials. This ratio was calculated by analyzing all the necessary parts for the 16 projects.
- There are 4 machines in the workshop, three of which are used for machining soft materials while other is used for pre-machining hard materials.
- The capacity of each machine is 1 and the set-up time varies according to the material type.
- There are 4 technicians and 1 supervisor working in the workshop, and 1 purchasing specialist working in the automation factory.
- Some of the parts produced in the workshop have some problems such as defect, design, quality problem. While some of them can be corrected by reprocessing, some are scrapped.
- Remanufacturing is required for certain parts, based on manufacturing defect, design error or material quality. 54% of the parts complete the process successfully, 38% require rework and 8% are scrapped. These ratios were calculated based on a historical analysis.
- The numbers of parts that come to the workshop first and need to be reprocessed after going to the factory are different from one another. Discarded items are also recorded, so the number of items that need to be reworked and the number of items that need to be discarded are calculated as a percentage. Parts that require rework always have priority in all queues, except for the truck and the procurement specialist.
- There are 4 trucks that transport materials between facilities. Loading and unloading times specified. When the process of each part is completed, the parts are transported in bundles as it will result in very high costs to transport one by one. Regardless of the type of part, once the number of waiting in the truck queue is 5, the transportation process is started. Each truck velocity is 30 km/h, it is specified in Arena as a 500 min/meter. Occupational safety and health are the most important matters to the company, and those speed limits are set by the company.
- If there are more than 20 pieces waiting in the queue of the purchasing specialist, the pieces are directed to the queue to be manufactured in workshop. It is assumed that these parts can be manufactured in workshop due to lack of data.

Table 1 shows the financial value of the projects and Table 2 demonstrates penalty costs which explain the magnitude of the penalty cost effect. In addition, project delays are essential to the company's profitability and reputation, which also has an impact on long-term sales.

Table 1. Project Values

Project Name	Project Value
Sub-project 1	$5,200,000
Sub-project 2	$3,500,000
Sub-project 3	$1,800,000
Sub-project 4	$7,500,000
Sub-project 5	$2,000,000
Line Project 1	**$20,000,000**
Sub-project 1	$9,000,000
Sub-project 2	$7,000,000
Sub-project 3	$4,000,000
Sub-project 4	$4,000,000
Sub-project 5	$2,500,000
Line Project 2	**$26,500,000**
Modification Project 1	$500,000
Modification Project 2	$800,000
Modification Project 3	$350,000
Modification Project 4	$350,000
Modification Project 5	$970,000
Modification Project 6	$650,000

Table 2. Project Penalty Costs and Effects

Project Name	Delay (Day)	Delay Cost/Day	Delay Cost Total	Penalty Cost Effect
Sub-project 1	40	$5,200	$208,000	4%
Sub-project 2	35	$3,500	$122,500	4%
Sub-project 3	22	$1,800	$39,600	2%
Sub-project 4	50	$7,500	$375,000	5%

(continued)

Table 2. *(continued)*

Project Name	Delay (Day)	Delay Cost/Day	Delay Cost Total	Penalty Cost Effect
Sub-project 5	15	$2,000	$30,000	2%
Line Project 1			**$775,100**	**4%**
Sub-project 1	45	$9,000	$405,000	5%
Sub-project 2	30	$7,000	$210,000	3%
Sub-project 3	50	$4,000	$200,000	5%
Sub-project 4	15	$4,000	$60,000	2%
Sub-project 5	15	$2,500	$37,500	2%
Line Project 2			**$912,500**	**3%**
Modification Project 1	12	$8,000	$96,000	19%
Modification Project 2	18	$12,800	$230,400	29%
Modification Project 3	16	$5,600	$89,600	26%
Modification Project 4	5	$5,600	$28,000	8%
Modification Project 5	9	$15,520	$139,680	14%
Modification Project 6	13	$10,400	$135,200	21%

4.2 Step 2: Process Mapping

The process mapping phase provides insight into the structure of existing and eventual process designs [4]. Process mapping offers numerous advantages, particularly for simulation modelling processes. The advantages can be listed as follows: Seeing the processes holistically, defining bottlenecks, minimizing non-value-added activities, and revealing alternative flows.

The process flow was drawn as shown in Fig. 1 with the help of professionals who have worked in the various departments of the organization for a long time. This step has significant importance for verification and validation process.

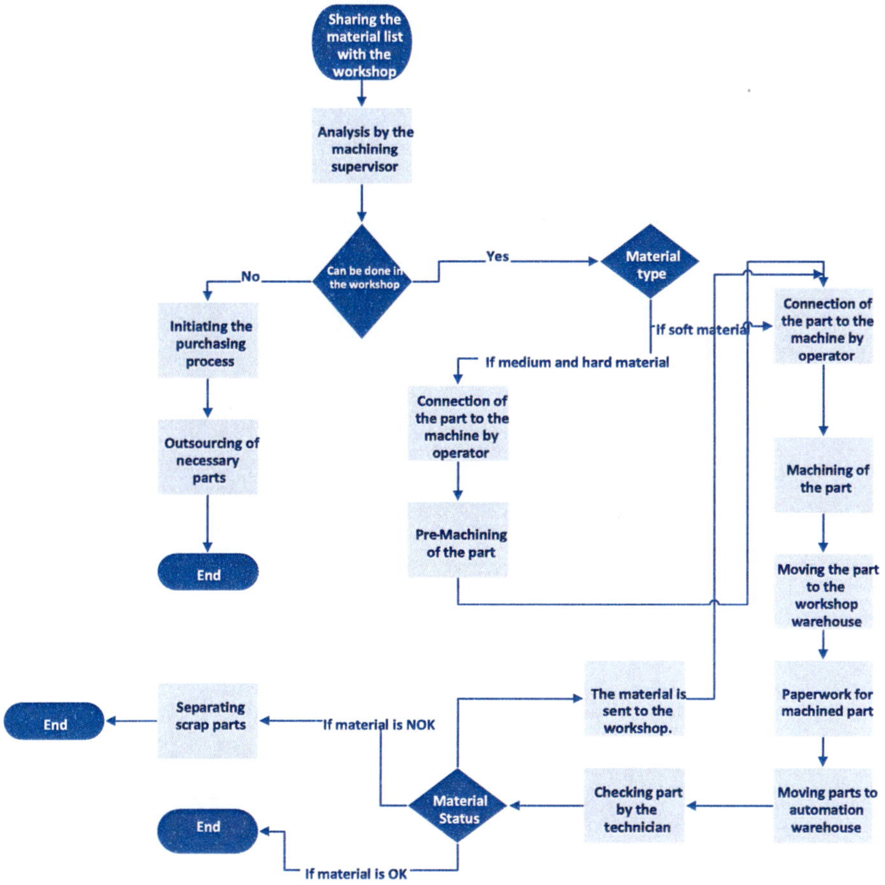

Fig. 1. Process Flow

4.3 Step 3: Simulation Modelling

Identical twin of the real system was created in a digital environment by using Arena Software. The system was run for 720 min with 50 repeats to achieve the most realistic results possible. Model verification and validation was conducted with the support of expert working in the company for a long time. Since verification and validation could be difficult for large and complex models, the model creation was started with small model which includes milestones and was added remaining element or blocks gradually.

4.4 Step 4: Output Analysis and Test Scenarios

Statistical results after running the simulation model are as in Table 3 and Table 4. Calculations were made regarding resource utilization rates, maximum, minimum, and average waiting times, and average queue length. Parameters to be measured were determined before modelling and related blocks and elements were created.

Table 3. Waiting rates for resources of current system

Tally Variables				
Waiting Times	Average	Minimum	Maximum	Observations
Machine1Q	0	0	0	46
Machine2Q	23.484	0	76,151	19
SupervisorQ	195.06	0	374,5	72
Technician1Q	0.31815	0	4,011	46
Technician2Q	0,11957	0	1,578	19
Technician3Q	64.424	0	100,39	73
Technician4Q	0.0224	0	0,44794	20
PurchaserQ	11.429	0	36,176	22
truckq	54.323	0	192,62	213

Table 4. Utilization rates for resources of current system

Utilization Rate	Average Utilization	Average Queue Length
Machine1	0,248	0
Machine2	0,692	0,619
Supervisor	0,986	39,4
Technician1	0,257	0,02
Technician2	0,218	0,00316
Technician3	0,765	6,8912
Technician4	0,137	6,2214
Purchaser	0,618	0,389
Truck	0,822	1,4482

As indicated in Table 3, Machine 1, Supervisor, Technician 3, Purchaser and Trucks have long wait times. Each item and improvement areas were discussed in detail below. As parallel with waiting times, Supervisor, purchaser, Machine 2, and trucks have high utilization rates (see Table 4). That raises some issues such as high occupancy rate, long cycle times, long waiting times, decreased quality and dependent system to specific people.

The interpretation of the results of the current system model is as follows.

- Supervisor has the longest waiting time and utilization rate. This is since a single supervisor who has high technical expertise and capability works in the automation plant and 60% of the parts are in-housed as company decision.
- There is a long waiting list for the purchasing specialist. When the number of parts on hold with the purchasing specialist exceeds 20, the parts are sent to the workshop for in-house machining. In this case, this part of the process should be improved as additional transport takes place and the workload of the workshop increases. To deal with this bottleneck, one more purchasing specialist can be hired, or the percentage of outsourced parts can be decreased with favoring workshop manufacturing. Parts are not outsourced solely for their technical feasibility, so favoring the in-house rather than outsourcing or increasing the number of procurement specialists can improve the process.
- There is a long waiting time for trucks. As the pieces are transported in groups, the wait time is not expected to be long, but the data indicates otherwise.
- One of the most important reasons for this is that 38% of the parts are returned to the workshop and reprocessed due to the quality problem. It can be regarded as a development area because it requires additional transport and machining. Another important reason is that the parts are waiting in the workshop because of the lack of communication after the parts are completed. As there is no automated equipment tracking system, finished parts cannot immediately enter the transportation queue, resulting in long queues.
- Technician 3 utilization and wait times in the queue are quite high. Since paperwork is a completely manual system, both compiling the incoming documents and entering the quality values into the system and working as a single person increase the waiting time of the parts. The improvements to be made here are thought to have a significant impact on the overall system.
- Considering the utilization rate of Machine 1 and Machine 2 and waiting times in the queue, a high waiting rate is seen in Machine 2. Due to the fact that machine 1 is used for soft and hard materials, there are 3 machines for soft materials but the long set-up time for hard materials does not consider. Therefore, it can be defined as improvement area.

Tables 5 and 6 shows the results for a test scenario where the number of machines for hard parts in the test scenario was increased to 2 by making the necessary configurations because Machine 2 has a high occupancy rate and Machine 1 does not have any waiting time. According to a company decision, the rate of in-house production of parts was set at 60%, but in the test scenario, this rate was raised to 70% because a significant bottleneck in the purchasing processes was established. An additional person was used in the test scenario due to the supervisor's high occupancy rate when performing the technical analysis of the parts.

When the waiting times of machine 1 and machine 2 in Table 5 meet with Table 3, it is seen that the machines are balanced. There was a 56% reduction in supervisor queue length. As demonstrated in Table 6, Machine 1 utilization was increased from

Table 5. Waiting rates for resources in test scenario

Tally Variables				
Waiting Times	Average	Minimum	Maximum	Observations
Machine1Q	9,45	0	17,93	29
Machine2Q	16,28	0	93,21	15
SupervisorQ	112,37	0	314,5	85
Technician1Q	0.3295	0	3,018	48
Technician2Q	0,1395	0	2,648	37
Technician3Q	52.342	0	91,39	62
Technician4Q	0.0128	0	0,446	28
PurchaserQ	5.46	0	19,176	22
truckq	54.323	0	192,62	213

Table 6. Utilization rates for resources of test scenario

Utilization Rate	Average Utilization	Average Queue Length
Machine1	0,429	0,344
Machine2	0,524	0,528
Supervisor	0,872	25,400
Technician1	0,214	0,090
Technician2	0,365	0,002
Technician3	0,695	4,836
Technician4	0,285	7,238
Purchaser	0,759	0,395
Truck	0,922	2,448

0,239 to 0,429. This 72% increase rate highly recommended to company since there is no additional cost for changing configuration settings. Supervisor utilization rate was decreased 0,986 to 0,872. The supervisor was overloaded before improvements of system. In terms of balancing the working hours, this change was also highly suggested to company provide sustainable working atmosphere. This table also proved that hiring supervisor was great opportunity to balance each supervisor utilization. The primary reason for the decrease in this rate was the employment of supervisors, and the increase in the percentage of domestic part production to 70% also affected the rate of beneficial use.

5 Conclusion

Simulation modelling is widespread and successful approach for system and sensitivity analysis. Digital software enables to create identical twin of real system to make any improvement in digital environment without physical changes and in a time-efficient manner [6]. However, this approach has some implementation difficulties, these can be solved by employing efficient tools such as problem analyzing and process mapping techniques.

Modelling project-based businesses, which is a time-consuming and complex process since each project has its own processes, as well as the drawing of the flow resulting from these variations and the gathering of data. However, given how simulation modelling affects operational excellence, it should unquestionably be regarded as a subject that requires research.

This study suggests process mapping to facilitate the time-consuming and complex process of simulation modelling. Process mapping offers a holistic overview by outlining each process's individual steps. The virtual system's use as a means of determining whether it accurately represents the real system is also a huge benefit. By providing a straightforward and usable methodology for resolving actual problems, this study seeks to improve the effectiveness of systems. Companies that produce project-based or mass-produced goods can use them for their own systems by following the guidelines in the study, regardless of the products they produce. By examining the impact of inputs and outputs on each other, the study's primary goal is to strategically identify improvement steps. When evaluating potential outcomes and selecting the better than current simulation, simulation modelling is very effective. According to the findings of this study, all conceivable scenarios have been tested, and system efficiency has been increased by selecting the scenario that best aligns with the business's strategic goals.

References

1. Ebert, C., Duarte, C.H.C.: Digital transformation. IEEE Softw. **35**(4), 16–21 (2018)
2. Colli, M., Cavalieri, S., Cimini, C., Madsen, O., Wæhrens, B.V.: Digital transformation strategies for achieving operational excellence: a cross-country evaluation (2020)
3. Greasley, A.: Using process mapping and business process simulation to support a process-based approach to change in a public sector organisation. Technovation **26**(1), 95–103 (2006)
4. Altiok, T., Melamed, B.: Simulation modeling and analysis with Arena. Elsevier (2010)
5. Law, A.M.: How to build valid and credible simulation models. In: 2019 Winter Simulation Conference (WSC), pp. 1402–1414. IEEE (2019)
6. Lugaresi, G., Matta, A.: Real-time simulation in manufacturing systems: Challenges and research directions. In: 2018 Winter Simulation Conference (WSC), pp. 3319–3330. IEEE (2018)

Prediction of Employee Turnover in Organizations Using Machine Learning Algorithms: A Decision Making Perspective

Zeynep Kaya(✉) and Gazi Bilal Yildiz

Hitit University/Industrial Engineering, Çorum, Türkiye
zeynep_k54@hotmail.com, bilalyildiz@hitit.edu.tr

Abstract. Digitalization can be defined as the transfer of activities performed in a field to digital environments. The application of digitalization in industry is revolutionary. The digitalization in industry can include applications such as collecting, analyzing, and managing company data with digital technologies, digitally monitoring and controlling the transfer of information between departments, and thus optimizing processes. In human resources management, digitalization can facilitate employee management in a variety of ways, increasing productivity and enabling better decisions. Human resources (HR) departments can develop more effective human resource management strategies by taking into account the amount of time employees are likely to work in the organization while making decisions such as incentives, bonuses, salary increases, and promotions. In this study, a decision support system is proposed to assist HR in determining the most appropriate departments for employees by predicting the potential working hours of current or new/to be hired employees in the organization. To estimate the potential work hours, we have used machine learning techniques that are widely used in the literature. We have adopted an assignment algorithm with work hour prediction to determine of the most suitable departments for employees. An application is carried out on a data set that has been published in the literature, and the results are discussed.

Keywords: Industry 4.0 · Human Resource Analytics · Machine Learning

1 Introduction

Employee turnover is a significant problem in organizations. It negatively impacts a wide range of issues, from morale and productivity to project continuity and long-term growth strategies. These problems result in a significant loss of time and money for the organization. In addition, an organization's production speed and quality can be negatively affected by the turnover of experienced and skilled employees. Therefore, predicting an employee's intention to resing gives the organization the opportunity to take preventive action. Predicting the period during which employees are likely to leave allows organizations to make decisions such as incentives, bonuses, promotions, etc. more effectively.

Z. Şen et al. (Eds.): IMSS 2023, LNME, pp. 118–127, 2024.
https://doi.org/10.1007/978-981-99-6062-0_12

Machine learning algorithms, have had great success in the prediction of future events based on historical data. In this study, a machine learning based prediction system is proposed for the solution of the employee turnover problem in organizations. Thus, the prediction of employee turnover period can be evaluated together with the performance of employees and their contributions to the organization, and can provide great benefits to organizations in determining policies according to the employees. In addition, enterprises can reconsider employees' career plans and reorganize working conditions to improve performance and productivity according to the prediction of employee turnover.

Another problem that is frequently encountered in the organizations is to determine the department in which the employees can make the greatest contribution to the organization. Performance assessment, which is considered in the framework, is a planned process that evaluates the development ability of the individual and his/her contribution to the success of the organization. It also reveals what training, reward, development and motivation the organization should provide to the employee [19]. It is not always easy to determine the appropriate department because employees have different skills, interests, and experiences. Since the determination of the appropriate departments for the employees has a direct impact on the success of the business, the solution of this problem will make a significant contribution to the business. The framework proposed in this study can be used as a decision-support system for determining the appropriate departments for employees.

In literature, turnover refers to the sum of intangible assets such as knowledge, skills, experience, creativity and other mental abilities of an employee who leaves the organization. The loss of such assets is an important factor that can reduce the value of the organization and at the same time reduce its competitive advantage [1]. The focus of this analysis is on the optimal utilization of employees. A meta-analysis review of human resource studies [2] found that the strongest predictors of retention were age, tenure, compensation, overall job satisfaction, and employee perceptions of fairness. Other similar research findings have suggested that personal or demographic variables, particularly age, gender, ethnicity, education, and marital status, are important factors in the prediction of voluntary employee turnover [3, 4]. Salary, working conditions, job satisfaction, supervision, promotion, recognition, growth potential, burnout, etc. are other characteristics that studies have focused on. [5, 6]. The frequent turnover of employees prevents the formation of a collective data base in the organization. It also reduces customer satisfaction because customers are constantly in contact with new employees. On the other hand, employee turnover leads to an undesirable situation that is the loss of employees may mean the loss of valuable knowledge with them, so it may cause the loss of competitive advantage [7]. Therefore, an organization should minimize employee turnover as much as possible to maintain its competitive advantage. Finding the reasons for employee turnover and preventing it is vital for an organization [8]. However, the use of heuristic methods by managers for this purpose can be difficult and time-consuming due to the consideration of many factors such as employee demographics and working conditions. The use of predictive analytical approaches can provide optimal combinations of employees and departments in the organization by giving managers a general idea of employee resignation rates [9].

The study involves selecting five different regression models for the dataset, comparing their performance, and selecting the best one. Based on this model, an infrastructure for a decision support system has been created. The proposed decision support system uses the results of the regression models as coefficients of the assignment problem to determine the appropriate departments for each employee.

This paper is organized in the following manner: Sect. 2 describes the algorithms used in this paper and their mechanisms. The characteristics of the data set, its preprocessing, and the exploratory data analysis are analyzed in Sect. 3. Section 4 presents the results of the study.

2 Methodology

Machine learning techniques are effective in making predictions. These techniques automatically identify patterns and links in data using statistical algorithms and computer methodologies, which can then be applied to forecast upcoming occurrences or outcomes. They rely on learning from historical/training data to map relevant dependent output variables to new input records based on appropriate independent variable values. Due to their capability to handle complicated correlated factors and their effectiveness in dealing with correlated variables, it is crucial to employ modern forecasting algorithms to obtain the best accuracy.

We can forecast staff turnover rates thanks to the benefits and predictive capabilities of machine learning algorithms. Using machine learning algorithms, a network of regression models was established, and its prediction outputs were utilized to generate an assignment problem. The predictions derived from the regression models were then considered as coefficients of the assignment problem. As a result, a decision-support system was created to identify the most appropriate departments for employees. It also provides information on employee turnover rates to human resources management.

Another problem faced by companies is determining the appropriate departments for employees. With the information obtained in the estimation of the employees' turnover rates, it is possible to predict in which departments the employees will work longer, and this information can be taken into account when determining the employees' departments. Therefore, this information can be used as a parameter in an assignment problem. The estimation of working hours can be used with objective coefficients of the assignment model to determine the most suitable department for employees.

2.1 Machine Learning Algorithms

Five prediction algorithms were used in this study. They are Extra Trees Regression, Random Forest Regression, Bagging Regression, LightGBM Regression, and XGBoost Regression. These five prediction algorithms are used to predict the values of the target variables using the characteristics of the samples. Extra Trees is built by combining many random trees to avoid overlearning. Random Forest is also an ensemble method of combining trees and is designed to produce low-variance and low-bias predictions. Bagging LightGBM is an ensemble method that combines many LightGBM trees and is designed to produce faster predictions. XGBoost is a gradient boosting method that adds new trees by focusing on the errors of previous trees and is popular in many machine learning applications.

Random Forest (RF) is a tree-based ensemble method that was developed to address the shortcomings of the traditional Classification and Regression Tree (CART) method. RF consists of a large number of simultaneously grown weak decision tree learners and is used to reduce both the bias and variability of the model [10]. RF uses bagging to increase the diversity of the trees, which in turn are grown from different training data sets, thus reducing the overall variability of the model. RF makes it possible to assess the relative importance of input features, which is useful for dimensionality reduction to improve model performance in high-dimensional data sets. RF changes an input variable while holding other input variables constant and measures the average reduction in the model's prediction accuracy, which is used to assign a relative importance score to each input variable [11].

The Extra Tree (ET) algorithm is a relatively new machine learning technique that was developed as an extension of the Random Forest algorithm and is less likely to overfit a data set [12]. ET uses the same principle as random forest and uses a random subset of features to train each base predictor. However, it randomly selects the best feature and the corresponding value to split the node. ET uses the entire training data set to train each regression tree [13].

Bagging regression is a parallel ensemble approach that deals with the propagation of a prediction model by including additional training data. This training data is added to the original set using a data imputation method. For each new set of training data, certain observations can be repeated during sampling. After bagging, the probability of each element in the reconstructed data set is the same. Increasing the size of the training data set has little effect on the predictive power. However, if the variation is adjusted to fit the desired result, the variation in the prediction can be significantly reduced. Each set of this dataset is automatically used to train new models [14].

XGBoost is a newly developed machine learning technique that has recently been widely used in many fields. It will be suitable for many applications because it is a well-organized, portable, and flexible approach [15]. As an efficient algorithm that combines the Cause Based Decision Tree (CBDT) and Gradient Boosting Machine (GBM) approaches, this technique has the ability to improve the boosting approach to process almost all types of data quickly and accurately. With these unique features, this algorithm can be efficiently used to develop predictive models by applying regression and classification to the target data set. XGBoost can also be used to process large data sets with many attributes and classifications. This algorithm provides practical and effective solutions to new optimization problems, especially when trade-offs between efficiency and accuracy are considered [16].

LGBM regression (Light Gradient Boosting Machine Regression) uses another innovative machine learning based data processing algorithm for more accurate residual value modeling and prediction. As a newly developed technique, it is designed by combining two novel data sampling and classification methods, namely Exclusive Feature Bundling (EFB) and Gradient-based One-Side Sampling (GOSS) [17]. With these combined features, data scanning, sampling, clustering and classification operations are performed properly and accurately in a short time compared to analogous techniques.

2.2 Assignment Problem

The assignment problem (AP) is defined as the assignment of m workers to n jobs. The classical assignment problem is a special case of transportation problems where the quantity of resources and demand are equal to 1 [20]. According to the quality of workers, assignment problems can be classified into three main categories: assignment models with at most one task per worker, assignment models with more than one task per worker, and multi-level assignment models [21, 22].

An assignment problem has been studied to ensure that workers are assigned to the appropriate departments in order to maximize the working time of workers in the organization under the current conditions. The developed model will identify departments that are likely to be more suitable for employees, thus providing decision support for human resource management. Human resource management can benefit from these suggestions when making decisions about salary, bonus, incentives, etc.

3 Application

3.1 Data Set

The basic knowledge and primary data on Predictive Human Resource Analytics was collected by William Walter [18] from Kaggle website. The data used in this proposed study contains 14,999 observations with each row representing a single employee. The fields in the dataset contain the following 10 variables:

- Satisfaction_level = The satisfaction level takes values between 0 and 1.
- Last_evaluation = The year elapsed since the last performance evaluation.
- Number_project = The number of projects completed while working.
- Average_montly_hours = The monthly average of hours spent at the workplace.
- Time_spend_company = The year(s) spent in the company.
- Work_accident = The employee's work accident status (0 'No Work Accident', 1 'Work Accident').
- Left = The employee's status at the workplace (0 'Not Leaving the Job', 1 'Leaving the Job').
- Promotion_last_5years = The employee's promotion status in the last five years (0 'Not Promoted', 1 'Promoted').
- Department = The employee's department. ("Sales": A, "Accounting": B, "HR": C, "Technical": D, "Support": E, "Management": F, "IT": G, "Product Manager": H, "Marketing": I, "R&D": J).
- Salary = Relative salary level (low, medium, high).

The preprocessing of filters focuses on two types of values in the data set, sample and attribute. The sample values are used for resampling. The samples are divided into two data sets, a training set and a test set. In the data set, 70% of the attribute time_spend_company is allocated for training and 30% for testing.

Figure 1 analyzes the time spent by employees in each department of the company and displays this data in a bar chart. This chart shows how many employees are in each department and the time spent by employees in the company is shown in different colors.

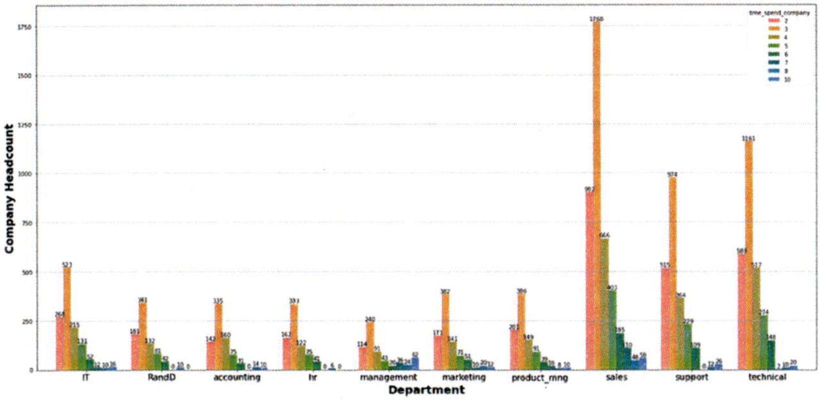

Fig. 1. Departments by Years Spent at the Company

This observation can also show the relationship between the number of employees and the department. Figure 1 shows 268 employees who have worked in the IT department for 2 years. It also shows that the number of employees in the sales, support and technical departments is high, and the number of employees in each department has worked for a maximum of 3 years.

3.2 Data Preprocessing

There are many difficulties in maintaining data in real life. There may be deficiencies in the data, incorrect entries may have been made, or extraordinary situations may have occurred. In such cases, it may be necessary to pre-process the data instead of using it directly in the model. The data preprocessing performed on the data in this study is as follows:

1. Detection of missing data.
2. Changing variable names.
3. Conversion of non-categorical variables into categorical variables.
4. Detection of outlier data.

4 Computational Results

In this section, the effectiveness of the implemented regression models was evaluated. Table 1 shows the performance of each machine learning algorithm in estimating employee turnover. R-Squared, RMSE and Time Taken were used as performance measures.

In Table 1, the performance of each machine learning algorithm is presented for predicting employee turnover rate. In the evaluated case study, the Extra Tree algorithm exhibits the lowest RMSE value based on the number of years spent in the company. Therefore, the Extra Tree algorithm is determined as the most suitable regression algorithm for predicting the duration of each employee's tenure in the company. A staffing model has been created using the predictions generated by this regression model.

Table 1. Performans measures of each ML techniques

		PERFORMANCE OUTPUTS		
		R-Squared	RMSE	Time Taken
MODEL	Extra Trees Regression	0,48	1,04	1,62
	Random Forest Regression	0,46	1,06	2,35
	Bagging Regression	0,40	1,11	0,25
	XGBoost Regression	0,34	1,17	1,56
	LightGBM Regression	0,25	1,24	0,08

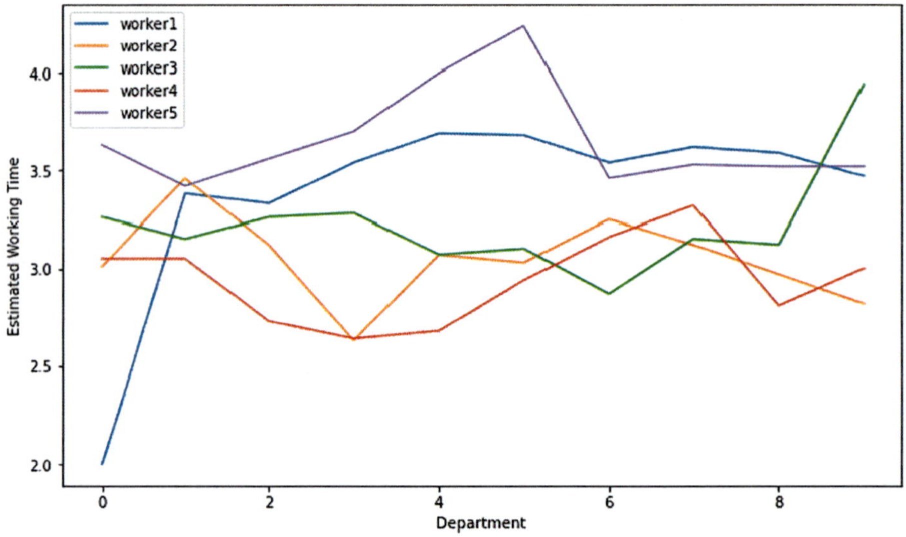

Fig. 2. Estimated work time of five employees according to the different departments

Figure 2 provides estimates of the turnover time of five different workers in different departments. While all other attributes are the same, only the departments were changed and differences in turnover times were observed. Thus, it was concluded that the departments have a significant impact on employee turnover times. For example, when the first employee works in the first department, it is estimated that he/she will leave the job within 2 years, while the estimated time to leave the job increases to 3.69 years when the employee is taken to the fifth department.

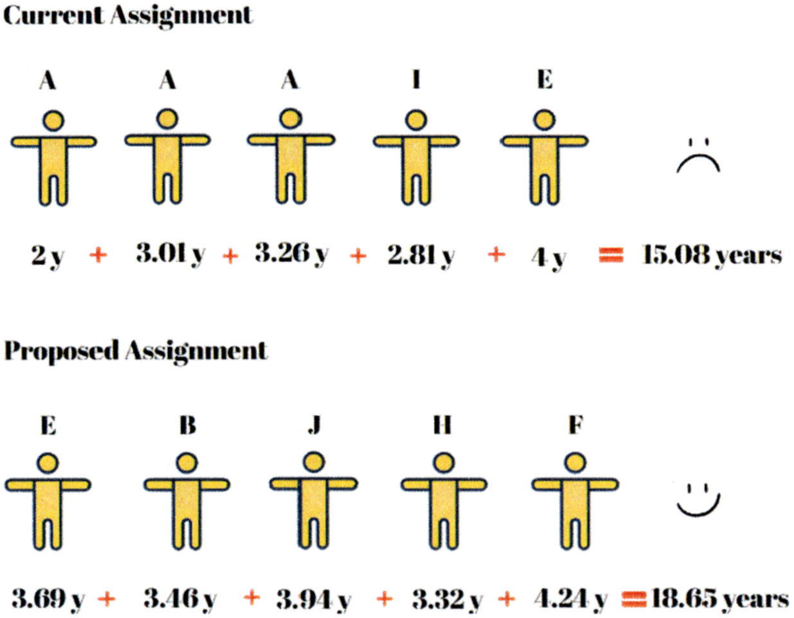

Fig. 3. Current status and recommended status for 5 employees

As an example, Fig. 3 shows the estimated work time for 5 workers in the departments where they currently have jobs. The total working time increases from 15.08 years to 18.65 years when an assignment problem is run to maximize the total working time of these workers. The model determined the appropriate department for each worker and recommended that the 1st worker be assigned to the E department, the 2nd worker to the B department, the 3rd worker to the J department, the 4th worker to the H department, and the 5th worker to the F department. Thus, it was observed that this could further increase the estimated working time of workers.

When assigning workers to departments, special constraints may be added to the assignment problem by taking into account the places where workers can work. Therefore, it is important to verify that the model's recommendations are consistent with the company's goals and priorities. Figure 4 shows the assignments suggested by the decision support system when we extend our example to 50 employees and 10 departments.

A significant difference in total employee work time was found between the current plan and the proposed plan in this example of 50 employees and 10 departments. In the case proposed by the decision support system, when all conditions were equal, the total work time of the employees was 34.52 years higher with only the changes in the departments.

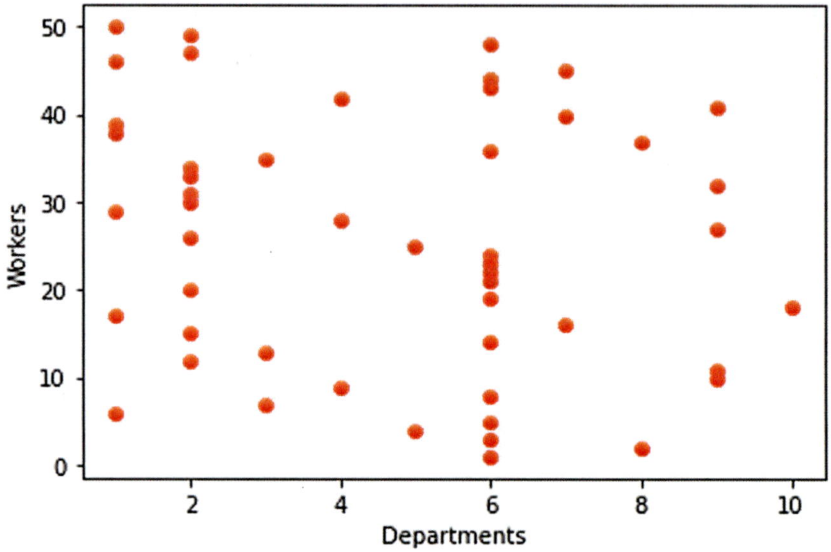

Fig. 4. Results of the assignment model

5 Conclusion

In this study, a decision support system has been created for human resources management. This decision support system estimates the time spent by a worker in the organization. In the same way, the working time of a newly applied candidate can be estimated. These results also serve as a reference for the future changes that HR will make in the working conditions of workers. On the other hand, the developed decision support system makes a suitable department estimation for both existing employees and a new employee. For this purpose, an assignment problem that maximizes the working time of employees is solved when choosing a department for each employee.

References

1. Stoval, M., Bontis, N.: Voluntary turnover: knowledge management – Friend or foe? J. Intellect. Cap. **3**(3), 303–322 (2002)
2. Cotton, J.L., Tuttle, J.M.: Employee turnover: a meta-analysis and review with implications for research. Acad. Manag. Rev. **11**(1), 55–70 (1986)
3. Finkelstein, L.M., Ryanand, K.M., King, E.B.: What do the young (old) people think of me? Content and accuracy of age-based metastereotypes. Eur. J. Work Organ. Psychol. **22**(6), 633–657 (2013)
4. Peterson, S.L.: Toward a theoretical model of employee turnover: a human resource development perspective. Hum. Resour. Dev. Rev. **3**(3), 209–227 (2004)
5. Liu, D., Mitchell, T.R., Lee, T.W., Holtom, B.C., Hinkin, T.R.: When employees are out of step with coworkers: how job satisfaction trajectory and dispersion influence individual-and unit-level voluntary turnover. Acad. Manag. J. **55**(6), 1360–1380 (2012)

6. Heckert, T.M., Farabee, A.M.: Turnover intentions of the faculty at a teaching-focused university. Psychol. Rep. **99**(1), 39–45 (2006)
7. Alao, D., Adeyemo, A.B.: Analyzing employee attrition using decision tree algorithms. Comput. Inf. Syst. Dev. Inform. J. **4**(1), 17–28 (2013)
8. Srivastava, D.K., Nair, P.: Employee attrition analysis using predictive techniques. In: 2017 International Conference on Information and Communication Technology for Intelligent Systems, Ahmedabad, India, pp. 293–300 (2017)
9. Raman, R., Bhattacharya, S., Pramod, D.: Predict employee attrition by using predictive analytics. Benchmarking: Int. J. **26**(1), 2–18 (2019)
10. Breiman, L.: Random forests. Mach. Learn. **45**, 5–32 (2001)
11. Rodriguez-Galiano, V., Sanchez-Castillo, M., Chica-Olmo, M., Chica-Rivas, M.J.O.G.R.: Machine learning predictive models for mineral prospectivity: an evaluation of neural networks, random forest, regression trees and support vector machines. Ore Geol. Rev. **71**, 804–818 (2015)
12. Geurts, P., Ernst, D., Wehenkel, L.: Extremely randomized trees. Mach. Learn. **63**(1), 3–42 (2006)
13. John, V., Liu, Z., Guo, C., Mita, S., Kidono, K.: Real-time lane estimation using deep features and extra trees regression. In: Image and Video Technology: 7th Pacific-Rim Symposium, PSIVT 2015, Auckland, New Zealand, November 25–27, 2015, Revised Selected Papers 7 (pp. 721–733). Springer International Publishing (2016). https://doi.org/10.1007/978-3-319-29451-3_57
14. Huang, J, Sun, Y, Zhang, J.: Reduction of computational error by optimizing SVR kernel coefficients to simulate concrete compressive strength through the use of a human learning optimization algorithm. Eng. Comput.1–18 (2021)
15. Friedman, J.H.: Greedy function approximation: a gradient boosting machine. Ann. Stat. 1189–1232 (2001)
16. Ramraj, S., Uzir, N., Sunil, R., Banerjee, S.: Experimenting XGBoost algorithm for prediction and classification of different datasets. Int. J. Control Theory Appl. **9**(40), 651–662 (2016)
17. Ke, G., et al.: Lightgbm: a highly efficient gradient boosting decision tree. Adv. Neural Inf. Process. Syst. **30** (2017)
18. Kaggle, "hr-comma-sep," Kaggle, Ed., ed (2019)
19. Armstrong, M.: Armstrong's Handbook of Performance Management: An Evidence Based Guide to Deliver High Performance, (4.Ed), Kogan Page, London (2009)
20. Kara, Doğrusal Programlama. Bilim Teknik, Ankara (2010)
21. Pentico, D.W.: Assignment problems: a golden anniversary survey. Eur. J. Oper. Res. **176**(2), 774–793 (2007)
22. Öncan, T.: A survey of the generalized assignment problem and its applications. INFOR: Inf. Syst. Oper. Res. **45**(3), 123–141 (2007)

Check for updates

Remaining Useful Life Prediction of Machinery Equipment via Deep Learning Approach Based on Separable CNN and Bi-LSTM

İbrahim Eke[1](✉) and Ahmet Kara[2]

[1] Graduate Education Institute, Foreign Trade and Supply Chain Management, Hitit University, Çorum, Turkey
ibrahimeke05@gmail.com.tr

[2] Department of Industrial Engineering, Hitit University, Çorum, Turkey
ahmetkara@hitit.edu.tr

Abstract. Predictive maintenance occupies a significant role to drop the operation and maintenance costs in production systems. Remaining useful life (RUL) prediction is one of the most preferred tasks in predictive maintenance decisions. Recently, deep learning techniques are extensively employed to accurately and effectively predict remaining useful life (RUL) by examining the past deterioration data of machinery and equipment failures. In this study, a deep learning approach that includes multiple separable convolutional neural networks (CNN), a bidirectional long short-term memory (Bi-LSTM) and fully-connected layers (FCL) are proposed to ensure more effective predictive maintenance planning. Separable CNN layers are applied to learn the nonlinear and sophisticated dependencies from the raw degradation data while the Bi-LSTM layer is employed to capture the long-short temporal characteristics. Besides, the dropout method and L2 regularization are used in the training stage of the proposed deep learning approach to achieve more accurate learning. The effectiveness of the proposed approach is verified by the popular FEMTO-bearing dataset presented by NASA. Finally, it is aimed that the experimental results provide better prognostic prediction compared with the benchmark models.

Keywords: Predictive maintenance · deep learning · prognostic prediction · separable convolution

1 Introduction

Maintenance of machine equipment is of paramount importance in the industrial and manufacturing sectors, as it directly impacts the efficiency, productivity, and profitability of an organization. Through the implementation of regular and systematic maintenance practices, machinery and equipment can operate at their optimal performance levels, reducing downtime and minimizing the risk of unexpected breakdowns [1].

Predictive maintenance is an advanced technology-based approach that focuses on predicting the future health and performance of equipment or systems, as well as detecting and diagnosing faults and failures in real-time [2]. It is an integrated process that

Z. Şen et al. (Eds.): IMSS 2023, LNME, pp. 128–137, 2024.
https://doi.org/10.1007/978-981-99-6062-0_13

involves the collection, analysis, and interpretation of data from various sources, such as sensors, diagnostics, and modelling, to provide insights into the condition of equipment or systems. Predictive maintenance is a proactive approach to predicting the future performance and health of a system, such as a machine, based on real-time data analysis. This approach involves the use of sensors, data analytics, and machine learning to monitor the health of a system and predict when maintenance is needed, which allows maintenance personnel to take corrective action before a failure occurs. The goal of predictive maintenance is to improve the reliability, availability, and safety of systems by detecting and diagnosing problems early before they result in downtime or failure, which can significantly reduce costs and increase efficiency [3].

Traditional predictive maintenance techniques rely on statistical and machine learning algorithms to analyze historical and real-time data to predict equipment failures and recommend maintenance actions. However, these techniques can be limited by the complexity and variability of data, which can make it difficult to identify patterns and relationships. Deep learning can overcome these limitations by automatically discovering patterns and relationships in complex data sets, including data from sensors, logs, and other sources. By training deep neural networks on large data sets, Deep learning algorithms can learn to identify patterns and relationships that are not easily detected by traditional machine learning techniques. This can lead to more accurate predictions of equipment failures and better recommendations for maintenance activities. Another advantage of deep learning is its ability to adapt to changing conditions. These algorithms can learn from new data as it becomes available, allowing them to adapt to changes in equipment performance and environmental conditions. This can help ensure that predictive models remain accurate and effective over time.

Predicting impending failure and estimating remaining useful life (RUL) is essential to avoid abrupt breakdown and schedule maintenance [4]. Increasing the accuracy of RUL prediction depends on determining the fundamental relationship between bearing deterioration progression and the current state of health. Therefore, the relationship between the two is also very important. To determine this relationship, effective feature compression and optimum feature selection are required. Similarly, it is difficult to determine a failure threshold since the health indicators of different machines are often different at the time of a failure [5]. Shen and Tang [3] proposed a novel data-driven method to address the challenge of data redundancy and initial prediction time in RUL prediction. This method involves extracting time-frequency features of vibration signals, constructing a nonlinear degradation indicator, and applying an attention mechanism called Multi-Head Attention Bidirectional-Long-Short-Term-Memory (MHA-BiLSTM). In the model proposed by Jiang et al. [6], a convolutional neural network (CNN) and an attention-based long short-term memory (LSTM) are used to partition a time series into multiple channels and improve performance by different deep learning approaches. Ren et al. [7] introduced a new method for the prediction of bearing RUL based on deep convolution neural network (CNN) and a new feature extraction method called the spectrum-principal-energy-vector. Yang et al. [8] addressed a new deep learning-based approach for predicting the remaining useful life (RUL) of rolling bearings based on long-short term memory (LSTM) with uncertainty quantification. The proposed method includes a fusion metric and an improved dropout method based on nonparametric kernel density to accurately estimate the RUL. Gupta et al. [9] addressed a deep learning approach for the real-time condition-based monitoring of bearings. A CNN-BILSTM model with attention mechanism for predicting an automatic RUL of bearings is developed by Xu et al. [10].

Furthermore, Sun et al. [11] presented a hybrid deep learning-based technique combining the convolutional neural network (CNN) and long short-term memory (LSTM) network to predict the short-term degradation of a fuel cell system used for commercial vehicles. Chang et al. [12] developed a LSTM network RUL prediction algorithm that is based on multi-layer grid search (MLGS) optimization, which integrates feature data and optimizes network parameters to ensure accuracy and effectively predict the non-stationary degradation of the bearing. A new deep learning framework called MSWR-LRCN for predicting the RUL of rolling bearings is presented by Chen et al. [13]. The framework incorporates an attention mechanism, a dual-path long-term recurrent convolutional network, and polynomial fitting to improve the RUL prediction accuracy. The DSCN proposed by Wang et al. [14] directly takes monitoring data acquired by different sensors as inputs, automatically learns high-level representations through separable convolutional building blocks, and estimates RUL through a fully-connected output layer. A hybrid approach based on deep order-wavelet convolutional variational autoencoder and a gray wolf optimizer for RUL prediction is proposed by Yan et al. [15].

In this research, a deep learning approach including multiple separable convolutional neural networks (CNNs), a bidirectional long short-term memory (Bi-LSTM) and fully connected layers (FCL) is adopted to accurately and efficiently estimate the remaining useful life (RUL) and enable more effective predictive maintenance planning. The separable CNN layers are deployed to learn non-linear and complex dependencies from raw distortion data, while the Bi-LSTM layer is used to capture long-short temporal features. Moreover, the dropout method and L2 regularization are used in the training phase of the proposed deep learning approach to achieve more accurate learning. The performance of the proposed approach is validated on the popular FEMTO Bearing dataset provided by NASA.

The organization of the research is as follows. Section 2 describes the technical background of the proposed approach for RUL prediction of machinery equipment via deep learning approach. In Sect. 3, the experimental setting and results are presented. Lastly, the conclusion is drawn in Sect. 4.

2 Materials and Method

In the proposed deep learning-based approach, separable CNN networks, Bi-LSTM, attention mechanism and full-connected layers are used to learn spatial-temporal and complex features in historical data of deterioration progressions. Detailed information about the deep learning model is presented below.

2.1 Convolution Neural Network (CNN)

Convolutional Neural Network (CNN) is a class of deep learning algorithms widely used to learn highly representative features from multi-sensor data. Recently, it has been used in image recognition, natural language processing, signal processing, and object detection [16]. The architecture of the CNN consists of multiple layers, including convolutional and pooling layers. The convolutional layers extract features from the input data by convolving filters over it, while the pooling layers reduce the features and reduce the number of parameters [17]. The convolution process is formulated as follows:

$$f_i = \delta(w_f \otimes x_i + b_f) \tag{1}$$

In the above formula, f_i stands for the features extracted by CNN, w_f for the kernel weights, b_f for the bias parameters and δ for the activation function. In addition, the operator $\otimes$ covers the retrieval process.

2.2 Separable Convolution Neural Network

Deeply separable convolution, also called separable convolution, aims to efficiently extract temporal and cross-channel relationships from different sensor data. Deeply separable convolutions have been widely applied in different fields because they reduce the computation time and the number of network parameters and avoid unnecessary learning correlations [18]. Unlike the traditional convolutional network, the depth separable convolution consists of two parts, including depth convolution and point convolution, as shown in Fig. 1. After deep convolution, the number of input channels remains the same [19].

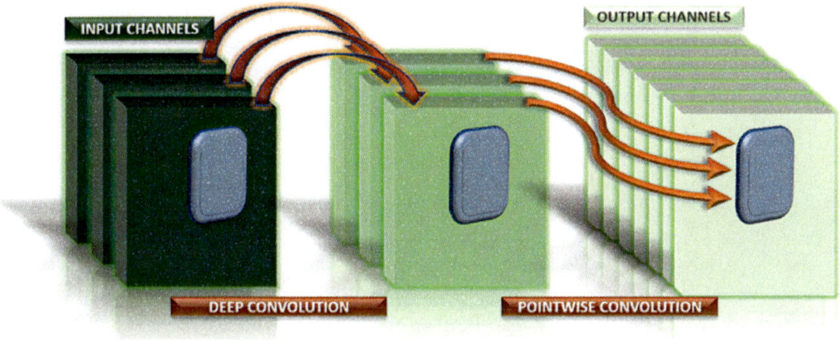

Fig. 1. Separable convolution network.

2.3 Bidirectional LSTM

LSTM uses a memory cell, an input gate, an output gate and a forget gate to control the flow of information through the network. The memory cell allows the network to selectively remember or forget information, while the gates help regulate the flow of information. In this study, unlike the traditional LSTM, bidirectional LSTM is used. Traditional LSTM can only utilize previous data for sequential input data. In other words, no future data of the sequential data are taken into account in the estimation of the current state. Bidirectional LSTM, on the other hand, utilizes the previous and future state of time series data simultaneously [20]. The final output of the network is obtained by combining the two hidden layers and can be calculated as follows:

$$\overrightarrow{h}_t = \tau(b_{\overrightarrow{h}} + \mathbf{W}_{\overrightarrow{h},\overrightarrow{h}} \cdot \overrightarrow{h}_{t-1} + \mathbf{W}_{\overrightarrow{h},x} \cdot x_t) \tag{2}$$

$$\overleftarrow{h}_t = \tau(b_{\overleftarrow{h}} + \mathbf{W}_{\overleftarrow{h},\overleftarrow{h}} \cdot \overleftarrow{h}_{t+1} + \mathbf{W}_{\overleftarrow{h},x} \cdot x_t) \tag{3}$$

$$h_t = g(\mathbf{W}_{h,\overleftarrow{h}} \cdot \overleftarrow{h}_t + \mathbf{W}_{h,\overrightarrow{h}} \cdot \overrightarrow{h}_t) \tag{4}$$

In the above equations, $\overrightarrow{h}_t$ and $\overleftarrow{h}_t$ represent the state information in the forward and backward layers, respectively. The operator $\tau(\cdot)$ represents the LSTM processing steps, while $g(\cdot)$ is the activation function.

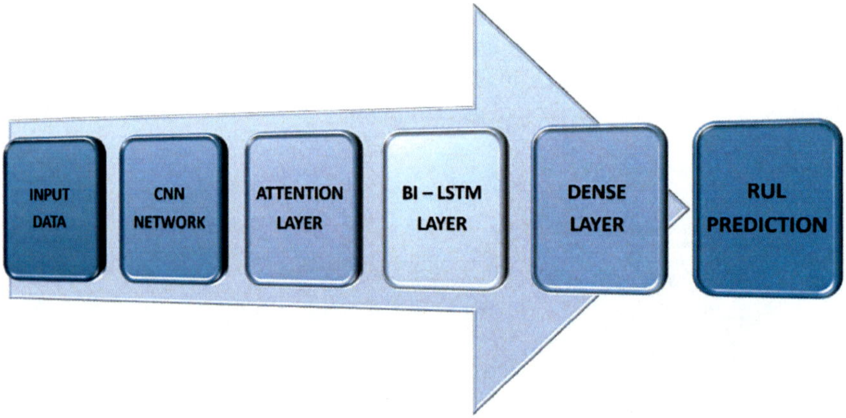

Fig. 2. General structure of the proposed approach.

The input data of the proposed approach is two-dimensional, $w_t \times f_t$. w_t, represents the time windows in the input data and f_t represents the predetermined number of features. The input data is first sent to two separable CNN networks with different kernel sizes. Through this process, it is planned to learn the complex and non-linear features in the input data. Then, the output of the CNN networks will be used by the self-attention mechanism. From the complex features and discriminative information to be obtained by CNN networks and attention mechanism, temporal dependencies will be extracted by Bi-LSTM network. Finally, the extracted features will be used by full-layer networks to predict the remaining lifetimes. Figure 2 shows the general structure of the proposed deep learning-based approach.

3 Experimental Setting and Results

3.1 Dataset

In this paper, we consider the FEMTO bearing dataset, which is widely used in the literature to predict the remaining life of machines and to evaluate the effectiveness of the proposed deep learning approach. The FEMTO dataset was collected by the PRONOSTIA test rig and made publicly available for the IEEE PHM 2012 prognostics competition [21]. The test rig consists mainly of an induction motor, a shaft, a speed controller, an assembly of two rollers and tested bearings. PRONOSTIA provides accelerated degradation of the bearings under three different operating conditions, and a total of up to seventeen failure operating datasets are provided, six training datasets and eleven test datasets.

3.2 Experimental Setting

The presented bearing RUL prediction approach deployed two separable CNN layers with kernel sizes of 5 and 3 as the first network component. The filter sizes of separable CNNs are set to 16 and 32, respectively. In addition, a Bi-LSTM with 16 units and two fully-connected layers with 32 and 1 units are used to accurately predict the bearing RUL. The dot product attention layer is adopted as the attention mechanism in the framework. To reduce the overfitting problem, a dropout method with a rate of 0.3 and an L2 regularization technique with a rate of 1e-4 are implemented. Mean Square Error (MSE) is handled as the loss function in this framework. The loss function minimization uses the Adam algorithm with a learning rate of 0.001.

A DNN with two fully-connected layers are used to verify the RUL prediction performance of the proposed method. Root mean square error (RMSE) is adopted as the assessment criterion. The experiments are carried out by means of Python v3.8.5 and TensorFlow v2.2.0.

3.3 Results

This section analysis the results of the proposed bearing RUL prediction approach based on deep learning by comparing with DNN benchmark. As an initial evaluation, the training loss curves of the proposed approach and DNN method are illustrated in Fig. 3. Considering the starting epochs, it is seen that the training loss of each model in the last epochs is at a low level. Therefore, the proposed framework and DNN produced less training loss at the end of the training period.

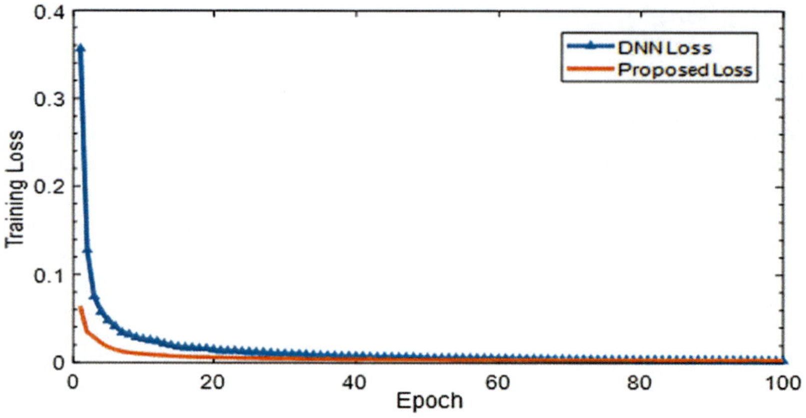

Fig. 3. Training loss curve derived by various techniques.

In this study, the effect of various time window sizes on the prediction accuracy of the proposed approach has been analyzed. In order to predict RUL of the bearings, the time window size is adjusted to 8, 16, and 32, respectively. Correspondingly, the box plots of the MAE score of the Bearing1_3 are demonstrated in Fig. 4. From this box plot, it is observed that MAE score of the proposed method at 16 gives better results compared with the different time window sizes. It was seen that both the mean and the variability of the MAE value were lower. The time window size in RUL prediction of bearings is set to 16 based on this result.

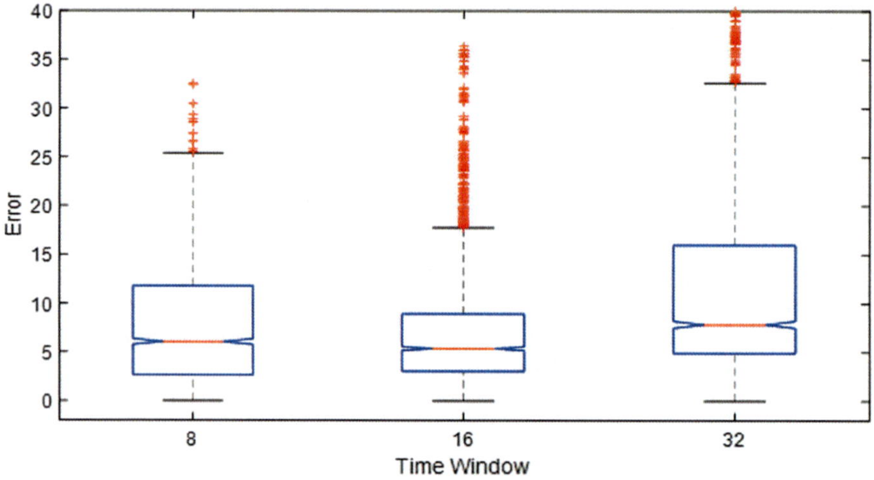

Fig. 4. Box plot of MAE scores under various time window sizes.

In Figs. 5(a) and (b), the RUL prediction results of the proposed and DNN approaches are compared with the actual RUL values of the Bearing1_3. In Fig. 5(a), it can be stated that, in spite of the local variations, the general pattern of degradation of the bearings can be represented by the proposed method. Moreover, compared to DNN method, the predictions of the proposed approach are very close to the actual values. On the other hand, it can be seen in both graphs that there is an increase in fluctuations towards the end of the time series.

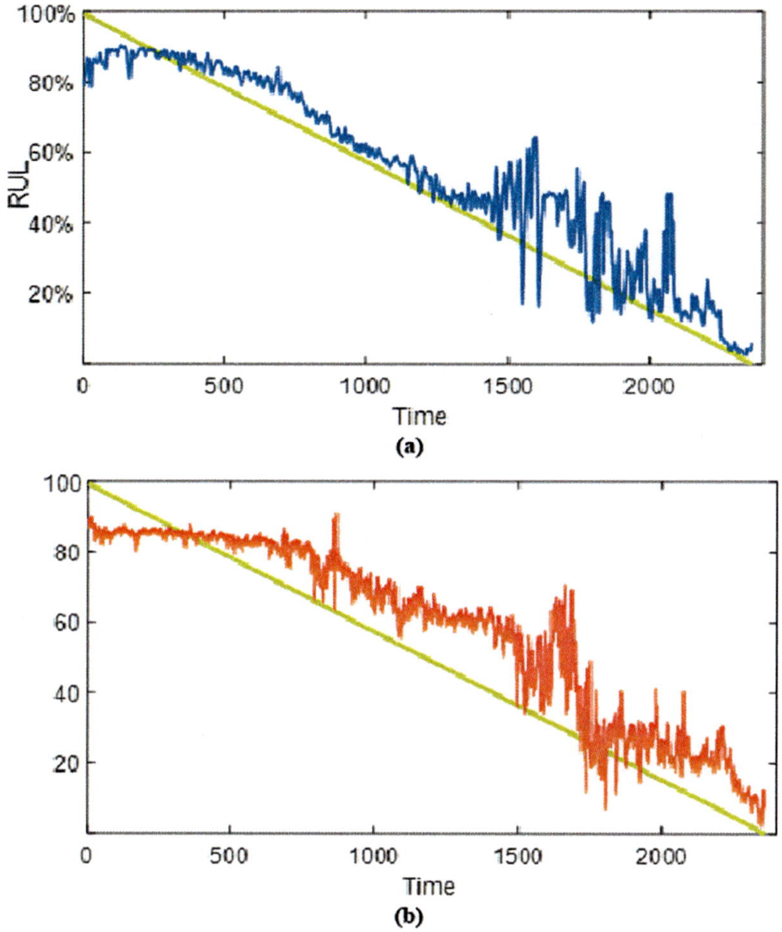

Fig. 5. RUL prediction results of the different methods.

Furthermore, Table 1 reported the comparison of results obtained by the proposed and DNN methods in terms of RMSE and MAE scores. According to these RMSE and MAE values, the proposed method provides more effective prediction performance in the Bearing1_3, Bearing2_7, and Bearing3_3 compared DNN method. For other bearings, DNN is better. In general, the proposed framework for RUL prediction is able to capture the degradation behavior of the bearings, but an effective hyper-parameter tuning is needed for better results.

Table 1. Comparison of the prediction errors of different methods.

Testing Bearing	Proposed		DNN	
	RMSE	MAE	RMSE	MAE
Bearing1_3	9.61	7.28	12.34	10.72
Bearing2_3	41.18	33.96	39.81	34.4
Bearing2_5	46.85	37.17	42.58	33.96
Bearing2_7	16.9	13.3	18.05	14.86
Bearing3_3	13.35	10.58	14.45	11.22

4 Conclusion

In this research, with the aim of prediction RUL using FEMTO bearing dataset, a hybrid approach based on deep learning has been introduced. To extract the effective patterns from the raw degradation data, the introduced framework consists of the combination of two separable CNN layers, a Bi-LSTM layer and the fully connected layers. Comparisons with the DNN model, was performed to evaluate the effectiveness of the proposed approach. Taking into account the experimental results, although the presented approach gives remarkable results for bearing prognostics, hyperparameter tuning with a meta-heuristic algorithm is required for more effective results.

References

1. Wei, Y., Wu, D., Terpenny, J.: Bearing remaining useful life prediction using self-adaptive graph convolutional networks with self-attention mechanism. Mech. Syst. Signal Process **188**, 110010 (2023)
2. Ouadah, A., Zemmouchi-Ghomari, L., Salhi, N.: Selecting an appropriate supervised machine learning algorithm for predictive maintenance. Int. J. Adv. Manuf. Technol. **119**(7–8), 4277–4301 (2022)
3. Shen, Y., Tang, B., Li, B., Tan, Q., Wu, Y.: Remaining useful life prediction of rolling bearing based on multi-head attention embedded Bi-LSTM network. Measurement **202**, 111803 (2022)
4. Ahmad, W., Khan, S.A., Islam, M.M.M., Kim, J.M.: A reliable technique for remaining useful life estimation of rolling element bearings using dynamic regression models. Reliab. Eng. Syst. Saf. **184**, 67–76 (2019)
5. Rathore, M.S., Harsha, S.P.: An attention-based stacked BiLSTM framework for predicting remaining useful life of rolling bearings. Appl. Soft. Comput. **131**, 109765 (2022)
6. Jiang, J.R., Lee, J.E., Zeng, Y.M.: Time series multiple channel convolutional neural network with attention-based long short-term memory for predicting bearing remaining useful life. Sensors **20**(1), 166 (2019)
7. Ren, L., Sun, Y., Wang, H., Zhang, L.: Prediction of bearing remaining useful life with deep convolution neural network. IEEE Access **6**, 13041–13049 (2018)
8. Yang, J., Peng, Y., Xie, J., Wang, P.: Remaining useful life prediction method for bearings based on LSTM with uncertainty quantification. Sensors **22**(12), 4549 (2022)

9. Gupta, M., Wadhvani, R., Rasool, A.: A real-time adaptive model for bearing fault classification and remaining useful life estimation using deep neural network. Knowl. Based Syst. **259**, 110070 (2023)
10. Xu, Z., et al.: A novel health indicator for intelligent prediction of rolling bearing remaining useful life based on unsupervised learning model. Comput. Ind. Eng. **176**, 108999 (2023)
11. Sun, B., Liu, X., Wang, J., Wei, X., Yuan, H., Dai, H.: Short-term performance degradation prediction of a commercial vehicle fuel cell system based on CNN and LSTM hybrid neural network. Int. J. Hydrogen Energy **48**(23), 8613–8628 (2023)
12. Chang, Z.H., Yuan, W., Huang, K.: Remaining useful life prediction for rolling bearings using multi-layer grid search and LSTM. Comput. Electr. Eng. **101**, 108083 (2022)
13. Chen, Y., Zhang, D., Zhang, W.: MSWR-LRCN: a new deep learning approach to remaining useful life estimation of bearings. Control Eng. Pract. **118**, 104969 (2022)
14. Wang, B., Lei, Y., Li, N., Yan, T.: Deep separable convolutional network for remaining useful life prediction of machinery. Mech. Syst. Signal Process **134**, 106330 (2019)
15. Yan, X., She, D., Xu, Y.: Deep order-wavelet convolutional variational autoencoder for fault identification of rolling bearing under fluctuating speed conditions. Expert Syst. Appl. **216**, 119479 (2023)
16. Hammad, M., Pławiak, P., Wang, K., Acharya, U.R.: ResNet-Attention model for human authentication using ECG signals. Expert Syst. **38**(6), e12547 (2021)
17. Yu, J., Zhang, C., Wang, S.: Multichannel one-dimensional convolutional neural network-based feature learning for fault diagnosis of industrial processes. Neural Comput. Appl. **33**(8), 3085–3104 (2021)
18. Shang, R., He, J., Wang, J., Xu, K., Jiao, L., Stolkin, R.: Dense connection and depthwise separable convolution-based CNN for polarimetric SAR image classification. Knowl. Based Syst. **194**, 105542 (2020)
19. Huang, G., Zhang, Y., Ou, J.: Transfer remaining useful life estimation of bearing using depth-wise separable convolution recurrent network. Measurement **176**, 109090 (2021)
20. Dong, S., Xiao, J., Hu, X., Fang, N., Liu, L., Yao, J.: Deep transfer learning based on Bi-LSTM and attention for remaining useful life prediction of rolling bearing. Reliab. Eng. Syst. Saf. **230**, 108914 (2023)
21. Nectoux, P., et al.: PRONOSTIA: an experimental platform for bearings accelerated degradation tests. In: IEEE International Conference on Prognostics and Health Management, PHM 2012, pp. 1–8 (2012)

Internet of Medical Things (IoMT): An Overview and Applications

Yeliz Doğan Merih[1,2](✉), Mehmet Emin Aktan[1,3], and Erhan Akdoğan[1,4]

[1] Health Institutes of Türkiye, İstanbul, Türkiye
{yeliz.merih,mehmetemin.aktan,erhan.akdogan}@tuseb.gov.tr
[2] Hamidiye Faculty of Nursing, University of Health Sciences, Hamidiye, İstanbul, Türkiye
[3] Department of Mechatronics Engineering, Bartın University, Bartın, Türkiye
[4] Department of Mechatronics Engineering, Yıldız Technical University, İstanbul, Türkiye

Abstract. Internet of Things (IoT) is an innovative technology that enables physical objects used in daily life to exchange data among themselves over the Internet. Healthcare is one of them. The basic idea behind the Internet of Medical Things (IoMT) applications is to detect and process patient data without any restrictions, and to provide remote communication with smart devices. Especially chronic disease follow-up and early warning applications are some of the current and effective usage of this technology. In addition, applications such as remote monitoring, telerehabilitation, smart sensors and medical device integration can be given as examples. The fact that it has the potential to increase the effectiveness and efficiency of not only patients but also healthcare professionals in the diagnosis and treatment processes is an encouraging factor for the use of the Internet of Things in the field of health.

When evaluated from the perspective of individuals, IoMT can enable better health self-management and personalization of the content of medical services. When evaluated from the perspective of the Ministry of Health and health institutions, it can reduce costs and ensure efficient use of the budget and increase the share allocated to other health services. For research institutions, smart healthcare can improve overall efficiency by reducing research costs and time.

In this study, it is aimed to examine the use and examples of IoMT technology in the health sector, to discuss the advantages, potential gains, and possible difficulties in the application of this innovative technology that can change the way of health service delivery, and to offer suggestions.

Keywords: Internet of Medical Things · Wearable technology · Healthcare

1 Introduction

In recent years, the rapid development of information and communication technologies has begun to change the working environment and service delivery of many sectors, especially the health sector. The Internet of Things (IoT), which is very popular and rapidly developing today, has great potential to help health professionals and improve people's quality of life, especially in the health sector. IoT is an innovative technology that can reduce healthcare interoperability challenges, increase the efficiency and effectiveness of diagnostic and treatment processes, and reduce healthcare costs [1–3].

Z. Şen et al. (Eds.): IMSS 2023, LNME, pp. 138–148, 2024.
https://doi.org/10.1007/978-981-99-6062-0_14

Individuals in today's developed societies tend to monitor and keep their health parameters under control, with the increasing awareness of healthy living. This phenomenon increases the demand for an informatics infrastructure where personal medical data can be transmitted electronically and for healthcare professionals who can communicate regardless of time and place. Internet of Medical Things (IoMT) technologies, which can collect and record personal health data in real-time, can provide connection and data exchange with remote servers over internet. When the mobile applications that we access via mobile phones and tablets, which have already entered our lives, are added to all these, the technological cycle is completed [3, 4].

One of the advantages of using IoMT is that it has tailor-made solutions and provides a globally accessible and regular database access. Successfully and broadly adapting IoMT technology to the healthcare industry will enable it to have better monitoring, sensing, communicating and controlling capabilities [2, 3]. Data types from different health management disciplines such as management, finance, logistics, stock control, diagnosis, therapy, treatment, meditation and daily activities can be collected through the IoMT. In this way, advantages such as providing simultaneous and reliable cost information, instant and accurate data transfer for diagnosis and treatment, making future predictions with the analysis of the collected data, and remote administration of health services especially during pandemics are obtained [4, 5].

This study aims to examine the use and examples of IoMT technology in the health sector, to discuss the advantages, potential gains and possible difficulties in the application of this innovative technology, and to offer suggestions.

2 Use of IoMT ın Healthcare

IoMT is a technology that connects physical and virtual objects in health and transfers information over the internet. These technologies have a wide application area and offer solutions with a service-oriented approach for planning health services and monitoring/detection in disaster management [1–6]. Despite limitations such as difficult working conditions and maintenance-deployment costs, IoMT is effectively used in real-time monitoring of disaster and epidemic scenarios, together with techniques such as artificial intelligence, machine learning and big data analysis integrated with web technologies. IoMT provides the digital transformation of the industry with wearable health technologies, remote patient monitoring systems, hospital information management systems, telerehabilitation and many similar applications (Fig. 1) [4, 7–9].

The biggest burden on the aging world population and global and national health systems is the rise of chronic diseases and the spread of diseases such as cancer, heart and diabetes. The increase in the elderly population and chronic disease has caused the size of IoMT technology in the healthcare industry market to increase from approximately $30 billion in 2015 to $140 billion in 2020. Many health institutions have already implemented applications that provide remote monitoring and management, and many people have already started using mobile applications to keep their health under control [1, 4, 8, 10].

Today, health-related measurements are costly and time-consuming. Patients need to go to health institutions, perform various examinations and take measurements in a

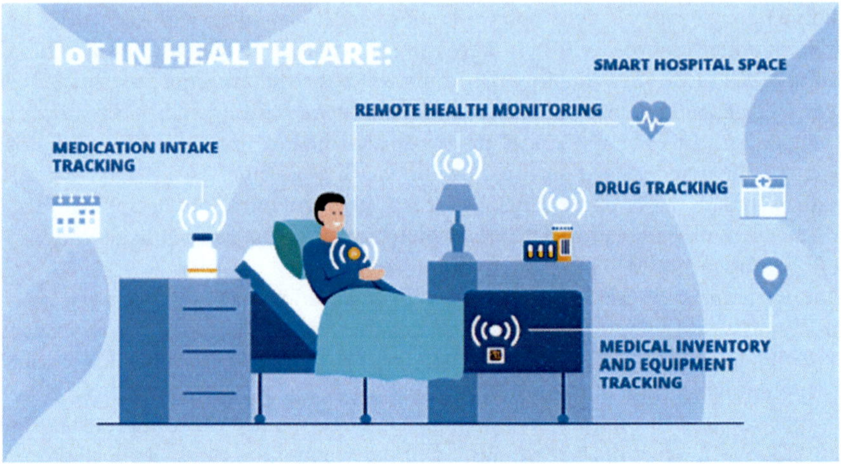

Fig. 1. IoMT in healthcare

laboratory. This causes patients with chronic diseases to spend a lot of money and time for routine measurements. In addition, transportation of elderly or bedridden patients to health institutions and intense human mobility during pandemics are other problems. IoMT technology reduces the need for direct patient-expert interaction to make the measurements and provides innovative ways to obtain and present the necessary data [1, 8–13]. When it comes to health, it doesn't matter whether the data collected is related to blood sugar level or heart rhythm, total steps taken and calories burned, it is vital for people to monitor all kinds of health information. [9–15].

Thanks to their features such as biocompatibility and small size, nano-devices promise various applications in many fields, especially in drug delivery, health monitoring, bio-hybrid implant, immune support systems and genetic engineering. It is seen that studies in this field generally focus on health monitoring. These applications aim to instantly detect hormonal and chemical disruptions that may occur for any reason by observing vital information such as body temperature and oxygen/cholesterol level through nano-sensors placed inside the body. [Table 1] [16–18].

One of the most important reasons for the increasing popularity of IoMT applications in recent years is that such applications increase the quality of life. In particular, the detection of cancer cells or pathological structures, which require very intensive processes for detection, can be done in a shorter time and at less cost thanks to nano-networks consisting of nano-devices. In addition, early diagnosis is the most important factor that increases the success of treatment in such diseases. In this context, thanks to the nano-devices in the body, it is possible to detect such diseases as soon as they occur [17–20].

The factors affecting the development of IoMT are given below.

- Increasing concerns about patient safety,
- The need to limit and reduce costs in healthcare,
- Expanding patient-centered care, diagnosis and treatment in health services,

- Spread of supporting technologies such as faster internet connection,
- Government initiatives promoting digital health,
- The popularity of personal mobile devices,
- Providing trainings that will ensure the adaptation of the stakeholders to the process.

The advantages of IoMT are given below.

- Reducing errors in medicine,
- Developing more effective methods in the treatment of diseases,
- Opportunity to make predictions for the future thanks to big data analysis,
- Reducing health costs,
- Increasing efficiency,
- Reducing the workload on health personnel,
- Improving inventory management.

When evaluated from the perspective of individuals, IoMT can enable better health self-management and personalization of the content of medical services. When evaluated from the perspective of government institutions and health institutions, it can reduce costs and provide a higher budget for other health expenditures. For research institutions, smart healthcare can increase the overall efficiency of research by reducing research cost and time. In a study, it was stated that health services will constitute 41% of the market share of IoT applications by 2025 [4–6].

In addition to many benefits of IoMT, there are also important limitations. High infrastructure costs are required to ensure effective and adequate use of this technology. In order to meet and reduce these costs, cost-effective methods of this technology should be investigated [1–21]. In addition, as in all information technologies in the field of health, data security is an important problem. The adequacy of cyber security measures in an important issue such as health data is still a question mark. Blockchain technology can provide a significant benefit in this regard. The lack of standards among the devices where IoMT technology is used is also an important problem. In addition, there are many legal problems regarding the production, sharing and use of health data between countries and health sectors. Stakeholders' lack of knowledge and motivation to use such technologies is also a major obstacle. This problem can be reduced by providing routine trainings to users and publishing informative documents and videos [1, 21–23].

3 Examples of IoMT in the Healthcare Sector

Adding components that support IoMT technology can provide more effective follow-up of patients. By using an IoMT device that has sensors that measure parameters such as body temperature, movement activity, blood pressure, glucose and oxygen in the blood, routine control and determination of the activity level of the patient can be done remotely. Patient data can be sent to doctors for examination, in case of a problem, the patient is called to the hospital or an ambulance is sent in case of an emergency [22, 24]. Thus, unnecessary hospital admissions and the spread of infectious diseases are prevented, more effective service can be provided to other patients and health expenditures are reduced.

Thanks to processing of the collected data and making inferences, it contributes to preventive health by detecting the diseases that did not show any symptoms before. In a study conducted in the USA, it was calculated that with the widespread use of IoMT applications, savings in health expenditures could increase up to 25% in the coming years [22–27].

Major applications of IoMT in healthcare are given below [Table 1]:

- Electronic health record.
- Telemedicine, telerehabilitation.
- Mobile healthcare applications (mhealth).
- Location tracking.
- Elderly care at home.
- Monitoring of chronic diseases.
- Preventive healthcare applications [25–29].

IoMT applications, especially solutions for patient safety and disease follow-up; health communication, medical equipment and drug management, field analysis, proactive treatment approaches and cost-effectiveness applications. In addition, services that produce analytical results such as remote monitoring systems, video health applications and mobile health technologies can be offered. For example, by monitoring the data on the increase in the use of painkillers or cough medicine and personal health data together in a region, inferences can be made about public health and possible negative factors in that region. Thus, alternative scenarios can be created about the course of the epidemic according to age, gender, geographical region, time and type of symptoms [1–3, 21–28].

Table 1. Examples of IoMT applications Source [1, 4, 21, 29–32].

Name	Working principle	Pictures
Wearable Health Technologies		
SugarBeat	SugarBeat, which measures blood sugar without using a needle with a tape attached to the arm, sends the data to the phone of the patient and the doctor.	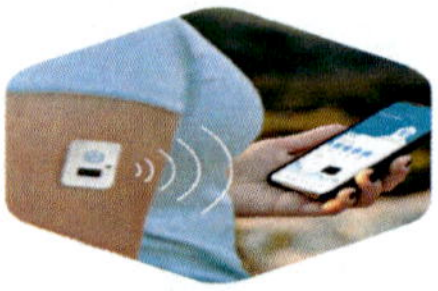
Alivecor	Cardial rhythm monitoring device that can be used in cases where long-term ECG monitoring is required. It can take six-channel ECG recording.	
Freestyle Libre	It is a product that provides continuous glucose measurement with a sensor attached to the body.	

(*continued*)

Table 1. (*continued*)

BodyGuardian	Heart rate, ECG, breathing frequency and physical activity obtained from a wearable sensor are sent to the center via smartphone and can be monitored by experts.	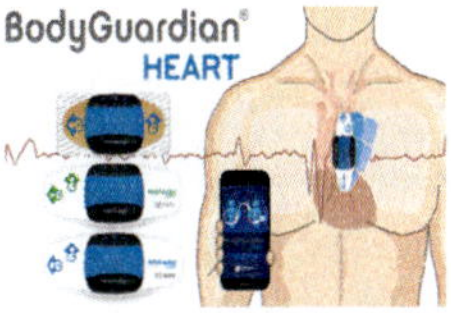
Abilify Mycite	When the drug is swallowed and reaches the stomach, it is detected by a sensor and a notification is sent to the phone.	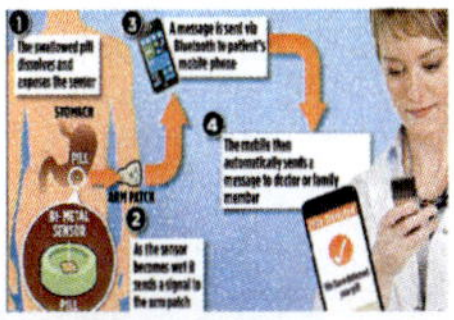
Sensatex	With the smart clothes, many values such as the user's body temperature, weight, heart rate, fat rate, daily activities are kept, and inferences are made about possible diseases.	
Smart Belt	The smart belt gives a warning when people eat more than they should. This warning is provided by measuring the space before eating and the tension after eating with a magnetic sensor.	
PillCamTM	It is an ingestible capsule technology that provides visualization of the esophagus, stomach and intestinal system.	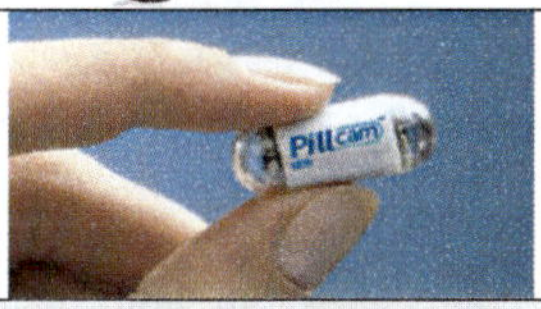
Remote Patient Monitoring Systems		
eVisit	It is a product that provides remote communication between healthcare professionals and patients and increases the effectiveness of home care services.	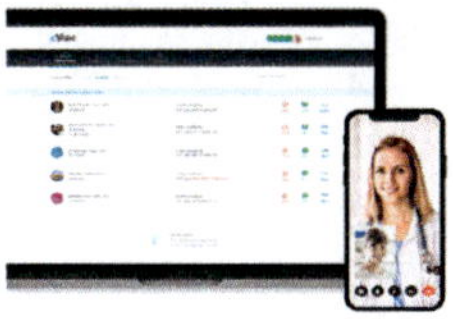
Pathway Genomics	With the application, individual health recommendations are made according to the genetic structure of the person.	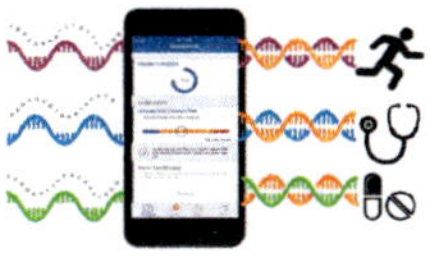

(*continued*)

Table 1. (*continued*)

Cardio Diagnostics	It was established with the vision of a proactive treatment approach from reactive treatment for cardiovascular diseases. The heart rhythms of the patients can be observed. Thus, early intervention can be made in critical situations.	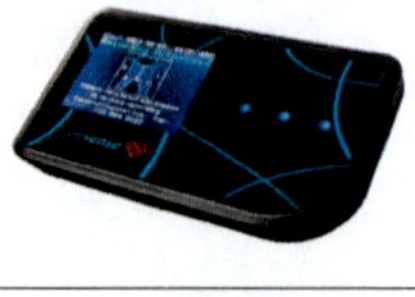
Amiko.io	It is an artificial intelligence-based application that provides effective and accurate inhaler treatment, which is frequently used in respiratory system diseases.	
AiCure	It is an application that monitors the long-term condition of patients and helps them to adhere to their drug treatments. With image processing, the face of the patients is recognized and it is confirmed that the drug has been taken.	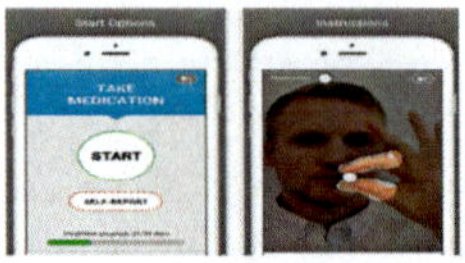
AdhereTech	When the patient forgets to take the medicine, it sends a warning and reminds the medicine.	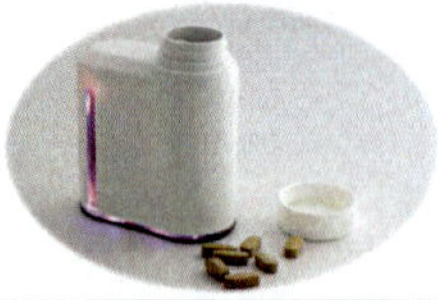
Hospital Information Management Systems		
GetWellNetwork	With the application, inpatients are reminded of the medications they need to take and the foods they should eat via IoT compatible devices such as smart TVs or tablets.	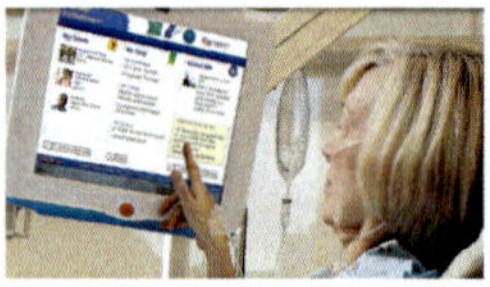
Stanley Healthcare	With the AeroScout Real-Time Location System application, it is possible to see the location and status of all employees, equipment and patients in the hospital.	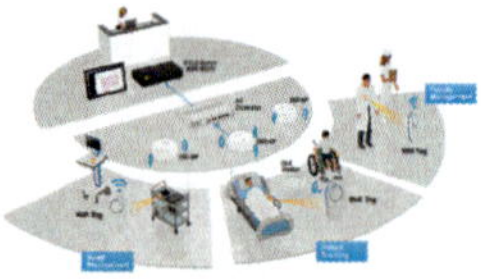
Rehub	ReHub is an online platform that connects the patient, the physiotherapist and the medical doctor. It facilitates the physical exercises from home and it is monitored by the professionals.	

In the research conducted with the help of IoMT-enabled wearable devices working over Bluetooth, the weight and blood pressure data of the patients were continuously followed up and cancer screenings were performed. With the help of treatment according to these data, it has been shown that cancer patients show much less severe symptoms. In another similar study, it was observed that the number of steps taken in a day by cancer patients using wearable IoMT devices increased from the first week. In the same study, two cancer patients with alarming data were identified and the additional precautions they needed could be taken in advance. In another study, the IoMT device (iTBra) detected temperature changes in the breast tissue and cancer was diagnosed at an early stage [9, 24–27].

The use of IoMT not only improves patient safety and quality of care but also reduces care costs. According to a study conducted in Singapore, the use of telephone therapy for diabetic radiculopathy alone saved $29.4 million. With the widespread use of IoMT in the field of health, many diseases can be detected at an early stage and deaths related to these can be reduced and financial savings can be achieved [10–13].

4 IoMT and Security

In-hospital devices such as drug pumps, patient follow-up, imaging and analysis devices and wearable medical devices carry the risk of being exposed to cyber-attacks during their communication over the internet. In addition to problems such as the failure of medical devices to receive updates or the need to make changes to the software, the diversity of devices also creates a wide attack surface. Millions of wearable medical devices are under threat due to security weaknesses in the data transmission processes of wearable devices over wireless communication protocols such as Bluetooth, Wi-Fi, NFC, Cellular, Zigbee [1, 4, 21–24].

Cyber security incidents that may occur on IoMT platforms lead to negative effects such as data loss, privacy violation, misuse of patient data. In addition, the deactivation of the devices on chronic patients who require continuous monitoring, the weaknesses that may arise in medical devices such as pacemakers, patient follow-up systems, infusion pumps make it necessary to consider IoMT devices in a special category [21, 33–35].

With the attack in 2015, a drug pump was seized and it was shown that various attack methods are possible, including injecting an overdose of drugs into the patient. In another study, it has been shown how the device can be hijacked by an attack on a device that performs the task of programming pacemakers before they are placed in the body. The passwords of the devices that monitor the devices of patients with pacemakers at home and send the data to the hospital have also been seized, and millions of devices around the world have become remotely accessible. Weaknesses in these devices have also attracted the attention of the academic world and various studies have been carried out on the subject. In the study titled "Know Your Enemy: Characteristics of Cyber-Attacks on Medical Imaging Devices" conducted in 2018, researchers found various vulnerabilities on tomography devices and showed the public the methods that could cause physical harm to the device and the patient. In a security study on patient follow-up systems in 2018, the patient data was displayed differently on the screen with the cyber-attack on the patient follow-up monitors in the intensive care unit. The most dangerous aspect of these attacks is that they are almost impossible to detect with classical cyber security solutions [1, 4, 24, 36–38].

Detection and prevention of such attacks is extremely critical, as they touch human life. The centers, which closely follow the developments in the field of health technologies, continue their efforts to develop a product focused on such attacks by starting a project called IoT-Medic in this context.

5 Conclusion

The primary purposes of IoMT are to reduce costs in care services, increase efficiency, early diagnosis and treatment and increase patient satisfaction. Technology is used in many areas related to health services such as protecting health, increasing the quality of life, predicting and preventing possible problems. With wearable technologies, medical data belonging to individuals can be monitored and shared with remote healthcare personnel. In this way, making remote patient care possible can reduce the length of stay in hospitals and help reduce costs.

When IoMT is examined in terms of healthcare institutions, it will increase the demands for the designing data center infrastructures. Increasing the security of wireless data communication, controlling access to medical data, determining potential usage rates and preparing strategic management plans with a good foresight are other critical issues. Finally, issues such as medical ethics, patient rights and privacy should be replanned within the framework of IoMT, and studies should be carried out to support the security. With the trainings to be carried out in this field, both health professionals and patients should be informed and the relevant regulations should be updated.

References

1. İleri, Y.Y.: Sağlık hizmetlerinde nesnelerin interneti (NİT): avantajlar ve zorluklar. J. Acad. Soc. Sci. **6**(67), 159–171 (2018)
2. Alemdar, H., Ersoy, C.: Wireless sensor networks for healthcare: a survey. Comput. Netw. **54**(15), 2688–2710 (2010)
3. Yao, W., Chu, C.H., Li, Z.: The adoption and implementation of RFID technologies in healthcare: a literature review. J. Med. Syst. **36**(6), 3507–3525 (2012)
4. Köse, G., Kurutkan, M.N.: Sağlık hizmetlerinde nesnelerin interneti uygulamalarının bibliyometrik analizi. Avrupa Bilim ve Teknoloji Dergisi. **27**, 412–432 (2021)
5. Tian, S., Yang, W., Grange, J.M., Wang, P., Huang, W., Ye, Z.: Smart healthcare: making medical care more intelligent. Global Health J. **3**(3), 62–65 (2019)
6. Asghari, P., Rahmani, A.M., Javadi, H.H.S.: Internet of things application: a systematic review. Comput. Netw. **148**, 241–261 (2019)
7. Bhatt, Y., Bhatt, C.: Internet of things in healthcare. In: Internet of Things and Big Data Technologies for Next Generation Healthcare. Springer International Publishing AG, pp. 13–33 (2017)
8. Xiang, G.Y., Zeng, Z., Shen, Y.J.: Present situation and development trend of China's intelligent medical construction. Chin. General Prac. **19**(24), 2998–3000 (2016)
9. Razdan, S., Sharma, S.: Internet of Medical Things (IoMT): overview, emerging technologies, and case studies. IETE Tech. Rev. **39**(4), 775–788 (2022)
10. Singh, R.P., Javaid, M., Haleem, A., Vaishya, R., Ali, S.: Internet of Medical Things (IoMT) for Orthopaedic in COVID-19 pandemic: roles, challenges, and applications. J. Clin. Orthopaedics Trauma. **11**(4), 713–717 (2020)

11. Tütüncü, D., Esen, M.F.: The use of Internet of Things in the management of epidemics: the case of COVID-19. Sağlık Akademisyenleri Dergisi. **8**(2), 169–177 (2021)
12. Hein, S., Bayer, S., Berger, R., Kraft, T., Lesmeister, D.: An integrated rapid mapping system for disaster management, the international archives of photogrammetry. Remote Sens. Spat. Inf. Sci. **42**, 499–504 (2017)
13. Perera, C., Liu, C.H., Jayawardena, S.: The emerging internet of things marketplace from an industrial perspective: a survey. IEEE Trans. Emerg. Top. Comput. **3**(4), 585–598 (2015)
14. Domingo, M.C.: An overview of the Internet of Things for people with disabilities. J. Netw. Comput. Appl. **35**(2), 584–596 (2012)
15. Xu, D.L., He, W., Li, S.: Internet of things in industries: a survey. IEEE Trans. Industr. Inf. **10**(4), 2233–2243 (2014)
16. Öztekin, A., Pajouh, F.M., Delen, D., Swim, L.K.: An RFID network design methodology for asset tracking in healthcare. Decis. Support Syst. **49**(1), 100–109 (2010)
17. Şahin, E., Dağdeviren, O., Akkaş, M.A.: Nano-nesnelerin internetinin gelecekteki uygulamalarına yönelik bir yol haritası. Avrupa Bilim ve Teknoloji Dergisi **26**, 174–179 (2021)
18. Balasubramaniam, S., Kangasharju, J.: Realizing the internet of nano things: challenges, solutions, and applications. Computer **46**(2), 62–68 (2013)
19. Al-Fuqaha, A., Guizani, M., Mohammadi, M., Aledhari, M., Ayyash, M.: Internet of Things: A survey on enabling technologies, protocols and applications. IEEE Commun. Surv. Tutorials. **17**(4), 2347–2376 (2015)
20. Manyika, J., Michael, C., Bughin, J., Dobbs, R., Bisson, P., Marrs, A.: Disruptive Technologies: Advances that will Transform Life, Business, and the Global Economy. CA, USA, San Francisco (2013)
21. Tekkeşin, A.İ.: IoMT Teknolojisi Sağlıkta Yeni Bir Devrim mi? http://dijitalsaglik.com.tr/IoMTTeknolojisi. Accessed: 16 Feb 2023
22. Segato, F., Masella, C.: Telemedicine services: how to make them last overtime. Health Policy Technol. **6**, 268–278 (2017)
23. Yang, F., et al.: Internet-of-Things-enabled data fusion method for sleep healthcare applications. IEEE Internet Things J. **8**(21), 15892–15905 (2021)
24. Yang, T., Gentile, M., Shen, C.F., Cheng, C.M.: Combining point-of-care diagnostic sand internet of medical things (IoMT) to combat the COVID-19 pandemic. Diagnostics **10**(4), 224 (2020)
25. Zafar, U., Shah, M.A., Wahid, A., Akhunzada, A., Arif, S.: Exploring IoT applications for disaster management: identifying key factors and proposing future directions, recent trends and advances in wireless and IoT-enabled networks. In: Recent Trends and Advances in Wireless and IoT-enabled Networks. Springer, pp. 291–309 (2019). https://doi.org/10.1007/978-3-319-99966-1_27
26. Zhang, J., Li, W., Han, N., Kan, J.: Forest fire detection system based on a ZigBee wireless sensor network. Front. For. China **3**, 369–374 (2008)
27. Fox, G., Connolly, R.: Mobile health technology adoption across generations: narrowing the digital divide. Inf. Syst. J. **28**, 995–1019 (2018)
28. Rueda, A., Krishnan, S.: Feature analysis of dysphonia speech for monitoring Parkinson's Disease. In: International Conference of Engineering in Medicine and Biology Society, pp. 2308–2311 (2017)
29. AlRasheed, A. Atkins, A.S., Campion, R.: Developing a guideline for hospital tracking and monitoring systems evaluation. In: International Conference on Computers in Management and Business, pp. 74–78 (2018)
30. Salama, D.: Smart life saver system for Alzheimer patients, down syndromes, and child missing using IoT. Asian J. Appl. Sci. **6**(1), 20–37 (2018)

31. Pinto, S., Cabral, J., Gomes, T.: We-Care: an IoT-based health care system for elderly people. In: International Conference on Industrial Technology, pp. 1378–1383 (2017)
32. Medikal Nesnelerin İnterneti Hayatımızı Kurtarabilir mi? Accessed 17 Feb 2023. https://thinktech.stm.com.tr/tr/medikal-nesnelerin-interneti-hayatimizi-kurtarabilir-mi?
33. Joyia, D.G.J., Liaqat, R.M., Farooq, A.,S.: Rehman Internet of Medical Things (IOMT): applications, benefits and future challenges in healthcare. J. Commun. **2**(4), 240–247 (2017)
34. Nalajala, P., Lakshmi, S.B.: A secured ıot based advanced health care system for medical field using sensor network. Int. J. Eng. Technol. **7**(2), 105–108 (2018)
35. Avaner, T., Avaner, E.: Yazılım teknolojileri ve sağlık yönetimi: HIMSS ya da dijital hastane hizmetleri üzerine bir değerlendirme". Yasama Dergisi. **37**, 5–28 (2018)
36. Qureshi, F., Krishnan, S.: Wearable hardware design for the Internet of Medical Things (IoMT). Sensors **18**, 3812 (2018)
37. Mustafaoğlu, A., Aktaş, F.: IoMT-based smart shoe design for healthy foot-flat feet gait analysis. Eur. J. Sci. Technol. **42**, 108–112 (2022)
38. Bozbuğa, N., Tekbaş, M., Gülseçen, S.: Tıbbi Nesnelerin İnterneti. İstanbul University Press, 451–478 (2021)

Using Social Media Analytics for Extracting Fashion Trends of Preowned Fashion Clothes

Noushin Mohammadian, Nusrat Jahan Raka, Meriel Wanyonyi, Yilmaz Uygun, and Omid Fatahi Valilai(✉)

School of Business, Social and Decision Sciences, Constructor University, Campus Ring 1, 28759 Bremen, Germany
OFatahiValilai@Constructor.University

Abstract. In recent years, the cloth industry has faced fast fashion trends. This has resulted in Fast fashion as a supply chain model for clothing and accessories that is supposed to respond quickly to the latest fashion trends by frequently updating the products already available in the inventory. However, Fast Fashion has created serious challenges for the sustainability of the clothing industry. This paper investigates the use of social media analytics to understand fashion trends in the preowned fashion industry. The study aims to establish a link between environmental pollution and fast fashion by investigating the preowned fashion industry from both consumer and business perspectives. To achieve this, the study proposes a social analytics (SA) approach to analyze social media posts and predict preowned fashion trends. By using SA techniques, the study hopes to provide valuable insights into consumer behavior and preferences in the preowned fashion industry, which can be used to promote sustainable fashion practices and reduce environmental pollution. Overall, the study demonstrates the potential of social media analytics in understanding and predicting fashion trends, with the goal of promoting sustainable fashion practices.

Keywords: Omnichannel strategy · Social media · Data analytics · Pre-owned cloth · Sustainability

1 Introduction

In today's world, most people are fashion-conscious and obsessed with fast fashion trends. From a related study done by the Waste and Resources Action Program (WRAP), it was found that in the UK 3.6 million tons of clothing were consumed in 2016 which was 16% more than the amount consumed in 2012 [1, 2]. WRAP also found out that in 2012, 24 million tons of carbon footprint was produced only from clothing items consumed in Britain which became 26.2 million tons in 2016 [3]. However, fast fashion trends can be responsible for these scenarios as people tend to buy clothing items or accessories frequently with the upcoming trends and get rid of the previous ones. Based on the research of the WRAP mentioned earlier, the lifetime of garment utilization in the UK is 3 years on average [2]. "*Fast fashion*" can be referred to as a supply chain model for clothing and accessories that is supposed to respond quickly to the latest fashion

Z. Şen et al. (Eds.): IMSS 2023, LNME, pp. 149–160, 2024.
https://doi.org/10.1007/978-981-99-6062-0_15

trends by frequently updating the products already available in the inventory. Usually, major clothing brands like Zara© and H&M© have embraced fast fashion by coming up with new fashion trends every three to five weeks. These trends tend to change frequently and henceforth shorten the practical service life of clothing items. Moreover, improper disposal of these clothes may lead to environmental pollution [4]. As most people are not willing to spend so much on fast-fashioned clothing items, the materials from which these are made tend to be cheap and short-lived. This means the higher the production the higher the tendency to abandon old clothes. Mostly, these items stay unused in people's closets for a while and as soon as that specific fashion trend is gone, they are simply thrown away without proper disposal. This kind of behavior keeps contributing to more environmental pollution. When these unused items are resold to new consumers, they are known as preowned items. Research studies with positive findings proved that using preowned clothing can sufficiently reduce environmental pollution. A similar study done in 2017 conducted life-cycle assessments on a cotton T-shirt, a pair of jeans, and a polyester dress and discovered that if the life span of these items were quadrupled, 75% of freshwater could be saved which was used in dyeing and other processes [4]. Another review study conducted in 2018 and published in the same journal investigated 41 studies and found all but one concluded that lengthening a garment's life by reusing it reduced its environmental impact [5].

H&M© and Zara© are among the big fashion brands existing, they are characterized by overproduction accompanied by lower prices [6, 7]. This has led to a quick turnover of clothes. While the US$2.4 trillion global fashion industry is growing at 5.5 percent, the fast-fashion sector has delivered growth at four times that rate since 2014 [8]. The fashion industry, however, calls for sustainability but a conflict of interest always exists because more consumers are inclined to a fast fashion mentality. Many producers try to change these perspectives, but consumers do not care about this unethical norm [9, 10]. However, different social media platforms can play an important role here to change the perspectives of people about preowned or recycled fashion items by promoting these items. Social media has been hailed as one of the most effective technologies for providing users with open communication and free markets [11]. The evolution of social media provides new opportunities for businesses to learn, innovate, and connect to open innovation platforms, which give them access to external sources of knowledge about technologies and markets [11–13]. Blogs and social media networking sites can be used for content sharing as well [14]. This study specifically aims to find how social media can be useful to influence people in buying preowned fashion items.

The main objective of this paper is to investigate social media post data on different platforms (Instagram©, Pinterest©) for recognizing user sentiments toward preowned fashion items. Exploiting the user's sentiment toward pre-owned fashion items on social media can be helpful to understand and predict certain fashion trends [15]. Therefore, this paper is focused to address how sentiment analysis can be applied to predict the preowned fashion item trend by using the data extracted from different social media platforms (Instagram©, Pinterest©) which is the primary research question of this study.

2 Literature Review

2.1 Environmental Aspect

As previously mentioned in the introduction, fast fashion has negative environmental impacts, while using pre-owned fashion items has positive environmental aspects. However, the environmental aspects of pre-owned fashion items have only been addressed from a motivational angle in this study and have not been investigated in detail. Criticism has emerged on the environmental impacts of fast fashion, as its effects are widespread. For instance, almost 8–10% of the world's carbon dioxide emissions are attributed to the fashion industry, which is nearly 5 billion tons worldwide. Additionally, the production of clothing uses approximately 79 trillion litres of water annually, with 20% of this water being used for dyeing and textile treatment. These figures illustrate the scale of the environmental impact that fast fashion has on the planet [16]. Previous researchers showed that 15% of clothing gets recycled while the rest gets into landfills. Big brands like H&M© offer recycling programs to their customers but it is noted if most of the buyers would return unused or idle clothing it would take around 12 years to recycle what was produced in 48 h [17]. Buying and selling of preowned items reduce global warming by 14% for a cotton T-shirt and 45% toxicity decline for polyester jeans [18]. Therefore, preowned clothing has a positive environmental impact and the best use for sustainable clothing is pre-owning it [17].

2.2 Preowned Fashion Items Concept

The preowned concept is related to secondhand shopping. According to Guiot and Roux (2010), secondhand shopping can be referred to as "*The acquisition of secondhand objects through methods and places of exchange that are generally distinct from those for new products*" [19, 20]. This means buying products, usually at discounted price rates, that were preowned or used previously by someone [19, 21, 22]. In recent times the secondhand online shopping market has boomed more than ever and during Covid-19 it expanded 21 times faster than the traditional textile industry [23, 24]. It can be assumed that as people are becoming more interested in sustainable living, they are gradually getting more inclined to secondhand or recycled fashion items. The present net worth of this global industry is around multi-billion-dollar which is expected to grow in the future [19, 25]. However, social media platforms can play an important role in promoting these preowned fashion items among consumers which have been addressed in this research.

2.3 Social Media Data Analytics

Nowadays, social media analytics is considered to be a powerful tool for establishing a product brand or creating a specific fashion trend. Although several data analytics methods are available for that, Sentiment Analysis is chosen to be used in this study to predict the preowned fashion trend. One of the main reasons behind using SA is that it is one of the most researched domains enabling researchers to analyze and categorize the true emotion of people. One of the main reasons behind using SA is that it is one of the most researched domains enabling researchers to analyze and categorize the true emotion of people [26]. Different social media platforms are easily accessible to people because of the revolutionary development of the internet and web 2.0 [12]. These large amounts of data can be extracted from different social media platforms to perform sentiment analysis. Especially for fashion-related items, people tend to express themselves more through these platforms by posting pictures or #posts. Moreover, according to Facebook data, fashion has been ranked as rank 4th among all Instagram© users' topmost interests [15]. However, the sentiments of the consumers are usually expressed through reviews, feedback posts, or simply by writing comments. These emotions or sentiments can mainly be categorized into positive or negative sentiments based on specific rating points (for example 4 or 5 stars) [26]. Although the challenge is to convert the sentiment into content and to access if the words of the reviewer display favorable or unfavorable emotions regarding the topic [27]. Mainly there are three types of machine learning strategies based on the data: supervised, semi-supervised, and unsupervised [28]. Using the unsupervised data extracted from different social media platforms also ensures that the results yielded from the analysis are non-biased. This method also applies to this study. These emotions or sentiments can mainly be categorized into positive or negative sentiments based on specific rating points (for example 4 or 5 stars) [26]. Although the challenge is to convert the sentiment into content and to access if the words of the reviewer display favorable or unfavorable emotions regarding the topic. Mainly there are three types of machine learning strategies based on the data: supervised, semi-supervised, and unsupervised [27]. Using the unsupervised data extracted from different social media platforms also ensures that the results yielded from the analysis are non-biased. This method also applies to this study.

2.4 Literature Gap Analysis

As illustrated in Table 1, the first theme mentioned in this study is the environmental aspects that serve as the motivational base behind this research. The negative environmental effects of fast fashion discussed in this paper prove that sustainable fashion trends must be introduced to save the environment. On the other hand, the positive environmental aspects mentioned earlier basically show how preowned clothing can reduce waste and save natural resources since the demand for new clothes will decline and save energy.

Similarly, the positive impact of preowned fashion items on the environment is one of the focused areas of this study. The preowned fashion industry context is discussed in detail in this paper to support and promote sustainability in the fashion industry. The aim of including this theme in this study is also to understand the behavior of consumers toward preowned fashion items which can be helpful in market analysis of the sustainable fashion industry.

Finally, the third theme of the study, the application of social media analytics is there to support the analysis of this research and to help in finding the answer to the research question. This theme specifically explains the part where a sentiment analysis approach is proposed to analyze the data extracted from different social media platforms to predict the preowned fashion item trends.

To conclude the research gap, the previous related studies in the field of sentiment analysis related to fashion industries have been considered. Several researchers [1, 28, 34]

Table 1. Comparison between the research themes of this research paper and other reviewed research papers.

Research Studies	Environmental aspect	Preowned fashion industry context	Social media analytics application
Jindal & Aron, 2021 [26]			Different SA approaches have been discussed
Xie et al., 2021 [1]	The relationship between sustainability and recycling	Different methods for recycling clothes have been addressed from technical, economic, and social perspectives	
Yuan and Lam, 2022 [15]			Used multimodal SA for analyzing social media posts related to fashion
Kerrice et al., 2022 [28]	Addressed the negative environmental impact of fast fashion	Abstract models recommended recycling clothing and establishing positive impacts on the environment	
S. Yi and X. Liu, 2020 [29]			Proposed MSVM approach for analyzing comments and sentiments in Twitter©
Sonali et al., 2019 [30]	Overconsumption of clothing effects on pollution Specific reasons relating to SCCBs		
Derwanz, 2021 [31]		Focused on preowned clothing promotion through digitalization	Digitalization and business models to promote preowned clothing
Yoonjae et al., 2022 [32]			Technology acceptance model (TAM) to promote e commerce
Hu et al.,2019 [33]			Text analytics framework to integrate different data about consumers from social media

(continued)

Table 1. (*continued*)

Research Studies	Environmental aspect	Preowned fashion industry context	Social media analytics application
Gajjar &Shah, 2022 [34]	How preowned clothing reduces environmental pollution	Consumer motivations towards preowned clothing	
This research	Establishes a link between environmental pollution and fast fashion	Investigation of the preowned fashion industry from both consumer and business perspectives	Proposes SA approach to analyze social media posts to predict preowned fashion trend

did their studies based on only the environmental aspects and the preowned or recycled clothing, but the social media analytics part was not included in these studies. Other research reviewed above [15, 26, 33] only focused on the last theme which is social media analytics. But none of the researchers investigated all these three aspects simultaneously. The closest match is found between this study and another reviewed study [15]where the similarity is that both researchers use a sentiment analysis approach for fashion-related social media posts. However, in this study, a sentiment analysis approach is proposed to predict the preowned fashion trend where the environmental aspects are also addressed from a motivational background. This makes it unique from all the reviewed literature.

3 Proposed Solution

3.1 Conceptual Model of a Social Media Analytics Enabled Sustainable Pre-owned Cloth System

The conceptual model (Fig. 1) below shows the comparison between pre-owned and fast fashion items with the help of different variables. The variables are raw materials costs, production costs, raw material, and retail inventory costs, and sales revenue. Due to a lack of necessary information about preowned clothing pros in the market, consumers are inclined more to fast fashion. The raw materials costs are negative because of factors like no cost for the raw materials, and no adequate labor for manufacturing the goods. Production costs are negatively affected because of low raw material costs. Raw material inventory costs are negative because less stock or no stock is stored in the warehouses awaiting customers' pick up or order. Retailer inventory costs are positive as they will stay in retailers' inventory awaiting pickup once needed. Sales revenue tends to be negative due to low sales and the inability to reach wider markets, however, CO2 emission is negative due to the nature of sustainable and eco-friendly preowned clothing.

Fast fashion is the opposite of preowned clothing. Their demand is too high and frequently produced. People are inclined to purchase more fast fashion due to different trends emerging on the market. Raw materials costs have increased due to high demand, leading to high production costs, this is shown in the positive relationship shown below.

Raw material and retail inventory costs have increased because of wide markets and the necessity for the product to reach the consumers in case of an order. High sales of these items lead to high profits in return.

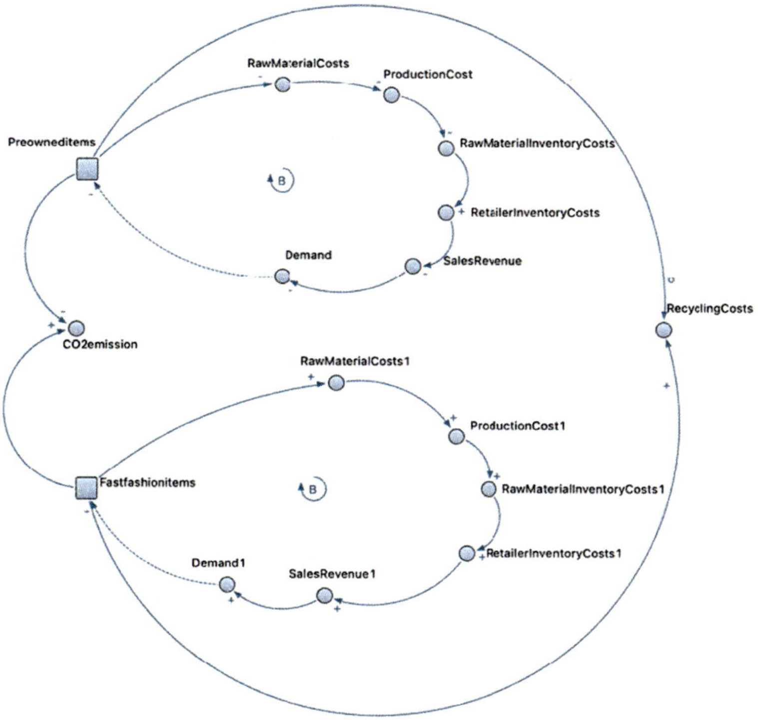

Fig. 1. System dynamic model of a conceptual model showing the comparison between preowned and fast fashion items without the influence of social media analytics.

However, in preowned clothing there are no raw materials and production costs as preowned garments are dealt with. This is an advantage to manufacturers as they save a lot, unlike fast fashion where assembling all raw materials is a must. Due to the nature of fast fashion raw materials are a must. Due to the nature of fast fashion they need to be recycled resulting in new costs which are avoided in preowned clothing. A move towards preowned clothing would save a lot of these costs. Thus, this can benefit both the manufacturers and consumers. The proposed model proposes the application of social media analytics capabilities by building a system dynamics model as shown in Fig. 2. The model explains the influence of social media analytics to promote preowned clothing and cut down CO2 emissions. As discussed above, different variables are affected by consumer choices of either fast fashion or preowned clothing. In recent years, social media has become an effective tool for booming businesses in the clothing industry. The quick rise of internet applications has led to comments, and reviews regarding daily activities [35]. Sentimental analysis is used here to analyze people's opinions towards the preowned fashion trend. People's opinions can be beneficial to businesses

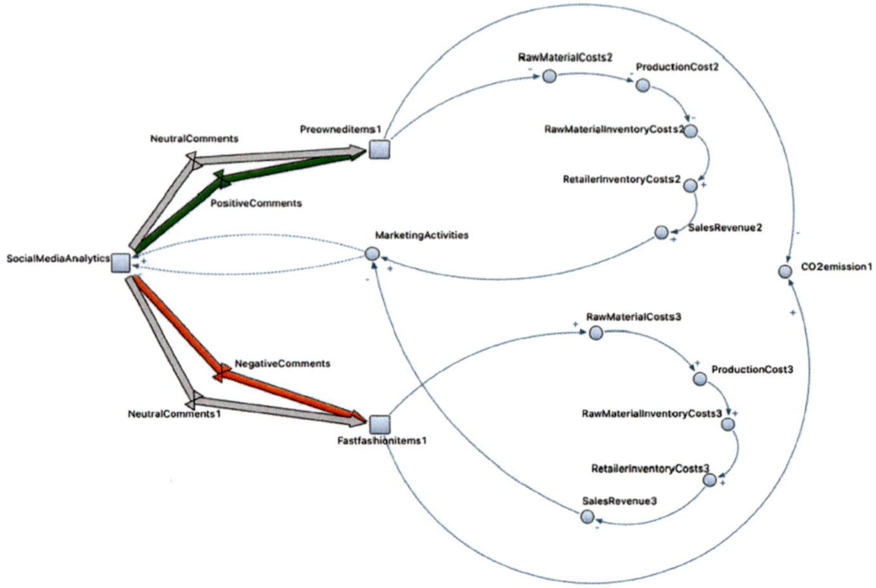

Fig. 2. System dynamic model of the proposed model showing the comparison between preowned and fast fashion items with the influence of social media analytics.

positively or negatively. Here social media analytics would imply either positive or negative comments. Positive comments are in support of moving towards preowned clothing while negative comments value fast fashion. It can be suggested that if the majority of the consumer group can be shifted towards this trend, then the environmental criteria can be positively impacted.

3.2 Framework for Detailed Solution

The proposed model involves analyzing comments extracted from Instagram© using sentiment analysis. The comments are categorized into three categories: positive for preowned, negative for fast fashion, and neutral for both. The first loop focuses on using social media analytics to influence consumer demand for preowned fashion items positively and for fast fashion items negatively. Each comment is analyzed to determine whether it inclines towards a certain choice. Preowned fashion items involve no raw materials or production costs since they are previously used items. The negative effect of raw material costs is avoided, and the revenue earned from sales can be reinvested in marketing activities via social media platforms. On the other hand, the second loop uses negative comments in line with social media analytics to promote preowned clothing and discourage people from buying fast fashion items. Fast fashion items are produced from scratch and thus have positive raw material and production costs. The low demand results in negative raw materials inventory and retailer inventory costs, which leads to low sales revenue. However, this would result in a significant reduction in CO2 emissions. The focus of the proposed model is on the influence of social media analytics on consumers' demand for preowned fashion items and fast fashion items, which is the base of the hypothesis tested in this study.

The impact of altering shoppers' behavior and transforming negative feedback into positive can lead to a decline in the fast fashion industry and a rise in pre-owned fashion trends. Consequently, this approach can reduce the amount of CO2 emissions generated during production. The fashion industry is responsible for approximately 10% of global carbon emissions each year, surpassing the combined carbon emissions of international flights and maritime shipping. To mitigate this problem, future efforts may utilize social media platforms such as Twitter© to measure the impact of individual tweets on reducing CO2 emissions.

4 Evaluation

4.1 Experiment Design Model

The research method chosen for this study is an inductive approach. This paper will analyze the sentiment of consumers' opinions toward pre-owned fashion items leveraging social media data (Instagram©). The workflow designed for this method is shown in Fig. 3 below:

Fig. 3. Analysis process.

In this research, the sentiment of certain text was obtained using the Python programming language. There are various methods to extract text sentiment, but to perform these algorithmic operations through machine learning and AI, the data needs to be pre-processed. This is because computers cannot understand human language directly, and instead rely on 0s and 1s. To extract emotions from text, the data must be fed into the algorithm. Social media posts often contain emotions, hashtags, incorrect spellings, and characters that pose a challenge for the machine to comprehend. Moreover, these texts often do not strictly follow grammatical rules. Therefore, it is necessary to pre-process the data and tailor it for the algorithm. There are several pre-processing steps to follow, and the approach used in this paper is to focus on techniques like lower case, removing punctuations, removing words and digits containing digits, removing stop words, stemming and lemmatization and tokenization.

In the Exploratory Data Analysis section, the data will be explored to extract insights. The analysis will examine patterns of likes and post distributions by location. Furthermore, sentiment analysis will be conducted to gain a deeper understanding of the data. Currently there is no existing dataset for the task of predicting the preowned fashion item trends using sentiment analysis. However, recently social media platforms like Instagram© have gained popularity among consumers where they can share personal images and posts related to fashion, lifestyles, or hobbies. These data are non-biased as they are shared by real consumers. These real-time data will be extracted from Instagram© to obtain the dataset and a set of pre-defined hashtags like #preowned, #secondhand, #thrift or #boycottfastfashion etc. While some of these # contain positive or neutral sentiment, some contain negative sentiment as well. While selecting the hashtags, they would be selected from the top-10 preowned fashion-related hashtags from the most popular Instagram© fashion hashtags released by best-hashtags.com. These data will be stored in our local repository for further analysis. The total number of positive or negative or neutral sentiments relating to each hashtag can be obtained. By collecting all the available data from Instagram© if preowned items positive posts are more than negative it means people are willing and able to shift to this sustainable trend. For fast fashion more positive sentiments shows that buyers are still inclined to this.

#Onlinestore can show the vast number of buyers accessing online platforms and their inclination towards either preowned or fast fashion items. #Resale shows number of items resold and if its market reach is wide. By using social media analytics, we can predict the trend and widespread use of social media can make consumers change their mindset about preowned fashion items. The next step involves creating a data repository that includes the top five selected hashtags: #preowned items, #second hand, #online store, #sustainablefashion, and #resale. To establish a solid ground for these posts, they will be grouped into three post categories: positive, negative, and neutral. The sentiment label for each post will be determined solely by the text. Each hashtag will be collected for the three post categories to assess their relevance. To comply with Instagram's terms and conditions, post IDs will be published in the data repository. Additionally, the pre-processing codes and sentiment labels associated with each post will also be published.

Subsequently, the results from the sentiment analysis will be analysed to determine consumer attitudes toward pre-owned fashion items. This analysis will have a marketing aspect, where the reasons for any positive, negative, or neutral sentiments will be investigated based on different criteria such as age, price, and quality concerns. Moreover, the grey area, which represents neutral comments, will also be investigated to understand what can be done to shift consumers more towards pre-owned fashion. This analysis can provide a good comparison between the pre-launch and post-launch phases of pre-owned fashion items, which can be effective in understanding the role of any promotion or influencers in marketing these kinds of fashion trends.

5 Conclusion

This study explores sentiment analysis on pre-owned fashion-related posts on Instagram©. The task is constructed based on a unimodal setting, whereby sentiment is determined by posts and associated texts. Due to the lack of existing datasets, a pre-owned fashion item-related sentiment analysis dataset was manually collected. Through careful analysis of consumer sentiment extracted from Instagram©, it is possible to predict a paradigm shift from fast fashion to pre-owned fashion, which can be helpful in creating new marketing strategies. It is important to note that all data extracted from Instagram© has been used solely for academic purposes in this research. The use of a framework can provide structure and guidance for marketing projects, helping to ensure that they are properly planned and executed. By incorporating the principles of system dynamics modelling, it is possible to gain a deeper understanding of the complex interrelationships between different factors that impact consumer behaviour, such as pricing, product features, and marketing messages. This, in turn, can lead to more effective marketing strategies that are better tailored to the needs and preferences of target audiences. Overall, the combination of a well-designed framework and a robust system dynamics model can be a powerful tool for marketers looking to create more effective and impactful marketing initiatives. By leveraging the insights gained from these tools, it is possible to better understand consumer behaviour, tailor marketing messages to specific audiences, and drive engagement through social media channels.

References

1. Xie, X., Hong, Y., Zeng, X., Dai, X., Wagner, M.: A systematic literature review for the recycling and reuse of wasted clothing. Sustainability **13**, 13732 (2021)
2. Peter Maddox. Why we should be focusing on clothing as well as plastics. https://wrap.org.uk/blog/2018/11/why-we-should-be-focusing-clothing-well-plastics. Accessed 20 Dec 2022
3. Peter Maddox. How to make the clothing sector fit for a net zero world? https://wrap.org.uk/blog/2020/01/how-make-clothing-sector-fit-net-zero-world. Accessed 19 Dec 2022
4. Zamani, B., Sandin, G., Peters, G.M.: Life cycle assessment of clothing libraries: can collaborative consumption reduce the environmental impact of fast fashion? J. Clean. Prod. **162**, 1368–1375 (2017)
5. Sandin, G., Peters, G.M.: Environmental impact of textile reuse and recycling – A review. J. Clean. Prod. **184**, 353–365 (2018)
6. Amatulli, C., Mileti, A., Speciale, V. Guido, G.: The Relationship between Fast Fashion and Luxury Brands: An Explanatory Study in the UK Market. In: Advertising and Branding: Concepts, Methodologies, Tools, and Applications, pp. 224–265 (2016). https://doi.org/10.4018/978-1-5225-1793-1.ch041
7. Olad, A.A., FatahiValilai, O.: Using of social media data analytics for applying digital twins in product development. In: 2020 IEEE International Conference on Industrial Engineering and Engineering Management (IEEM), pp. 319–323 (2020). https://doi.org/10.1109/IEEM45057.2020.9309834
8. Stringer, T., Mortimer, G., Payne, A.R.: Do ethical concerns and personal values influence the purchase intention of fast-fashion clothing? JFMM **24**, 99–120 (2020)
9. McNeill, L., Moore, R.: Sustainable fashion consumption and the fast fashion conundrum: fashionable consumers and attitudes to sustainability in clothing choice: sustainable fashion consumption and the fast fashion conundrum. Int. J. Consum. Stud. **39**, 212–222 (2015)
10. Mohammadian, N., Mechai, N., Fatahi Valilai, O.: Social media product data integration with product lifecycle management; insights for application of artificial intelligence and machine learning. In: 2022 IEEE International Conference on Industrial Engineering and Engineering Management (IEEM), pp. 1469–1473 (2022). https://doi.org/10.1109/IEEM55944.2022.9989781
11. Björk, J., Magnusson, M.: Where do good innovation ideas come from? Exploring the Influence of network connectivity on innovation idea quality. J. of Prod. Innov. Manage. **26**, 662–670 (2009)
12. Adebanjo, D., Michaelides, R.: Analysis of Web 2.0 enabled e-clusters: A case study. Technovation **30**, 238–248 (2010)
13. Tuyishime, A.-M., Fatahi Valilai, O.: Sustainable last mile delivery network using social media data analytics, pp. 840–874 (2022). https://doi.org/10.15480/882.4696

14. Fuchs, C.: From digital positivism and administrative big data analytics towards critical digital and social media research! Eur. J. Commun. **32**, 37–49 (2017)
15. Yuan, Y., Lam, W.: Sentiment analysis of fashion related posts in social media. In: Proceedings of the Fifteenth ACM International Conference on Web Search and Data Min-ing, pp. 1310–1318. ACM (2022). https://doi.org/10.1145/3488560.3498423
16. Niinimäki, K., et al.: The environmental price of fast fashion. Nat. Rev. Earth Environ. **1**, 189–200 (2020)
17. Michele, C.H.: Used Clothing vs. Fast Fashion: Why Vintage is Always Greener
18. Farrant, L., Olsen, S.I., Wangel, A.: Environmental benefits from reusing clothes. Int. J. Life Cycle Assess. **15**, 726–736 (2010)
19. Evans, F., Grimmer, L., Grimmer, M.: Consumer orientations of secondhand fashion shoppers: the role of shopping frequency and store type. J. Retail. Consum. Serv. **67**, 102991 (2022)
20. Guiot, D., Roux, D.: A second-hand shoppers' motivation scale: antecedents, consequences, and implications for retailers. J. Retail. **86**, 355–371 (2010)
21. Padmavathy, C., Swapana, M., Paul, J.: Online second-hand shopping motivation – Conceptualization, scale development, and validation. J. Retail. Consum. Serv. **51**, 19–32 (2019)
22. Fernando, A.G., Sivakumaran, B., Suganthi, L.: Comparison of perceived acquisition value sought by online second-hand and new goods shoppers. Eur. J. Mark. **52**, 1412–1438 (2018)
23. Secondhand clothing is in fashion - and it's helping the environment. https://www.weforum.org/agenda/2020/12/secondhand-clothing-environment-fashion-style-climate-change-sustainability/. Accessed 26 Dec 2022
24. Gulnaz, K.: The Secondhand Market Is Growing Rapidly, Can Challengers Like Vinokilo Thrive And Scale?
25. Mohammad, J., Quoquab, F., Mohamed Sadom, N.Z.: Mindful consumption of second-hand clothing: the role of eWOM, attitude and consumer engagement. J. Fashion Mark. Manage.: Int. J. **25**, 482–510 (2021)
26. Jindal, K., Aron, R.: WITHDRAWN: a systematic study of sentiment analysis for social media data. Mater. Today: Proc. (2021). https://doi.org/10.1016/j.matpr.2021.01.048
27. Yue, L., Chen, W., Li, X., Zuo, W., Yin, M.: A survey of sentiment analysis in social media. Knowl. Inf. Syst. **60**, 617–663 (2019)
28. Bailey, K., Basu, A., Sharma, S.: The environmental impacts of fast fashion on water quality: a systematic review. Water **14**, 1073 (2022)
29. Biradar, S.H., Gorabal, J.V., Gupta, G.: Machine learning tool for exploring sentiment analysis on twitter data. Mater. Today: Proc. (2021). https://doi.org/10.1016/j.matpr.2021.11.199
30. Diddi, S., Yan, R.-N., Bloodhart, B., Bajtelsmit, V., McShane, K.: Exploring young adult consumers' sustainable clothing consumption intention-behavior gap: a behavioral reasoning theory perspective. Sustainable Prod. Consum. **18**, 200–209 (2019)
31. Derwanz, H.: Digitalizing Local Markets: The Secondhand Market for Pre-owned Clothing in Hamburg, Germany, pp. 135–161 (2021). https://doi.org/10.1108/S0190-128120210000041007
32. Bae, Y., Choi, J., Gantumur, M., Kim, N.: Technology-based strategies for online secondhand platforms promoting sustainable retailing. Sustainability **14**, 3259 (2022)
33. Hu, Y., et al.: Generating business intelligence through social media analytics: measuring brand personality with consumer-, employee-, and firm-generated content. J. Manag. Inf. Syst. **36**, 893–930 (2019)
34. Shah, P., Gajjar, C.: Secondhand shopping: understanding consumer behavior toward pre-owned clothing in India. WRIPUB (2021). https://doi.org/10.46830/wripn.20.00035
35. Wankhade, M., Rao, A.C.S., Kulkarni, C.: A survey on sentiment analysis methods, applications, and challenges. Artif. Intell. Rev. **55**, 5731–5780 (2022)

An Ordered Flow Shop Scheduling Problem

Aslıhan Çakmak[1(✉)], Zeynep Ceylan[2], and Serol Bulkan[3]

[1] Kocaeli Health and Technology University, Kocaeli, Turkey
aslihancakmak.ie@gmail.com
[2] Samsun University, Samsun, Turkey
zeynep.ceylan@samsun.edu.tr
[3] Marmara University, Istanbul, Turkey
sbulkan@marmara.edu.tr

Abstract. The subject of the study is ordered flow shop scheduling problems which were first seen in the literature in the 1970s. The main objective is to get a fast and good solution. For the purpose of the study, firstly, the ordered flow shop scheduling problems were defined. Then, a heuristic method was suggested for the ordered flow shop scheduling problems and a sample problem was solved and discussed. A genetic algorithm (GA) based on the complexity property was developed. Full enumeration was applied to determine the optimal makespan values based on Smith's rule because Smith had specified the conditions under which circumstances the permutation can be optimum.

This study is one of the few studies in the literature to obtain optimum solutions up to 15 jobs ordered flow shop problems in a very short amount of time. The developed GA heuristic can also solve large size ordered flow shop scheduling problems very fast. The significant advantage of the proposed method is that, while Smith's method rule not does not work with large size problems since it requires full enumeration to identify best solution to follow a convexity property where GA algorithm can find a good solution very fast.

Keywords: Ordered flowshop · Scheduling · Genetic Algorithms · Makespan · Convexity property

1 Introduction

The problem of flow shop scheduling was first introduced by Johnson (Johnson 1954) 60 years ago, and since then, numerous studies have been conducted on this topic. Some of the variations of this problem have been formulated by researchers such as Dudek et al. (Dudek et al. 1992), Elmaghraby (Elmaghraby 1968), Gupta and Stafford (Gupta & Stafford, 2006) and Hejazi and Saghafianb (Hejazi and Saghafian 2005).

Scheduling involves allocating resources to different activities that have varying characteristics within a specific timeframe (Torkashvand et al. 2017). In scheduling problems, there is a set of jobs that require either single or multiple operations for completion. This is the essence of scheduling problems. Additionally, there is one workstation for each

Z. Şen et al. (Eds.): IMSS 2023, LNME, pp. 161–173, 2024.
https://doi.org/10.1007/978-981-99-6062-0_16

type of operation, meaning that if there are m-operation jobs, there must be m workstations. In summary, there are two types of jobs: single station and multi-station. Single and parallel machine problems fall under the category of single station shops, while flow shop and open shops belong to the multi-station shop type.

Many studies utilize a processing time matrix, in which the values are randomly selected from a predetermined distribution. To evaluate the effectiveness and efficiency of solution procedures, randomly generated problems are often employed. These problems feature processing times that are drawn from the same distribution used in job sequencing research.

There is no difference in difficulty between hypothetical and physical problems. Additionally, many industries encounter ordered flow shop problems. Therefore, instead of hypothetical problems, utilizing problems with actual processing times is more beneficial in practical situations.

The ordered flow shop scheduling problem was introduced by Smith (Smith 1968). This problem is also known as the ordered matrix problem and can be considered a subcategory of the flow shop problem with ordered processing times.

The primary objective of this study is to solve ordered flow shop problems with the makespan objective. Firstly, the flow shop scheduling problem and the ordered flow shop scheduling problem were explained. Then, the use of genetic algorithms for solving the permutation flow shop problem was outlined. Finally, the proposed method and its results were discussed.

The proposed method utilized the convexity property to obtain a feasible solution, and modifications were made to the crossover and mutation processes in this method.

2 Flow Shop Scheduling Problem

Assume that there are n-jobs and m-machines, and the jobs need to be scheduled across these machines. For a problem to be classified as an m-machine flow shop problem, it needs to satisfy three requirements. Firstly, each must consist of m-operations. Secondly, each operation must require a different machine. Finally, all jobs must be processed in the same order through the machines. If these three criteria are met, that problem can be referred to as an m-machine flow shop problem (Aldowaisan and Allahverdi 2004).

Flow shop scheduling problems can be classified into two groups based on the machine characteristics: machines can have buffer spaces, or the processing is continuous without any interruptions between machines. When the processing is continuous and there are no interruptions, the problem is referred to as the no-wait flow shop problem. The no-wait flow shop problem arises when discontinuous processing is not allowed, as in the case of industries such as metal or food production, where the process must continue from start to end without interruption.

This study utilized the m-machine flow shop problem with the objective of minimizing the total completion time, which is also known as the makespan objective.

3 Ordered Flow Shop Scheduling Problem

Ordered flow shop problems must satisfy two properties. Firstly, given two jobs, A and B, with known processing times, if the processing time of A on any machine is smaller than the processing time of B, then A must also have a smaller or equal processing time on all machines compared to B.

Secondly, for ordered flow shop problems, if the processing times of several jobs are known, and if any job has its jth smallest processing time on any machine, then every job must have their j^{th} smallest processing time on the same machine.

Based on these 2 properties, jobs can be sequenced in the ascending order according to their processing times, thus transforming the problem into an ordered flow shop problem with the minimum makespan criterion and permutation schedules.

There are various flow shop sequencing problems in the literature (Panwalkar and Woollam 1980), with the most interesting ones, according to the researchers, including the classical (n × m) problem, the (n × m) problem with no waiting, the (n × m) ordered problem, and the (n × m) ordered problem with no waiting.

In the classical problem, there are infinite intermediate storages between machines, meaning that there is no limitation on the number of jobs waiting in the queue between machines. Redid and Ramamoorthy (Reddy & Ramamoorthy) and Wismer (Wismer) introduced a restriction of no waiting with some justification.

The classical (n × m) problem and (n × m) problem with no waiting are considered for not only the makespan but also the mean flowtime criteria. Efficient solution procedures are not obtainable for all of these problems, but there are some exceptions. Nonetheless, both of these problems are considered unsolvable in computational complexity. However, some specialized techniques can be developed that are applicable to these problems (Coffman).

Special characteristics of processing times in flow shops have been considered in the development of the n × m ordered problem. Smith (Smith et al. 1975) and Panwalkar (Panwalkar et al. 1973) provided the practical basis for this problem.

The (n × m) ordered problem with no waiting has been considered with the makespan criterion by Panwalkar and Woollam (Panwalkar & Woollam). They have developed simple and efficient procedures for certain cases of the problem.

Smith (Smith, Dudek, & Panwalkar, 1973) demonstrated that jobs ordered in ascending or descending order based on their processing times are optimal for ordered flow shop problems if the maximum processing time for each job occurs on the first or last machine. Therefore, if the maximum processing time for each job occurs on the first machine, the jobs can be arranged in descending order. Conversely, if the maximum processing time occurs on the last machine, the jobs can be ordered in ascending order. However, if the maximum processing time occurs on an intermediate machine, Smith (Smith, Dudek, & Panwalkar, 1973) proposed an algorithm to minimize the makespan, where the minimum makespan sequence is arranged in ascending order of processing times followed by the remaining jobs in descending order.

Smith's method generates 2^{n-1} alternative sequences in 4 steps as follows:

Step 1: the jobs need to be ordered in terms of ascending processing times on the first machine.

Step 2: in each partial sequence, the lowest ranking job which has not been arrayed should be placed in the leftmost and rightmost unfilled array position.

Step 3: Step 2 should be repeated until the first n-1 jobs are placed into every array. The highest-ranking job has to be put only into the unfilled position.

Step 4: for makespan, the 2^{n-1} sequences are evaluated. The sequences with the lowest makespan are the optimum ones.

4 Genetic Algorithm for Permutation Flow Shop Problem

Genetic Algorithms (GAs) are heuristic search techniques and intelligent randomized search strategies that can find near-optimal solutions for complex problems (Iyer and Barkha 2004). These algorithms are highly effective in solving problems, especially when the search space is vast and standard techniques are unable to efficiently solve the problem. GAs are based on the natural genetics and mechanics of natural selection. In GA, chromosomes represent solutions in the search space. The crossover and mutation operators are determined in each GA. How they are defined is significant.

Several steps need to be followed while applying GAs to problems. Firstly, the feasible solutions of the problems must be transformed into chromosomes, which are a string-type structures.

A standard GA starts by generating a set of presumed or randomly produced solutions, which are called chromosomes, as the initial population. Over a series of generations or iterations, it evolves different sets of solutions, i.e., chromosomes, that are better than the previous ones, with the aim of finding the optimal solution to a problem. The objective function specifies the suitability of each chromosome in each generation. A subset of the chromosomes is selected for reproduction based on their fitness value. The fitness value of each chromosome for the flow shop scheduling problems is taken as the commutual of the makespan. The number of offspring that an individual parent produces is directly proportional to its fitness value, and those with lower fitness values are eliminated via the natural selection procedure. New chromosomes, or offspring, are generated by the applying of genetic operators such as mutation and crossover to the reproduced chromosomes. The new chromosomes constitute the subsequent generation, and the process continues iteratively until a termination criterion is met.

A permutation flow shop problem involves multiple jobs and machines that need to be processed in a specific order on each machine. The primary objective of this problem is to determine a sequence for the jobs that minimizes the overall completion time.

The permutation flow shop scheduling problem is a type of assembly line problem in which there are m distinct jobs that need to be processed on n different machines. All jobs have to be processed in the same order on each machine, and the processing times for each job are predetermined and fixed. The main objective of this problem is to find a job sequence that results in the minimum completion time.

Processing time matrix of the problem:

$P = (p_{ij}), p_{ij} > 0, i = 1,\ldots,m\ j = 1,\ldots,n$

At any given time, each machine processes exactly one job. Moreover, each job is processed on exactly one machine. The problem is to find a sequence of jobs with the minimum makespan, where the objective is to minimize the maximum completion time among all jobs. That is, $\max_i C_i$ is minimized, where C_i is the completion time for job *i*.

A standard implementation exists for solving the permutation flow shop problem using a GA. The components of a standard GA can be described as follows.

The Scheme of Encoding: job sequences can be viewed as chromosomes. For example, in a five job problem, a chromosome would be represented as [12543], which also represents the alignment. Job1 is processed first on all machines, followed by job2, job5, job4 and finally job3 on all machines. This is a natural choice for this particular problem.

Initial Population: the initial population consists of N randomly generated job arrays, where N represents the size of the population.

Fitness Evaluation Function: the fitness evaluation function aims to simulate the natural process of survival of the fittest. To achieve this, the fitness evaluation function assigns a value to each member of the population that reflects their relative superiority. If r_i, where $i = 1,\ldots,N$, represents the mutual of the makespan of the strings in the population, the fitness value assigned to string i would be proportional to $f_i = r_i / \sum_j r_j$. Our objective is to minimize the makespan.

Reproduction: based on their fitness values, individual chromosomes from the current population are replicated. That replication process is performed by randomly selecting chromosomes with a probability proportional to their fitness value. To generate the next generation, $N \times f_i$ copies of string i are expected to be present in the gene pool. This creates the mating pool, to which crossover and mutation operators are applied in order to generate the next generation.

Crossover: the primary objective of crossover is to swap information between randomly selected parental chromosomes to generate improved offspring. The crossover process combines the genetic material of two parental chromosomes to generate offspring for the next generation. This offspring retains the desirable characteristics of the parent chromosomes. This exchange is also intended to investigate superior genes.

For the crossover implementation of crossover, two parents are randomly selected from the mating pool. The parents are duplicated with a probability $1\text{-}p_c$ where p_c is the crossover probability. The following process is then performed.

In the standard genetic algorithm, a single point crossover is performed between two parents. This is done by randomly selecting a number *k* between 1 and *l*-1, where *l* is the length of the string and is greater than 1. The result of the interchanging all characters from position $k + 1$ to *l* is the creation of two new strings. It is important to note that this coding scheme can produce infeasible strings. For example, in an eight job problem with $k = 3$ and the crossing strings are [12345678] and [58142376].

This problem can produce [12342376] and [58145678], but these are infeasible solutions due to the recurrence of jobs. Therefore, a modification needs to be made to the standard crossover procedure. The first modification is to replicate all characters of the first parent's chromosome until location *k*.

Mutation: two distinct locations of the chromosome are randomly selected, and the jobs at these locations are swapped during the crossover process. Then, the mutation operator is independently applied to each child obtained from the crossover. A small probability p_m is used when applying the mutation operator. Mutation expands the search space, preventing the selection and crossover from focusing on a narrow area of the search space and preventing the GA from getting stuck in a local optimum.

Criteria of Termination: that criterion determines the number of iterations at which the algorithm will terminate in the GA.

5 Proposed Method

This section provides a detailed explanation of the proposed method for scheduling an ordered flow shop. The method involves applying crossover and mutation operations and takes into account the requirement for a feasible solution with convexity. Each step of the approach is described in sequence.

The first step of the proposed method involved creating an ordered matrix that represented the processing times of jobs on different machines. To form the matrix, two key properties of ordered flow shop problems were taken into account and random values were assigned to its elements. The software was only provided with information on the number of jobs and machines available, such as a problem with 10 jobs and 5 machines or one with 30 jobs and 20 machines. The proposed method considers problems with a range of 10 to 30 jobs and 5 to 20 machines.

Setting the population size was another important criterion that had to be determined. In this method, the population size was fixed as 2n, where n denotes the number of jobs.

In addition, specific stopping criteria, also known as termination criteria, were defined for each distinct ordered matrix. The stopping criteria employed in the problem include 500 iterations, 1000 iterations and 2000 iterations.

The next step was to form the parents. While the parents were selected randomly, first the peak was found and then they were aligned in ascending order from the starting point to the peak, in descending order from the peak to the end. It was crucial to ensure that there was only one peak in the matrix.

The third step involved applying crossover to generate offspring from the parents. The parents were randomly divided into two parts, with the division not necessarily at the middle point. The first part of the first parent was then copied, and the second part of the second parent was inserted according to the rule described in the second step, which involved arranging the jobs in ascending order up to the highest point, and in descending order thereafter. It should be noted that ensuring the matrix had only one peak was also critical during this crossover operation.

In the fourth step, mutation was performed by randomly selecting two elements and leaving the other unchanged. The selected elements were then repositioned according to the rule described in the second step, which involved arranging the jobs in ascending order up to the highest point, and in descending order thereafter.

The parent selection process employed the Roulette Wheel Selection method. This involved randomly selecting two numbers and picking the parents whose cumulative fitness values were closest to them. It should be noted that the fitness values were calculated by dividing 1 to the C_{max} of parents.

In the final step, elitism was applied to generate a new population that also represented the solution. This involved forming 50% of the new population by selecting the best 25% of the current population and 25% of the child population. The remaining 50% of the new population was filled with individuals from the current and child populations that were not selected in the previous step.

6 Solutions

The population size was set to 2n, where n represented the number of jobs. The proposed method was tested for different mutation rates, stopping criteria, and varying numbers of jobs and machines. The method was implemented using the C programming language. An example problem and its corresponding results are provided below.

Table 1 displays the ordered matrix generated by the computer using random values that were generated based on certain rules explained in this study. For this problem, the mutation probability was set to 0.1 the crossover probability was set to 0.6 and the number of iterations was set to 500.

Table 1. Ordered Matrix.

J/M	M1	M2	M3	M4	M5
J_1	2	34	9	18	14
J_2	28	62	40	56	52
J_3	3	41	13	28	17
J_4	39	90	53	79	65
J_5	48	104	68	94	70

The iteration that yielded the best fitness value was the 5th iteration, and it took 4773849 ns to obtain the result. In this context, the best fitness value refers to the lowest value achieved. For this particular problem, the best fitness value was 525, and the best alignment associated with that fitness value was [J_1, J_2, J_5, J_4, J_3].

The convexity of the result is demonstrated in Fig. 1, where the processing times for each machine are plotted according to the best alignment. The figure shows that the processing times initially increase and then decrease, indicating a convex relationship.

Tables 2 to 3 and Figs. 2–5 provide summary results for different problem sizes. The mutation probability, crossover probability, and number of iterations were set to 0.1, 0.6 and 500 respectively for all problems shown in the tables and figures. The elements of the matrix are between 1 and 800. Each matrix was executed 10 times and the minimum, maximum, and average makespan values as well as average execution

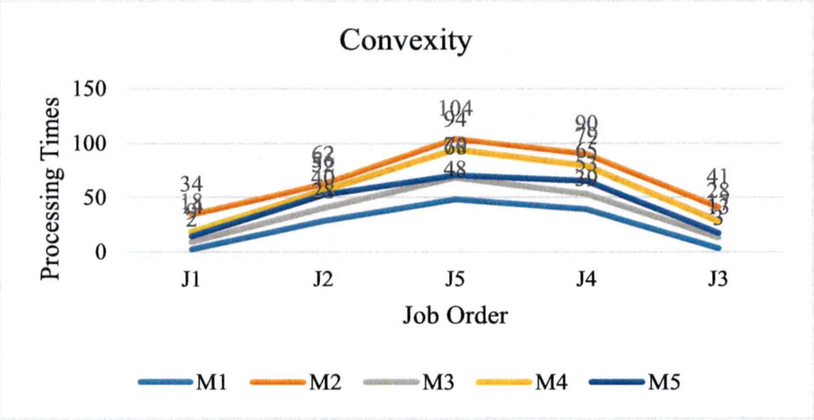

Fig. 1. Convexity Property.

times were recorded. For instance, in a 10 jobs and 5 machines problem, 10 different matrixes were created and each matrix was executed 10 times.

A new C programming language code was developed for comparison purposes. In this proposed algorithm, the jobs were sorted in ascending order based on their processing times and were then placed sequentially in the first and last empty locations of the array. For a problem with n jobs, there are 2^{n-1} arrays generated. The makespan and processing time for each array were determined, and the best arrays were selected.

Table 2. Proposed GA Results for 20 Job Problems.

Problem Rate	Makespan			Time (Seconds)
	Minimum	Maximum	Average	Average
CB_20J-5M_01	4776	4781	4777	4.77
CB_20J-10M_01	7730	7733	7730	4.94
CB_20J-15M_01	12256	12261	12257	11.41
CB_20J-20M_01	15530	15545	15533	36.68

Table 3. Proposed GA Results for 30 Job Problems.

Problem Rate	Makespan			Time (Seconds)
	Minimum	Maximum	Average	Average
CB_30J-5M_01	9430	9430	9430	1.03
CB_30J-10M_01	16154	16161	16157	53.54
CB_30J-15M_01	8267	18267	17263	69.03
CB_30J-20M_01	23275	23304	23290	122.79

* Others are not included in the table due to page limit.

7 Result and Discussion

The proposed method improved a genetic algorithm based on the complexity property and implemented the full elimination technique based on Smith's rule. Smith's rule provides the conditions under which the permutation can be optimal.

The proposed method successfully generated results for problems ranging from 10 to 30 jobs and 5 to 20 machines.

On the other hand, Smith's method could not be executed for 20 and 30 job problems due to their size and insufficient computer memory. The algorithms were executed on a computer with Intel i7 processor, 8 GB of RAM, a 500 GB hard disk, and the Windows10 operating system. The computer's specifications can be considered a limitation for solving the problem. The results of the proposed method for 10 to 30 job problems are presented in Fig. 5.

Figure 2 compares the time performance of the proposed algorithm and Smith's method for scheduling 10 jobs. As shown in the figure, the execution times are identical for both methods when there are up to 15 jobs to be scheduled. However, for more than 15 jobs, the proposed algorithm shows a significant improvement in execution time, thanks to its elimination approach. Time comparison between the proposed algorithm and Smith's method for 15 jobs is shown in Fig. 3. Although the execution times are comparable for 10 jobs, the difference becomes more apparent as the number of jobs increases to 15. Notably, even though the makespan values are the same for both methods when scheduling 10 jobs, the proposed algorithm can still save time compared to Smith's method.

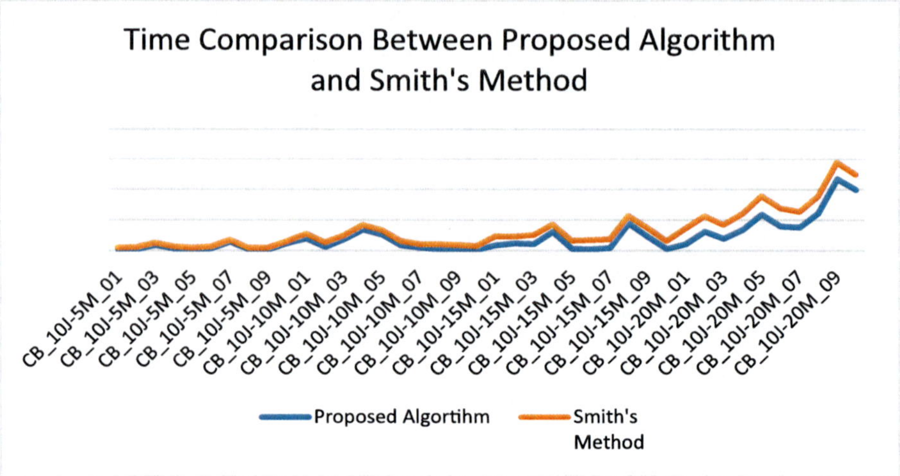

Fig. 2. Time Comparison Between two Methods (10 jobs).

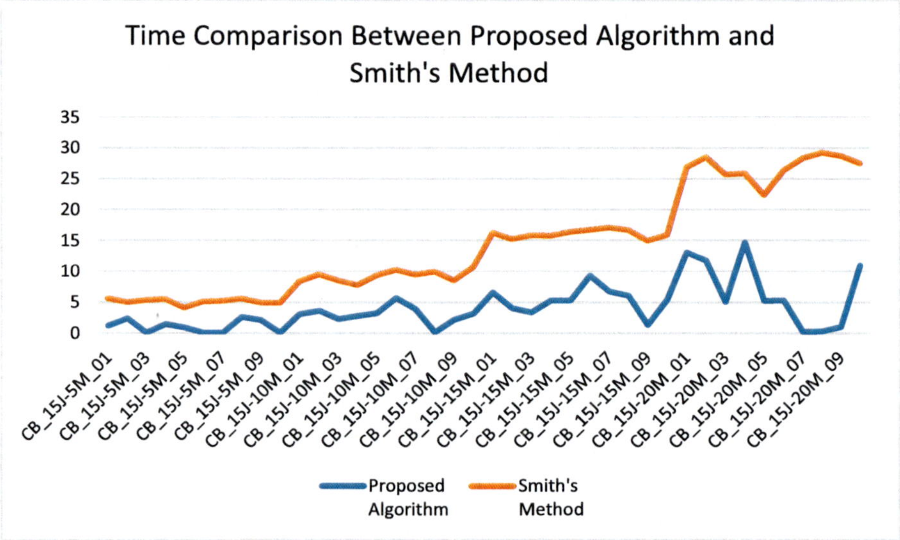

Fig. 3. Time Comparison Between two Methods (15 jobs).

Figure 4 depicts a time comparison between the two methods, for all problem sizes. As demonstrated in the figure, Smith's method could not produce results for problems with more than 15 jobs, whereas the proposed algorithm was able to solve all problem sizes.

Figure 5 shows the makespan comparison of the two methods for all data in the study. As the figure clearly indicates, Smith's method fails to provide solutions for problems that involve more than 15 jobs and 20 machines.

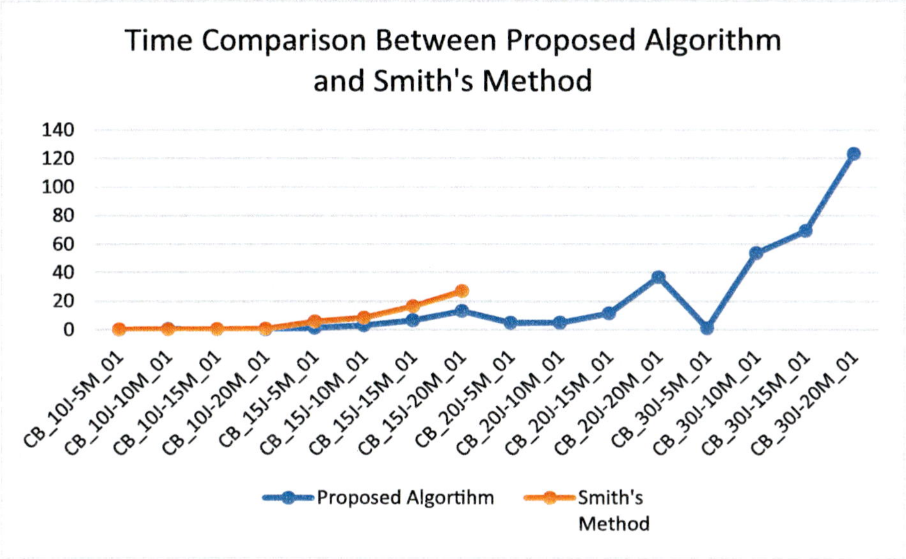

Fig. 4. Time Comparison Between two Methods (10 to 30 jobs).

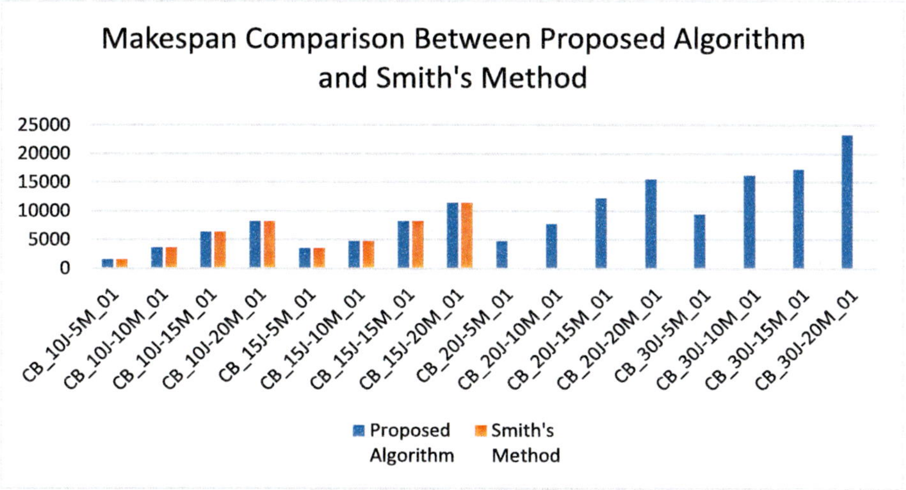

Fig. 5. Makespan Comparison Between two Methods (10 to 30 jobs).

8 Conclusion and Future Research

Although the optimum sequencing characteristic is evident in ordered flow shop problems, finding the optimal solution becomes impossible as the problem size increases due to computational explosion.

The present study found that the proposed algorithm can achieve the optimum solution for scheduling problems with up to 15 jobs. We obtained the same optimum solutions that Smith's method achieved for up to 15 jobs by using a genetic algorithm. While it may be possible to solve problems with 20 jobs using more powerful computers, the proposed algorithm can still produce good solutions for problems with more than 20 jobs. However, we cannot guarantee that full elimination will always yield the optimum solution on our computers beyond 15 jobs. Nevertheless, since the makespans obtained by the proposed algorithm and Smith's method are identical, we can be confident that our algorithm can solve the problem. The proposed algorithm can quickly reach a good solution, and the fact that we found the optimum solutions for 10 and 15 jobs suggests that the algorithm should also work well for 20 and 30 jobs. Smith's method provides a rule for achieving the optimum solution based on the convexity principle and by performing full elimination.

In summary, as the problem size increases, the proposed method outperforms Smith's method by finding an optimal solution in less time. In addition, Smith's method fails to provide any solution for problems with more than 15 jobs, while the proposed algorithm can handle up to 30 jobs. Overall, the proposed algorithm is superior to Smith's method as it achieves a better makespan in less time.

References

Aldowaisan, T., Allahverdi, A.: New Heuristics for M-Machine No-Wait Flowshop to Minimize Total Completion Time. Kuwait University, Kuwait (2004)

Coffman, E. (n.d.). Preface. Opns. Res. (26), 1–2

Dudek, R., Panwalkar, S., Smith, M.: The lessons of flow scheduling research. Oper. Res. **40**, 7–13 (1992)

Elmaghraby, S.: The machine sequencing problem-review and extensions. Nav. Res. Log. Q. **15**, 205–232 (1986)

Gupta, J., Stafford, E.: Flow shop scheduling research after five decades. Eur. J. Oper. Res. **169**, 375–381 (2006)

Hejazi, S., Saghafian, S.: Flow shop scheduling problems with makespan criterion: a review. Int. J. Prod. Res. **43**, 2895–2929 (2005)

Iyer, S., Barkha, S.: Improved genetic algorithm for the permutation flowshop scheduling problem. Comput. Oper. Res. **31**, 593–606 (2004)

Johnson, S.: Optimal two and three stage production schedules with setup times included. Nav. Res. Log. Q. **1**, 61–68 (1954)

Panwalkar, S., Woollam, C.: Ordered flow shop problems with no in-process waiting: further results. J. Opl. Res. Soc. **31**, 1039–1043 (1980)

Panwalkar, S., Woollam, C. (n.d.).: Flow shop scheduling problems with no in-process waiting: a special case. J. Opl. Res. **30**, 661–664 (1979)

Panwalkar, S., Dudek, R., Smith, M.: Sequencing research and industrial scheduling problem. In: proceedings of the Symposium on the Theory of Scheduling and Its Applications. Berlin (1973)

Reddy, S., Ramamoorthy, C. (n.d.).: On the flowshop sequencing problem with no wait in process. Opl. Res. (23), 323–331

Smith, M.: A Critical Analysis of Flow Shop Sequencing. Texas Tech University, Lubbock (1968)

Smith, M., Dudek, R., Panwalkar, S.: Job sequencing with ordered processing time matrix, pp. 481–486. Wisconsin: 43rd Meeting of ORSA (1973)

Smith, M., Panwalkar, S., Dudek, R.: Flowshop sequencing problem with ordered processing times matrices. Manage. Sci. **21**, 544–549 (1975)

Torkashvand, M., Naderi, B., Hosseini, S.: Modelling and scheduling multi-objective flow shop problems with interfering jobs. Appl. Soft. Comput. **54**, 221–228 (2017)

Wismer, D. (n.d.).: Solution of the flowshop scheduling problem with no intermediate queue. Op. Res. **20**, 689–697

Fuzzy Logic Based Heating and Cooling Control in Buildings Using Intermittent Energy

Serdar Ezber[1](✉), Erhan Akdoğan[1], and Zafer Gemici[2]

[1] Mechatronics Eng. Department, Yıldız Technical University, Istanbul, Turkey
serdarezber@gmail.com, eakdogan@yildiz.edu.tr
[2] Mechanical Eng. Department, Yıldız Technical University, Istanbul, Turkey
zgemici@yildiz.edu.tr

Abstract. Energy-saving technologies for heating and cooling systems in buildings have received much attention. Energy saving is even more important in the heating and cooling systems of intermittent use places such as mosques. Because the stop-start operation of the mechanical systems causes the overall efficiency to decrease even more, and the thermal comfort of the environment cannot be provided. In order to save energy and adapt to variable environmental conditions, this study investigates the fuzzy logic control structure of the intermittent radiant heating and cooling system. Fuzzy logic control can handle imprecise or uncertain information, making it an effective solution in complex systems where mathematical models are difficult to derive. The control strategies included actual ambient air temperature, outdoor temperature, return water temperature and system on/off timing. Expert knowledge and observations of system performance were used to create fuzzy logic rules. Studies were carried out on the radiant floor heating and cooling system with ground source heat pumps and thermal energy storage of a mosque.

Keywords: Fuzzy Logic · Energy Efficiency · Intermittent Operation · Operation Strategy · Radiant Heating and Cooling System

1 Introduction

Energy demand is increasing day by day with the effect of increasing population and developing technology. This brings serious problems in energy supply and environmental problems. To overcome these problems and challenges, decision-makers are making more radical decisions and imposing restrictions on societies. Some of these constraints are sanctions for more efficient use of available energy. Therefore, considerable efforts are being made to reduce energy consumption and increase efficiency in industry and households. Buildings are responsible for 30% of total global energy consumption, and there are many different reasons for the increasing energy demand in buildings [1]. However, different intended uses of buildings lead to different energy consumption behaviors. A dwelling that consumes energy continuously and a dwelling that needs energy intermittently do not have the same energy efficiency. Spaces that consume energy intermittently are generally less efficient.

Z. Şen et al. (Eds.): IMSS 2023, LNME, pp. 174–187, 2024.
https://doi.org/10.1007/978-981-99-6062-0_17

Mosques are a good example of places that consume energy intermittently due to their periodic use. They are used for worship five times a day, for about one hour each time. At these times, only a small part of the capacity of the mosque is utilized. On the other hand, there are times when the capacity is fully utilized, such as Friday prayers, tarawih prayers, and Eid prayers. For this reason, mechanical installations in mosques are designed according to full capacity usage. However, due to economic reasons, an adequate air conditioning system cannot always be provided, and a solution is developed by adding split-type air conditioners. However, in crowded prayers where the cooling capacity is insufficient, the congregation tries to cool the interior by opening the windows for natural ventilation, especially in hot weather. In cold weather, an electric heating system is chosen to heat a few rows. These methods lead to higher energy consumption and deteriorate the thermal comfort inside the mosque. In addition, the short-term use of the space, the inability to fully utilize the capacity during prayer times and the long intervals between prayer times cause inefficient operation of the mechanical installations and excessive energy consumption. Generally, the interior surfaces of mosques in Turkey are covered with materials such as marble or plaster and the floors are covered with carpets.

Attia et al. simulated the control of air conditioning systems in buildings using fuzzy logic and conventional PID control and showed that fuzzy logic control is more efficient [2]. Budaiwi et al. developed a mosque model and created a simulation environment to investigate the effect of the operational strategies of the air conditioning system on this model. In this study, the mosque was divided into different zones and the operation of the air conditioning system with different strategies at certain times was simulated and a significant reduction in annual energy consumption was observed [3]. Krzaczek et al. developed a heating and cooling system called Thermal Barrier (TB) in pipes placed inside the wall. The system is controlled using fuzzy logic and the developed fuzzy logic control system can be implemented without the need for any building model and topology [4]. Ulpiani et al. compared different control methods for heating systems in buildings in terms of energy efficiency. In the application, electric radiators and sensors measuring temperature at different points were used. In fuzzy logic control, outdoor and indoor temperatures and the derivative of indoor temperature are used as parameters. They showed that the fuzzy logic control method outperforms both PID and on-off controllers and reduces energy consumption by 30–70% [5].

There are many studies aiming to improve the intermittent operation efficiency of ventilation and heating systems. A systematic review on this topic was conducted by Hu et al. [6]. Different energy-saving technologies have been used, such as optimizing wall thermal performance [7], improving energy use efficiency [8], optimizing building form [9], and using new energy sources and materials [10]. Intermittent operation and night setback systems are generally considered for energy savings in buildings. The night setback system saves energy by reducing the temperature at unnecessary times in the heating system control [11]. Ling et al. used an air-source heat pump for the heating system of a school and compared the effects of different operating strategies to provide thermal comfort in the building while achieving energy efficiency. To find the optimal operating conditions, they simulated 64 different operating conditions by correlating 8 start times and 8 return water temperature values, and the results showed that operating at a lower water temperature with an earlier start time is more energy efficient than operating at a higher water temperature with a later start time while meeting the thermal comfort requirement [12]. Kim et al. optimized the control of the central heating system of university buildings operating according to the outdoor temperature in a specified time period and achieved a 25% energy savings over the period. In the proposed control strategy, an on-off control is implemented and is simulated using a dynamic simulation programme. It was shown that with a simple on-off control, excessive temperature rise can be prevented and the indoor temperature can be kept within a comfortable range with less energy use [13]. Cho and Zaheer-uddin compared two different air conditioning strategies. In the conventional control method, a heating program table is created based on the outdoor temperature and the heating system is operated accordingly. The predictive control method consists of four steps: estimating the outdoor temperature, estimating the boiler heat output, determining the heating hours during the day, and determining the total heating hours during the day. Simulation and experimental studies were carried out, and the predictive control strategy provided an energy savings between 10% and 20% compared to the conventional control strategy [14].

Gwerder et al. applied PWM control of thermal active building systems in intermittent operation. Two different PWM control methods were implemented in a testbed, the first one is derived based on a simple TABS model. For the second method, a simpler solution was developed to reduce the tuning efforts. The basis of this solution is that an equal amount of energy must be delivered to the thermally activated structure in the PWM operation phase as in the continuous operation mode. PWM control enables energy efficient control solutions by intermittently operating the TABS zone pump [15]. Mustafa Fatih et al. used a microcontroller to control the room temperature with fuzzy logic. They determined the operating states of the cooling and heating fans based on the created rule table. It was observed that the result obtained from the comparison of the simulated system in computer environment and the real physical system was consistent [16]. The low-temperature hot water underfloor heating system does not affect indoor thermal comfort conditions during intermittent operation. Chao et al. developed a control strategy that can adjust the on/off time by considering the average temperature value of the feed water and the return water temperature value. Using the control strategy used in urban residential buildings, they determined different operation times between rooms according to the characteristics of indoor temperature variation [17]. Mustafa et al. performed greenhouse system control using fuzzy logic structure with PLC. They controlled the environmental temperature and humidity inputs and heating, fogging, and ventilation outputs using a fuzzy logic structure in greenhouse systems where it is difficult to extract the mathematical model. With the rule base created, the environmental conditions required for the greenhouse were prepared by the PLC [18].

In this study, the control of the radiant heating and cooling system of a mosque, which is one of the buildings using intermittent energy, is provided with a fuzzy logic structure using PLC. In the control of the system, which is non-linear and has different variable ambient values, the operating time of the heating-cooling system is determined with the inputs of the indoor temperature value, outdoor temperature value, and return water temperature value of the mosque. The pump is operated with PLC by considering the working time calculated with the fuzzy logic structure according to the prayer times.

All input and output signals of the system are collected in the PLC and all operations in the fuzzy logic structure are performed by PLC.

2 Material and Method

There are different methods to save energy without affecting comfort conditions in heating and cooling systems of intermittent energy consuming buildings. Fuzzy logic is a suitable method for the control of systems where it is difficult to obtain mathematical models. Fuzzy logic can produce consistent results even in uncertain and approximate situations by imitating human decision-making mechanism.

In this study, a control system using a fuzzy logic structure and operated by a PLC is developed to control the radiant heating and cooling system of a mosque located in the YTU Davutpasa campus. A ground-source heat pump is used as the heat source in the radiant heating and cooling system.

Expert knowledge is utilized to create a control system that adapts to varying environmental conditions and saves energy, considering the usage periods of the mosque. The operating time of a circulation pump is determined by a fuzzy logic algorithm that takes the outdoor temperature, the indoor temperature of the mosque, and the return water temperature as reference to save energy.

2.1 Physical System

The radiant heating and cooling system consists of two ground-source heat pumps, circulation pumps, thermal energy storage system, two boilers, valves, temperature sensors, buried pipes, collectors, and a control panel. Heating and cooling of the floor are achieved by circulating water through pipes placed on the mosque floor. Due to these features, radiant systems can be used both in summer and winter according to the needs. Some of the advantages of these systems are that they take up less space, operate quietly, require less cleaning, can be divided into zones and have the potential to save energy.

The experimental building is a mosque located in YTU Davutpasa campus, which has a transitional climate between the Black Sea and the Mediterranean. The mosque architecture is a rectangular concrete structure. The floor is carpeted and the underfloor heating system is divided into two zones, front and back rows.

2.2 Radiant Heating and Cooling System

The radiant heating and cooling system consists of a ground-source heat pump, an energy storage tank, underground pipes, and auxiliary equipment. The ground-source heat pump can be used for both heating and cooling modes. The maximum outlet water temperature in heating mode is 55 °C, and the minimum outlet water temperature in cooling mode is 7 °C. The operating principle of the heat pump unit is to transfer the energy it takes from the ground to the circulating water and thus, to heat or cool the water. The water at the outlet of the heat pump unit is sent to the boiler and from here to the energy storage tank or the underfloor pipes. The water at the outlet of the underfloor pipes is returned to the supply source back. The operation of the system is selected one of the energy storage tank or heat pump units and sent to the underfloor pipes in the mosque floor.

2.3 Measurement Setup

The control of the system is carried out depending on the indoor temperature value of the mosque, outdoor temperature value, and return water temperature value. In addition to the measurement data used in the control system, the supply water temperature value, supply water flow rate, and return water flow rate are transferred to the system as data. T-type thermocouples are used to measure the ambient temperature value, while PT1000 is used to measure water temperatures. Ultrasonic flow meter is used for flow measurement. The data are instantaneously transferred to the PLC system. All data, including the operating time and circulation pump operating status data output by the fuzzy logic algorithm running in the PLC, are transferred to the cloud system and continuously recorded at one-minute intervals.

2.4 Control System

A PLC, commonly used in industrial automation systems, was preferred for the control of the system. The necessary signals were collected on the PLC panel and the actuators in the system were operated according to the algorithm output created with fuzzy logic structure.. The system output is based on the calculation of the heating time and the activation of the circulation pump a certain time before the prayer times calculated according to this time. The circulation pump continues to run for a fixed time period of time, which is set when the prayer times are entered. The data processing of the temperature sensors, determination of the current prayer times according to the coordinates, all numerical operations in the fuzzy logic structure, and the operation of the actuators were carried out using the PLC program. C language and Ladder programming technique were used as the PLC programming language.

Fuzzy Logic Controller Configuration. The fuzzy logic controller consists of fuzzification, rule base, fuzzy inference, and defuzzification. A fuzzy logic-based controller structure is shown in Fig. 1.

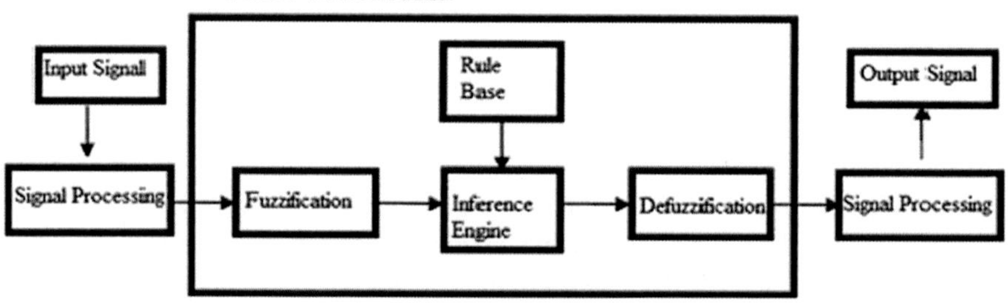

Fig. 1. Fuzzy Logic Control Structure

Fuzzification. The fuzzification process allows us to express numerical input and output signals in linguistic terms and membership values. Each parameter is defined using linguistic expressions with expert knowledge. The membership functions of all input and output parameters to be used in the fuzzy system are shown in the table and figures below (Tables 1, 2, 3, 4 and Figs. 2, 3, 4, 5).

Table 1. Membership functions for Indoor Temperature (IT)

Membership Function	Shape	Points
cold	trapezoid	0:0:15:16
comfort	triangle	15:17:19
warm	triangle	17:19:22
Hot	trapezoid	19:22:40:40

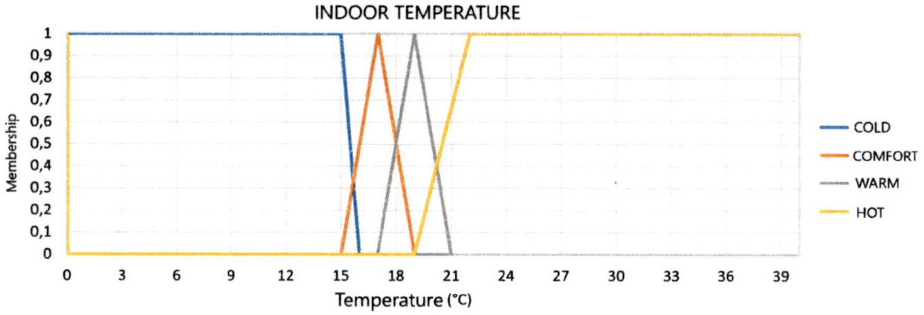

Fig. 2. Membership functions for Indoor Temperature

Table 2. Membership functions for Outdoor Temperature (OT)

Membership Function	Shape	Points
cold	trapezoid	0:0:14:16
normal	trapezoid	14:16:17:19
warm	trapezoid	17:19:20:22
hot	trapezoid	20:22:40:40

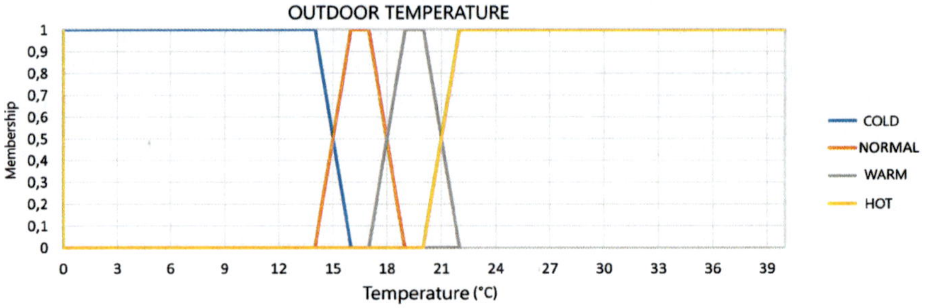

Fig. 3. Membership functions for Outdoor Temperature

Table 3. Membership functions for Return Water Temperature (RWT)

Membership Function	Shape	Points
cold	trapezoid	0:0:15:20
normal	trapezoid	15:20:25:30
hot	trapezoid	25:30:40:40

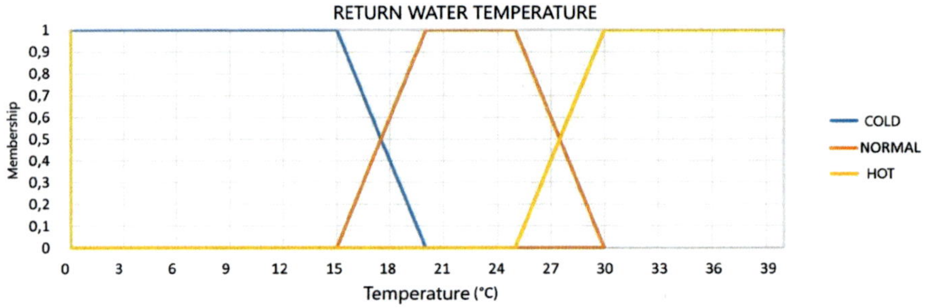

Fig. 4. Membership functions for Return Water Temperature

Fuzzy Rule Base and Inference Mechanism. When it is decided to design a fuzzy system, the first step is to obtain a table of IF-THEN rules. These rules are usually created by

Table 4. Membership functions for Working Time (WT)

Membership Function	Shape	Points
close	singleton	0
latest	triangle	0:20:40
late	triangle	20:40:60
normal	triangle	40:60:80
early	triangle	60:80:100

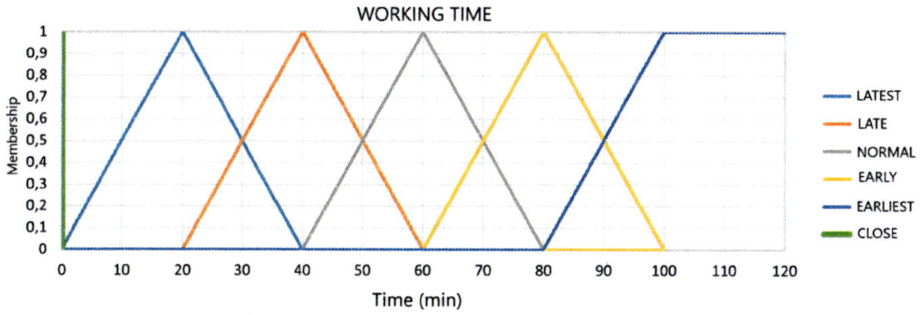

Fig. 5. Membership functions for Working Time

consulting an expert. When creating the rules, the input parameters are connected to each other using the AND operator. The membership degree of the output of each rule is calculated by taking the minimum of the inputs since the "AND" operator is used (Tables 5 and 6).

Defuzzification. The results obtained from the inference mechanism cannot be used directly in physical systems, so the fuzzy result obtained must be defuzzified and converted into numerical values. Therefore, the obtained fuzzy result needs to be defuzzified. There are many different defuzzification methods available and the Weighted Plateau Average Method-WPA method weights the midpoint (m) of all output membership function plateaus (cores). These weights are either the height (h) of each plateau or the membership degrees ($\mu(x)$) of the membership function. In this application with PLC, the Weighted Plateau Average Method (WPA) is used in the defuzzification process [19]. The defuzzification is performed using the fuzzy values and membership degrees of all fired rules. The mathematical expression of the defuzzification method is shown in Eq. (1) and the curve is shown in Fig. 6.

$$WPA = \frac{\sum_{i=1}^{n} m_i x h_i}{\sum_{i=1}^{n} h_i} \tag{1}$$

Table 5. Rules for Heating

No	IT	OT	RWT	WT
1	cold	cold	cold	earliest
2	cold	cold	normal	earliest
3	cold	cold	hot	earliest
4	cold	normal	cold	earliest
5	cold	normal	normal	earliest
6	cold	normal	hot	early
7	cold	warm	cold	earliest
8	cold	warm	normal	early
9	cold	warm	hot	early
10	cold	hot	cold	early
11	cold	hot	normal	early
12	cold	hot	hot	early
13	comfort	cold	cold	early
14	comfort	cold	normal	early
15	comfort	cold	hot	early
16	comfort	normal	cold	normal
17	comfort	normal	normal	normal
18	comfort	normal	hot	late
19	comfort	warm	cold	normal
20	comfort	warm	normal	late
21	comfort	warm	hot	latest
22	comfort	hot	cold	late
23	comfort	hot	normal	latest
24	comfort	hot	hot	latest
25	warm	cold	cold	normal
26	warm	cold	normal	normal
27	warm	cold	hot	normal
28	warm	normal	cold	normal
29	warm	normal	normal	late
30	warm	normal	hot	late
31	warm	warm	cold	late
32	warm	warm	normal	latest

(continued)

Table 5. *(continued)*

No	IT	OT	RWT	WT
33	warm	warm	hot	close
34	warm	hot	cold	latest
35	warm	hot	normal	latest
36	warm	hot	hot	close
37	hot	cold	cold	normal
38	hot	cold	normal	normal
39	hot	cold	hot	late
40	hot	normal	cold	late
41	hot	normal	normal	latest
42	hot	normal	hot	close
43	hot	warm	cold	latest
44	hot	warm	normal	close
45	hot	warm	hot	close
46	hot	hot	cold	latest
47	hot	hot	normal	close
48	hot	hot	hot	close

Table 6. Rules for Cooling

No	IT	OT	RWT	WT
1	hot	hot	cold	late
2	hot	hot	normal	normal
3	hot	hot	hot	early
4	hot	warm	cold	latest
5	hot	warm	normal	late
6	hot	warm	hot	normal

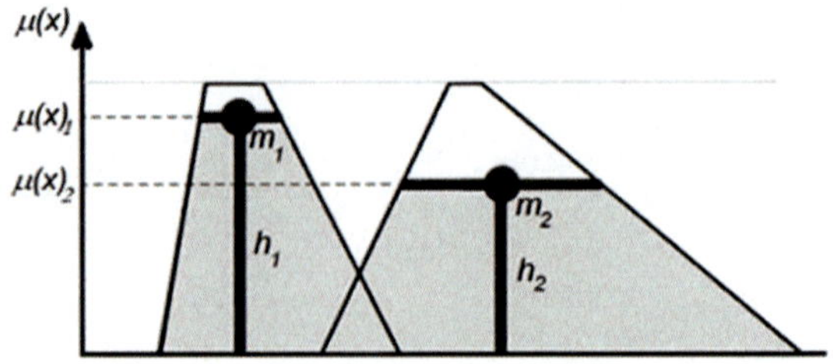

Fig. 6. Weighted Plateau Average Method- WPA [19]

3 Results

The time that determines how long before the prayer times the system will be activated in heating and cooling modes is calculated with a fuzzy logic structure. All these calculations are performed with codes written in the PLC program. The system was prepared and used in the mosque and automatically calculated the working time according to the changing air temperature values and activated the necessary actuators. The studies related to the heating mode were carried out during the winter season. The manual operation of the system is started approximately two hours before the prayer times. In this way, depending on the person operating the system, different operating times are decided every day or even before every prayer time. With the implemented fuzzy logic structure, the operating times are automatically calculated without the need for such an intervention and the necessary controls are provided by the PLC. According to the changing outdoor temperatures, the opperating time of the system before prayer times has decreased to around 50 min, even in winter months. Accordingly, when these values are taken into consideration, it is seen that energy savings between 10%-25% are achieved due to the system not being operated unnecessarily. In Fig. 7, the calculated operating time before the morning prayer time on 22 February 2023, is around 100 min on average. In Fig. 8, the calculated operating time before the afternoon prayer time varies between 80 min and 40 min, depending on the values. Some operating times calculated according to different temperature values are shown in Table 7.

Table 7. Fuzzy Logic Controller

No	IT (°)	OT (°)	RWT (°)	WT (min)
1	16.2	14.1	29.4	76
2	16.2	15.1	29.5	60
3	16	9.2	32.1	80
4	15	11.9	29.4	107
5	16.3	16.1	29.7	41

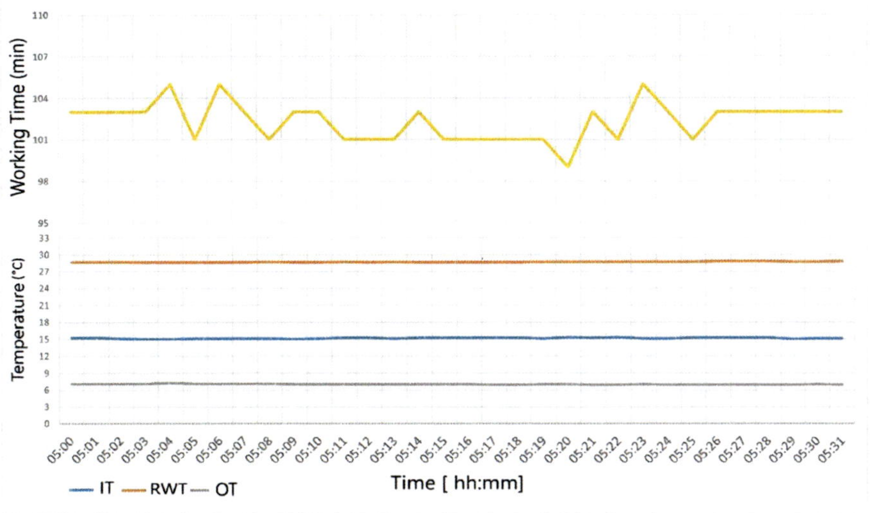

Fig. 7. Fuzzy Logic Controller for morning

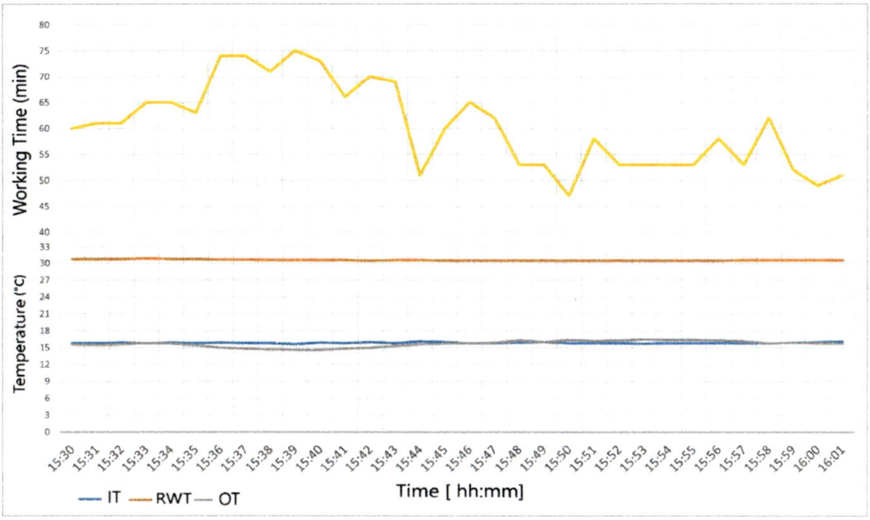

Fig. 8. Fuzzy Logic Controller for afternoon

4 Conclusion

The optimal control method for energy savings and automatic operation in places that use intermittent energy due to the presence of variable parameters that can affect the environmental conditions, and the unpredictability of when they will occur, is a PLC-based fuzzy logic control structure. It has been observed that the structure that controls the radiant heating cooling control system adapts to all kinds of working conditions due to its dynamic structure. The system evaluates data instantaneously and produces an output accordingly to ensure that the radiant heating system runs for a sufficient amount of time. When the results obtained in the winter season are evaluated, the system prevents unnecessary operation by optimizing the working times on the days when the daily air temperature differences are high.

Furthermore, it is important in terms of optimum energy consumption that all heating and cooling systems of buildings that consume intermittent energy, such as mosques, are controlled from a single center. In order to achieve this, the radiant cooling-heating system can work in coordination with the ventilation systems that control the ambient temperature, taking into account the values such as the ambient temperature value, humidity value, CO2 level, etc. inside the mosque. Additionally, more effective control systems can be developed using machine learning methods by utilizing information about the density of people in the mosque.

References

1. Buildings A source of enormous untapped efficiency potential. https://www.iea.org/topics/buildings. Accessed 10 Mar 2023
2. Attia, A., Rezeka, S.F., Saleh, A.M.: Fuzzy logic control of air-conditioning system in residential buildings. Alexandria Eng. J. **54**(3), 395–403 (2015)
3. Budaiwi, I., Abdou, A.: HVAC system operational strategies for reduced energy consumption in buildings with intermittent occupancy: the case of mosques. Energy Convers. Manag. **73**, 37–50 (2013)
4. Krzaczek, M., Florczuk, J., Tejchman, J.: Improved energy management technique in pipe-embedded wall heating/cooling system in residential buildings. Appl. Energy **254** (2019). https://doi.org/10.1016/j.apenergy.2019.113711
5. Ulpiani, G., Borgognoni, M., Romagnoli, A., Di Perna, C.: Comparing the performance of on/off, PID and fuzzy controllers applied to the heating system of an energy-efficient building. Energy Build. **116**, 1–17 (2016)
6. Hu, C., Xu, R., Meng, X.: A systemic review to improve the intermittent operation efficiency of air-conditioning and heating system. J. Build. Eng. **60** (2022). https://doi.org/10.1016/j.jobe.2022.105136
7. Tunçbilek, E., Arici, M., Krajčík, M., Nižetić, S., Karabay, H.: Thermal performance-based optimization of an office wall containing PCM under intermittent cooling operation. Appl. Therm. Eng. **179** (2020). https://doi.org/10.1016/j.applthermaleng.2020.115750
8. Ge, J., Li, S., Chen, S., Wang, X., Jiang, Z., Shen, C.: Energy-efficiency strategies of residential envelope in China's Hot Summer–Cold Winter zone based on intermittent thermal regulation behavior. J. Build. Eng. **44** (2021). https://doi.org/10.1016/j.jobe.2021.103028
9. Wang, J., Liu, S., Liu, Z., Meng, X., Xu, C., Gao, W.: An experimental comparison on regional thermal environment of the high-density enclosed building groups with retro-reflective and high-reflective coatings. Energy Build. **259** (2022). https://doi.org/10.1016/j.enbuild.2022.111864
10. Liu, F., Yan, L., Meng, X., Zhang, C.: A review on indoor green plants employed to improve indoor environment. J. Build. Eng. **53** (2022). https://doi.org/10.1016/j.jobe.2022.104542
11. Jacquet, S., Bel, C.L., Monfet, D.: In situ evaluation of thermostat setback scenarios for all-electric single-family houses in cold climate. Energy Build. **154**, 538–544 (2017)

12. Ling, J., Tong, H., Xing, J., Zhao, Y.: Simulation and optimization of the operation strategy of ASHP heating system: a case study in Tianjin. Energy Build. **226** (2020). https://doi.org/10.1016/j.enbuild.2020.110349
13. Kim, M.S., Kim, Y., Chung, K.S.: Improvement of intermittent central heating system of university building. Energy Build. **42**, 83–89 (2010)
14. Cho, S.H., Zaheer-uddin, M.: Predictive control of intermittently operated radiant floor heating systems. Energy Convers. Manag. **44**, 1333–1342 (2003)
15. Gwerder, M., Tödtli, J., Lehmann, B., Dorer, V., Güntensperger, W., Renggli, F.: Control of thermally activated building systems (TABS) in intermittent operation with pulse width modulation. Appl. Energy **86**, 1606–1616 (2009)
16. Tosun, M.F., Gençkal, A.A., Şenol, R.: Fuzzy logic based room temperature control with modern control methods. Süleyman Demirel Univ. J. Nat. Appl. Sci. **23**, 992–999 (2019)
17. Ma, C., Liu, Y., Song, C., Wang, D.: The intermittent operation control strategy of low-temperature hot-water floor radiant heating system. In: Li, A., Zhu, Y., Li, Y. (eds.) Proceedings of the 8th International Symposium on Heating, Ventilation and Air Conditioning. Lecture Notes in Electrical Engineering, vol. 263, pp. 259–268. Springer, Heidelberg (2014). https://doi.org/10.1007/978-3-642-39578-9_28
18. Ayan, M., Şenol, R.: Fuzzy logic based - remote access greenhouse automation. Düzce Üniversitesi Bilim ve Teknoloji Dergisi **4**, 734–746 (2016)
19. Erdun, H.: Fuzzy logic defuzzification methods with examples, September 2020. https://doi.org/10.13140/RG.2.2.19014.09282

Generating Linguistic Advice for the Carbon Limit Adjustment Mechanism

Fatma Şener Fidan[1](✉), Sena Aydoğan[2], and Diyar Akay[3]

[1] Abdullah Gul University, 38090 Kayseri, Turkey
fatma.senerfidan@agu.edu.tr, fatmasener@gmail.com
[2] Gazi University, 06570 Ankara, Turkey
[3] Hacettepe University, 06800 Ankara, Turkey

Abstract. Linguistic summarization, a subfield of data mining, generates summaries in natural language for comprehending big data. This approach simplifies the incorporation of information into decision-making processes since no specialized knowledge is needed to understand the generated language summaries. The present research employs linguistic summarization to examine the circumstances surrounding the Carbon Border Adjustment Mechanism, one of the most significant regulations confronting exporting nations to the European Union, and will be adopted to support sustainable growth. In this paper, associated with several attributes of the countries and product flow from exporting countries to European countries were defined as nodes and relations, respectively. Before the modeling phase, fuzzy c-means automatically identified fuzzy sets and membership degrees of attributes. During the modeling phase, summary forms were generated using polyadic quantifiers. A total of 1944 linguistic summaries were produced between exporting countries and European countries. Thirty-five summaries have a truth degree greater than or equal to the threshold value of 0.9, which is considered reasonable. The provision of natural language descriptions of the Carbon Border Adjustment Mechanism is intended to aid decision-makers and policymakers in their deliberations.

Keywords: Linguistic summarization · EU CBAM · Climate policy · Carbon leakage · Sustainable Development · Data Mining · Fuzzy Set Theory

1 Introduction

The Carbon Limit Adjustment Mechanism (CBAM), a component of the "Fit for 55" climate policy package, is one of the organizations that make up the European Green Consensus, which was founded by the European Union (EU) to reduce greenhouse gas (GHG) emissions by 2050. This technology, implemented as a part of the EU Emissions Trading System, will attempt to minimize the danger of carbon leakage (EU ETS). As a measure of compensating for the price disparity between EU-made and non-EU-made goods, CBAM stipulates the introduction of import levies. This disparity is caused by the differing environmental regulations in exporting nations [1]. Although CBAM is

Z. Şen et al. (Eds.): IMSS 2023, LNME, pp. 188–199, 2024.
https://doi.org/10.1007/978-981-99-6062-0_18

applied to energy-intensive industries such as steel, cement, electricity, fertilizer, and aluminium, it is evident that it will significantly impact the industries and policies of countries. It is because CBAM will cover the subsectors involved in the supply chain and the enterprises that supply these industries with products. Consequently, nations and industries must establish plans for this transformation and settle on substantial courses of action.

The number of studies on CBAM is growing daily in the academic literature. Most research has addressed CBAM analysis, regulatory compliance, and the carbon leakage link. Mehling et al. (2019), for example, conducted a thorough review of CBAM and recommended a new approach after analyzing the legal, administrative, and environmental aspects of regulation [2]. Using data from the Global Trade Analysis Project (GTAP) and pollution from carbon leakage, Naegele and Zaklan (2019) evaluated the issue of carbon leakage in European manufacturing of EU ETS emission costs [3].

The literature also investigates, albeit in a limited capacity, the effects of CBAM on countries [4]. Tang et al. (2015) analyzed the impact of carbon-based border tax adjustments on the Chinese trade and concluded that they would be detrimental to Chinese exports [4]. Zhong and Pei (2022) evaluated the effects of CBAM on changes in competitiveness and welfare, including in various regions of China. It was concluded that the EU's growth in local supply-demand could motivate China to work on climate targets [5]. Overland and Sabyrbekov determined which nations whose industries rely on fossil fuels might effectively withstand CBAM [6]. According to the CBAM Opposition Index, the most significant competitors are Iran, Ukraine, the United States, the United Arab Emirates, Egypt, China, India, Kazakhstan, Russia, and Belarus. Perdana and Vielle (2022) assessed the impact of EU-CBAM and explored eight countermeasures for the Least Developed Countries [7]. It was stated that welfare would decline due to the decline in exports of less developed nations. Most CBAM research has focused on studying the system and how it will affect countries, while only a few have surveyed all nations. In addition, the literature lacks methods, such as data mining, for incorporating the research outcomes into decision-making.

Knowledge discovery, also known as data mining, involves a variety of approaches that may be generally categorized into two groups: descriptive and predictive methods [8, 9]. The goal of applying these approaches to studying massive datasets is to extract knowledge. Typically, descriptive data mining techniques involve statistical summarizing, in which the features of a database are statistically summarized using metrics such as the median, mean, and standard deviation. Although these approaches are less computationally costly, statistics summaries such as "the average daily sales for the firm are $10,425.4" might be challenging to recall. In comparison, statements such as "the average daily sales of the company are around a hundred thousand dollars" are more thorough and simpler to grasp. To construct such summaries, Yager's notion of linguistic summarization has received considerable attention under several names [10], including fuzzy quantification [11, 12], semi-fuzzy quantifier [13, 14], fuzzy association rules [15, 16], and fuzzy rules [17, 18]. Fuzzy sets may be used to represent linguistic words such as "about a hundred thousand dollars," which is a key step in linguistic summarization. In other words, fuzzy sets can be used to designate database properties during the linguistic summarizing process. Linguistic summarization has applications in diverse fields, such as computer systems [19], investment funds [20], Kansei Engineering [21], human behavior modeling [22], sensor data for elderly care [23], human resources [24], oil price forecasting [25], fall detection [26], social networks [27], and transportation [28], due to its effectiveness in extracting knowledge from large datasets. Please refer to references for further information [29, 30].

Although it has been applied in various fields, its application in sustainable development and CBAM is not yet documented. This study utilized linguistic summarization to investigate the CBAM, which was implemented to assure sustainable development, one of the most pressing concerns facing exporting countries to the EU today. It has contributed to the development of the approach by providing decision-makers and policymakers with relevant linguistic summaries and by adapting the linguistic summary method to a different subject.

2 Materials and Methods

2.1 Preliminaries

A fuzzy set A on universe X, is defined $A = \{\langle x, \mu_A(x)\rangle | x \in X\}$ where $\mu_A(x) : X \to [0, 1]$ is the membership degree of x. Let Y be the set of objects, $Y = \{y_1, y_2, \ldots, y_M\}$, S be the set of attributes $S = \{s_1, s_2, \ldots, s_K\}$ and X_k be the domain of $s_k (k = 1, \ldots, K)$. A linguistic summary in the form of "$Q\ B\ Y$ s are/have A. . [T]" consists of four components as (i) a linguistic quantifier Q, , (ii) a linguistic summarizer A, , (iii) a linguistic pre-summarizer B, , and (iv) truth degree of the summary T. Following the extraction of potential summaries from a dataset, the truth degree of each sentence is calculated by Eq. (1).

$$T = Q\left(\frac{\sum_{m=1}^{M} (\mu_A(y_m) \wedge \mu_B(y_i))}{\sum_{m=1}^{M} \mu_B(y_i)}\right) \tag{1}$$

2.2 Implementation

This research aims to provide linguistic summaries for decision-makers by employing several variables for countries exporting to EU member states under the scope of CBAM. The steps of the investigation are outlined below.

Step 1: Determining Countries
The initial step is to decide which countries will be included in the study. As a result, 177 countries selling products to any EU country were identified as input. There were 86 countries with exports worth more than \$1 billion from these countries. Due to a lack of data, nine countries were dropped from the study, leaving 77 exporting countries as the basis for the research.

Table 1 lists the 27 EU nations included in the study and the 77 countries that export to the EU.

Table 1. Countries

EU Country	Exporter Countries (EC)
Austria, Belgium, Bulgaria, Croatia, Cyprus, Czech, Denmark, Estonia, Finland, France, Germany, Greece, Hungary, Ireland, Italy, Latvia, Lithuania, Luxembourg, Malta, Netherlands, Poland, Portugal, Romania, Slovakia, Slovenia, Spain, Sweden.	Albania, Algeria, Angola, Argentina, Australia, Azerbaijan, Bahrain, Bangladesh, Belarus, Bosnia And Herzegovina, Brazil, Burkina Faso, Cambodia, Cameroon, Canada, Chile, China, Colombia, Congo, Costa Rica, Cuba, Dominican Republic, Ecuador, Egypt, Ethiopia, Georgia, Ghana, Guatemala, Iceland, India, Indonesia, Iran, Iraq, Israel, Japan, Jordan, Kazakhstan, Kenya, Korea, Republic of, Kuwait, Lebanon, Liberia, Macedonia, North, Malaysia, Mali, Mexico, Moldova, Republic of, Montenegro, Morocco, New Zealand, Nigeria, Norway, Oman, Pakistan, Panama, Peru, Philippines, Qatar, Russian Federation, Saudi Arabia, Senegal, Serbia, Singapore, South Africa, Sri Lanka, Switzerland, Thailand, Togo, Tunisia, Türkiye, Ukraine, United Arab Emirates, United Kingdom, United States of America, Uruguay, Uzbekistan, Viet Nam.

Step 2: Determining Attributes
This step identified the attributes of network data employed in the study. Nodes have the attributes of the Environmental Performance Index (EPI) index and population, while relations have the attributes of import value (USD thousands) and distance (km). The first attribute is the EPI, which reveals the environmental performance of countries to safeguard human health and support ecosystem vitality [31]. EPI was developed by Yale University and Columbia University in collaboration with the World Economic Forum and the European Commission Joint Research Center to compare countries' environmental performance. The second attribute determined was the country's population. In studies of countries, the population is a crucial factor [32]. The third attribute is the import value of exporting countries to the EU nations. The last attribute was selected as the distance between countries (km). Also, the current CBAM restriction targeted carbon-intensive industries primarily. To evaluate CBAM with the LS method, GTIP code 68: Objects with stone, plaster, cement, asbestos, mica, and similar materials were selected. Therefore, import value values were collected for this sector.
Step 3: Data Preparation with Fuzzy Sets
A fuzzy c-means approach was used before the modelling phase to automatically identify fuzzy sets and membership degrees of features and quantifiers by referencing fuzzy sets [33]. MATLAB was employed in the identification process [34]. Low, medium, and high are linguistic summarizers used to label fuzzy collections. Fuzzy sets extracted by the fuzzy c-means algorithm are depicted in Fig. 1.

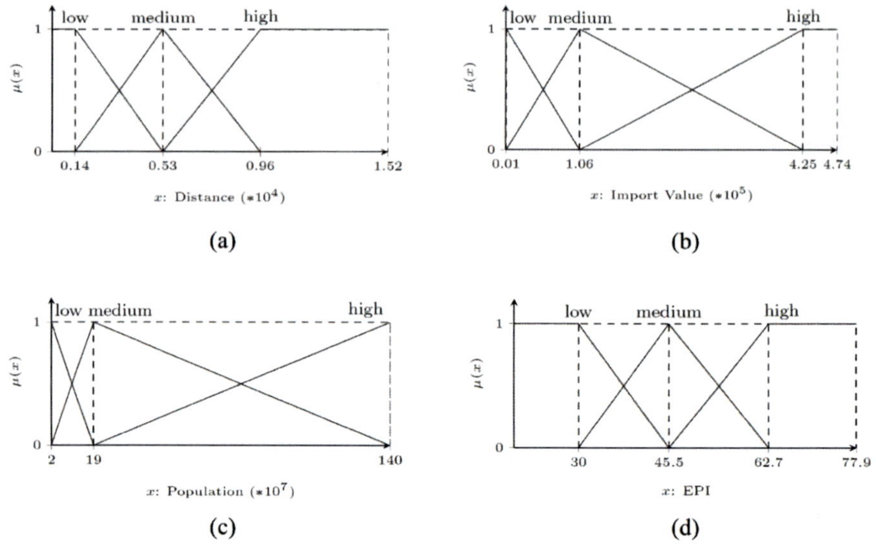

Fig. 1. The membership degrees of the variables: (a) Distance, (b) Import value, (c) Population, (d) EPI

Step 4: Modelling

During the modelling phase, polyadic quantifiers were utilized to obtain summary forms [35]. The generated summaries were assessed using the semi-fuzzy quantifiers-based evaluation method [36]. Genç et al. (2020) proposed semi-fuzzy quantifiers to evaluate the summaries in the form of polyadic quantification [37]. The semi-fuzzy iteration operator in the summary form "Q A Y s are in R with $Q\prime$ B Y s" is defined as $It(Q, Q\prime)[A, B, R] \Leftrightarrow Q[A, \{a|Q\prime[B, R_{(a)}]\}]$, where Q and $Q\prime$ are semi-fuzzy quantifiers, A and B are the fuzzy subsets of the universe X for the attributes v_1 and v_2, R is a fuzzy relation, $R_{(x_i)} = \{x_j|R(x_i, x_j)\}$, and F is a Quantifier Fuzzification Mechanism (QFM). When QFM is applied to F^I for the finite case, the fuzzy value of the linguistic summary's truth is delivered. For more information, please refer to Genc et. al (2020) [37]

For example, "Most countries export small quantities of the products to EU countries." is a polyadic quantification, and iteration may be used to express the meaning in terms of its constituents. The "export" is a relation between the sets of "ECs" and "products". The sentence is valid under one interpretation if and only if a set contains most ECs, each of whom exports a few products.

3 Results and Discussion

Linguistic summaries for all possible combinations of quantifiers and summarizers were generated and evaluated with MATLAB. A total of 1944 linguistic summaries were generated between ECs and EU countries. Thirty-five summaries have a truth degree greater than or equal to the threshold value of 0.9, which is considered reasonable. In the obtained summaries, important features were examined. Summaries with a truth degree higher than 0.9 are given in Table 2.

Table 2. Generated Summaries

Summaries	*T*
Few of the EC with medium EPI import high to a few of the EU countries with high EPI	0.99999
Few of the EC with low EPI import high to a few of the EU countries with high EPI	0.99999
Few of the EC with low population import high to a few of the EU countries with high EPI	0.99999
Few of the EC with low EPI import high to a few of the EU countries with low population	0.99998
Few of the EC with medium EPI import high to a few of the EU countries with low population	0.99998
Few of the EC with low population import high to a few of the EU countries with low population	0.99998
Few of the EC with low EPI import medium to a few of the EU countries with low population	0.98452
Few of the EC with low EPI import medium to a few of the EU countries with high EPI	0.98444
Few of the EC with low population import medium to a few of the EU countries with low population	0.97952
Few of the EC with low population import medium to a few of the EU countries with high EPI	0.97944
Few of the EC with medium population import high to a few of the EU countries with high EPI	0.97906
Few of the EC with medium population import high to a few of the EU countries with low population	0.97905
Few of the EC with medium EPI import medium to a few of the EU countries with low population	0.96616
Few of the EC with medium EPI import medium to a few of the EU countries with high EPI	0.96608
Few of the EC with high EPI high medium to a few of the EU countries with high EPI	0.94355
Few of the EC with high EPI import high to a few of the EU countries with low population	0.94354
Half of the EC with high population import high to a few of the EU countries with high EPI	0.94064
Half of the EC with high population import high to a few of the EU countries with low population	0.94063

(*continued*)

Table 2. (*continued*)

Summaries	*T*
Few of the EC with low population import low to half of the EU countries with low population	0.92547
Few of the EC with low EPI import low to half of the EU countries with low population	0.92547
Few of the EC with low populations have a high distance to half of the EU countries with low population	0.92547
Few of the EC with high populations have a low distance to half of the EU countries with low population	0.92546
Few of the EC with medium EPI import low to half of the EU countries with low population	0.92544
Most of the EC with low population import low to half of the EU countries with low population	0.92538
Few of the EC with high EPI have a high distance to half of the EU countries with low population	0.92533
Most of the EC with low EPI import low to half of the EU countries with low population	0.92524
Half of the EC with low population import low to half of the EU countries with low population	0.92447
Most of the EC with medium EPI import low to half of the EU countries with low population	0.92436
Half of the EC with high population import medium to a few of the EU countries with low population	0.92319
Half of the EC with low EPI import low to half of the EU countries with low population	0.92290
Half of the EC with medium EPI import low to half of the EU countries with low population	0.92153
Few of the EC with medium EPI have a high distance to half of the EU countries with low population	0.91614
Few of the EC with high EPI import low to half of the EU countries with low population	0.91024
Half of the EC with high population import medium to a few of the EU countries with high EPI	0.90824
Few of the EC with medium population import low to half of the EU countries with low population	0.90129

In this section, the results were illustrated using graphs. Figure 2 highlights the relationship between ECs and EU nations based on linguistic summaries with an accuracy level exceeding the criteria. According to the findings, comparisons were made between exporter nations and only EU nations with a high EPI and a low population. These summaries indicated that EU countries have a higher EPI and a smaller population than exporter nations. As a result, it was anticipated that implementing CBAM would boost the sustainability of exporting countries and, thus, the value of EPI.

Fig. 2. Acquired Network with the LS

Figures 3, 4, and 5 depict the networks of the EU countries' qualifiers concerning those of the exporting countries, as determined by the study. In addition, the importance of the accuracy degrees between 0.9 and 1 in the constructed networks ranges from 1 to 9. One represents the highest degree of precision, while 9 represents the lowest degree of precision. This application supported both visualization and analysis.

The statement in Figure 3 with the best level of accuracy among the summaries derived for the ECs' population is that a few ECs with low populations export a substantial proportion of their goods to a few EU countries. However, no summary is given for EU countries with the most inclusive linguistic qualifier "most." The only summary for ECs with "Most" is that numerous low-population ECs under-export to around half of the EU countries with low populations. According to the summary produced with the second inclusive qualifier "half," ECs with medium or low populations export very little to half of the EU nations with a low population. The summary containing EPI values reveals that only a few ECs with high sustainable value export to EU nations. According to the collected results, countries selling to half of the EU countries with a low population tend to have low or moderate EPI levels. For exporting nations with high EPI values, no summary with the qualifiers "most" or "half" was obtained. A sample of the EC nations with high export levels for the given product code. High EPI EU countries are average or

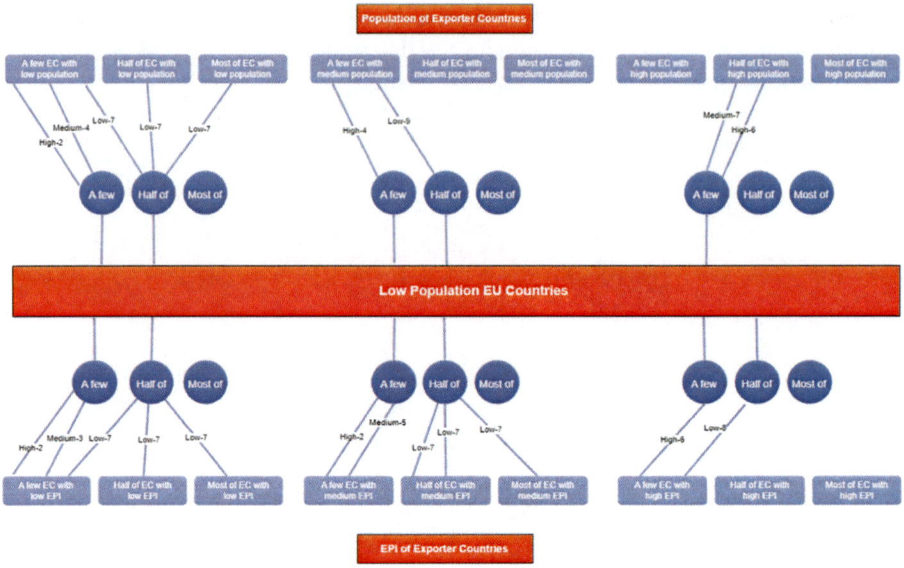

Fig. 3. Low Population EU Countries and ECs Network for Import Value

low EPI countries. According to the collected results, the EPI values for EC are generally low. The majority of ECs have a large population.

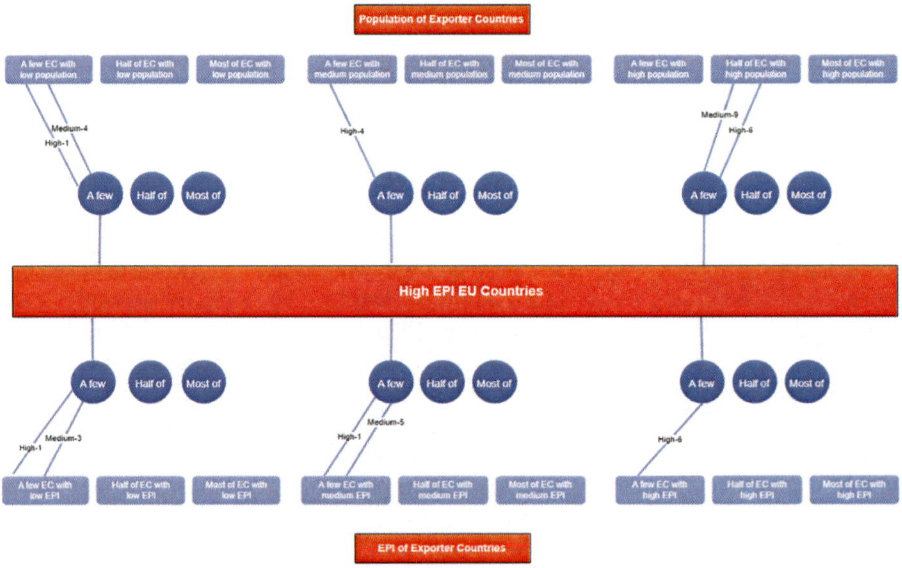

Fig. 4. High EPI EU Countries and ECs Network for Import Value

Figure 4 illustrates the derived summaries of import value between EC countries and EU members. A small number of ECs with a low or medium EPI export to a small number of EU members with a high EPI. A few CEs with a high EPI export fewer goods to a few EU countries with a high EPI. The conclusion that can be drawn from these summaries is that EC countries must enhance their sustainability. Half of the population-dense EC nations export mainly to a few EU nations. Few EU nations with significant EPI export the most, whereas those with low populations export the least.

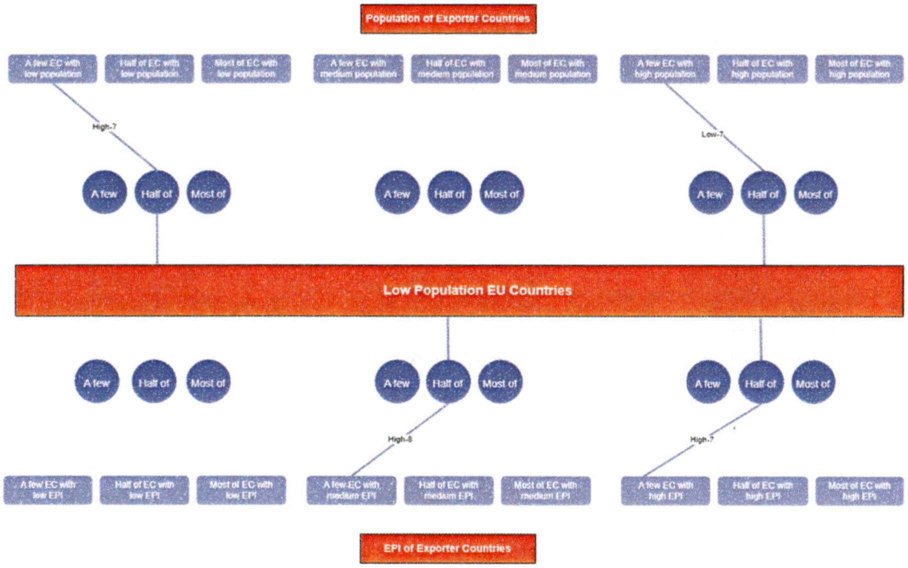

Fig. 5. High EPI EU Countries and ECs Network for Distance

Figure 5 illustrates the derived summaries of distances between EC countries and EU members. A few ECs with high or moderate EPI are distant from most EU members with low populations. Moreover, ECs with a large population have a shorter distance than ECs with a small population. It illustrates that even if ECs with a large distance have high EPI values, they will face adverse conditions in CBAM.

4 Conclusion

This study examined the application of LS, which aims to give decision-makers supplementary data through the use of datasets, in the fields of the green deal and CBAM, which have recently been on the agenda of many countries. Uniquely, the LS discipline has adapted its usage to CBAM, which has only recently begun to be explored in the literature. Consequently, 1944 unique linguistic summaries were obtained, and 35 of those above the threshold were examined.

It can be expanded in future research by examining all product groupings within the purview of CBAM and integrating the regions, development levels, and energy mixes of these countries as attributes to facilitate efficient decision-making and provide an overview of exporting nations.

References

1. Sato, S.Y.: EU's carbon border adjustment mechanism: will it achieve its objective (s)? J. World Trade. **56**, 383–404 (2022)
2. Mehling, M.A., Van Asselt, H., Das, K., Droege, S., Verkuijl, C.: Designing border carbon adjustments for enhanced climate action. Am. J. Int. Law. **113**, 433–481 (2019)
3. Naegele, H., Zaklan, A.: Does the EU ETS cause carbon leakage in European manufacturing? J. Environ. Econ. Manag. **93**, 125–147 (2019)
4. Tang, L., Bao, Q., Zhang, Z., Wang, S.: Carbon-based border tax adjustments and China's international trade: analysis based on a dynamic computable general equilibrium model. Environ. Econ. Policy Stud. **17**, 329–360 (2015)
5. Zhong, J., Pei, J.: Beggar thy neighbor? On the competitiveness and welfare impacts of the EU's proposed carbon border adjustment mechanism. Energy Policy. **162**, 112802 (2022)
6. Overland, I., Sabyrbekov, R.: Know your opponent: which countries might fight the European carbon border adjustment mechanism? Energy Policy **169**, 113175 (2022)
7. Perdana, S., Vielle, M.: Making the EU carbon border adjustment mechanism acceptable and climate friendly for least developed countries (2022)
8. Han, J., Kamber, M., Pei, J.: Data Mining Concepts and Techniques, Massachusetts, vol. 3 (2012)
9. Vercellis, C.: Business Intelligence: Data Mining and Optimization for Decision Making. Wiley, Hoboken (2011)
10. Yager, R.R.: A new approach to the summarization of data. Inf. Sci. **28**, 69–86 (1982)
11. Barro, S., Bugarín, A.J., Carinena, P., Díaz-Hermida, F.: A framework for fuzzy quantification models analysis. IEEE Trans. Fuzzy Syst. **11**, 89–99 (2003)
12. Delgado Calvo-Flores, M., Sánchez Fernández, D., Vila Miranda, M.A.: Fuzzy cardinality based evaluation of quanti® ed sentences (1999)
13. Díaz-Hermida, F., Bugarín, A.: Semi-fuzzy quantifiers as a tool for building linguistic summaries of data patterns. In: 2011 IEEE Symposium on Foundations of Computational Intelligence (FOCI), pp. 45–52. IEEE (2011)
14. Dıaz-Hermida, F., Bugarın, A., Cariñena, P., Barro, S.: Voting-model based evaluation of fuzzy quantified sentences: a general framework. Fuzzy Sets Syst. **146**, 97–120 (2004)
15. Dubois, D., Hüllermeier, E., Prade, H.: A systematic approach to the assessment of fuzzy association rules. Data Min. Knowl. Discov. **13**, 167–192 (2006). https://doi.org/10.1007/s10618-005-0032-4
16. Martin, T., Shen, Y., Majidian, A.: Discovery of time-varying relations using fuzzy formal concept analysis and associations. Int. J. Intell. Syst. **25**, 1217–1248 (2010)
17. Dubois, D., Prade, H.: What are fuzzy rules and how to use them. Fuzzy Sets Syst. **84**, 169–185 (1996)
18. Rasmussen, D., Yager, R.R.: Finding fuzzy and gradual functional dependencies with SummarySQL. Fuzzy Sets Syst. **106**, 131–142 (1999)
19. Niewiadomski, A.: Cylindric extensions of interval-valued fuzzy sets in data linguistic summaries. J. Ambient Intell. Humaniz. Comput. **4**, 369–376 (2013). https://doi.org/10.1007/s12652-011-0098-3
20. Kacprzyk, J., Zadrożny, S.: Fuzzy logic-based linguistic summaries of time series: a powerful tool for discovering knowledge on time varying processes and systems under imprecision. Wiley Interdiscip. Rev. Data Min. Knowl. Discov. **6**, 37–46 (2016)

21. Akgül, E., Delice, Y., Aydoğan, E.K., Boran, F.E.: An application of fuzzy linguistic summarization and fuzzy association rule mining to Kansei Engineering: a case study on cradle design. J. Ambient Intell. Humaniz. Comput. **13**, 2533–2563 (2022). https://doi.org/10.1007/s12652-021-03292-9
22. Alvarez-Alvarez, A., Alonso, J.M., Trivino, G.: Human activity recognition in indoor environments by means of fusing information extracted from intensity of WiFi signal and accelerations. Inf. Sci. **233**, 162–182 (2013)
23. Wilbik, A., Keller, J.M., Alexander, G.L.: Linguistic summarization of sensor data for eldercare. In: 2011 IEEE International Conference on Systems, Man, and Cybernetics, pp. 2595–2599. IEEE (2011)
24. Tré, G., Dziedzic, M., Britsom, D., Zadrożny, S.: Dealing with missing information in linguistic summarization: a bipolar approach. In: Angelov, P., et al. (eds.) Intelligent Systems'2014. AISC, vol. 322, pp. 57–68. Springer, Cham (2015). https://doi.org/10.1007/978-3-319-11313-5_6
25. Boran, F.E., Akay, D.: A generic method for the evaluation of interval type-2 fuzzy linguistic summaries. IEEE Trans. Cybern. **44**, 1632–1645 (2013)
26. Anderson, D., Luke, R.H., Keller, J.M., Skubic, M., Rantz, M., Aud, M.: Linguistic summarization of video for fall detection using voxel person and fuzzy logic. Comput. Vis. Image Underst. **113**, 80–89 (2009)
27. Yager, R.R.: Concept representation and database structures in fuzzy social relational networks. IEEE Trans. Syst. Man Cybern.-Part Syst. Hum. **40**, 413–419 (2009)
28. Alvarez-Alvarez, A., Sanchez-Valdes, D., Trivino, G., Sánchez, Á., Suárez, P.D.: Automatic linguistic report of traffic evolution in roads. Expert Syst. Appl. **39**, 11293–11302 (2012)
29. Boran, F.E., Akay, D., Yager, R.R.: An overview of methods for linguistic summarization with fuzzy sets. Expert Syst. Appl. **61**, 356–377 (2016)
30. Delgado, M., Ruiz, M.D., Sánchez, D., Vila, M.A.: Fuzzy quantification: a state of the art. Fuzzy Sets Syst. **242**, 1–30 (2014)
31. Esty, D.C., Levy, M.A., Srebotnja, T., De Sherbinin, A., Kim, B., Anderson, B.: Pilot Environmental Performance Index. Yale Centre for Environmental Law & Policy, NewHaven (2006)
32. Barnes, D.F., Floor, W.M.: COUNTRIES: a challenge for economic developmento. Annu. Rev. Energy Env. **21**, 497–530 (1996)
33. Bezdek, J.C., Ehrlich, R., Full, W.: FCM: the fuzzy c-means clustering algorithm. Comput. Geosci. **10**, 191–203 (1984)
34. Matlab, S.: Matlab. MathWorks Natick MA (2012)
35. Szymanik, J., Szymanik, J.: Complexity of Polyadic Quantifiers. Quantifiers Cogn. Log. Comput. Perspect. 101–121 (2016)
36. Díaz-Hermida, F., Bugarín, A., Barro, S.: Definition and classification of semi-fuzzy quantifiers for the evaluation of fuzzy quantified sentences. Int. J. Approx. Reason. **34**, 49–88 (2003)
37. Genc, S., Akay, D., Boran, F.E., Yager, R.R.: Linguistic summarization of fuzzy social and economic networks: an application on the international trade network. Soft Comput. **24**, 1511–1527 (2020). https://doi.org/10.1007/s00500-019-03982-9

Autonomous Mobile Robot Navigation Using Lower Resolution Grids and PID-Based Pure Pursuit Controller

Ahmed Al-Naseri and Erkan Uslu(✉)

Computer Engineering Department, Yildiz Technical University, İstanbul, Turkey
ahmed.naseri@std.yildiz.edu.tr, euslu@yildiz.edu.tr

Abstract. In the modern era, mobile robots are gaining special attention in various intralogistics operations such as warehousing, manufacturing, electrical high-voltage substations, roads, etc. These vehicles should be capable of effectively recognizing their routes, avoiding singularities and obstacles, and making a decision according to the environment. Hence an advanced control mechanism is required for path planning and navigation to work effectively in that dynamic environment. Keeping insight, into the challenges faced by the mobile robot this study aims to develop path planning using global (A* and Dijkstra) and local planners (Pure Pursuit) in a 2D navigation system, utilizing g-mapping for Simultaneous localization and mapping (SLAM) and Adaptive Monte Carlo Localization (AMCL) for probabilistic localization, to solve this issue. The system is designed to be compatible with the Robot Operating System (ROS) ecosystem. Path planning is carried out on a lower-resolution grid covering the navigable areas and the Pure Pursuit approach is enriched with a Proportional-integral-derivative (PID) controller. The results have shown that proposed schemes give superior performance in challenging obstacle-based warehouse systems, compared to publicly available ROS navigation planners.

Keywords: Lower Resolution Grids · PID Control · Pure Pursuit · Navigation · ROS · G-mapping · AMCL · SLAM · Localization

1 Introduction

Lately, the emergence of autonomous robots has revolutionized a myriad of fields, propelling significant advancements in efficiency and productivity. The realm of robotics technology has undergone a paradigm shift, with robots evolving to become intelligent and autonomous entities, far beyond the mere impact of software applications and bionics. This cutting-edge technology is also commonly referred to as a mobile autonomous robot, marking a new era of unparalleled innovation and scientific progress. Any robot will be autonomous and intelligent if it can move around without being constrained and has the intelligence to get around any obstacles that may be in its path. The Robot Institute of America provided the most well-known definition of a robot, stating that it is "a multifunctional, reprogrammable manipulator capable of moving materials, tools,

Z. Şen et al. (Eds.): IMSS 2023, LNME, pp. 200–213, 2024.
https://doi.org/10.1007/978-981-99-6062-0_19

specialized devices or components, through movements incorporated with parameter programming for the performance of various tasks" [1]. A mobile robot is a device that is controlled by a computer and equipped with sensor devices and other technologies that allow it to identify its surroundings and perform specific tasks. Utilizing an artificial intelligence software, a mobile robot is capable of executing pre-defined tasks. The autonomous mobile robot typically completes a set job in three steps: perception (sense), planning, interpreting (process), and motion (action). While mobile vehicles can perform tasks that are typically associated with humans, mobile robots and humans often share similar functional elements when it comes to executing similar tasks. However, it's important to note that a mobile robot doesn't necessarily have to resemble or behave like a human to perform its designated tasks [2]. The robot's capacity to create simultaneous maps of the surroundings and situate itself accordingly to the environment is the key component for mobile robots. Various algorithms can be utilized to develop a solution for SLAM (Simultaneous Localization and Mapping). In recent years, practical applications of probabilistic SLAM have become increasingly impressive, covering larger areas in more challenging environments. The investigation of SLAM procedures is an expanding field within robotics. The laser scanner detects a relatively small displacement in the form of a cloud point between the robot and any given landmark. Odometer and inertial measurement unit (IMU) sensor to adjust the mobile robot's position and orientation. Without knowing the environment's feature information, it is impossible to pinpoint the exact locations of landmarks. For non-linear models with their original sensor data, the G-Mapping-based SLAM technique can be applied [3]. Therefore, it is possible to use the resulting map with localization systems. One popular localization system is Adaptive Monte Carlo Localization (AMCL) which implements a particle filter approach [4]. ROS is a middleware designed for robot programming, which is free and open-source. It facilitates the development and reuse of code across different robotics applications, thereby assisting researchers and developers. ROS offers a range of packages, drivers, frameworks, and resources that can be utilized in the creation and development of robot applications. The fundamental idea behind ROS is to provide a framework that can be incorporated into other robotics systems by making minimal changes to the code. Furthermore, Gazebo and RVIZ are programs that simulate 3D visualization and enable sensor data to be viewed in the environment [5]. Robotics solutions are being used in an increasing variety of industries one of the applications is an automated guided vehicle (AGV), which is a transportation automation tool that is widely employed in manufacturing, intelligent warehouses, and other industries. Yet, the primary task of AGV is to trace the pre-defined path due to non-holonomic restrictions and the system's nonlinearity. AGVs can only travel a certain, constrained path since there is an insufficient place to use obstacle avoidance; this is known to be a major challenge for AGVs. In order to carry out their designated functions, AGVs have sensors, mapping, and localization technologies [6]. Our system integrates Pure Pursuit with a Proportional, Integrated, and Derivative controller, which employs a low-resolution grid to address some of the drawbacks observed. The aim of this paper is to outline why mobile robots face difficulty with obstacle avoidance in narrow pathways within confined spaces, particularly in warehouses. Our focus is on ensuring that the mobile robot stays on a defined path, where one of the challenges is the issue of unoptimized parameter sets that cause the robot to move backward until it collides with an obstacle, a problem commonly encountered in classical Pure Pursuit. Another limitation is the issue of overshooting with velocity and acceleration when the robot approaches the goal, which triggers the activation of recovery behavior for an extended period. To overcome these challenges and simultaneously increase the speed of the robot's trajectory toward the target, we have investigated a Pure Pursuit model based on a Proportional-Integral-Derivative (PID) controller.

2 Methodology

To address the challenge of navigating through narrow and restricted paths within the warehouse map, a grid of markers has been placed on the map. These markers are selected by the path planner, and the robot turns its orientation to reach them. The markers are separated 0.2 m from adjacent ones. This approach, known as "point-based control," is a common method for precise navigation in robotics. The PID-based Pure Pursuit as well as the local planners used in this study can use those markers placed on the map. The PID-based Pure Pursuit controller enhances navigation, therefore, validating the concept that overshoot and oscillations in trajectory tracking may be efficiently handled utilising PID and Pure Pursuit. These solutions aim to enhance the stability, precision, and effectiveness of trajectory monitoring.

2.1 ROS and Gazebo

Robot Operating System (ROS) is a software architecture that is used to create and execute robot software applications. ROS is a flexible, modular, and distributed framework that is hardware and software agnostic, making it versatile [7]. It's a "node," basis system which is an executable file crucial to the ROS network. The ROS network continues to distribute messages when this node is running, and any programs that have subscribed to ROS be able to submit the necessary. Other than messages, ROS supports service-based and action-based communication approaches. Server-based communication is useful when a result is also requested for the given inputs in an asynchronous manner. On the other hand, action-based communication arises from the need for feedback, besides the end result, throughout the process itself. Also, ROS provides packages on subjects such as SLAM, localization, path planning, navigation control, Gazebo, and RVIZ. Gazebo is a standalone simulation environment that can generate simulated sensor data and can provide interaction with the simulated robot. The modelling of robots is represented in the URDF format, whereas representations of environments are provided in the SDF format.

2.2 SLAM

A computational SLAM technique is used by mobile robots and autonomous vehicles to construct a map of an unknown environment while simultaneously estimating the robot's location within that environment. The robot uses an onboard LIDAR sensor to collect data for the surrounding environment and then processes this data to build a map of that environment. SLAM algorithms can be based on a range of techniques, including feature-based, grid-based, graph-based, and filter-based approaches. The most commonly used approach for solving the problem of SLAM by G-Mapping algorithm which is a specific implementation of the SLAM problem to incorporate probabilistic methods to estimate the robot's pose and map the environment. In mobile robotics G-Mapping tries to find the best match between the laser scan and the existing map by starting from an initial robot point given by the odometry and moving in a hill-climbing manner. G-Mapping is utilized to construct a grid-based representation of the environment, which is saved for later use in localization [8].

Simultaneous Localization and Mapping (SLAM) based on G-Mapping algorithm is given in Alg. 1.

Algorithm 1. G-Mapping

```
Output: Updated Map at time t M_t, Estimated Robot Pose
at time t X_t
Initialize map (M_0)
X_t, Z_t, t ← X_0, Z_0, 0
Main loop
do {
    Increment time step
    t ← t + 1
    Receive sensor data (Z-t)
    Perform scan matching
    Update robot pose (X_t) and map probabilities p(m_i))
    Check if the loop should continue
} while (continueLoop)
Output updated map (M_t)
```

2.3 AMCL

The methodology tackles the challenges encountered by mobile robots in warehouse environments through the deployment of AMCL and G-Mapping (Fig. 1).

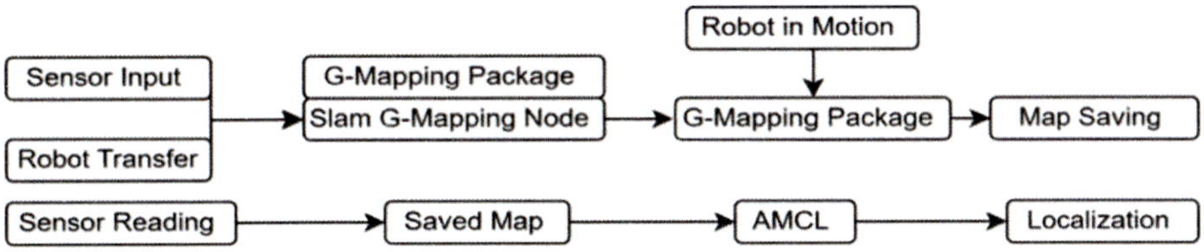

Fig. 1. Mapping and localization flow

AMCL employs a particle filter to gauge the robot's pose based on sensor readings and a known map generated by G-Mapping. The particle filter selectively resamples particles in accordance with the likelihood of the sensor readings, securing a dependable estimate of the robot's pose. Adopting these advanced techniques to ensure accurate localization, therefore estimate robot pose represents the robot position at any time given as (x, y, θ), namely the position and the orientation.

2.4 Navigation

Robot navigation is a fundamental aspect of robotics that enables a robot to move autonomously in an environment. Navigation has two sub-tasks namely path planning and navigation control. Path planning can be performed in ROS by using global planners, which generate a high-level plan that spans the entire environment. On the other hand, navigation control is implemented with local planners in ROS, which generates low-level motion commands and focuses on short-term movements, taking into account the sensor data and the created path.

Path Planning. The global planners, Dijkstra, and A* are used to generate a path from the robot's starting position to the goal position. The path planning is carried out on a lower resolution grid covering the navigable areas to reduce computational complexity as a proposed method. Overall, this methodology enables the robot to navigate efficiently in warehouse environments while avoiding obstacles and singularities and also, utilizes a global cost map. The path-planning method used by the Dijkstra algorithm is greedy [9]. The Dijkstra algorithm does not take into account the network's spatial distribution properties, making the path search non-directional. A* algorithm introduces a heuristic as the cost estimation reaching the target from the current position added to the cost reaching to current position from the start. If the heuristic estimation is less or equal to the actual traversable cost, then A* can find the optimal path. Dijkstra and A* exploration behaviors are given in Fig. 2.

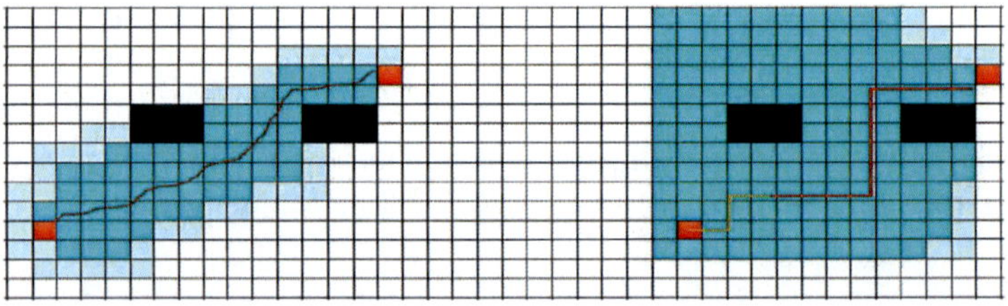

Fig. 2. Dijkstra exploring behavior (on the right), A* exploring behavior (on the left)

Navigation Controller. Following the local goal determination, the local planner combines sensory data with the global path and plans the motion commands, bringing the robot closer to the destination. The intended course can avoid obstructions and stays within the confines of the planned path. Well-known path controllers namely TEB, DWA [10], and Pure Pursuit are given in the following sections.

Time Elastic Band (TEB). TEB planner tries to optimize velocity and acceleration concerning both desired robot position and times between each position as constraints. Based on the global planner plan, it generates a local plan that consists of an array of n robot positions. Q consists of n consecutive robot positions that are (x_i, y_i, θ_i) values and τ consists of $n-1$ time interval values ΔT_i. Q and τ are concatenated together to form B, to be fed into a global objective function to be optimized. The goal function maximizes B. The dynamic limits placed on the robots and the separation between obstacles are only two examples of the many variables that make up the goal function. The course that incurs the lowest cost will be chosen for the robot to follow. Based on the two successive configurations and the intervals of two consecutive postures, translational and rotational velocities are calculated, thus accelerating the velocity as it varies over time. In Fig. 3a constraints (positions and time intervals) to be met during optimization are given for the TEB planner [11].

DWA (Dynamic Window Approach). The planning process in this algorithm is aided by the robot's dynamics. It will compute several estimates of trajectories over some time while sampling the robot's velocities. A two-dimensional search space will be produced using the approximated trajectory distribution. Circular trajectories, dynamic windows, and permissible velocities are the three criteria on which the trajectories in the search space are based. The velocity vector is computed at a specific time interval and sent to the robot for it to achieve the destination using the intended paths. In Fig. 3b simulated trajectories to be evaluated are represented for the DWA method [12].

Pure Pursuit. The local planner, Pure Pursuit, is used to follow the generated path by considering the robot's kinematic constraints. The Pure Pursuit algorithm is a common method based on peek route tracking control and is frequently used as a control algorithm for agricultural machinery's automatic navigation system. It describes the geometric relationship between the vehicle's front wheel angle and the destination point (look ahead point) on the planned path using a straightforward mathematical statement. Look ahead distance plays an important role in algorithms path tracking behavior. In Fig. 3c Pure-Pursuit approach is presented with L look ahead distance to the goal point (x_g, y_g), $\Delta\phi$ heading deviation and $(1/\gamma_{sp})$ instantaneous rotation radius [13]. Pure Pursuit approach is enriched with a PID controller as a proposed method to improve the tracking performance.

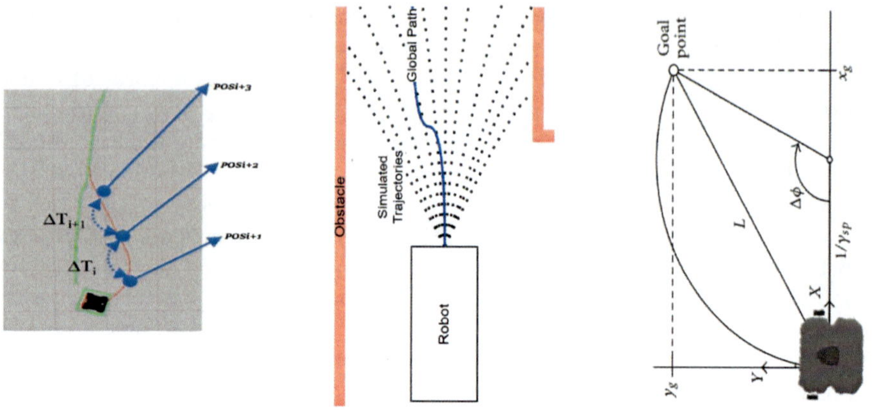

(a) (b) (c)

Fig. 3. Local planner approaches, a) TEB, b) DWA, c) Pure Pursuit

3 Proposed Method

3.1 Low-Resolution Grids

The most popular low-level sensor-based environment model used in robotics is an occupancy grid. The occupancy grid approach, which was initially created for the fusion of sonar data, is particularly robust for the fusion of noisy data. It is desirable to have a high-resolution map for better environmental representation. However, as the map resolution increases path calculation and execution times increase exponentially. Therefore, apart from the map grids we introduced a lower-resolution grid representation of the created map and used it for navigation purposes. Low-resolution grids also enabled us to force path calculation and execution to be on pre-defined waypoints.

3.2 PID Controller

PID controller is frequently used for mechanical control because it is straightforward, reliable, and simple to install. Three components comprise the PID controller: differential, integral, and proportional. The PID controller's proportional control component is used to modify the value of the fundamental error; in the integral control component, the constant error that develops during process control is removed; and the differential control component specifically states how quickly errors are eliminated. PID controller flow is given in Fig. 4a.

3.3 Pure Pursuit with PID

Further to the pure-pursuit strategy, we add a PID controller to the Pure Pursuit algorithm to help follow the low-resolution grid points more accurately [14]. The PID controller uses the current error between the robot's position and the desired position on the path to calculate a control signal that adjusts the robot's steering (ω) and velocity (ν). In classical pure-pursuit, the linear velocity is a fixed value whereas, in the implementation, we propose to adjust the linear velocity as well. The linear velocity control signal $\nu(t)$ is given by Eq. 1. Where $e_{lin}(t)$ is the error signal, $K_{P_{lin}}$, $K_{I_{lin}}$, and $K_{D_{lin}}$ are the proportional, integral, and derivative gains, for linear velocity respectively. The integral and derivative terms are computed using numerical integration and differentiation.

$$\nu(t) = K_{P_{lin}} \cdot e_{lin}(t) + K_{I_{lin}} \cdot \int e_{lin}(t)dt + K_{D_{lin}} \cdot \frac{de_{lin}(t)}{dt} \tag{1}$$

By tuning the gains of the PID controller, we can adjust the response of the robot to the error signal and improve its ability to accurately follow the low-resolution grid points. In addition to the linear PID controller, we also incorporate an angular PID controller to adjust the robot's orientation toward the next point. The Overall flow for Pure-Pursuit with PID control is given in Fig. 4b.

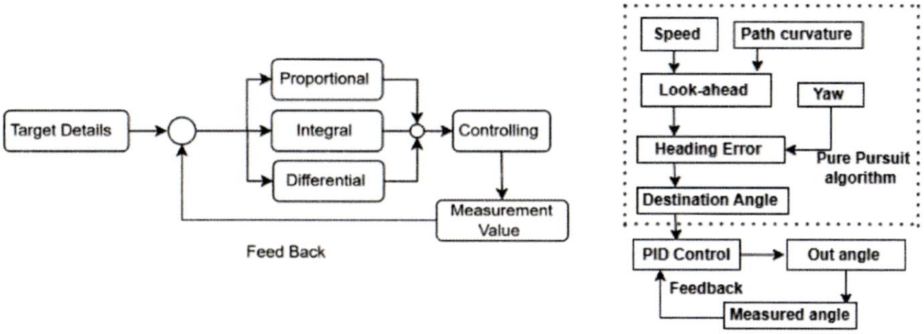

(a) (b)

Fig. 4. PID Controller Chart, b) Proposed Pure Pursuit with PID Control Chart

The PID control signal for the angular velocity control signal $\omega(t)$ is given by Eq. 2. Where $e_{ang}(t)$ is the error signal, $K_{P_{ang}}$, $K_{I_{ang}}$, and $K_{D_{ang}}$ are the proportional, integral, and derivative gains, for angular velocity respectively. The integral and derivative terms are computed using numerical integration and differentiation.

$$\omega(t) = K_{P_{ang}} \cdot e_{ang}(t) + K_{I_{ang}} \cdot \int e_{ang}(t)dt + K_{D_{ang}} \cdot \frac{de_{ang}(t)}{dt} \tag{2}$$

The combination of both linear and angular PID controllers can help the robot navigate more accurately and efficiently toward the pre-determined points. By utilizing the PID point-based controller, the robot navigation behavior is given in Fig. 5 and Alg. 2.

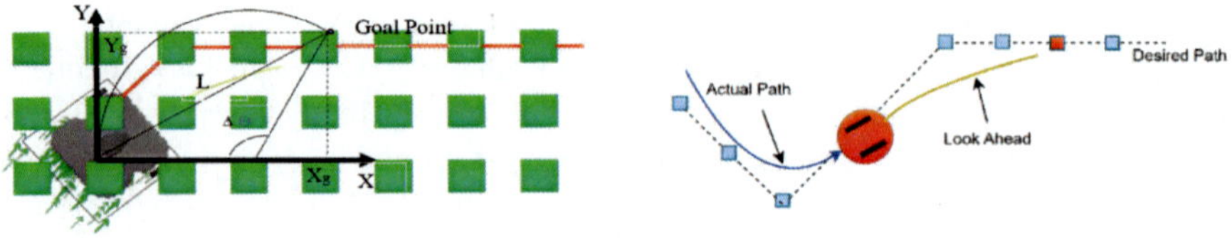

Fig. 5. Coordinate System of A* with Pure Pursuit Algorithm and PID controller

Algorithm 2. Pure Pursuit with PID controller

```
Input: RobotPose, RobotGoal, GlobalPath
Output: v and w
While(true)
  Average_cost = calculate_cost(costmap)
  Goal = getTargetPoint(GlobalPath, RobotPose, LookAhead)
  Error_distance, Error_orientation = calcu-
late_errors(RobotPose, Goal)
  If(Average_cost < 80.0) then
    If(Error_distance > 0.5) then
      W = saturate(PID(Error_orientation))
      V = saturate(PID(Error_distance))
    Else
      Goal reached
      Return true
  Else
    Return false

calculate_errors
Input: RobotPose, RobotGoal
Output: Error_distance, Error_orientation
Error_distance = get euclidean dist(RobotGoal, RobotPose)
Error_orientation = get orientation err from quaterni-
ons(RobotGoal, RobotPose)

getTargetPoint
Input: GlobalPath, RobotPose, LookAhead
Output: TargetPoint
N = path_size(GlobalPath)
distance = 0
Get
For(i<N and distance < LookAhead)
  distance = get euclidean distance(RobotGoal, RobotPose)
  if(distance>LookAhead)
    Target_point = GlobalPath[i]
```

4 Experimental Setup

Software Setup: ROS (Robot Operating System) served as the software platform for the mobile-wheeled robot testing experiment.

Environment and Robot Model: A simulated warehouse environment using Gazebo with the RVIZ tool was chosen to visualize the map and markers (Fig. 6). The model of the robot has a 2D LIDAR sensor with a resolution of 0.2 degrees and coverage of 10 m in all directions.

Mapping and Localization: The G-Mapping algorithm is used for map creation and AMCL is used to localize the robot on the saved G-Mapping map.

Path Planning: We used the ROS navigation stack's path planning modules namely, A* and Dijkstra algorithms, that utilize the global cost map. PID-based Pure Pursuit, TEB, and DWA algorithms are used as navigation controllers that utilize a local cost map to generate collision-free speed controls for the robot [14, 15]. To optimize the robot's navigation performance, PID controller parameters were tuned.

(a)

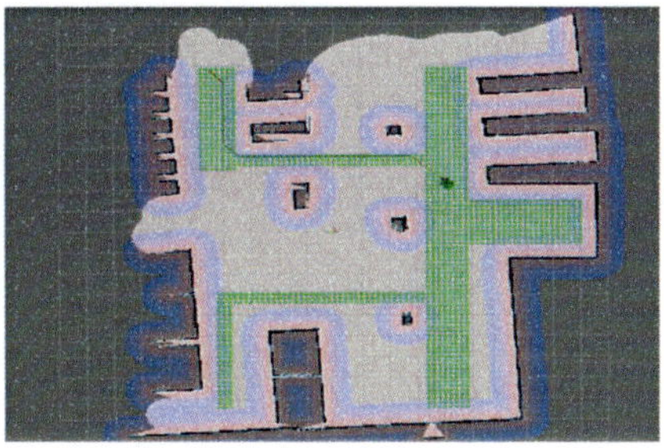

(b)

Fig. 6. Warehouse a) environment, b) map

Measurements: Three performance metrics are used for evaluation, the average time, position, and orientation errors to reach the goal position. To calculate the position and orientation errors, the Euclidean distance between the robot's final position and the goal position, and the absolute contrast of the robot's final orientation, and the goal orientation respectively have been utilized. Multiple trials were conducted to calculate the average values of the performance metrics, through a cluttered environment finding the average of 5 trials for 10 different located targets with fixing the initial pose.

Parameters: In this study, to evaluate the performance of the proposed controller by tuning linear and angular gains, Ziegler-Nichols (ZN) heuristic method was applied. ZN method is a popular heuristic approach for tuning PID controllers based on step-response experiments. The best PID parameter setting for Pure Pursuit PID control in the hypothesis is given in Table 1.

Several experiments are conducted to evaluate the robot's performance. In each experiment, the starting point is fixed, and the goal's location is altered to evaluate the robot's ability to navigate in a variety of locations. Experiments' outcomes were analyzed using descriptive statistics and graphical representations. The table of results compares

Table 1. Linear and Angular PID Parameters

Parameter	$K_{P\,lin}$	$K_{D\,lin}$	$K_{I\,lin}$	$K_{P\,ang}$	$K_{D\,ang}$	$K_{I\,ang}$	v_{max}	ω_{max}	Look Ahead
Value	0.7	0.2	0.25	1.9	0.97	0.45	0.26	1.28	12.90

the average time required to reach the target position error and the average orientation error across various experiments.

5 Results and Discussion

After a lengthy investigation with a simulated mobile robot navigation, the implications of combining A* with PID-based Pure Pursuit revealed no collisions or goal overshoots. The results for the aforementioned metrics were obtained as follows: the competition between the global planners Dijkstra/A* and the competition between the local planners after driving tests is given in Table 2.

Table 2. Global and Local Planners Comparison

	Dijkstra& PP+PID	A* & PP+PID	A* & PP	A* & DWA	A* & TEB
Average time to arrive at goal [s]	75.091	74.125	83.847	94.884	79.756
Average position error [meters]	0.187	0.152	0.163	0.466	0.467
Average orientation error [degrees]	1.434	1.534	0.146	0.135	0.318
Collision	1	0	1	0	0
Overshoot the goal	0	0	0	0	16

It's clear from Table 2 that A* is faster than Dijkstra in competition time as mentioned in the literature [14], while A* Pure Pursuit PID-based model has the shortest implemented time. Local planner comparisons based on execution times, the position error and the orientation error with respect to the target points are further detailed and given in Fig. 7.

The combination of the PID controller and Pure Pursuit algorithm allows for real-time adjustments and precise control, minimizing the risk of collisions. The PID controller continuously adjusts the robot's motion parameters, while Pure Pursuit ensures that the robot follows a collision-free trajectory.

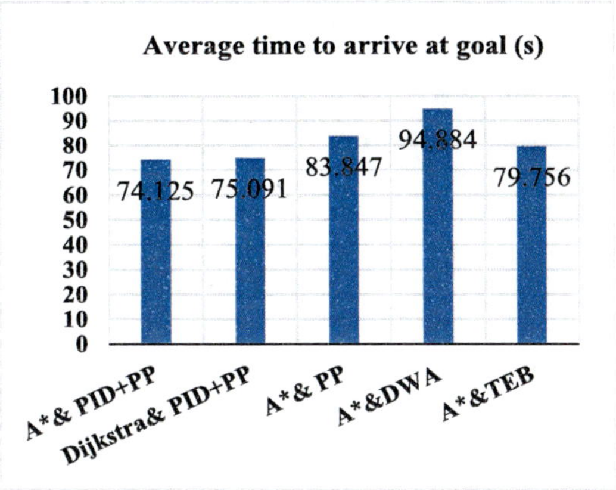

(a)

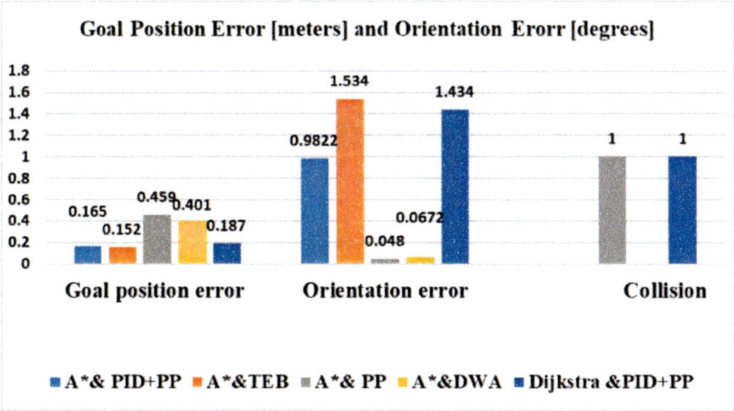

(b)

Fig. 7. Local planner comparisons, a) time comparison, b) position error and orientation error

6 Conclusion

This paper examines the impact of Proportional Integral Derivative (PID) on Pure Pursuit control, integrates the curvature information of the tracking path into the Pure Pursuit model, describes several scenarios in which AGVs may be useful, and then applies a PID controller to achieve global optimisation. AGV uses its own driving linear and angular velocities as well as the upcoming road curve to perform adaptive Pure Pursuit control, which improves tracking accuracy and ensures stable steering. The efficiency of traditional Pure Pursuit has been enhanced, and with the aid of this straightforward approach, the navigation of any robot in similar narrow warehouse environments has been resolved. The proposed solution hypothesizes that the implementation of PID controllers can improve trajectory tracking accuracy and prevent track system oscillation failure in traditional Pure Pursuit. Nevertheless, it is crucial to acknowledge the limitations of combining the A* global planner with PID controllers as local trajectory controllers, especially when dealing with complex trajectory heading changes or dynamic trajectories. First, if the A* heuristic function uses an overestimate of the actual path distance, then non-optimal paths can be encountered. Secondly selection of the look-ahead distance changes the behaviour of the controller dramatically and should be fine-tuned based on the working environment's properties. Lastly, PID parameters should be fine-tuned according to the calculated path's properties. Future efforts may concentrate on the creation of more sophisticated control techniques capable of overcoming these complexities.

References

1. Considine, D.M., Considine, G.D.: Robot technology fundamentals. In: Considine, D.M., Considine, G.D. (eds.) Standard Handbook of Industrial Automation, pp. 262–320. Springer, Boston (1986). https://doi.org/10.1007/978-1-4613-1963-4_17
2. Baek, E.-T., Im, D.Y.: ROS-based unmanned mobile robot platform for agriculture. Appl. Sci. **12**, 4335 (2022)
3. Zhao, J., Li, J., Zhou, J.: Research on two-round self-balancing robot SLAM based on the gmapping algorithm. Sensors **23**, 2489 (2023)
4. Peavy, M., Kim, P., Oyediran, H., Kim, K.: Integration of real-time semantic building map updating with adaptive monte carlo localization (AMCL) for robust indoor mobile robot localization. Appl. Sci. **13**, 909 (2023)
5. Mattila, J., Ala-Laurinaho, R., Autiosalo, J., Salminen, P., Tammi, K.: Using digital twin documents to control a smart factory: simulation approach with ROS, gazebo, and twinbase. Machines **10**, 225 (2022)
6. Xing, Z., Liu, H., Wang, T., Chew, E.P., Lee, L.H., Tan, K.C.: Integrated automated guided vehicle dispatching and equipment scheduling with speed optimization. Transp. Res. Part E: Logist. Transp. Rev. **169**, 102993 (2023)
7. Cherubin, S., Kaczmarek, W., Daniel, N.: Autonomous robot project based on the robot operating system platform. Problemy Mechatroniki: uzbrojenie, lotnictwo, inżynieria bezpieczeństwa, **13**(50), 85–108 (2022)
8. Zhou, L., Zhu, C., Su, X.: SLAM algorithm and navigation for indoor mobile robot based on ROS. In: 2022 IEEE 2nd International Conference on Software Engineering and Artificial Intelligence (SEAI) (2022)
9. Luo, M., Hou, X., Yang, J.: Surface optimal path planning using an extended Dijkstra algorithm. IEEE Access **8**, 147827–147838 (2020)
10. Ugalde Pereira, F., Medeiros de Assis Brasil, P., de Souza Leite Cuadros, M.A., Cukla, A.R., Drews Junior, P., Tello Gamarra, D.F.: Analysis of local trajectory planners for mobile robot with robot operating system. IEEE Latin Am. Trans. **20**, 92–99 (2022)
11. Wu, J., Ma, X., Peng, T., Wang, H.: An improved timed elastic band (TEB) algorithm of autonomous ground vehicle (AGV) in complex environment. Sensors **21**, 8312 (2021)
12. Yang, H., Teng, X.: Mobile robot path planning based on enhanced dynamic window approach and improved A* algorithm. J. Robot. **2022** (2022)
13. Yang, Y., et al.: An optimal goal point determination algorithm for automatic navigation of agricultural machinery: Improving the tracking accuracy of the Pure Pursuit algorithm. Comput. Electron. Agric. **194**, 106760 (2022)

14. Nawawi, S.W., Abdeltawab, A.A.A., Samsuria, N.E.N., Sirkunan, N.A.: Modelling, simulation and navigation of a two-wheel mobile robot using pure pursuit controller. ELEKTRIKA-J. Electr. Eng. **21**, 69–75 (2022)
15. Looi, C.Z., Ng, D.W.K.: A study on the effect of parameters for ROS motion planer and navigation system for indoor robot. Int. J. Electr. Comput. Eng. Res. **1**, 29–36 (2021)

A Digital Twin-Based Decision Support System for Dynamic Labor Planning

Banu Soylu[1](✉) and Gazi Bilal Yildiz[2]

[1] Industrial Engineering Department, Erciyes University, 38020 Kayseri, Türkiye
bsoylu@erciyes.edu.tr
[2] Industrial Engineering Department, Hitit University, 19030 Çorum, Türkiye
bilalyildiz@hitit.edu.tr

Abstract. The digital twin technology coordinates digital and physical spaces in order to improve the current and future actions in the system based on the real-time data. It allows organizations to follow and optimize their systems in a virtual environment before performing actions in reality. Digital twin can play a role in labor planning within an organization as well as in smart manufacturing environments such as performing the simulation of different scenarios, helping to determine the most efficient use of multi-skilled workers. In case of unexpected absences or changes in labor resource during a shift, organizations need to reconsider labor assignments to reduce downtime and inefficiencies. Traditionally, these actions are performed by the shift supervisor. In industry 4.0 concept, we design a digital twin-based decision support system with simulation capabilities for dynamic labor planning. The proposed system allows the unit to adapt to new conditions and also provides performance measures for the future state of the system. Additionally, the operator can simulate different scenarios and evaluate their performances. We present the results and performance of the proposed system on a case example.

Keywords: Digital twin · workforce planning · cross trained workers · absenteeism · job shop

1 Introduction

Manufacturing systems are often faced with unexpected absences or fluctuations in labor resources during a shift, which can have a significant impact on productivity, efficiency, and profitability. On average, unplanned absences consume 2.0–2.3% of all scheduled work hours in the U.S. service sector and up to 5% in certain industries [9]. To address this issue, planners can consider applying an automatic labor allocation system that can adapt quickly and easily to changes in labor resource levels. Dynamic labor planning involves assigning cross trained operators to different jobs in real-time [1]. Cross-trained workers help reduce the impact of absenteeism on production output and overtime cost [8]. The key issue here is for organizations to have multi-skilled workforce that are ready to accomplish the new mission. Easton [9] examined the performance of full and partial cross-training policies with that of dedicated specialists and found that the cross-trained

Z. Şen et al. (Eds.): IMSS 2023, LNME, pp. 214–223, 2024.
https://doi.org/10.1007/978-981-99-6062-0_20

workforce often, but not always, dominated the performance of a specialized workforce. In recent years, cross-training has attracted the interest of researchers and practitioners due to its potential to increase productivity and improve workers' socioeconomic welfare [2–4].

While long-run labor scheduling has been a widely studied problem in the operations management literature (see e.g., review articles [5, 6]), short-run dynamic labor planning has received relatively less attention. There is considerable research on staffing and scheduling decisions, while a few articles address real time labor allocation, where cross-trained workers are redeployed in real time to adjust the labor supply and demand [1]. Simulation is a widely used technique for making staffing and operational decisions. Feng and Fan [10] studied the dynamic multi-skilled workforce planning problem by exploring the effect of worker pool size and cross-training level on the performance of the production line through simulation. Mou and Robb [1] designed a simulator for typical grocery store operations such as in-store shopping, checkouts, shelf inventory management, and evaluate effects of reallocation decisions. Annear et al. [11] applied approximate dynamic programming to schedule multi-skilled technicians throughout the job shop. Easton [9] used two-stage stochastic model to schedule cross-trained workers and then simulate the system under uncertain demand and employee attendance to reallocate available cross-trained workers.

A manufacturing cell is an arrangement of machines in a job shop environment to produce families of parts with similar processing steps [7]. A *seru* production system is a novel variant of the cellular manufacturing system (CMS). It merges the flexibility of job shops and conveyor assembly lines [3]. One of the critical elements of the *seru* production is multi-skilled workforce. Ertay and Da Ruan [7] presented a decision-making approach for multi-skilled labor assignment problem in a CMS. Ferjani et al. [12] studied the dynamic assignment problem of multi-skilled workers and presented an online heuristic approach.

According to the above, the current methods in the literature do not adequately reflect the flexibility of labor planning problem. Smart technologies such as digital-twin can help collect and analyze the necessary data in real-time, contributing to the dynamic nature of the problem. In this study, we present a decision-making framework based on digital twin technology. The inputs of the decision-making approach and the abstraction level of the digital twin technology can be adapted according to the technological advancement of the organization. To show the implementation and performance of the system, we performed an example case from a cellular manufacturing system in job shop environment.

2 Digital Twin Technology-Enabled Decision Making

The digital twin is a critical technology for industrial digital transformation in the era of Industry 4.0 [17]. The shop floor has always been an important application object for the digital twin [18–20]. The digital twin can be described by three main components [16]: (1) A physical reality, (2) a virtual representation, and (3) a process of decision making and interpretation between physical reality and virtual representation as represented in Fig. 1.

Physical systems, physical processes and the physical environment are the building blocks of physical reality [16]. In our study, physical system refers to the labor working

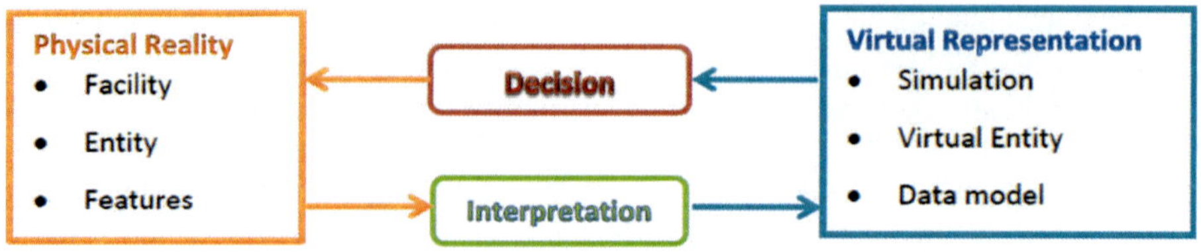

Fig. 1. Description of digital twin components

place, which can range from a single unit to the entire working place. In the physical system, the tasks are performed by human laborers or automated operators like robots and cobots, which are called the entities. The performance of the physical system is measured based on the inputs and outputs, including the use of resources and production results.

The physical processes in this study include factors which affect the state of entities. The states of the entities could be absenteeism or availability at some level while the factors, which cause state changes over time, could be training, rotating, resignation, maintenance, and fluctuations in demand etc.

The physical environment covers the management information systems as well as the processes where the data is collected. The physical system is surrounded by a management environment using (primitive or advanced) information systems. The databases, resource tracking systems, and resource planning software help to manage the physical environment. In our system, amount of resources, their abilities, resource requirements of units, constraints, restrictions, capacities are collected from the environment for a given time horizon.

The virtual representation of physical reality depends on the level of abstraction. Abstraction is the process of simplifying a complex system by focusing on its essential features and ignoring irrelevant details. A virtual representation with a shallow level of abstraction will have a more general and simplified view of the physical system, while a deeper level of abstraction will result in a more complex and detailed representation. This tradeoff between accuracy and complexity is important to consider when creating a digital twin or other virtual representations, as the level of detail required for an accurate model may not be feasible or cost-effective. Ultimately, the level of abstraction used will depend on the specific needs and requirements of the application.

The main components of the virtual representation are the construction of the model, the characterization of the entity and the simulation of the system as shown in Fig. 1.

Construction of the model refers to the process of formulating the system behavior using mathematical equations, models, algorithms or other computational methods. Here we design a decision support system which process the data from physical environment and returns the resource allocation plan. Digital twin technology enabled these decisions to move from the tactical level to the operational level. In this way, it is possible to obtain the best possible resource allocation plan in real-time, depending on the current state of the system. The main inputs of the decision support system are related to resources and their characteristics, as well as resource requirements of production cells and specifications. These inputs are usually available either digitally or on a simple worksheet in any physical environment. The most critical part of the proposed methodology is the determination of which inputs are necessary and at what level of detail. Depending on these answers, the level of abstraction of the virtual representation will be formed. Unnecessary input data increases the cost of data collection and storage, and reduces the computational performance of the system. On the other hand, missing a required input can result in loss of information and reduced accuracy.

The interpretation element pre-processes data to provide input to the model in an appropriate format, and also provides warnings about features when necessary. It also helps to get quick responses in case of abnormal situations.

Once the model and attributes of the entity are established, the system can be simulated. Simulation allows users to observe how the system responds to the resource allocations provided by the model and to evaluate alternative scenarios, either user-specified or system-generated. With the help of virtual representation of the labor planning system, it is possible to take reasonable actions in emergency situations, evaluate proposals, and predict future performance without executing the solution in real-life.

Based on this general outline, we next present the details of a digital twin technology-enabled decision support system (TW-DSS) for dynamic labor planning. Although we show the implementation on a cellular manufacturing system, it is possible to adapt it to flow shops and production lines.

2.1 The TW-DSS Framework

In a cellular manufacturing system, the machines, equipment and labor are grouped into a cell to produce families of similar products [13]. The labor force and their cross-training abilities are essential inputs to a cellular manufacturing system. We propose a digital-twin framework integrated with a decision support system (DSS), which is capable of evaluating labor planning strategies at the shop level. Figure 2 illustrates the main components of the framework.

A cell can consist of machines, assembly tables, equipment, etc. This content is allowed to change, but we assume that we know the final configuration. Other assumptions regarding the cell are as follows.

- The production plan for a given cell is ready before the start of the shift.
- Preventive maintenance interruptions are also known and considered in the production plan.
- Some jobs may have urgency.

There are several means of disturbances, either external or internal, in a manufacturing system. External ones are the variations in demand and supply, while the internal disruptions are variations in processing times, worker absenteeism, machine failures and defects. The responding mechanism against these disruptions is either using buffering strategies such as safety stock, excess capacity, rescheduling, or flexibility strategies such as labor and machine [14]. Buffering strategies are often used to address variations in demand and supply, but these strategies may not be as effective for dynamic labor planning for a short period of time. The unexpected changes in labor resource generally occur within a short period of time, such as hours or a shift, and may limit the ability to adjust schedules. Therefore, labor and machine flexibility provide efficient solutions compared to costly buffering strategies.

The labor flexibility corresponds to the number of machines or stations that a worker has qualification. Operators have varied skill sets and levels depending on their experience at different stations. The machine flexibility refers to the number of multi-purpose machines performing different operations [15]. In this study, we only consider the labor flexibility and the allocation of available operators in order to maximize the utility of the unit, which is measured as the expected number of jobs completed and the operator utilization. The assumptions regarding the resource level are as follows.

- Operator skills and levels are known, and unused skills do not expire.
- The working hour capacity of each operator is known.
- There are several positions in each cell, and each position requires single qualified operator.
- Operators may not be available on a given shift due to health problems, training exercises, breaks etc.

There can be some constraints such as the number of qualified workers necessary for each cell, or the set of partner cells that are able to support a cell. All of the above are inputs for the proposed DSS. There are daily work hour limits for each operator. These limits are determined by the characteristics of the job assigned to the operator. For example, if the job requires the operator to work while standing, this may be detrimental to his/her health. Therefore, we have included daily limits depending on the characteristics of the jobs.

The assignment of operators to cells and positions based on their skills is performed using the DSS. The DSS module first determines the *capacity* of each cell in terms of available machining hours. This information is usually available in advance and is not expected to change in the short term, but it is reduced by (un)scheduled maintenance. The *lack_of_labor* procedure calculates the difference between the required labor level and the available labor level for each cell and a given shift, depending on the capacity and the production schedule. However, this is not enough to determine the *performance* of a worker. The experiences a worker has, called history, and his/her displeasure from these practices are used to determine his/her performance for a given position. The *limit* of an operator refers to the restriction of working hours according to the characteristics of the jobs.

The assignment procedure determines the allocation plan of operators in a shift. The aim of this procedure is to assign operators to positions in order to maximize the expected number of completed jobs and sum of operator utilization. For this purpose, we suggest two strategies as *fastest-operator-first* or minimum *deviation* from average workload (workload balance of operators). The procedure evaluates the availability, production plan, limit and performance of operators, and additional constraints.

The workforce simulator module is used to evaluate the labor-job allocation plan returned from the DSS module and also to make real-time reallocations of workers and observe its effect on the system. The digital twin gets the real time availability of workforce and can monitor their assignment plan. The user can change the positions of operators using the menu. When the workers are overloaded or underskilled, the alerts warn the user. After running the simulator, performance graphs and tables are produced.

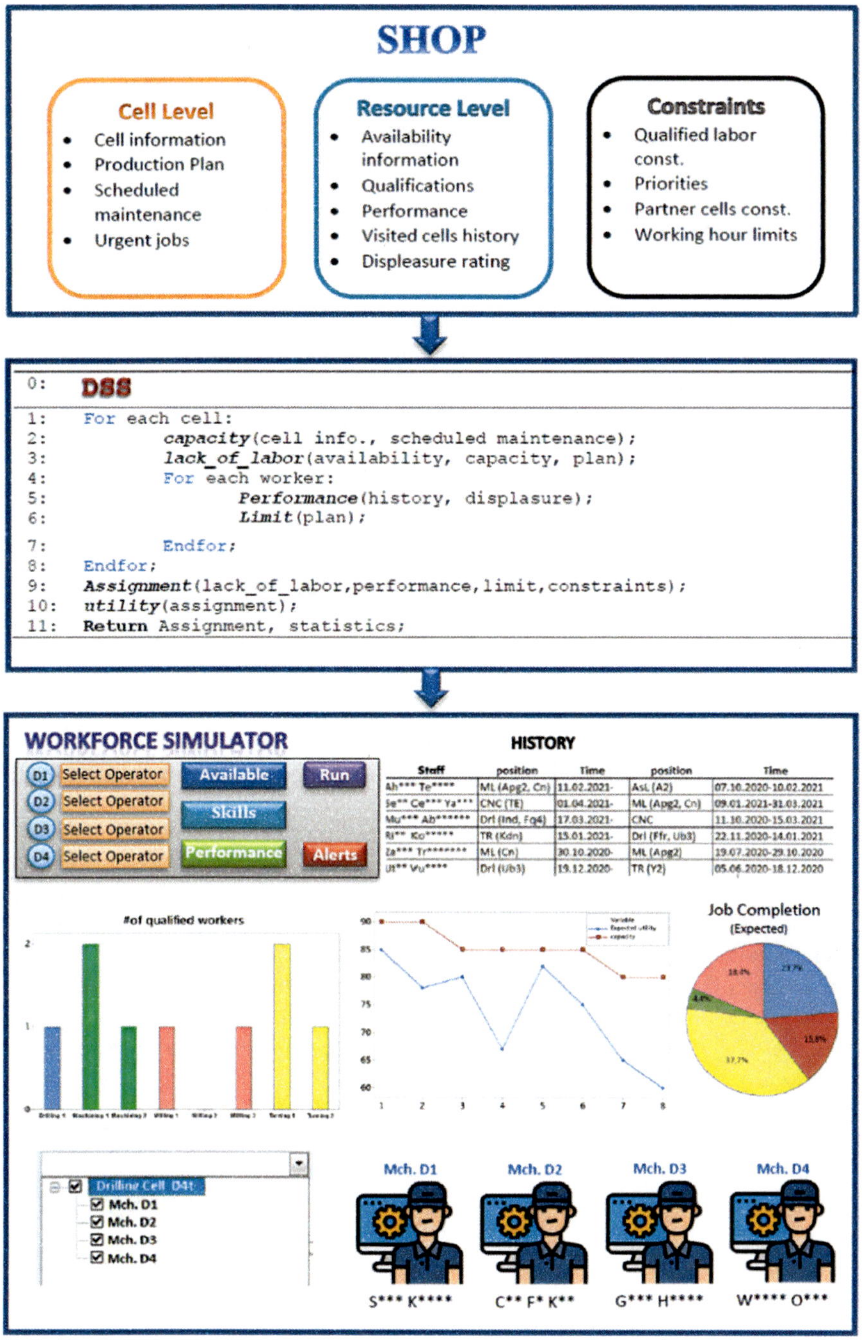

Fig. 2. Illustration of the TW-DSS Framework

3 Computational Results

In the case example considered in this section, we simulate a manufacturing cell and worker reallocation scenarios under fluctuations on worker resource and machine breakdowns. There are 5 operators and each operator is assigned to a single machine in the cell. There are 6 different jobs in a weekly production plan with different sizes such as j1:25, j2:30, j3:38, j4:13, j5:45, j6:30 units. The expertise and skill level of each operator to process these jobs is different and is shown in Table 1. There are 4 skills coded as A, B, C, and D. The skill level 1.0 refers to the expertise on this skill, while a skill level less than 0.4 is considered unskilled. The skill requirement of jobs and standard time are as follows: j1(A)[80min], j2(B,D)[40min.], j3(B)[100min.], j4(A,C)[60min.], j5(C,D)[120min.], and j6(D)[20min.]. The daily capacity of an operator is 480min. - machine breakdown duration. The processing time p_{ij} of a job j by operator i depends on the standard time s_j and the skill level requirement of the job j, i.e., $p_{ij} = \max_k \left\{ \frac{s_j}{l_{ik}} \right\}$ where l_{ik} is the skill level of operator i for skill k.

Table 1. Skill levels of operators

Operator i	Skill A	Skill B	Skill C	Skill D
1	1.0	0.8	0.8	0.7
2	0.8	0.9	0.8	0.7
3	0.7	0.6	0.9	0.8
4	0.9	0.7	0.8	0.7
5	0.8	0.9	0.3	0.9

The probability that a worker is unavailable in a day is assumed to be uniformly distributed and less than 0.05. The probability of a machine breakdown is also less than 0.05. The breakdown duration t (min.) is from U[0, 48].

The DSS explained above determine the worker assignments according to two strategies. The *fastest-operator-first* strategy is called the *Strategy*1, the balancing strategy is called the *Strategy*2, and the current assignment determined by the cell supervisor is called the *current*.

3.1 Results and Discussion

All algorithms were coded in Python 3.8, and experiments were conducted on a personal computer with Intel(R) Core (TM) i5-11400H 2.7 GHz 16GB RAM.

Table 2 shows the average processing time (min.) of jobs in each strategy. In the simulation, Operator 5 was absent on one day of the week. In addition, the machine that Operator 3 operates breaks down for 30 min. According to this table, Strategy 1 helps to complete jobs faster since it assigns an incoming job to the fastest available worker. The processing times in Strategy 2 are not better than Strategy 1, even close to the current situation; however, it presents a balanced workload, as can be seen in Fig. 3.

Table 2. Average processing time (min.) of jobs in each strategy

Job	*Current*	*Strategy*1	*Strategy*2
J1	98.2	90.1	96.7
J2	53.9	47.8	54.0
J3	129.5	116.1	127.3
J4	78.3	75.0	75.8
J5	164.3	151.4	161.7
J6	25.4	22.4	25.3

Figure 3 shows the average (daily) working and idle times for each operator. As shown in the figure, Operator 5 worked less on average because he was absent one day. However, Strategy 2 loads more jobs to Operator 5 in the remaining days to balance the workload of operators. On the other hand, the current system and Strategy 1 do not perform reallocations by taking into account absenteeism and downtime, i.e., they are memoryless.

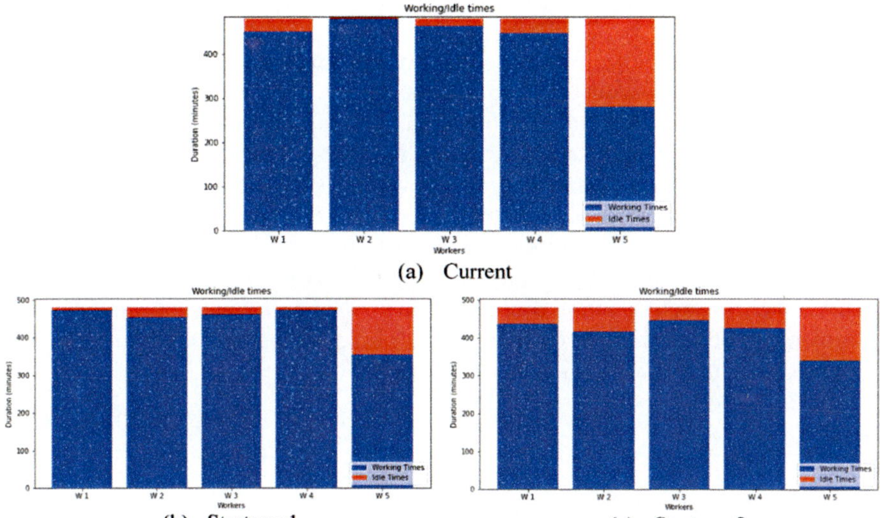

(a) Current

(b) Strategy 1

(c) Strategy 2

Fig. 3. Average working and idle times of each operator

Figure 4 illustrates the percentage of completed and incomplete jobs for a weekly schedule. According to this figure, a higher percentage of jobs are completed when strategy 1 is applied. Strategy 2 is close to the current situation.

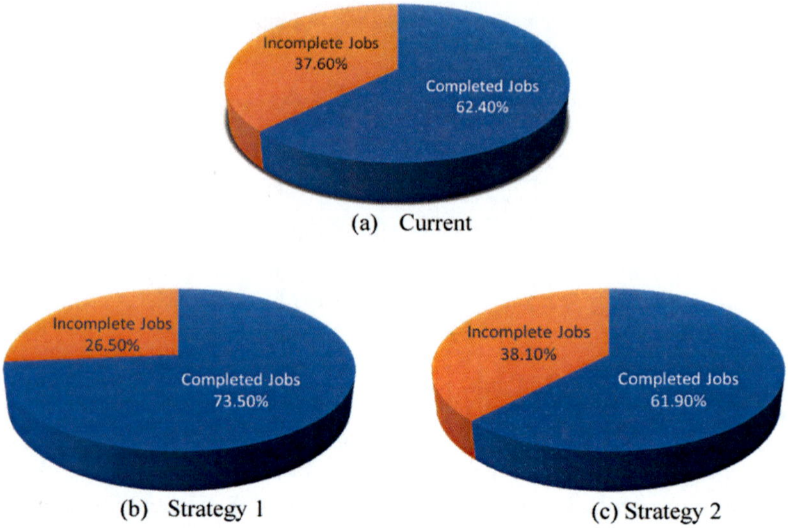

Fig. 4. Ratio of completed/uncomplete jobs (weekly) for each strategy

4 Conclusion

This study presents a digital twin-based decision support system for reallocation of operators in real-time under absenteeism and downtimes. The DSS reallocates jobs based on one of two strategies. According to our simulation results, in the fastest-worker-first strategy, more jobs are completed as expected. The TW-DSS framework has the ability to simulate the system in real-time, and also allows the user to make real-time changes in the positions of operators and observe the performance measures. Since the framework takes into account the real-world dynamics, such as alerts for overload or unskilled worker assignments, worker availability, limits and performance, etc., it is applicable.

The further research direction could be to integrate green strategies into the DSS of this framework. For example, prioritizing energy-efficient machines, while performing jobs, can contribute to the sustainability of the system.

References

1. Mou, S., Robb, D.J.: Real-time labour allocation in grocery stores: a simulation-based approach. Decis. Support Syst. **124**, 113095 (2019)
2. Wang, H., Alidaee, B., Ortiz, J., Wang, W.: The multi-skilled multi-period workforce assignment problem. Int. J. Prod. Res. **59**(18), 5477–5494 (2021)
3. Liu, C., Yang, N., Li, W., Lian, J., Evans, S., Yin, Y.: Training and assignment of multi-skilled workers for implementing seru production systems. Int. J. Adv. Manuf. Technol. **69**, 937–959 (2013). https://doi.org/10.1007/s00170-013-5027-5
4. Hopp, W.J., Tekin, E., Van Oyen, M.P.: Benefits of skill chaining in serial production lines with cross-trained workers. Manage. Sci. **50**(1), 83–98 (2004)
5. Özder, E.H., Özcan, E., Eren, T.: A systematic literature review for personnel scheduling problems. Int. J. Inf. Technol. Decis. Mak. **19**(06), 1695–1735 (2020)

6. Van den Bergh, J., Beliën, J., De Bruecker, P., Demeulemeester, E., De Boeck, L.: Personnel scheduling: a literature review. Eur. J. Oper. Res. **226**(3), 367–385 (2013)
7. Ertay, T., Ruan, D.: Data envelopment analysis based decision model for optimal operator allocation in CMS. Eur. J. Oper. Res. **164**(3), 800–810 (2005)
8. Slomp, J., Suresh, N.C.: The shift team formation problem in multi-shift manufacturing operations. Eur. J. Oper. Res. **165**(3), 708–728 (2005)
9. Easton, F.F.: Cross-training performance in flexible labor scheduling environments. IIE Trans. **43**(8), 589–603 (2011)
10. Feng, Y., Fan, W.: A hybrid simulation approach to dynamic multi-skilled workforce planning of production line. In: Proceedings of the Winter Simulation Conference 2014, pp. 1632–1643. IEEE (2014)
11. Annear, L.M., Akhavan-Tabatabaei, R., Schmid, V.: Dynamic assignment of a multi-skilled workforce in job shops: an approximate dynamic programming approach. Eur. J. Oper. Res. **306**(3), 1109–1125 (2023)
12. Ferjani, A., Ammar, A., Pierreval, H., Elkosantini, S.: A simulation-optimization based heuristic for the online assignment of multi-skilled workers subjected to fatigue in manufacturing systems. Comput. Ind. Eng. **112**, 663–674 (2017)
13. Singh, N.: Design of cellular manufacturing systems: an invited review. Eur. J. Oper. Res. **69**(3), 284–291 (1993)
14. Rocky Newman, W., Hanna, M., Jo Maffei, M.: Dealing with the uncertainties of manufacturing: flexibility, buffers and integration. Int. J. Oper. Prod. Manag. **13**(1), 19–34 (1993)
15. Palominos, P., Quezada, L., Moncada, G.: Modeling the response capability of a production system. Int. J. Prod. Econ. **122**(1), 458–468 (2009)
16. VanDerHorn, E., Mahadevan, S.: Digital twin: generalization, characterization and implementation. Decis. Support Syst. **145**, 113524 (2021)
17. Liu, X., et al.: A systematic review of digital twin about physical entities, virtual models, twin data, and applications. Adv. Eng. Inform. **55**, 101876 (2023)
18. Jia, W., Wang, W., Zhang, Z.: From simple digital twin to complex digital twin part I: a novel modeling method for multi-scale and multi-scenario digital twin. Adv. Eng. Inform. **53**, 101706 (2022)
19. Jia, W., Wang, W., Zhang, Z.: From simple digital twin to complex digital twin part II: multi-scenario applications of digital twin shop floor. Adv. Eng. Inform. **56**, 101915 (2023)
20. Fang, X., Wang, H., Liu, G., Tian, X., Ding, G., Zhang, H.: Industry application of digital twin: from concept to implementation. Int. J. Adv. Manuf. Technol. **121**(7–8), 4289–4312 (2022). https://doi.org/10.1007/s00170-022-09632-z

An Active Learning Approach Using Clustering-Based Initialization for Time Series Classification

Fatma Saniye Koyuncu(✉) and Tülin İnkaya

Bursa Uludag University, 16059 Nilüfer Bursa, Turkey
fatmasaniyekoyuncu@gmail.com, tinkaya@uludag.edu.tr

Abstract. The increase of digitalization has enhanced the collection of time series data using sensors in various production and service systems such as manufacturing, energy, transportation, and healthcare systems. To manage these systems efficiently and effectively, artificial intelligence techniques are widely used in making predictions and inferences from time series data. Artificial intelligence methods require a sufficient amount of labeled data in the learning process. However, most of the data in real-life systems are unlabeled, and the annotation task is costly or difficult. For this purpose, active learning can be used as a solution approach. Active learning is one of the machine learning methods, in which the model interacts with the environment and requests the labels of the informative samples. In this study, we introduce an active learning-based approach for the time series classification problem. In the proposed approach, the k-medoids clustering method is first used to determine the representative samples in the dataset, and these cluster representatives are labeled during the initialization of active learning. Then, the k-nearest-neighbor (KNN) algorithm is used for the classification task. For the query selection, uncertainty sampling is applied so that the samples having the least certain labels are prioritized. The performance of the proposed approach was evaluated using sensor data from the production and healthcare systems. In the experimental study, the impacts of the initialization techniques, number of queries, and neighborhood size were analyzed. The experimental studies showed the promising performance of the proposed approach compared to the competing approaches.

Keywords: Machine Learning · Active Learning · Clustering · Initialization · Time Series

1 Introduction

With the increase of digitalization, data collection and storage activities have become widespread in production and service systems in recent years. The use of artificial intelligence techniques become prominent to improve performance of these systems. A sufficient amount of labeled data is needed for the training process of the artificial intelligence models. Although data storage opportunities increase, large amounts of data are unlabeled. In some of the systems, it is possible to annotate data with labels automatically

Z. Şen et al. (Eds.): IMSS 2023, LNME, pp. 224–235, 2024.
https://doi.org/10.1007/978-981-99-6062-0_21

or at a low cost. On the other hand, in some systems, resource allocation is needed for annotation, and the annotation task is costly or difficult. Therefore, data samples should be selected for annotation so that both high accuracy and low cost are ensured in the development of artificial intelligence models. Motivated by this, active learning, a branch of machine learning, is considered in this study.

Active learning enables the training of artificial intelligence models with a small number of labeled data. It includes the selection of the unlabeled data for annotation by an expert or oracle to improve the performance of the model [1]. It aims to increase the accuracy of the artificial intelligence models, and in the meanwhile, keep the annotation cost at a minimum [2].

Figure 1 shows the active learning process. There is an unlabeled data pool, and the class labels of these data are not known a priori. Labeled data are needed to train a classifier model. For this reason, query samples are selected from the unlabeled dataset, and the selected samples are annotated by an expert. The expert who annotates data with labels can be a person, a machine, or a database system. The query strategy aims to identify the most informative regions and data samples for querying [2]. The query sample is added to the labeled dataset after annotation. This process continues until the stopping criterion is satisfied or the labeling process is finished [1].

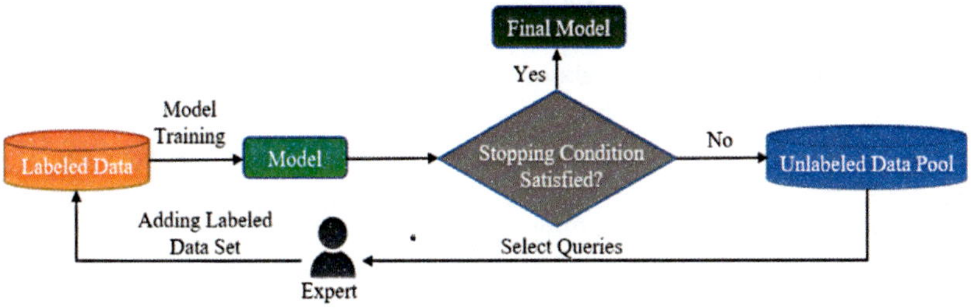

Fig. 1. Active learning process

This study focuses on the time series classification problem. A time series represents values from time-dependent consecutive measurements, and the time series classification aims to predict the class label of a new time series. It is a widely encountered problem in real-life applications such as healthcare, energy, finance, manufacturing, and so on [3], and machine learning techniques are used as a solution approach. When the labeled data are scarce or not available, active learning approaches are used to increase the performance of the machine learning models with minimum annotation costs [4].

In this study, we propose a clustering-based initialization for active learning when there is no labeled sample in time series classification. During initialization, the k-medoid clustering is implemented for selecting the most representative samples to be labeled. Then, the k-nearest-neighbor (KNN) algorithm is used for the classification task. In the active learning process, the query selection is performed according to uncertainty sampling. The proposed approach is evaluated using sensor data from production and healthcare systems.

In summary, this work contributes to the literature by introducing a KNN-based active learning approach with the k-medoid clustering initialization for time series classification. In particular, the proposed approach yields reasonable classification accuracy with a small number of labeled samples. Moreover, it can be applied to real-life problems easily.

The organization of this paper is as follows. In Sect. 2, we present the literature review. In Sect. 3, we explain the proposed method. The experimental study is given in Sect. 4. Finally, Sect. 5 includes the conclusion and future research directions.

2 Literature Review

Recently active learning has attracted the interest of many researchers, and several studies are proposed in this field. Kumar and Gupta [1], Aggarwal [2] and Settles [4] compiled the main studies in this field and provided taxonomies of active learning strategies.

A group of studies applied active learning approaches to time series problems. Gweon and Yu [5], He et al. [6], Saaedi et al. [7] and Peng et al. [8] developed neighborhood-based active learning approaches to time series classification. For instance, Gweon and Yu [5] proposed an active learning strategy and a nearest-neighbor-based method. In the proposed approach, the Euclidean distance was calculated and ranked for each unlabeled sample. The informativeness score showed the uncertainty of an unlabeled time series for utility. A high informativeness score was assigned to a sample that was close to the current decision boundary with unlabeled samples nearby and had a different pattern from the labeled samples so far. They used uncertainty and utility metrics to calculate the informativeness score. The updated 1-nearest-neighbor algorithm was applied as the classification algorithm.

He et al. [6] studied active learning and semi-supervised learning in time series classification problems. They proposed a sampling strategy by ranking the informativeness of unlabeled samples according to their uncertainty and local data density. They used cosine similarity to calculate the uncertainty. They considered local data density using the reverse k-nearest neighborhood method. They also calculated a score for the unlabeled data considering uncertainty, local data density, and weight. Moreover, they integrated the proposed approach with a semi-supervised learning method. Saeedi et al. [7] proposed a multi-expert and cost-sensitive active learning strategy to monitor mobile health data. They aimed to minimize the total cost of annotation by different experts. Mobile health data often start with unlabeled data. For this reason, they proposed an initial learning algorithm, an entropy and ensemble-based query strategy working with a random forest classifier, and an expert selection algorithm for the case of multiple experts.

Peng et al. [8] introduced a score based on uncertainty and utility to select the most informative samples in the time series. Uncertainty indicated the difficulty of prediction with the current classifier, and utility measured the correlation between time series samples. They used the nearest neighborhood algorithm as the classifier. They proposed a greedy algorithm for selecting query samples that maximized the overall knowledge of the training set.

Another group of studies integrated clustering methods with active learning to increase efficiency. For example, Zhang and Dai [9] proposed an active learning method based on support vector regression for the time series prediction problem. An efficient, cost-sensitive active learning model, namely a multi-criteria active weighted model, was proposed to solve the time series prediction problem in unbalanced datasets. First, weighted support vector regression was used as a predictor to assign different costs to samples with cost-sensitive learning. Then, the weighted support vector regression was trained with a small initial training data to obtain the set of support vectors and non-support vectors. Then, the kernel k-means algorithm was applied to the unlabeled samples along with the training samples of non-support vectors.

Shim et al. [10] applied active learning to the semiconductor manufacturing sector. They used a k-means clustering algorithm for grouping the faulty wafer maps into clusters and performed cluster-level labeling according to the similarity of wafer maps. Clustering was first performed for the unlabeled wafer maps. Then, a convolutional neural network (CNN) was trained with the labeled dataset so that the wafer maps were classified according to their fault types. CNN's cluster-level uncertainties were calculated for the unlabeled wafer maps. Considering uncertainties, query samples were selected and labeled at the cluster level by an engineer. Using this method, a high-performance CNN with a low explanation cost was obtained.

Qi and Ting [11] proposed an active semi-supervised proximity propagation clustering algorithm, considering the local outlier factor. The local outlier factor algorithm was used to reduce the impact of outliers in the data. Then, a semi-supervised clustering algorithm was used to obtain better performance.

Mai et al. [12] proposed the Act-DBSCAN method that combined the density-based clustering algorithm (DBSCAN) and active learning. Binary lower-bound similarities were used to initialize the cluster structure. The results showed that Act-DBSCAN outperformed other competing techniques such as active spectral clustering in real datasets.

In the literature, some studies addressed the initialization (initial labeling) problem in active learning. For instance, Grimova et al. [13] studied the classification of records obtained during the stages of sleep. The focus was on the problem of the initialization of active learning. For the classification of unlabeled samples, they performed the classification using the 1-nearest-neighbor algorithm with the Euclidean distance. The proposed method was compared with the initialization with the k-means algorithm. They observed that the proposed method showed better results.

Nguyen and Smeulders [14] applied the k-medoid algorithm for initial clustering. The selection criterion has been chosen as the most representative training samples in the same cluster. The Gaussian noise model was used to infer the class labels for non-representative samples.

To sum up, there exist a limited number of studies that consider the initialization of unlabeled data in time series classification. To fill this gap, we propose an active learning approach with clustering-based initialization. In particular, the most representative time series are labeled using the k-medoid clustering during initialization, and the classification task is performed using KNN. During active learning, time series having the least certain class labels are annotated.

3 Proposed Approach

In this study, all samples are unlabeled initially, and the aim is to select the most representative samples using clustering during initialization. The flowchart of the proposed approach is given in Fig. 2. The proposed approach consists of three stages: data normalization, initialization with the k-medoid clustering algorithm, and query selection, labeling and training. The details of each stage are explained as follows.

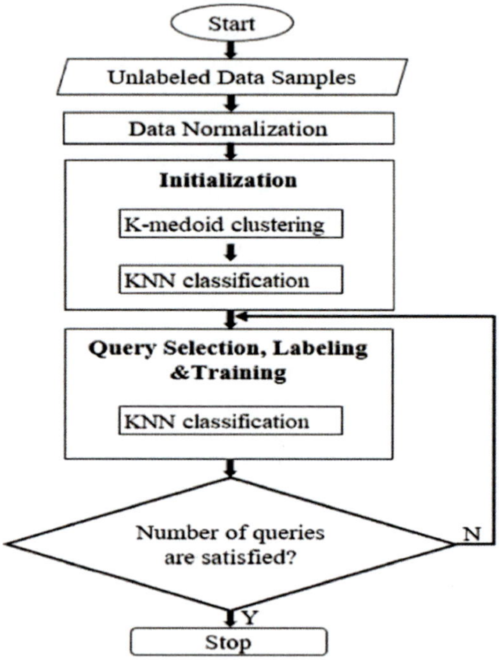

Fig. 2. The proposed active learning algorithm

Stage 1 - Data normalization: Z-score normalization is used so that the impact of scale differences among the attributes is removed [15].

Stage 2 - Initialization with the k-medoid clustering algorithm: In this stage, clustering is used to determine the samples to be labeled during initialization. That is, among the unlabeled samples, the most representative samples are selected for labeling using the k-medoid clustering algorithm.

Clustering algorithms form groups of data samples such that similar samples are assigned to the same cluster whereas dissimilar samples are assigned to different clusters. The k-medoid clustering method is one of the partitioning methods, and its pseudocode is given in Fig. 3 [15]. In Fig. 3, given a dataset D with n data samples, the aim is to determine k representative samples of the k clusters. For this purpose, the sum of the dissimilarities between each data sample and its representative is minimized. The clustering error, E, is calculated according to Eq. (1).

$$E = \sum\nolimits_{i=1}^{k} \sum\nolimits_{x \in C_i} dist(x, o_i) \quad (1)$$

where k denotes the number of clusters, x indicates the data sample, o_i shows the representative sample of cluster C_i, and $dist(.)$ is the dissimilarity function.

When a new representative sample is selected for swapping, the clustering error, E, is recalculated. The total cost of swapping, S, is the difference in the clustering error values when one of the representative samples is replaced with a non-representative sample. If the total cost is negative ($S < 0$), then the representative sample, o_j, is swapped with the sample o_{random}. If the total cost is positive, no change is performed. The iterations continue until there is no improvement in the total cost.

Algorithm 1: The k-medoid Clustering Algorithm

Input: D: dataset with n samples, k: number of clusters
Output: Set of k clusters

(1) choose k samples in dataset D arbitrarily as initial representatives.
(2) **repeat**
(3) assign each sample to the cluster with the nearest representative.
(4) select a non-representative sample, o_{random}, randomly.
(5) compute the total cost, S, of swapping the representative sample, o_j, with o_{random}.
(6) **if** $S < 0$ **then**, swap o_j with o_{random} and update the set of k representatives.
(7) **until** no change.

Fig. 3. The k-medoid clustering algorithm

After the execution of the k-medoid algorithm, k representative samples are annotated. Using these labeled samples, the KNN algorithm is trained for time series classification. The KNN algorithm is one of the supervised learning methods, and it is widely used in time series classification. It is a distance-based algorithm, and the classification is performed according to the labels of the neighboring points.

The pseudocode of the KNN algorithm is given in Fig. 4 [16]. Let x_i and y_i denote data sample i and its class label, respectively. Also, let x_{ij} denotes the jth attribute of data sample i, and each sample has d attributes. The training dataset includes all data samples and their class labels, i.e. $D = \{(x_1, y_1), (x_2, y_2),..., (x_n, y_n)\}$. The neighboring points *of test sample a are determined* according to a dissimilarity measure, and Euclidean distance is used for this purpose in this study. The number of neighbors, k, is an input parameter. The class label of the test sample a is classified according to the most frequent class label among its k nearest neighbors.

Stage 3 - Query selection, labeling and training: In this stage, uncertainty sampling is applied as the query selection method. Uncertainty sampling is one of the heterogeneity-based methods. In these methods, the samples in the heterogeneous regions, i.e. close to the decision boundary, are considered to be more valuable for learning the decision boundary [2]. Hence, uncertainty sampling selects the least certain sample from the unlabeled dataset for labeling. The least certain sample x^* is determined according to

Algorithm 2: The K-Nearest-Neighbor (KNN) Algorithm

Input: $D = \{(x_1, y_1), (x_2, y_2), \ldots, (x_n, y_n)\}$: training dataset, k: number of nearest neighbors, a: test sample.
Output: Class label of test sample a.
(1) Find the k nearest neighbors of sample a using Euclidean distance, $dist(a, x_i) = \sqrt{\sum_{j=1}^{d} (a_j - x_{ij})^2}$.
(2) Estimate the class label of sample a using the most frequent class label among k neighbors.

Fig. 4. The k-nearest-neighbor (KNN) algorithm

Eq. (2) and Eq. (3) [17].

$$x^* = \underset{x}{\operatorname{argmax}}(1 - P(\hat{y}|x)) \quad (2)$$

$$\hat{y} = \underset{y}{\operatorname{argmax}}(P(y|x)) \quad (3)$$

where $P(\hat{y}|x)$ denotes the probability of classifying sample x to class label y, and $\hat{y}$ denotes the most likely class label.

4 Experimental Study

In this section, the experimental study is explained in detail.

4.1 Datasets and Implementation Details

The performance of the proposed approach was tested on four datasets for time series classification, namely ECG5000, Wafer, FordA and FordB [18]. Table 1 shows the properties of the datasets.

The ECG5000 dataset contains 20 hours of ECG data downloaded from the Physionet [18]. There are 5000 samples in this dataset. Firstly, the data were extracted. Then, the heartbeats were reduced to equal lengths using the interpolation technique. There are five classes and 5500 samples in this dataset.

The Wafer dataset contains data related to semiconductor production [18]. The dataset has been created by storing in-line process data recorded from various sensors during the processing of wafer silicones. There are two classes in the dataset, namely normal and abnormal. There is a class imbalance between normal and abnormal class labels. That is, 10.7% of the training data and 12.1% of the test data belong to the abnormal class. There is a total of 7164 samples in the dataset.

The FordA and FordB datasets contain measurements of engine noise in the automotive production [18]. The aim is to find whether a specific symptom exists or not. The training and test datasets of FordA have been created in normal operating conditions, under minimal noise contamination. There are 4921 samples and two classes in the FordA dataset. The training dataset for FordB has been collected in normal operating conditions whereas its test dataset has been collected under noisy conditions. In the FordB dataset, there is a total of 4446 samples, and there are two classes.

In all four datasets, the training and test sets are provided separately. Hence, in the experiments, the unlabeled samples were selected from the training set. After labeling, they were used in the training of the algorithm. The test dataset was used for evaluation purpose.

Table 1. Dataset properties

Dataset	Number of Training Samples	Number of Test Samples	Number of Features	Number of Classes
ECG5000	500	4500	140	5
Wafer	1000	6164	152	2
FordA	3601	1320	500	2
FordB	3636	810	500	2

The performance of the proposed method was compared with random selection for the initialization. In the random selection, five randomly selected samples were labeled during initialization. All experimental studies were performed on the Python platform. Pandas, Numpy, Sklearn, Scipy and ModAL Python libraries were used in the experiments. For each initialization method, five repetitions were performed, and their averages were reported.

To determine the labels during initialization, we set the number of clusters to 5 ($nc = 5$) in the k-medoid algorithm. In KNN, we considered three values for the number of nearest neighbors, i.e. $k = \{1, 3, 5\}$, and they were abbreviated as 1NN, 3NN, and 5NN, respectively. In active learning, the number of queries was increased to 150. That is, 150 samples were selected and labeled.

The performance of the proposed approach was evaluated using an accuracy score. The accuracy score is calculated based on the confusion matrix as shown in Table 2.

Table 2. Confusion matrix

	Predicted Positive (1)	Predicted Negative (0)
Actual Positive (1)	True Positive (TP)	False Negative (FN)
Actual Negative (0)	False Positive (FP)	True Negative (TN)

In Table 2, True Positive (TP) denotes the number of positive samples that the model correctly classified as positive, True Negative (TN) shows the number of negative samples that the model correctly classified as negative, False Positive (FP) indicates the number of negative samples that the model incorrectly classified as positive, and False Negative

(FN) is the number of positive samples that the model incorrectly classified as negative. The accuracy score is the ratio of TP and TN samples to all samples as shown in Eq. (4). An accuracy score close to one indicates a better classification.

$$Accuracy = \frac{\mathrm{TP} + \mathrm{TN}}{TP + TN + FP + FN} \quad (4)$$

4.2 Results and Discussion

The accuracy scores of the proposed approach and random selection are given in Table 3. The best values are bolded in Table 3.

The following implications can be inferred from Table 3:

- For $k = 1$, the k-medoid algorithm outperforms the random selection for all query samples in the ECG5000 dataset. On the other hand, the k-medoid algorithm provides better accuracy scores up to the 50th query in the Wafer dataset. After the 50th query, the random selection is slightly better than the k-medoid algorithm in this dataset. In the FordA dataset, the k-medoid algorithm is slightly better than the random selection for 150th query only. For the other number of queries, the random selection yields better accuracy scores. In the FordB dataset, the k-medoid algorithm is prominent up to the 100th query whereas the random selection shows higher performance for the rest of the queries.
- For $k = 3$, the accuracy scores of the k-medoid algorithm are higher than the random selection up to the 100th and 25th queries in the ECG5000 and Wafer datasets, respectively. However, as the number of queries increases, the random selection gives slightly better results. In the FordA and FordB datasets, the k-medoid algorithm has higher performance during initialization and when the number of queries is small. On the other hand, the performance of the k-medoid algorithm fluctuates back and forth as the number of queries increase. In addition, the best performances in the 150th query for both datasets are obtained with the k-medoid algorithm.
- For $k = 5$, in both ECG5000 and Wafer datasets, the k-medoid algorithm and random selection yield very close accuracy values when the number of queries is less than 75. When the number of queries increases, the k-medoid algorithm results in slightly better accuracy values compared to random selection. In the FordA dataset, the k-medoid algorithm initially has higher accuracy values, while the random selection performs slightly better than the k-medoid algorithm with the increase in the number of queries. In the FordB dataset, the k-medoid algorithm outperforms the random selection except for the initialization and the 75th query.
- For all datasets, the average computational time for the k-medoid clustering algorithm is 2.7 s. Also, the query selection, labeling and training stage takes an average of 118.5 s for all 150 queries in all datasets.

As a result, for the small neighborhood sizes ($k = 1$, $k = 3$), the k-medoid algorithm achieves almost 90% of accuracy scores during initialization of active learning in the ECG5000 and Wafer datasets. However, the accuracy values are initially around 50% in the FordA and FordB datasets. Even the number of labeled queries increases, the accuracy

Table 3. The comparison of the accuracy scores for the proposed approach and random selection

1NN		Number of Queries							
Dataset	Method	Initialization	1	25	50	75	100	125	150
ECG5000	Random	0.800	0.800	0.797	0.795	0.795	0.794	0.794	0.792
	K-medoid	**0.904**	**0.906**	**0.897**	**0.894**	**0.894**	**0.893**	**0.894**	**0.893**
Wafer	Random	0.850	0.850	0.961	0.974	**0.982**	**0.982**	**0.989**	**0.989**
	K-medoid	**0.892**	**0.892**	**0.962**	**0.976**	0.981	0.981	0.988	0.988
FordA	Random	**0.530**	**0.539**	**0.577**	**0.558**	**0.599**	**0.607**	**0.614**	0.616
	K-medoid	0.516	0.516	0.554	0.545	0.592	0.595	0.608	**0.617**
FordB	Random	0.508	0.519	0.525	0.516	0.530	**0.554**	**0.549**	**0.542**
	K-medoid	**0.559**	**0.558**	**0.527**	**0.522**	**0.535**	0.546	0.537	0.533
3NN		Number of Queries							
Dataset	Method	Initialization	1	25	50	75	100	125	150
ECG5000	Random	0.712	0.714	0.916	0.925	0.926	0.933	**0.936**	**0.936**
	K-medoid	**0.891**	**0.893**	**0.920**	**0.934**	**0.934**	**0.934**	0.933	0.933
Wafer	Random	0.856	0.888	0.958	**0.993**	**0.994**	**0.994**	**0.994**	**0.994**
	K-medoid	**0.892**	**0.892**	**0.960**	0.991	0.993	0.993	0.993	0.993
FordA	Random	0.510	0.509	**0.558**	**0.600**	**0.609**	**0.637**	**0.646**	0.637
	K-medoid	**0.516**	**0.516**	0.538	0.574	0.595	**0.637**	0.639	**0.643**
FordB	Random	0.497	0.496	0.506	0.505	**0.526**	**0.530**	**0.511**	0.527
	K-medoid	**0.533**	**0.523**	**0.514**	**0.510**	0.517	**0.530**	0.505	**0.532**
5NN		Number of Queries							
Dataset	Method	Initialization	1	25	50	75	100	125	150
ECG5000	Random	**0.584**	**0.584**	0.922	**0.931**	**0.936**	0.937	0.938	0.938
	K-medoid	**0.584**	**0.584**	**0.927**	**0.931**	0.934	**0.939**	**0.939**	**0.939**
Wafer	Random	**0.892**	**0.892**	0.947	0.983	**0.993**	0.993	0.993	0.993
	K-medoid	**0.892**	**0.892**	**0.953**	**0.990**	**0.993**	**0.994**	**0.994**	**0.994**
FordA	Random	0.510	0.507	**0.557**	**0.596**	**0.615**	**0.629**	**0.640**	**0.654**
	K-medoid	**0.516**	**0.516**	0.547	0.586	0.611	0.620	0.626	0.647
FordB	Random	**0.501**	0.493	0.496	0.500	**0.524**	0.531	0.533	0.539
	K-medoid	0.495	**0.526**	**0.506**	**0.511**	0.501	**0.541**	**0.544**	**0.551**

values become almost 65% and 55% in these datasets, respectively. These performances can be attributed to the weakness of KNN in the noisy datasets. Moreover, for the small neighborhood sizes, the k-medoid algorithm outperforms the random selection in the ECG5000, Wafer and FordB datasets. For the large neighborhood sizes ($k = 5$) and a high number of queries, the performances of both approaches are very close. To sum up, when the annotation cost is high, coupling the k-medoid algorithm with 1NN and 3NN yields good performance in the ECG5000, Wafer and FordB datasets.

5 Conclusion

In this study, we addressed the time series classification problem, which is widely encountered in production and service systems. Initially, the dataset is unlabeled, and labeling each sample is a costly task. For this reason, the selection of samples for annotation becomes a challenging issue.

A clustering-based initialization is proposed for active learning in this study. During initialization, the proposed approach uses the k-medoid algorithm to determine the representative samples to be labeled. The labeled samples are input to KNN, one of the widely used methods for time series classification. Also, uncertainty sampling is used for the query selection.

The experimental studies with production and healthcare systems showed that, even with a small number of labeled samples, the proposed method increases the accuracy in small neighborhood sizes. Hence, the proposed method can be used in reducing labeling costs and improving the model performance in real-world datasets. On the other hand, the proposed approach has limitations in the presence of noise.

As a future study, different clustering methods and query selection mechanisms can be studied to improve the performance. Also, noise handling mechanisms can be integrated to the proposed approach.

References

1. Kumar, P., Gupta, A.: Active learning query strategies for classification, regression, and clustering: a survey. J. Comput. Sci. Technol. **35**(4), 913–945 (2020)
2. Aggarwal, C.C.: Data Mining: The Textbook, 1st edn. Springer, New York (2015). https://doi.org/10.1007/978-3-319-14142-8
3. Esling, P., Carlos, A.: Time-series data mining. ACM Comput. Surv. **45**(1), 1–34 (2012)
4. Settles, B.: Active learning literature survey. Technical report, University of Winconsin-Madison Department of Computer Sciences (2009)
5. Gweon, H., Yu, H.: A nearest neighbor-based active learning method and its application to time series classification. Pattern Recogn. Lett. **146**, 230–236 (2021)
6. He, G., Li, Y., Zhao, W.: An uncertainty and density based active semi-supervised learning scheme for positive unlabeled multivariate time series classification. Knowl.-Based Syst. **124**, 80–92 (2017)
7. Saeedi, R., Sasani, K., Gebremedhin, A.H.: Co-meal: cost-optimal multi-expert active learning architecture for mobile health monitoring. In: Proceedings of the 8th ACM International Conference on Bioinformatics, Computational Biology, and Health Informatics, pp. 432–441 (2017)
8. Peng, F., Luo, Q., Ni, L.M.: ACTS: an active learning method for time series classification. In: IEEE 33rd International Conference on Data Engineering (ICDE), San Diego, CA, USA, pp. 175–178 (2017)
9. Zhang, J., Dai, Q.: A cost-sensitive active learning algorithm: toward imbalanced time series forecasting. Neural Comput. Appl. **34**(9), 6953–6972 (2022)

10. Shim, J., Kang, S., Cho, S.: Active cluster annotation for wafer map pattern classification in semiconductor manufacturing. Expert Syst. Appl. **183**, 115429 (2021)
11. Qi, L., Ting, L.: Active semi-supervised affinity propagation clustering algorithm based on local outlier factor. In: 37th Chinese Control Conference (CCC), Wuhan, China, pp. 9368–9373 (2018)
12. Mai, S.T., He, X., Hubig, N., Plant, C., Böhm, C.: Active density-based clustering. In: IEEE 13th International Conference on Data Mining, Dallas, TX, USA, pp. 508–517 (2013)
13. Grimova, N., Macas, M., Gerla, V.: Addressing the cold start problem in active learning approach used for semi-automated sleep stages classification. In: IEEE International Conference on Bioinformatics and Biomedicine (BIBM), Madrid, Spain, pp. 2249–2253 (2018)
14. Nguyen, H.T., Smeulders, A.: Active learning using pre-clustering. In: Proceedings of the 21st International Conference on Machine Learning, Banff, Canada (2004)
15. Han, J., Kamber, M., Pei, J.: Data Mining: Concepts and Techniques, 3rd edn. Morgan Kaufmann, Waltham (2012)
16. Almomany, A., Ayyad, W.R., Jarrah, A.: Optimized implementation of an improved KNN classification algorithm using Intel FPGA platform: Covid-19 case study. J. King Saud University Comput. Inf. Sci. **34**(6), 3815–3827 (2022)
17. Prudencio, R.B.C., Carlos, S., Ludermir, T.B.: Uncertainty sampling methods for selecting datasets in active meta-learning. In: The 2011 International Joint Conference on Neural Networks (IJCNN), pp. 1082–1089. IEEE, San Jose (2011)
18. The UEA & UCR time series classification repository. http://www.timeseriesclassification.com. Accessed 14 June 2023

Finger Movement Classification from EMG Signals Using Gaussian Mixture Model

Mehmet Emin Aktan[1,2(✉)], Merve Aktan Süzgün[3], Erhan Akdoğan[1,4], and Tuğçe Özekli Mısırlıoğlu[5]

[1] Health Institutes of Türkiye, 34718 İstanbul, Turkey
[2] Department of Mechatronics Engineering, Bartın University, 74110 Bartın, Turkey
maktan@bartin.edu.tr
[3] Department of Neurology, Istanbul University-Cerrahpaşa, 34098 İstanbul, Turkey
[4] Department of Mechatronics Engineering, Yıldız Technical University, 34349 İstanbul, Turkey
[5] Department of Physical Medicine and Rehabilitation, Istanbul University-Cerrahpaşa, 34098 İstanbul, Turkey

Abstract. Hands are the most used parts of the limbs while performing complex and routine tasks in our daily life. Today, it is an important requirement to determine the user's intention based on muscle activity in exoskeletons and prostheses developed for individuals with limited mobility in their hands due to traumatic, neurologic injuries, stroke etc. In this study, 5-finger movements were classified using surface electromyography (EMG) signals. The signals were acquired from forearm via the 8-channel Myo Gesture Control Armband. EMG signals from three participants were analyzed for the movements of each finger, and the activity levels of the channels were compared according to the movements. Following, movement classification was performed using the Gaussian mixture network, a statistical artificial neural network model. According to the experimental results, it was seen that the model achieved an accuracy of 73.3% in finger movement classification.

Keywords: finger movement classification · sEMG · Gaussian Mixture Model · artificial neural network

1 Introduction

Hands are used the most in daily life and are exposed to the biggest strain and trauma. Complete or partial loss of function in the hands may occur due to ageing, traumatic injuries and neurologic diseases. These movement limitations are tried to be eliminated as much as possible with various surgical interventions and rehabilitation processes. In amputation cases, solutions are found for the problems of patients with individually designed prostheses. Evaluation of muscle activities is extremely important both in treatment processes and in prosthesis

Z. Şen et al. (Eds.): IMSS 2023, LNME, pp. 236–246, 2024.
https://doi.org/10.1007/978-981-99-6062-0_22

applications. Measuring muscle contraction levels according to finger movements during hand rehabilitation is important in determining target position and force values. Especially in robotic rehabilitation, while the control parameters are determined by the doctor according to the patient's condition or automatically selected by the system, the evaluation of muscle contraction levels and providing optimum contraction of the muscles increase the effectiveness of the treatment. In exoskeleton robots, which are also used for therapeutic purposes or as motion supporters, movement classification is made with the signals received from the muscles, and the limbs are moved with various actuation systems. In upper limb prosthesis applications, the intended motion of the patient is determined by the muscle contraction signals received from the forearm or upper arm, and motors connected to the prosthetic fingers are moved accordingly. In order to achieve high success in these applications, it is necessary to determine the muscles and activity levels in each finger movement separately. After finding the critical features in the raw EMG signal, finger movements can be detected from muscle contraction levels with various classification algorithms [1].

There is a bundle of studies in the literature on movement classification from muscle contraction levels. These studies differ in terms of the target limb, the number of channels and the classification method preferred. While the number of channels is one of the most important parameters affecting classification accuracy, it creates a negative effect in terms of cost and complexity [2]. In the study conducted by Caesarendra et al. [3], the movement of the five fingers was tried to be estimated using the adaptive neuro-fuzzy input system method over an 8-channel EMG. In the system where the general classification accuracy is 72%, still, the accuracy of the thumb movement has a lower value of 20%. In the study conducted by Lee et al. [4], the signals received from the three-channel EMG device and nine different hand movements were classified with ANN-based classifiers and the results were compared. In the study conducted with 10 different subjects, accuracy values ranging from 54.4% to 67.5% were obtained. Bhattachargee et al. [5] used the dataset containing EMG data of 10 different hand movements to classify movements with the Gradient Boosting method. In this study, where an accuracy value of 98.5% was obtained, no experiment was done with real subjects. In the study published by Tuncer et al. [6], using the EMG dataset containing 15 different hand movements, classification was made with the multi-centred binary pattern method and an accuracy value of 99% was obtained. However, when the results are analysed in detail, it is seen that lower accuracy values are obtained in real-time motion classification applications with subjects.

In this study, activity levels of forearm muscles were determined during five different finger movements. Using the Gaussian Mixture Model developed by Tsuji et al. [7], the signals received from the 8-channel EMG device and finger movements were classified. It was tested with three participants and an average accuracy of 73.3% was obtained.

The paper is organized as follows. Section 2 contains a brief discussion of the EMG device, signal pre-processing steps and details of the Log-Linearized

Gaussian Mixture Network (LLGMN) model. Section 3 presents the experimental results along with relevant discussion. Lastly, Sect. 4 contains concluding remarks and some recommendations regarding future developments.

2 Material and Method

In this study, the finger movements are detected using muscle contraction signals from the forearm. LLGMN model [7] is used for the movement classification process. MYO Armband device was used to measure muscle contraction levels (Fig. 1).

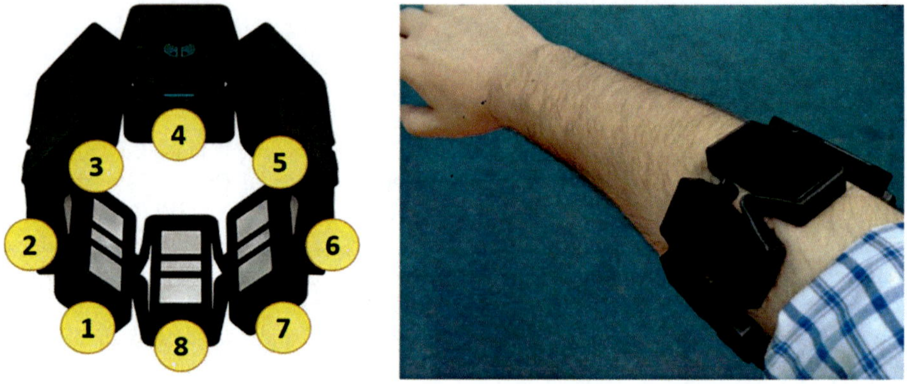

Fig. 1. MYO Armband channels and placement on the arm

Raw signals were received through the 8-channel electrodes on the MYO Armband and subjected to various pre-processing steps. In the first step, the raw EMG signals from the 8 electrodes were amplified, rectified and filtered, respectively. Since the amplitude of the raw EMG signals was very low; first, the 20 dB amplification was performed. Then, the values in negative alternance were converted into positive by rectification. Finally, filtering was done to eliminate unwanted noise in the EMG signal. For this, a 2nd order low-pass Butterworth filter was used. The corner frequency of the filter was chosen as 3 Hz. These filtered signals were sampled. The sampled signals were identified as $EMG_i(t)(i = 1, 2, 3...8)$. The $EMG_i(t)$ parameter was normalized so that the sum of the signals from the 8 channel electrodes was 1. The normalized EMG signal was defined in Eq. 1.

$$EMG_i'(t) = \frac{EMG_i(t) - EMG_i^{rest}}{\sum_{i=1}^{L}(EMG_i(t) - EMG_i^{rest})} \tag{1}$$

Here, $EMG_i'(t)$ represents normalised EMG signals, EMG_i^{rest} represents the mean value of $EMG_i(t)$ in the rest position of the relevant limb. The detailed diagram for the processing of EMG signals was given in Fig. 2.

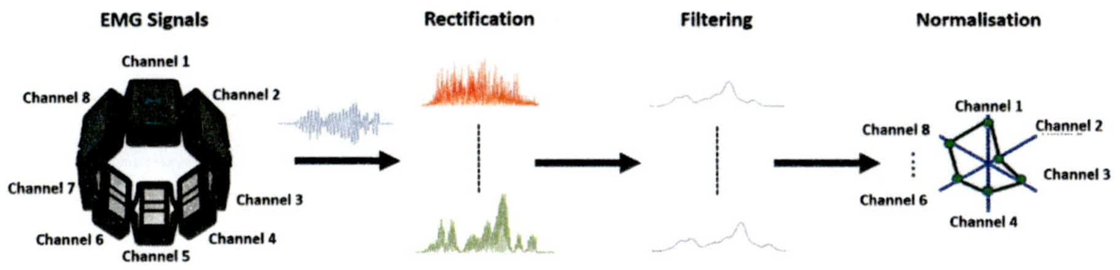

Fig. 2. EMG signal processing steps

In order to perform the movement classification, the muscular contraction level (MCL) must be calculated using the processed EMG signals. The equation used for the MCL calculation is given in Eq. 2.

$$MCL(t) = \frac{1}{2}\sum_{n=1}^{N}\left(\frac{EMG_n(t) - EMG_n^{rest}}{EMG_n^{max} - EMG_n^{rest}}\right) \tag{2}$$

Here, EMG_n^{rest} and EMG_n^{max} represent the muscular contraction level at rest and at maximum contraction, respectively. n is the number of channels (n = 8). For rest and maximum contraction, a 30 kg adjustable hand grip strengthener was used. Pictures taken during maximum contraction and rest are shown in Fig. 3.

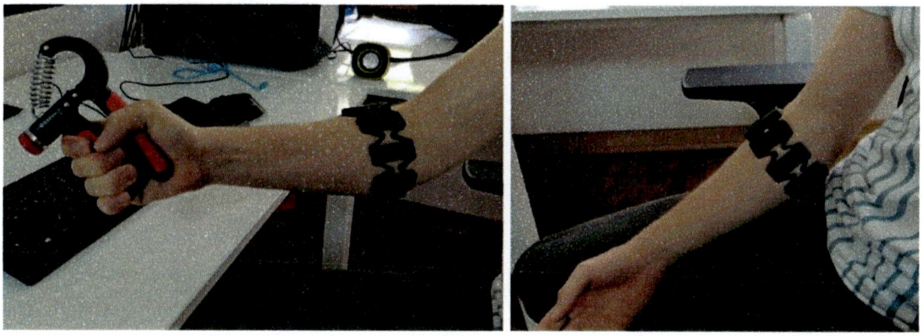

Fig. 3. EMG measurement at maximum contraction and rest position

After signal processing, normalization and MCL calculation, motion classification was performed. Motion classification is the detection of human limb motion using EMG signals. After the signal processing stages, the data was given to LLGMN, a statistical artificial neural network model, where motion classification was performed. With the movement information obtained at the output of the network, it was decided which finger to move. The structure of the LLGMN network is demonstrated in Fig. 4.

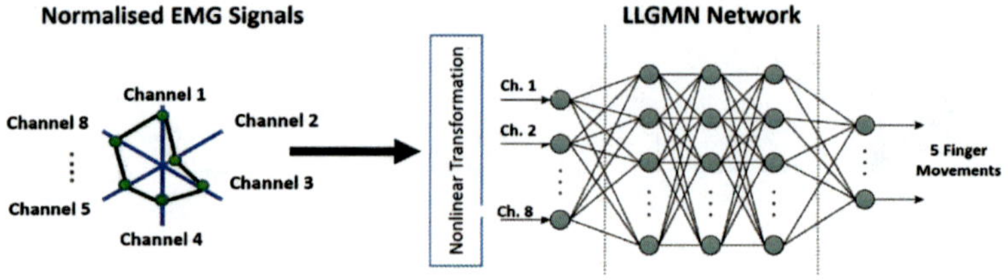

Fig. 4. LLGMN network structure

3 Results and Discussion

EMG signals obtained via the defined system were analysed for the movement of each finger, and which channels were more active in which movement, and movements that could affect each other and negatively affect the result of movement classification were determined. Data collection was done with 3 healthy subjects with the permission of the Istanbul University-Cerrahpasa Ethical Committee (ID: E-83045809). The exclusion criteria were having history of upper limb fracture, upper limb nerve injury, upper limb peripheral neurophaty, diabetes mellitus, hypo or hyperthyroidism, cervical radiculopathy, rheumatologic diseases, and kidney or liver failure. The personal information of the subjects is given in Table 1.

Table 1. Personal information of the subjects

	Subject A	Subject B	Subject C
Gender	Male	Male	Male
Age (year)	34	26	21
Height (cm)	175	170	180
Weight (kg)	78	72	69
Dominant Hand	Right	Right	Right

The 8-channel EMG signals from Subject A for each finger movement are given in Figs. 5, 6, 7, 8 and 9.

When the figures were examined, it was seen that channels 3, 4 and 8 were dominant in thumb movement. Also, channels 3, 4 and 8 are active in the index finger, middle finger and ring finger movements, and channel 7 was also active in addition to these. In the ring finger movement, channels 7 and 8 seemed more dominant than the other fingers. In addition to channels 3, 4 and 8, channels 1, 2 and 5 were also dominant in the little finger movement. Root mean square-RMS values of EMG signals from three subjects were calculated and averaged. Thus, the activity levels of 8-channels were expressed numerically in each finger movement. The results are in Table 2.

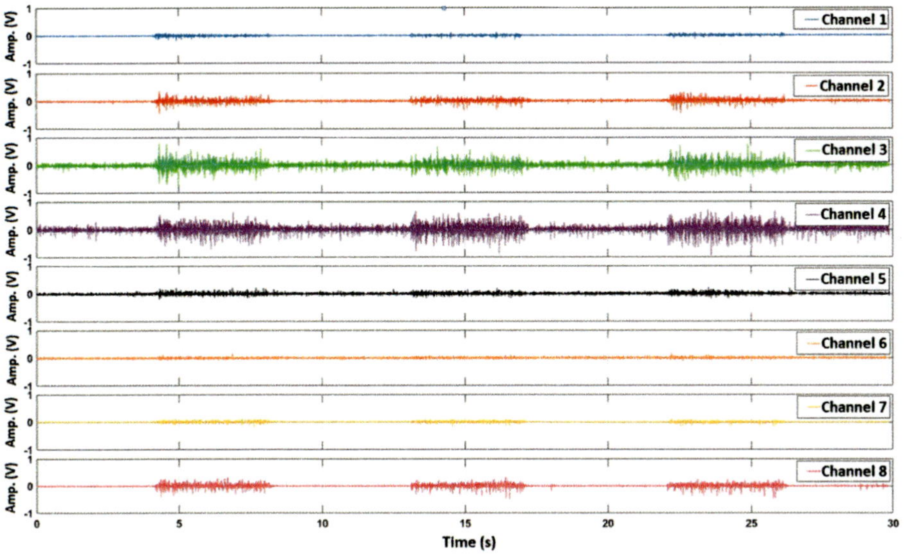

Fig. 5. Muscle contraction levels during thumb flexion and extension movement

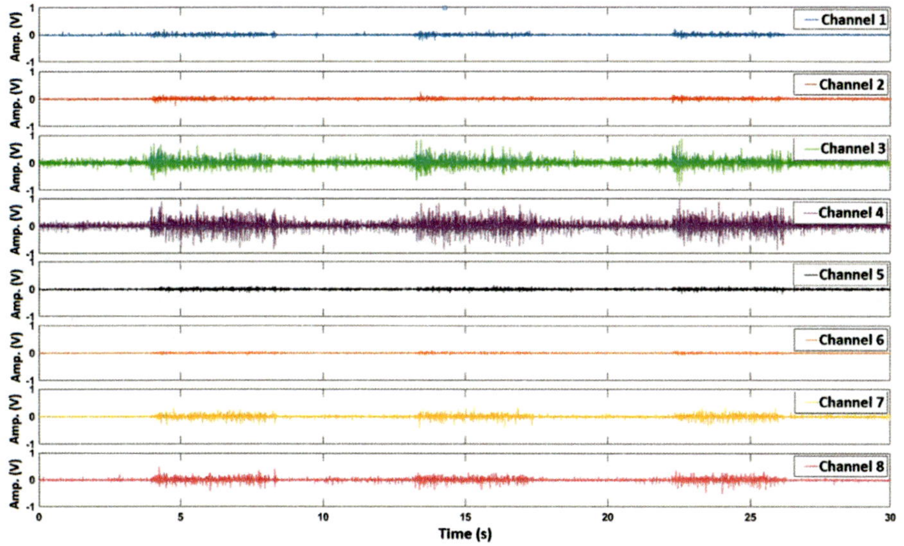

Fig. 6. Muscle contraction levels during index finger flexion and extension movement

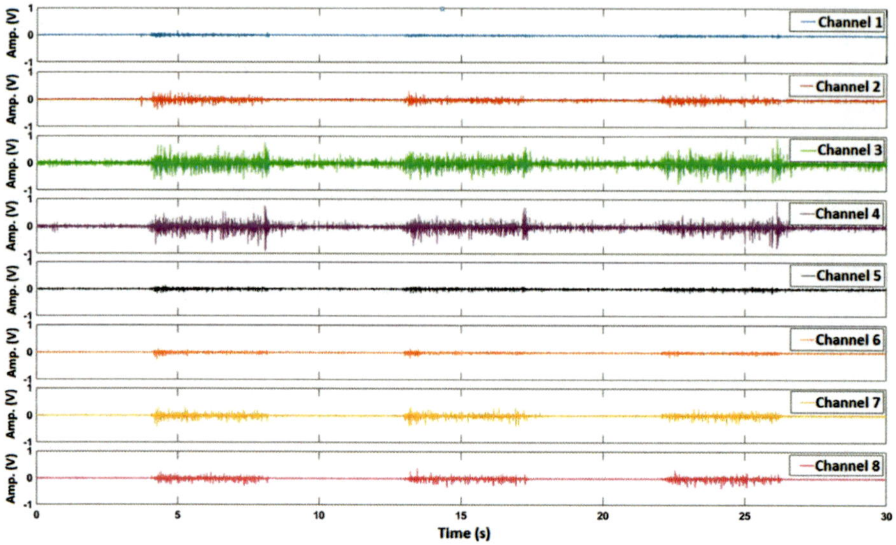

Fig. 7. Muscle contraction levels during middle finger flexion and extension movement

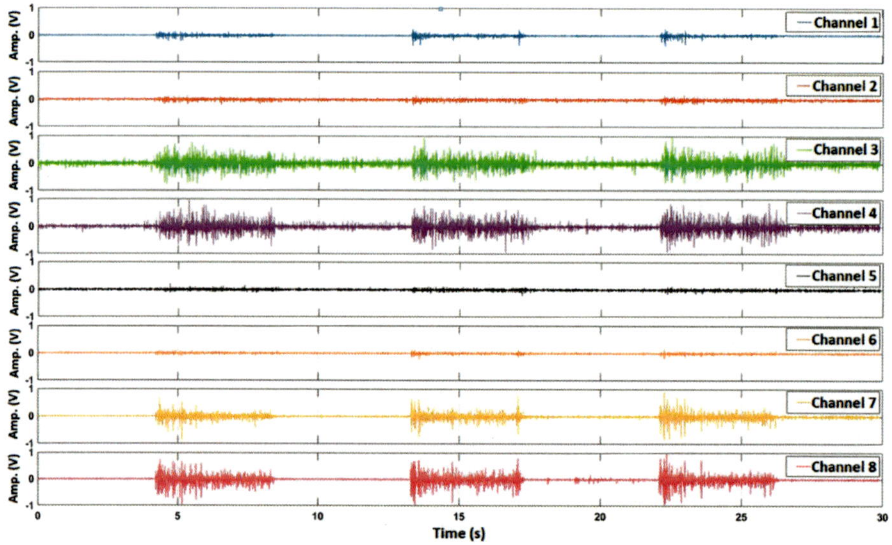

Fig. 8. Muscle contraction levels during ring finger flexion and extension movement

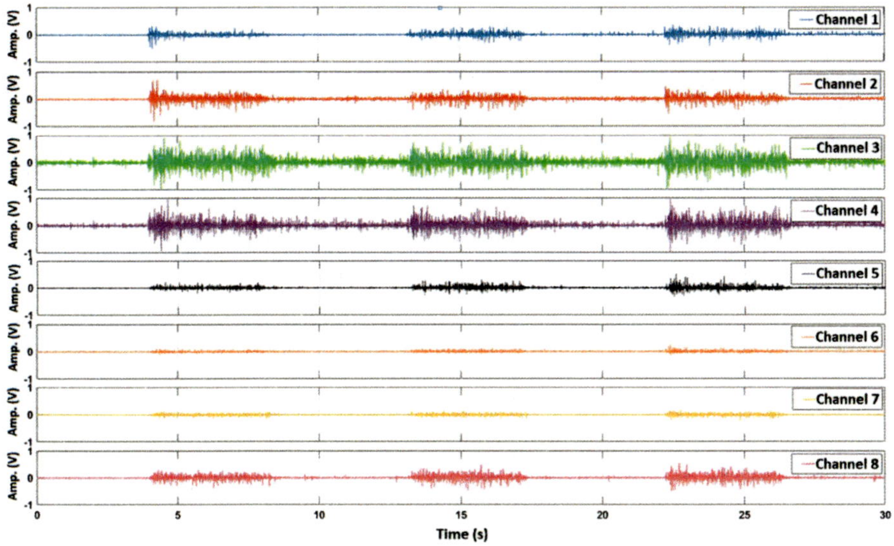

Fig. 9. Muscle contraction levels during little finger flexion and extension movement

Table 2. RMS values of EMG signals according to finger movements

	Thumb	Index Finger	Middle Finger	Ring Finger	Little Finger
Channel 1	0,0222	0,0332	0,0178	0,0329	0,0528
Channel 2	0,0546	0,0338	0,0494	0,0352	0,0816
Channel 3	0,1235	0,1337	0,1403	0,1557	0,1724
Channel 4	0,1624	0,1833	0,1186	0,1603	0,1432
Channel 5	0,0476	0,0327	0,0310	0,0333	0,0517
Channel 6	0,0284	0,0216	0,0259	0,0245	0,0251
Channel 7	0,0254	0,0588	0,0534	0,0928	0,0308
Channel 8	0,0579	0,0654	0,0457	0,1298	0,0761

In order to determine the movement classification performance of the developed system, subjects were asked to perform some targeted experiments. In the test process, the subjects were informed about the project and the operation of the system was explained. The MYO Armband was placed on the right forearm. Each subject was asked to make 20 independent finger movements randomly. It was ensured that each finger movement was applied for 3 s, waiting for 10 s between movements. The timing was performed by the subjects themselves. During this test, the output of the LLGMN model for each movement of the subject was noted. It was planned to record the movements of the subject with the flex sensor, but it was not applied since it was thought that attaching any element (sensor, glove, etc.) to the subject's hand could affect the movement and muscle contractions [8,9], and it was recorded by observation. Accordingly, the movement classification results obtained from the three subjects are given in Table 3. The results shown in red in the table indicate the movements that the system detected incorrectly.

Table 3. Movement classification test resultsm

	Subject A		**Subject B**		**Subject C**	
	Movement	**Output**	**Movement**	**Output**	**Movement**	**Output**
1	Ring	Ring	Thumb	Thumb	Thumb	Thumb
2	Index	Middle	Little	Little	Middle	Middle
3	Thumb	Thumb	Middle	Index	Little	Little
4	Thumb	Index	Index	Index	Thumb	Thumb
5	Middle	Middle	Thumb	Thumb	Index	Middle
6	Little	Little	Ring	Ring	Ring	Ring
7	Ring	Ring	Index	Middle	Middle	Middle
8	Index	Middle	Little	Little	Little	Little
9	Thumb	Thumb	Thumb	Thumb	Thumb	Middle
10	Little	Little	Middle	Middle	Middle	Middle
11	Little	Index	Ring	Ring	Little	Little
12	Thumb	Thumb	Thumb	Thumb	Thumb	Thumb
13	Index	Index	Little	Little	Index	Index
14	Middle	Middle	Index	Index	Middle	Middle
15	Ring	Little	Thumb	Thumb	Thumb	Index
16	Little	Little	Ring	Ring	Ring	Ring
17	Middle	Middle	Little	Ring	Little	Little
18	Thumb	Middle	Index	Index	Middle	Ring
19	Index	Middle	Middle	Middle	Ring	Ring
20	Little	Little	Thumb	Middle	Index	Middle

When the movement classification results in Table 3 were examined, it revealed 13 correct, 7 wrong results for Subject A; 16 correct, 4 wrong results for Subject B; 15 correct and 5 wrong results for Subject C. When the test performance checked according to the movements for all subjects, 11 correct, 5 incorrect results for thumb; 5 correct, 6 incorrect results for index finger; 9 correct, 3 incorrect results for middle finger; 8 correct, 1 incorrect results for ring finger and 11 correct, 1 incorrect results for little finger has been obtained. When the results were examined according to the movements, it was seen that the most incorrect output was in the index finger. The most incorrect results for the index finger were given inappropriately as the middle finger. The reason for this can be shown that the dominant channel numbers in Table 2 were similar for the index and middle fingers. The most successful results were obtained for the little finger. Again, when Table 2 was examined, it was seen that the most distinctive movement according to EMG channels was in the little finger. The same was true for the ring finger. Accordingly, the similarity of dominant signals in the EMG channels affected the results of the LLGMN model. Similar movements could be confused with each other. As a solution, the number of channels should be increased, and different muscles should be evaluated. When the overall performance of the system regarding motion classification was calculated, it was seen that an accuracy rate of 73.3% was obtained.

4 Conclusion

In this study, EMG signals from the forearm were examined according to different finger movements, and the dominant muscle groups in each finger movement were determined and compared. By measuring muscular contraction levels, movement classification of the five fingers of the hand was made. According to the results obtained in the experiments with three subjects, it was seen that the system achieved an accuracy rate of 73.3% in the relevant classification. A high accuracy rate could not be obtained in the classification of index finger movements due to the nonlinearity of the human musculature, EMG measurements being made from the forearm, and the inability to differentiate the muscle groups responsible for the index and middle finger movements by the current system. In the next study, measurements can be taken from points close to the hand by increasing the number of EMG channels in order to increase the accuracy of movement classification of the index finger. The classifier results have a big potential to be transferred to an exoskeleton mechanism and used for therapeutic purposes at the clinical site.

Acknowledgement. This work has been supported by the Scientific Research Projects Coordination Unit of Bartin University (Project Number: 2019-FEN-A-014).

References

1. Phinyomark, A., Phukpattaranont, P., Limsakul, C.: Feature reduction and selection for EMG signal classification. Expert Syst. Appl. **39**, 7420–7431 (2012). https://doi.org/10.1016/j.eswa.2012.01.102
2. Li, G., Schultz, A.E., Kuiken, T.A.: Quantifying pattern recognition- based myoelectric control of multifunctional transradial prostheses. IEEE Trans. Neural Syst. Rehabil. Eng. **18**, 185–192 (2010). https://doi.org/10.1109/TNSRE.2009.2039619
3. Caesarendra, W., Tjahjowidodo, T., Nico, Y., Wahyudati, S., Nurhasanah, L.: EMG finger movement classification based on ANFIS. J. Phys.: Conf. Ser. **1007** (2018). https://doi.org/10.1088/1742-6596/1007/1/012005
4. Lee, K.H., Min, J.Y., Byun, S.: Electromyogram-based classification of hand and finger gestures using artificial neural networks. Sens. (Basel) **22**, 225 (2021). https://doi.org/10.3390/s22010225

5. Bhattachargee, C.K., Sikder, N., Hasan, M.T, Nahid, A.A.: Finger movement classification based on statistical and frequency features extracted from surface EMG signals. In: International Conference on Computer, Communication, Chemical, Materials and Electronic Engineering (IC4ME2), Rajshahi, pp. 1–4 (2019). https://doi.org/10.1109/IC4ME247184.2019.9036671
6. Tuncer, T., Dogan, S., Subasi, A.: Novel finger movement classification method based on multi-centered binary pattern using surface electromyogram signals. Biomed. Signal Process. Control **71**, 103153 (2022). https://doi.org/10.1016/j.bspc.2021.103153
7. Tsuji, T., Fukuda, O., Ichinobe, H., Kaneko, M.: A log-linearized Gaussian mixture network and its application to EEG pattern classification. IEEE Trans. Syst. Man Cybern.-Part C: Appl. Rev. **29**, 60–72 (1999). https://doi.org/10.1109/5326.740670
8. Riddle, M., MacDermid, J., Robinson, S., Szekeres, M., Ferreira, L., Lalone, E.: Evaluation of individual finger forces during activities of daily living in healthy individuals and those with hand arthritis. J. Hand Ther. **33**, 188–197 (2020). https://doi.org/10.1016/j.jht.2020.04.002
9. Woods, S., et al.: Effects of wearing of metacarpal gloves on hand dexterity, function, and perceived comfort: a pilot study. Appl. Ergon. **97**, 103–119 (2021). https://doi.org/10.1016/j.apergo.2021.103538

Calculation of Efficiency Rate of Lean Manufacturing Techniques in a Casting Factory with Fuzzy Logic Approach

Zeynep Coskun[1(✉)], Adnan Aktepe[2], Süleyman Ersöz[2], Ayşe Gül Mangan[1], and Uğur Kuruoğlu[1]

[1] Akdaş Döküm A.Ş, Ankara, Turkey
zeynepcoskun.98@hotmail.com, {amangan,ugurk}@akdas.com.tr
[2] Industrial Engineering, Kırıkkale University, Kırıkkale, Turkey
sersoz@kku.edu.tr

Abstract. Sectoral growth is increasing day by day and the competition market is growing with it. At the same time, customer awareness is also increasing. As customer awareness increases, the quality of service provided should also increase. One of the ways that companies will apply in order to maintain their existence in this competitive environment and to prevent customer loss is to make the lean manufacturing philosophy a corporate culture. It is a production approach that does not contain any unnecessary elements in the lean manufacturing structure, minimizes waste and aims to increase efficiency in production. When moving to the lean manufacturing philosophy, it is of great importance for companies to draw a correct road map. This study was applied to the product/product group produced in a foundry. Value stream mapping (VSM), which is one of the lean manufacturing techniques for the determined product/product group, was made and the current situation value stream map was created. With value stream mapping, bottlenecks and losses in the process were determined and a future situation value stream map was created. Lean manufacturing techniques were applied at these determined points, problems were eliminated and productivity increase was achieved in production. Fuzzy logic was used to clearly determine the productivity increase. Fuzzy logic creates numerical models by imitating the human mind many vague, non-numerically expressed terms that we use daily. With fuzzy logic, the efficiency rate was modeled numerically and the contribution of the lean manufacturing techniques applied to productivity was determined.

Keywords: Lean Manufacturing · Efficiency · Artificial Intelligence · Fuzzy Logic

1 Introduction

In today's conditions, the competitive environment is increasing. Every day, a new company is included in the market and customer demands are increasing rapidly. At the same time, companies are struggling with increasing cost and quality defects. One of the ways companies will resort to in order to survive in this competitive environment, to make profits and to increase their efficiency is to make lean manufacturing a corporate culture.

Z. Şen et al. (Eds.): IMSS 2023, LNME, pp. 247–258, 2024.
https://doi.org/10.1007/978-981-99-6062-0_23

Lean manufacturing is developed to meet business processes and increasing customer demands. The methods used in lean manufacturing aim to improve product quality and reduce costs by eliminating all waste in the production system [1]. The main purpose in the philosophy of lean manufacturing is to focus on value stream processes. Only in this way can wastes be eliminated and plans can be developed to eliminate these wastes [2].

There are eight basic wastes (muda) in lean manufacturing: defects, overproduction, inventory, waiting, transportation, motion, extra-processing and unused talent. These types of waste occur in many areas in the production processes. By eliminating these wastes, an increase in productivity can be achieved in production.

Considering that the concepts of efficiency, quality and competition have gained more importance recently, the promotion of lean manufacturing practices in both service and manufacturing enterprises contributes to a significant gain in industrial zones and therefore in the field of production. With the increase in efficiency, an increase is achieved in customer satisfaction, product quality and production processes.

Due to the requirements of the era, the understanding of industry 4.0 in production is increasing. Industry 4.0, known as the fourth industrial revolution, is the industrial revolution that includes automation systems, data exchanges, and production technologies. Industry 4.0 represents progress in three points [3]:

- Digitization of production: information systems for management and production planning,
- Automation: Systems using machines and data collection from production lines,
- Automatic data exchange: linking production facilities in a comprehensive supply chain

With Industry 4.0, an increase in productivity is foreseen in production. One of the important developments brought by Industry 4.0 is artificial intelligence technology. Thanks to artificial intelligence, it processes information quickly, contributing to the acceleration of decision-making processes and drawing a roadmap for future studies.

In this study; Lean manufacturing concept, value stream mapping method and fuzzy logic are explained. The application part of this study was done in Akdaş Döküm A.Ş. Losses in production were determined by value stream mapping. These losses were minimized by applying lean production methods and increased productivity was achieved. Fuzzy logic approach was used to determine the rate of increase in productivity.

2 Literature Research

In this study, a literature review on lean manufacturing practices and artificial intelligence integration was conducted. In particular, studies using fuzzy logic approach were examined. There are various studies on lean manufacturing and artificial intelligence studies in the literature.

Lean manufacturing is defined as a production system that does not contain any unnecessary elements in its structure and that factors such as error, stock, workmanship, development processes, production area, waste, cost, customer dissatisfaction are minimized. The lean manufacturing method basically aims to eliminate or reduce the waste that is frequently experienced in businesses. Various techniques and methods have been developed for this [4].

Value stream mapping-VSM: For this, first of all, process analyses are made and the processes are standardized. While performing these analyses, activities that do not create value and are defined as waste are identified [5, 6].

Total Productive Maintenance-TPM: With this approach, all employees are encouraged to continuously make small improvements and preventive maintenance and be involved in the system. With this approach, it aims at zero failure and minimum production loss [7]. The aim is not to fix the error, but to prevent the error.

POKA-YOKE: It is known as error caused by carelessness. The main goal is to ensure product quality. The flow of the faulty product in the system should be stopped and should not proceed to the next process. For this, each employee should be trained to check and detect potential defects/errors.

Single Minute Exchange of Dies-SMED: In order to reduce waiting times, it is tried to shorten the setup-set times. It is the elimination of non-value-added activities by reducing set-up times [8].

5S: It is a workplace organization. Every employee is included and the work environment is organized. In this way, providing an efficient and productive working environment. In addition, the business environment and business leaks are made more visible in this way [9]. The 5S, which starts with the letters of the week, includes five steps: sorting, organizing, cleaning, standardizing and discipline.

KAIZEN: In Japanese, “Kai” means change, “Zen” means good, and kaizen means continuous improvement. It is a technique applied in every field. It aims to eliminate activities that do not create added value for the customer, to continuously improve production processes and to eliminate waste [10].

In this section, a literature review was conducted on the studies made with lean manufacturing and artificial intelligence integration.

[11] developed a fuzzy logic advisory system to be used for decision support purposes in lean manufacturing applications. The aim of the study was to assist SME practitioners in estimating the impact of the lean manufacturing application during the implementation phase. Data were collected from 10 SMEs. Three input variables and eight rules are defined to the fuzzy logic system. With the developed fuzzy logic system, businesses can predict the possible relative cost of implementing lean manufacturing, can predict the return on investment of lean manufacturing practices, can also be used as a standard business tool to evaluate the performance status of the business, and the need area can be determined by analyzing factors such as resource availability. [12] presents a new fuzzy logic-based metric tool to measure lean storage performance, facilitating goal setting, performance monitoring, and lean storage implementation. This modelled integrates seven indicators from the intersection of seven key lean management variables, eight lean waste, and four key warehousing activities. It helps to reconcile the storage of improvement goals and key activities based on lean principles with the current situation and the goals and policies of the organization. The suggested scorecard also allows to measure the improvements in development of a lean project. [13], on the other hand, shows the structure of the decision-supported tool developed to optimize the procurement and shipment processes. He emphasized the importance of good management of the stock level in a hospital and its effect on the workflow process. It will allow the main performance criteria such as quality, cost, delivery time to be combined with social,

societal and environmental aspects to improve logistics performance. [14] report a study conducted to evaluate the level of leanness in their study. A lean measurement model was designed and the lean index was calculated. A computerized decision support system has been developed and this decision support system has been determined as a decision support system for fuzzy logic-based simplicity evaluation. The fuzzy simplicity index has been a guide in identifying weak areas that need improvement. [15] on the other hand, aimed to deal with the multidimensional concept, unavailability criterion and uncertainty arising from subjective and uncertain human reasoning for the measurement of the degree of weakness in the method proposed in their study. The reason for using fuzzy logic is that it is associated with the uncertainty of human judgment regarding the degree of implementation of lean implementations, the length of lean implementation and an additional assessment regarding the use of multiple evaluators. Lean implementation grade is scored using the value stream method. To test the proposed method, a survey was conducted in the manufacturing industry and the results of the survey were presented. [16] firstly determined the types of information and created database management for SMED. Then, the SMED intelligent decision support system was established and rules were created for the textile industry. By applying SMED techniques in a company's production line, the production system has been made efficient and some of the determined rules have been integrated into the company with an intelligent decision support system. The effects of the intelligent decision support system on SMED and the situations that may occur in the future are emphasized.

3 Method

Lean manufacturing is a process that started when two engineers, Taiichi Ohno and Eiji Toyoda, who worked for Toyota company in the 1950s, went to America to visit Ford company. Ford, which was a pioneer in the sector at that time, adopts the concept of mass-type production. As a result of their observations, they came to the conclusion that mass-type production includes too much waste in production. Deciding that the mass production approach is not a suitable production concept for Japan, they returned to Japan and laid the foundations of lean manufacturing at Toyota.

3.1 Value Stream Mapping

Value stream mapping is one of the lean manufacturing techniques. "Value stream mapping" is the set of value-added and non-value-added activities essential to each product and needed to create a product along the main streams. The production flow from raw material to the customer and the product development process can be defined as the basic flows applicable to each product [17]. The value stream map shows both information and material flows. In this way, it identifies waste and opportunities to improve value [18]. The value stream mapping process starts with the order request from the customer and continues until the product is shipped to the customer. Value stream mapping is not focused on a single process. It provides an overall view of the manufacturing process from start to finish. In the value stream mapping method, two maps are prepared, namely the current state and the future state. It is created with the help of a standardized set of shapes.

The baseline value stream is created using the collected data when creating the map. It shows the bottlenecks, stocks, waste, cycle time, supply time and information flow that occur in the process.

Future State Value Stream Map, on the other hand, reveals undetected inefficiencies in a value properties, builds to create a VSM. Once hidden inefficiencies in the process have been identified, a future state map can be presented showing predictions for how the system will be improved.

3.2 Artificial Intelligence and Fuzzy Logic

Every person has a certain intelligence and intelligence has the ability to improve itself by learning, training and experiences over time. It has the ability to be emulated by intelligence, software or onboard chips. In this case, intelligence is called 'Artificial Intelligence' [19]. It has many sub-branches such as Artificial Intelligence, Expert Systems, Fuzzy Logic, Genetic Algorithm and Artificial Neural Networks, and it has found a wide research and application area, especially in recent years [19]. Fuzzy means unclear or unclear. Many terms we use in daily life have a fuzzy structure. For example; Terms such as old-young, long-short, hot-cold, fast-slow contain blur. Fuzzy logic can correspond to any value in the range 0–1. The fuzzy logic approach gains the power of machines to process the private data of people and to work by benefiting from their experiences and intuitions. Contrary to classical logic, fuzzy logic is not two-level, but very advanced operations. While gaining this power, it uses verbal expressions instead of numerical expressions [19].

4 Experiments

The application started with the determination of the product/product group. Then, current state and future state maps were created. The efficiency ratio was determined with the MATLAB Fuzzy application by applying the lean manufacturing techniques suggested in the future state map.

4.1 Identification of the Product/Product Group

The application was made in Akdaş Döküm A.Ş. Akdaş Döküm A.Ş produces according to the order. For this reason, while choosing the part, the criterion of being frequently produced before and having a high order rate afterwards was taken into consideration. The reason for this is to collect clearer data and to make comparisons before and after the application.

Selected track: It is a part that acts as a moving jaw in jaw crusher machines. Jaw crushers are a compression type crusher machine consisting of a fixed jaw and a movable jaw placed inside. With the help of the movable jaw, the material is compressed towards the fixed jaw, crushed and brought to the desired size. It is mostly used in the cement and mining industry.

4.2 Creation of the Current Situation Map

While creating the current situation map, the shading method was used. The shading method is the method of observing the production during the production process and noting the operations, postures, losses, wastes and times. The process from the production of the part to its shipment was monitored with the shading method. Since the operation times are long, hours are written on the map and activities that do not create value are indicated in red.

The production plan is prepared after the order is received from the customer. Production plan is made monthly and weekly. Raw materials are supplied in bulk, not on a piece basis. The firm has a pull type production system. In the pull type production, the next process is realized by demanding the parts needed from the previous process at the right time and in the required amount [20]. In pull type production, production is made according to demand, there is a zero-stock target, information flow between processes is slow, quality control is done at every stage, timing and quality are key performance criteria, and machine setup times are generally short.

Value add and non-value add processes were examined. The effect of these processes on production was calculated.

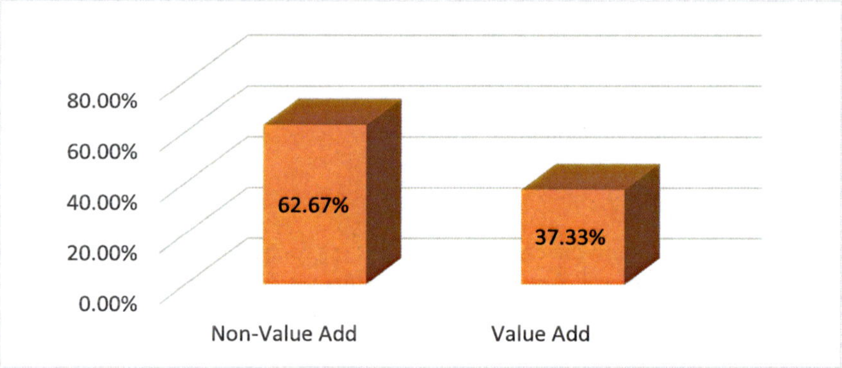

Fig. 1. Percentage of Value Added and Non-Value-Added Activities

The current situation map created is as follows (Fig. 2);

The cycle time in Fig. 1 refers to the time it takes to complete a job. The number of operators represents the number of employees doing that job.

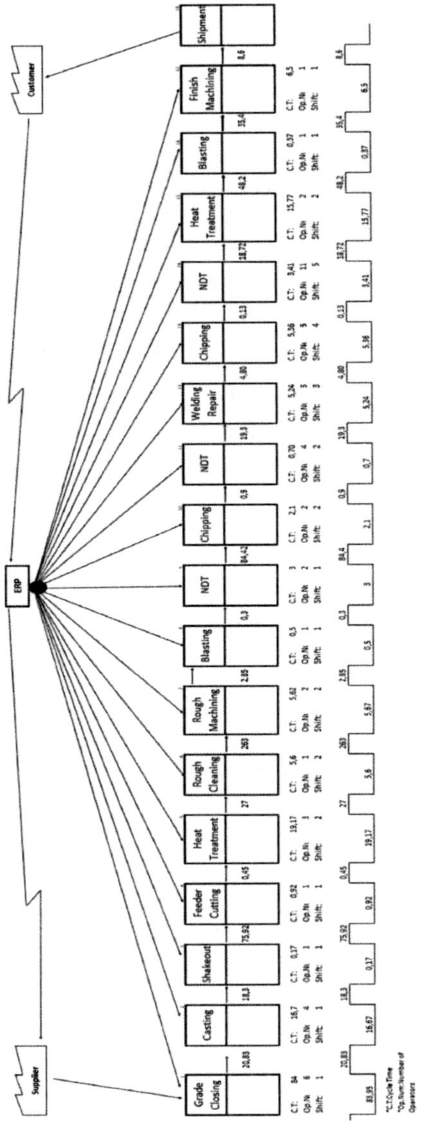

Fig. 2. Current State Value Stream Map

4.3 Creating a Future State Map

After the current situation map was created, the problems that emerged were determined and the future situation map was created by making suggestions for improvement in the places where these problems were experienced. Since the production times are long in the processes, the times are calculated in hours and the operations that do non-value add are shown in red.

The future state map created is as follows (Fig. 3);

4.4 Application of Lean Manufacturing Techniques

Three improvement suggestions were made in the studies carried out. First, it was determined that some operations' data entries were not made to the ERP system. For this problem, the planning unit and the ERP consultancy company carried out a joint study. Missing operations were defined in the ERP system and opinions were exchanged with the personnel working in these operations. Information was given about the ERP entry, and in the new case, the data entries of these operations began to be made to the ERP.

It was observed that there was a lot of irregularity in welding and grinding operations and as a result of this irregularity, the working personnel moved too much to search for materials-equipment. 5S studies were carried out to ensure orderliness and cleanliness in these areas. The work area was cleaned. The work was disciplined by performing weekly 5S field inspections.

There was a lot of transportation between the part workstations and it was determined that unnecessary waiting came to the fore. Kobetsu kaizen work was done in these areas. The target has been determined as meeting the transportation and waiting times. Especially since there is transportation between the grinding and welding stations, the locations of these two workstations were changed and the transportation distance was reduced. The order of processing parts is done randomly. Here, FIFO (First in First Out) application was made. FIFO is a method often used in material flow and inventory management. The first product to arrive is the first product to leave the process. Here, the first piece that comes in with FIFO is processed first and waiting times are reduced.

4.5 MATLAB Application

In this study, a comparison of the current situation and the future situation was made based on 5 of the 8 basic mudas. Fuzzy logic was used in this study, since the data in operations such as blasting, rough cleaning, rough machining, finish machining are not clear and are given approximately. Fuzzy logic problems can be solved with MATLAB application. MATLAB is a programming language and many mathematical operations can be easily performed. It uses a graphical interface. The combination of fuzzy logic and MATLAB application provides an effective solution method in solving problems involving uncertainty.

A muda evaluation form was created for the current situation and the future situation. The evaluations were made by the factory manager and the unit supervisor (Table 1).

5 mudas were defined as inputs and efficiency was obtained as output (Fig. 4).

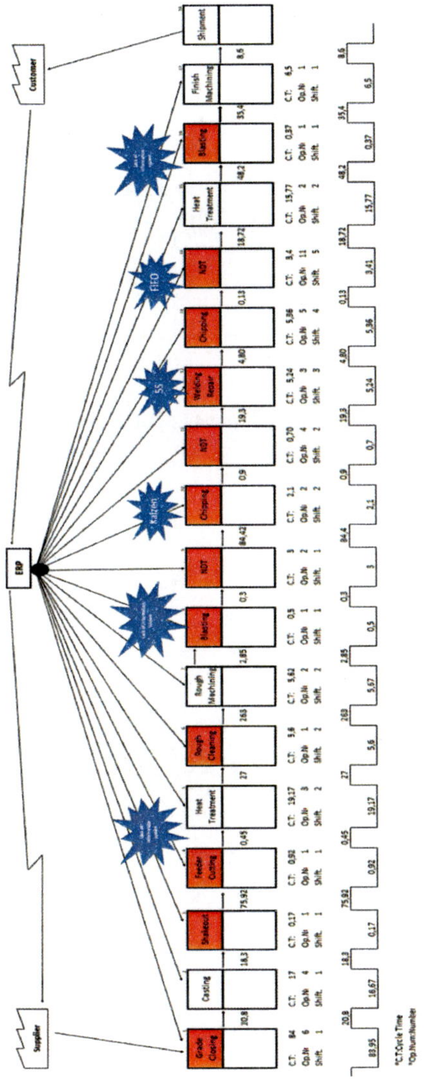

Fig. 3. Future State Value Stream Map

Table 1. Muda Evaluation Form

	The current situation	The future status
1. Waiting	Good	Very Good
2. Defects	Good	Good
3. Motion	Good	Very Good
4. Transportation	Average	Good
5. Unused talent	Good	Very Good

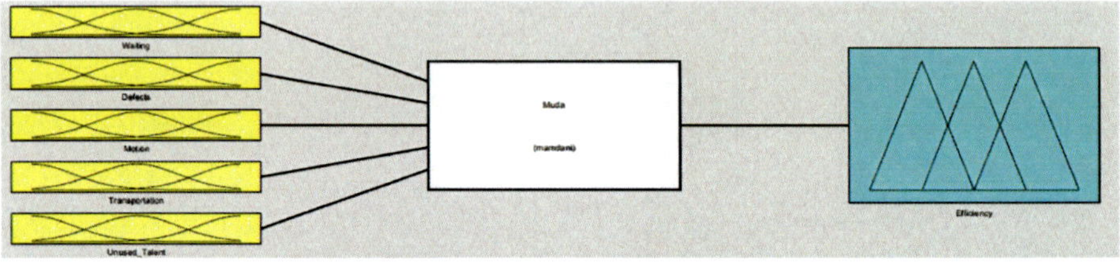

Fig. 4. MATLAB Fuzzy Input-Output Variables

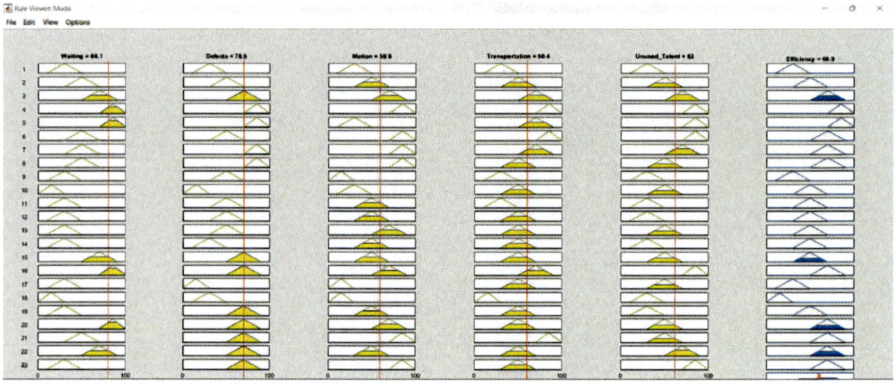

Fig. 5. Current Situation Efficiency Determination

When the current status information was entered, it was determined that the efficiency was 60.9% (Fig. 5).

The techniques and suggestions found for the future situation were applied. As a result, the new situation was evaluated and the data were entered. In the new case, it was observed that the efficiency increased to 71.9% (Fig. 6).

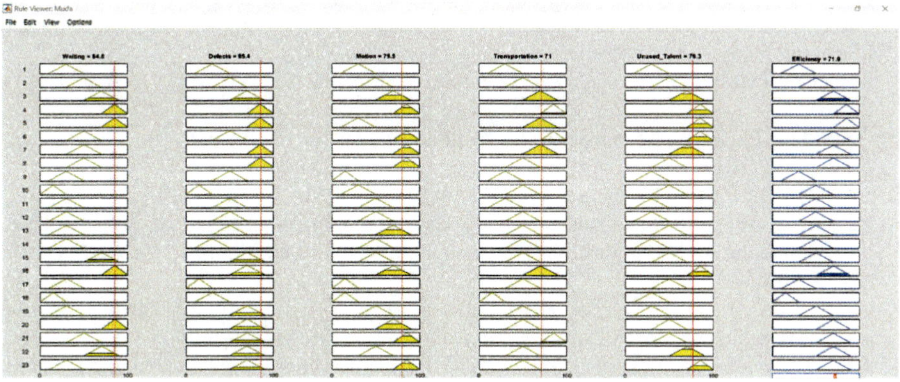

Fig. 6. Future Situation Efficiency Determination

5 Conclusion

Lean manufacturing techniques and fuzzy logic were used in this study. It started to work with value stream mapping method, which is one of the lean manufacturing techniques. As part of the value stream mapping study, a current state and future state map was created. While this mapping was being done, work was followed by the shading method. With the value stream mapping study, the wastes, losses, value-creating and non-value-creating activities in the process were identified. Five of the eight basic mudas were based: waiting, defects, motion, transportation and unused talent. Suggestions for improvement were made to increase the efficiency of the process. The suggestions made were put into practice in the production area. The five basic mudas based on the application were evaluated by the factory manager and the unit supervisor. Fuzzy logic was used to calculate the effect of the studies on productivity increase. The main reason for using fuzzy logic was that in some operations the data was entered not clearly but approximately. Fuzzy logic is used in situations that are unclear, ambiguous, or non-linear.

Before and after evaluation data were evaluated with MATLAB Fuzzy. According to the results obtained, it was determined that while the efficiency of lean manufacturing techniques was 60.9% before the application, the efficiency increased to 71.9% after the techniques with improvement suggestions were applied. Here, we have observed that lean manufacturing practices are solutions to the problems we experience in our daily life, production, working life and in many areas, and that they have a serious impact on productivity increase. Data were collected for a product group determined in this study. The same application can be applied by collecting data on other product groups produced in the company. In addition, this study will be the decision maker in the selection of lean manufacturing techniques to be applied in other units throughout the company.

References

1. Sivaslı, E.: A study on using lean techniques in business processes. Master thesis, Institute of Social Sciences, Department of Total Quality Management, Dokuz Eylül University (2006)

2. Maskell, B., Baggaley, B.: Practical Lean Accounting: A Proven System for Measuring and Managing the Lean Enterprise. Productivity Press, New York (2004)
3. Roblek, V., Meško, M., Krapež, A.: A complex view of industry 4.0. Sage Open **6**(2), 1–11 (2016)
4. Öztürk, H., Elevli, B.: Lean manufacturing philosophy in the mining industry. J. Eng. Brains **2**(1), 24–32 (2017)
5. Baggaley, B.: Costing by value stream. J. Cost Manag. **17**(3), 24–30 (2003)
6. Birgün, S., Gülen, K.G., Özkan, K.: Using value stream mapping technique in transition to lean manufacturing: an application in the manufacturing industry. Commer. Univ. J. Sci. **5**(9), 47–59 (2006)
7. Chan, F.T.S., Lau, H.C.W., Ip, R.W.L., Chan, H.K., Kong, S.: Implementation of total productive maintenance: a case study. Int. J. Prod. Econ. **95**, 71–94 (2005)
8. Tanık, M.: Improving die setting times with SMED methodology: a lean six sigma application. J. Muğla Univ. Inst. Soc. Sci. **25**, 117–140 (2010)
9. Michalska, J., Szewieczek, D.: The 5S methodology as a tool for improving the organisation. J. Achieve. Mater. Manuf. Eng. **24**(2), 211–214 (2017)
10. Gürdal, K.: Current approaches in cost management. Politik Kitapevi, Ankara (2007)
11. Achanga, P., Shehab, E., Roy, R., Nelder, G.: A fuzzy-logic advisory system for lean manufacturing. Int. J. Comput. Integr. Manuf. **25**(9), 839–852 (2012)
12. Buonamico, N., Muller, L., Camargo, M.: A new fuzzy logic-based metric to measure lean warehousing performance. Int. J. **18**(2), 96–111 (2017)
13. Jordon, K., Dossou, P.-E., Junior, J.C.: Using lean manufacturing and machine learning for improving medicines procurement and dispatching in a hospital. Flexible Autom. Intell. Manuf. **38**, 1034–1041 (2019)
14. Balaji, S.R., Vinodh, S.: Fuzzy logic based leanness assessment and its decision support system. Int. J. Prod. Res. **49**(13), 4027–4041 (2011)
15. Susilawati, A., Tan, J., Bell, D., Sarwar, M.: Fuzzy Logic based method to measure degree of lean activity in manufacturing industry. J. Manuf. Syst. **34**, 1–11 (2015)
16. Kemalbay, V.: Mold change in single minutes intelligent decision support system and its application in the textile industry. Master thesis, Graduate School of Natural and Applied Sciences, Istanbul Technical University, İstanbul (2012)
17. Rother, M., Shook, J.: Learning to See, Versiyon 1.2. The Lean Enterprise Institute Inc., Brookline (1998)
18. Adalı, M., Kiraz, A., Akyüz, U., Halk, B.: The use of value stream mapping technique in the transition to lean manufacturing: application in a large-scale tractor business. Sakarya Univ. J. Sci. Inst. **21**(2), 242–251 (2016)
19. Elmas, Ç.: Fuzzy Logic Controllers Seçkin Yayıncılık. Ankara (2003)
20. Gökşen, Y.: From traditional manufacturing to flexible manufacturing: a comparative review. Dokuz Eylül Univ. J. Soc. Sci. Inst. **5**(4), 32–48 (2003)

Simulated Annealing for the Traveling Purchaser Problem in Cold Chain Logistics

Ilker Kucukoglu[1] (✉), Dirk Cattrysse[2], and Pieter Vansteenwegen[2]

[1] Industrial Engineering Department, Bursa Uludag University, Bursa, Turkey
ikucukoglu@uludag.edu.tr

[2] KU Leuven Institute for Mobility – CIB, KU Leuven, Leuven, Belgium
{dirk.cattrysse,pieter.vansteenwegen}@kuleuven.be

Abstract. Transportation of perishable food in cold chain logistics systems is crucial in order to preserve the freshness of the products. Due to the extended traveling times and frequent stops, planning the distribution operations in cold chain logistics plays a vital role in minimizing the deterioration cost of the products. In order to minimize the total cost of cold chain logistics activities related to the purchase of perishable products, the route and procurement operations have to be well-planned. In this context, this paper addresses the well-known traveling purchaser problem (TPP) and extends the TPP by considering the procurement of perishable products. This is called the traveling purchaser problem in cold chain logistics (TPP-CCL). In the TPP-CCL, the demand for a number of perishable products is provided from a number of markets, where the products purchased at markets are transported by a temperature-controlled vehicle. In addition to the transportation and procurement cost, the deterioration cost of the products is taken into account in the problem. The problem is formulated as a non-linear mixed-integer programming model in which the objective is to find the best procurement and route plan for the purchaser that minimizes the total cost. Considering the complexity of the problem, a simulated annealing (SA) algorithm is proposed to solve the TPP-CCL. The SA is formed by using a number of local search procedures, where the procedures are randomly selected to find a new solution in each iteration. The proposed SA is performed for a TPP-CCL problem set that includes different-sized instances. The results of the SA are compared to the GUROBI solver results. A better result is obtained by the SA for most of the instances. The computational results show that the proposed SA outperforms the GUROBI results by finding better results in shorter computational times.

Keywords: Traveling Purchaser Problem · Cold-Chain Logistics · Mathematical Modeling · Simulated Annealing

1 Introduction

The traveling purchaser problem (TPP) is the generalization of the well-known traveling salesperson problem (TSP) and aims to satisfy a number of product demands from a number of markets with minimum total cost [1]. Distinct from the TSP, TPP includes three operational decisions: the selection of the markets to be visited, the number of products purchased from each market, and the route of the purchaser.

Z. Şen et al. (Eds.): IMSS 2023, LNME, pp. 259–274, 2024.
https://doi.org/10.1007/978-981-99-6062-0_24

The TPP was first introduced by Ramesh [2] in 1981 and extended with different assumptions. One of the widely studied variants of the TPP is the uncapacitated TPP, in which the amount of a product is sufficient to satisfy demand if it is available in a market [3]. Voß [4] considered a fixed market visiting cost for the uncapacitated TPP. Another variant of the TPP is the bi-objective TPP, in which the objective function includes different goals [5]. Coi and Lee [6] extended the TPP with a budget constraint for multiple vehicles. In addition to the cost-based or delivery-based constraints, a number of studies addressed the TPP with environmental concerns, such as green TPP [7], sustainable TPP [8], and solid green TPP [9]. Since the TPP belongs to the class of combinatorial optimization problems and has been shown to be NP-hard, many heuristic and meta-heuristic solution approaches have been introduced in the literature [1]. The early heuristic approaches are the solution construction methods, such as the generalized saving heuristic [10], the tour reduction heuristic [11], and the commodity adding heuristic [12]). In addition to the solution construction methods, a number of local search methods were developed to solve TPP. Recently, meta-heuristic solution approaches have been used to find better results, such as ant colony optimization [13], transgenetic algorithm [14], and variable neighborhood search [15].

In this paper, a new variation of the TPP called the traveling purchaser problem in cold chain logistics (TPP-CCL) is introduced for the procurement plan of perishable foods in cold chain logistics. The TPP-CCL takes into account the release time of the perishable foods in the markets and their additional deterioration costs during transportation. The aim of the problem is to minimize the total cost of the purchaser. The TPP-CCL is formulated as a non-linear mixed-integer programming model. To solve the considered problem, a simulated annealing (SA) algorithm is proposed. The proposed SA employs an advanced local search (LS) procedure to efficiently search the solution space. This paper contributes to the literature by introducing a new variant of the TPP by considering cold chain operational activities. Based on a non-linear deterioration cost function, which is commonly used in the cold chain literature, a non-linear mixed-integer mathematical model formulation is introduced for the problem. In addition to the deterioration costs, the mathematical formulation considers the release time of the perishable products in each market. To the best of our knowledge, the TPP has not been addressed for perishable foods with product release times. With this assumption, a more realistic operational plan can be provided for the procurement of perishable products, which are available at different times in the markets due to the fact that they are produced at different times of the day. In addition to the new variant of the TPP, this study proposes a simulated annealing algorithm in which a problem-specific local search operator is used to generate new solutions. With the help of an advanced search mechanism, the proposed SA is capable of finding effective results for the TPP-CCL.

The remainder of the paper is organized as follows. Section 2 introduces the details of the TPP-CCL and its mathematical formulation. The proposed SA is presented in Sect. 3. The computational results and performance analysis of the proposed algorithm are given in Sect. 4. Finally, the conclusions are given in Sect. 5.

2 Problem Definition and Model Formulation

In TPP-CCL, the demand for a set of perishable products is provided from a set of capacitated markets, where the available quantity of a product at a market can be less than the product demand or even zero. For each product at a market, there exists a product release time. Therefore, the purchaser can only buy a product at a market after its release time. The tour of the purchaser starts and ends at the depot node. The purchaser collects the perishable products by using a temperature-controlled vehicle, where the capacity of the vehicle is greater than the total product demand. Each market can be visited once at most. The aim of the problem is to minimize total traveling, procurement, and product deterioration costs. Deterioration of the perishable products depends on the time they stand in the vehicle and the opening time of the vehicle door for loading. The loading time of the purchaser at any market is proportional to the number of products purchased at the market. The loading operation can be carried out when all products to be purchased from the market are available. Regarding the assumptions given above, the mathematical model of the TPP-CCL is formulated by using the parameters and decision variables presented in Table 1.

$$Min \sum_{i \in V} \sum_{j \in V} c_{ij} x_{ij} + \sum_{k \in K} \sum_{i \in M_k} p_{ik} z_{ik} + \sum_{k \in K} \sum_{i \in M_k} p_{ik} z_{ik} \left(1 - e^{-\theta_1 (r_0 - r_i - s_i - w_i)}\right)$$
$$+ \sum_{i \in M} l_i \left(1 - e^{-\theta_2 s_i}\right) + \sum_{i \in M} \vartheta (s_i + w_i) \tag{1}$$

S.t.

$$\sum_{i \in M_k} z_{ik} = d_k \quad k \in K \tag{2}$$

$$z_{ik} \le q_{ik} y_i \quad k \in K, \quad i \in M_k \tag{3}$$

$$o_{ik} \le z_{ik} \le q_{ik} o_{ik} \quad k \in K, \quad i \in M_k \tag{4}$$

$$\sum_{\substack{j \in V \\ i \neq j}} x_{ij} = y_i \quad i \in M \tag{5}$$

$$\sum_{\substack{i \in V \\ i \neq j}} x_{ij} = y_j \quad j \in M \tag{6}$$

$$\sum_{i \in M} x_{i0} = 1 \tag{7}$$

$$\sum_{j \in M} x_{0j} = 1 \tag{8}$$

Table 1. Parameters and decision variables used in the mathematical model formulation.

Parameters			
K	Set of products	$\{0\}$	Depot
M	Set of markets	V	Set of markets and depot ($M \cup \{0\}$)
M_k	Set of markets in which the product k is available ($M_k \subseteq M$), $k \in K$		
d_k	Demand amount of product k, $k \in K$		
q_{ik}	Available amount of product k at market i, $k \in K, i \in M_k$		
p_{ik}	Price of product k at market i, $k \in K, i \in M_k$		
rl_{ik}	Release time of product k at market i, $k \in K, i \in M_k$		
c_{ij}	Traveling cost from node i to node j, $i, j \in V$		
t_{ij}	Traveling time from node i to node j, $i, j \in V$		
h_i	Unit service time to purchase any product at market i; $i \in M$		
θ_1	Coefficient of deterioration of products per unit of time while products are in the vehicle		
θ_2	Coefficient of deterioration of the products in a unit of time due to opening the vehicle door		
ϑ	The cost of cooling the products in one unit of time at the market		
γ	Large number		
Decision Variables			
x_{ij}	Binary variable: 1 if the purchaser travels from node i to node j, otherwise 0; $i, j \in V, i \neq j$		
y_i	Binary variable: 1 if market i is visited by the purchaser, otherwise 0; $i \in M$		
z_{ik}	Amount of product k purchased from market i; $k \in K, i \in M_k$		
o_{ik}	Binary variable: 1 if product k is purchased at market i, otherwise 0; $k \in K$, $i \in M_k$		
r_i	Arrival time of purchaser to node i; $i \in V$		
s_i	Time spent at market i; $i \in M$		
w_i	Waiting time of vehicle at market i; $i \in M$		
l_i	The total purchasing cost of the products in the vehicle before it reaches market i; $i \in M$		

$$t_{0i} \leq r_i + \gamma(1 - x_{0i}) \quad i \in M \tag{9}$$

$$t_{0i} \geq r_i - \gamma(1 - x_{0i}) \quad i \in M \tag{10}$$

$$r_i + s_i + w_i + t_{ij} \leq r_j + \gamma\left(1 - x_{ij}\right) \quad i \in M, \quad j \in V, \quad i \neq j \tag{11}$$

$$r_i + s_i + w_i + t_{ij} \geq r_j - \gamma\left(1 - x_{ij}\right) \quad i \in M, \quad j \in V, \quad i \neq j \tag{12}$$

$$s_i = \sum_{k \in K} h_i z_{ik} \qquad i \in M \tag{13}$$

$$w_i \geq r l_{ik} - r_i - \gamma(1 - o_{ik}) \qquad k \in K, \qquad i \in M_k \tag{14}$$

$$l_i + \sum_{k \in K} p_{ik} z_{ik} \leq l_j + \gamma\big(1 - x_{ij}\big) \qquad i, j \in M, \qquad i \neq j \tag{15}$$

$$x_{ij} \in \{0, 1\} \qquad i, j \in V, \qquad i \neq j \tag{16}$$

$$y_i \in \{0, 1\} \qquad i \in M \tag{17}$$

$$o_{ik} \in \{0, 1\} \qquad k \in K, \qquad i \in M_k \tag{18}$$

$$z_{ik} \geq 0 \qquad k \in K, \qquad i \in M_k \tag{19}$$

$$r_i \geq 0 \qquad i \in V \tag{20}$$

$$s_i, l_i, w_i \geq 0 \qquad i \in M \tag{21}$$

The objective function (1) aims to minimize total traveling, procurement, and deterioration cost. In detail, each sub-term in the objective function determines the transportation cost of the vehicle on the move (f_1), the procurement cost (f_2), the deterioration cost due to the waiting of products in the vehicle (f_3), the deterioration cost due to the door-opening (f_4), and the refrigeration cost of the vehicle while waiting at market nodes (f_5). Here, f_1 and f_2 are the cost functions considered in the original TPP. f_3–f_5 are formed regarding the existing literature related to cold chain logistics (i.e., [16–18]). In case there is a degree of importance among these cost items, f_1–f_5 may be weighted by using importance coefficients. Constraints (2) ensure that the demand for each product is satisfied by the purchaser. Constraints (3) and (4) provide the capacity restrictions of the markets where the purchased amount of a product at any market cannot exceed the available stock. Constraints (5) and (6) guarantee that the purchaser enters and leaves each visited market exactly once. Constraints (7) and (8) assure that the route of the purchaser starts and ends at the depot node. Constraints (9)–(12) determine the arrival time of the purchaser to the visited nodes. The service time of the purchaser at market nodes is determined by constraints (13). The waiting time of the purchaser at the visited markets is computed by constraints (14). Constraints (15) determine the cumulative price of the products in the vehicle (l_i) before it reaches a market, where l_i is used to determine the deterioration cost of products while waiting in the vehicle at market nodes. Constraints (16)–(21) identify the decision variables of the model.

Based on the formulation given above, called hereafter Model 1, the purchaser is allowed to load the vehicle at any market point after all purchased products are available at the market. On the other hand, the model does not limit the purchaser for extra waiting at any market even though all products are ready. In this context, the purchaser has an opportunity to reduce deterioration costs by waiting at the markets in the route before loading the vehicle. In this way, the transportation time of the products can be reduced in case there is a potential waiting time at further market nodes. As an alternative to this situation, the extra waiting option of the purchaser can be avoided by using the constraints (22) and (23), where w'_{ik} is a free variable and equals to difference between the release time of product k at market i and the arrival time of the purchaser to the corresponding market; $k \in K$, $i \in M_k$. The extended version of the model is called Model 2.

$$w'_{ik} = rl_{ik}o_{ik} - r_i \qquad k \in K, \qquad i \in M_k \tag{22}$$

$$w_i = max\{0, max_{k \in K}\{w'_{ik}\}\} \qquad i \in M \tag{23}$$

3 Solution Methodology

SA is a stochastic method for solving combinatorial problems proposed by Kirkpatrick, Gelatt, and Vecchi in 1983 [19]. The main feature of the SA is the solution acceptance mechanism (known as the Metropolis acceptance criterion) that allows accepting worse solutions in the search with a probability in order to escape from the local optimal. Thanks to the Metropolis criterion, SA has been successfully employed to solve many routing problems, such as capacitated vehicle routing problem with loading constraints [20], two-echelon vehicle routing problem [21], traveling salesperson problem [22], green vehicle problem [23], location routing problem [24], inventory routing problem [25], etc. Motivated by its efficiency, this study proposes a simulated annealing algorithm to solve the TPP-CCL in which the solution generation mechanism is formed by using an advanced local search procedure.

The proposed SA starts with generating an initial solution ($\boldsymbol{X}_0$) by iteratively selecting one of the product types and purchasing from the unvisited/visited markets according to unit insertion price. The markets with smaller unit prices are selected until the demand for the selected product is satisfied. In the main loop of the SA, a new solution ($\boldsymbol{X}'$) is generated through the existing solution ($\boldsymbol{X}$) by using a number of search methods, where these methods are categorized into two groups: procurement-based search methods and route-change-based search methods. The procurement-based search methods are utilized to make a change in the procurement plan of the purchaser. In this context, an unvisited market insertion, a visited market removal, and product exchange between the visited markets are the used moves. In the route-change-based methods, the order of the visited markets is changed by using four different moves: exchanging a pair of markets, moving a market to a forward position, moving a market to a backward position, and reversing a sub-path in the route. The moves are employed in the local search procedure in a nested structure. The procedure initially starts with $\boldsymbol{X}' = \boldsymbol{X}$ and generates a new solution ($\boldsymbol{X}^{LS}$) by applying one of the procurement-based search methods. Following the procurement-based search method, one of the randomly selected route-change-based searches is carried out to improve $\boldsymbol{X}^{LS}$. As long as an improvement is provided by the selected move, the route-change-based search is repeated. At the end of the route-change-based search, if the $\boldsymbol{X}^{LS}$ is better than the $\boldsymbol{X}'$, the $\boldsymbol{X}'$ is updated as $\boldsymbol{X}' = \boldsymbol{X}^{LS}$ and the procedure continues by randomly selecting a procurement-based search to be carried out. Otherwise, the local search procedure is terminated. The methods used in the local search procedure are implemented by using the best-improvement strategy. After the local search procedure, X' is accepted according to the SA solution acceptance criterion where $X = X'$ if $f\left(X'\right) < f(X)$ or $e^{\Delta/T} \geq rnd$. Here, Δ is the total cost difference between the new solution and the existing solution ($\Delta = f\left(X'\right) - f(X)$), T is the temperature, and rnd is the standard uniform distributed random number. At the end of the main loop of the SA, the temperature is decreased by using a cooling coefficient (c). The SA is terminated if the algorithm reaches a maximum number of iterations. Based on the definition of the SA given above, the pseudo-code of the algorithm is presented in Fig. 1.

```
Set the algorithm parameters T = T_0, c
Generate an initial solution X = X_0, X^Best = X
Do //Main loop of the algorithm
    Do
        X' = X
        Randomly select a procurement-based search and generate a new solution X^LS
        Do
            Randomly select a route-change-based search and apply to X^LS
        Loop Until (X^LS is not improved by the selected move)
        If f(X^LS) < f(X') Then
            X' = X^LS
        End If
    Loop Until (X' is not improved)
    Δ= f(X') − f(X)
    If Δ< 0 Or e^(−Δ/T) ≥ rnd Then
        X = X'
        If f(X) < f(X^Best) Then
            X^Best = X
        End If
    End If
    T = T × c
Loop Until (Stopping criterion is met)
```

Fig. 1. Pseudo-code of the proposed SA.

4 Computational Results

The performance of the proposed SA is analyzed by using a well-known capacitated TPP benchmark problem set introduced by Laporte et al. [26]. The TPP instances are adapted to the TPP-CCL by adding a release time for each available product in the markets. In addition to the product release time, a service time for each market is identified. The release times and service times are determined using a uniform random distribution, where the lower and upper limits of the distribution are identified according to a feasible solution obtained for the TPP instance. Finally, the deterioration costs and the refrigeration cost of the vehicle while waiting at market locations are specified. For the traveling time, it is assumed as $t_{ij} = c_{ij}$. By using this scheme, 315 different-sized TPP-CCL instances are generated, where each instance characteristic is identified by three parameters: the number of nodes ($|V| = \{10, 20, 30\}$), the number of product types ($|K| = \{10, 20, 30\}$) in the instance, and λ parameter ($\lambda = \{0.1, 0.5, 0.7, 0.8, 0.9, 0.95, 0.99\}$) that controls the number of markets in a feasible solution through the product demand. The benchmark problem set includes five different instances for each parameter combination. The generated problems are labeled as "TPP.CCL.$|V|$.$|K|$.λ.[*#Instance*]".

The proposed algorithm is carried out for each instance with 10 independent runs with a 10,000 iteration limit. The SA parameters are identified as $T_o = 1000$, $c = 0.92$, and $L = 20$. In order to evaluate the performance of the proposed SA, the LS without SA and the GUROBI solver are used as competitor algorithms. As in the SA computations, the LS is run 10 times for each instance, and each run is terminated at 10,000 iterations. On the other hand, the GUROBI results are obtained with a time limit of one hour. Each computation is carried out on a personal computer with an Intel® Core™ i7-8665U CPU and 16GB of memory.

The SA, LS, and GUROBI are implemented for both Model 1 and Model 2. Table 2 and Table 3 show the summary of the computational results based on Model 1 and Model 2, respectively. The details of the results are given in Appendix Tables A1, A2, A3, A4, A5 and A6, in which each row shows the averages of the results obtained for five instances belonging to a parameter combination ($\{|V|, |K|, \lambda\}$).

In Table 2 and Table 3, the GUROBI results are presented through the best-found integer solution (f_G), the optimality gap of the GUROBI solver ($G_{opt}\%$), and the solution time in second (t_G). The notations used for the LS results are the best-found solution of LS over 10 runs (f_{LS}^B), the average solution of LS over 10 runs (f_{LS}^A), the average computational time of LS over 10 runs (t_{LS}), the percentage gap between the f_{LS}^B and f_G ($G_{LS}^B\%$), and the percentage gap between the f_{LS}^A and f_G ($G_{LS}^A\%$). Similar notations are used to show SA results as follows: the best-found solution of SA over 10 runs (f_{SA}^B), the average solution of SA over 10 runs (f_{SA}^A), the average computational time of SA over 10 runs (t_{SA}), the percentage gap between the f_{SA}^B and f_G ($G_{SA}^B\%$), and the percentage gap between the f_{SA}^A and f_G ($G_{SA}^A\%$). The percentage gaps are calculated by using the Eq. (24) where a negative gap value indicates that a lower total cost is obtained by LS or SA compared to the GUROBI solution. Regarding the results presented in Table 2 and Table 3, it can be expressed that the proposed SA outperforms GUROBI by finding better results in less computational time. Particularly, for the large-sized instances, the average percentage gaps between the SA result and GUROBI results are greater than 20%. Similar results are provided by the LS, where a better average result is obtained for each problem size excepting the instances with the size $\{|V| = 10, |K| = 10\}$. On the other hand, when the SA is compared to the LS, better results are obtained by SA for almost all problem sizes.

$$G_{LSorSA}^{BorA}\% = \frac{f_{LSorSA}^{BorA} - f_G}{f_G} \times 100\% \tag{24}$$

Table 2. Computational results for Model 1.

Problem Size		GUROBI			LS					SA				
					Best Solution		Average Solution			Best Solution		Average Solution		
$\|V\|$	$\|K\|$	f_G	$G_{opt}\%$	t_G	f_{LS}^B	$G_{LS}^B\%$	f_{LS}^A	t_{LS}	$G_{LS}^A\%$	f_{SA}^B	$G_{SA}^B\%$	f_{SA}^A	t_{SA}	$G_{SA}^A\%$
10	10	2874.62	0.61	2380.37	2874.04	−0.02	2878.46	0.76	0.13	2874.04	−0.02	2876.49	0.85	0.07
10	20	3997.62	10.14	3600.00	3910.36	−1.96	3919.71	1.11	−1.72	3910.36	−1.96	3917.07	1.18	−1.79
10	30	4851.21	16.15	3600.00	4664.82	−3.38	4680.30	1.34	−3.08	4664.79	−3.38	4676.89	1.39	−3.14
20	10	3808.71	17.77	3600.00	3580.96	−5.39	3632.08	1.24	−4.01	3574.72	−5.55	3625.41	1.37	−4.21
20	20	6026.63	29.63	3600.00	5298.64	−11.58	5357.47	2.07	−10.62	5292.14	−11.70	5348.79	2.12	−10.76
20	30	8241.95	38.00	3600.00	6885.38	−16.10	6982.94	2.83	−14.91	6884.37	−16.11	6960.02	2.96	−15.18
30	10	5471.45	39.06	3600.00	4338.26	−20.06	4451.90	2.04	−17.93	4323.62	−20.36	4428.11	2.07	−18.33
30	20	8583.46	44.63	3600.00	6530.52	−23.03	6660.89	4.07	−21.47	6521.51	−23.12	6634.04	4.13	−21.73
30	30	11551.10	45.38	3600.00	8787.53	−23.70	8978.70	6.27	−22.05	8778.45	−23.78	8956.62	6.46	−22.26
Average		6156.31	26.82	3464.49	5207.84	−11.69	5282.49	2.41	−10.63	5202.67	−11.78	5269.27	2.50	−10.81

Table 3. Computational results for Model 2.

Problem Size		GUROBI			LS					SA				
					Best Solution		Average Solution			Best Solution		Average Solution		
$\|V\|$	$\|K\|$	f_G	$G_{opt}\%$	t_G	f_{LS}^B	$G_{LS}^B\%$	f_{LS}^A	t_{LS}	$G_{LS}^A\%$	f_{SA}^B	$G_{SA}^B\%$	f_{SA}^A	t_{SA}	$G_{SA}^A\%$
10	10	2875.67	2.29	3600.00	2875.39	−0.01	2879.08	0.83	0.09	2875.39	−0.01	2877.43	0.95	0.08
10	20	3934.86	7.53	3600.00	3911.20	−0.58	3919.40	1.08	−0.42	3911.20	−0.58	3916.40	1.24	−0.39
10	30	4751.13	13.41	3600.00	4667.80	−1.68	4680.69	1.28	−1.42	4667.75	−1.68	4678.73	1.50	−1.46
20	10	3692.86	12.92	3600.00	3589.45	−2.43	3640.19	1.33	−1.01	3584.60	−2.56	3625.63	1.37	−1.42
20	20	5707.91	23.38	3600.00	5302.93	−6.83	5358.41	2.13	−5.86	5298.13	−6.92	5348.82	2.14	−6.04
20	30	7691.00	31.66	3600.00	6891.22	−9.88	6985.75	2.94	−8.64	6888.73	−9.90	6965.61	2.95	−8.88
30	10	5665.52	37.49	3600.00	4347.15	−21.66	4465.04	2.04	−19.36	4336.52	−21.83	4438.59	2.05	−19.96
30	20	8936.06	44.86	3600.00	6547.01	−25.96	6664.47	4.10	−24.65	6539.35	−26.05	6643.49	4.13	−24.87
30	30	11118.71	41.01	3600.00	8798.96	−19.98	9037.45	6.27	−17.86	8781.82	−20.14	8966.00	6.41	−18.46
Average		6041.52	23.84	3600.00	5214.57	−9.89	5292.28	2.44	−8.79	5209.28	−9.96	5273.41	2.53	−9.04

Another finding of the computational experiments is the effect of restricting the extra waiting before vehicle loading on the total cost. In order to compare Model 1 and Model 2, five small-sized instances ($|V| = 10$ and $|K| = 10$) are selected and resolved by GUROBI without a time limitation. For each instance, an optimal solution is found for both model types. Table 4 presents the selected instances and the results based on the cost items of the objective function ($f_1 - f_5$). In this context, the results given in Table 4 show that avoiding extra waiting for the purchaser directly affects the total cost of the purchaser. As a result, it should be expressed that less cost can be provided by allowing waiting for the purchaser at markets if waiting is allowed.

Table 4. Comparison of Model 1 and Model 2 based on the cost items.

Problem	Model 1						Model 2					
	f_1	f_2	f_3	f_4	f_5	Total Cost	f_1	f_2	f_3	f_4	f_5	Total Cost
TPP.CCL.10.10.5.1	1649	240	63.25	20.99	16.06	1989.30	1649	240	64.26	20.99	16.06	1990.31
TPP.CCL.10.10.5.3	1612	611	118.50	17.83	20.54	2379.87	1612	611	132.86	17.83	20.54	2394.23
TPP.CCL.10.10.95.2	2151	354	64.49	34.44	18.58	2622.52	2151	354	65.53	34.44	18.58	2622.56
TPP.CCL.10.10.95.3	1902	511	111.46	32.67	19.22	2576.35	1902	521	108.28	31.31	19.40	2581.99
TPP.CCL.10.10.99.2	1362	295	38.29	12.38	19.83	1727.50	1362	295	38.55	12.38	19.83	1727.76

5 Conclusion

In this study, the traveling purchaser problem in cold chain logistics (TPP-CCL) is introduced, considering the deterioration cost of perishable foods in cold chain logistics operations. TPP-CCL considers a number of perishable products to be purchased from a set of markets, where the products at markets become available at a specific time. In this context, the aim of the problem is to find the procurement and route plan for the purchaser that minimizes total traveling, procurement, and deterioration cost. Regarding the exponential computations of the deterioration rate, the problem is formulated as a non-linear mixed-integer programming model. To solve the TPP-CCL, a simulated annealing (SA) algorithm is proposed. The algorithm is formed by using a problem-specific local search (LS) procedure in which route-based and purchase-based moves are employed. In computational studies, the performance of the proposed algorithm is tested on a benchmark problem set, which includes different-sized instances. The results of the SA are compared to GUROBI results and LS results used in the SA. A better result is obtained by the proposed SA for most of the instances. As a result of the experiments, the proposed SA outperforms both the competitor algorithms. Furthermore, it is concluded from the computational experiments that allowing extra waiting for the purchaser at markets may reduce the total cost of the purchaser.

For future research, this study can be extended by considering additional restrictions of cold chain logistic activities. In this context, multiple vehicles with different cooling technologies or incompatible products can be taken into account for the purchaser. On the other hand, a maximum traveling time for each product can be considered for the perishable products. In addition to the new assumptions for the problem, problem-specific exact solution approaches can be implemented to find the optimal solution for the TPP-CCL.

Acknowledgement. This work is supported by the Commission of Scientific Research Projects of Bursa Uludag University, Project Number FUİ-2022-1042.

Appendix

Detailed computational results for Model 1 and Model 2

Table A1. Results for the TPP-CCL instances of size $|V| = 10$ for Model 1.

Problem Type		GUROBI			LS					SA				
					Best Solution		Average Solution			Best Solution		Average Solution		
$\lvert K\rvert$	λ	f_G	$G_{opt}\%$	t_G	f_{LS}^B	$G_{LS}^B\%$	f_{LS}^A	t_{LS}	$G_{LS}^A\%$	f_{SA}^B	$G_{SA}^B\%$	f_{SA}^A	t_{SA}	$G_{SA}^A\%$
10	0.1	3120.43	1.24	2419.12	3117.86	−0.08	3131.76	0.73	0.34	3117.86	−0.08	3124.94	0.88	0.13
10	0.5	2479.53	0.73	1784.57	2478.04	−0.04	2479.35	0.71	0.01	2478.04	−0.04	2478.88	0.77	−0.01
10	0.7	2899.42	0.87	3052.95	2899.60	0.01	2900.21	1.01	0.03	2899.60	0.01	2900.21	0.86	0.03
10	0.8	2756.98	0.40	2102.57	2756.89	0.00	2756.89	0.67	0.00	2756.89	0.00	2756.89	0.80	0.00
10	0.9	3287.69	0.61	2945.02	3287.64	0.00	3290.80	0.66	0.10	3287.64	0.00	3289.26	0.80	0.05
10	0.95	3051.94	0.18	2487.60	3051.96	0.00	3058.47	0.81	0.20	3051.96	0.00	3056.75	1.00	0.15
10	0.99	2526.31	0.22	1870.76	2526.26	0.00	2531.75	0.74	0.27	2526.26	0.00	2528.47	0.86	0.10
20	0.1	3679.73	10.08	3600.00	3598.05	−1.90	3613.16	1.07	−1.44	3598.05	−1.90	3612.38	1.10	−1.49
20	0.5	3994.98	4.92	3600.00	3975.37	−0.47	3985.61	0.99	−0.22	3975.37	−0.47	3983.34	1.16	−0.27
20	0.7	3915.53	9.32	3600.00	3793.92	−2.80	3800.96	1.07	−2.63	3793.92	−2.80	3795.13	1.24	−2.77
20	0.8	4127.13	12.67	3600.00	3976.95	−3.13	3981.40	1.30	−3.03	3976.95	−3.13	3978.48	1.14	−3.10
20	0.9	4146.09	6.90	3600.00	4103.37	−0.96	4113.22	1.12	−0.71	4103.37	−0.96	4111.15	1.29	−0.77
20	0.95	3913.13	10.71	3600.00	3860.51	−1.21	3869.53	1.06	−1.01	3860.51	−1.21	3867.27	1.22	−1.04
20	0.99	4206.76	16.38	3600.00	4064.36	−3.26	4074.07	1.13	−3.03	4064.36	−3.26	4071.77	1.15	−3.09
30	0.1	4851.38	17.32	3600.00	4709.48	−2.83	4742.23	1.22	−2.21	4709.48	−2.83	4731.56	1.32	−2.39
30	0.5	4927.72	17.90	3600.00	4609.96	−5.85	4624.31	1.17	−5.56	4609.96	−5.85	4619.81	1.36	−5.66
30	0.7	5112.91	17.98	3600.00	4754.79	−5.51	4765.17	1.30	−5.36	4754.79	−5.51	4757.06	1.51	−5.48
30	0.8	4488.92	12.05	3600.00	4405.15	−1.68	4420.47	1.55	−1.36	4404.97	−1.69	4421.34	1.35	−1.35
30	0.9	4728.30	14.66	3600.00	4613.70	−2.28	4629.39	1.31	−1.94	4613.70	−2.28	4625.07	1.57	−2.03
30	0.95	4862.75	14.57	3600.00	4791.49	−1.53	4802.27	1.29	−1.30	4791.49	−1.53	4801.50	1.33	−1.32
30	0.99	4986.49	18.57	3600.00	4769.17	−4.00	4778.23	1.52	−3.82	4769.17	−4.00	4781.87	1.25	−3.75

Table A2. Results for the TPP-CCL instances of size $|V| = 20$ for Model 1.

Problem Type		GUROBI			LS					SA				
					Best Solution		Average Solution			Best Solution		Average Solution		
$\lvert K\rvert$	λ	f_G	$G_{opt}\%$	t_G	f_{LS}^B	$G_{LS}^B\%$	f_{LS}^A	t_{LS}	$G_{LS}^A\%$	f_{SA}^B	$G_{SA}^B\%$	f_{SA}^A	t_{SA}	$G_{SA}^A\%$
10	0.1	4192.49	23.23	3600.00	3786.28	−9.05	3856.56	1.30	−7.58	3786.28	−9.05	3849.35	1.52	−7.69
10	0.5	3689.01	15.31	3600.00	3566.36	−3.07	3632.39	1.22	−1.29	3556.12	−3.37	3620.68	1.46	−1.59
10	0.7	3824.59	16.86	3600.00	3545.47	−6.67	3587.10	1.32	−5.48	3545.47	−6.67	3570.80	1.50	−5.96
10	0.8	3543.27	18.32	3600.00	3415.77	−3.26	3449.13	1.18	−2.32	3384.83	−4.02	3447.04	1.19	−2.40
10	0.9	4343.93	19.38	3600.00	4098.89	−5.57	4147.59	1.35	−4.40	4095.91	−5.65	4140.29	1.56	−4.54
10	0.95	3489.09	16.73	3600.00	3202.33	−6.94	3270.50	1.16	−4.72	3202.83	−6.92	3275.87	1.16	−4.86
10	0.99	3578.56	14.60	3600.00	3451.61	−3.21	3481.27	1.11	−2.30	3451.61	−3.21	3473.82	1.20	−2.43
20	0.1	6629.70	33.90	3600.00	5739.58	−13.24	5799.88	2.18	−12.34	5736.28	−13.29	5776.10	2.35	−12.71
20	0.5	6456.10	31.26	3600.00	5602.83	−12.76	5646.57	2.13	−12.12	5599.11	−12.82	5643.11	2.26	−12.16
20	0.7	5840.25	29.92	3600.00	5115.47	−12.27	5192.06	2.14	−10.95	5104.39	−12.46	5175.65	1.98	−11.25

(continued)

Table A2. (*continued*)

Problem Type		GUROBI			LS					SA				
					Best Solution		Average Solution			Best Solution		Average Solution		
$\|K\|$	λ	f_G	$G_{opt}\%$	t_G	f_{LS}^B	$G_{LS}^B\%$	f_{LS}^A	t_{LS}	$G_{LS}^A\%$	f_{SA}^B	$G_{SA}^B\%$	f_{SA}^A	t_{SA}	$G_{SA}^A\%$
20	0.8	6037.68	28.64	3600.00	5404.54	−10.06	5477.13	2.03	−8.92	5393.64	−10.28	5462.88	2.24	−9.12
20	0.9	6265.40	28.37	3600.00	5497.90	−12.24	5548.42	2.27	−11.41	5497.93	−12.24	5544.65	2.15	−11.48
20	0.95	6080.99	33.29	3600.00	5258.76	−13.02	5325.27	2.01	−11.99	5241.26	−13.27	5327.07	2.06	−11.89
20	0.99	4876.31	22.07	3600.00	4471.39	−7.49	4512.94	1.72	−6.62	4472.34	−7.50	4512.06	1.75	−6.73
30	0.1	7841.78	36.09	3600.00	6630.43	−15.60	6736.32	2.70	−14.29	6621.12	−15.68	6695.94	2.88	−14.68
30	0.5	7981.06	37.21	3600.00	6694.05	−15.33	6812.02	2.63	−13.69	6694.05	−15.33	6806.28	2.68	−13.98
30	0.7	8768.70	37.70	3600.00	7345.52	−15.47	7438.17	2.85	−14.47	7349.94	−15.43	7389.69	3.03	−14.95
30	0.8	7894.64	41.16	3600.00	6407.17	−18.71	6513.57	2.73	−17.35	6407.17	−18.71	6505.05	2.79	−17.44
30	0.9	8382.25	38.13	3600.00	6984.79	−16.47	7070.95	2.87	−15.51	6984.79	−16.47	7046.71	3.05	−15.81
30	0.95	8931.52	38.43	3600.00	7550.57	−15.27	7641.52	3.23	−14.34	7550.57	−15.27	7619.90	3.29	−14.56
30	0.99	7893.69	37.30	3600.00	6585.13	−15.87	6668.03	2.83	−14.73	6582.92	−15.90	6656.55	2.97	−14.83

Table A3. Results for the TPP-CCL instances of size $|V| = 30$ for Model 1.

Problem Type		GUROBI			LS					SA				
					Best Solution		Average Solution			Best Solution		Average Solution		
$\|K\|$	λ	f_G	$G_{opt}\%$	t_G	f_{LS}^B	$G_{LS}^B\%$	f_{LS}^A	t_{LS}	$G_{LS}^A\%$	f_{SA}^B	$G_{SA}^B\%$	f_{SA}^A	t_{SA}	$G_{SA}^A\%$
10	0.1	5566.38	37.20	3600.00	4453.11	−19.47	4552.57	1.85	−17.70	4438.60	−19.72	4514.35	1.85	−18.37
10	0.5	5864.88	36.19	3600.00	4731.61	−17.29	4864.21	2.32	−14.57	4729.41	−17.30	4835.52	2.34	−14.93
10	0.7	5481.15	41.88	3600.00	4184.95	−23.36	4308.30	2.00	−21.10	4171.45	−23.64	4292.94	2.03	−21.36
10	0.8	5224.29	39.28	3600.00	4074.45	−21.56	4228.34	2.15	−18.53	4055.00	−21.92	4213.97	2.08	−18.84
10	0.9	5271.33	40.19	3600.00	4143.62	−20.73	4180.54	1.96	−20.10	4139.36	−20.79	4178.16	2.09	−20.11
10	0.95	5890.40	40.55	3600.00	4616.13	−21.40	4784.16	2.18	−18.48	4611.59	−21.54	4738.48	2.22	−19.20
10	0.99	5001.71	38.15	3600.00	4163.97	−16.60	4245.20	1.85	−15.06	4119.94	−17.64	4223.38	1.86	−15.49
20	0.1	7403.38	44.07	3600.00	5806.60	−21.43	5909.79	3.04	−20.01	5790.78	−21.62	5897.98	3.16	−20.17
20	0.5	9877.00	48.93	3600.00	6918.68	−26.62	7082.36	4.20	−24.93	6906.50	−26.70	7027.43	4.40	−25.26
20	0.7	8324.20	41.94	3600.00	6490.15	−21.77	6638.90	4.07	−19.85	6495.43	−21.70	6606.12	4.10	−20.26
20	0.8	8578.89	48.20	3600.00	6291.49	−26.47	6402.23	4.02	−25.18	6281.18	−26.60	6377.40	4.05	−25.45
20	0.9	8278.45	40.39	3600.00	6591.03	−19.89	6715.26	4.21	−18.42	6584.48	−19.95	6664.00	4.36	−18.95
20	0.95	8416.62	44.54	3600.00	6549.18	−22.11	6696.04	4.32	−20.32	6549.18	−22.11	6694.50	4.33	−20.41
20	0.99	9205.69	44.32	3600.00	7066.55	−22.90	7181.63	4.61	−21.57	7043.05	−23.13	7170.88	4.52	−21.63
30	0.1	13025.45	50.03	3600.00	9215.54	−28.96	9418.18	5.85	−27.41	9219.09	−28.93	9404.66	5.76	−27.55
30	0.5	10701.92	42.53	3600.00	8372.23	−21.77	8614.90	5.68	−19.56	8364.34	−21.85	8606.23	5.70	−19.62
30	0.7	11358.03	47.23	3600.00	8452.13	−25.56	8571.64	6.31	−24.52	8457.97	−25.51	8551.52	6.64	−24.68
30	0.8	11485.16	45.83	3600.00	8805.68	−23.36	8996.88	6.57	−21.67	8787.23	−23.46	8952.57	7.13	−22.05
30	0.9	11526.76	45.03	3600.00	8980.64	−21.94	9125.99	6.62	−20.73	8963.67	−22.09	9129.20	6.93	−20.68
30	0.95	10583.14	44.35	3600.00	8004.36	−24.21	8220.72	5.88	−22.18	7987.31	−24.39	8159.03	6.06	−22.80
30	0.99	12177.24	42.65	3600.00	9682.14	−20.09	9902.63	6.99	−18.31	9669.52	−20.21	9893.14	6.97	−18.40

Table A4. Results for the TPP-CCL instances of size $|V| = 10$ for Model 2.

Problem Type		GUROBI			LS					SA				
					Best Solution		Average Solution			Best Solution		Average Solution		
$\|K\|$	λ	f_G	$G_{opt}\%$	t_G	f_{LS}^B	$G_{LS}^B\%$	f_{LS}^A	t_{LS}	$G_{LS}^A\%$	f_{SA}^B	$G_{SA}^B\%$	f_{SA}^A	t_{SA}	$G_{SA}^A\%$
10	0.1	3119.03	3.12	3600.00	3118.23	−0.02	3127.34	0.81	0.12	3118.23	−0.02	3122.89	0.89	0.25
10	0.5	2481.69	2.52	3600.00	2481.35	−0.01	2481.79	0.83	0.00	2481.35	−0.01	2481.52	0.94	0.01
10	0.7	2901.41	2.36	3600.00	2901.28	0.00	2906.03	0.87	0.14	2901.28	0.00	2902.49	1.05	0.04
10	0.8	2757.26	1.91	3600.00	2757.15	0.00	2757.15	0.83	0.00	2757.15	0.00	2757.15	0.96	0.00
10	0.9	3290.08	2.69	3600.00	3289.52	−0.02	3291.25	0.83	0.04	3289.52	−0.02	3291.16	0.99	0.04
10	0.95	3053.32	2.24	3600.00	3053.32	0.00	3058.34	0.84	0.15	3053.32	0.00	3057.32	0.99	0.13
10	0.99	2526.92	1.22	3600.00	2526.92	0.00	2531.71	0.78	0.22	2526.92	0.00	2529.50	0.85	0.12
20	0.1	3618.99	7.76	3600.00	3599.48	−0.50	3615.30	1.14	−0.37	3599.48	−0.50	3605.03	1.24	−0.03
20	0.5	3997.02	7.45	3600.00	3977.50	−0.47	3986.62	0.98	−0.28	3977.50	−0.47	3984.88	1.16	−0.24
20	0.7	3813.64	7.23	3600.00	3793.94	−0.49	3797.13	1.03	−0.42	3793.94	−0.49	3796.78	1.22	−0.42
20	0.8	4001.99	7.73	3600.00	3978.15	−0.55	3988.26	1.09	−0.32	3978.15	−0.55	3986.97	1.29	−0.34
20	0.9	4126.15	7.27	3600.00	4103.48	−0.57	4108.92	1.14	−0.42	4103.48	−0.57	4107.48	1.27	−0.46
20	0.95	3883.60	6.60	3600.00	3860.51	−0.54	3866.74	1.05	−0.39	3860.51	−0.54	3862.66	1.21	−0.49
20	0.99	4102.64	8.68	3600.00	4065.36	−0.90	4072.80	1.12	−0.72	4065.36	−0.90	4070.98	1.32	−0.77
30	0.1	4872.18	16.66	3600.00	4716.80	−3.03	4742.04	1.25	−2.53	4716.80	−3.03	4745.18	1.49	−2.50
30	0.5	4659.72	12.02	3600.00	4610.77	−1.06	4626.52	1.14	−0.72	4610.77	−1.06	4618.76	1.37	−0.89
30	0.7	4839.24	13.37	3600.00	4755.57	−1.63	4767.07	1.24	−1.42	4755.57	−1.63	4762.80	1.49	−1.51
30	0.8	4504.67	13.15	3600.00	4407.73	−2.01	4419.07	1.34	−1.76	4407.73	−2.01	4421.94	1.54	−1.71
30	0.9	4670.78	12.24	3600.00	4616.82	−1.14	4632.39	1.33	−0.81	4616.51	−1.15	4628.74	1.54	−0.88
30	0.95	4874.65	13.45	3600.00	4795.63	−1.58	4802.53	1.38	−1.44	4795.63	−1.58	4798.77	1.58	−1.51
30	0.99	4836.67	13.00	3600.00	4771.25	−1.32	4775.22	1.30	−1.24	4771.25	−1.32	4774.92	1.51	−1.24

Table A5. Results for the TPP-CCL instances of size $|V| = 20$ for Model 2.

Problem Type		GUROBI			LS					SA				
					Best Solution		Average Solution			Best Solution		Average Solution		
$\|K\|$	λ	f_G	$G_{opt}\%$	t_G	f_{LS}^B	$G_{LS}^B\%$	f_{LS}^A	t_{LS}	$G_{LS}^A\%$	f_{SA}^B	$G_{SA}^B\%$	f_{SA}^A	t_{SA}	$G_{SA}^A\%$
10	0.1	4075.29	19.60	3600.00	3820.37	−5.68	3866.27	1.21	−4.57	3806.17	−5.99	3844.79	1.28	−5.09
10	0.5	3639.34	13.83	3600.00	3592.83	−1.06	3637.06	1.29	0.12	3592.83	−1.06	3633.56	1.48	0.05
10	0.7	3598.21	8.08	3600.00	3545.47	−1.37	3587.88	1.55	−0.20	3545.47	−1.37	3570.37	1.30	−0.68
10	0.8	3516.78	12.89	3600.00	3388.77	−3.18	3450.77	1.23	−1.49	3385.16	−3.27	3445.10	1.41	−1.69
10	0.9	4224.85	13.88	3600.00	4100.33	−2.96	4157.67	1.55	−1.51	4100.33	−2.96	4133.29	1.35	−2.12
10	0.95	3246.74	9.62	3600.00	3223.20	−0.55	3296.75	1.23	1.80	3207.11	−1.05	3278.42	1.36	1.18
10	0.99	3548.78	12.57	3600.00	3455.21	−2.22	3484.91	1.23	−1.25	3455.11	−2.22	3473.84	1.39	−1.59
20	0.1	6111.60	24.72	3600.00	5739.28	−5.93	5801.33	2.40	−4.85	5738.86	−5.96	5784.57	2.26	−5.21

(continued)

Table A5. (*continued*)

Problem Type		GUROBI			LS					SA				
					Best Solution		Average Solution			Best Solution		Average Solution		
$\|K\|$	λ	f_G	$G_{opt}\%$	t_G	f_{LS}^B	$G_{LS}^B\%$	f_{LS}^A	t_{LS}	$G_{LS}^A\%$	f_{SA}^B	$G_{SA}^B\%$	f_{SA}^A	t_{SA}	$G_{SA}^A\%$
20	0.5	6074.93	25.56	3600.00	5609.00	−7.61	5672.86	2.24	−6.57	5607.71	−7.63	5667.32	2.12	−6.67
20	0.7	5511.46	23.64	3600.00	5130.07	−7.00	5200.33	2.10	−5.75	5116.87	−7.27	5166.99	1.97	−6.34
20	0.8	5778.10	22.97	3600.00	5398.68	−6.09	5473.11	2.10	−4.77	5400.38	−6.06	5475.11	2.24	−4.79
20	0.9	5789.89	19.63	3600.00	5501.96	−5.03	5548.55	2.14	−4.21	5501.96	−5.03	5538.71	2.26	−4.37
20	0.95	5825.40	27.08	3600.00	5260.57	−9.65	5296.14	2.12	−9.09	5242.91	−9.91	5297.08	2.25	−9.08
20	0.99	4863.98	20.09	3600.00	4480.96	−6.53	4516.57	1.79	−5.79	4478.22	−6.59	4511.99	1.89	−5.85
30	0.1	7270.14	28.24	3600.00	6634.47	−7.97	6758.08	2.78	−6.31	6623.38	−8.09	6679.46	2.75	−7.31
30	0.5	7578.96	32.38	3600.00	6696.12	−10.33	6780.07	2.75	−9.17	6702.71	−10.21	6781.80	2.87	−9.13
30	0.7	8128.72	31.35	3600.00	7356.02	−8.63	7443.37	3.07	−7.57	7350.18	−8.69	7424.75	2.96	−7.74
30	0.8	7660.93	37.98	3600.00	6414.36	−16.19	6504.43	2.91	−15.04	6414.36	−16.19	6480.38	2.80	−15.34
30	0.9	8027.33	33.11	3600.00	6991.83	−12.79	7080.69	2.99	−11.72	6991.83	−12.79	7065.53	3.08	−11.91
30	0.95	8094.05	30.68	3600.00	7557.75	−6.52	7651.99	3.14	−5.42	7553.05	−6.57	7642.77	3.30	−5.49
30	0.99	7076.90	27.90	3600.00	6587.95	−6.70	6681.62	2.92	−5.21	6585.56	−6.74	6684.61	2.91	−5.21

Table A6. Results for the TPP-CCL instances of size $|V| = 30$ for Model 2.

Problem Type		GUROBI			LS					SA				
					Best Solution		Average Solution			Best Solution		Average Solution		
$\|K\|$	λ	f_G	$G_{opt}\%$	t_G	f_{LS}^B	$G_{LS}^B\%$	f_{LS}^A	t_{LS}	$G_{LS}^A\%$	f_{SA}^B	$G_{SA}^B\%$	f_{SA}^A	t_{SA}	$G_{SA}^A\%$
10	0.1	6148.98	41.09	3600.00	4451.36	−25.85	4549.10	1.83	−24.09	4442.46	−25.96	4554.60	1.93	−24.00
10	0.5	6244.18	35.94	3600.00	4752.67	−20.17	4888.46	2.28	−17.11	4733.30	−20.52	4849.70	2.34	−18.27
10	0.7	5671.69	40.64	3600.00	4215.11	−25.20	4318.61	1.99	−23.34	4172.52	−25.89	4285.77	1.92	−23.89
10	0.8	5134.55	34.19	3600.00	4078.76	−19.67	4198.21	2.16	−17.33	4081.59	−19.61	4188.52	2.06	−17.51
10	0.9	5421.38	37.86	3600.00	4149.28	−22.52	4207.11	1.97	−21.49	4149.02	−22.52	4203.26	2.04	−21.61
10	0.95	5982.14	38.71	3600.00	4618.24	−22.09	4791.77	2.13	−19.08	4620.67	−22.08	4746.72	2.18	−19.87
10	0.99	5055.74	34.02	3600.00	4164.66	−16.08	4302.05	1.96	−13.11	4156.06	−16.24	4241.54	1.89	−14.56
20	0.1	8325.64	47.31	3600.00	5818.22	−29.47	5967.89	3.15	−27.70	5799.74	−29.59	5937.00	3.16	−28.04
20	0.5	9705.19	48.46	3600.00	6912.41	−27.69	7027.87	4.52	−26.40	6920.43	−27.66	7017.31	4.55	−26.59
20	0.7	8824.69	43.06	3600.00	6534.30	−25.68	6640.99	4.11	−24.52	6510.98	−25.99	6625.51	4.00	−24.70
20	0.8	9243.40	49.28	3600.00	6314.08	−30.73	6414.33	3.82	−29.63	6308.25	−30.80	6386.67	4.07	−29.94
20	0.9	8321.82	37.82	3600.00	6602.61	−19.87	6693.44	4.41	−18.78	6609.80	−19.80	6661.02	4.26	−19.16
20	0.95	8372.15	41.82	3600.00	6564.36	−21.09	6649.25	4.28	−20.09	6554.51	−21.22	6668.55	4.31	−19.80
20	0.99	9759.55	46.31	3600.00	7083.11	−27.16	7257.49	4.41	−25.41	7071.74	−27.29	7208.35	4.56	−25.85
30	0.1	12119.73	44.31	3600.00	9245.62	−22.64	9474.35	5.69	−20.70	9228.56	−22.75	9454.04	5.73	−20.81
30	0.5	10472.12	39.80	3600.00	8376.07	−19.86	8610.68	5.56	−17.64	8353.73	−20.07	8547.13	5.94	−18.23
30	0.7	10564.36	40.90	3600.00	8467.11	−19.66	8685.76	6.44	−17.54	8465.14	−19.68	8627.19	6.49	−18.15
30	0.8	11268.45	41.14	3600.00	8817.32	−20.84	9048.57	6.89	−18.77	8774.18	−21.21	8973.62	7.14	−19.46

(*continued*)

Table A6. *(continued)*

Problem Type		GUROBI			LS					SA				
					Best Solution		Average Solution			Best Solution		Average Solution		
$\|K\|$	λ	f_G	$G_{opt}\%$	t_G	f_{LS}^B	$G_{LS}^B\%$	f_{LS}^A	t_{LS}	$G_{LS}^A\%$	f_{SA}^B	$G_{SA}^B\%$	f_{SA}^A	t_{SA}	$G_{SA}^A\%$
30	0.9	11203.16	41.57	3600.00	9007.16	−18.36	9219.63	6.56	−16.53	8988.16	−18.54	9154.57	6.75	−17.07
30	0.95	9909.30	38.07	3600.00	7993.05	−18.73	8248.24	6.02	−16.29	7989.93	−18.83	8149.58	5.86	−17.19
30	0.99	12293.85	41.28	3600.00	9686.42	−19.78	9974.96	6.75	−17.55	9673.05	−19.91	9855.84	6.97	−18.34

References

1. Manerba, M., Mansini, R., Riera-Ledesma, J.: The traveling purchaser problem and its variants. Eur. J. Oper. Res. **259**(1), 1–18 (2017)
2. Ramesh, T.: Traveling purchaser problem. Opsearch **18**(1–3), 78–91 (1981)
3. Boctor, F.F., Laporte, G., Renaud, J.: Heuristics for the traveling purchaser problem. Comput. Oper. Res. **30**(4), 491–504 (2003)
4. Voß, S.: Dynamic tabu search strategies for the traveling purchaser problem. Ann. Oper. Res. **63**(2), 253–275 (1996)
5. Almeida, C.P., Gonçalves, R.A., Goldbarg, E.F., Goldbarg, M.C., Delgado, M.R.: An experimental analysis of evolutionary heuristics for the biobjective traveling purchaser problem. Ann. Oper. Res. **199**(1), 305–341 (2012)
6. Choi, M.J., Lee, S.H.J.E.S.: The multiple traveling purchaser problem for maximizing system's reliability with budget constraints. Expert Syst. Appl. **38**(8), 9848–9853 (2011)
7. Hamdan, S., Larbi, R., Cheaitou, A., Alsyouf, I.: Green traveling purchaser problem model: a bi-objective optimization approach. In: 7th International Conference on Modeling, Simulation, and Applied Optimization (ICMSAO), Sharjah, United Arab Emirates (2017)
8. Cheaitou, A., Hamdan, S., Larbi, R., Alsyouf, I.: Sustainable traveling purchaser problem with speed optimization. Int. J. Sustain. Transp. 1–20 (2020)
9. Roy, A., Gao, R., Jia, L., Maity, S., Kar, S.: A noble genetic algorithm to solve a solid green traveling purchaser problem with uncertain cost parameters. Am. J. Math. Manage. Sci. 1–15 (2020)
10. Golden, B., Levy, L., Dahl, R.: Two generalizations of the traveling salesman problem. Omega **9**(4), 439–441 (1981)
11. Ong, H.L.: Approximate algorithms for the travelling purchaser problem. Oper. Res. Lett. **1**(5), 201–205 (1982)
12. Pearn, W. L.: On the traveling purchaser problem. Technical Note (1991)
13. Bontoux, B., Feillet, D.: Ant colony optimization for the traveling purchaser problem. Comput. Oper. Res. **35**(2), 628–637 (2008)
14. Goldbarg, M.C., Bagi, L.B., Goldbarg, E.F.G.: Transgenetic algorithm for the traveling purchaser problem. Eur. J. Oper. Res. **199**(1), 36–45 (2009)
15. Ochi, L., Silva, M., Drummond, L.: Metaheuristics based on GRASP and VNS for solving the traveling purchaser problem. In: IV Metaheuristic International Conference (MIC 2001), Porto, Portugal (2001)
16. Wang, S., Tao, F., Shi, Y., Wen, H.: Optimization of vehicle routing problem with time windows for cold chain logistics based on carbon tax. Sustainability **9**(5), 694 (2017)
17. Qin, G., Tao, F., Li, L.: A vehicle routing optimization problem for cold chain logistics considering customer satisfaction and carbon emissions. Environ. Res. Public Health **16**(4), 576 (2019)

18. Qi, C., Hu, L.: Optimization of vehicle routing problem for emergency cold chain logistics based on minimum loss. Phys. Commun. **40**, 101085 (2020)
19. Kirkpatrick, S., Gelatt, C.D., Vecchi, M.P.: Optimization by simulated annealing. Science **220**, 671–681 (1983)
20. Wei, L., Zhang, Z., Zhang, D., Leung, S.C.H.: A Simulated annealing algorithm for the capacitated vehicle routing problem with two-dimensional loading constraints. Eur. J. Oper. Res. **265**, 843–859 (2018)
21. Redi, A.A.N.P., Jewpanya, P., Kurniawan, A.C., Persada, S.F., Nadlifatin, R., Dewi, O.A.C.: A simulated annealing algorithm for solving two-echelon vehicle routing problem with locker facilities. Algorithms **13**(9), 218 (2020)
22. Zhan, S.-H., Lin, J., Zhang, Z.-J., Zhong, Y.-W.: List-based simulated annealing algorithm for traveling salesman problem. Comput. Intell. Neurosci. **2016**, 1–12 (2016)
23. Karagul, K., Sahin, Y., Aydemir, E., Oral, A.: A simulated annealing algorithm based solution method for a green vehicle routing problem with fuel consumption. In: Paksoy, T., Weber, G.-W., Huber, S. (eds.) Lean and Green Supply Chain Management. ISORMS, vol. 273, pp. 161–187. Springer, Cham (2019). https://doi.org/10.1007/978-3-319-97511-5_6
24. Ferreira, K.M., Queiroz, T.A.: Two effective simulated annealing algorithms for the location-routing problem. Appl. Soft Comput. **70**, 389–422 (2018)
25. Alvarez, A., Munari, P., Morabito, R.: Iterated local search and simulated annealing algorithms for the inventory routing problem. Int. Trans. Oper. Res. **25**, 1785–1809 (2018)
26. Laporte, G., Riera-Ledesma, J., Salazar-González, J.-J.: A branch-and-cut algorithm for the undirected traveling purchaser problem. Oper. Res. **51**(6), 940–951 (2003)

A Machine Vision Algorithm Approach for Angle Detection in Industrial Applications

Mehmet Kayğusuz[1] (✉), Barış Öz[1,2], Ayberk Çelik[1,3], Yunus Emre Akgül[1,3], Gözde Şimşek[1,2], and Ebru Gezgin Sarıgüzel[1,2]

[1] Ar-Ge Merkezi, Mamur Teknoloji Sistemleri, Istanbul, Turkey
{mehmet.kaygusuz,baris.oz,emre.akgul,gozde.simsek, ebru.gezgin}@mamurtech.com, ayberk.celik@teknomuhendislik.com
[2] Graduate School of Science and Engineering, Yildiz Technical University, Istanbul, Turkey
[3] Graduate School of Science and Engineering, Kocaeli University, İzmir, Kocaeli, Turkey

Abstract. In automatic feeding systems, feeding of characteristic workpieces by mechanical tools causes accuracy and cost difficulties. For this reason, in systems where special workpieces are fed, image processing applications are necessary to obtain characteristic features of a product. In this study, a novel image processing algorithm is proposed for feeding a workpiece which has characteristic geometrical structures. The proposed algorithm is based on obtaining geometrical and rotational properties of the product and the gradient-based analysis as follows. The first step is to extract features from the shape of the workpiece, this step includes noise reduction, filtering, and edge detection operations. The gradient values of the edge information are used to create the angle-length vector pair in the second step. The workpiece rotation information is derived from length values indexed with angle information. The last step involves determination of the workpiece position in the 2D coordinate system. The coordinate information is used to determine the position of the gripper holder. The coordinates and angle are transmitted to the feed control. The proposed algorithm is applied on the 800 images that are collected from manufactured products. Rotation angle of the workpiece is determined by a tolerance of 1.5°. It is seen that results have sufficient accuracy for industrial applications.

Keywords: Visual Inspection · Computer Vision · Image Processing · Angle Detection · Feeding Systems First Section

1 Introduction

In industrial applications, detecting angles accurately and efficiently is crucial for maintaining high quality and efficiency in production processes. Machine vision algorithms have become increasingly popular for this purpose due to their ability to provide fast and accurate measurements in a non-contact manner.

One approach for angle detection is to analyse the geometric properties of the image matrix in the machine vision system. Wen et al. [1] developed a system using video frames to detect the movement angle of a test object attached to a rotating disk relative

Z. Şen et al. (Eds.): IMSS 2023, LNME, pp. 275–283, 2024.
https://doi.org/10.1007/978-981-99-6062-0_25

to a marked point on the disk. In systems where the axes of movement change, additional parameters are needed to calculate the product angle. Jie et al. [2] developed a system for faucet quality control that determines the rotation trajectory and centre of rotation of the faucet handle tail using different images. Then, the product centre that can be clearly extracted from a single image of the faucet mechanism is used to calculate the angle between these two points. Shu et al. [3] aim to detect the steering angle of a vehicle using a camera placed on the vehicle. They used the difference between the right and left wheels to determine the direction of the steering. Another approach is to use machine learning algorithms to analyse images and classify them according to their angles. This approach has been used in applications such as detecting the angle of a weld joint [4] and measuring the angle of a drill bit [5].

The accuracy and efficiency of machine vision algorithms for angle detection depend on various factors, such as the quality of the image, the resolution of the camera, the lighting conditions, and the choice of algorithm. Therefore, it is important to carefully select and optimize the parameters of the machine vision system for each specific application.

Vibratory bowl feeders are commonly used in industrial part feeding systems. This system accurately positions parts using traps to determine the product movement behaviour. In this method, the assembly parts encounter traps, which are a mechanical separation method in the direction of movement, ensuring that the product obtains the necessary position for assembly [6]. Traditionally, vibration-based feeding methods may not meet feeding needs in terms of efficiency and accuracy. In this case, machine vision systems that can be integrated into feeding systems are required to transfer product information to the control system. Therefore, the transportation information of the feeding product has been solved by integrating visual control systems into the vibratory feeders, transferring the product's characteristics to the feeding system at specific positions, thereby bridging the gap between the movement system and the feeding system [7]. Haugaard et al. [8] integrated the motion trap with machine vision systems to develop a flexible and fast part feeding system, allowing for the feeding of a wide range of parts.

This study involves a non-contact measurement method to obtain information about the angularity and position of assembly parts. The product positioning information is obtained by analysing the geometric properties of the image matrix in the machine vision system. The angle of rotation of the product is the output of the cosine theorem between the centre point and the product target point relative to a fixed reference point.

To achieve the objectives of this study, a machine vision-based algorithm will be developed for detecting angles in industrial applications. The proposed approach will consist of two main steps, the design of illumination and separation traps for the extraction of product contour regions and the application of geometric analysis of the image matrix for obtaining product position information. The developed algorithm will be integrated into a vibration-based feeding system to obtain accurate and efficient feeding. The details of the proposed approach will be explained in the following section.

2 Material and Method

The feeding system consists of three parts: image acquisition, image processing algorithm, and motion control system dependent on product information. For image acquisition, a 10 MP, 10 FPS area scanning camera and a 35 mm lens were used. The product was illuminated by a 24V LED strip surrounding the product, except for the feeding inlet, to ensure optimal illumination and obtain precise angle information. The motion control system is composed of servo mechanisms and components. Figure 1 shows the outer frame of the product obtained under the machine vision system.

Fig. 1. Machine vision system

The workflow of the visual error detection system is shown in Fig. 2. The product's movement is provided by the vibration unit, and a sensor that checks for its presence at the pick-up point creates one of the image acquisition conditions. When the product reaches its position, the PLC sends a signal to the camera, and image acquisition is carried out by turning on the illumination. The generated image matrix is used to obtain the angle information in the image processing step, and this angle information is transferred to the motion control system. Thus, the control system feeds the products to the feeding system by keeping them in the correct angle.

The first step in the image processing algorithm is to detect the contour of the product. This is achieved by thresholding the image to separate the product from the background. The threshold value is determined by analyzing the histogram of the image. After thresholding, the image is processed using morphological operations to remove noise and fill in any gaps in the product contour. The resulting binary image is then used to extract the product contour using the edge detection algorithm. The product contour is represented as a set of ordered points in the image coordinate system. From these points, the centroid of the product contour is computed, which gives the position of the center of the product. The orientation of the product is then computed by fitting a line through the product contour points and calculating the angle between the line and the horizontal axis. The resulting angle provides the required information for the control system to adjust the product feeding system accordingly.

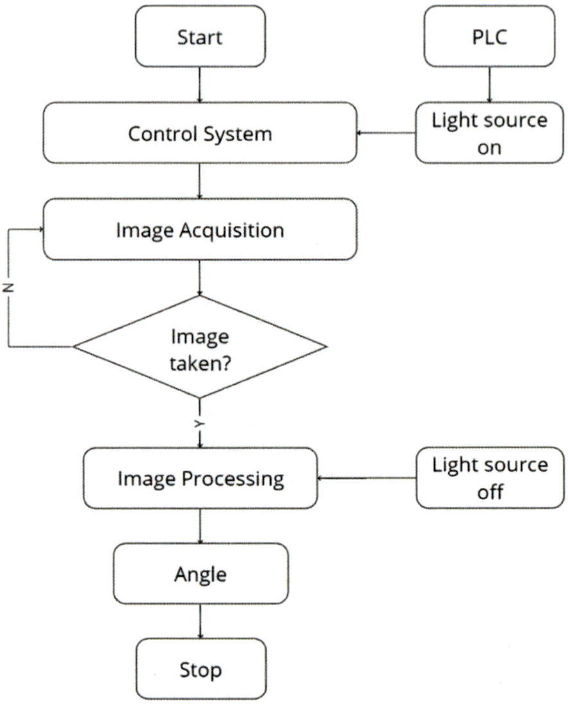

Fig. 2. Workflow of the proposed method

The visual inspection algorithm consists of four main steps. The first step is feature extraction, where the necessary information about the product is extracted from the captured image. In this study, the desired feature is the angle of the straight section of the product with respect to the centre position. The second step is to obtain the angle-length relation vector with respect to the centre of the product, which will be used to determine the angle of the straight section.

The third step involves estimating the coordinates of the product's centre point. Finally, the extracted features are transferred to the motion control system for accurate feeding of the product. The image processing algorithm flowchart is shown in Fig. 3.

The position and angle information of the surrounding points relative to the center of the product create the desired features. Therefore, the outer contour of the product must be clearly defined. Angles that vary towards contour points at a fixed interval from the outer contour center enable the analysis of the image matrix. As a result of the analysis, information is obtained by indexing the angle value with the angle-length mapping vector information. To obtain the product positioning information with high accuracy, the product feature points created at fixed intervals must be analyzed. Thus, the analysis frequency should be increased in the region where the position information of the product is desired to be obtained. In addition, the analysis frequency should be reduced in the remaining regions to increase the speed performance. The images obtained as a result of the analysis are as in Fig. 4.

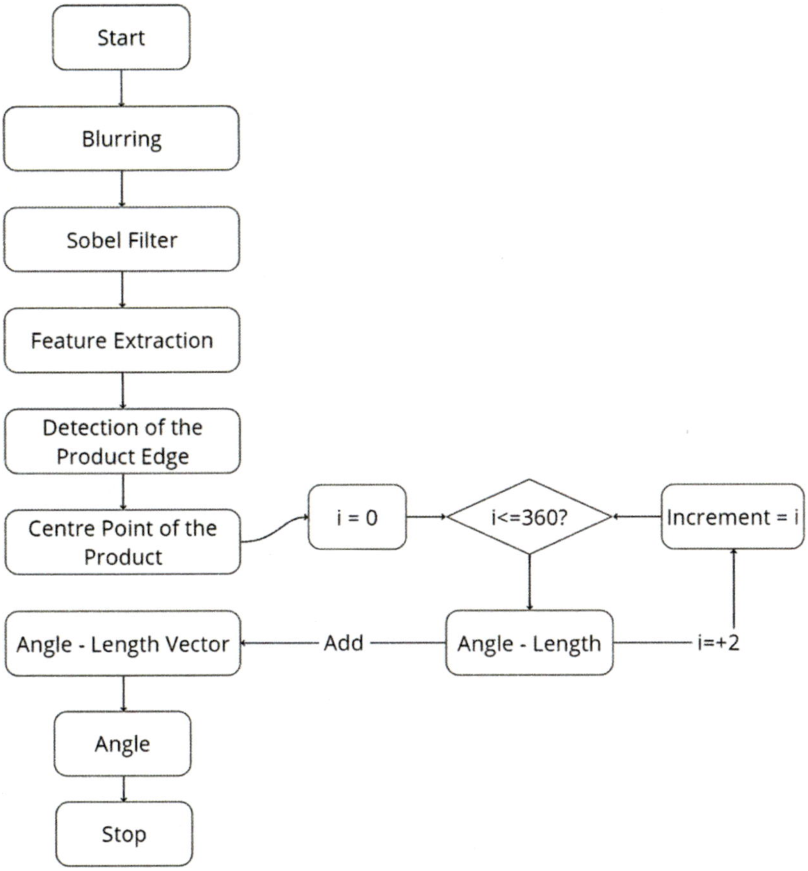

Fig. 3. Workflow of the image processing algorithm

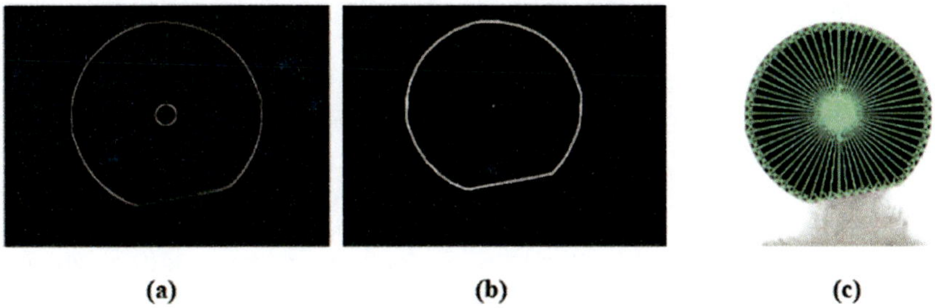

Fig. 4. (a) Sobel filter output (b) Outer contour and center point detection (c) Feature points detection

Therefore, firstly, in Fig. 4a, the Sobel filter and in Fig. 4b, contour-based feature extraction methods are used for shape extraction and product detection. After the product shape extraction, the remaining areas are removed from the image to obtain the feature vector by eliminating the external noise from the image matrix. In the third image in Fig. 4, feature points obtained around the product at fixed intervals from the product center are used for product positioning. Each point has four pieces of information: x and y coordinates of the point, distance of the point from the center, and angle value between the point and the center. The output for product positioning is the angle value, and a fixed center information is required for precise determination of the product's location within the placement compartment. This center information is provided by the circle located in the center of the product due to the trapezoidal shape of the product outer contour. The vectors that are created from the center of a product to the contour of the product with equal angle spacing can be expressed as follows:

Let (x_c, y_c) be the center coordinates of the product, and (x_i, y_i) be the contour points at equal angle spacing with respect to the center of the product. Then, the angle θ_i between the x-axis and the line connecting (x_c, y_c) and (x_i, y_i) can be calculated as follows:

$$\theta_i = \text{atan2}(y_i - y_c, x_i - x_c) \tag{1}$$

where atan2 is the four-quadrant inverse tangent function. The distance r_i between the center of the product and the contour point (x_i, y_i) can be calculated using the Euclidean distance formula:

$$r_i = \sqrt{(x_i - x_c)^2 + (y_i - y_c)^2} \tag{2}$$

Thus, each contour point can be represented by a vector with magnitude r_i and direction θ_i. These vectors can be used to obtain the angle-length mapping vector information required for product feature analysis. The obtained feature vector, sorted by length values according to the angle information, gives the product orientation angle of the point closest to the center position. At the same time, the line formed by the selected point from the center position and the line obtained from the flat part of the product create two perpendicular lines.

To increase the angle precision and reduce the overall processing time, optimization study was carried out to determine the correct frequency of points. Figure 5 shows equally spaced angle/length data points. Correct feeding angle is the minimum value of these data points. As there are missing data between two measured distances, a search algorithm is applied to measure missing points and find optimal result.

In this study, line search algorithm is used to optimize decreasing step size in a 1D vector. Line search is an optimization algorithm used to find the minimum of a function along a given direction. It is a popular method in optimization and machine learning, where it is often used in conjunction with gradient descent or other optimization methods.

The basic idea of line search algorithm is to iteratively search along a given direction for the minimum of the function, using a one-dimensional search. At each iteration, the algorithm takes a step in the given direction, and then computes the function value at the new point. It then searches along the line connecting the old and new points for the

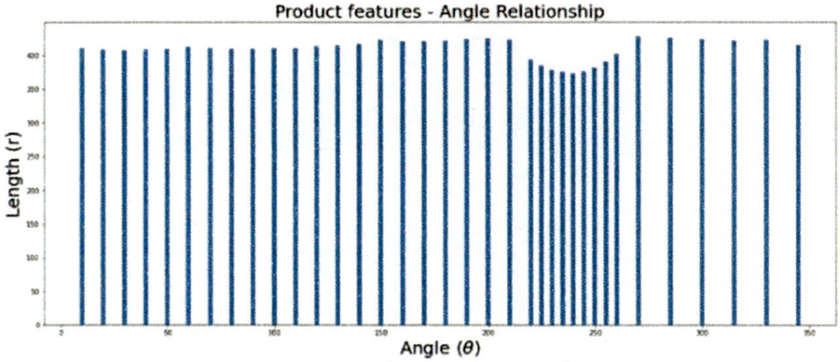

Fig. 5. Measured data

minimum of the function and takes a step in that direction. This process is repeated until convergence is reached. Wolfe conditions provide additional criteria for step acceptance and help to ensure convergence to the minimum of the function. They are a set of necessary conditions for convergence that ensure that the step size is not too small or too large. They are defined as:

$$f\left(x_k + \alpha^k d_k\right) \leq f(x_k) + c_1\alpha^k \nabla f(x_k)^T d_k \tag{3}$$

$$\left|\nabla f\left(x_k + \alpha^k d_k\right)^T d_k\right| \leq c_2\left|\nabla f(x_k)^T d_k\right| \tag{4}$$

where c_1 and c_2 are constant parameters, α_k a positive scalar value that determines the step size in the search direction. d_k the search direction vector, which is a descent direction, meaning that the directional derivative of the objective function in the search direction is negative. T a closed interval containing zero, which is used in the line search procedure to find a suitable value for alpha.The interval is usually defined as [0, t_{max}], where t_{max} is a maximum step size. In this study, line search algorithm is applied where all major decreasing criteria is met. Resulting image is shown on Fig. 6.

The desired point is the point where the length value depending on the angle is minimum. As a result of the analysis, the points shown in blue show the regions where the step size decreases, and the points shown in green show the regions where the step size increases. Between these points, line search algorithm is carried on both directions to ensure finding minimum point. Minimum distance represents angle of the product.

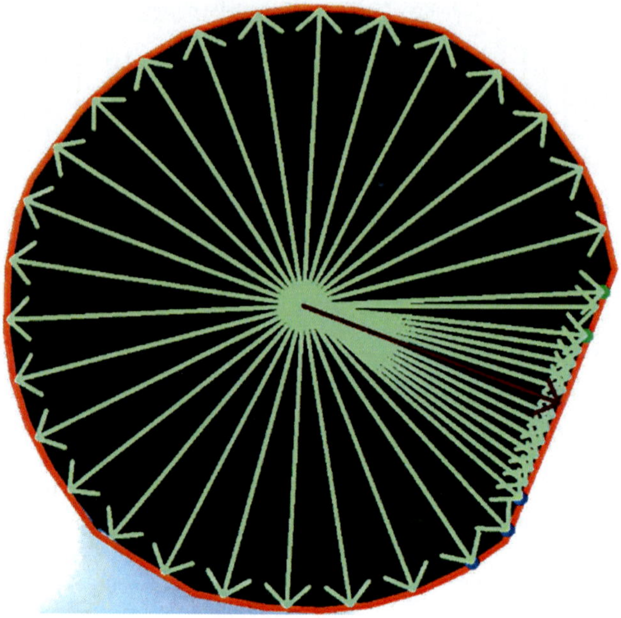

Fig. 6. Resulting image.

3 Conclusion

In this study, a visual control system integrated into the vibratory feeding system was designed. In this system, the product position and rotation information are used in the system work, thus creating a sensitive and efficient feeding system. In our study, in which we perform gradient-based feature vector extraction, the analysis points are increased or decreased by utilizing the geometric features of the product contour. Higher frequency analysis points were created in the part where the product for which the orientation information is wanted to be obtained is flat. In other parts, the analysis points have been reduced and the speed and precision have been increased. The central position of the assembly piece was used as the starting parameter of the analysis process against the problem of the product not being placed in the desired position. The proposed algorithm is applied on 800 images that are collected from manufactured products. Rotation angle of the workpiece is determined by a tolerance of 1.5°. It is seen that results have sufficient accuracy for industrial applications.

References

1. Wen, F., Zou, X., Liu, Y., Li, J.: Detection of movement angle based on machine vision. J. Phys. Conf. Ser. **1168**(1), 012071 (2019)
2. Jie, Z., Li, J., Zeng, J., Li, Y.: Application of machine vision technology in faucet quality control. J. Phys. Conf. Ser. **1756**(1), 012013 (2021)

3. Shu, W., Zhang, L., Xue, S., Wu, Z.: Vehicle steering angle detection based on machine vision. IEEE Access **8**, 88688–88695 (2020)
4. Li, Y., Tian, Z., Li, J., Li, Z.: Automated angle detection of weld joint based on machine learning. J. Intell. Manuf. **32**, 1481–1491 (2021)
5. Kwon, O., Jeon, Y.: Measuring drill bit angles using machine vision and deep learning. J. Mech. Sci. Technol. **33**(2), 759–766 (2019)
6. Mathiesen, S., et al.: Optimisation of trap design for vibratory bowl feeders. In: International Conference on Robotics and Automation (2018)
7. Malik, A., et al.: Advances in machine vision for flexible feeding of assembly parts. In: International Conference on Flexible Automation and Intelligent Manufacturing (2019)
8. Haugaard, R., et al.: A flexible and robust vision trap for automated part feeder design. In: International Conference on Intelligent Robots and Systems (2022)

Integrated Infrastructure Investment Project Management System Development for Mega Projects Case Study of Türkiye

Hakan Inaç[1](✉) and Yunus Emre Ayözen[2]

[1] Directorate for Strategy Development, Head of Investment Management & Control Department, Ministry of Transport and Infrastructure, Ankara, Turkey
hakan.inac@uab.gov.tr

[2] Head of Directorate for Strategy Development, Ministry of Transport and Infrastructure, Ankara, Turkey
yunusemre.ayozen@uab.gov.tr

Abstract. Making investment decisions for infrastructure projects, monitoring and controlling investments is one of the critical issues that policy makers need decision support. Successful integrated project management is needed to evaluate the projects structurally and economically from a holistic perspective. It is necessary to design a framework to manage factors such as financing, technical capacity, contract management, equipment planning, known as limited resources in infrastructure projects. A roadmap guide is needed to manage these projects with limited resources. The use of digital systems in this regard provides an important advantage. Especially in public administration, monitoring the planned, ongoing, and completed projects on a portal with digitalisation opportunities are possible. GIS-based UYS developed by The Ministry of Transport and Infrastructure for digital monitoring of infrastructure projects is described in this article. Within the scope of system engineering, the UYS system has modules such as contract, performance, and planning. Program and cost performance indices can be calculated using the UYS methodology. Thanks to the UYS system, it's possible to keep track of the goals set for contract management and constantly monitor the most critical tasks. The developed data health approach provides to the monitoring of the current data condition on the UYS platform. In addition, other projects related to the mega-project that followed can be monitored via UYS with integrated management. Approaches to the development of these modules are discussed in this paper. With the concept of earned value, the planned projection and actual follow-up for the projects are carried out, and activities with an increased risk level can be closely followed. It is also explained as a case study through the work of "Ankara-Sivas HSR Project", a mega-project in which the earned value approach of the UYS system is followed.

Keywords: Transportation Management System · Cost Performance Index · Earned Value

Z. Şen et al. (Eds.): IMSS 2023, LNME, pp. 284–297, 2024.
https://doi.org/10.1007/978-981-99-6062-0_26

1 Introduction

Project management has become one of the essential topics in recent years that helps all stakeholders of the project duration to succeed. Project life cycle effectively leads to more likelihood of success with project delivery. This duration is composed of different phases. According to Project Management Institute's (PMI) PMBOK (Project Management Body of Knowledge), each project process consists of five basic steps, including initiation, planning, executing, monitoring - controlling and closing. These steps detail every aspect of a project, from minor elements such as dependencies and individual responsibilities, to the basic structure of a project with critical milestones, such as timing and budget.

Understanding the different phases is helpful to ensure that everything that should be considered deeply during the project's lifecycle. It allows the project manager to plan, execute, monitor, and control the project in a structured manner, ensuring that it stays on a monitor and meets the goals and objectives set out in the project charter. Additionally, by following a project lifecycle, the project team can be more efficient and effective in their work, and stakeholders can better understand the project's progress and outcome.

2 Literature Review

Earned value management (EVM) is a project management technique used to measure the performance of a project in terms of its progress and performance. It provides a method for measuring project performance by comparing the amount of work that has been completed to the amount of work planned to be completed. It allows the calculation of cost and schedule variances and performance indices and forecasts the project cost and schedule duration [1].

Earned Value Management (EVM) is commonly used in construction projects and other types of projects such as engineering, manufacturing, and IT projects. In construction projects, EVM can be used to measure work progress and compare it to the project plan and budget. This allows project managers, decision makers to monitor progress, identify differences from the plan, and forecast future project performance. With increasing the project budget, different parts and details should be considered for the rest of the project duration. Mega projects, which have a financial threshold as EUR 75 million [2] need efficient risk management during construction because if there are any possible deviations from meeting their objectives, mega projects are forced to deliver behind schedule and over budget. While managing such risks, each project's stakeholders has some responsibilities [3]. To manage and control these risks, value engineering (VE) has become one of the vital methods throughout the years.

VE was developed during World War II (WWII) by Lawrence Miles of General Electric. Seeking a way to make the most efficient use of war-limited funds and raw materials, Miles devised a team-oriented technique that determines the objective of a project, service, or process; analyses functions; and examines each step for ways to increase efficiency and cut costs and completion time [4]. However, Miles and GE used this method in their studies as "value analysis" instead of value engineering [5]. Over time, value engineering became a method to gather processes and product development.

After WWII, Toyota identified the eight types of waste that should be eliminated from not only their production facilities, but also supplier's facilities in disposing of this waste. Toyota's production system's success in minimizing costs and increasing productivity was accepted as a success of value engineering [6]. Success in value engineering has been significant in other disciplines and provides systematic approaches for the project duration. It is used in various industries and disciplines, including construction, manufacturing, service, healthcare, and government.

All stakeholders in project management progress have issued a series of regulatory policies on their own. To deal with various factors during this process, intelligent services become vital to find a connection and estimate trends of the project among cumbersome and uncertain mass information, providing the most detailed information and suitable solutions for green financial risk, favoring early detection and precise decision-making of green financial risk [7]. Intelligent service management refers to managing and coordinating activities related to intelligent gathering and analysis. This includes collecting, analyzing, and disseminating information relevant to national security and decision-making processes. The goal of intelligence service management is to provide timely and accurate information to decision-makers in order to support informed decision-making. To gather and analyze effectively, it may be necessary to coordinate the efforts of multiple intelligence agencies. This could involve bringing together information from various sources. Once this information has been consolidated, the next step is to develop and implement effective strategies for collecting and analyzing it. Intelligence service management also involves ensuring the protection of classified information and the legality and ethical appropriateness of intelligence activities.

Project management tools and intelligent agents can be used together to improve project management processes. Project management tools are software applications designed to help teams plan, organize, and track the progress of their projects. They typically include task management, time tracking, resource allocation, and project tracking. Intelligent agents, on the other hand, are computer programs that can perform tasks on behalf of a user or autonomously. In project management, intelligent agents can be used to automate routine tasks, such as monitoring project status and sending notifications, freeing up time for project managers to focus on other tasks. When integrated with project management tools, intelligent agents can provide valuable insights and improve the project's overall efficiency.

Construction projects are usually separated into several management phases, such as decision, design, implementation, and transfer [8]. Therefore, Project management is an essential aspect of construction projects, as it helps to ensure that the project is completed on time, within budget, and to the desired level of quality. Effective project management helps to keep all stakeholders informed and on the same page, which can help to minimize delays, reduce costs, and improve overall project outcomes [9].

Effective planning and control methods have a significant impact on the management of the projects [10]. It composes of developing a detailed project plan that includes a budget and schedule, as well as identifying the work breakdown structure (WBS) and the associated cost and schedule estimates. The project plan should be detailed enough to accurately track and report progress and performance. The criteria of time, cost, and quality have become some of the parameters to measure the performance of construction projects [11]. It requires regularly measuring the progress of the work and comparing it to the project plan and budget. This allows project managers to identify variances from the plan and make data-driven decisions to address any issues. Also, while identifying and focusing on the different categories of project stakeholders, defined a successful project as one that is able to meet the needs and expectations of the project stakeholders [12]. Therefore, the project management process should be understandable for all parts of the project and ensure that everyone has the same level of knowledge about the project.

AI related developments have been used at project management related projects. Data selection, data transformation, data mining and using patterns to have an estimate on the relationships between the duration of a previous construction works development at project have been dealt with predictive methods [13]. AI technology, and specifically the area of machine learning, enables machines to improve themselves automatically by using past experiences to make decisions in new situations [14].Through the creation and use of smart analysis techniques to aid decision-making, it is possible to prevent construction process failures before they occur and anticipate possible thresholds requirements during their operational phase [14].

Digitalization in project management can bring many benefits to construction projects, such as improved communication and collaboration among parts of projects, better tracking of project progress, and increased efficiency and cost savings. In the area of digitalization, lifecycle of the project could be monitored easily to monitor project duration prediction, cost prediction, software effort prediction, risk prediction and so on [15]. Digitalization in project management can also help improve communication with stakeholders and clients, and increase transparency and accountability throughout the project. From a project management perspective, digital applications have very extensive areas for infrastructure construction projects because infrastructure works from the point of view of construction operation compose of transport networks like roads, railways, or waterways. Therefore, development in this area has been closely followed by the construction sector.

A data-driven project management methodology is one of the crucial developments in digitalization in project management. This method uses data and analytics to inform and guide decision-making throughout the project management. This can include using data to set project goals, track progress, identify and mitigate risks, and make adjustments to ensure the project is completed on time and within budget. A methodology for project management that is based on data enables project managers to effectively plan, monitor, and control projects, ensuring that they are completed on schedule and within budget. The use of data is widely recognized as a means of improving decision-making by professionals. Data analysis is essential in carrying out risk analysis and in determining appropriate actions to take when uncertainty poses a threat to the project. It involves using data intelligently to facilitate better decision-making [16].

Three vital aspects could be mentioned for project management. One of them is a baseline schedule, which is a project plan that serves as a benchmark for assessing the risk of the project and monitoring its performance as it is carried out [16]. Activities as stated at the schedule gains more importance to follow continuation of the project. Otherwise, it may cause possible delays and costs. Therefore schedule risk analysis could be accepted another aspect for project management. Earned value analysis has gained importance during risk analysis in order to prevent possible delays in resource and time management [17]. Also, project control is another prominent element during project management. The purpose of project control is to ensure that a project achieves its objectives, but currently it's only being considered in terms of measuring performance indicators [18].

3 Methodology

It has become necessary in the digitalized world to monitor many planned, ongoing and completed infrastructure projects, especially with today's technologies. To implement new technologies in infrastructure projects and increase the project management knowledge within the changes in the digital world, Transportation Management System (UYS) has been implemented by the Minister of Transportation and Infrastructure in Türkiye since 2020.

Up to the 2000s, the Ministry of Transport has been followed the projects of its responsibility (see Fig. 1). All developments in the project are reported by the consultant and the contractor to the responsible Minister's members to control and approve the developments in field. These developments are detail-reported to the region and Minister Head Departments to follow the field's cost and physical performance progress. In traditional project management methods, all these information transfer about project progress is done not only via mail or in word/excel type of documents.

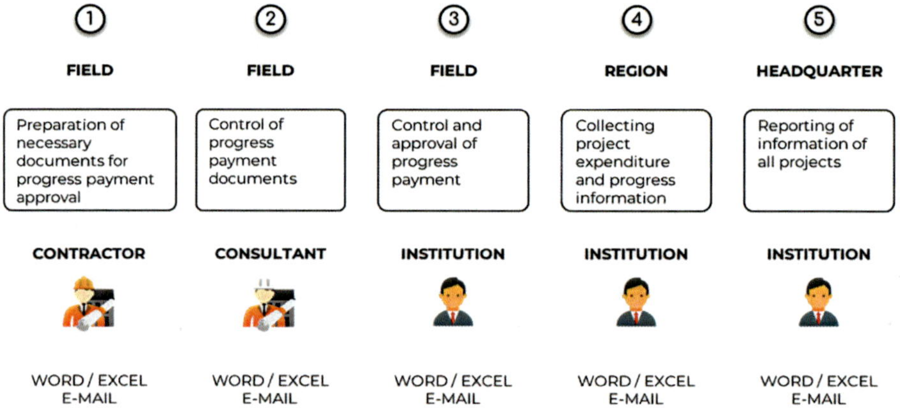

Fig. 1. Change of organizational relations - project reporting process-traditional methodology

UYS is a system in which all project stakeholders are involved. Contractors and consultancy firms have been a part of the system. After preparation of necessary document-related payment is completed, system data entry is expected from Consultancy firms. If there is no consultancy firm on project, Minister's members are duty to give data to system. They must periodically enter data related to performance (physical and cost progress) planning, reports, and photos. The Institutions of Ministry do sustainability of the system. The data entrance process of the system depends on the hierarchy of the institution and if the institution coordinator approves the entered data, system will launch the information. This helps control the data and allows it to be corrected in case of a possible error, before it is reported to the ministry. When each data entry process is completed, information is provided from the system as mail or pop-ups for the approval and control of the relevant units. This data-driven system can be easily connected from anywhere with internet access. Figure 2 represents the data-driven UYS system working principles between all stakeholders.

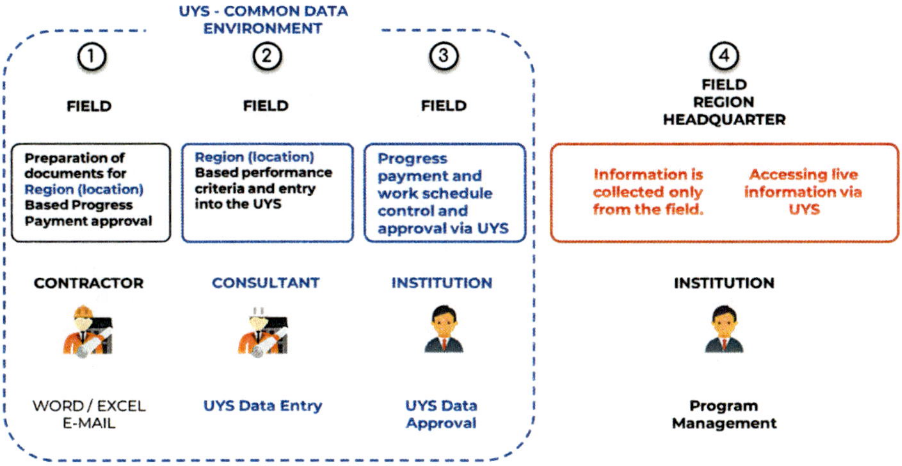

Fig. 2. Change of organizational relations of project monitoring process with UYS methodology

UYS systematically holds past data about the project. Depending on the period information, physical and cost developments, critical activities and opinions of field managers can be accessed. Also, project management must examine the status of projects that are followed up regularly and systematically within the terms of the agreements. Schedules about the budget and physical development at the site have been offered to the Minister at the beginning of the projects. For each of the activity, payment durations for the project have been determined. Still, changes in conditions (problems related construction, related economic factors, etc.) cause the project not to be suitable as stated in time and budget schedule. This force the project managers if these delays are not predictable, so possible delays and budget exceeding conditions could be estimated for Minister of Transport. UYS is data driven based and with the help of periodic monitoring, system provides information about the future estimations about project. Earn value analysis have been done at the system. This helps to see actual physical and budget that has been done and compare the actual and planned situation, which is determined at the contract and schedule of the project. CPI (Cost Performance Index) and SPI (Schedule Performance Index) are used as a parameter for earn value analysis at the system. Parameters are calculated based on the equations below.

$$EV = PPxCV \tag{1}$$

$$\mathrm{CPI} = \frac{\mathrm{EV}}{\mathrm{AC}} \tag{2}$$

$$\mathrm{SPI} = \frac{\mathrm{EV}}{\mathrm{PV}} \tag{3}$$

where,

PP: Physical Progress, which defines the completed work with respect to total scheduled work at the project duration.

CV: Contract Value is a value of the total scope of the project, which is accepted by authority.

PV: Planned Value, which is determined by the cost and schedule baseline.

AC: Actual Cost, which is determined by the actual cost incurred on the project.

EV: Earned Value is the amount earned based on the work completed and is determined by multiplying the percent completed with the budgeted amount.

If everything goes on the same at the schedule of project, CPI and SPI values become 1. However, different situations may affect CPI and SPI values of the project at different periods differently. If CPI and SPI is lower than 1, it tells that the project is above the expected budget and that the project is behind the work schedule. Otherwise, It explains project conditions like that the project is below the expected cost and ahead of the work schedule. Also, earn value analysis provides some information about cost and budget expectation about project based on CPI and SPI.

$$\mathrm{EAC} = \frac{\mathrm{BAC}}{\mathrm{CPI}} \tag{4}$$

$$\mathrm{EPD} = \frac{\mathrm{TD}}{\mathrm{SPI}} \tag{5}$$

where,

EAC: Estimate at Completion, which is determined by periodically or when a significant change happens to the project.

BAC: Budget of Completion, which is sum of all budgets allocated to a project scope.

EPD: Estimated Project Duration which is determined by periodically or when a significant change happens to the project.

TD: Total Duration of the project, which is determined at the contract and time schedule of project.

UYS not only provides information about budget and time schedule performance about projects but also helps to predict the critical activities at project on the map, which may cause the extension of the project duration. The system is working on GIS-based, so project activities could be seen on the map.

4 Case Study of Ankara Sivas High Speed Railway Project

The Ankara-Sivas railway project, located in the East-West development corridors in Türkiye, connects the cities of Ankara and Sivas. It is a significant infrastructure project and the National Transport & Logistics Master Plan of the country defines a project as one of the visionary projects of 2023, which is the 100th anniversary of the republic. The railway is expected to reduce travel time between the two cities from approximately 10 h to just over 3 h, making it a much more convenient and faster mode of transportation for passengers. Additionally, the railway is expected to positively impact the local economy by stimulating economic growth, creating job opportunities, and increasing tourism. Therefore, under construction status of the project has been regularly monitored by the Ministry of Transport.

UYS has been designed as a system where the vision projects of Türkiye can be seen within the scope of the Ministry, in the tender process, in the investment program or within the scope of the Transport & Logistic Master Plan. The current status of the project, field reports and critical activities, and cost and physical progress in all works can be monitored through different modules at the system. Like many other essential and vision projects of Türkiye, Ankara- Sivas High-Speed Railway Project is monitored at UYS.

4.1 Project Identification Information

The project was first included in the investment program in 2007. Until today, a total of 44 different contracts have been signed with different contractors regarding infrastructure, bridges & tunnels and electric-mechanic works (see Fig. 3). The region has become a problematic condition for construction works due to the challenging climate conditions and soil profile of that area.

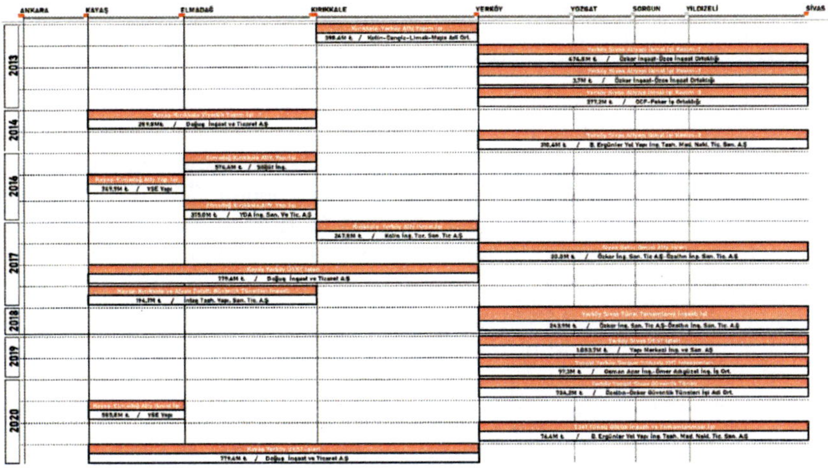

Fig. 3. Contracts made on the line of project over the years

Ankara - Sivas High-Speed Railway line consists of 7 stations. The line from Ankara to Sivas at the end of the production will also make services to Yozgat and Kırıkkale. It will be an essential corridor on the west-east axis of the Turkish railway systems and provides services to 4 cities in Türkiye. Also, the High Speed Rail line is composed of 8 stations (see Fig. 4).

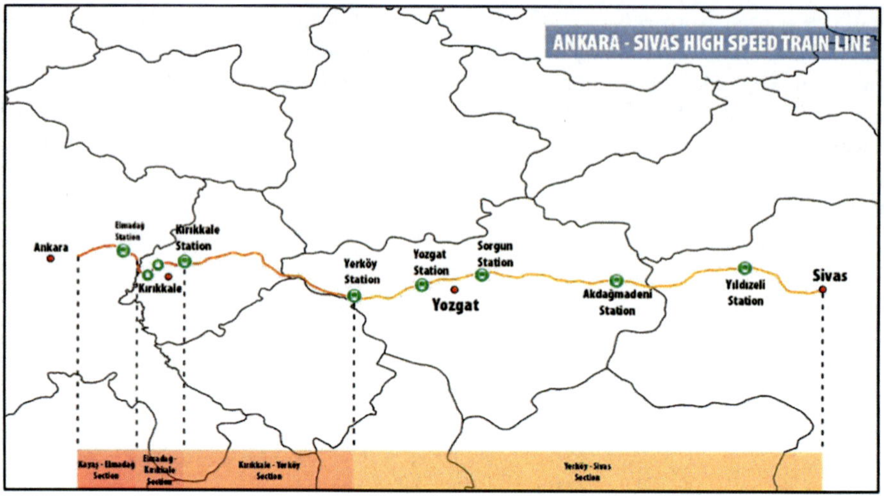

Fig. 4. General view of the Ankara-Sivas HSR line

Construction progress development based on the infrastructure and electric and signalization works is close to complete based on January 2023 site physical progress report. General information for the project and the site's physical developments are summarized below.

Line composed of different construction activities and critical activities. However, critical activity for the line was the construction process at T15 and T16 tunnels at the Elmadağ-Kırıkkale section. For the line to serve, it was necessary to closely monitor the critical activities on the line (T15–T16 tunnels). The ministry has carried out the rough construction processes and then the electrical signaling processes. The stops and critical productions that will be opened when the line is completed are shown in the figure below.

As a result of the close following of current construction activities at site by the Ministry and working with the consultant company to accelerate this process, the construction activities at T15 and T16 tunnels have been completed at the end of 2022. The remaining activities, which are electrical and signalization works, are close to be completed and Ankara Sivas HSR line is expected to be operated on April 23, 2023. One of the critical success of the Ministry of Transport during project management of the project is that digitalization programs (UYS) were used closely in the process. Following of the construction activities, their status according to the work schedule and earned value analyzes were made through the digital platform UYS system. In this way, the monitoring of the construction on site could be handled by the Ministry all over Türkiye easily. This gave a chance to make quick decision-making in the solution of the existing problems and prevented possible threads to extend construction active is the duration at the site (Fig. 5) and (Table 1).

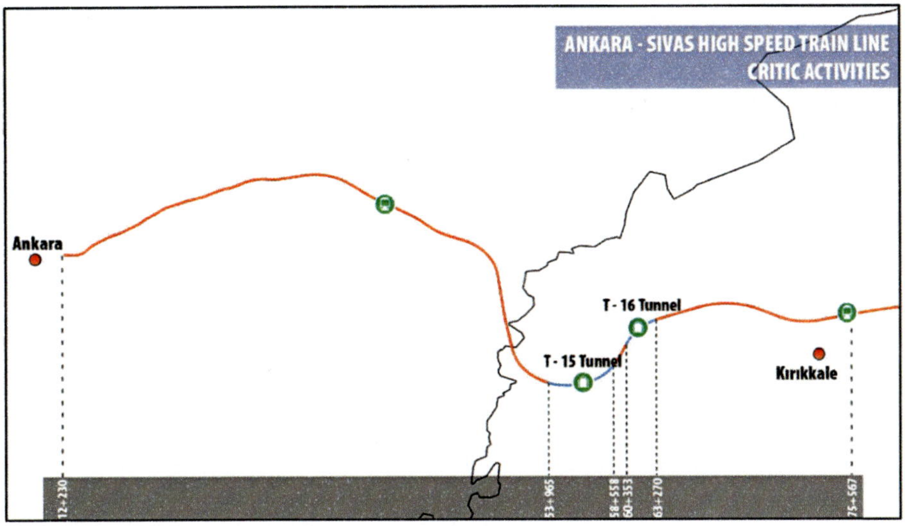

Fig. 5. Critical activities at Ankara Sivas HSR line

Table 1. General information and physical site progress at Ankara-Sivas HSR

General Information	
Length (km)	393 km (151 km: Kayaş-Yerköy/242 km: Yerköy-Sivas
Operational speed (km/h)	250 km/h
Type of operation	Electrical and signalized HSR
Number of station	8 (Elmadağ, Kırıkkale, Yerköy, Yozgat, Sorgun, Akdağmadeni, Yıldızeli and Sivas)
Total number of tunnel	49
Total tunnel length (km)	66,081
Completed tunnel number	49
Completed tunnel construction length (km)	66,081
Most extended tunnel length (m)	5,125
Total Number of Viaduct	49
Completed Viaduct Number	49
Completed viaduct length	27,211
Longest viaduct length	2,220

(*continued*)

Table 1. (*continued*)

Total volume of Cut (m^3)	114 million
Completed Cut volume	114 million m^3
Total volume of Fill (m^3)	30.9 million
Completed Fill volume	30.9 million m^3
Physical Progress	
Physical Progress at Kayas-Sivas Section (%)	100
Electric & Signal Works Progress	
Signalization and Telecommunication Works Physical Progress at Kayas-Sivas (%)	97
Signalization and Telecommunication Works Physical Progress at Kayas-Sivas (%)	95
Electrification Physical Progress at Ballışeh-Sivas (%)	97
Signalization and Telecommunication Works Physical Progress at Ballışeh-Sivas (%)	95

4.2 UYS Processes in Project Management

Ongoing construction works at related contracts at Ankara-Sivas HSR project are periodically followed up on the UYS system by Ministry of Transportation. Yerköy-Sivas section of the HSR project whose production status is in progress is shown on the system. Regarding the period for which data is monitored on the system, the contract page contains general information about the contract, as well as the details of the contract and importance milestones of the contract, which could be seen as a timeline in this screen as shown in Fig. 6.

The cash and physical progress of the project is updated periodically. The cash progress status according to the physical progress in the field and the physical conditions according to the cash status are included in the system with earned value analysis. The analysis could be seen as UYS systems' graphs and historic bar charts. Estimations based on the previous periods are used for possible extensions of project cost and durations. Using this approach, the project duration situation and possible extensions of the project budget could be viewed at the performance screens at UYS (Fig. 6(a)). As a result of the studies carried out in the section of Ankara-Sivas HSR, cost and physical condition of the construction works is approximately 90% levels. After October 2020, when digitalization works accelerated, cost and schedule performance progresses coincide with what was planned in the relevant business. Studies at Ankara-Sivas HSR that continues in accordance with the work program, goes well according to schedule of the project, so the SPI and CPI value of the is seen as 1 at speedometer charts at UYS.

Critical activities can be seen on the panels. As shown in UYS screen of Ankara-Sivas HSR, critical activities is represented as related icons and condition of them could be followed with the color. All activities at the project is seen as green, which means all

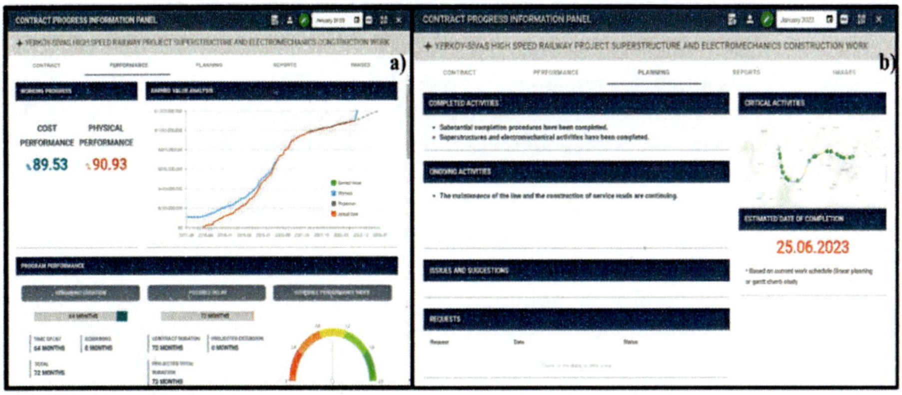

Fig. 6. a) Performance evaluation screen based on CPI, b) Evaluation of critic activities of one of the sections at Ankara-Sivas HSR.

activities goes well according to schedule; otherwise, system represents critic activities as red. During the construction process of Ankara-Sivas HSR, the developments in the T15 and T16 tunnels were followed by authorities on the UYS screens. If the icons It is also possible to have information about the completed, ongoing productions or bottlenecks at the site at the executive summary's panels at UYS planning screens as shown in Fig. 6(b). Based on developments at a construction site, suggestions, demands screens could be these screens can also be displayed in existing or suggestion requests and requests. The system comprises essential documents that have been delivered, which include the approved work schedule by the administration, reports submitted to the administration, and related progress documents. Maintaining these documents in the system enhances accessibility to past records. The digitization of information and documents in the system facilitates decision-making and administrative processes in the project, as it allows for access from various environments. This is a crucial outcome of the digitalization process and UYS allows authorities to see key project reports.

5 Conclusion

The perspective of Ministry's 2053 Türkiye targets, it is essential to follow up and coordinate the ongoing and survey project works. Therefore, studies have been done to lead equipment and employees at site and to follow critical activities. For the effective management of time and money during project schedule, digital project management tools are used all over the world as well as in our country.

UYS, a digital project management tool, has been improved for the Ministry of Transport and Infrastructure and is still used to new versions to meet demands and increase institutional capacity to project management. As seen in the case of Ankara Sivas HSR observations stated in relevant sections in this study, the digital project management tool, UYS, becomes beneficial for efficient project planning. UYS allows project managers and Ministry to plan projects more efficiently by organizing tasks, assigning responsibilities, and setting deadlines based on the earn value analysis. Also, as stated in most of the digital technologies advantage, UYS provides real-time communication for all project stakeholders. UYS facilitates real-time communication between Ministry, clients, consultants and other related stakeholders. This increases collaboration among all stakeholders in the projects. The system could be used anywhere via the internet by computer or phones regardless of geographic location. This enables stakeholders to work together seamlessly and leads to faster project delivery, so any issues can be resolved or be possible to deal with quickly. Moreover, Ankara-Sivas HSR As seen in the case of Ankara-Sivas HSR, activities that may cause delays in time and cost performances of the project timeline are accepted as critical activity. Delays in critical activities such as T15 and T16 cause prolongation in the duration and budget of the project. At this point, UYS has been important for Ministry to take some precautions and quickly make decisions, so UYS is used to relocate resources properly. This ensures that resources, such as time, budget, and personnel, are utilized efficiently, leading to cost savings and improved project outcomes. Overall, digital project management tools are essential for institutions, businesses and organizations to manage projects, collaborate efficiently, and ensure successful project outcomes.

References

1. Naderpour, A., Mofid, M.: Improving construction management of an educational center by applying earned value technique. In: The Twelfth East Asia-Pacific Conference on Structural Engineering and Construction, pp. 1945–1952 (2011)
2. European Commision, Guide to Cost-Benefit Analysis of Investment Projects (2014)
3. Nabawy, M., Khodeir, L.M.: A systematic review of quantitative risk analysis in construction of mega projects. Ain Shams Eng. J. **11**(4), 1403–1410 (2020)
4. Borkenhagen, K.: Value Engineering: An Incredible Return on Investment (1999)
5. Adem Ö.Ğ.Ü.T., Rıfat, İ. R. A. Z., Zerenler, M.: Değer Mühendisliği (Value Engineering) Uygulamalarinin Fonksiyonel Etkinlik Açisindan Işletmelerin Somut Ve Soyut Varliklarina Yönelik Olasi Etkileri. SÜ İİBF Sosyal ve Ekonomik Araştırmalar Dergisi, pp. 51–68 (2007)
6. Örnek, Ş.: Bir Yönetim Tekniği Olarak Değer Mühendisliği. Dokuz Eylül Üniversitesi Sosyal Bilimler Enstitüsü Dergisi (2003)
7. Chen, H., Zhao, X.: Green financial risk management based on intelligent service. J. Clean. Prod. **364**, 132617 (2022)
8. Hu, W.: Information lifecycle modeling framework for construction project lifecycle management. In: 2008 International Seminar on Future Information Technology and Management Engineering (2008)
9. Zwikael, O., Chih, Y.Y., Meredith, J.R.: Project benefit management: Setting effective target benefits. Int. J. Proj. Manage. **36**, 650–658 (2018)
10. Gao, M., Wu, X., Wang, Y.H., Yin, Y.: Study on the mechanism of a lean construction safety planning and control system: An empirical analysis in China. Ain Shams Engineering Journal **14**(2), 101856 (2023)
11. Al-Tmeemy, S.M.H.M., Abdul-Rahman, H., Harun, Z.: Future criteria for success of building projects in Malaysia. Int. J. Proj. Manage. **29**(3), 337–348 (2011)
12. Unegbu, H.C.O., Yawas, D.S., Dan-Asabe, B.: An investigation of the relationship between project performance measures and project management practices of construction projects for the construction industry in Nigeria. J. King Saud Univ.-Eng. Sci. **34**(4), 240–249 (2022)
13. Sheoraj, Y., Sungkur, R.K.: Using AI to develop a framework to prevent employees from missing project deadlines in software projects-case study of a global human capital management (HCM) software company. Adv. Eng. Softw. **170**, 103143 (2022)

14. Guenther, R., Beckschulte, S., Wende, M., Mende, H., Schmitt, R.H.: AI-based failure management: value chain approach in commercial vehicle industry. In: 32nd CIRP Design Conference (2022)
15. Li, H., Cao, Y., Lin, Q., Zhu, H.: Data-driven project buffer sizing in critical chains. Autom. Constr. **135**, 104134 (2022)
16. Vanhoucke, M.: The Data-Driven Project Manager: A Statistical Battle Against Project Obstacles (2018)
17. Song, J., Martens, A., Vanhoucke, M.: Using earned value management and schedule risk analysis with resource constraints for project control. Eur. J. Oper. Res. **297**(2), 451–466 (2022)
18. Kivila, J., Martinsuo, M., Vuorinen, L.: Sustainable project management through project control in infrastructure projects. Int. J. Proj. Manage. **35**(6), 1167–1183 (2017)

Municipal Solid Waste Management: A Case Study Utilizing DES and GIS

Banu Çalış Uslu[1(✉)], Vahit Atakan Kerçek[1], Enes Şahin[1], Terrence Perrera[2], Buket Doğan[3], and Eyüp Emre Ülkü[3]

[1] Department of Industrial Engineering, Engineering Faculty, Marmara University, Istanbul, Turkey
bcalis@marmara.edu.tr

[2] Department of Engineering and Mathematics, Sheffield Hallam University, Sheffield, England
T.D.Perera@shu.ac.uk

[3] Department of Computer Engineering, Faculty of Technology, Marmara University, Istanbul, Turkey
{buketb,emre.ulku}@marmara.edu.tr

Abstract. This research aims to compare two well-known solution methodologies, namely Geographical Information Systems (GIS) and Discrete Event Simulation (DES), which are used to design, analyze, and optimize the solid waste management system based on the locations of the garbage bins. A significant finding of the study was that the application of the simulation methodology for a geographical area of a size of 278 km^2 was challenging in that the addition of the geographical conditions to the developed model proved to be time-consuming. On the other hand, the simulation model that was developed without adding geographical conditions revealed that the number of bins could be reduced by 60.3% depending on the population size and garbage density. However, this model could not be implemented since the required walking distance was higher than 75 m, which is greater than the distance the residents could be reasonably expected to travel to reach a bin. Thus, using a cutoff value of 75 m, the total number of bins can be reduced by 30% on average with regard to the result obtained from the GIS-based solution. This can lead to an annual cost reduction of 93.706 € on average in the collection process and carbon dioxide release reduction of 18% on average.

Keywords: Location Optimization · Municipal Solid Waste Management · Simulation · DES · GIS

1 Introduction

The proper allocation of garbage bins has become an important issue in the efficient management of solid wastes in cities (Rathore et al., 2020). Many cities of developing countries, including the Turkish city of Istanbul, face the problem of an insufficient

This work is supported by the Newton-Katip Celebi Fund, the Scientific and Technological Research Council of Turkey (TUBITAK), and the Royal Academy of Engineering (RAEng) via the Transforming Systems through Partnership Program under Grant Number TSP1307.

Z. Şen et al. (Eds.): IMSS 2023, LNME, pp. 298–311, 2024.
https://doi.org/10.1007/978-981-99-6062-0_27

number of garbage bins in suitable locations. Inappropriate and incomplete positioning of bins causes irregular garbage clusters in the streets, leading to the accumulation of garbage at different points, a large increase in the number of waste collection points, and eventually, to greater labor and costs. Moreover, the presence of multiple collection points adds to the carbon emission and cost in the environment (Das et al., 2020). The current problems in solid waste threaten human health and affect crucial human life issues such as budget losses and environmental problems, which are exacerbated by the increasing population and uncontrolled consumption. A critical issue in solid waste management is the collection and transportation of solid wastes to appropriate locations. The reduction of the costs through the optimization of solid waste disposal points can guarantee resident satisfaction. In consideration of these practical issues, the optimization of the position and rotation of solid waste disposal points has become a requirement for municipalities. Thus, this study aims to solve the garbage accumulation problem in the streets by exploring the optimal location points using simulation and a GIS-based approach for municipal solid waste management. Residents of a neighborhood may have different expectations since some residents may be unwilling to have bins that are too close to their homes in consideration of its certain undesirable consequences such as bad smells, disturbing noises, and heavy traffic while some residents want closely located bins to their residences. The main purpose of the municipality is to satisfy the expectations of its citizens rather than the realization of the aim of this project, which constitutes a challenge to this study.

This research paper aims to optimize the allocation of solid waste bins using a comparative analysis. Although Maltepe is a relatively big area to implement such optimal designs, there are comparable studies focusing on different districts. The overall goals of the optimization of the location of solid waste bins are summarized below:

1. Determining the number of solid waste bin need;
2. Reducing the environmental impacts that are likely caused by regional waste generation;
3. Reducing the costs of the waste collection process through optimization;
4. Increasing social welfare by reducing pollution.
5. Comparison of the effectiveness of DES and GIS methodologies in the location of solid waste bins

Article consists of five sections, which introduce the research objective, review related literature, implement the methodology within the framework of the case handled in this study, discussions of the results, and elaborate on the conclusions of the study, respectively.

2 Related Literature

The effective design of solid waste management ensures the planning of processes that are oriented to reduce operational costs and harmful waste-caused environmental impact (Anvar et al., 2018). Numerous studies have addressed this problem utilizing discrete event simulation (DES), soft computing, mixed integer linear programing and geographical information systems (GIS).

Lyoo et al. (2020) developed a discrete event simulation model that measures residents' satisfaction levels according to their occupations to analyse waste generation and municipal waste management. Ochoa et al. (2017) developed a behavior-based solid waste management model through dynamic multiple simulation. Ramos et al. (2018) developed a dynamic rotation model with stochastic discrete-event simulation approach, taking into account the occupancy rates of solid waste bins via sensors.

Yadav and Karmakar (Yaday and KArmakar, 2020) explained different Collection and Transportation (C&T) techniques worldwide such as backyard, door-to-door, and pneumatic pre-applied. Municipality officials can choose the most appropriate method with reference to their research. Adeleke and Ali (2021) studied the optimization of the waste collection points. Unlike most papers, the sources of waste generation were handled separately. A Lagrangian relaxation (L.R.) and a Simple Linear Heuristic were developed. The location of the sites, the appointment of customers to the sites, and the assignment of bins to the collection sites were included in the model as decision variables. The researchers suggested a course that considers optimal threshold distance and heterogeneous population distribution for future research. Rathore and Sarmah (2019) focused on a location-allocation problem and proposed a mixed-integer linear programming model as a 2-phase methodology. The researchers also contributed a perspective to the literature by considering safety and rag-picking factors.

Badran and El-Haggar (2006) proposed mixed-integer programming with the objective of finding the most appropriate locations of collection sites in Egypt. They aimed to minimize the costs including fixed and operating costs. Samanlioglu (2013) developed a multiobjective mixed-integer model for the collection and transportation of industrial hazmat wastes in the Marmara region of Turkey, which can also be referred to as a multi-criteria decision-making problem. Author tried to minimize the total cost, transportation risk, and site risk. Toutouh et al. (2020) proposed single and multi-objective heuristics based on the PageRank method and two multi-objective evolutionary algorithms (MOEA). Researchers considered three demand levels comprising low, normal, and high for waste generations, which enabled symbolizing the potential variances of waste generation rate throughout the year. Their outcomes demonstrated that the applied MOEAs improved PageRank in all scenarios.

Rossit et al. (2019) developed bi-objective integer programming. Authors aimed to determine the appropriate "garbage accumulation points". And minimizing the investment cost to install the bins and minimizing the routing cost related to the collection frequency. The augmented ε-constraint method (AUGMECON) was employed to solve the problem. Blazquez and Paredes-Belmar (2020) built a MILP to solve bin location allocation. The model's objective was to minimize bin location cost and the model was used to determine the number of bins at each collection site. Then, the researchers built another MILP to determine waste collection routes for the waste collection vehicles, depicting objective function as minimum travel distance with several assumptions.

Geographical Information Systems (GIS), especially ArcGIS, are also another widely used method to optimally allocate solid waste bins (Rathore and Sarmah, 2019; Khan and Samadder, 2016; Erfani et al., 2016). ArcGIS provides comprehensive route optimization tools when routing is of concern. Khan and Samadder (2020) built a model using uniform distances between waste bins with easy accessibility using ArcGIS Network Analyst, thus allowing an easier VRP (Vehicle Routing Problem) approach. One of the functions of ArcGIS, namely Minimize Facilities, can be used in the optimization of bin locations while assuming bins are facilities with coverage areas (Erfani et al., 2016).

Although numerous article on GIS, the majority of these studies diversely employed GIS in location-allocation and vehicle routing problems. Some of them are; Illeperuma and Samarakoon (2010) solved a greenfield-type problem for the bin location-allocation problem. Hemmelmayr et al. (2014) used GIS for processing the data through regression analysis, they formalized linear equations between the areas (m^2) of households. Nithya et al. (2012) studied on the waste coverage of the district. To achieve the optimal coverage, they first determined the number of bins to be allocated through a basic formulation that includes waste generation per day, waste density, the size of bins after determining the type, the filling rate of the bin, and frequency of collection. After necessary information was acquired, a geospatial database was created. Using one of the many built-in objective functions of GIS for maximum coverage, a selection between predetermined coverage radii was made to achieve optimum coverage. Similar to the previous work, Khan and Samadder (2016) determined the number of bins through basic formulation. In continuation, ArcGIS network service area solver and Euclidean distance tools were used to allocate waste bins efficiently. After the allocation of the bins, the ArcGIS network analyst tool was used to solve the vehicle routing problem using the default algorithm of the tool, namely Dijkstra's algorithm.

Erfani et al. (2017) calculated the population distribution in the buildings that were located in their study areas to create a low-bias density map and obtain GIS results in a very straightforward way. The coverage percentages of bins were used to track performance to obtain the lowest number of bins with optimum efficiency. In another article, Vijay et al. (2005) proposed a Triangulated Irregular Network (TIN) to use GIS for the calculation of slopes in the allocation of bins. They aimed to allocate the bins so that the collection process would be mostly descending sloping with regard to the prevalence of collection by cart pullers in a given district at a given time. Then, service areas were determined to maximize the coverage of wastes.

Kallel et al. (2016) solved a vehicle routing problem using the Network Analyst tool of ArcGIS. They compared the initial scenario to three different scenarios, with the final scenario being the reallocation of bins. Although the objective of the problem was the determination of the optimum vehicle route, the final scenario, which was the optimized route using the modified collection method, yielded the best results by far and offered an adequate, if not optimal, bin network. From a different standpoint than that of MSW-related articles, Bolouri et al. (2018) allocated fire stations with respect to the service area. Using the Minimize Facilities tool of ArcGIS, they found a solution in which all study areas were covered by a fire station's 5-min radius. After obtaining results from GIS, they entered the results into a specified Genetic Algorithm (G.A.), which speeds up the G.A. process while finding the near-optimal solution. G.A. works on the sets of current facility locations and potential facility locations. The same processes were also done using the Simulated Annealing algorithm. Their results revealed that after 150 generations per algorithm, G.A. yielded better results, culminating in its acceptance as the solution. The results indicated that the use of GA to find optimal or near-optimal solutions after determining bin locations using ArcGIS could be appropriate in the present study.

Paul et al. (2017) examined the application of the Geographic Information System (GIS) to analyze the garbage bin problem in Kolkata, India, and optimize solid waste collection. Using the projection coordinate system, they geographically referenced and digitized in the ArcGIS environment. The road networks in the Google Earth Environment were updated and added to ArcGIS. The researchers assumed a maximum distance of 500 m between two bins. During the analysis, the entire study area was covered with as little bin space as possible. Das and Bhattacharyya (2015) proposed an optimal waste collection and waste transport route to minimize the length of transport routes for waste collection and garbage trucks. The problem of waste collection and transport was modeled on the hash integer program, which yielded a heuristic solution that can provide the shortest and most convenient way for waste collection and transport. They showed that the proposed solution shortened the current length of the waste collection path by more than 30%.

Ebrahimi et al. (2017) identified two case studies that revealed the change in waste production changes in time, predicted the increase in the amount of recycling through infrastructure changes, and were supported by GIS and geospatial methods, which enable accurate decision-making. They used GIS to spatially identify and analyze solutions for resource estimation and production flows. The kernel density estimation (KDE) method was used for the determination of the ratio of solid waste accumulation in garbage bins. Location allocation analyses were carried out to examine the locations of recycling and bins. The researchers found that waste production was directly or indirectly associated with food sales.

3 Methodology

This research aims to determine the optimal number of solid waste bins and their locations to increase the efficiency of the collection process. The optimization of this process contributes to sustainable municipal management by increasing social welfare, reducing collection costs, and reducing CO2 emissions during collection operations. Stochastic modeling, probability, statistics, simulation, and GIS-based optimization were used to determine the optimal number of solid waste bins and their best locations.

The model developed within the scope of this research was focused on a case study of the Municipality of Maltepe, one of the largest municipalities in Istanbul. A metropolis, Istanbul receives immigration and grows rapidly, leading to unplanned urbanization. The city has 18 districts and a population of 515,021 as of 2020. Although more waste collection shifts are needed, unplanned urbanization obstructs the efficient route planning for waste collection and transportation. For example, the disorganization of the streets leads to a more complex route due to the incorrect positioning of bins. This research aims to determine the optimal number of solid waste bins and their locations to increase the efficiency of the collection process. The optimization of this process contributes to sustainable municipal management by increasing social welfare, reducing collection costs, and reducing CO2 emissions during collection operations. Stochastic modeling, probability, statistics, simulation, and GIS-based optimization were used to determine the optimal number of solid waste bins and their best locations.

In this research paper, the location optimization problem for municipal solid waste management (MSWM) is carried out. The model is referred to as the maximized capacitated coverage problem (MCCP) in reference to the capacity limit of a bin. The methodology of the study consists of four main stages that are:

1. Data Collection
2. Data Analysis
3. Estimation
4. Optimization

3.1 Data Collection

The following information was obtained from Maltepe Municipality as historical data:

- Annual waste amounts per district (Ton), • Daily Waste amount (Kg),
- Number of Metal & Plastic Bins,
- Waste Amount per Bin and
- Bin Fullness Rate for the Districts

Figure 1 shows the distribution of the districts of Maltepe Municipality.

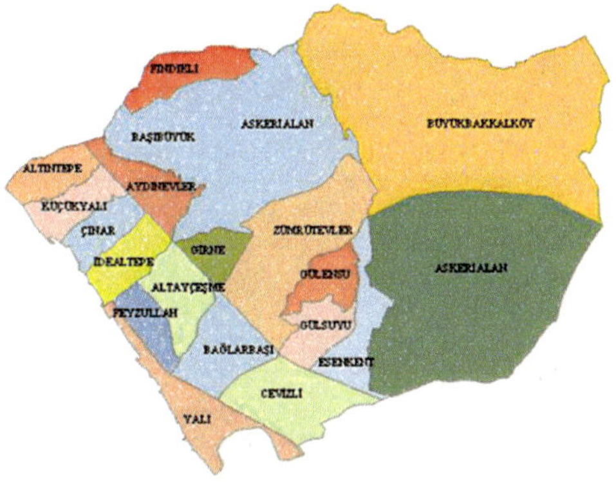

Fig. 1. Districts of Maltepe Municipality

The following information was obtained from ENDEKSA commercial software.

- Each neighborhood's population density (see Fig. 2).
- Characteristics of each neighborhood (education level, gender distribution, etc.

The following information was gathered through observation.

- Time lost per stop

The following information was gathered from the literature.

- Key performance parameters
- Implementation methodology
- CO_2 emission rate

The unit cost of fuel consumption.

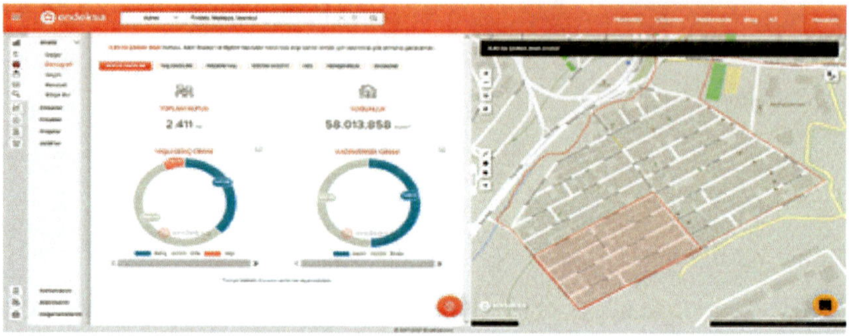

Fig. 2. Population Density in each Neighborhood

3.2 Data Analysis

The data was analysed using the following statistical methods (Fig. 3).

- Frequency analysis (See Fig. 4).
- Data consistency check
- Correlation tests

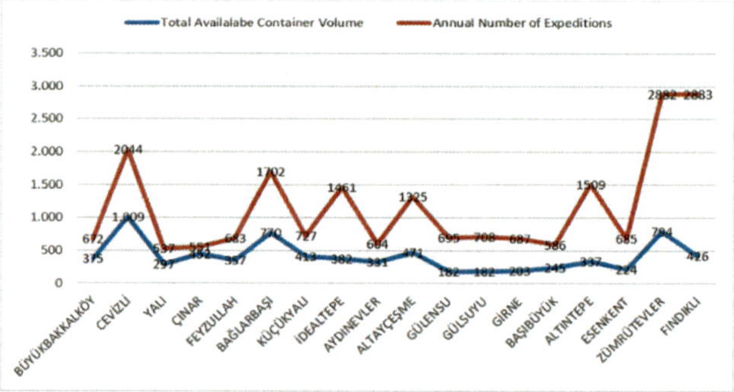

Fig. 3. An example of statistical analysis: Total available bin volume versus the annual number of the expeditions

3.3 Estimation

ARENA simulation was used for the determination of the expected number of solid waste bins that are needed. ARENA input analyzer is used to distribution fitting for neighborhood-based waste density: (See Table 1).

ARENA simulation software is used to process modeling:

- Area-based population
- The volume of waste bins

- Waste density constant (0.12 kg/L)
- Waste per capita distributions

The required number of bins was calculated by dividing the daily waste amount per district by 73.92, which is the desired volume of waste for each bin, i.e., 80% fullness rate. The daily waste amount is the main output of the simulation model. The number of required bins is the processed output to satisfy our requirements.

The model was implemented by Arena Simulation Software 16.10.00001 on Lenovo Legion with 9. Generation Intel® CoreTM i7-9750H CPU and 16 GB RAM.

Table 1. Waste generation probability of each district, acquired via ARENA input analyzer

District	*Waste Amount Expression*	*District*	*Waste Amount Expression*
Altayçeşme	0.3 + WEIB(0.054, 4.27)	Feyzullah	0.25 + WEIB(0.331, 2.83)
Altıntepe	0.16 + 1.53*BETA(10.6, 19.3)	Fındıklı	0.34 + 1.11*BETA(3.8, 4.44)
Aydınevler	0.19 + GAMM(0.032, 11.3)	Girne	NORM(0.713, 0.0926)
Bağlarbaşı	NORM(0.721, 0.0993)	Gülensu	0.24 + LOGN(0.499, 0.153)
Başıbüyük	0.1 + LOGN(0.347, 0.0906)	Gülsuyu	0.01 + 0.03*BETA(10.6, 13)
Büyükbakkalköy	0.71 + GAMM(0.0856, 8.13)	İdealtepe	0.29 + 1.53*BETA(10.9, 16.1)
Cevizli	NORM(0.926, 0.107)	Küçükyalı	0.04 + LOGN(0.421, 0.18)
Çınar	0.14 + LOGN(0.345, 0.125)	Yalı	NORM(0.764, 0.184)
Esenkent	NORM(0.52, 0.096)	Zümrütevler	0.3 + ERLA(0.0262, 12)

3.4 Optimization

Since the values obtained by the simulation method do not include the geographical conditions, the results that were yielded by the simulation were accepted as the initial candidate and the problem was optimized using the ArcGIS software.

3.5 Additional Parameters

Following additional parameters are used to analyze the system.

- ENDESA could provide information for a minimum of 5 hectares per hub. Thus, each neighborhood was divided into hubs with respect to this value and 75 m were accepted as the cutoff for each bin (See Fig. 5).

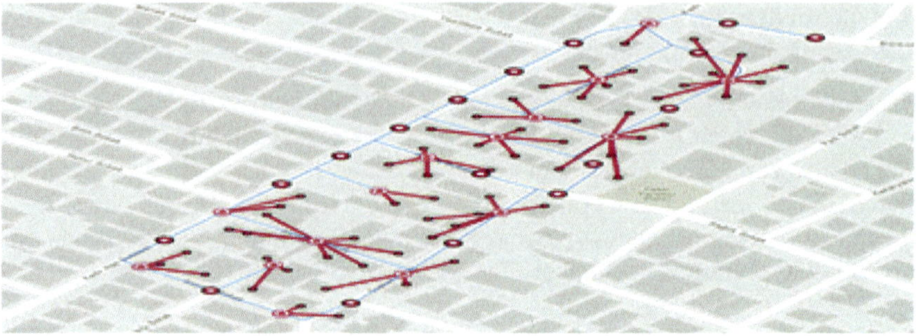

Fig. 4. An example solution using 75 m cutoffs

- Boundaries of the waste collection areas, namely hubs (Endeksa).
- Road networks of the hubs (ArcGIS Pro Map, Google Maps).
- Locations of waste sources namely demand points (ArcGIS Pro Map, Google Maps).
- Population and area of each hub (Endeksa).
- Daily waste production rate per capita for each district (Maltepe Municipality).
- Waste bin capacity (Maltepe Municipality).
- The initial solution to input location-allocation analysis (optional, previous simulation results).
- The road network of each hub is evaluated in its hub only.
- Every bin has the same capacity of 92.4 kg of waste.
- Each demand point has a predetermined weight value according to its hub. This value is the same for all demand points within a particular hub.
- Each demand point can be allocated to one bin only.
- Initial solutions are taken from the results from the simulation model for each hub of each district.
- Solutions are made concerning 100% coverage, meaning no demand point is left unallocated.
- The cost function that calculates the distance between bins and demand points was set to a power function and its parameter was set to 2.

4 Analysis of Results

One of the significant findings obtained from the research is that the application of the discrete event simulation methodology to a geographical area of this size (278 km^2) was challenging and time-consuming during the insertion of the geographical conditions into the developed model. On the other hand, the simulation model that was developed without adding geographical conditions revealed that the number of bins that were needed could be reduced by 60.3% depending on the population and garbage density. However, this model could not be implemented considering the walking distance of above 75 m, which was set as the upper limit of distance the residents could travel.

Moreover, the results of the GIS solution revealed that the current number of bins that were placed by the municipality in Başıbüyük, Esenkent, and Zümrütevler neighborhoods were insufficient at the rates of 81%, 9%, and 23%, respectively. However, the number of bins in the remaining 15 neighborhoods exceeded the needed number (See Table 2). The results of optimization revealed that a total of 1408 extra bins were used throughout the Municipality.

Table 2. Comparison of the current garbage bins in each district and the results obtained from DES and GIS

District	*Current Bins*	*DES Bins*	*GIS Bins*	*Difference*	*Difference (%)*
Altayçeşme	487	287	390	−97	−20%
Altıntepe	423	238	321	−102	−24%
Aydınevler	325	179	239	−86	−26%
Bağlarbaşı	907	460	463	−444	−49%
Başıbüyük	201	117	363	162	81%
Büyükbakkalköy	315	231	291	−24	−8%
Cevizli	1084	450	520	−564	−52%
Çınar	540	189	217	−323	−60%
Esenkent	267	150	292	25	9%
Fethullah	339	184	266	−73	−22%
Fındıklı	365	559	607	242	66%
Girne	170	166	222	52	31%
Gülensu	237	148	333	96	41%
Gülsuyu	229	147	275	46	20%
İdealtepe	407	274	331	−76	−19%
Küçükyalı	513	152	226	−287	−56%
Yalı	378	116	254	−124	−33%
Zümrütevler	747	738	916	169	23%
Total	**7934**	**4785**	**6526**	**−1408**	**−18%**

The environmental and economic benefits of the reductions are as follows: (See Table 3).

- A reduction of approximately 342.9 tons in CO2 emissions can be achieved,
- Fuel consumption can be reduced by about 127,932.5 L, And, hence,
- A cost reduction of 930,069.275 TL (93,127.81 Euros) can be achieved from the share that is allocated by the municipality to the solid waste collection system.

Furthermore, the results indicated that sustainable solid waste management could be achieved to a greater degree. The improvement in social welfare through the ease of access (75 m cutoff for each bin) that is provided to the residents by placing bins at the right points and improvements in services through the increased share from the municipal budget.

Table 3. CO_2 and cost reduction of each district

District	*Co_2 Emission Difference (Ton) in a Year*	*Reduced Diesel Consumption (L) in a Year*	*Reduced Cost (TL) in a Year*
Altayçeşme	−23,7	− 8.851,25	− 64.348,59
Altıntepe	−24,9	− 9.307,50	− 67.665,53
Aydınevler	−21,0	− 7.847,50	− 57.051,33
Bağlarbaşı	−108,6	− 40.515,00	− 294.544,05
Başıbüyük	39,6	14.782,50	107.468,78
Büyükbakkalköy	−5,9	− 2.190,00	− 15.921,30
Cevizli	−137,9	− 51.465,00	− 374.150,55
Çınar	−77,5	− 28.926,25	− 210.293,84
Esenkent	6,1	2.281,25	16.584,69
Fethullah	−17,9	− 6.661,25	− 48.427,29
Fındıklı	59,2	22.082,50	160.539,78
Girne	12,7	4.745,00	34.496,15
Gülensu	23,5	8.760,00	63.685,20
Gülsuyu	11,2	4.197,50	30.515,83
İdealtepe	−18,6	− 6.935,00	− 50.417,45
Küçükyalı	−70,2	− 26.188,75	− 190.392,21
Yalı	−30,3	− 11.315,00	− 82.260,05
Zümrütevler	41,3	15.421,25	112.112,49
Total	**−342,9**	**−127,932.5 L**	**−930,069.275**

Maximized capacitated coverage method was used to solve the problem via ArcGIS software, and the location of each bin was determined for each neighbourhood. Figure 5 and Fig. 6 show the areas of the bins in the Fındıklı and Altayçeşme districts as examples.

5 Conclusions

An environmentally friendly, green, and intelligent solid waste management is of great importance, especially considering the limited resources of developing countries. In this study, a simulation study was carried out to obtain the initial results, which were then introduced as input variables to the GIS system for different threshold values and analysis. The number of solid waste bins was estimated using vehicle activity data, bin types, numbers, location information, waste amounts, and population settlement information to achieve optimal solid waste management in Maltepe, Istanbul.

Based on the GIS-based solution model, the number of bins was reduced from 7934 to 6526. The results showed that CO2 emissions could be reduced up to 342.9

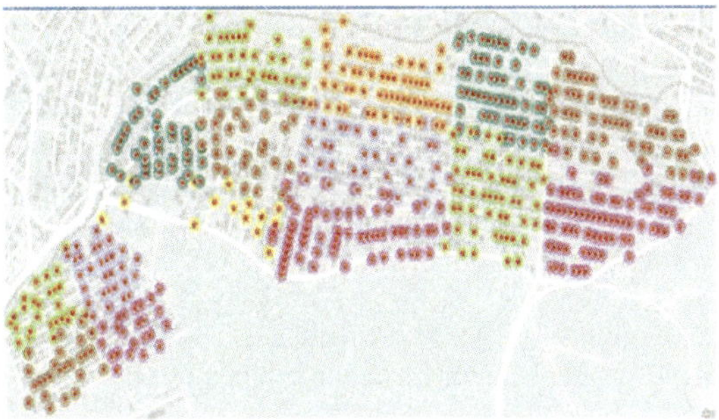

Fig. 5. Bin locations for Fındıklı

Fig. 6. Bin locations for Altayçeşme

tons annually. This reveals that our primary goals to improve environmental health and establish balanced distribution and resource use were achieved. In addition, according to the results of this study, an annual fuel saving of 127,932.5 L is likely, which corresponds to approximately 930,069.275 TL (93,706 EURO).

The municipality can promptly implement the study results and achieve considerable improvements in solid waste management. Furthermore, a separate study is planned to calculate the optimum route for the optimized bin locations.

References

Auth Adeleke, O.J., Ali, M.M.: An efficient model for locating solid waste collection sites in urban residential areas. Int. J. Prod. Res. **59**(3), 798–812 (2021)

Anwar, S., Elagroudy, S., Razik, M.A., Gaber, A., Bong, C.P.C., Ho, W.S.: Optimization of solid waste management in rural villages of developing countries. Clean Technol. Environ. Policy **20**(3), 489–502 (2018)

Arebey, M., Hannan, M., Basri, H., Abdullah, H.: Solid waste monitoring and management using RFID, GIS and GSM. In: 2009 IEEE Student Conference on Research and Development (SCOReD), pp 37–40. IEEE (2009)

Badran, M., El-Haggar, S.: Optimization of municipal solid waste management in port said-Egypt. Waste Manage. **26**(5), 534–545 (2006)

Blazquez, C., Paredes-Belmar, G.: Network design of a household waste collection system: a case study of the commune of Renca in Santiago, Chile. waste Manage. **116**, 179–189 (2020)
Bolouri, S., Vafaeinejad, A., Alesheikh, A.A., Aghamohammadi, H.: The ordered capacitated multi-objective location-allocation problem for fire stations using spatial optimization. ISPRS Int. J. Geo Inf. **7**(2), 44 (2018)
Chalkias, C., Lasaridi, K.: A GIS based model for the optimisation of municipal solid waste collection: the case study of Nikea, Athens, Greece. WSEAS Trans. Environ. Dev. **5**(10), 640–650 (2009)
Coutinho-Rodrigues, J., Tralhão, L., Alçada-Almeida, L.: A bi-objective modeling approach applied to an urban semi-desirable facility location problem. Eur. J. Oper. Res. **223**(1), 203–213 (2012)
Das, R., Shaw, K., Irfan, M.: Supply chain network design considering carbon footprint, water footprint, supplier's social risk, solid waste, and service level under the uncertain condition. Clean Technol. Environ. Policy **22**(2), 337–370 (2020)
Das, S., Bhattacharyya, B.K.: Optimization of municipal solid waste collection and transportation routes. Waste Manage. **43**, 9–18 (2015)
Di Felice, P.: Integration of spatial and descriptive information to solve the urban waste accumulation problem. Procedia Soc. Behav. Sci. **147**, 182–188 (2014)
Ebrahimi, K., North, L., Yan, J.: GIS applications in developing zero-waste strategies at a mid-size American university. In: 2017 25th International Conference on Geoinformatics, pp 1–6. IEEE (2017)
Erfani, S.M.H., Danesh, S., Karrabi, S.M., Shad, R.: A novel approach to find and optimize bin locations and collection routes using a geographic information system. Waste Manage. Res. **35**(7), 776–785 (2017)
Ghiani, G., Laganà, D., Manni, E., Triki, C.: Capacitated location of collection sites in an urban waste management system. Waste Manage. **32**(7), 1291–1296 (2012)
Hemmelmayr, V.C., Doerner, K.F., Hartl, R.F., Vigo, D.: Models and algorithms for the integrated planning of bin allocation and vehicle routing in solid waste management. Transp. Sci. **48**(1), 103–120 (2014)
Illeperuma, I., Samarakoon, L.: Locating bins using GIS. Int. J. Eng. Technol. IJET-IJENS **10**(02), 75–84 (2010)
Kallel, A., Serbaji, M.M., Zairi, M.: Using GIS-based tools for the optimization of solid waste collection and transport: case study of Sfax City, Tunisia. J. Eng. 2016 (2016)
Kao, J.J., Lin, T.I.: Shortest service location model for planning waste pickup locations. J. Air Waste Manag. Assoc. **52**(5), 585–592 (2002)
Karadimas, N.V., Loumos, V.G.: GIS-based modelling for the estimation of municipal solid waste generation and collection. Waste Manage. Res. **26**(4), 337–346 (2008)
Khan, D., Samadder, S.: Allocation of solid waste collection bins and route optimisation using geographical information system: a case study of Dhanbad City India. Waste Manage. Res **34**(7), 666–676 (2016)
Lyoo, C.H., Jung, J., Choi, C., Kim, E. Y.: Modeling and simulation of a municipal solid waste management system based on discrete event system specification. In: Proceedings of the 11th Annual Symposium on Simulation for Architecture and Urban Design, pp. 1–8 (2020)
Nithya, R., Velumani, A., Kumar, S.: Optimal location and proximity distance of municipal solid waste collection bin using GIS: a case study of Coimbatore city. WSEAS Trans. Environ. Dev. **8**(4), 107–119 (2012)
Ochoa, A., Rudomin, I., Vargas-Solar, G., Espinosa-Oviedo, J.A., Pérez, H., Zechinelli-Martini, J.L.: Humanitarian logistics and cultural diversity within crowd simulation. computacion sistemas, **21**(1), 7–21 (2017)
Paul, K., Dutta, A., Krishna, A.: Location/allocation of waste bins using GIS in Kolkata municipal corporation area. Int. J. Emerg. Technol. **8**(1), 511–520 (2017)

Ramos, T.R.P., de Morais, C.S., Barbosa-Povoa, A.P.: The smart waste collection routing problem: alternative operational management approaches. Expert Syst. Appl. **103**, 146–158 (2018)

Rathore, P., Sarmah, S.: Modeling transfer station locations considering source separation of solid waste in urban centers: a case study of Bilaspur City, India. J. Clean. Prod. **211**, 44–60 (2019)

Rathore, P., Sarmah, S.P., Singh, A.: Location-allocation of bins in urban solid waste management: a case study of Bilaspur City, India. Environ. Dev. Sustain. **22**(4), 3309–3331 (2020)

Rossit, D.G., Nesmachnow, S., Toutouh, J.: A Bi-objective integer programming model for locating garbage accumulation points: a case study. Revista Facultad de Ingeniería Universidad de Antioquia **93**, 70–81 (2019)

Samanlioglu, F.: A multi-objective mathematical model for the industrial hazardous waste location-routing problem. Eur. J. Oper. Res. **226**(2), 332–340 (2013)

Toutouh, J., Rossit, D., Nesmachnow, S.: Soft computing methods for multiobjective location of garbage accumulation points in smart cities. Ann. Math. Artif. Intell. **88**(1), 105–131 (2020)

Vijay, R., Gupta, A., Kalamdhad, A.S., Devotta, S.: Estimation and allocation of solid waste to bin through geographical information systems. Waste Manage. Res. **23**(5), 479–484 (2005)

Yadav, V., Karmakar, S.: Sustainable collection and transportation of municipal solid waste in urban centers. Sustain. Cities Soc. **53**, 101937 (2020)

A Development of Imaging System for Thermal Isolation in the Electric Vehicle Battery Systems

İlyas Hüseyin Güvenç[1(✉)] and H. Metin Ertunç[2]

[1] Robo Automation and Engineering Company, Gölcük, Kocaeli, Turkey
ilyashguvenc@gmail.com

[2] Mechatronics Engineering Department, Kocaeli University, İzmit, Kocaeli, Turkey
hmertunc@kocaeli.edu.tr

Abstract. As the intelligent technologies progress, the control systems and designing of efficient process systems has become more important. Besides that, analysis of data in the production lines play crucial role in the process. One of the essential parts of the systems in electric vehicles is battery systems and cooling of these batteries can be implemented by thermal isolators. In this study, an industrial imagining control system using profiler laser sensor for thermal mastic surface mapping was developed. For this, 3D image data were acquired and analyzed by a Laser Profiler Sensor that attached on a robot. In that case, the size of the mastic lines was measured and checked whether the sizes remained within tolerances which are defined by the car producers. Location data are sent to the robot and sealing process begins as defined. When processing is done, the operator is visually informed about the results. The operator can see mistakes on the mastic lines that can't be seen by eyes. Thus, applications that out of tolerances can be determined.

Keywords: Profiler Sensor · Thermal Isolation · 3D Visualization · Electric Vehicles

1 Introduction

'Industrial 4.0' is a new concept in industrial engineering that was first introduced in an article published by the German government in November 2011 as a high-tech strategy for 2020. This concept does not only cover industrial revolution stages but also future industry development trends to achieve smarter manufacturing processes such as reliance on Cyber-physical systems (CPP), construction of Cyber-physical manufacturing systems (CPP), and implementation and operation of smart factories [1]. On the other hand, one of the most important technological developments in the automotive sector have been defined as CASE (Connectivity, Autonomy, Shared Mobility, Electrification) [3]. As can be seen in Fig. 1, the ages of the industrial revolution are illustrated.

In this paper, the thermal sealant process was applied to the coolers that are placed between the battery cells in accordance with the standards. Then a computer software has been prepared to process the data to be collected with precise measurement and control. The simulation of the thermal sealant application process in the factory environment is given in Fig. 2.

Z. Şen et al. (Eds.): IMSS 2023, LNME, pp. 312–319, 2024.
https://doi.org/10.1007/978-981-99-6062-0_28

Fig. 1. The Ages of Industrial Revolution

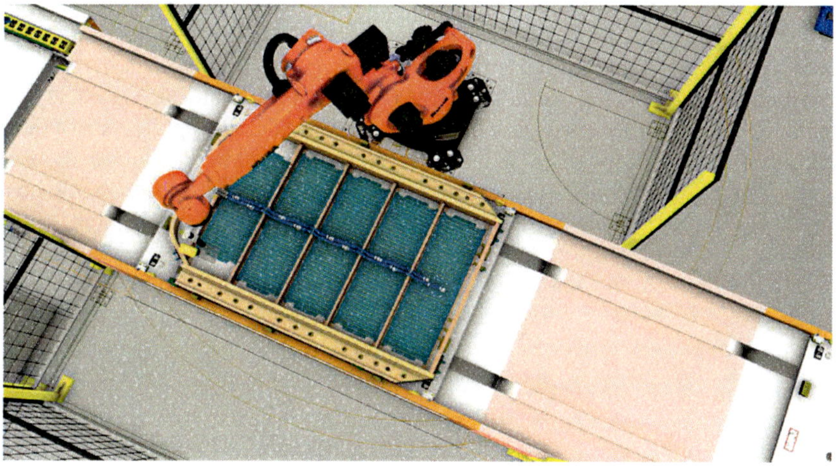

Fig. 2. Simulation of Thermal Sealant Application Process

2 Material and Method

One of the most important factors determining the life and stable operation of battery cells is cooling in accordance with the standards. To perform this process in tolerances, it is necessary to apply thermal sealer between the coolers and the battery cells with a standard and optimal application. The software prepared within the scope of the study, which the mastic application was performed in a stable and high precision control was performed after the application was made.

The robot sealant application trajectory, which is written according to the reference part in the current system, creates problems such as the wrong amount of sealant and incorrect application trajectory in parts that come in different geometries and positions. As a result of the values obtained from the sensor, the calibration of the part to be applied sealant on the X-axis has been ensured. According to the calibration values, the working principle to be performed during the thermal sealant application has been arranged in the robot program.

The heat transfer sealant process applied to the battery unit, which is the most important system component in electric vehicles, is a critical operation for the life and efficiency of the battery. The quality problems that may occur in the mastic operation are among the important factors of the battery life. The software and system of the cooling will be provided appropriately in this study. Then a healthy battery system will have been developed by minimizing possible problems.

2.1 Profiler Sensor

It is the sensor that creates the x and z coordinates of each point on this line according to the sensor by sending laser light in the form of a line. The profile sensor is attached to the end of the robot that will apply the sealant. While the robot is applying the sealant, the profile sensor positioned just behind it measures the height and width of the applied sealant. Considering the characteristics of the sensor, the profile sensor can make a measurement in 250 ms. UDP (User Datagram Protocol) communication system is used to quickly retrieve and process these values. It is expected that the application of thermal sealant to the battery cooling plate will be 7.2 mm (±0.6 mm) high, 3.6 mm (±0.6 mm) wide, 272 mm (±3 mm) long. At this sensitivity, the laser profiler sensor should be able to make measurements with a maximum resolution of 49 microns in height and 123 microns in width. The profile sensor has no impact, contamination, etc. it is placed at a height of 130 mm from the application plate so as not to be affected by environmental factors.

2.2 Robot and PLC

The speed of the robot to be used in the system will be 300 mm/s during the thermal sealant application. This makes it possible to process data at 75 microns, when the aforementioned 250 microsecond data transfer rates are exceeded. The algorithm showing the general structure of the robot program is given in Fig. 3.

The PLC program communicates between robot and computer software. The commands from the computer software are transferred to the PLC program, from the PLC program to the Robot program. In this context, the whole system is working in relation to each other. The working algorithm of the PLC program is given in Fig. 4.

2.3 Computer Application

Computer software is the main component of the applied mastic scanning and control system. The software works interactively with the PLC and the robot to manage the system.

The UDP connection protocol is used to read the data quickly. However, MSSQL database is preferred for the storage of the read data. With the acquisition of big data, procedures have been written in the database and in the software to organize the data and keep its size low. At the same time, a single line of data is recorded for each 'sealer' data read from the profile sensor. In this context, the conversion of the data to the 'JSON' format and the increase in the size of the database has been prevented.

The instant data received from the profile sensor has been filtered and the recording of meaningless data has been prevented. The software adds an error log when the width and height values are outside the tolerance values. In this context, because of recording the error, mastic number and layer number on a battery plate, the location of the error can

Step-1 : Start
Step-2: Recieve the command from PLC
Step-3: Go to starting position
Step-4: Receive the command from PLC
Step-5: Go to endpoint
Step-6: Receive the command from PLC
Step-7: Profile sensor reads mastic
Step-8: Send reading start signal to PLC
Step-9: Send reading finish signal to PLC
Step-10: Finish mastic read
Step-11: If mastic read is small than 30,
y= y+5 and go to step-7.
Step-12: Go to starting position
Step-13: Wait the command from PLC
Step-14: Finish

Fig. 3. Robot Off-line Path Algorithm

Step-1 : Start
Step-2: Recieve value from system
Step-3: Send the command to Robot
Step-4: Receive the command from Robot
Step-5: Send the value to system
Step-6: Finish

Fig. 4. PLC Algorithm

also be traced through the real mastic. The home screen of the prepared software includes connectivity features. This display includes multiple PLCs, profile sensors and features that can be connected to a computer. There are also features that database connection information can change from the home screen interface. Figure 5 illustrates the working principle of computer software.

Within the general features of the software, the reading and monitoring functions are added. The format of the data requested to be read from the profile sensor can be set in this method. The necessary functions are also written for calibration settings of the profile sensor. The necessary settings for reading and integrating different types of mastic systems have been added. The main screen in the software is given in Fig. 6.

The Display Profile screen is designed to provide a snapshot of data from the profile sensor. Thanks to this interface, data exchange is checked instantly. The form size of the data is given in Fig. 7.

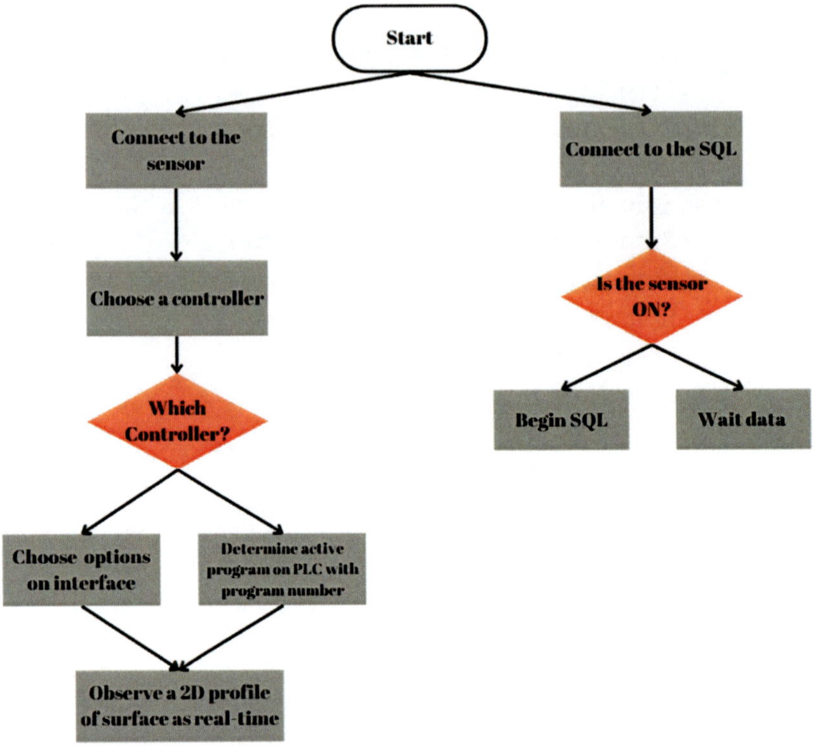

Fig. 5. Flow Chart of Computer

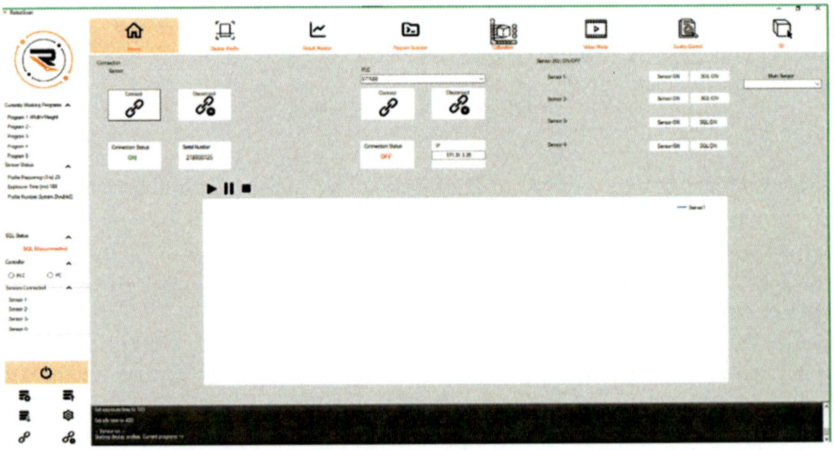

Fig. 6. Interface Design of Computer

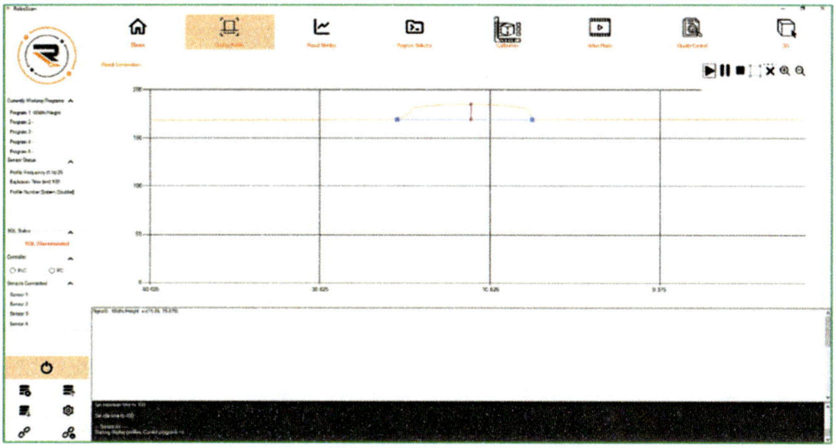

Fig. 7. Display of Profile of Seal

The software has four reading functions. Thanks to these functions, the data formats that are desired to be analyzed are defined from the interface. According to the selected reading function, the values of the data received by sending the profile command to the sensor also change. In this case, the data formats to be controlled can also be edited from the interface screen. In Fig. 8, the Program Selector screen design is given.

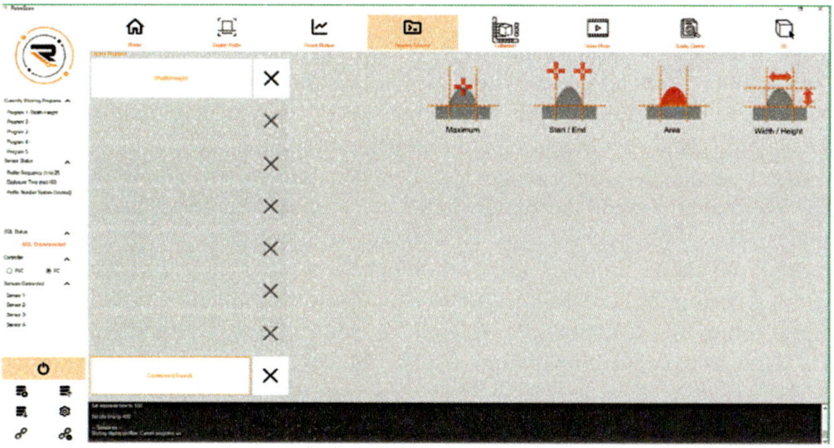

Fig. 8. Program Selector Interface

Additional features of the software include screens with calibration, 3D visualization, video recording and quality control parameters. The software has been designed to be used by the operator in the factory environment and has been expanded with additional features. The images of the other screens are given in Fig. 9.

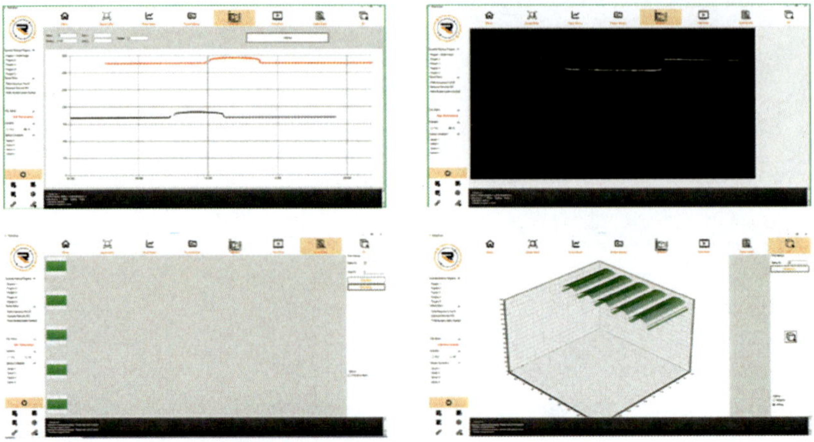

Fig. 9. Additional Features of Interface

3 Conclusion and Discussion

In conclusion, the development of intelligent technologies has brought about significant advancements in control systems and efficient process designs, particularly in the realm of electric vehicles and battery systems. The focus on data analysis in production lines has become crucial for ensuring high-quality outputs. This study presents an industrial imaging control system that utilizes a profiler laser sensor to map the thermal mastic surface, an essential component for cooling the batteries. By acquiring and analyzing 3D image data using a Laser Profiler Sensor attached to a robot, the size of the mastic lines can be measured and checked against predefined tolerances set by car producers. This allows for the detection of any deviations beyond the acceptable limits.

Maintaining proper cooling in accordance with standards is vital for the longevity and stable operation of battery cells. Achieving this within tolerances requires the application of a thermal sealer between the coolers and battery cells with precise and standardized techniques. The developed software ensures stable and highly accurate control over the mastic application process. After the application is completed, a comprehensive evaluation is performed, and the operator is visually informed of the results. This enables the identification of any mistakes or deviations that may not be apparent to the naked eye. Consequently, applications falling outside the defined tolerances can be promptly detected and rectified.

The study addresses the challenges associated with applying sealant using a robot sealant application trajectory that is pre-determined based on a reference part in the existing system. The variations in part geometries and positions often lead to incorrect sealant amounts and application trajectories. Through sensor data analysis, the calibration of the part to be sealed on the X-axis is achieved. The robot program is then adjusted based on the calibration values to ensure an appropriate working principle during the thermal sealant application process.

The heat transfer sealant process applied to the battery unit, as the critical component in electric vehicles, plays a vital role in their overall performance and efficiency. Quality issues in the mastic operation can significantly impact the battery's lifespan. By incorporating the developed software and system for cooling, this study aims to provide optimal cooling conditions and minimize potential problems. The successful implementation of these techniques will result in the development of a robust battery system that ensures both the longevity and efficiency of electric vehicles.

References

1. Zhou, K., Liu, T., Zhou, L.: Industry 4.0: towards future industrial opportunities and challenges. In: 2015 12th International Conference on Fuzzy Systems and Knowledge Discovery (FSKD), Zhang-jiajie, China, pp. 2147-2152 (2015). https://doi.org/10.1109/FSKD.2015.7382284
2. Young, G.O.: Synthetic structure of industrial plastics. In: Peters, J. (ed.) Plastics, 2nd ed., vol. 3, pp. 15-64, New York: McGraw-Hill (1964)
3. Yeni Sanayi Devrimi Akıllı Üretim Sistemleri Yol Haritası. TÜBİTAK Bilim. Teknoloji ve Yenilik Politikaları Daire Başkanlığı (2016)
4. Haug, K., Pritschow, G.: Robust laser-stripe sensor for automated weld-seam-tracking in the shipbuilding industry. In: Proceedings of the 24th Annual Conference of the IEEE Industrial Electronics Society, IECON 1998, (Cat. No.98CH36200), Aachen, Germany, vol. 2, pp. 1236–1241 (1998).https://doi.org/10.1109/IECON.1998.724281
5. Chang, D., et al.: A new seam-tracking algorithm through characteristic-point detection for a portable welding robot. Rob. Comput. Integr. Manuf. **28**(1), 1–13 (2012). ISSN 0736–5845. https://doi.org/10.1016/j.rcim.2011.06.001
6. Luo, H., Chen, X.: Laser visual sensing for seam tracking in robotic arc welding of titanium alloys. Int. J. Adv. Manuf. Technol. **26**, 1012–1017 (2005). https://doi.org/10.1007/s00170-004-2062-2
7. Gao, M., Li, X., Yang, Y., He, Z., Huang,J.: An automatic sealing system for battery lid based on machine vision. In: 2017 IEEE 26th International Symposium on Industrial Electronics (ISIE), Edinburgh, UK, pp. 407-411 (2017). https://doi.org/10.1109/ISIE.2017.8001281
8. Huang, Y., Xiao, Y., Wang, P., et al.: A seam-tracking laser welding platform with 3D and 2D visual information fusion vision sensor sys-tem. Int. J. Adv. Manuf. Technol. **67**, 415–426 (2013). https://doi.org/10.1007/s00170-012-4494-4
9. Hang, K., Pritschow, G.: Reducing distortions caused by the welding arc in a laser stripe sensor system for automated seam tracking. In: Proceedings of the IEEE International Symposium on Industrial Electronics, ISIE 1999, (Cat. No.99TH8465), Bled, Slovenia, vol. 2, pp. 919–924 (1999).https://doi.org/10.1109/ISIE.1999.798737
10. Rout, A., Deepak, B.B.V.L., Biswal, B.B.: Advances in weld seam tracking techniques for robotic welding: A review. Rob. Comput. Integr. Manuf. **56**, 12–37 (2019). ISSN: 0736–5845. https://doi.org/10.1016/j.rcim.2018.08.003

Resolving the Ergonomics Problem of the Tailgate Fixture on the Robotic Production Line

Abdullah Burak Arslan(✉)

ROBO Automation and Engineering, Gölcük, Kocaeli, Turkey
abdullah.arslan@robo.com.tr

Abstract. In this study, a novel technique was developed to solve ergonomic problems that were encountered in a robotic production line. Ergonomic is essential in production lines both for physical health of operators and for the improvement of the cycle time. In this process of tailgate in the robotic production line, the operator in the field bends over more than it should. This situation eventually impairs the health of the operator. The fixture must be horizontal for this sealing process, and it must have a safe angle for the operator. The operator starts the process after positioning a part and a sealer robot starts sealing process. This ergonomic problem was solved by designing a piston and bearing mechanism. The main problem is the positioning of the piston in the design. For this, position of the piston was calculated and determined. Then an appropriate CAD design was developed. Designed parts were manufactured and assembled with standard equipment. The ergonomic analysis was successfully completed and assembled to the production process.

Keywords: Ergonomics · fixture · robotic lines · design · process

1 Introduction

Robotic processes are known to be fast and precise. Workpieces that are processed in this robotic line must be hold and stable during the process. This type of tooling equipment is known as fixtures. Operators set the workpieces, which is based on the specifications of that fixture and the process sheets [1]. One of the crucial topics regarding fixtures is their design. The design of a fixture has a significant effect on product quality and complexity. The complexities in the automotive industries are increasing, leading to the development of advanced tools and techniques to provide support. Furthermore, there is always a necessity for these tools to improve ergonomics while keeping costs low [2].

Presently, industries are attempting to unify innovative products and modern technologies to maintain the performance and health of employees. It is essential to correctly understand the requests made by employees pertaining to their work areas to make the changes required for maintaining a healthy workforce [3]. The operators working on the station lines are exposed to significant ergonomic risks due to inverted postures, environmental factors, and repetitive motions. Therefore, industries are studying and

Z. Şen et al. (Eds.): IMSS 2023, LNME, pp. 320–325, 2024.
https://doi.org/10.1007/978-981-99-6062-0_29

adapting numerous methods to evaluate work situations and postures in order to overcome ergonomic problems. Along with ergonomic studies, the existing processes must continue to work as usual [4].

There are positioners used in industrial areas. They are used to rotate and process position to suit the operator's needs. Positioners contribute several advantages, including improved ergonomics and adjustable lifting capabilities [5]. Ghasem et al., reported that the common musculoskeletal disorders was observed to be immense among body-in-white assembly line workers [6]. Jones et al., studied musculoskeletal disorders marked for that 30% of total claims cost and represents 40% of total time loss more than any other injury category. Neck, shoulders, and low back are the most frequently affected body regions [7].

Narayanan et al., conducted a study on a flexible fixture and a spot weld cell. The work cell is developed to handle multiple body-in-white part varieties and also for a human-robot cooperative cell. They proposed concept to shorten the design and fabrication time of fixture. The fixture mount was chosen based on the ergonomics for operator loading and to withstand the payload of the fixture [8].

2 Ergonomics and Process Problems

2.1 Ergonomic Problem

Operators have limitations on their ability to bend and lift. When a part is lifted and set on the fixture, the operator bends over almost 45° angle with the part. In short term this brings health issues for the operator. Figure 1 shows the fixture height position, which must remain at that angle during the sealing process. Figure 2 represents the ergonomics problem of the operator.

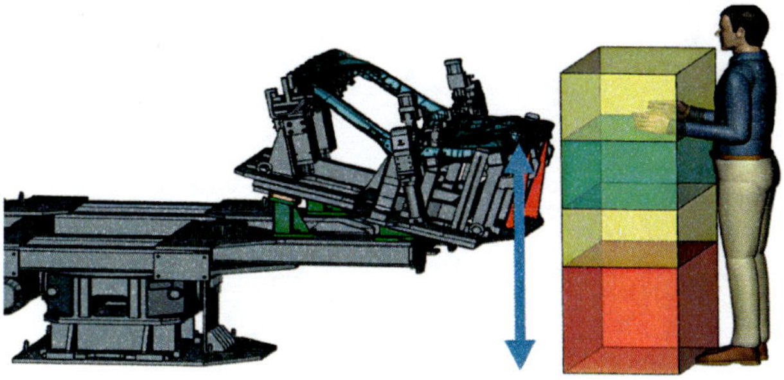

Fig. 1. The fixture height.

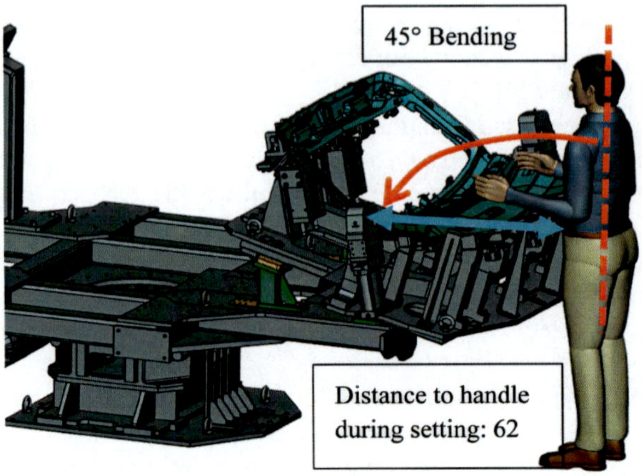

Fig. 2. Setting position.

2.2 Robot Sealing Process Problem

Sealing is a glue that provides bonding between materials while also preventing rust. It must be applied before materials are merged. The robot has a tool that provides sealing process. This tool has 3 cameras on it. These cameras have to see the part vertically from a nearly 90° angle. Figure 3 shows the horizontal position in which the operator cannot set the part that is necessary for this process.

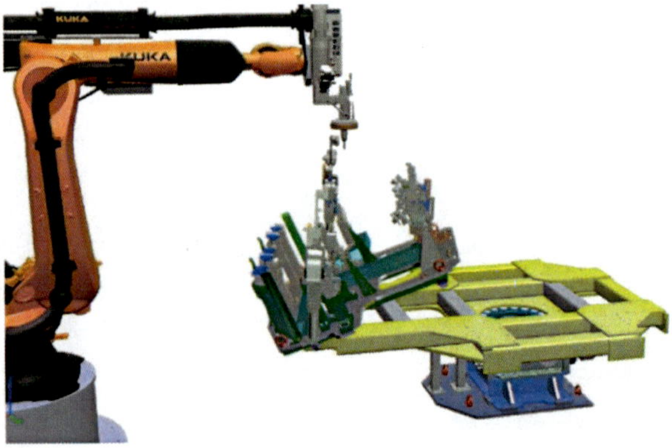

Fig. 3. Horizontal position for sealing.

Figure 4 shows an optimal position for the operator to set the part. The operator can lift the part and set to the fixture with this suitable angle. However, at this particular angle, the robot's access to the sealing points is limited, and the cameras are unable to provide an accurate measurement of the process.

Fig. 4. Vertical position for the operator.

2.3 Equipments

Figure 5 shows the position of lifting. Lifting is supported by 6 bar pressured air and distance to bearing axis is 750 mm. You can see the used equipment in Table 1. Also, it can be seen the piston can lift and pull easily without any impact in Table 2.

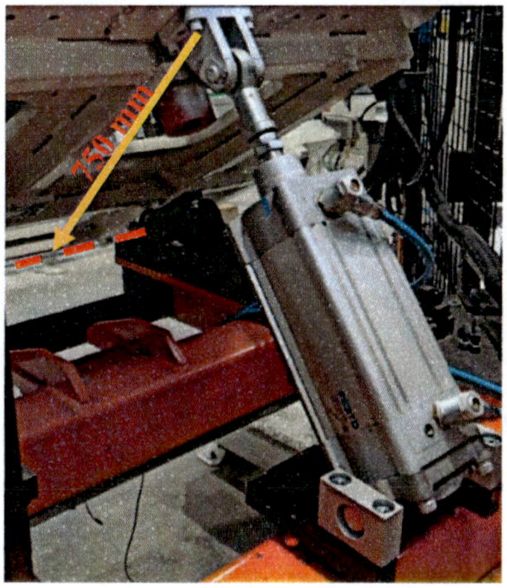

Fig. 5. Lifting position of the piston.

3 Conclusion

Through this modification, both ergonomic and process problems have been effectively resolved. The piston provides sufficient lift power, and the fixture's position enables proper sealing with the robot. After testing the modified setup several times, no problem with the system was observed. This has been approved by the client as well.

- This setup is simple to apply to most of fixtures, and its cost is low,
- By manipulating the fixture, we were able to quickly solve both ergonomic and sealing problems,
- Based on Table 2, piston can lift and pull total mass for a long time,
- It was proven that fixtures can be modified and manipulated after they had been manufactured.

Table 1. Table captions should be placed above the tables.

Equipment	**Type**	Pieces
Piston	DSBC 100–150-PA-Z0	1
Bearings	RASEY 30 XL-N	6
Rod eye	SGS M20x1,5	1
Trunnion support	CRLNZG 100–125	2
Clevis foot	LBG-100	1

Table 2. Calculation of Impulse

Unit	Value
Mass of the fixture (kg)	500
Impact speed (m/s)	0,888
Thrust torque (Nm)	800
Radius of mass (mm)	800
Total reaction power (N)	4417

References

1. Zhou, Y., Li, Y., Wang, W.: A feature-based fixture design methodology for the manufacturing of aircraft structural parts. Rob. Comput. Integr. Manuf. **27**, 986–993 (2011)
2. Rowan, M.P., Wright, P.C.: Ergonomics is good for business. Work Study **43**, 7–12 (1994)
3. Varaprasada, M., Kampurath, V., Ananda, R., Chaitanya, M.: Innovative industrial and workplace ergonomics in modern organizations.Int. J. Eng. Res. Technol. **4** (2015)

4. Annarumma, M., Pappalardo, M., Naddeo, A.: Methodology development of human task simulation as PLM solution related to OCRA ergonomic analysis. In: Cascini, G. (eds.) Computer-Aided Innovation (CAI), The International Federation for Information Processing, vol. 277, pp. 19–29 . Springer, Boston, (2008). https://doi.org/10.1007/978-0-387-09697-1_2
5. Shinde, S.N., et al.: Design of welding fixtures and positiners. Int. J. Eng. Res. Gener. Sci. **2**, 681–689 (2014)
6. Ghasemkhani, M., Aten, S., Azam, K.: Musculoskeletal symptoms among automobile assembly line VVorkers. J. Appl. Sci. **6**, 35–39 (2006)
7. Jones, T., Kumar, S.: Six years of injuries and accidents in the sawmill industry of Alberta. Int. J. Ind. Ergon. **33**, 415–427 (2004)
8. Elangovan, M., Thenarasu, M., Narayanan, S., Shankar, P.S.: Design of flexible spot welding cell for Body-In-White (BIW) assembly. Period. Eng. Nat. Sci. **6**, 23–38 (2018)

Digital Transformation with Artificial Intelligence in the Insurance Industry

Samet Gürsev(✉)

Agesa Life and Pension A.Ş., İstanbul, Turkey
sametgursev@gmail.com

Abstract. In today's world artificial intelligence studies are applied in many sectors. There are companies in the insurance field that use data-driven and innovative technologies. The insurance industry depends on the ability to be competitive in the market with the right analysis of customer data and the right products. The insurance industry is a challenging area where customer data needs to be analyzed correctly. Artificial intelligence-supported analysis of customer behavior provides a competitive and more profitable structure. All steps from actuarial calculation to customer claim transactions can become much more effective with artificial intelligence. In the research, artificial intelligence products used by today's insurance companies were examined. An alternative model based on real practice data, much more detailed than other studies in the literature, is recommended in the research. The research offers innovative approaches to artificial intelligence and digital transformation studies.

Keywords: First Artificial intelligence · Digital Transformation · Industry 4.0 · Insurance

1 Introduction

Digital transformation leads to major changes in insurance. Instead of the traditional insurance model, an approach where all business processes are digital. Many concepts such as Artificial Intelligence, Internet of Things, Blockchain, machine learning, and cloud systems are involved in insurance processes [1]. New generation technology is used in subjects such as information security, virtual assistants, risk calculation, customer data analytics [2]. After the Covid-19 epidemic, all companies have seen the importance of digitalization overnight.

Every area where the insurance company comes into contact with its customers is affected by digital transformation [3]. Customers can use agencies, banks, and direct sales specialist channels to purchase insurance products [4]. In the new generation approach, channels such as online sales, mobile application sales, Interactive Voice Response, Chatbot are added to these channels. Customer service channels increase similarly [5]. Video chat, complaint notification via mobile application, online transaction channels are provided.

Z. Şen et al. (Eds.): IMSS 2023, LNME, pp. 326–335, 2024.
https://doi.org/10.1007/978-981-99-6062-0_30

In the research, the insurance sector is examined as a whole. Examples of digital transformation of institutions working in this field are clustered. Researches are presented on the direction in which current processes can go with the philosophy of industry 4.0. The industry 4.0 approach is not just using artificial intelligence and similar new generation technologies [6]. To be able to personalize the customer experience, to make it perfect and to adapt to the market demands in the fastest way.

Customer satisfaction is of great importance in the insurance industry [7]. Companies in this field survive with customer loyalty and satisfaction [8]. Artificial Intelligence models achieve success by examining the correct data collected with new generation technological tools [9]. The literature includes reviews for artificial intelligence applications in the insurance industry. The research presents a model based on real applications, covering all these studies and providing more detailed content.

2 Insurance and Artificial Intelligence Applications

The goal of the value chain is to provide the customer with optimum cost [10]. The big data in question is the innovation of insurance companies' opportunities for them to apply technologies to their benefit [11]. Another area developed and transformed by artificial intelligence is the insurance industry [12]. The world insurance market is dominated by big brands that have not changed significantly for a long time [13]. However, with the developing technology, change is inevitable and AI foreshadows these changes. The wealth of data provided by artificial intelligence to the insurance industry is enormous [14]. Insurers need AI and machine learning tools to take advantage of this wealth [15]. The insurance industry has witnessed many changes in information technologies in recent years [16]. Variables such as developments in hardware, software and internet, lower cost and data processing time created high profit potential [17]. New technologies are collected under seven main headings in the literature research and sector examples review (Fig. 1).

Machine Learning; It is the first main title in the industry. It is a structure that can learn from examples and gives high performance over time [19]. This method is used effectively in matters such as the detection of financial crimes, autonomous task performance, cost-increasing fake damage detection, and fraud analysis.

Natural Language Processing (NLP); Obtains data from texts. Interface implementation that speaks to the customer, known as a chatbot [20]. It is an application that can answer customer questions and provide guidance. It provides great benefits in after-sales processes.

Internet of Things (Iot); electronics and internet combination of features. Smart watches, home devices, wearables are examples [21]. It provides support in collecting accurate data about the insured. Example; Driver performance measurement and pricing accordingly with an IoT device that measures car usage.

Predictive Analytics; It is to predict the future period by analysing past data. It is the important value that keeps the insurance industry alive [22]. It is critical for market analysis, customer segmentation, and new product studies.

Machine Vision; It is to provide machine learning with image, video analysis [23]. Risk analysis is made by analysing the satellite images of the place where disaster insurance will be made (machine vision algorithm). For home insurance, accurate premium calculation is provided with picture and video analysis.

Robotic Process Automation, (RPA); data extraction, form filling, file migration, etc. Automation technology that performs office tasks done by humans [24]. The operational workload for insurance companies is reduced and the policy acceptance phases are accelerated. Manual work processes are eliminated, personnel are used more effectively.

Deep Learning; Deep learning algorithm based on data; it is a specialized type among machine learning algorithms in terms of learning [25]. The analyses here provide performance increase for decision support systems.

Artificial intelligence applications in the insurance sector have found less place in the literature. Although the sector is very suitable for customer data analytics and related results, there has not been enough research. According to Riikkinen et al. [1] looked at artificial intelligence applications in the insurance industry with a value-oriented approach. Park and et al. [3] made recommendations on AI-assisted decision-making tools. Kum et al. [2], Ho et al. [4], Eling et al. [5] examined the applications made on big data, especially risk analysis processes, in terms of value management in the insurance industry. Gramegna and Giudici [6] focused on insurance sales processes. Kelly et al. [26] examined the practices of risk management processes. Johnson et al. [7] examined examples of artificial intelligence in health insurance. Paul et al. [8] reviewed customer data forecasting. Stern et al. [9], Mullis et al. [10], Hermann and Masawi [12], Sakthivel and Rajita [27], Keller [14] examined different insurance industry practices and finance examples. Lior [28], Ceylan [15], Sinha and Sohakak [16] cite different examples of artificial intelligence projects in Sufriyana [29] countries.

Gupta et al. [17], Yoshikawa [18], Murray et al. [30] Ejiyi et al. [20] comparatively discusses AI-supported solutions in payment systems and insurance operations. Amerriad et al. [31], Shi et al. [32], Sharma and Shood [33], Paruchuri [34] examined machine learning applications, new generation algorithms and innovative models in the field of insurance. Massaro [35] has researched insurance industry decision support systems. Kaushik et al., [36] Lee and Oh [37], Fung and Polania [38] examined the effects of artificial intelligence in the industry from a financial perspective. Cioffi et al. [39] has conducted research on artificial intelligence, machine learning, smart processes and change areas of all sectors. All technologies in Fig. 1 work much more effectively with successful artificial intelligence examples arranged with new generation algorithms. Many of the models and tools in the artificial intelligence literature can be applied very successfully in the insurance industry. The model suggestions in the research were applied by making all these studies special. World-renowned global insurance brands are now managing their customer journeys through this model.

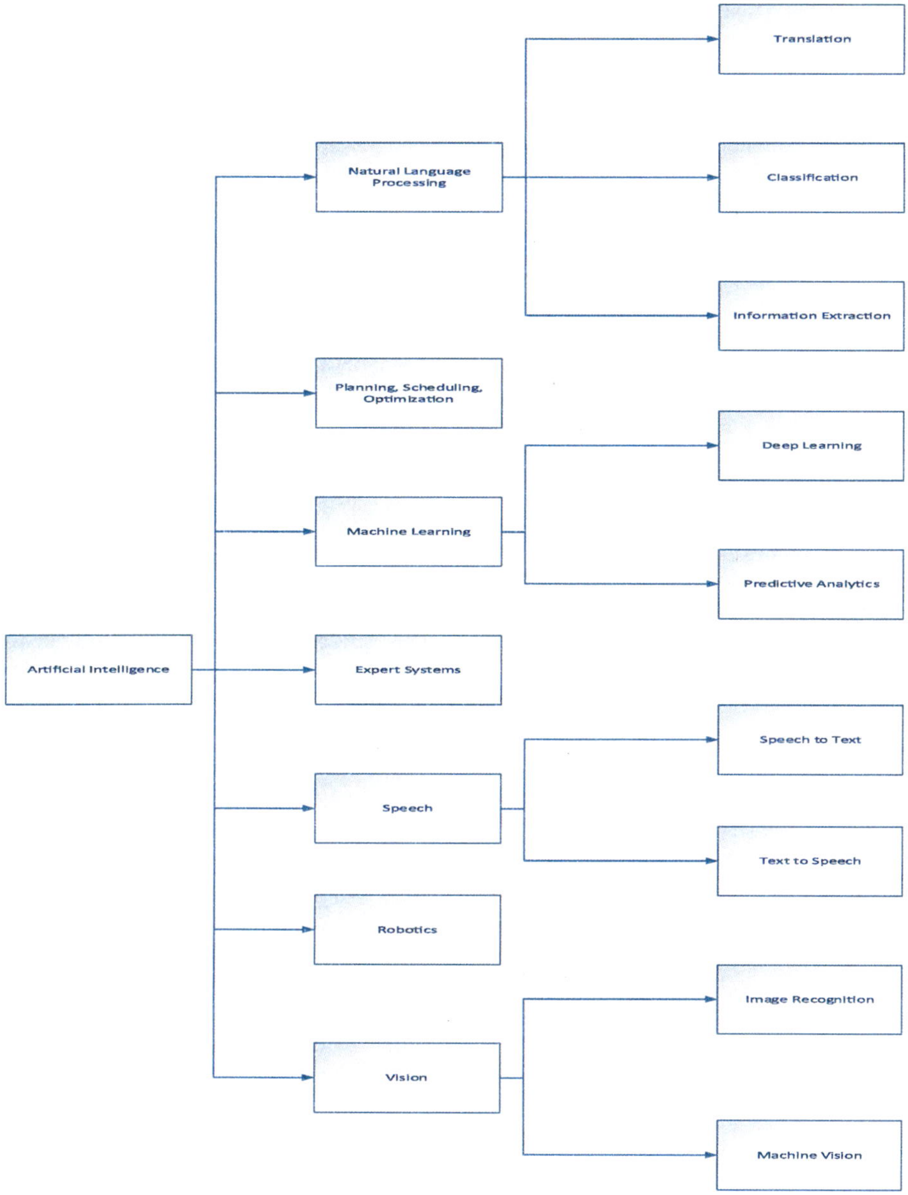

Fig. 1. Artificial Intelligence in Insurance Company [18]

3 Predictive Analytic Process

Predictive Analytics can predict future events and behaviours using statistical methods. Forecasting success is related to the accuracy of the retrospective data, to have sufficient amount, to contain different cases [40]. The insurance process itself is a data estimate. The risk that the insurer buys, the probability of the customer making an accident. Each insurance company optimizes its premium and coverage accounts using statistical calculations [41].

The digitalization process provides benefits for low-cost and fast product operation [42]. The insurance company analyses the data and offers a policy price offer [43]. However, there are more estimated data about the customer than actual data [44]. When customer behaviours and habits are not analysed in detail, forecasts can often bring high costs. Bringing fast and innovative products to market is a long and difficult process for most companies. Artificial intelligence technology promises a world where a driver can analyse his/her bad behaviours recently and update his/her current insurance policy according to the renewed accident risk.

Customers report damage to their insurance in certain periods. For example, people who have health insurance report their hospital expenses to the company when they are sick. The process here is reviewed and approved by expert people [45]. Customers who wait for a long time to make simple transactions on the insurance policy become unhappy. It is also a negative customer experience for a customer with car insurance to wait for a long time for damage after a car accident, to determine the damage late by experts, and not to use the car for a long time.

Customers can leave pre-sales and after-sales services happily with artificial intelligence support [46]. First of all, data must be collected from the right channels. The data obtained must be clean and organized. With data mining methods, the process can be turned into a correct process (Fig. 2).

In Fig. 2, a chatbot and RPA supported structure proposal is presented. An early damage warning system was established with data estimation method [46]. The feedback provided via Chatbot (NLP) analyses customer demands in the most accurate way. Claim applications, complaint applications, customer requests, these data are collected and analysed by artificial intelligence and the most appropriate sales strategies are provided [47]. Analysis of current customers' data contributes to potential product sales. In terms of information security and Fraud, artificial intelligence contributes to the data security of the company. As a result of all these improvements, customer demands continue to be resolved as quickly as possible.

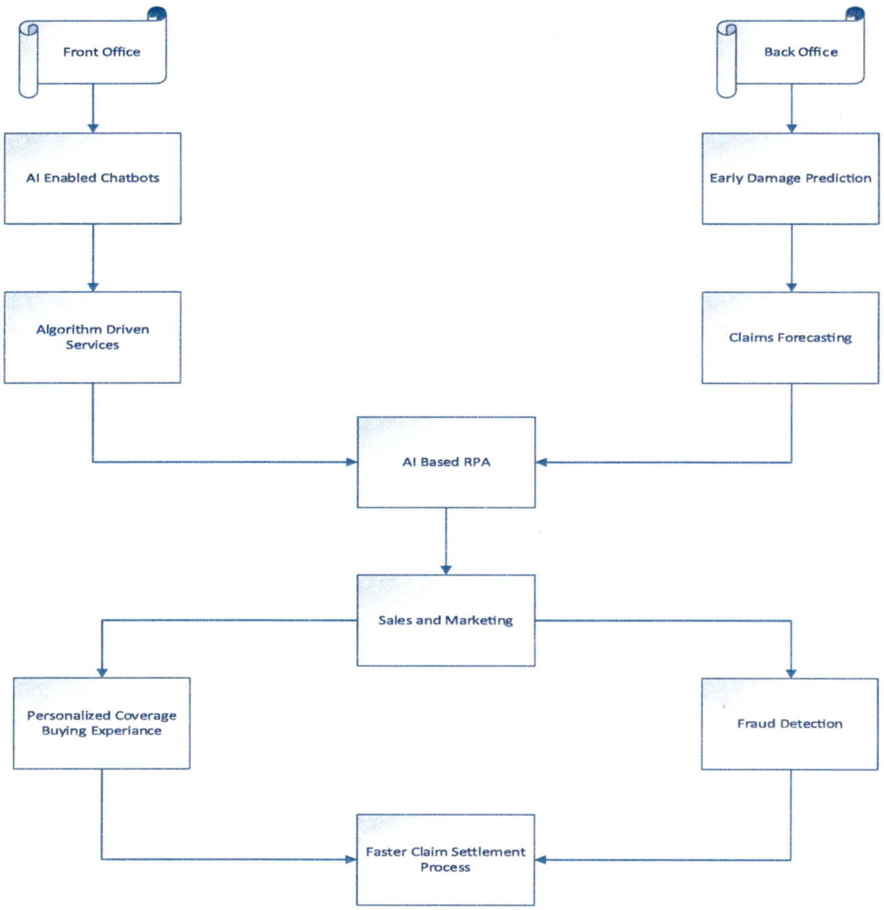

Fig. 2. Insurance Industry Example Artificial Intelligence Flow

4 Customer Experience-Oriented AI Model

Insurance companies analyse their customers' risks and purchase them. The customer pays for the risk. If the client risk is large, coverage increases and it becomes an expensive underwriting process [48]. Artificial intelligence is used in many areas such as insurance sales, risk calculation, after-sales operations. The structure proposed in the third part of the research is for operations, and the gains here are examined for one year [49]. Many technological methods are used in the recommended artificial intelligence model. It is a long-term project for all of these technologies to come together and work together.

In the research, improvement measurement was made based on the common gains of the company in the sector. Customer Demand Fulfilment Rate; the rate of meeting and responding to all requests submitted by customers. This ratio is found as the ratio of incoming customer requests/completed customer requests. Customer Satisfaction Rate;

It is calculated over the result of the satisfaction survey score that the company periodically transmits to its customers. Number of Insurance Policy Sales; the monthly increase in the number of policies for all insurance products of the company. Fraud Registration Numbers; the number of all faulty transactions and illegal processes within the scope of fraud in the total application [50]. Efficiency Increase in Operational Business Processes is the amount of improvement in the output/input ratios of operational processes across the company [51].

Table 1. Increase Rates After Artificial Intelligence Implementation

	INCREASE RATE
The Customer Demand Fulfilment Rate	40%
Customer Satisfaction Rate	90%
Number of Insurance Policy Sales	15%
Operational Business Processes	70%

During the one-year observation period (Table 1); great improvements have been achieved with the recommended model structure and artificial intelligence effect. The Customer Demand Fulfilment Rate increased by 40%. Customer Satisfaction Rate increased by 90%. Number of Insurance Policy Sales increased by 15%. Fraud Registration Numbers decreased by 100%. Efficiency Increase in Operational Business Processes resulted in 70% improvement.

5 Conclusion

Our world is changing rapidly. Industry 4.0 and the approaches it brings are changing the production and service processes. The financial industry is changing rapidly. The competition is so brutal. Customers change companies very easily. Customer retention and sustainable growth are as important as the number of sales. Banks are closing branches. The insurance industry is growing in the field of online sales. This change process proceeds through digital transformation and quickly involves the customer with a simple application on the mobile phone.

The research clusters artificial intelligence tools in the insurance industry. It gives industry application examples of these tools. Suggests an exemplary model for a successful artificial intelligence setup. Measures benefits from this model with a one-year follow-up value. Artificial intelligence tools, chatbot, RPA, Machine Learning are very expensive and long-term investments. Firms do not make these investments without a break-even analysis. If the model proposed by the research is implemented with a correct integration, it can make great contributions.

All businesses in the field of Insurance and Finance have to provide digital transformation in all their business processes. The changing world after Industry 4.0 provides the customer with a completely different experience with personalized vehicles. If the literature review studies are examined, it is seen that while conceptual algorithms guide the digital return journey, artificial intelligence does not take into account real uses. Based on industry examples, the research proposes a recommendation model that synthesizes artificial intelligence models and next-generation algorithms with Industry 4.0 technology tools. The improvement data in a single company has changed noticeably in a period of one year, with the analysis made. The improvement data in a single company has changed noticeably in a period of one year, with the analysis made.

Artificial intelligence is a rapidly developing field that makes great contributions. Researchers who want to work in this field are recommended to try artificial intelligence tools and models in different sectors. Each step of the model proposed in the research can be passed through the productivity analysis processes separately. Forecasts for Industry 4.0 and beyond are largely dependent on the development of artificial intelligence. Solution proposals brought by artificial intelligence in this field will bring companies to a further stage.

References

1. Riikkinen, M., Saarijärvi, H., Sarlin, P., Lähteenmäki, I.: Using artificial intelligence to create value in insurance. Int. J. Bank Mark. (2018)
2. Kumar, N., Srivastava, J.D., Bisht, H.: Artificial intelligence in insurance sector. J. Gujarat Res. Soc. **21**(7), 79–91 (2019)
3. Park, S.H., Choi, J., Byeon, J.S.: Key principles of clinical validation, device approval, and insurance coverage decisions of artificial intelligence. Korean J. Radiol. **22**(3), 442 (2021)
4. Ho, C.W., Ali, J., Caals, K.: Ensuring trustworthy use of artificial intelligence and big data analytics in health insurance. Bull. World Health Organ. **98**(4), 263 (2020)
5. Eling, M., Nuessle, D., Staubli, J.: The impact of artificial intelligence along the insurance value chain and on the insurability of risks. Geneva Papers Risk Insur.-Issues Pract. **47**, 1–37 (2021). https://doi.org/10.1057/s41288-020-00201-7
6. Gramegna, A., Giudici, P.: Why to buy insurance? An explainable artificial intelligence approach. Risks **8**(4), 137 (2020)
7. Johnson, M., Albizri, A., Harfouche, A.: Responsible artificial intelligence in healthcare: Predicting and preventing insurance claim denials for economic and social wellbeing. Inf. Syst. Front. 1–17 (2021). https://doi.org/10.1007/s10796-021-10137-5
8. Paul, L.R., Sadath, L., Madana, A.: Artificial intelligence in predictive analysis of insurance and banking. In: Artificial Intelligence, pp. 31–54. CRC Press (2021)
9. Stern, A.D., Goldfarb, A., Minssen, T., Price II, W.N.: AI insurance: how liability insurance can drive the responsible adoption of artificial intelligence in health care. NEJM Catal. Innov. Care Deliv. **3**(4), CAT-21 (2022)
10. Mullins, M., Holland, C.P., Cunneen, M.: Creating ethics guidelines for artificial intelligence and big data analytics customers: the case of the consumer European insurance market. Patterns **2**(10), 100362 (2021)
11. Islam, M.M., Yang, H.C., Poly, T.N., Li, Y.C.J.: Development of an artificial intelligence–based automated recommendation system for clinical laboratory tests: Retrospective analysis of the national health insurance database. JMIR Med. Inform. **8**(11), e24163 (2020)
12. Herrmann, H., Masawi, B.: Three and a half decades of artificial intelligence in banking, financial services, and insurance: a systematic evolutionary review. Strateg. Chang. **31**(6), 549–569 (2022)
13. Owens, E., Sheehan, B., Mullins, M., Cunneen, M., Ressel, J., Castignani, G.: Explainable Artificial Intelligence (XAI) in Insurance. Risks **10**(12), 230 (2022)

14. Keller, B.: Promoting responsible artificial intelligence in insurance. Geneva Association-International Association for the Study of Insurance Economics (2020)
15. Erem Ceylan, I.: The Effects of Artificial Intelligence on the Insurance Sector: Emergence, Applications, Challenges, and Opportunities. The Impact of Artificial Intelligence on Governance, Economics and Finance **2**, 225–241 (2022). https://doi.org/10.1007/978-981-16-8997-0_13
16. Sinha, K.P., Sookhak, M., Wu, S.: Agentless insurance model based on modern artificial intelligence. In: 2021 IEEE 22nd International Conference on Information Reuse and Integration for Data Science (IRI), pp. 49–56. IEEE (2021)
17. Gupta, S., Ghardallou, W., Pandey, D.K., Sahu, G.P.: Artificial intelligence adoption in the insurance industry: evidence using the technology–organization–environment framework. Res. Int. Bus. Financ. **63**, 101757 (2022)
18. Yoshikawa, J.: Sharing the costs of artificial intelligence: universal no-fault social insurance for personal injuries. Vand. J. Ent. Tech. L. **21**, 1155 (2018)
19. Singh, S.K., Chivukula, M.: A commentary on the application of artificial intelligence in the insurance industry. Trends Artif. Intell. **4**(1), 75–79 (2020)
20. Ejiyi, C.J., et al.: Comparative analysis of building insurance prediction using some machine learning algorithms (2022)
21. Winston, P.H.: Artificial Intelligence. Addison-Wesley Longman Publishing Co., Inc., Boston (1984)
22. Boden, M.A. (ed.): Artificial Intelligence. Elsevier, Amsterdam (1996)
23. Ramesh, A.N., Kambhampati, C., Monson, J.R., Drew, P.J.: Artificial intelligence in medicine. Ann. R. Coll. Surg. Engl. **86**(5), 334 (2004)
24. Fetzer, J.H.: What is artificial intelligence? In: Fetzer, J.H. (ed.) Artificial Intelligence: Its Scope and Limits Studies in Cognitive Systems, vol. 4, pp. 3–27. Springer, Dordrecht (1990). https://doi.org/10.1007/978-94-009-1900-6_1
25. Gunning, D., Stefik, M., Choi, J., Miller, T., Stumpf, S., Yang, G.Z.: XAI—Explainable artificial intelligence. Sci. Robot. **4**(37), eaay7120 (2019)
26. Kelley, K.H., Fontanetta, L.M., Heintzman, M., Pereira, N.: Artificial intelligence: implications for social inflation and insurance. Risk Manag. Insur. Rev. **21**(3), 373–387 (2018)
27. Sakthivel, K.M., Rajitha, C.S.: Artificial intelligence for estimation of future claim frequency in non-life insurance. Glob. J. Pure Appl. Math. **13**(6), 1701–1710 (2017)
28. Lior, A. (2022). Insuring AI: The role of insurance in artificial intelligence regulation. Harvard Journal of Law and Technology, 1
29. Sufriyana, H., Wu, Y.W., Su, E.C.Y.: Artificial intelligence-assisted prediction of preeclampsia: development and external validation of a nationwide health insurance dataset of the BPJS Kesehatan in Indonesia. EBioMedicine **54**, 102710 (2020)
30. Murray, N.M., et al.: Insurance payment for artificial intelligence technology: methods used by a stroke artificial intelligence system and strategies to qualify for the new technology add-on payment. Neuroradiol. J. **35**(3), 284–289 (2022)
31. Amerirad, B., Cattaneo, M., Kenett, R.S., Luciano, E.: Adversarial artificial intelligence in insurance: from an example to some potential remedies. Risks **11**(1), 20 (2023)
32. Shi, Y., Sun, C., Li, Q., Cui, L., Yu, H., Miao, C.: A fraud resilient medical insurance claim system. In: Proceedings of the AAAI Conference on Artificial Intelligence, vol. 30, no. 1 (2016)
33. Sharma, V., Sood, D.: The role of artificial intelligence in the insurance industry of India. In: Big Data Analytics in the Insurance Market, pp. 287–297. Emerald Publishing Limited (2022)
34. Paruchuri, H.: The impact of machine learning on the future of insurance industry. Am. J. Trade Policy **7**(3), 85–90 (2020)

35. Massaro, A.: Implementation of a decision support system and business Intelligence algorithms for the automated management of insurance agents activities. Int. J. Artif. Intell. Appl. (IJAIA), **12**(3) (2021)
36. Kaushik, K., Bhardwaj, A., Dwivedi, A.D., Singh, R.: Machine learning-based regression framework to predict health insurance premiums. Int. J. Environ. Res. Public Health **19**(13), 7898 (2022)
37. Lee, J., Oh, S.: Analysis of success cases of insurtech and digital insurance platform based on artificial intelligence technologies: focused on ping an insurance group ltd. in China. J. Intell. Inf. Syst. **26**(3), 71–90 (2020)
38. Fung, G., Polania, L.F., Choi, S.C.T., Wu, V., Ma, L.: Artificial intelligence in insurance and finance. Front. Appl. Math. Stat. **7**, 795207 (2021)
39. Cioffi, R., Travaglioni, M., Piscitelli, G., Petrillo, A., De Felice, F.: Artificial intelligence and machine learning applications in smart production: progress, trends, and directions. Sustainability **12**(2), 492 (2020)
40. Kusiak, A.: Smart manufacturing. Int. J. Prod. Res. **56**(1–2), 508–517 (2018)
41. Kusiak, A.: Intelligent manufacturing: bridging two centuries. J. Intell. Manuf. **30**, 1–2 (2019). https://doi.org/10.1007/s10845-018-1455-2
42. Zhang, L., Zhou, L., Ren, L., Laili, Y.: Modeling and simulation in intelligent manufacturing. Comput. Ind. **112**, 103123 (2019)
43. Banjanović-Mehmedović, L., Mehmedović, F.: Intelligent manufacturing systems driven by artificial intelligence in Industry 4.0. In: Handbook of Research on Integrating Industry 4.0 in Business and Manufacturing, pp. 31–52. IGI global (2020)
44. Yang, S., Wang, J., Shi, L., Tan, Y., Qiao, F.: Engineering management for high-end equipment intelligent manufacturing. Front. Eng. Manag. **5**(4), 420–450 (2018)
45. Zhong, R.Y., Xu, X., Klotz, E., Newman, S.T.: Intelligent manufacturing in the context of Industry 4.0: a review. Engineering, **3**(5), 616–630 (2017)
46. Liang, S., Rajora, M., Liu, X., Yue, C., Zou, P., Wang, L.: Intelligent manufacturing systems: a review. Int. J. Mech. Eng. Robot. Res. **7**(3), 324–330 (2018)
47. McFarlane, D., Sarma, S., Chirn, J.L., Wong, C., Ashton, K.: Auto ID systems and intelligent manufacturing control. Eng. Appl. Artif. Intell. **16**(4), 365–376 (2003)
48. Wan, J., Yang, J., Wang, Z., Hua, Q.: Artificial intelligence for cloud-assisted smart factory. IEEE Access **6**, 55419–55430 (2018)
49. Gray-Hawkins, M., Lăzăroiu, G.: Industrial artificial intelligence, sustainable product lifecycle management, and internet of things sensing networks in cyber-physical smart manufacturing systems. J. Self-Gov. Manag. Econ. **8**(4), 19–28 (2020)
50. Zhou, G., Zhang, C., Li, Z., Ding, K., Wang, C.: Knowledge-driven digital twin manufacturing cell towards intelligent manufacturing. Int. J. Prod. Res. **58**(4), 1034–1051 (2020)
51. Yuan, C., Li, G., Kamarthi, S., Jin, X., Moghaddam, M.: Trends in intelligent manufacturing research: a keyword co-occurrence network based review. J. Intell. Manuf. **33**(2), 425–439 (2022). https://doi.org/10.1007/s10845-021-01885-x

Development of Rule-Based Control Algorithm for DC Charging Stations and Simulation Results

Furkan Üstünsoy[1(✉)] and H. Hüseyin Sayan[2]

[1] Gazi University Institute of Natural and Applied Sciences, 06560 Ankara, Turkey
ustunsoy.furkan@gmail.com

[2] Faculty of Technology, Gazi University, 06560 Ankara, Turkey
hsayan@gazi.edu.tr

Abstract. The rapid increase in electrical appliances and systems today means an increasing load on the main power grid. This rising will increase much faster in the near future, since especially electric vehicle load can create a local or global peak load on the grid. This will cause power quality problems such as frequency or voltage instabilities on the main power grid. However, topologies that will reduce the load that the electric vehicle will create on the grid have been suggested by researchers recently. To solve this problem, it is recommended to use the batteries of electric vehicles connected to the grid. In this context, the operation of V2G (Vehicle-to-Grid), V2H (Vehicle-to-Home) or V2V (Vehicle-to-Vehicle) topologies using vehicle batteries has begun to be seen as an opportunity. The most critical issue in the operation of these topologies is to control the energy flow in a coordinated manner, considering the losses. For this reason, a smart charge control algorithm is needed for the operation of this charge topologies. In this study, a rule-based control algorithm has been developed for these topologies, which also considering user satisfaction. This developed algorithm was simulated in real time according to a scenario prepared in the Matlab/Simulink program and the results were analyzed.

Keywords: Charging Strategy · V2G · G2V · V2V · V2H

1 Introduction

The majority of individual vehicles currently used work with fossil-based fuels (diesel, gasoline, LPG, etc.). However, countries shift production to hybrid electrical or full electric vehicles due to the fact that fossil-induced fuels will be exhausted and CO2 emissions. As a matter of fact, the share of electric vehicles in vehicle sales worldwide is increasing. This shows that electric vehicles will increase rapidly in the near future and ultimately replace completely internal combustion vehicles. In this context, the use of electric vehicles will bring many advantages and disadvantages.

Today, the rapid increase in electrical devices or systems means an increasing load on the main grid. This increase will rise much faster in the near future, especially since

Z. Şen et al. (Eds.): IMSS 2023, LNME, pp. 336–346, 2024.
https://doi.org/10.1007/978-981-99-6062-0_31

electric vehicles can create local or general peak loads on the grid. As a matter of fact, the topologies that will reduce this load to be created by the electric vehicle on the grid are recently recommended by the researchers. Researchers find the solution of this problem again in electric vehicles. In this context, the use of vehicle batteries for the operation of V2G, V2H or V2V topologies has begun to be seen as an opportunity. For example, an integrated control scheme is presented in [1] to realize the V2G topology in a distribution grid with renewable energy sources. In a similar study [2], the energy management strategy was evaluated with a two-way V2G topology to enable the penetration of hybrid electric vehicles (PHEV). In [3], a real-time energy sharing method that uses Grid-to-Vehicle (G2V) and Vehicle-to-Vehicle (V2V) topologies simultaneously is proposed. Similarly, a new decentralized power management algorithm is proposed in [4] to keep the voltage profiles within their limits and to reprogram the charging and discharging of the PEV. In [5], the V2V energy sharing concept is used, which proposes a smart and comprehensive framework for managing and allocating energy between EVs. Another study considers a V2G programming approach that can provide power balancing services to the grid while mitigating the battery aging problem [6]. In [7], within the framework of a research project, many possible applications were categorized and an open source dynamic simulation was developed to identify the most promising conditions. In [8], an algorithm was developed to control the energy management of the campus area of a public institution in line with the optimal economic target and simulated in Matlab/Simulink environment.

In this study, a rule-based control algorithm has been developed to operate V2G, V2H and V2V topologies on DC microgrid. This control algorithm, unlike the literature, is designed by considering driver behavior. The developed algorithm is simulated in the Simulink environment as discrete time (DT). In the simulation, the behavior of 3 different topologies at different times was examined and the results were presented.

2 Operating Topologies and DC Charging Station Architecture

2.1 V2G Operating Topology

V2G topology is that the vehicles provide electrical energy to the grid by hierarchical control at the charging stations to which they are connected. In this context, the power given can be active or reactive characteristic. Therefore, electric vehicles can also be used as a reactive power compensator. In short, if we accept the main grid as a microgrid, charging stations can also be considered as distributed battery storage units. In order to implement this hierarchical management, the data of a large number of active charging points should be examined and the most appropriate control method should be applied by the central grid operator.

2.2 V2H Operating Topology

V2H Topology is a control method recommended for charging points built on the existing grid infrastructure. With this topology, it is aimed to feed the existing residential load to which the charging point is connected to reduce grid stress with vehicle batteries. Since it is a more local solution than V2G, it is easy and efficient to operate. For this reason, researchers frequently use this topology in smart charging algorithms.

2.3 V2V Operating Topology

V2V topology is a new management principle that can be used for all charging station architectures. The main objectives of the energy exchange between electric vehicles at a charging station designed according to microgrid principle is to reduce of the grid stress and to fulfill the functions that battery storage systems can do. The most critical issue in the operation of this topology is to control the energy given and the energy taken by considering the losses. If this issue is not taken into consideration, frequency and voltage instability for the common AC charging area and voltage instability for the common DC charging area will occur. However, V2V and V2H topologies should not be considered only as energy exchange within the same station. On Blockchain-like platforms, there may be a commercial contract between a user who wants to sell energy from the battery and another user in different stations who want to get more affordable energy than the network tariff. The same can be performed between housing users and vehicles too. In recent years, researchers have been working on this issue with the optimum cost targets and reducing grid stress. Figure 1 is given symbolic architectures for these topologies in which the flow of energy is shown.

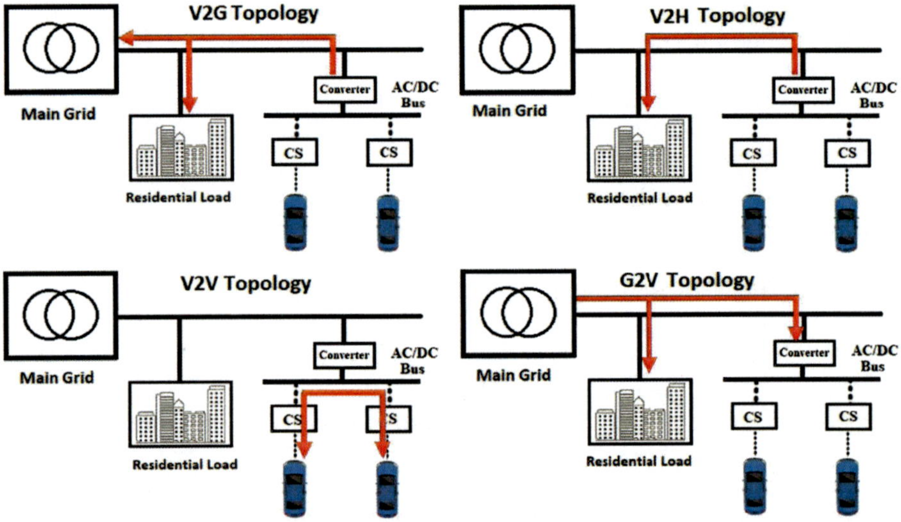

Fig. 1. Symbolic architectures for operating topologies of EV charge

In the literature, there are two basic approaches used to develop the smart charging algorithm within the charging station [9]. The first is a rule-based approach. In this approach, algorithm is developed with experience-based rules or formulas. In this way, dynamic results can be developed at any time according to variable inputs. The two are optimization-based algorithms. In such approaches, the charging station is managed by optimizing to find the best result of a particular goal.

2.4 DC Charging Station Architecture

Microgrid configurations must be used in order to integrate renewable energy sources into charging areas and to apply smart charging algorithms. A microgrid is a system that can operate in island mode with existing energy sources and energy storage units, and that can reduce the grid load. In this context, designing each of the electric vehicle charging points with this principle is very meaningful in terms of controllability and design flexibility.

Microgrids are designed to have a common DC, AC or hybrid bus. Those with DC character are called DC microgrid, those with AC character are called AC microgrid, and those with AC-DC character are called hybrid AC-DC microgrids. Charging points should also be designed based on these topologies. As a matter of fact, this issue is frequently emphasized in studies conducted in the literature [10–15]. DC charging infrastructure for both fast charging stations and charging points from the grid is more prominent in both literature studies and charging infrastructure applications due to the following advantages [16].

- The energy flow of PV panels and battery storage units is DC content.
- The energy flow to the batteries of electric vehicles is DC content.
- More suitable for smart charging algorithms.
- Making operating topologies easier with DC charging.

In this study, a charging station architecture with a common DC bus was designed and analyzes were carried out on this infrastructure in the Simulink environment. An example DC charging station architecture for residential areas is given in Fig. 2.

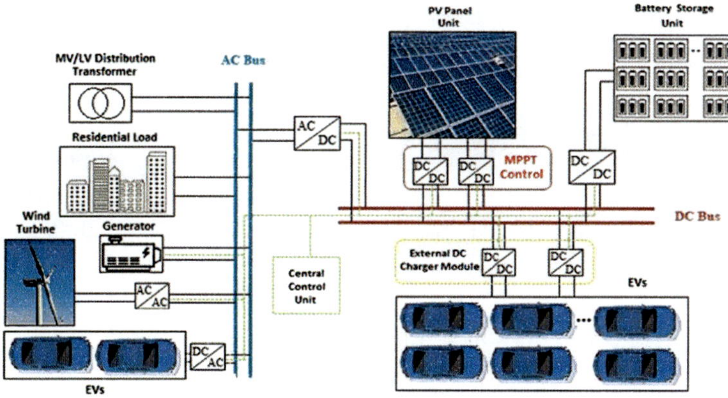

Fig. 2. DC microgrid based charging station

3 Rule-Based Control Algorithm and Simulation Results

3.1 Rule-Based Control Algorithm

In this study, a rule-based algorithm has been developed that takes into account user behaviors. First of all, specific information such as state of charge (SOC) level, battery charge voltage (V_{ev}), battery capacity (C_p), maximum and minimum charging power (P_{max}, P_{min}), maximum amount of energy that can be used (E_{max}) and the actual amount of energy used (E_{real}) for the integrated vehicles into the charging station is taken as the input of the algorithm on the local grid. Since the control architecture proposed in this study is centralized control method, it is assumed that the topology mode selection is determined by Distribution System Operator (DSO). Vehicles are labeled according to the received data and the mode selection of the DSO. This labeling determines the producer or consumer mode of operation of the vehicles. Finally, the usage frequency index (f_j) of each vehicle is calculated and the charging scheme is created for station.

The flow chart of the developed rule-based algorithm is given in Fig. 3. When the flow chart is examined, if a new situation occurs by checking the vehicle that leaves the station or newly integrated at any time during the cycle, the whole algorithm is taken from the beginning and a new charging scheme is created in real-time.

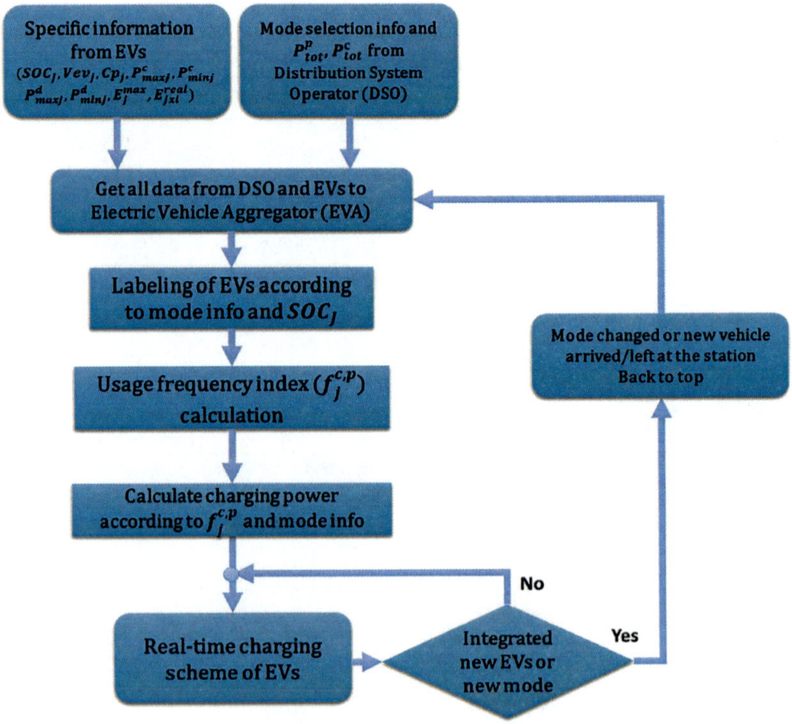

Fig. 3. Real-Time Rule-based control algorithm flowchart

In the developed algorithm, the behavior of the users is considered by calculating the usage frequency index and creating the charging scheme accordingly. Here, it is assumed that E_{real} and E_{max} values are calculated by the battery control system in the vehicle. These data, which are calculated in real time, are taken into the system as the input of the algorithm with the station local network. E_{real} and E_{max} are calculated by the battery control system according to Eq. 1 and Eq. 2.

$$\mathrm{E}_j^{\max(c,p)} = \left[\mathrm{V}_j^{ev(c,p)}.I_j^{cmax}.\mathrm{T_H} + \mathrm{V}_j^{ev(c,p)}.|I_j^{dmax}|.\mathrm{T_L}\right].\frac{24}{(\mathrm{T_H} + \mathrm{T_L})} \tag{1}$$

$$\mathrm{E}_{jxi}^{real(c,p)} = \int_0^{t_c} P_{jxi}^c(t)dt + \int_0^{t_d} |P_{jxi}^p(t)|dt \tag{2}$$

According to the calculated E_{real} and E_{max} values, the daily average frequency index (μ_{av}) is calculated in Eq. 3 and Eq. 4. In Eq. 5, the instantaneous usage frequency index (f_j) is calculated for each vehicle in the Electric Vehicle Aggregator (EVA).

$$\mu_{jxi}^{c,p} = \frac{\mathrm{E}_{jxi}^{real(c,p)}}{\mathrm{E}_j^{\max(c,p)}} \tag{3}$$

$$\mu_j^{av(c,p)} = \frac{\sum_{\mathrm{i}=1}^{\mathrm{n}} \mu_{jxi}^{c,p}}{\mathrm{n}}, \ (\mu_{\mathrm{J0}} = 0) \tag{4}$$

$$f_j^{c,p} = 1 + \mu_j^{av(c,p)} \tag{5}$$

According to the calculated for each vehicle instantaneous frequency index, the label status of the vehicles and the topology mode, the charging scheme is created and the system is started. The final charge pattern is calculated according to Eq. 6 and Eq. 7. The limitations of the study are that the reference power values are within the maximum and minimum power values ($P_{minj}^d \leq P_j^p \leq P_{maxj}^d, P_{minj}^c \leq P_j^c \leq P_{maxj}^c$).

$$P_j^p = \frac{P_{tot}^p}{u.f_j^p}, P_j^p = \mathrm{I.V}_j^{ev(p)}, I_j^{refp} = \frac{P_j^p}{\mathrm{V}_j^{ev(p)}} \tag{6}$$

$$P_j^c = \frac{P_{tot}^c.f_j^c}{v}, P_j^c = \mathrm{I.V}_j^{ev(c)}, I_j^{refc} = \frac{P_j^c}{\mathrm{V}_j^{ev(c)}} \tag{7}$$

Here, P_j^p is the reference power value of the vehicles with the prosumer label and P_j^c is the reference power values of the vehicles with the consumer label.

3.2 Simulation Results

In order to show the real-time operating characteristics of the rule-based charge management algorithm proposed in this study, a charging station with a common DC bus is designed in Matlab/Simulink environment. This charging station is designed in discrete time analysis format. Half-bridge bidirectional non-isolated DC/DC converter circuit architecture is used for electric vehicle charging. In order to operate the V2G topology at the DC charging station, a 3-phase grid connected inverter has been designed. A 3-phase full-wave uncontrolled rectifier is used for the energy need of the charging station. The charging schemes produced as a result of the algorithm were used as a reference for these charging stations. Since the voltage changes of electric vehicle batteries during charging/discharging are negligible, they are assumed to be constant. For this reason, the power flow is provided by controlling the current values in the simulation. In the simulation, 4 EVs were used and the vehicles were modeled as 73 Ah Li-ion batteries at 300 V voltage. The Simulink model of the designed DC charging station is given in Fig. 4.

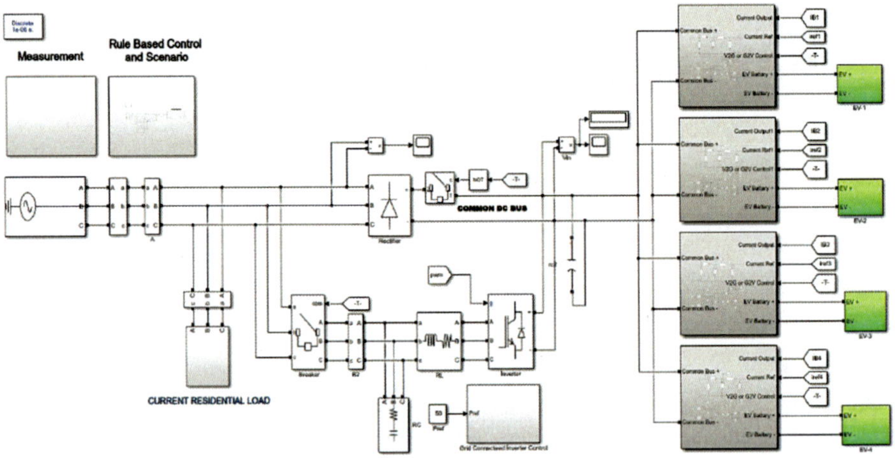

Fig. 4. Simulink model of DC charging station

In order to operate the rule-based charging algorithm in the designed Simulink model, different types of topologies were run at separate times during the simulation period. Here, the charging scheme is determined according to the calculated usage frequency index, the selected topology mode and the vehicle tag type.

In the designed DC charging station, 3 different scenarios were operated in 4 different time intervals. The G2V topology was operated in the 0–0.3 s and 0.7–0.9 s interval. V2V topology was operated in the range of 0.3–0.7 s. In the 0.9–2 s interval, V2G and V2H topologies were operated. All vehicles in the G2V timeframe are consumer-labeled. Therefore, the load of electric vehicles has been added to the current residential load. Since the %SOC values are higher in the V2V time interval, EV-3 and EV-4 are producer-labeled. EV-1 and EV-2 are consumer-labeled as they have a lower %SOC. Since energy transfer takes place between vehicles in this time period, the load of electric vehicles loaded on the grid has been reduced. In the V2G time interval, all vehicles supported the main grid and the current residential load. In Fig. 5, the power changes of the current residential load and total grid load during the simulation are given.

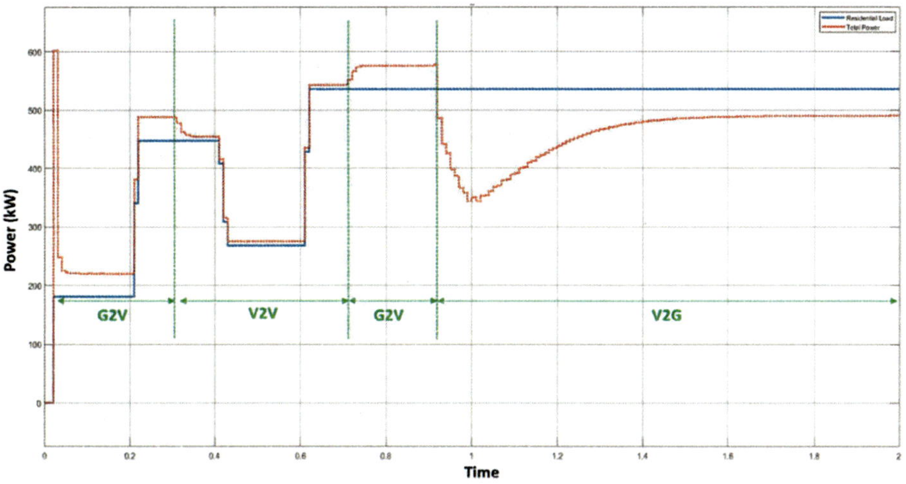

Fig. 5. Total load and residential load during the simulation

When the changes are examined, it is seen that the total power reaches the desired values with very short delays in the V2V and G2V topology transitions. Since vehicles are charged with nominal charging power in the G2V time interval, approximately 40 kW of additional load is included in the grid. In the V2V time interval, the energy of the vehicles with the consumer label was provided through the vehicles with the producer label. In this way, very little additional load was included in the grid during this time period. In order to ensure the stability of the DC bus voltage in the V2V time interval, the total consumption power is always kept higher than the total generation power. If this rule was not applied, the DC bus voltage would start to rise rapidly and unstable behavior would occur, since more energy would start to accumulate on the DC bus. When the graph is examined, it will be understood that the total load is always slightly higher than the current residential load in the V2V time interval. Here, u is the number of vehicles with the producer-labeled and v is the number of vehicles with the consumer-labeled;

$$\text{If} \sum_{j=1}^{u} P_j^p \geq \sum_{j=1}^{v} P_j^c;\ P_j^p = \frac{P_{tot}^p}{u.f_j^p} - \frac{P_{cal}}{u} \tag{8}$$

According to Eq. 8, a new charging scheme has been created for vehicles with producer labels. Here, P_{cal} is used as a constant and very little calibration power. In Fig. 6, the graph of the current changes during the simulation is given.

When Fig. 6 is examined, it is understood that the vehicles are discharged with current value of approximately 4 times the nominal value for 300 ms from the moment the V2G topology is started to operate. In the V2G time interval, all vehicles reached their nominal discharge current values within 0.9 s. In addition, when the current values reach the nominal values, it is seen that the energy flow to the grid is inversely proportional to the usage frequency index (f_j). In Fig. 7, %SOC changes of EVs during the simulation are given. Because the %SOC values are higher, the producer-labeled EV-3 and EV-4 supplied power to other vehicles in the V2V time interval.

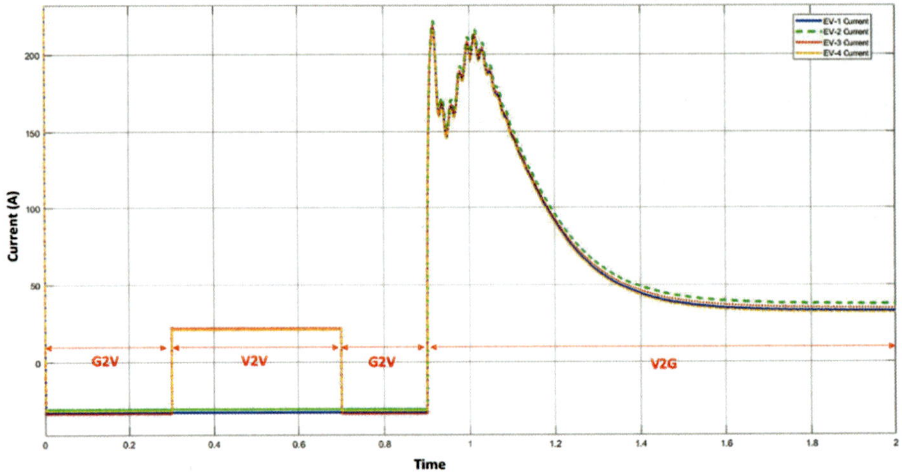

Fig. 6. Current of EVs during the simulation

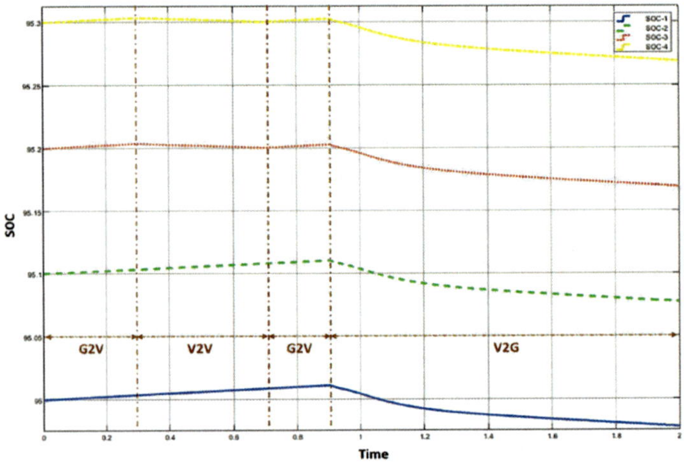

Fig. 7. %SOC of EVs during the simulation

Common DC bus voltage and grid voltage-current changes during the simulation are given in Fig. 8. In Fig. 8 (b), it is observed that the voltage remains constant, but fluctuations occur in the current amplitude at the topology change moments. The power fluctuation in Fig. 5 has also occurred due to the fluctuation in the current amplitude.

When Fig. 8(a) is examined, it is observed that there is fluctuation in the DC bus voltage at the topology transition moments. However, the busbar voltage exhibited stable behavior during the simulation.

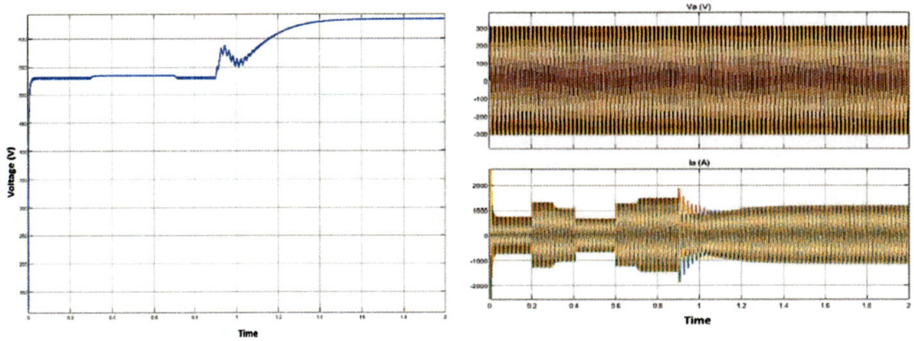

(a) (b)

Fig. 8. a) DC bus voltage during the simulation b) Grid current and voltage during the simulation

4 Conclusion and Future Works

Harmonics, grid frequency and voltage instability can be shown as the most important detrimental effects of grid integration of EVs. As a result of the energy supply-demand imbalance due to the peak load of electric vehicles on the grid, grid frequency and voltage instability is likely to occur. For this reason, grid integration of electric vehicles has been frequently studied by researchers recently. A lot of work has been done recently to overcome the possible supply-demand imbalance caused by the grid integration of EVs. Among these, the integration of renewable energy sources into the grid and the development of smart algorithms for EV charging have come to the fore. Developed smart charging algorithms have to consider user behavior. However, it is almost impossible to model user behavior as it exhibits stochastic behavior.

For this reason, in this study, a rule-based algorithm was developed and simulation results were presented in order to consider the user behavior and to show the operating characteristics of the charging operation topologies. In the rule-based algorithm, the usage habits of each driver are modeled as a usage frequency index and charging topologies are operated. In this way, at the end of the charging period, the grid load stress is reduced and user satisfaction is increased.

In future studies, the usability of the proposed rule-based algorithm in real applications can be tested by applying small powers. In addition, it is predicted that the user behavior model proposed in this study will be a source of inspiration for future studies.

Acknowledgements. This study (BAP Project Number: FDK-2021-7261) was supported by Gazi University Scientific Research Projects Unit.

References

1. Gao, S., Chau, K.T., Liu, C., Wu, D., Chan, C.C.: Integrated energy management of plug-in electric vehicles in power grid with renewables. IEEE Trans. Veh. Technol. **63**(7), 3019–3027 (2014)
2. Wang, X., Liang, Q.: Energy management strategy for plug-in hybrid electric vehicles via bidirectional vehicle-to-grid. IEEE Syst. J. **11**(3), 1789–1798 (2017)
3. Üstünsoy, F., Sayan, H.H.: Real-time realization of network integration of electric vehicles with a unique balancing strategy. Electr. Eng. **103**, 2647–2660 (2021). https://doi.org/10.1007/s00202-021-01259-9
4. Al Essa, M.J.M.: Power management of PEV using linear programming with solar panels and wind turbines in smart grids. Electr. Eng. **105**, 1761–1773 (2023)
5. Shurrab, M., Singh, S., Otrok, H., Mizouni, R., Khadkikar, V., Zeineldin, H.: A stable matching game for V2V energy sharing–a user satisfaction framework. IEEE Trans. Intell. Transp. Syst. **23**(7), 7601–7613 (2022)
6. Li, S., Gu, C.: Multi-objective bi-directional V2G behavior optimization and strategy deployment. In: Cao, Y., Zhang, Y., Gu, C. (eds.) Automated and Electric Vehicle: Design, Informatics and Sustainability. Recent Advancements in Connected Autonomous Vehicle Technologies, vol. 3, pp. 135–152. Springer, Heidelberg (2023). https://doi.org/10.1007/978-981-19-5751-2_8
7. Villante, C., Ranieri, S., Duronio, F., De Vita, A., Anatone, M.: An energy-based assessment of expected benefits for V2H charging systems through a dedicated dynamic simulation and optimization tool. World Electr. Veh. J. **13**(6), 99 (2022)
8. Üstünsoy, F., Serkan Gönen, H., Sayan, H., Yılmaz, E.N., Karacayılmaz, G.: Microgrid design optimization and control with artificial intelligence algorithms for a public institution. In: Hemanth, J., Yigit, T., Patrut, B., Angelopoulou, A. (eds.) ICAIAME 2020. LNDECT, vol. 76, pp. 418–428. Springer, Cham (2021). https://doi.org/10.1007/978-3-030-79357-9_41
9. Turker, H., Bacha, S.: Smart charging of Plug-in Electric Vehicles (PEVs) in residential areas: Vehicle-to-Home (V2H) and Vehicle-to-Grid (V2G) concepts. Int. J. Renew. Energy Res. **4**(4), 859–871 (2014)
10. Mouli, G.C., Bauer, P., Zeman, M.: Comparison of system architecture and converter topology for a solar powered electric vehicle charging station. In: 9th International Conference on Power Electronics and ECCE Asia. IEEE, Seoul, Korea (South) (2015)
11. Hadero, M., Khan, B.: Development of DC microgrid integrated electric vehicle charging station with fuzzy logic controller. Frontiers İn Energy Research **10**, 922984 (2022)
12. Locment, F., Sechilariu, M.: Modeling and simulation of DC microgrids for electric vehicle charging stations. Energies **8**(5), 4335–4356 (2015)
13. Zheng, Y., Niu, S., Shang, Y., Shao, Z., Jian, L.: Integrating plug-in electric vehicles into power grids: a comprehensive review on power interaction mode, scheduling methodology and mathematical foundation. Renew. Sustain. Energy Rev. **112**, 424–439 (2019)
14. Chen, Z., Liu, Q., Xiao, X., Liu, N., Yan, X.: Integrated mode and key issues of renewable energy sources and electric vehicles' charging and discharging facilities in microgrid. In: 2nd IET Renewable Power Generation Conference (2013)
15. Can, E.: The design and experimentation of the new cascaded DC-DC boost converter for renewable energy. Int. J. Electron. **106**(9), 1374–1393 (2019)
16. Eltoumi, F.: Charging station for electric vehicle using hybrid sources. Ph.d. thesis, Bourgogne Franche-Comté (2020)

LCL Filter Design and Simulation for Vehicle-To-Grid (V2G) Applications

Sadık Yildiz[1](✉) and Hasan Hüseyin Sayan[2]

[1] Institute of Natural and Applied Sciences, Gazi University, 06560 Ankara, Turkey
sadikyildiz06@gmail.com

[2] Faculty of Technology, Electrical Electronic Engineering Department, Gazi University, 06560 Ankara, Turkey
hsayan@gazi.edu.tr

Abstract. There is a remarkable increase in the number of electric vehicles (EV) with the increase in the demand for renewable energy sources. The integration of EVs into the grid has become an important issue with the widespread use of EVs. The grid integration of EVs has detrimental effects on power quality. The charging topologies such as vehicle-to-grid energy transfer (V2G), grid-to-vehicle energy transfer (G2V), vehicle-to-vehicle energy transfer (V2V) have been developed in order to overcome this problem. In this study, a microgrid using V2G and G2V topologies has been designed for a building and the EVs in this building. In this designed microgrid, 120 kw energy flow is provided for the loads in the building and the charging of EVs. When an extra load is added to the grid in the building, the energy above 120 kw is supplied from the EVs in the microgrid (V2G) to support the grid. The LCL filter design has been carried out for the grid connected inverter used in the V2G topology in the microgrid. After determining the output power, switching frequency, busbar voltage, etc. values of the three-phase inverter for the designed V2G topology, the LCL filter parameters have been calculated. The total harmonic distortion (THD) has been determined according to the calculated parameter values. It has been observed that the THD performance of the LCL filter is better than the THD performance of the LC and L filters. The designed microgrid simulations and LCL filter analyses have been carried out in the MATLAB 2020b Simulink program. With the simulations, G2V and V2G topologies have been analysed and LCL filter design has been carried out for the grid connected inverter used in the V2G topology.

Keywords: LCL Filter · V2G · G2V · Smart Grid

1 Introduction

The demand for renewable energy sources is increasing day by day, and there is a remarkable increase in the number of electric vehicles (EV) with the increasing use of renewable energy sources. As this remarkable increase will disrupt the supply-demand balance, the integration of EVs into the grid has become an important issue. The integration of EVs into the grid will place unpredictable over loads on the existing grid. The studies have

Z. Şen et al. (Eds.): IMSS 2023, LNME, pp. 347–358, 2024.
https://doi.org/10.1007/978-981-99-6062-0_32

been carried out in recent years in order to overcome this network capacity problem. The charging topologies such as vehicle-to-grid energy transfer (V2G), grid-to-vehicle energy transfer (G2V), vehicle-to-vehicle energy transfer (V2V) have been developed with the studies carried out [1].

In the V2G topology, when there is a power demand above the power capacity of the grid to which the EV is connected, this demand is supplied from the EV group which connected to the grid. The demand for three-phase grid-connected inverters with pulse width modulation (PWM) that control for grid connection and power control of EVs has been increasing in recent years [2, 3]. The harmonics occur due to the switching elements in the inverters which used for the connection of EVs to the grid [4]. The total harmonic distortion (THD) value of the fundamental frequency current transferred to the grid in grid-connected inverters must comply with international standards [5]. A filter must be used at the output of the inverter in order to transfer the current to the grid by complying with these standards. There are many filtering methods for grid-connected inverters. The most commonly used filter type is the LCL filter. The design of the LCL filter is smaller than other filters and it has lower cost. But the determining the parameters of the LCL filter is more complicated. Therefore, the parameters must be accurately calculated and analysed in order for the system to remain in a steady state. The LCL filter parameters should be calculated by determining the output power, switching frequency, busbar voltage, etc. values of the designed three-phase inverter [6].

In this study, a microgrid which consisting of a building and EVs and which using V2G and G2V topologies is designed. In this designed microgrid, 120 kW energy flow is provided for the loads in the building and the charging of the EVs. When an extra load is added to the microgrid inside the building, more than 120 kW of energy is provided from the electric vehicles in the grid to support the grid. In other words, V2G topology is operated. In this study, the filter design has been carried out for grid connected inverters that used in V2G topology. According to the determined parameter values, THD performance of LCL filter has been shown to be better by comparing it with LC and L filters. The G2V and V2G topologies have been examined with the simulations and The LCL filter design has been carried out for the grid connected inverter that used in the V2G topology.

2 EV Charge Topologies

The charge topologies have three components. The first of these components is the current conventional grid. This grid is our main energy source. The other component is the batteries of the EVs. The batteries are other energy source that can store energy. The third component is the energy management system that controls the energy flow between these two energy sources. The different charging topologies have been developed by using these three components. These topologies are G2V, V2G and V2V (Fig. 1). With these topologies, the bidirectional energy flow is carried out.

The charging stations are becoming more widespread and the charging units are set up in the car parks of the owners' homes with the increase in the use of EVs. At these charging stations and parking lots, the current conventional grid is used to charge the batteries of the EVs. This topology is called G2V.

With the development of control methods, the bidirectional energy flow control is carried out in AC and DC grid [7, 8]. The parking lot or charging station where EVs charge would act as a microgrid. When there is a distortion in power quality or an increase in demand in this microgrid, the batteries of the vehicles connected to the microgrid would provide energy to the microgrid with independent inverters (DC to AC). This topology is called V2G.

As the smart grid perspective develops, the vehicles in the same microgrid use each other's batteries to charge their batteries (V2V topology) [1].

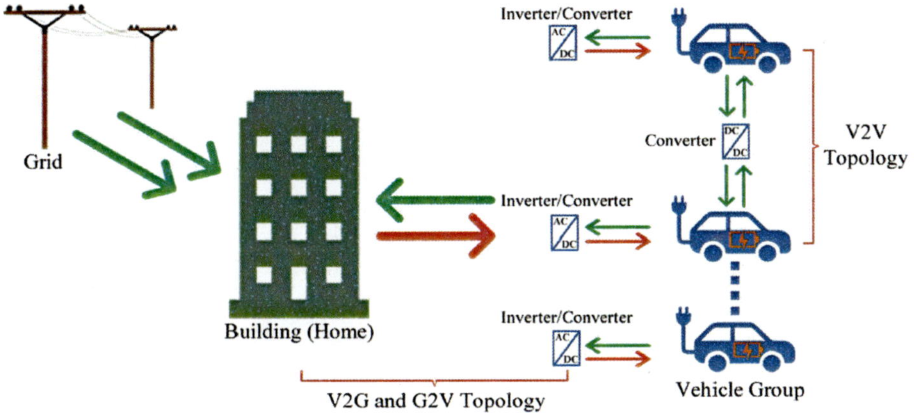

Fig. 1. EV Charge Topologies

3 Basic Power Filters

The power quality of the inverter output is very important in the V2G topology. The signal at the inverter output contains harmonics at different levels due to the frequency of the switching signal. For this reason, the use of filters at the inverter output is necessary [6, 9]. The filter types L, LC and LCL (Fig. 2) are used in grid-connect inverters. The most commonly used filter in grid-connected inverters is the L filter. It must have a high inductance value to reduce current ripple. For this reason, the dimensions of the L filter are very large, so the cost is also high [10]. In order to reduce the larger dimensions of the L filter, the LC filter that another filter type is used. In the LC Filter, the resonance frequency is not constant due to the uncertainty of the grid inductance. Therefore, the LC filter is not suitable for grid-connected inverters [10–12]. The LCL filters have a higher harmonic suppression ratio [12]. The harmonics negatively affect the network and the loads in the system. Also, it could even cause serious damage. According to IEEE-519 harmonic standards; The THD value of the current drawn from the grid should be less than 5% [12–14]. The switching frequency of the inverters is greatly reduced by using the LCL filter. LCL filter could be designed with a smaller dimension, so cost savings could also achieve [12].

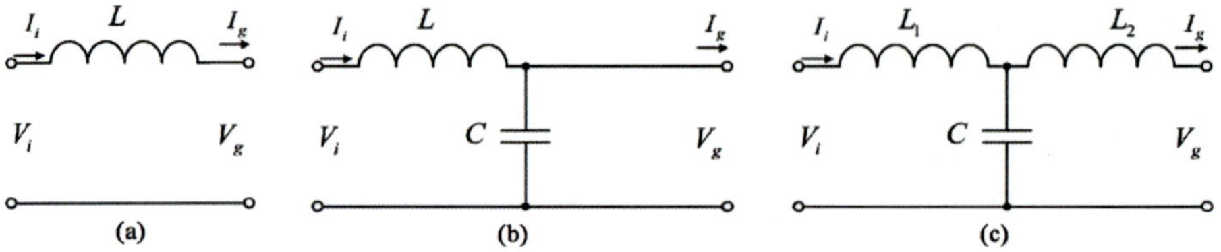

Fig. 2. Basic Power Filters: (a) L Filter (b) LC Filter (c) LCL Filter

3.1 LCL Filter Design

The LCL filter has been designed for the three-phase grid-connected inverter in V2G topology (Fig. 3). The LCL filter could suppress high-order harmonics at its output. But the design should be done very carefully. The incorrect design values could cause increased distortion at the filter output. Therefore, LCL filter should be designed with appropriate values.

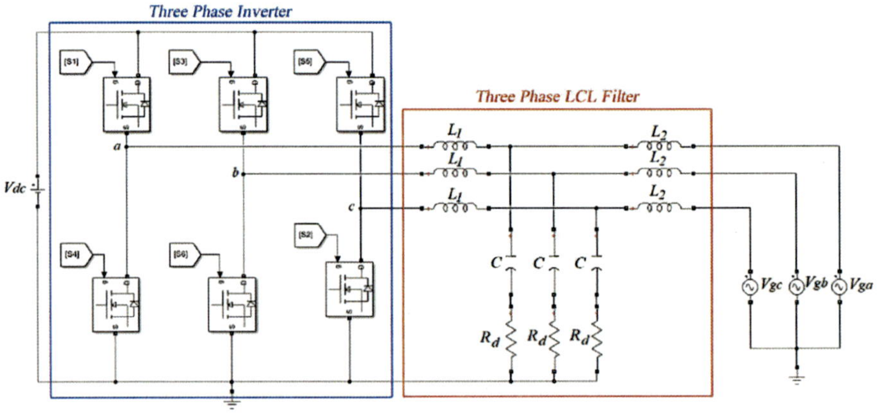

Fig. 3. Three-Phase Grid-Connected Inverter with LCL Filter

The transfer function of LCL filter is given in Eq. 1.

$$\frac{I_g(S)}{V_i(S)} = \frac{1}{(L_1L_2C)S^3 + (R_1L_2C + R_2L_1C)S^2 + (R_1R_2C + L_1 + L_2)S + (R_1 + R_2)} \tag{1}$$

In LCL filter design, the resonant frequency should be far from the mains frequency. The resonance frequency of the LCL Filter is calculated as in Eq. 2.

$$f_r = \frac{1}{2\pi}\sqrt{\frac{L_1 + L_2}{L_1L_2C}} \tag{2}$$

The performance of the LCL filter is affected by the selected resonance frequency, the value of the capacitor C, and the inductance values of the L1 and L2 coils. For this reason, Eq. 3 should be considered in the selection of the resonant frequency.

$$10f_g \leq f_r \leq (f_{sw}/2) \tag{3}$$

The reactive power requirements can cause resonance of the capacitor interacting with the grid. Therefore, passive or active damping must be added by including a resistor in series with the capacitor. The passive damping solution is chosen in this study, but active solutions could also be applied [14, 15]. The value of the damping resistor (R_d) connected in series with the filter capacitor could be calculated as follows:

$$R_d = \frac{1}{3w_r C} \tag{4}$$

In order to perform the filter design, the following parameters have to be known:

Vn: Line to Line RMS Voltage at the inverter output
Vp: Phase voltage at the inverter output
Pn: Nominal Active Power
Vd: DC line voltage
fg: Grid Frequency
fsw: Switching Frequency
fr: Resonance Frequency

Filter values are obtained by using base values [14]:

$$Z_b = \frac{V_n^2}{P_n} \tag{5}$$

$$C_b = \frac{1}{w_g Z_b} \tag{6}$$

Z_b: Base Impedance
C_b: Base Capacitance

During the design of the filter capacitance, the power correction factor was considered to be a maximum of 5%. This calculated value corresponds to 5% of the base capacity value.

$$C = 0.05C_b \tag{7}$$

The maximum amount of ripple in the current at the inverter output is given in Eq. 8.

$$\Delta I_{Lmax} = \frac{2V_{DC}}{3L_1}(1-m)mT_{sw} \tag{8}$$

T_{sw}: Switching time of the inverter
m: Modulation factor of the inverter

It is seen that the highest peak-to-peak current harmonic occurs at m = 0,5. The maximum amount of ripple of the inductance current is given in Eq. 12.

$$\Delta I_{Lmax} = \frac{V_{DC}}{6f_{sw}L_1} \tag{9}$$

L_I is the inductance on the inverter side. The 10% ripple in current is considered for the maximum nominal current. The maximum nominal current is shown in Eq. 10.

$$\Delta I_{Lmax} = 0.1 I_{max} \tag{10}$$

$$I_{max} = \frac{P_n\sqrt{2}}{V_p} \tag{11}$$

With these equations, the L1 value is calculated as in Eq. 12.

$$L_1 = \frac{V_{dc}}{6f_{sw}\Delta I_{Lmax}} \tag{12}$$

The LCL filter equivalent circuit is analysed as a current source to calculate the ripple reduction for each harmonic frequency. The LCL filter limits its own value to 20% to reduce the expected 10% current ripple and creates a 2% ripple in the output current [15, 16]. The relation between the harmonic current of the inverter and the grid and the simplified version of this relation are given in Eq. 13.

$$\frac{i_g}{i_i} = \frac{1}{\left|1 + r\left|1 - L_1 C_b w_{sw}^2 x\right|\right|} = k_a \tag{13}$$

The inductance value (L_2) on the grid side can be calculated by using the k_a value obtained in Eq. 13.

$$L_2 = \frac{\sqrt{\frac{1}{k_a^2}} + 1}{C_f w_{sw}^2} \tag{14}$$

The constant *ka* in Eq. 13 and 14 is expressed as the desired reduction ratio. Displayed as $C = 0.01 / 0.05Cb$.

The constant r is defined as the ratio of the inductances on the grid and inverter side to each other. In this case, the resulting equation is given in Eq. 15.

$$L_2 = rL_1 \tag{15}$$

3.2 Simulation and Analysis Results

Three-phase grid-connected inverter model used in V2G topology (Fig. 4) has been designed in MATLAB 2020b/Simulink program. The Grid values and LCL filter values for the simulation of the designed model are given in Table 1. The simulations of the model have been performed using these values as a reference.

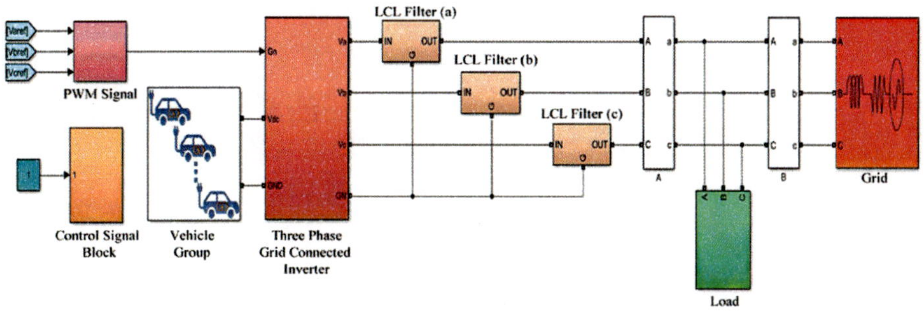

Fig. 4. Three-Phase Grid-Connected Inverter Model used V2G Topology

As it is known, the inverter has a DC signal at its input and an AC signal at its output. The output signal of the inverter has to be a pure sine wave. But the output voltage of an inverter contains harmonics and ripple in practice. These ripples and harmonics are suppressed using the filter. In addition, the phase angles of the grid and the inverter output voltages and currents must overlap in grid-connected inverters. By using the filter, the phase angles of the inverter signal and the grid signal are overlapped in a short time. The three-phase V-I (Voltage-Current) graph of the grid-connected inverter is given in Fig. 5. In Fig. 5a, the V-I graph of the grid-connected inverter with filter is given. As seen in Fig. 5a, harmonics have been suppressed and phase angles have been adjusted. The simulation has been repeated without the filter and the unfiltered V-I graph is given in Fig. 5b. In the current graph, it is seen that the phase angles do not overlap and there are harmonics.

The grid has a total energy of 120 kW in the simulation model. It feeds a load with 75 kW of this energy and charges the EVs in the charging station with 45 kW of it. Here, the G2V topology is operated. A second load of 100 kW is connected to the grid. The V2G topology is starting to operate since the grid cannot supply the demanded 175 kW of energy. The 55 kW energy portion (175 kW–120 kW = 55 kW) required by the grid is provided from the batteries of the EVs in the charging station.

Table 1. System Parameters

fg	Grid Frequency	50 Hz
f_s	Switching Frequency	10 kHz
V_g	Grid Voltage	380 V
V_{dc}	DC Link (Battery) Voltage	800 V
L_1	Inverter Side Inductor of LCL Filter	1 mH
L_2	Grid Side Inductor of LCL Filter	500 µH
C	Capacitor of LCL Filter	100 µf
R_d	Damping Resistor	10 Ω

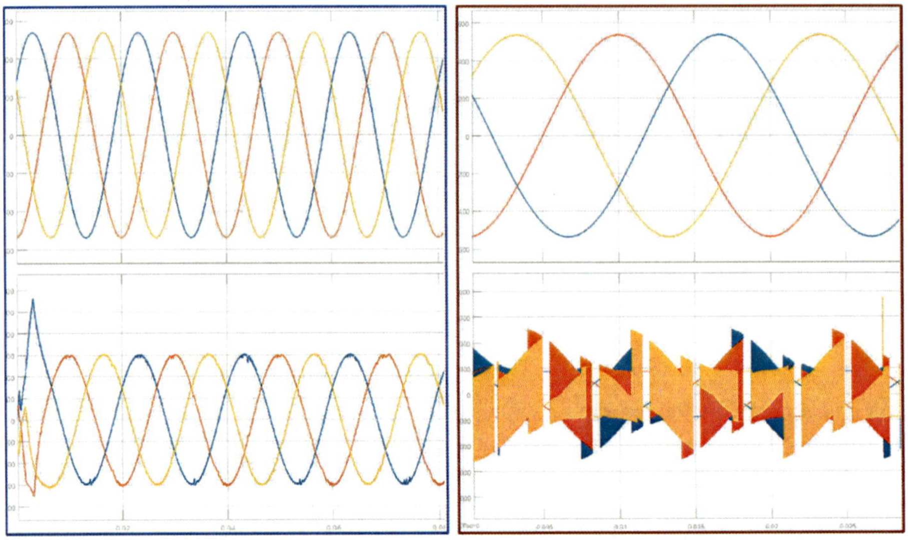

(a) (b)

Fig. 5. Three-Phase V-I Graph a) with LCL Filter b) without Filter

In Fig. 6, the energy flow graph of the grid is given. In Fig. 7, the energy flow graph of EVs is given. EVs charge their batteries by getting 45 kW of energy from the grid in the G2V topology. During the V2G topology, EVs provide 55 kW of energy support to the grid for feeding a total of 175 kW of load.

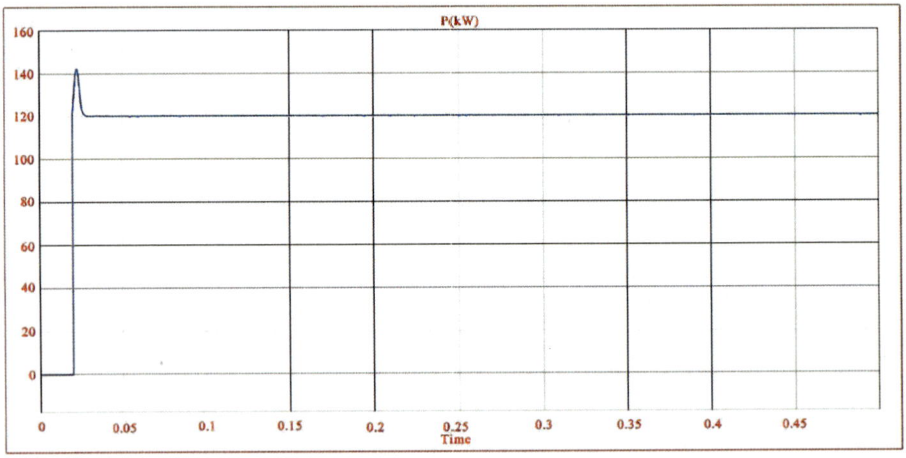

Fig. 6. Energy Flow Graph of the Grid

In the three-phase grid-connected inverter model used for the V2G topology, simulations have been performed for L, LC, LCL filters, respectively, and THD values have

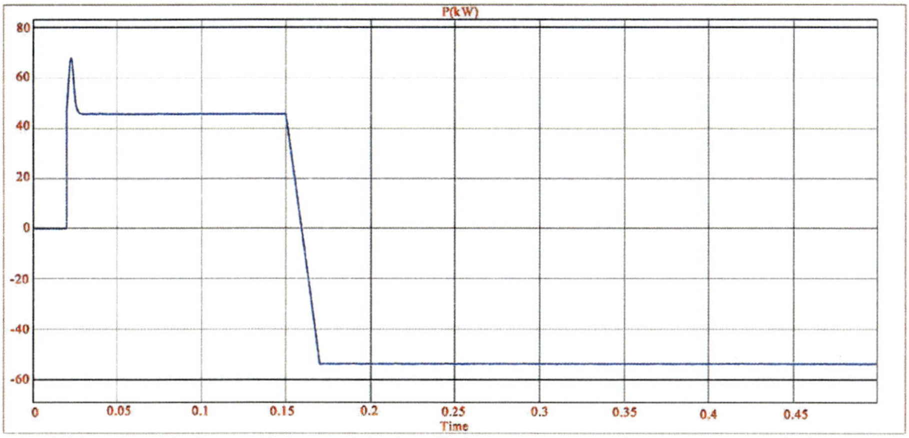

Fig. 7. Energy Flow Graph of EV Battery Group

been measured. According to IEEE-519 harmonic standards; It has been emphasized that the THD value of the current drawn from the grid should be less than 5%. THD value is less than 5% for all three filters. FFT analyses have been performed for L, LC, LCL filters, respectively, and the results are given in Fig. 8, Fig. 9 and Fig. 10.

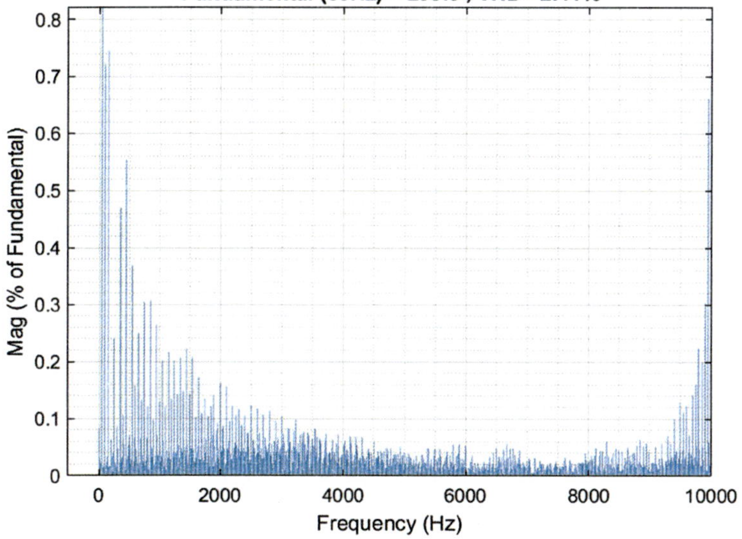

Fig. 8. L Filter FFT Analysis

THD values for L, LC, LCL filters have been measured as 2.41%, 2.38% and 1.52%, respectively when looking at the results of the FFT analysis. It is seen that the LCL filter gives better results than the others. Also, the FFT analysis of the simulation has

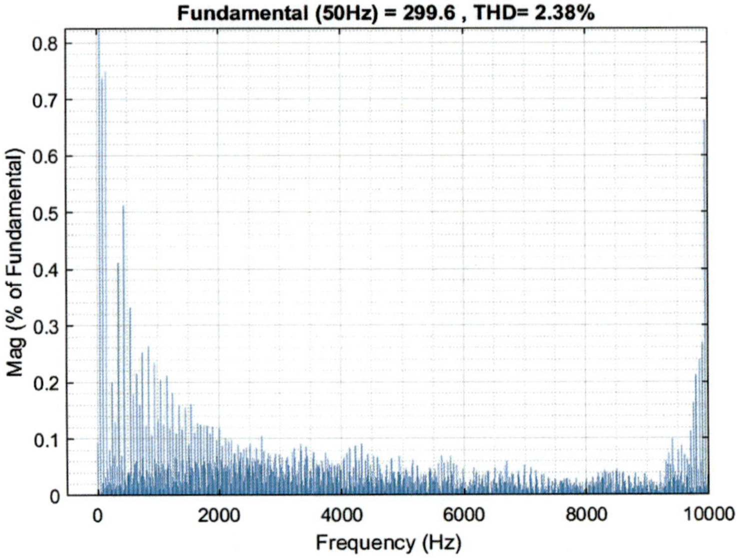

Fig. 9. LC Filter FFT Analysis

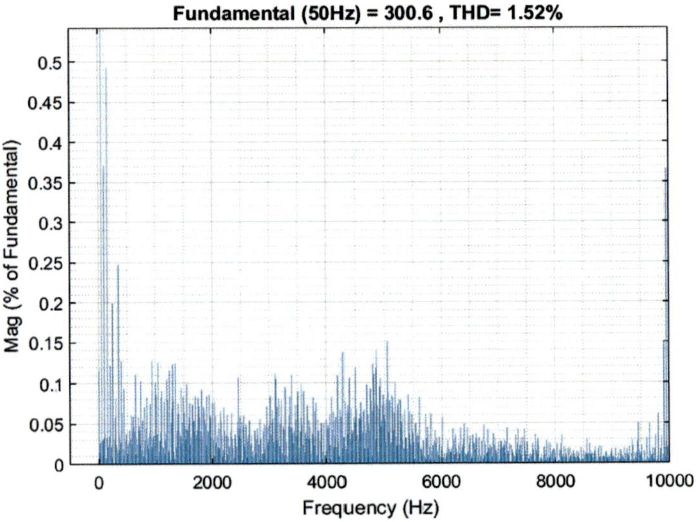

Fig. 10. LCL Filter FFT Analysis

been performed by removing the filter from the designed three-phase grid-connected inverter. In the FFT analysis, the THD value has been measured as 1038.28% (Fig. 11). According to this result, the necessity of using filters is seen.

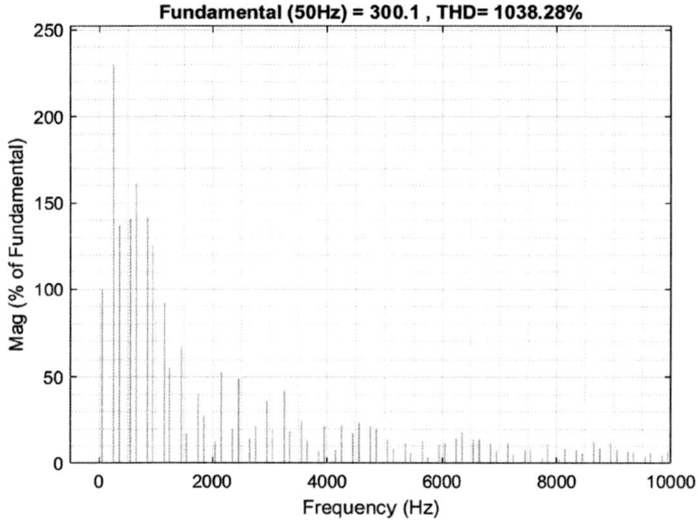

Fig. 11. FFT Analysis without Filter

4 Conclusion

The number of electric vehicles is increasing day by day. This increase disrupts the supply-demand balance in the existing electricity grid. For this reason, the demand for renewable energy sources is also increasing. In order to offer a solution to this problem, charging topologies such as vehicle-to-grid energy transfer (V2G), grid-to-vehicle energy transfer (G2V), vehicle-to-vehicle energy transfer (V2V) have been developed. These topologies are foreseen to support renewable energy sources. V2G topology has been studied in this paper. V2G topology has been used to support the micro-grid in the current grid.

Inverters are used to connect EVs to the grid. The filter at the output of the inverter gains importance in order to suppress the harmonics caused by the switching elements. In this paper, grid-connected inverter filters which are used in V2G topology have been studied. The design of the LCL filter has been explained. it has been shown that the LCL filter is more efficient with the simulation results. The THD value of the fundamental frequency current transferred to the grid in grid-connected inverters should be below 5% according to international standards. According to the simulation results of the designed system, THD value for L, LC, LCL filters has been measured as 2.41%, 2.38% and 1.52%, respectively. The suggested model has been found to be efficient with the simulation results. The power quality at the inverter output of the designed model has been significantly improved with LCL filter.

Acknowledgements. This study (BAP Project Number: FDK-2023-8335) has been supported by Gazi University Scientific Research Projects Unit.

References

1. Üstünsoy, F., Sayan, H.H.: Real-time realization of network integration of electric vehicles with a unique balancing strategy. Electr. Eng. **103**, 2647–2660 (2021). https://doi.org/10.1007/s00202-021-01259-9
2. Can, E., Sayan, H.H.: Development of fractional sinus pulse width modulation with β gap on three step signal processing. Int. J. Electron. **110**(3), 527–546 (2023)
3. Jeong, H.-G., Lee, K.-B., Choi, S., Choi, W.: Performance improvement of LCL-filter-based grid-connected inverters using PQR power transformation. IEEE Trans. Power Electron. **25**(5), 1320–1330 (2010)
4. Üstünsoy, F., et al.: Autonomous operation of microgrid and minimization of fault in case of failure in high-voltage lines. Politeknik Dergisi **23**(4), 1371–1377 (2020)
5. Alamri, B., Alharbi, Y.M.: A framework for optimum determination of LCL-filter parameters for N-level voltage source inverters using heuristic approach. IEEE Access **8**, 209212–209223 (2020)
6. Dursun, M., DÖŞOĞLU, M. K.: LCL filter design for grid connected three-phase inverter. In: Özseven, T., Karaca, V., 2nd International Symposium on Multidisciplinary Studies and Innovative Technologies 2018, ISMSIT, Ankara, Turkey, pp. 1–4 (2018)
7. Mittal, S., Singh A., Chittora, P.: EV control in G2V and V2G modes using SOGI Controller. In: IEEE 3rd Global Conference for Advancement in Technology (GCAT), Bangalore, India, pp. 1–6 (2022)
8. Kumar, S., Likassa, K., Ayenew, E., Sandeep, N., Udaykumar, R.Y.: Architectural framework of on-board integrator: an interface for grid connected EV. In: IEEE AFRICON, Cape Town, South Africa, pp. 1167–1172 (2017)
9. Jiang, S., Liu, Y., Liang, W., Peng, J., Jiang, H.: Active EMI filter design with a modified LCL-LC filter for single-phase grid-connected inverter in vehicle-to-grid application. IEEE Trans. Veh. Technol. **68**(11), 10639–10650 (2019)
10. Renzhong, X., Lie, X., Junjun, Z., Jie, D.: Design and research on the LCL filter in three-phase PV grid-connected inverters. Int. J. Comput. Electr. Eng. **5**(3), 322–325 (2013)
11. Choi, W., Lee, W., Han, D., Sarlioglu, B.: Shunt-series-switched multi-functional grid-connected inverter for voltage regulation in vehicle-to-grid application. In: IEEE Transportation Electrification Conference and Expo (ITEC), Long Beach, CA, USA, pp. 961–965 (2018)
12. Fidan, İ, Dursun, M., Fidan, Ş: Üç Fazlı Eviriciler İçin LCL Filtre Tasarımı ve Deneysel Analizi. Düzce Üniversitesi Bilim ve Teknoloji Dergisi **7**(3), 1727–1743 (2019)
13. Karabacak, M., Kılıç, F., Saraçoğlu, B., Boz, A.F., Ferikoğlu, A.: Şebeke Bağlantılı Eviriciler için LLCL Filtre Tasarımı. Detaylı Bir Performans Analizi. Politeknik Dergisi **19**(3), 251–260 (2016)
14. Kahlane, A.E.W.H., Hassaine, L., Kherchi, M.: LCL filter design for photovoltaic grid connected systems. In: Revue des Energies Renouvelables SIENR'14, Ghardaïa, pp. 227–232 (2014)
15. Reznik, A., Simões, M.G., Al-Durra, A., Muyeen, S.M.: LCL filter design and performance analysis for grid-interconnected systems. IEEE Trans. Ind. Appl. **50**(2), 1225–1232 (2014)
16. Kim, Y.-J., Kim, H.: Optimal design of LCL filter in grid-connected inverters. IET Power Electron. **12**, 1774–1782 (2019)

Airline Passenger Planes Arrival and Departure Plan Synchronization and Optimization Using Genetic Algorithms

Süraka Derviş[1] and Halil Ibrahim Demir[2](✉)

[1] Toyota Motor Europe, Brussels, Belgium
[2] Industrial Engineering Department, Sakarya University, Sakarya, Turkey
hidemir@sakarya.edu.tr

Abstract. Although aviation has been hit hard by the pandemic in the last years, air transport is expected to continue to increase, even if some companies go into crisis. Although some companies went bankrupt during the pandemic, new companies were established, and the planes in the bankrupt companies started to fly in other airlines. The number of employees in the aviation industry is also increasing. Road, maritime, and rail transport cannot keep up with the increase in air transport. By 2037, the annual passenger capacity will be predicted to exceed 8 billion or even 9 billion. There are unique airline connections between more than 20 thousand city-pairs worldwide. Despite this, many passengers fly via transit as there are no direct flights from many cities to many cities or because direct flights are sometimes not economical or common. Since, naturally, passengers do not like to wait in transfers, passengers are lost to other flights and airline companies if the airline does not have a suitable flight at the appropriate time for the transfer. Therefore, synchronization and optimization of international and domestic flights are important at major hub airports and major airlines. In this study, we will try to solve this problem by using the genetic algorithm to maximize the number of total passengers for Istanbul Airport, an important global hub airport. The genetic algorithm is a search heuristic that was inspired by Charles Darwin's theory of natural evolution. This algorithm reflects the process of natural selection, where the fittest individuals are selected for reproduction to produce offspring of the next generation.

Keywords: Airline arrival and departure synchronization · Airline arrival and departure optimization · Genetic algorithms

1 Introduction

The airline industry has been crucial to global transportation for over a century. It has provided a fast and efficient mode of travel for people and goods, connecting cities and countries worldwide. However, the industry has also faced numerous challenges, including economic downturns, pandemics, fuel price fluctuations, and environmental concerns. Despite these challenges, the airline industry has continued to evolve and

Z. Şen et al. (Eds.): IMSS 2023, LNME, pp. 359–367, 2024.
https://doi.org/10.1007/978-981-99-6062-0_33

adapt, introducing new technologies, services, and business models to meet the changing needs of travelers. Our study aims to reduce passenger loss resulting from transit time in a prominent airline company. The transit point is Istanbul airport, and both domestic and international destinations are included. The original plan involves arrivals and departures scheduled with a 5-min interval for one week, and this plan is repeated weekly. In this problem two dimensional chromosomes are used for both arrivals and departures. Two chromosomes with two axes are used and both algorithms are applied these two two-dimensional chromosomes in parallel. In a week there are 2016 5-min time slots for both arrivals and departures. 2016 is the gene number in Y-axis. For every 5-min time slots maximum 7 arrivals and maximum 7 departures are possible. These 7 arrivals or departures represent X-axis. In total, we have 2016 x 7 = 14112 genes for both arrival and departure flights in Y-axis and X axis. To evaluate the effectiveness of different approaches, we will compare the results of the original plan with genetic algorithms, and evolutionary strategy.

2 Literature Survey

While there is no specific study on optimizing airline arrival and departure synchronization, numerous studies have been conducted on other problems in the airline industry. For instance, some studies have focused on optimizing runway and air traffic to enhance efficiency. Montlaur and Delgado[1] developed an optimization model using statistical models for passenger allocation, minimum return time, and noise considerations. Sáez et al. [2] created a model to assign optimal routes and arrival times while Mondoloni and Rozen [3] developed a prediction model based on orbital synchronization. Zhang et al. [4] proposed a criterion selection method for the aircraft landing problem based on the single machine scheduling problem. Similarly, Çiftçi and Özkır [5] developed a mathematical model to minimize connection times for transfer passengers using annealing and taboo search algorithms, and Gu et al. [6] created a dynamic optimization model for fuel consumption. Jungai and Hongjun [7] developed a dynamic optimization model to reduce air traffic flight delays, and Torres [8] presented an approach inspired by swarm theory for air traffic flow management. Finally, Borhani [9] used a multi-objective genetic algorithm to analyze the air transport network's structure and optimize it for fewer airlines, passenger collection, reduced route changes, and shorter travel time.

Several studies have focused on crew planning and assignment. Medard and Sawhney [10] developed a model using shortest path algorithms that incorporated simple tree search and more complex column generation and solution techniques. Yaakoubi et al. [11] utilized a machine learning-based model to perform crew planning for 50,000 flights, while Deveci and Demirel [12] used genetic algorithms and evolutionary strategies for team planning.

Repko and Santos [13] proposed a multi-period modeling approach that utilized a scenario tree to solve the airline demand planning problem under demand uncertainty. The tree comprised decision points at multiple time stages of the planning horizon, with branches representing demand variation scenarios.

Regarding maintenance planning, Deng et al. [14] presented a practical dynamic programming-based methodology that optimizes the long-term maintenance control schedule for a heterogeneous fleet of aircraft. Lin et al. [15] developed a fleet maintenance decision-making model based on the component business model for fatigue structures. This model aimed to minimize airline maintenance costs and maximize fleet availability.

Studies on revenue management aimed to maximize transfer revenue with other airline partners by using a simulation-based optimization model developed by Graf and Kimms [16]. Lawhead and Gosavi [17] utilized the reinforcement learning algorithm to optimize revenue management. Aslani et al. [18] developed a simulator that examines the decision-making process of network revenue management by associating observations with revenue management metrics such as bid price or adjustment cost.

3 Problem Definition

The issue under examination is that a prominent turkish airline company is experiencing a significant loss of passengers due to longer transit times. After examining demand and booking data, the airline discovered that passengers are willing to wait for flights with a transit duration of 1–3 h and prefer to fly with their airline. For flights with a transit duration of 3–5 h, only 50% of the passengers wait, while the remaining prefer to book with another airline. For flights with a transit duration of 5–7 h, only 20% of the passengers wait, and for flights with a transit duration of 7–10 h, only 10% of the passengers wait. None of the passengers are willing to wait for flights with a transit duration exceeding 10 h.

Therefore, the airline aims to maximize the total number of passengers by optimizing the synchronization of international and domestic flights. However, with over 54,000 possible transit combinations, it is not feasible to manually test all combinations to find the optimal solution. To address this challenge, we propose the use of genetic algorithms and evolutionary strategy algorithms to maximize the number of passengers without exhaustively testing all possible combinations. Derviş et al. [19] used evolutionary strategies before. In this research genetics algorithms are used and compared with previous research results.

4 Methods

4.1 Original Plan (OP)

This is the plan being used by the airline company. According to this plan, the total number of passengers gained during transfers is 65659.

4.2 Genetic Algorithms (GA)

Genetic algorithms are a type of search and optimization technique that takes inspiration from the natural process of evolution. GA searches for the most optimal solution in a complex multidimensional space, with the guiding principle of "survival of the fittest." John Holland first proposed these algorithms at the University of Michigan, and they work by evaluating multiple solutions in the solution space rather than a single one. By considering many points at once, genetic algorithms increase the likelihood of finding a better overall solution. They have proven to be very effective in solving complex problems with very large search spaces, where other optimization methods may struggle. While genetic algorithms do not guarantee the absolute best solution, they can find good, acceptable solutions within a reasonable timeframe Demir and Phanden [20]. Flowchart of GA is given in Fig. 1.

The airline company plans to have a weekly flight. In this study, each flight will represent one gene. In the plan, flights are indicated by the destination airport. There are 401 destinations in total. A maximum of 7 planes can take off at the same 5-min time interval, similarly a maximum of 7 planes can land on at the same 5-min time interval. The schedule starts at 00:00 on Monday and continues until 23:55 on Sunday, every 5 min.

We used GA and obtained the following results as in Table 1,2 and 3: we used one-point, two-point, and three-point crossover. We obtained the best results in GA with one-point crossover as in Table 1.

Table 1. Gained transfer passengers using GA with one-point crossover

Iter #	seed1 Best	seed1Avg	seed1Worst	Seed2Best	Seed2Avg	Seed2Worst	Seed3Best	Seed3Avg	Seed3Worst	OP
0	73149	51804	20631	78983	51830	33771	65659	51348	36840	65659
10	73149	64086	56119	78983	60588	49454	69927	61715	54717	65659
20	73149	64086	56119	78983	60715	50722	69927	61715	54717	65659
30	73149	64439	59183	78983	60715	50722	69927	65052	62266	65659
40	73149	64439	59183	78983	60715	50722	69927	65052	62266	65659
50	73149	64439	59183	78983	60715	50722	69927	65052	62266	65659

Table 2. Gained transfer passengers using GA with two-point crossover

Iter #	seed1Best	seed1Avg	seed1Worst	Seed2Best	Seed2Avg	Seed2Worst	Seed3Best	Seed3Avg	Seed3Worst	OP
0	70276	50512	18013	65659	53704	36580	72408	56135	27790	65659
10	70407	61532	49093	65659	60928	57132	75752	68789	63245	65659
20	70407	61532	49093	65659	60928	57132	75752	68789	63245	65659
30	70407	61532	49093	65659	60928	57132	75752	68789	63245	65659
40	70407	61719	50970	65659	61455	58717	75752	68789	63245	65659
50	70407	62436	52510	65659	61455	58717	75752	68789	63245	65659

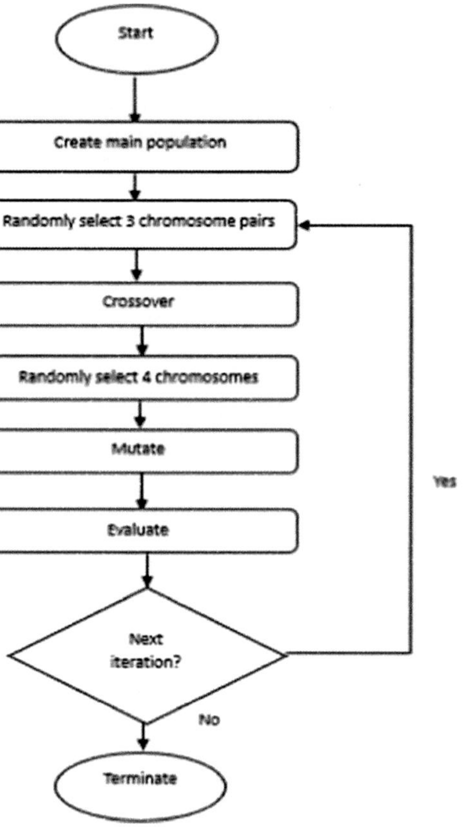

Fig. 1. Flowchart of Genetic Algorithms

Table 3. Gained transfer passengers using GA with three-point crossover

Iter #	seed1Best	seed1Avg	seed1Worst	Seed2Best	Seed2Avg	Seed2Worst	Seed3Best	Seed3Avg	Seed3Worst	OP
0	72367	55994	34971	67171	53956	37495	68460	50387	28795	65659
10	72372	65871	57474	67173	62135	53864	71314	62363	53846	65659
20	72372	65871	57474	67173	62135	53864	71314	62363	53846	65659
30	72372	65871	57474	67173	62490	57414	71314	62363	53846	65659
40	72372	65871	57474	67173	62490	57414	71314	62363	53846	65659
50	72372	65871	57474	67173	62490	57414	71314	62363	53846	65659

4.3 Evolutionary Strategy

Evolution Strategies (ES's) belong to the Evolutionary Algorithms (EA's) class, which are a type of direct search and optimization technique inspired by nature. ES's use mutation, recombination, and selection to evolve increasingly better solutions from a population of individuals containing candidate solutions. The roots of ES's date back to the mid-1960s when researchers at the Technical University of Berlin developed the first schemes for evolving optimal shapes of minimal drag bodies using Darwin's evolution principle.

ES's can be applied to a wide range of optimization problems, including continuous, discrete, combinatorial search spaces with or without constraints and mixed search spaces. The main difference between ES and genetic algorithms is the operators used. ES only uses the mutation operator, while genetic algorithms use both mutation and crossover operators. Each iteration uses the same number of new offspring in all search techniques.

This study will use ES with a mutation operator for 100 iterations. The mutation will be applied by randomly selecting 100 genes from the original plan, shuffling them, and replacing them in the same locations. In each iteration, the passenger number of the new solution is compared with the passenger number of the current best solution. If the new solution has a higher passenger number, the best solution is updated accordingly.

5 Experiments

We coded the program using Python programming language in Google Colab platform. Genetic algorithms and evolutionary strategy were run for 50 iterations. Total passenger number in the original plan was 65659.

Average, max and min total passenger numbers for each algorithm are shown in tables 4 and 5. Best, average and worst comparisons between algorithms are shown in Figs. 2, 3 and 4.

Table 4. Best, average & worst chromosome results for GA over each 10 iterations

Iter #	Best	Avg	Worst
10	78983	51830	33771
20	78983	60588	49454
30	78983	60715	50722
40	78983	60715	50722
50	78983	60715	50722

Table 5. Best, average & worst chromosome results for ES over each 10 iterations

Iter #	Best	Avg	Worst
10	82964	81731	80949
20	87222	85884	85156
30	91599	90647	89972
40	94982	94749	94348
50	102922	100564	99716

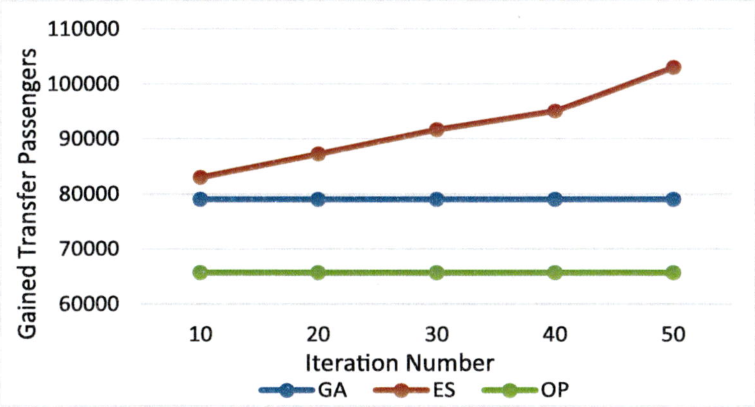

Fig. 2. Best chromosome comparison of each algorithm

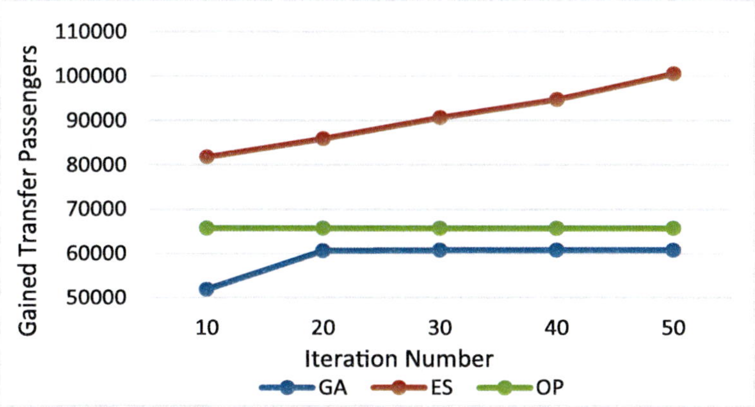

Fig. 3. Average chromosome comparison of each algorithm

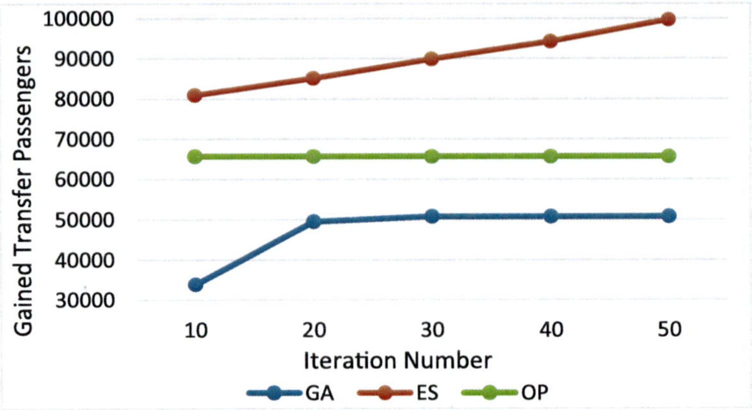

Fig. 4. Worst chromosome comparison of each algorithm

6 Conclusion

We studied airline flights optimization using genetic algorithms and compared with evolutionary strategy. One point-cross over gave the best result in genetic algorithms.

After 50 iterations we can see that the evolutionary strategy gave us better solutions than the original plan and the genetic algorithms. The total passenger number in the original plan was 65659, While the best passenger number in the evolutionary strategy was 102922, the performance in genetic algorithms was 78983. As an overview we can see that evolutionary strategy gave the best performance compared to the genetic algorithms specifically for this problem.

References

1. Montlaur, A., Delgado, L.: Flight and passenger delay assignment optimization strategies. Transp. Res. Part C: Emerg. Technol. **81,** 99–117 (2017). https://doi.org/10.1016/j.trc.2017.05.011
2. Sáez, R., Prats, X., Polishchuk, T., Polishchuk, V.: Traffic synchronization in terminal airspace to enable continuous descent operations in trombone sequencing and merging procedures: An implementation study for Frankfurt airport. Transp. Res. Part C: Emerg. Technol. **121**, 102875 (2020). https://doi.org/10.1016/j.trc.2020.102875
3. Mondoloni, S., Rozen, N.: Aircraft trajectory prediction and synchronization for air traffic management applications. Prog. Aerosp. Sci. **119**, 100640 (2020). https://doi.org/10.1016/j.paerosci.2020.100640
4. Zhang, J., Zhao, P., Zhang, Y., Dai, X., Sui, D.: Criteria selection and multi-objective optimization of aircraft landing problem. J. Air Transp. Manage. **82**, 101734 (2020). https://doi.org/10.1016/j.jairtraman.2019.101734
5. Çiftçi, M.E., Özkır, V.: Optimising flight connection times in airline bank structure through simulated annealing and tabu search algorithms. J. Air Transp. Manage. **87**, 101858 (2020). https://doi.org/10.1016/j.jairtraman.2020.101858

6. Gu, J., Tang, X., Hong, W., Chen, P., Li, T.: Real-time optimization of short-term flight profiles to control time of arrival. Aerosp. Sci. Technol. **84**, 1164–1174 (2019). https://doi.org/10.1016/j.ast.2018.09.029
7. Jungai, T., Hongjun, X.: Optimizing arrival flight delay scheduling based on simulated annealing algorithm. Phys. Procedia **33**, 348–353 (2012). https://doi.org/10.1016/j.phpro.2012.05.073
8. Torres, S.: Swarm theory applied to air traffic flow management. Procedia Comput. Sci. **12**, 463–470 (2012). https://doi.org/10.1016/j.procs.2012.09.105
9. Borhani, M.: Evolutionary multi-objective network optimization algorithm in trajectory planning. Ain Shams Eng. J. **12**(1), 677–686 (2021). https://doi.org/10.1016/j.asej.2020.07.001
10. Medard, C.P., Sawhney, N.: Airline crew scheduling from planning to operations. Eur. J. Oper. Res. **183**(3), 1013–1027 (2007). https://doi.org/10.1016/j.ejor.2005.12.046
11. Yaakoubi, Y., Soumis, F., Lacoste-Julien, S.: Machine learning in airline crew pairing to construct initial clusters for dynamic constraint aggregation. EURO J. Transp. Logist. **9**(4), 100020 (2020). https://doi.org/10.1016/j.ejtl.2020.100020
12. Deveci, M., Demirel, N.Ç.: Evolutionary algorithms for solving the airline crew pairing problem. Comput. Ind. Eng. **115**, 389–406 (2018). https://doi.org/10.1016/j.cie.2017.11.022
13. Repko, M.G., Santos, B.F.: Scenario tree airline fleet planning for demand uncertainty. J. Air Transp. Manage. **65**, 198–208 (2017). https://doi.org/10.1016/j.jairtraman.2017.06.010
14. Deng, Q., Santos, B.F., Curran, R.: A practical dynamic programming based methodology for aircraft maintenance check scheduling optimization. Eur. J. Oper. Res. **281**(2), 256–273 (2020). https://doi.org/10.1016/j.ejor.2019.08.025
15. Lin, L., Wang, F., Luo, B.: An optimization algorithm inspired by propagation of yeast for fleet maintenance decision making problem involving fatigue structures. Appl. Soft Comput. **85**, 105755 (2019). https://doi.org/10.1016/j.asoc.2019.105755
16. Graf, M., Kimms, A.: Transfer price optimization for option-based airline alliance revenue management. Int. J. Prod. Econ. **145**(1), 281–293 (2013). https://doi.org/10.1016/j.ijpe.2013.04.049
17. Lawhead, R.J., Gosavi, A.: A bounded actor–critic reinforcement learning algorithm applied to airline revenue management. Eng. Appl. Artif. Intell. **82**, 252–262 (2019). https://doi.org/10.1016/j.engappai.2019.04.008
18. Aslani, S., Modarres, M., Sibdari, S.: On the fairness of airlines' ticket pricing as a result of revenue management techniques. J. Air Transp. Manage. **40**, 56–64 (2014). https://doi.org/10.1016/j.jairtraman.2014.05.004
19. Derviş, S., Demir, H.I., Phanden, R.K., Kökçam, A.H., Erden, C.: Optimization of arrival and departure plans of airline passenger aircraft using evolutionary strategy. Int. Symp. Intell. Manuf. Serv. Syst., 36–42 (2021)
20. Demir, H.I., Phanden, R.K.: Due-date agreement in integrated process planning and scheduling environment using common meta-heuristics. In: içinde Integration of Process Planning and Scheduling: Approaches and Algorithms, pp. 161–182. CRC Press/Taylor & Francis Group, Boca Raton (2020)

Exploring the Transition from "Contextual AI" to "Generative AI" in Management: Cases of ChatGPT and DALL-E 2

Samia Chehbi Gamoura[1](✉), Halil İbrahim Koruca[2], and Kemal Burak Urgancı[2]

[1] HuManis Laboratory of EM Strasbourg Business School, Strasbourg, France
samia.gamoura@em-strasbourg.eu

[2] Industrial Engineering Department, Süleyman Demirel University, Isparta, Turkey
{halilkoruca,kemalurganci}@sdu.edu.tr

Abstract. The transition towards Generative Artificial Intelligence (GAI) is rapidly transforming the digital realm and providing new avenues for creativity for all humanity. In the past two years, several generative models have disrupted worldwide, including ChatGPT and DALL-E 2, developed by OpenAI, which are currently receiving significant media attention. These models can generate new content, respond to prompts, and automatically create new images and videos. Nevertheless, despite this progress of GAI, research into its application in business and industry is still in its infancy. Generative AI is bringing ground-breaking innovations that go beyond the limitations of conventional Contextual AI. This new type of AI can generate novel patterns in human-like creativity, encompassing various forms of content such as text, images, and media. It transforms how people communicate, create, and share content, taking organizations by surprise. Unfortunately, these organizations were not fully prepared as they were focused on the advancements and impacts of Contextual AI. Given the significant organizational-societal opportunities and challenges posed by generative models, it is crucial to comprehend their ramifications. However, the excessive hype surrounding GAI currently makes it difficult to determine how organizations can effectively utilize and regulate these powerful algorithms. In research, the primary question is how organizations can manage the intersection of human creativity and machine creativity, and how can they leverage this intersection to their advantage? To address this question and mitigate concerns related to it, a comprehensive understanding of GAI is essential. Therefore, this paper aims to provide technical insights into this paradigm and analyze its potential, opportunities, and constraints for business and industrial research.

Keywords: Generative Artificial Intelligence · Contextual Artificial Intelligence · business · industry · ChatGPT · DALL-E 2

Z. Şen et al. (Eds.): IMSS 2023, LNME, pp. 368–381, 2024.
https://doi.org/10.1007/978-981-99-6062-0_34

1 Introduction

The advancements in artificial intelligence (AI) have led to the creation of several innovative tools, including ChatGPT. With its ability to generate coherent and contextualized responses to a given prompt, ChatGPT has garnered considerable attention from both individuals and organizations. Many individuals have already tested ChatGPT by introducing at least one simple prompt, and some have even started using it extensively for various purposes, such as writing essays, solving problems, correcting, or translating texts (Korzynski, et al., 2023). However, organizations are still exploring the potential of this tool and questioning how it can be leveraged to generate profit and improve overall productivity.

First and foremost, it is important for organizations to understand that not all can easily deploy and integrate emerging Generative AI tools such as ChatGPT, Dall-E 2, MidJourney, Jasper, Stable Diffusion, and others due to the required maturity level of transformation. As per (Ivančić, et al., 2019), digital transformation is an ongoing process of enhancing digital maturity in organizations by leveraging both digital technologies and organizational practices (Reix, 2004). The same author suggests that an appropriate level of maturity would enable the integration of advanced technologies. However, according to the authors of a report published by the MIT Sloan Management Review (Kane, et al., 2015), digital maturity lies in how organizations integrate appropriate digital technologies to transform their activities, following a digital strategy that fosters digital maturity (Reix, 2004). Hence, organizations must avoid chasing the hype and carefully analyze their strategies and the maturity of their transformation before deploying such technologies. Additionally, selecting the appropriate Generative model for a specific use case is a significant challenge for organizations, as the field is emerging and evolving rapidly (Castelli & Manzoni, 2022). These advanced AI tools require a certain level of maturity and cannot be deployed without proper preparation and guidance (Park, et al., 2023). Thus, the main objective of this paper is to provide guidance and recommendations on the effective use of Generative AI tools in organizations.

This paper is divided into three main parts. In Sect. 1, we provide a context for the transition from contextual AI to Generative AI. Section 2 illustrates the impact of Generative AI on management theories. In Sect. 3, we present our proposed Business-Value driven model for Generative AI applications. Finally, a brief discussion and conclusion are presented.

2 From Contextual AI to Generative AI in Management

Over the years, "Contextual AI" has undergone significant evolution, leading to the emergence of "Generative AI". "Contextual AI" primarily focuses on pre-determined tasks, such as classification, regression, and prediction, where the model is trained to analyze and make decisions based on a set of input data (Tang, et al., 2021). However, with increasing demand for more innovative and creative solutions, Generative AI has gained popularity. Unlike Contextual AI, Generative AI not only analyzes pre-existing data but also generates new content, such as images, text, and even music (Castelli & Manzoni, 2022). Generative AI employs complex algorithms that learn from patterns in data to generate novel solutions (Holmgard, et al., 2014). The rise of Generative AI has brought forth a new era of AI applications for organizations, ranging from creating text, media, images, art, and music to developing novel solutions for complex problems.

Generative AI is a branch of AI that focuses on the creation and generation of new content. Generative models include GPT-3 (Generative Pre-trained Transformer-3), Generative Adversarial Networks (GAN), and Variational Auto-Encoders (VAE) (Olmo, et al., 2021). Recent advancements in machine learning and deep learning techniques, particularly GAN networks and their variants, have resulted in several emerging systems in this field, such as Text-to-Text models like ChatGPT, Text-to-Images like DALL-E 2, Text-to-Music, Text-to-Video, and even Text-to-Code like Alpha Code (Marcus, et al., 2022). These models enable the generation of novel content by learning patterns in existing data and using those patterns to generate new content. That is probably caused by the extensive advancements in Machine Learning and Deep Learning techniques in recent years, and particularly the GAN networks and their variants, which are nowadays popular techniques (Castelli & Manzoni, 2022).

One of the most popular generative models in Text-to-Text is ChatGPT, which is a variant of the GPT language model developed by OpenAI® and released on November 30, 2022 (Zhou, et al., 2023). ChatGPT is an interactive tool that belongs to the textual Generative AI category and has the ability to generate new comprehensible text by using prompts as inputs (Korzynski, et al., 2023). Additionally, ChatGPT is a Chatbot technology that falls under the category of "Conversational AI", which is currently experiencing a surge in popularity among both society and organizations (Gkinko & Elbanna, 2022). According to the Deloitte digital® report, the global market size of Conversational AI is expected to reach €145.78 billion by 2024, with an annual growth rate of 30.2% (Deloitte digital, 2019). Research on this topic, as per the specialized analytical application for scientific production, Dimensions.ai, published in management and management journals, has nearly doubled in the last four years (2018 to 2022), increasing from 784 to 1,512. The ChatGPT system utilizes the transformer architecture, which is one of the recent key advancements in AI. This type of architecture belongs to Artificial Neural Networks (ANN) and was introduced by Vaswani et al. in their paper (Vaswani, et al., 2017) for Natural Language Processing (NLP) tasks. The transformer architecture is guided by "self-attention" layers, which prioritize the weight assigned to words and phrases based on their importance in the textual input. Additionally, the system includes "feed-forward" layers with residual connections to enhance the model's ability to comprehend complex patterns in the data (Zhou, et al., 2023). While ChatGPT is a well-known textual Generative AI tool, others such as Bard, recently developed by Google, are emerging (Korzynski, et al., 2023).

In the realm of generative models for Text-to-Image, DALL-E 2 is a noteworthy system created by OpenAI® in April 2022 [2]. DALL-E 2 can generate new synthetic images that correspond to input text (captions) (Marcus, et al., 2022). The system is based on a Deep Learning model for Text-to-Image generation (OpenAI, Jun 28, 2022). DALL-E 2 holds tremendous potential for image generation and augmentation in several management fields, such as brand creation, visual marketing, product design, and other applications (Tuomi, 2023) (Fig. 1).

The literature on Generative AI in management and business is currently in its early stages, with most publications being preprints. However, two contrasting views have emerged in the literature: the optimistic view and the pessimistic view. According to (Korzynski, et al., 2023), younger managers tend to be more accepting of new technologies than older generations, and this could affect their views on Generative AI. In the optimists' camp, authors argue that Generative AI will undoubtedly serve management research and academia, as this new technology has the potential to impact managerial work at the strategic, functional, and administrative levels (Korzynski, et al., 2023). On the other hand, the pessimistic view is that Generative AI may generate false knowledge and be prone to biases, leading to incorrect decisions and actions (McGee, 2023.). Fears about the capabilities of AI are not new, as many studies have expressed concerns about the socio-organizational problems caused by automation and autonomy, such as the dehumanization of work, loss of social connections, and loss of the value of professional gestures (Akerkar, 2019).

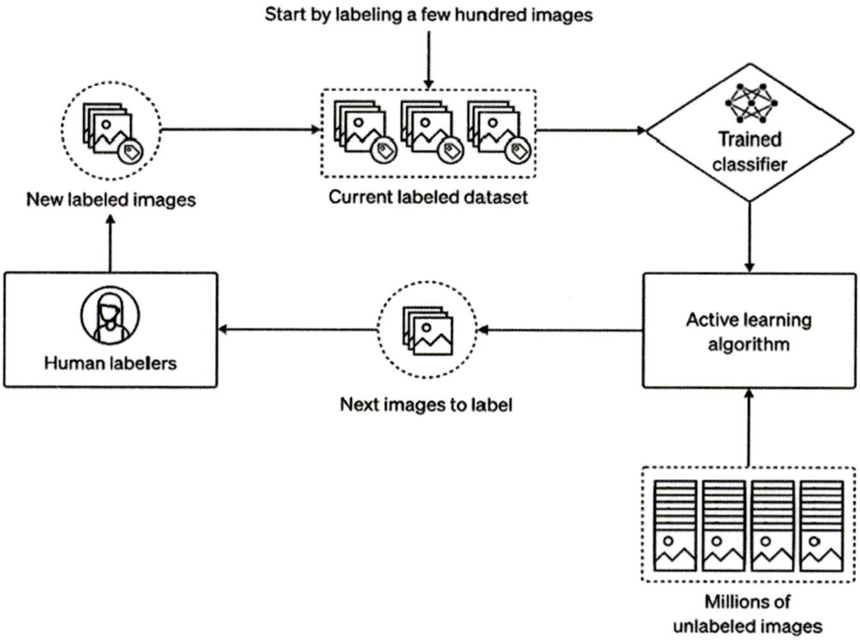

Fig. 1. The training data process of DALL-E 2 system by OpenAI® (OpenAI, Jun 28, 2022)

The McKinsey Global Survey report (McKinsey, 2019) has revealed that although investments in AI are rapidly increasing, its effective adoption is slowed down for various reasons, including socio-organizational factors. This report mentions reluctance towards the changes that AI implies as one of these reasons. According to (Shin, 2021), these problems have significantly contributed to the emergence of movements of fear and reluctance towards AI, thus slowing down its acceptance. However, with Generative AI, fears are now directed towards machines' capacity for creativity, such as generating new essays, journalistic speeches, product designs, ideas, and more. This raises several questions about acceptance, as well as ethics, cyber criminality (Mijwil & Aljanabi, 2023), patent use, and intellectual property (Strowel, 2023) among other concerns.

3 Position of Generative AI in Management Theories

The field of management science is concerned with understanding and improving organizational performance through the application of scientific methods and principles. One key aspect of this field is the development and application of organizational theories, which provide frameworks for understanding how organizations function and how they can be managed effectively (Simon, 1987). These theories draw on various disciplines of management, including technology management, to explain why organizations behave the way they do, what factors contribute to their success or failure, and how managers can intervene to improve their performance using various tools and theories (Korzynski, et al., 2023). As Generative AI is emerging in management, we aim to explore the relationship between this emerging paradigm and key organizational theories in management science to highlight their impact and relevance. In the following sections, we will discuss some argumentation points about the most common management theories and how they relate to Generative AI.

In 1887, Herbert Simon introduced the concept of Bounded Rationality Model (BRM) (Simon, 1987). This theory explains how human cognitive abilities are limited by various constraints, including time constraints, which can affect the ability to make optimal decisions. This raises the question of the optimality of our decision-making abilities as humans. One way to address the theory of bounded rationality is by developing procedural rationality, as suggested in (Dean Jr & Sharfman, 1993). Considering the advancements made by ChatGPT today, one might wonder what Herbert Simon would think of them. Generative AI has shown great proficiency in procedural rationality, particularly in areas such as inferencing, decision-making, and problem-solving (Korzynski, et al., 2023).

When considering the ability of ChatGPT to generate new and innovative ideas and creative solutions, the Innovation Theory may come to mind (Brock & Von Wangenheim, 2019). This theory is based on Schumpeter's concept of "creative destruction", which explores processes related to "disruptions" and "mutations" (Schumpeter, 2007). In fact, generative AI can play an important role in helping organizations stay ahead of the curve and adapt to rapidly changing market conditions. The ability of generative AI, such as ChatGPT, to generate new and innovative ideas is a key aspect of its potential utility. The Innovation theory is particularly relevant in this context, as it emphasizes the importance of disruptive processes in driving progress and innovation (Brock & Von Wangenheim, 2019) Schumpeter's concept of "creative destruction" highlights the need for organizations to continually innovate and adapt to changing market conditions, which can be facilitated by the use of generative AI (Schumpeter, 2007).

The Change Theory, which views digital transformation as a process of change, is supported by numerous studies, including (Panenkov, et al., 2021). Generative AI can play a crucial role in facilitating this process of change. Firstly, it can enable the implementation of change by providing new content easily and more automatically, along with insights and recommendations based on vast amounts of data. It can also assist organizations in developing and testing new ideas and strategies for change management. For instance, generative AI can be used to simulate the impact of various change initiatives, enabling decision-makers to evaluate the potential outcomes of different strategies and select the most effective one (Holmgard, et al., 2014). By leveraging generative AI, organizations can benefit from the insights and recommendations provided by advanced algorithms, which can help them to identify areas for improvement and implement changes more effectively. The Change Theory provides a framework for understanding the process of digital transformation as an ongoing process of change, which can be facilitated by the use of generative AI (Panenkov, et al., 2021).

Integrating AI algorithms into management processes involves making changes to these processes through the flow of information systems, which are essentially information processing procedures. As such, we can draw a connection to the Organizational Information Processing Theory. This theory is based on the idea of an organization evolving within a system that integrates various internal and external processes characterized by their complexity and uncertainty (Nikitaeva & Salem, 2022). The use of appropriate generative AI technology can help to reduce the uncertainty and complexity of these processes, as noted in (Sensoy, et al., 2020).

The utilization of generative AI within organizations holds immense potential as a valuable tool for improving business operations and enhancing customer experiences (Korzynski, et al., 2023). However, to fully realize this potential, it is crucial that this concept be grounded in Organizational Theories (Brown, et al., 2014). Organizational theories within the field of management science provide a focused study on how organizations operate, why they behave the way they do, and how they can be managed effectively. Addressing this challenge is essential for organizations seeking to effectively leverage generative AI tools such as ChatGPT and others.

Understanding the factors that influence the acceptance or rejection of new technological paradigms or technologies has become an increasingly important topic in management (Momani & Jamous, 2017). This is because technology adoption affects not only the success of the technology itself, but also the success of the broader organizational transformation process. Recent research has focused on the Technology Acceptance Theories (TAM) framework, which has gained momentum in the wake of the introduction of new generative tools like ChatGPT (Sensoy, et al., 2020) The term "acceptance" refers to users' intention or willingness to use a new concept (Davis, et al., 1989). Among these new concepts, generative AI and smart chatterbots like ChatGPT rank among the most disruptive and challenging for organizations to accept (Shin, 2021).

4 Proposed BV-Driven Model for Generative AI in Management

The concept of Business Value (BV) has been a subject of research in management for many years, and has been associated with various contexts (Mooney, et al., 1996). Initially, much of this research linked BV to information technology (IT), which characterized the last two decades. However, with the advent of data analytics systems and the Big Data paradigm, much of this research has shifted towards these systems, which are considered a form of technological evolution that includes IT (Shanks, et al., 2011). Several theoretical frameworks have been proposed in these researches, such as those presented as empirical models. Recently, there has been an increasing interest in Business Values derived from AI systems and approaches, as evidenced by Davenport's book published by MIT (Davenport, 2018). The main focus of these researches is on how to derive Business Values in the context of AI, which is typically done through an inductive approach (Wetering, et al., 2018).

In their work (Enholm, et al., 2022), the authors distinguish between two types of Business Value that can be generated by AI: internal values that use AI to improve internal managerial processes, and external values that are used for commercializing products and services, in direct contact with the outside world (clients, suppliers, etc.) (Chehbi Gamoura, et al., 2020). Our research aims to cover all types of Business Values that can be derived by AI, without specialization in any particular type. *Wamba-Taguimdje et al.* (Wamba-Taguimdje, et al., 2020) define the capabilities of AI, called "AICAP (AI capabilities)", in a company as its ability to create an effective model for the creation and capture of Business Value. This model is based on three types of resources: AI management capability (AIMC), personal AI expertise (AIPE), and AI infrastructure flexibility (AIIF). The model captures a significant influence on organizational performance and process innovation through digital transformation projects based on AI at two levels: the process level and the organizational level. At the process level, there are three categories of effects related to Business Values: automation effect, informational effect, and transformational effect, which are also found in the conceptual model of *Mooney et al.* (Mooney, et al., 1996). At the organizational level, there are three categories of performance (interpreted as Business Values in our paper): financial performance, marketing performance, and administrative performance.

Building on the work of *Wamba-Taguimdje et al.* (Wamba-Taguimdje, et al., 2020) and *Mooney et al.* (Mooney, et al., 1996), we propose a new conceptual model that identifies seven Business Values that can be derived from Generative AI in management. We refer to this model as the "7-BV model" (see Fig. 2, Fig. 3), and Table 1 provides a brief overview of each of the Business Values. The 7-BV model expands on the previous models by specifically focusing on Generative AI and its potential impact on management. The seven Business Values identified in the model are based on a thorough review of the relevant literature and empirical evidence. They are as follows: [insert the seven Business Values with brief descriptions.

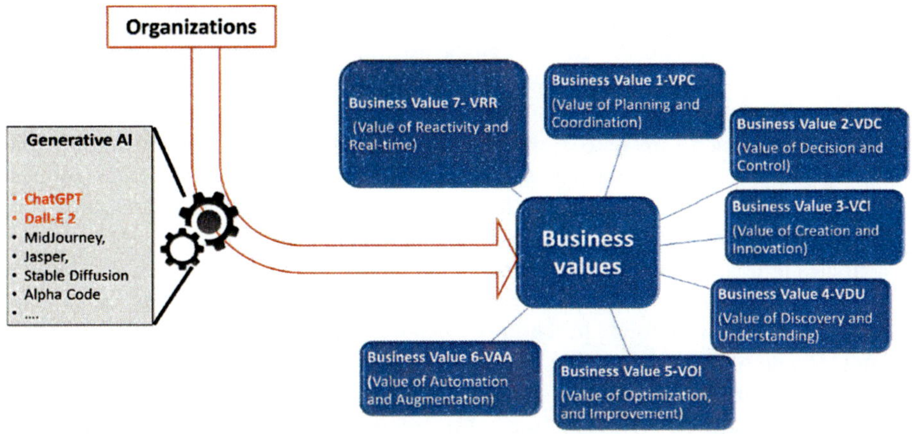

Fig. 2. Proposed model (7-BV) of Business Values categorizations for Generative Artificial Intelligence models in Business applications

1. **Business Value 1-VPC (Value of Planning and Coordination):** Generative AI can significantly contribute to production planning by optimizing resource allocation in terms of time and quantity (Park, et al., 2023). Additionally, it can aid in task and employee scheduling by leveraging real-time data and considering constraints such as availability, skills, and deadlines. Generative models can also enhance team coordination by providing automated real-time updates, alerts, and notifications to team members (Korzynski, et al., 2023).
2. **Business Value 2-VDC (Value of Decision and Control):** Generative AI models, such as Generative Adversarial Networks (GANs) and Variational Autoencoders (VAEs), can aid enterprises in decision making by generating new data based on existing data (Korzynski, et al., 2023). This can be particularly valuable in situations where data is limited or incomplete, or when enterprises want to explore various scenarios or possibilities without conducting experiments or collecting new data. For example, ChatGPT can suggest recommendations by interacting with explicit prompts (such as sales or customer service inquiries), or can assist in decision making by utilizing implicit prompts from customer behavior and experience (Olmo, et al., 2021).
3. **Business Value 3-VCI (Value of Creation and Innovation):** Generative AI is a powerful tool for fostering innovation in organizations by enabling the creation of new products, services, and business models that can disrupt existing markets and create new ones (OpenAI, Jun 28, 2022) The contribution of Generative AI to this value can be divided into four directions. First, it can facilitate the generation of new ideas by identifying patterns and insights in large amounts of data (Holmgard, et al., 2014). Second, it can be used to develop and refine concepts by simulating different scenarios and experimenting with various parameters to assess feasibility (Mijwil & Aljanabi, 2023.). Third, Generative AI can aid in the design and prototyping of new products and services by leveraging machine learning and computer vision techniques (McGee, 2023.). Finally, it can enable personalization of products and services based

on customer preferences and behavior, by creating customized experiences that resonate with individual customers through the use of feedback data (Zhou, et al., 2023). This can lead to increased customer satisfaction and loyalty.

4. **Business Value 4-VDU (Value of Discovery and Understanding):** Generative AI can be highly useful for pattern discovery, including information retrieval and text mining (McGee, 2023.). For instance, ChatGPT can be trained to recognize entities (such as people, organizations, and locations) within text data in documents, and it can also perform clustering analysis to group together entities that share common attributes or characteristics (Korzynski, et al., 2023). This can help identify relationships between entities and the patterns that connect them. Applications for Generative AI in this value can be found in sentiment analysis, topic modeling, complex network analysis, and other domains (McGee, 2023.).
5. **Business Value 5-VOI (Value of Optimization and Improvement):** Generative AI can play a critical role in optimizing operations by analyzing large datasets and identifying patterns and business insights (Holmgard, et al., 2014). This can help organizations reduce costs, increase efficiency, and improve overall performance. Generative AI can also automate repetitive and time-consuming tasks, freeing up employees to focus on higher-value activities. For instance, ChatGPT can help users improve their creativity by providing advice on new ideas, supporting brainstorming sessions, or offering enhanced writing prompts (Tuomi, 2023).
6. **Business Value 6 - VAA (Value of Automation and Augmentation):** Generative AI can contribute to automation in customer service, enabling automated customer interactions, and in human resource management (HRM) by streamlining HR-related processes, such as resume screening and employee on boarding (Korzynski, et al., 2023). ChatGPT can also be applied to automate repetitive tasks, such as scheduling actions or generating business documents (Zhou, et al., 2023). Additionally, ChatGPT can provide emotional support to users by listening to their concerns, offering advice, providing help and support, and providing a sympathetic ear, which can be particularly helpful for employees in organizations. Integrating generative transformer models of AI with AR/VR can provide enterprises with a powerful tool for creating new and immersive experiences for customers or employees (Castelli & Manzoni, 2022).
7. **Business Value 7 - VRR (Value of Reactivity and Real-time):** Generative transformer models can create realistic 3D models for use in AR/VR applications, or produce new virtual environments in real-time (OpenAI, Jun 28, 2022). In addition, ChatGPT can build conversational interfaces that respond to users' (employees, customers, suppliers, etc.) requests in real-time, and automate tasks to respond to events in real-time without human intervention (Wu, et al., 2019).

For example, ChatGPT could be used to monitor social media for mentions of a brand and automatically respond to users in real-time (Tuomi, 2023). This value can also be applied to other domains, such as finance, where real-time data analysis and decision-making can be crucial (Ali & Aysan, 2023).

Table 1. Business Values and related applications in Contextual and Generative AI

Business Value Categories		Applications in management	Examples of AI algorithms used to extract the corresponding BV	Related references	
				Contextual AI	Generative AI
Business Value 1	Value of Planning and Coordination *(VPC)*	Production planning, scheduling of tasks, employees, etc., team coordination, job shop scheduling, etc.	K-Nearest Neighbors (KNN), Ant Colony Optimization algorithm (ACO), Particle Swarm Optimization (PSO)	(Chehbi Gamoura, et al., 2018)	(Park, et al., 2023)
Business Value 2	Value of Decision and Control (VDC)	Detection of new patterns (e.g. markets, customers, etc.), detection of new relationships (e.g. associations of behaviours, objects, etc.)	Expert Systems, Artificial Neural Networks (ANN), Decision trees	(Kucharavy, et al., 2021-[notre contr. 40])	(Korzynski, et al., 2023)
Business Value 3	Value of Creation and Innovation (VCI)	Design of new products, new services, etc	knowledge-based system (KBS), Multi-Arm Bandit algorithlms (MAB)	(Sharma, et al., 2021-[notre contr. 45])	(Holmgard, et al., 2014)
Business Value 4	Value of Discovery and Understanding (VDU)	New markets, new products, new customer segments, behavior detection, etc	K-Means, clustering Random Forests, Decision Trees	(Kucharavy, et al., 2019-[notre contr. 28])	(Tang, et al., 2021)
Business Value 5	Value of Optimization, and Improvement (VOI)	Market positioning, acquisition of competitive advantage, international expansion, etc	Genetic Algorithms, PSO	(Chehbi Gamoura, 2021)	(Wu, et al., 2019)
Business Value 6	Value of Automation and Augmentation (VAA)	Warehouse robots, trading agents, etc	Deep Learning, Multi Agent Systems (MAS)	(Chehbi Gamoura, 2022)	(Korzynski, et al., 2023)
Business Value 7	Value of Reactivity and Real-time (VRR)	Real-time tracking in logistics, web responsiveness with intelligent agents, etc	MAS, Deep learning	(Arslan Kazan, et al., 2022-[notre contr. 49])	(Olmo, et al., 2021)

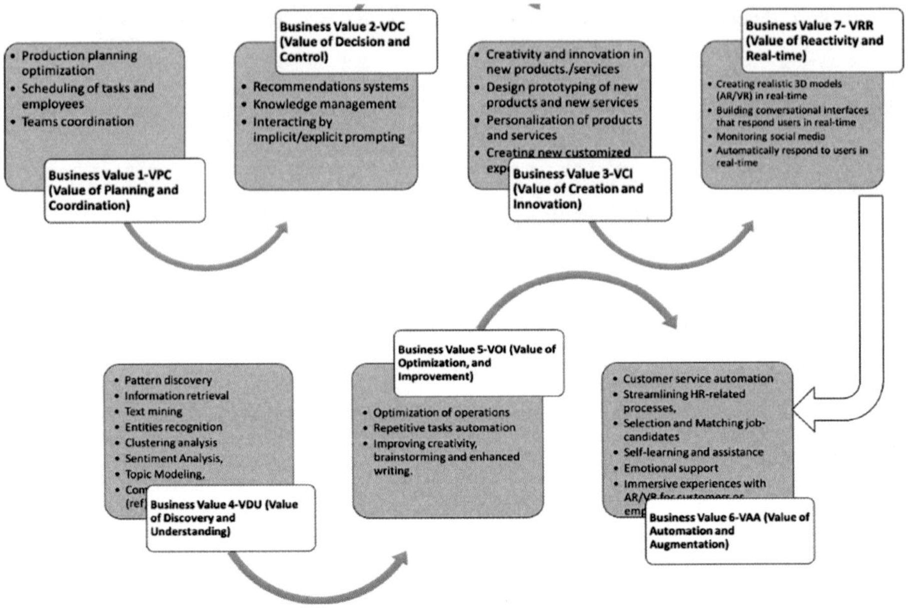

Fig. 3. Potentials of Generative Artificial Intelligence models in Business applications according to our proposed model of 7BV categorization

5 Conclusion

In conclusion, Generative AI has emerged as a more advanced form of AI that focuses on creating new content rather than just analyzing existing data. This has led to the development of advanced machine learning models such as ChatGPT and DALL-E 2, which can generate text and images based on input data. The popularity of Generative AI is expected to continue growing, with Conversational AI being one of the key areas of interest for both organizations and society. In this paper, we have discussed the position of Generative AI regarding the most common management theories and proposed a new conceptual model, the 7-BV, for the use of Generative AI in management. The 7-BV model is Business-Value-Driven and provides a framework for organizations to leverage Generative AI to achieve their business objectives.

Generative AI holds immense promise for revolutionizing the field of management and generating substantial business value for organizations. However, as the technology is still in its nascent stage, there is a need for extensive research to comprehend its implications, limitations, as well as the technical, organizational, and ethical impact on organizations. To fully exploit the potential of generative AI, it is imperative that we delve deeper into these aspects and identify ways to address potential challenges while leveraging the benefits of this transformative technology.

References

Akerkar, R.: Artificial Intelligence for Business. Springer, Cham (2019)

Ali, H., Aysan, A.: What will ChatGPT revolutionize in financial industry?. SSRN, p. 4403372 (2023)

Arslan Kazan, C., Koruca, H., Chehbi Gamoura, S.: Dynamic data-driven failure mode effects analysis (FMEA) and fault prediction with real-time condition monitoring in manufacturing 4.0. In: Hemanth, D.J., Kose, U., Watada, J., Patrut, B. (eds.) ICAIAME 2021. Engineering Cyber-Physical Systems and Critical Infrastructures, vol. 1, pp. 773–790. Springer, Cham (2022). https://doi.org/10.1007/978-3-031-09753-9_60

Brock, J., Von Wangenheim, F.: Demystifying AI: what digital transformation leaders can teach you about realistic artificial intelligence. Calif. Manage. Rev. **61**(4), 110–134 (2019)

Brown, A., Fishenden, J., Thompson, M.: Digitizing Government. Palgrave Macmillan, London (2014)

Castelli, M., Manzoni, L.: Generative models in artificial intelligence and their applications. Appl. Sci. **12**(9), 4127 (2022)

Chehbi Gamoura, S.: Predictive reinforcement learning algorithm for unstructured business process optimisation: case of human resources process. Int. J. Spatio-Temp. Data Sci. **1**(2), 184–214 (2021)

Chehbi Gamoura, S.: Processus «Achat 5.0» et «Acheteurs Augmentés»: l'intelligence artificielle collective pour l'automatisation de la sélection multifournisseurs via des chat-bots dotés d'aversion au risque: cas d'un constructeur automobile français en post-COVID-19. Rev. Française Gestion Industrielle **36**(1), 83–111 (2022)

Chehbi Gamoura, S., Derrouiche, R., Malhotra, M., Damand, D.: Predictive cross-management of disaster plans in big data supply chains: fuzzy cognitive maps approach. In: Acts of International Conference of Modelling, Optimization and SIMulation (MOSIM), Toulouse, France (2018)

Chehbi Gamoura, S., Koruca, H.I., Sharma, S.: Blockchain adoption in business models: what is the business value? In: ICEBM, Singapour, pp. 1–12 (2020)

Davenport, T.: The AI Advantage: How to Put the Artificial Intelligence Revolution to Work. MIT Press, Cambridge (2018)

Davis, F., Bagozzi, R., Warshaw, P.: User acceptance of computer technology: a comparison of two theoretical models. Manage. Sci. **35**(8), 982–1003 (1989)

Dean, J., Jr., Sharfman, M.: The relationship between procedural rationality and political behavior in strategic decision making. Decis. Sci. **24**(6), 1069–1083 (1993)

Deloitte digital: Conversational AI: the next wave of customer and employee experiences. Deloitte digital Australie, Canberra (2019)

Enholm, I., Papagiannidis, E., Mikalef, P., Krogstie, J.: Artificial intelligence and business value: a literature review. Inf. Syst. Front. **24**(5), 1709–1734 (2022)

Gkinko, L., Elbanna, A.: The appropriation of conversational AI in the workplace: a taxonomy of AI chatbot users. Int. J. Inf. Manage. 102568 (2022)

Holmgard, C., Liapis, A., Togelius, J., Yannakakis, G.: Generative agents for player decision modeling in games (2014)

Ivančić, L., Vukšić, V., Spremić, M.: Mastering the digital transformation process: Business practices and lessons learned. Technol. Innov. Manag. Rev. **2**, 9 (2019)

Kane, G., et al.: Strategy, Not Technology, Drives Digital Transformation. MIT Sloan Management Review and Deloitte University Press, New York (2015)

Korzynski, P., et al.: Generative artificial intelligence as a new context for management theories: analysis of ChatGPT. Cent. Eur. Manage. J. (2023)

Kucharavy, D., et al.: Using map of contradiction for decision support within warehouse design process, Berlin (Allemagne), pp. 1–28 (2019)

Kucharavy, D., Damand, D., Chehbi Gamoura, S., Barth, M.: Supporting strategic decision-making in manufacturing 4.0 with mix of qualitative and quantitative data analysis. In: 13ème Conférence Internationale de Modélisation, Optimisation et Simulation (MOSIM 2020). MOSIM, Rabat (2021)

Marcus, G., Davis, E., Aaronson, S.: A very preliminary analysis of Dall-E 2. Cornel University Press. arXiv preprint arXiv:2204.13807 (2022)

McGee, R.: Ethics committees can be unethical: the ChatGPT response (2023)

McKinsey: Driving impact at scale from automation and AI. McKinsey, Chicago (2019)

Mijwil, M., Aljanabi, M.: Towards artificial intelligence-based cybersecurity: the practices and ChatGPT generated ways to combat cybercrime. Iraqi J. Comput. Sci. Math. **4**(1), 65–70 (2023)

Momani, A., Jamous, M.: The evolution of technology acceptance theories. Int. J. Contemp. Comput. Res. **1**(1), 51–58 (2017)

Mooney, J., Gurbaxani, V., Kraemer, K.: A process oriented framework for assessing the business value of information technology. ACM SIGMIS Database: DATABASE Adv. Inf. Syst. **27**(2), 68–81 (1996)

Nikitaeva, A., Salem, A.: Institutional framework for the development of artificial intelligence in the industry. J. Inst. Stud. **13**(1), 108–126 (2022)

Olmo, A., Sreedharan, S., Kambhampati, S.: GPT3-to-plan: extracting plans from text using GPT-3. arXiv preprint arXiv:2106.07131 (2021)

OpenAI. DALL·E 2 pre-training mitigations, CA, USA: Technical report (2022)

Panenkov, A., Lukmanova, I., Kuzovleva, I., Bredikhin, V.: Methodology of the theory of change management in the implementation of digital transformation of construction: problems and prospects. In: E3S Web of Conferences, s.l.:EDP, vol. 244, p. 05005 (2021)

Park, J., et al.: Generative agents: interactive simulacra of human behavior. arXiv preprint arXiv: 2304.03442 (2023)

Reix, R.: Systèmes d'information et management des organisations. Vuibert (2004)

Schumpeter, J.: Prophet of innovation (2007)

Sensoy, M., Kaplan, L., Cerutti, F.S.M.: Uncertainty-aware deep classifiers using generative models. In: Proceedings of the AAAI Conference on Artificial Intelligence, vol. 34, no. 04, pp. 5620–5627 (2020)

Shanks, G., Bekmamedov, N., Sharma, R.: Creating value from business analytics systems: a process-oriented theoretical framework and case study (2011)

Sharma, S., Chehbi Gamoura, S., Prasad, D., Aneja, A.: Emerging legal informatics towards legal innovation: current status and future challenges and opportunities. Legal Inf. Manage. **21**, 218–235 (2021)

Shin, D.: The effects of explainability and causability on perception, trust, and acceptance: Implications for explainable AI. Int. J. Hum Comput Stud. **146**, 102551 (2021)

Simon, H.: Making management decisions: the role of intuition and emotion. Acad. Manag. Perspect. **1**(1), 57–64 (1987)

Strowel, A.: ChatGPT and generative AI tools: theft of intellectual labor? IIC-Int. Rev. Intellect. Property Compet. Law **54**, 1–4 (2023)

Tang, B., Ewalt, J., Ng, H.: Generative AI models for drug discovery. In: Saxena, A.K. (ed.) Biophysical and Computational Tools in Drug Discovery. Topics in Medicinal Chemistry, vol. 37, pp. 221–243. Springer, Cham (2021). https://doi.org/10.1007/7355_2021_124

Tuomi, A.: AI-generated content, creative freelance work and hospitality and tourism marketing. In: Ferrer-Rosell, B., Massimo, D., Berezina, K. (eds.) ENTER 2023. Springer Proceedings in Business and Economics, pp. 323–328. Springer, Cham (2023). https://doi.org/10.1007/978-3-031-25752-0_35

Vaswani, A., et al.: Attention is all you need. In: Advances in Neural Information Processing Systems, p. 30 (2017)

Wamba-Taguimdje, S.L., Wamba, S.F., Kamdjoug, J.R.K., Wanko, C.E.T.: Influence of artificial intelligence (AI) on firm performance: the business value of AI-based transformation projects. Bus. Process Manage. J. **26**(7), 1893–2192 (2020)

van de Wetering, R., Mikalef, P., Krogstie, J.: Big data is power: business value from a process oriented analytics capability. In: Abramowicz, W., Paschke, A. (eds.) BIS 2018. LNBIP, vol. 339, pp. 468–480. Springer, Cham (2019). https://doi.org/10.1007/978-3-030-04849-5_41

Wu, J., Qian, X., Wang, M.: Advances in generative design. Comput. Aided Des. **116**, 102733 (2019)

Zhou, C., et al.: A comprehensive survey on pretrained foundation models: a history from BERT to ChatGPT. arXiv preprint arXiv:2302.09419 (2023)

Arc Routing Problem and Solution Approaches for Due Diligence in Disaster Management

Ferhat Yuna(✉) and Burak Erkayman

Department of Industrial Engineering, Engineering Faculty, Ataturk University, 25240 Erzurum, Turkey
{ferhat.yuna,erkayman}@atauni.edu.tr

Abstract. Due to the increasing number of disasters in the world, the number of studies in the field of disasters is increasing. The planning, implementation, management and coordination of disaster management activities are very important. Search and rescue, humanitarian assistance, evacuation operations, etc. special infrastructures must be provided to manage situations. One of the most important issues in disaster management is establishing due diligence immediately after the disaster occurs. Quick detection of debris, especially during earthquakes, is extremely important and necessary to reduce the number of casualties. As infrastructures such as the internet, telephone and power lines are damaged in major disasters such as earthquakes, due diligence becomes difficult. Identifying wreckage locations is essential to support search and rescue efforts. In this study, an arc routing problem is considered to determine the condition of buildings in a region affected by an earthquake. The objective is to determine the locations as quickly as possible by checking every road and path in the disaster area at least once. For disaster management, however, it is of great importance to obtain a quick solution to arc routing problems rather than optimal results. Therefore, the solution was sought by the heuristic method of nearest neighbor search, which is widely used in the literature, and the results were recorded.

Keywords: Arc routing · heuristic approach · disaster management · due diligence

1 Introduction

Disasters are events where it is uncertain when they will occur. Therefore, it is very important to always be prepared for these events. Once a disaster occurs, it is important to respond as quickly as possible. An important phase of good disaster management is to plan search and rescue operations quickly and properly. Quick planning plays a major role in reducing the loss of life.

The earthquake is the focus of this study. Communication and transportation infrastructures are severely damaged during major disasters such as earthquakes. This situation affects communication with the disaster area, transportation to the region, and due diligence operations in the region. It is necessary to quickly identify the wreckage on which

Z. Şen et al. (Eds.): IMSS 2023, LNME, pp. 382–387, 2024.
https://doi.org/10.1007/978-981-99-6062-0_35

search and rescue operations will be conducted. This situation plays an important role in reducing casualties.

During major disasters, infrastructures such as electricity, internet, etc. are severely damaged. For this reason, teams responding to the disaster are also affected by this situation. Identifying heavily damaged or destroyed buildings during an earthquake and directing search and rescue teams to the debris is difficult. Because of the damage to infrastructure, it is more difficult to determine where the destruction lies in the first moments of the disaster. In this study, a nearest neighbor search heuristic is proposed to perform due diligence studies in a region devastated by earthquakes. The problem is an arc routing problem. It aims to control every street in the affected area by passing them at least once. The best solutions to the related problem can be found through solvers. However, the focus of this study is on rapid damage assessment at a time when there are major deficits in solvers, computers, the internet, electricity, etc.

2 Literature

Since the costs associated with the edges may vary in real-life applications, Keskin and Yılmaz have proposed a formulation for the postman problem that takes this into account [1]. The arc-cycle formulation, an extended version of the formulation in Wang and Wen's [2] pioneering work that directly models time-varying CPP, was proposed by Sun et al. [3]. The cost of travel on a bow varies depending on travel time and time. Vincent and Lin proposed an iterated greedy heuristic for the time-dependent prize-collecting arc routing problem [4]. Tagmouti et al. investigated an arc routing problem with capacity constraints and time-dependent service costs [5]. In another study, Tagmouti et al. proposed a variable neighborhood descent heuristic method to solve the problem under capacity constraints and time-dependent service costs [6]. Street sweeping [7], electric meter reading [8], refuse collection [9] and snow removal [10] are real-life problems modeled as CARP (capacitated arc routing problem). Another of these problems has been winter gritting. Tagmouti et al. discussed the dynamic capacity arc routing problem for winter gritting applications [11]. Black et al. have described a new problem called TD-PARP (Time-Dependent Prize-Collecting Arc Routing Problem). A solution to this problem was sought using the Variable Neighborhood Search and Tabu Search meta-heuristics [12].

Post-disaster damage assessment studies have recently been an important area of study. For example, Nex et al. introduced a new approach for real-time UAV mapping of building damage [13]. Again, a rapid approach to disaster response using UAV and aircraft was presented by Sugita et al. [14]. Another study is the deep learning-based damage map estimation proposed by Tran et al. [15]. Many of these studies are not suitable for use immediately after major disasters. Because, in order to carry out these works, there is a need for qualified personnel to carry out the related works besides electrical power, internet, and computers. The focus of this study is a due diligence proposal to be used in cases where there is no infrastructure such as telephone, internet, computer, electrical power etc. For this purpose, a nearest neighbor search heuristic based on the arc routing problem is proposed.

3 Methodology

The arc routing problems (ARP) is one of the well-studied problems in the literature [1]. The goal of the Chinese Postman problem (CPP) is to find the shortest closed path that visits each edge of an undirected network. CPP is a variant of ARP. It is classified as undirected and directed depending on the state of the edges. Another issue to consider is cost. For the problem in this study, the cost is considered as distance. Therefore, the edges in this study are undirected. Moreover, the problem is symmetric since the distances are independent of the transition direction.

The heuristic proposed in the study aims to visit all arcs at least once without returning to the initial arc. The purpose of the heuristic is not to create a tour that visits the arcs. It is sufficient to visit the arcs at least once in any direction. By passing through each street or avenue at least once, it is ensured that the situation there is determined.

The stages of heuristics are as follows:

1. Initially save the number of passes from the edges as 0 ($n_i = 0$). n_i represents the number of passes of the edge i.
2. Select the starting node. Rank the number of passes and costs of the edges that are connected to the node from smallest to largest.
3. Select the edge with the least cost from the edges with the least number of passes. If the number of passes of the edges is equal, choose the least costly edge. If the number of passes of the edges is equal and the costs are equal, make the selection at random. Increase the number of passes of the selected edge by 1 ($n_i = n_i + 1$).
4. When the new node is reached, update the starting node with the new node and return to Step 2.
5. Continue until the number of passes of all edges is at least 1.

With the proposed heuristic, a search and rescue team with only a network and the costs of the edges can quickly complete debris detection with this information. One of the most important points of the heuristic is that it can be easily calculated by people without the need for any technological infrastructure. It is a very useful heuristic in environments where infrastructures are damaged, such as phones, computers, and the internet.

4 Case Study

The proposed nearest neighbor search heuristic is run in a network consisting of 5 nodes and 7 edges. Details of each step of the heuristic solution are given. Cost refers to the individual cost of all edges that can be traveled from the starting node. The number of passes is the number of times that each edge that can be traveled from the starting node has been used separately. The feasible edge refers to the edges that can be used in the next step according to the proposed heuristic (Table 1).

The route formed according to the result of the proposed heuristic is as follows: 1-3-2-5-3-1-2-4-5. The cost of the relevant route is: $4 + 6 + 3 + 7 + 4 + 5 + 10 + 8 = 53$ units. As can be seen, all edges have been visited at least once. The twice visited edge is only 1–3 arcs.

Table 1. An Example of Proposed Heuristics

The Heuristic Solution	
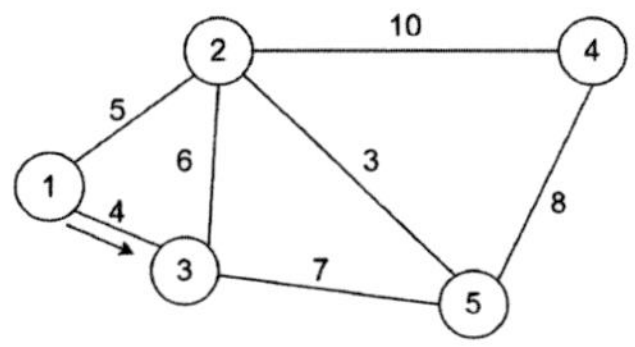	Start Node: 1 Costs: 1-2 (5), 1-3 (4) Number of Passes: 1-2 (0), 1-3 (0) Feasible Edges: 1-2 (5), 1-3 (4) Selected Edge: 1-3 (4)
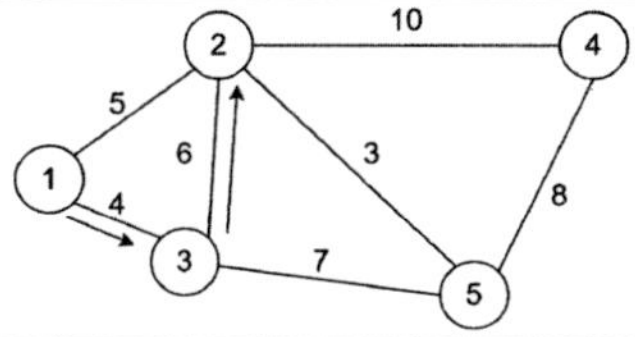	Start Node: 3 Costs: 3-1 (4), 3-2 (6), 3-5 (7) Number of Passes: 3-1 (1), 3-2 (0), 3-5 (0) Feasible Edges: 3-2 (6), 3-5 (7) Selected Edge: 3-2 (6)
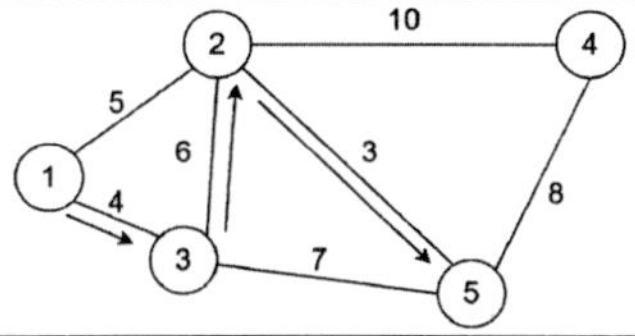	Start Node: 2 Costs: 2-1 (5), 2-3 (6), 2-5 (3), 2-4 (10) Number of Passes: 2-1 (0), 2-3 (1), 2-5 (0), 2-4 (0) Feasible Edges: 2-1 (5), 2-5 (3), 2-4 (10) Selected Edge: 2-5 (3)
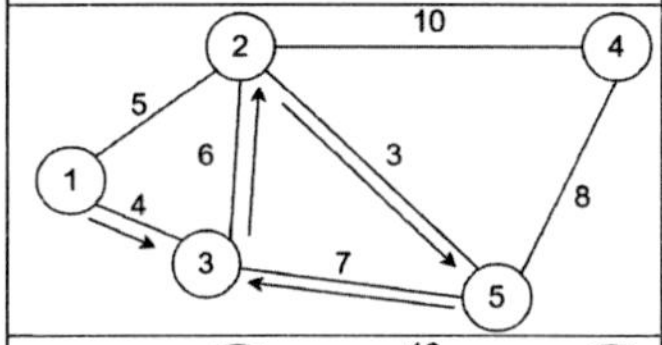	Start Node: 5 Costs: 5-3 (7), 5-2 (3), 5-4 (8) Number of Passes: 5-3 (0), 5-2 (1), 5-4 (0) Feasible Edges: 5-3 (7), 5-4 (8) Selected Edge: 5-3 (7)
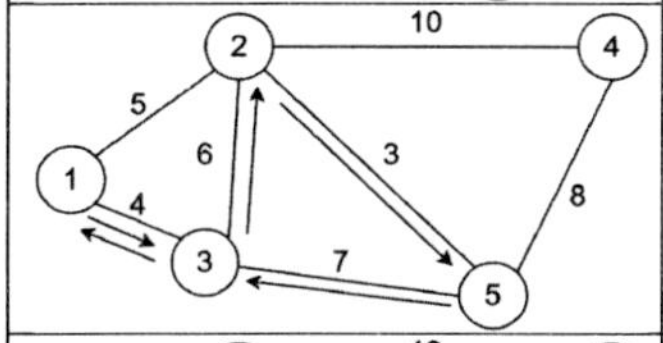	Start Node: 3 Costs: 3-1 (4), 3-2 (6), 3-5 (7) Number of Passes: 3-1 (1), 3-2 (1), 3-5 (1) Feasible Edges: 3-1 (4), 3-2 (6), 3-5 (7) Selected Edge: 3-1 (4)
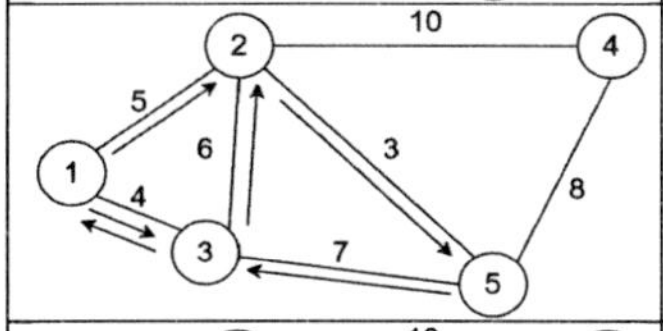	Start Node: 1 Costs: 1-2 (5), 1-3 (4) Number of Passes: 1-2 (0), 1-3 (2) Feasible Edges: 1-2 (5) Selected Edge: 1-2 (5)
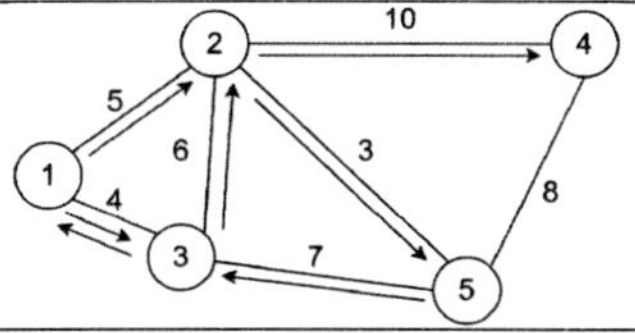	Start Node: 2 Costs: 2-1 (5), 2-3 (6), 2-5 (3), 2-4 (10) Number of Passes: 2-1 (1), 2-3 (1), 2-5 (1), 2-4 (0) Feasible Edges: 2-4 (10) Selected Edge: 2-4 (10)
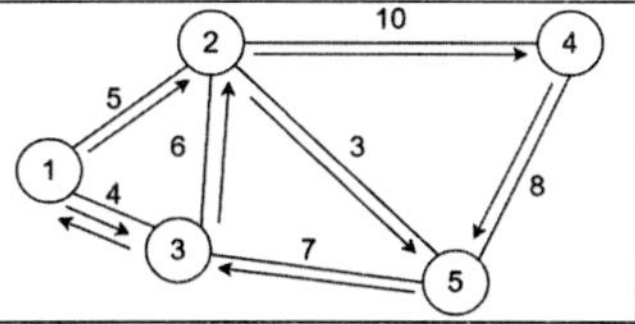	Start Node: 4 Costs: 4-2 (10), 4-5 (8) Number of Passes: 4-2 (1), 4-5 (0) Feasible Edges: 4-5 (8) Selected Edge: 4-5 (8)

5 Conclusion

After the disaster has occurred, it is necessary to establish a detailed damage assessment in a timely manner. This is of great importance in accelerating search and rescue activities and assisting survivors. Vehicles such as unmanned aerial vehicles have accomplished great things in this field in recent years. However, there may be such great disasters that it becomes very difficult to reach the disaster site. In addition to this transportation problem, various infrastructures are damaged. Electrical power, computer, telephone, communication tools, etc. types of equipment are vital needs for first response and disaster management. In cases where these infrastructures cannot be provided, it is necessary to know where the debris is, especially in the hours immediately after the disaster occurs. With the nearest neighbor search heuristic proposed for this purpose, debris detection can be performed without the need for any infrastructure, technology or qualified personnel.

The only information needed for the proposed heuristic is a detailed map of the relevant disaster area and the distance to the streets. In light of this information, the damage assessment team, which starts working from any point, will have passed all the ways as soon as possible. Thus, a great contribution will be made to the disaster response and disaster management stages. Because search and rescue efforts are dynamic and complex. The more consciously the work is started, the faster the chance of responding to the disaster occurs. Therefore, detection studies have importance to direct all disaster management and response studies.

References

1. Keskin, M.E., Yılmaz, M.: Chinese and windy postman problem with variable service costs. Soft Comput. **23**(16), 7359–7373 (2019)
2. Wang, H.F., Wen, Y.P.: Time-constrained Chinese postman problems. Comput. Math. Appl. **44**(3–4), 375–387 (2002)
3. Sun, J., Meng, Y., Tan, G.: An integer programming approach for the Chinese postman problem with time-dependent travel time. J. Comb. Optim. **29**, 565–588 (2015)
4. Vincent, F.Y., Lin, S.W.: Iterated greedy heuristic for the time-dependent prize-collecting arc routing problem. Comput. Ind. Eng. **90**, 54–66 (2015)
5. Tagmouti, M., Gendreau, M., Potvin, J.Y.: Arc routing problems with time-dependent service costs. Eur. J. Oper. Res. **181**(1), 30–39 (2007)
6. Tagmouti, M., Gendreau, M., Potvin, J.Y.: A variable neighborhood descent heuristic for arc routing problems with time-dependent service costs. Comput. Ind. Eng. **59**(4), 954–963 (2010)
7. Bodin, L.D., Kursh, S.J.: A computer-assisted system for the routing and scheduling of street sweepers. Oper. Res. **26**(4), 525–537 (1978)
8. Stern, H.I., Dror, M.: Routing electric meter readers. Comput. Oper. Res. **6**(4), 209–223 (1979)
9. Bodin, L., Fagin, G., Welebny, R., Greenberg, J.: The design of a computerized sanitation vehicle routing and scheduling system for the town of Oyster Bay, New York. Comput. Oper. Res. **16**(1), 45–54 (1989)
10. Haslam, E., Wright, J. R.: Application of routing technologies to rural snow and ice control. Transp. Res. Rec. (1304) (1991)

11. Tagmouti, M., Gendreau, M., Potvin, J.Y.: A dynamic capacitated arc routing problem with time-dependent service costs. Transp. Res. Part C: Emerg. Technol. **19**(1), 20–28 (2011)
12. Black, D., Eglese, R., Wøhlk, S.: The time-dependent prize-collecting arc routing problem. Comput. Oper. Res. **40**(2), 526–535 (2013)
13. Nex, F., Duarte, D., Steenbeek, A., Kerle, N.: Towards real-time building damage mapping with low-cost UAV solutions. Remote Sens. **11**(3), 287 (2019)
14. Sugita, S., Fukui, H., Inoue, H., Asahi, Y., Furuse, Y.: Quick and low-cost high resolution remote sensing using UAV and aircraft to address initial stage of disaster response. In: IOP Conference Series: Earth and Environmental Science, vol. 509, no. 1, p. 012054. IOP Publishing (2020)
15. Tran, D.Q., Park, M., Jung, D., Park, S.: Damage-map estimation using UAV images and deep learning algorithms for disaster management system. Remote Sens. **12**(24), 4169 (2020)

Integrated Process Planning, Scheduling, Due-Date Assignment and Delivery Using Simulated Annealing and Evolutionary Strategies

Onur Canpolat[1](✉), Halil Ibrahim Demir[1], and Caner Erden[2]

[1] Department of Industrial Engineering, Faculty of Engineering, Sakarya University, Sakarya, Turkey
{onurcanpolat,hidemir}@sakarya.edu.tr
[2] Department of International Trade and Finance, Faculty of Applied Science, Sakarya University of Applied Science, Sakarya, Turkey
cerden@subu.edu.tr

Abstract. Process planning, scheduling, and due date assignment functions are the three fundamental manufacturing functions. Traditionally, these functions were examined independently in production systems. The integrated communication of these functions is one of the most efficient ways to ensure high customer satisfaction, nonetheless, in the technological and competitive climate of today. Although these functions have been integrated in academic research over the past few decades, in practice, they are still generally performed sequentially and independently. Although limited research exists that integrates the three functions, this study introduces delivery as a fourth function. The objectives of this study are to make a significant contribution to the literature by demonstrating the integrated nature of four functions in manufacturing systems, with a view to increasing efficiency compared to traditional solutions. The study also seeks to investigate the impact of incorporating the delivery function. Customers are not viewed as being equal, as is the case with many of the other integrated studies that can be found in the literature; rather, each client is given special consideration. Each of the four job shops is unique and has a different number and location of customers. The study solves the complex problem by utilizing simulated annealing and evolutionary strategies algorithms. Both method-based and job shop-based comparisons are made, and the results show which methods perform better in each job shop. The results demonstrate that the integrated system offers an improvement of approximately 50% compared to independent systems. Furthermore, the study found that, across all four job shops, the evolutionary strategies (ES) outperformed simulated annealing (SA) in terms of results.

Keywords: Integrated process planning · scheduling and due-date assignment · simulated annealing · evolutionary strategies · delivery scheduling · distribution · transportation

Z. Şen et al. (Eds.): IMSS 2023, LNME, pp. 388–401, 2024.
https://doi.org/10.1007/978-981-99-6062-0_36

1 Introduction

Process planning, scheduling and due date assignment are three important functions. In classical planning, these three functions worked separately, even though they strongly affect each other [1]. This produces poor inputs for later phases and degrades the effectiveness of the overall solution. As a result, there can be significant losses in performance measures and shop floor (SF) productivity decreases. Besides, customers may become dissatisfied when given unnecessarily long due dates. Non-integrated production functions can cause losses across multiple different areas. Therefore, the integration of manufacturing functions is crucial. The integration will provide benefits such as the elimination of conflicting objectives between functions and the improvement of function-to-function communication, which will greatly enhance productivity, performance, and quality.

The task of integration is crucial but also challenging. Even the scheduling problem on its own is categorized as non-polynomial hard (NP-Hard) [2], which means exact solutions are only feasible for minor issues. When we expand our focus to encompass integrated problems, the difficulty level increases significantly. Research reveals that meta-heuristic algorithms are commonly used to solve analogous problems. This study applies simulated annealing and evolutionary strategies techniques tailored for this structure to evaluate their effectiveness against traditional approaches by comparing performance on global metrics.

This research defines the IPPSDDAD problem, which incorporates the delivery function as a fourth component into the integration of process planning, scheduling, and due date assignment functions. Initially, an independent structure with non-integrated functions is employed to assess performance with increasing levels of functional integration. While investigations on integrating three functions have been conducted in recent years and are documented in the literature, there remains significant potential for further inquiry within this domain. The integration of four functions constitutes an unprecedented area that warrants exploration and analysis by researchers.

The punishment of tardiness in scheduling has been approached in various ways in the literature, including punishing earliness and tardiness, maximum absolute lateness, or the number of tardy jobs. However, this research adopts a different approach by penalizing the sum of weighted tardiness, earliness, and due date-related costs. This approach is adopted to ensure that realistic due dates are set and to prevent unnecessary delays, particularly for important customers. The penalization of weighted tardiness aims to prevent late deliveries, which can lead to customer dissatisfaction, loss of customers, a damaged reputation, and price reductions. While tardiness has traditionally been the only aspect penalized, the present study recognizes that earliness can also be problematic in a JIT environment. Early deliveries can result in additional costs such as stock holding, storage, and spoilage, all of which are considered earliness costs. Therefore, the present study also penalizes weighted earliness to address this issue.

This research examines traditional job shop manufacturing and evaluates four distinct environments with varying numbers of jobs: 25, 50, 75, and 100. Additionally, the study acknowledges that customers may not carry equal significance. Many previous studies have neglected to consider customer weights while scheduling or establishing due dates. Important customers are also prioritized in the delivery phase. By privileging the important customer at every stage of production, this study aims to investigate how the integration of the four functions impacts the global solution in comparison to the non-integrated solution (SIRO-RDM). Additionally, the study examines the performance of the heuristic algorithms used and the effect of utilizing varying levels of customer importance on overall performance.

Section 2 will provide by giving an overview of the sub-problems explored in existing literature, followed by describing the methods and modeling used in Sects. 3 and 4. The obtained findings are presented in Sect. 5, with a final comprehensive evaluation and commentary in the last section.

2 Integration Studies

The task of assigning jobs to machines is informed by process planning, which determines the sequence of machines for each job. Scheduling problems can be studied in isolation or in integration with other factors such as process planning, due date assignment or delivery.

2.1 Integrated Process Planning and Scheduling (IPPS)

Regarding IPPS, Hutchison et al. [3] proposed two offline and one real-time scheduling plan. Although the plan gives the general optimum solution, it can only be applied to small problems. Jiang and Hsiao [4] proposed an analytical solution to the problem, but their 0–1 binary programming can only be applied to small problems. Zhang and Mallur [5] developed an integrated model consisting of three modules. These modules are process planning, production scheduling and decision-making module.

Brandimarte [6] used a multi-objective approach to utilize process plan flexibility in scheduling. Kim and Egbelu [7] studied the scheduling problem involving alternative process plans. Research has shown that mathematical models are useful in solving small problems, but they have not been as successful at solving larger problems. In most of the studies from the 90s, the problems were generally small and discrete, while in the early 2000s, larger problems and alternative routes were studied. Yang et al. [8] considered an IPPS problem that aims to minimize the total completion time for a single-machine parallel batching with disparate job families. Recent studies have used artificial intelligence and heuristic algorithms (and simulation in a few studies) to study more realistic integration problems where dynamic process plans, uncertainties and multiple objectives are realized rather than just achieving integration. In the literature, heuristic solutions are generally used for these problems. When the studies are analyzed, it is observed that batch manufacturing has been studied in a limited number of studies. The number of studies that attribute different importance to customers is extremely small.

2.2 Scheduling with Due Date Assignment (SWDDA)

Gordon et al. [9] conducted a comprehensive literature review in the field of SWDDA and noted that there is a continuous interest in SWDDA studies. The authors noted that in the context of Just in Time (JIT) production, the completion of jobs is expected to match the delivery date precisely rather than before the delivery date, as is typical in traditional production environments. This is since completing jobs too early or too late can result in additional costs associated with either late or early completion. According to Cheng et al. [10], earliness leads to unnecessary inventory holding and tardiness leads to customer dissatisfaction and contract non-compliance costs. Zhao et al. [11] investigated single-machine scheduling and due date assignment, where the processing time of a job depends on both its start time and its position in a queue. Xiong et al. [12] consider a single-machine SWDDA problem in an environment where a machine breaks down randomly at a given time with a certain probability.

The performance functions in SWDDA problems can be composed of factors such as earliness, tardiness, number of tardy jobs, due date-related costs and due window-related costs. In most of the studies in the literature, due dates are given in terms of process times and the number of operations, but customer weights are not considered. It is notable that the integration of single machine scheduling and common due date is widespread in most of the studies conducted in the 90s. Generally, the cost of earliness and tardiness has been considered, and due date-related costs are mentioned in very few studies. The number of studies that mention important customers is very limited.

2.3 Integrated Process Planning, Scheduling and Due Date Assignment (IPPSDDA)

Although the integration of the three functions has the potential to produce very efficient results, it has not yet found a large place in the literature, probably because it is a difficult and complex topic. There are only a few studies on the IPPSDDA problem. Demir and Taskin [13] studied this issue as a Ph.D. thesis. Then, Çeven and Demir [14] studied performance improvement by integrating the due date with the IPPS problem in their Master thesis. In the following years, studies such as Demir et al. [15–17] continue to cover the topic. Erden [18] dynamized the integration of three functions with stochastic and dynamic arrivals. Jobs can arrive at the SF at any time according to an exponential distribution. Demir and Erden [19] solved the dynamic IPPSDDA problem with an ant colony algorithm. Erden et al. [20], Demir et al. [21], Demir and Phanden [22], Demir et al. [23] and Demir et al. [24] solved the IPPSDDA problem using hybrid evolutionary strategy, tabu search and annealing simulation, particle swarm optimization.

2.4 Integrated Production and Delivery Scheduling (IPDS)

In studies where scheduling is integrated with delivery, the concept of delivery is sometimes expressed with different concepts such as vehicle routing, delivery, distribution, and transportation. These studies also try to optimize delivery in different ways. For example, in some studies, delivery is done by the manufacturer and the products are delivered to the customer's doorstep (Tonizza Pereira and Seido Nagano [25]; Garcia and Lozano [26]. In some studies, delivery optimization is performed until the moment when the products are loaded onto the vehicle and the vehicle's path is not examined [27]. In some studies, third-party logistics (3PL) companies are used for delivery planning [28]. Zografos and Androutsopoulos [29] proposed a method for routing and scheduling trucks carrying hazardous materials using heuristic algorithms. Chen et al. [30] presented a nonlinear mathematical model for scheduling the production of perishable food products and routing delivery vehicles. Fu et al. [31] developed a two-stage heuristic algorithm integrating scheduling and vehicle routing in a company in the metal packaging industry.

In the context of integrated production and delivery, previous studies have shown that scheduling and delivery functions are typically combined, while process planning, and due date assignment are often disregarded. Additionally, some studies have only focused on optimizing delivery up until the loading of products onto the vehicle, without considering the vehicle's route. Most of the integrated studies have concentrated on straightforward delivery operations, such as direct shipments to customers or predetermined/fixed routes, with an emphasis on minimizing transportation costs. However, no prior research has explored the importance of customers in the delivery phase. Moreover, unlike previous literature, this research will investigate how the distance traveled by the vehicle during the delivery phase influences performance.

3 Methods

This study employs a combination of ES and SA methods to achieve integration. ES is regarded as the forerunner of GA, differing from the latter in that it solely employs mutation operators, eschewing crossover operators altogether. The initial step of ES involves adding the chromosomes from the extant population, initially comprising ten chromosomes, into an array and subsequently calculating and recording their performance values, equivalent to the population size. The performance values are then ranked in ascending order, and the selection probabilities for each chromosome are computed accordingly. A mutation is then performed by randomly selecting a chromosome based on the selection probability. A total of 10 chromosomes are altered, and their mutated counterparts are retained in the mutation array. Performance values for the current and mutational populations are computed, and the chromosomes are sorted in ascending order based on their performance. The best chromosomes and population sizes are carried forward to the next iteration, and the process is repeated for several iterations.

Simulated annealing (SA) is a heuristic algorithm that circumvents the issue of being trapped in a local optimum by incorporating stochasticity, which allows it to explore solutions in a larger search space [32]. SA is a prevalent method for solving combinational NP-hard problems [33]. Initially, SA generates new solutions using a cooling rule from an initial temperature and evaluates them against an initial solution. If the new solution is better, SA transitions to the new solution [34]. However, to avoid becoming trapped in local optima, SA must occasionally accept new solutions that are inferior to the current solution. The selection of which suboptimal solutions to accept is determined stochastically using a probability function. The distinguishing characteristic of SA from other neighboring search algorithms is its ability to evade local minima. This ability is solely due to the algorithm's willingness to accept suboptimal solutions to a certain extent [35].

4 The IPPSDDAD Problem

In the study, a job shop-type manufacturing environment with one vehicle is discussed. In this environment, it is aimed that process planning, scheduling, due date assignment and delivery operations work in an integrated manner. When an order is received, the due date is determined by considering the customer, operation, and route information of this order, then this order is scheduled according to the machine densities, and after the production is completed, it is loaded on the vehicle and delivered to the customer.

4.1 Definition and Modeling of the Problem

The problem involves numerous customers and requires consideration of two distinct production routes for each job, with every job having three operations. The significance of each customer varies based on their level of importance categorized as very important, important, moderately important, or slightly important. Each classification is assigned a corresponding value: 2.5, 1, 0.5 and 0.33, respectively; higher values are given to more significant customers who take priority in scheduling and due date assignment when solving the problem using either individual delivery or batch-type delivery methods.

Integration into an SF must first be supported by process strategies. The process plan comprises information such as the jobs on the SF, their operations, the machines on which they will be performed, their times of operation, etc. Table 1 provides an illustration of a process plan.

Table 1. Sample Process Plan

Jobs	Routes	Op. 1 (Machine)		Op. 2 (Machine)		Op. 3 (Machine)		Customer Importance
J1	R0	6	1	5	2	9	2	2.50
	R1	8	2	6	2	5	1	
J2	R0	9	2	8	1	3	1	0.50
	R1	8	1	4	1	3	2	
J3	R0	3	1	3	1	9	2	2.50
	R1	4	1	7	2	6	1	
J4	R0	9	2	5	2	7	1	1.00
	R1	6	2	7	1	5	1	

After determining the information about the jobs, the assignment of jobs to batches is performed. Each job belongs to only one batch and each batch has five jobs (customers). When allocating jobs to batches, the processing time in the SF (p_j), the importance of the customer (w_j) and the distance of the customer to the SF (d_{0j}) are calculated as in Eq. (1) below. After the batching values (BV) of the jobs are determined, the jobs are divided into batch groups of five by sorting the BV from smallest to largest.

$$BV = \left(\sum p_j + d_{0j}\right) * \frac{1}{w_j} \tag{1}$$

Having knowledge about the location of each customer is essential to ensure successful delivery. In this study, customer locations are defined as random points on a coordinate system. The SF is considered the center (0, 0) in the coordinate system. The distances to each other and to the SF are calculated using Eq. (2) where d_{ij} represents the distance and x and y represent the coordinates. Figure 1 shows the distance matrix by the determination of all distances.

$$d_{ij} = \left|(y_j - y_i)\right| + \left|(x_j - x_i)\right| \tag{2}$$

Distance	Depot	J1	J2	J3	J4	J5	J6	J7	J8	J9
Depot	-1	23	23	23	15	7	5	22	16	11
J1	23	-1	46	26	38	24	28	13	35	18
J2	23	46	-1	22	14	28	18	45	11	28
J3	23	26	22	-1	16	30	18	39	11	12
J4	15	38	14	16	-1	14	10	37	9	20
J5	7	24	28	30	14	-1	12	23	23	18
J6	5	28	18	18	10	12	-1	27	11	10
J7	22	13	45	39	37	23	27	-1	34	27
J8	16	35	11	11	9	23	11	34	-1	17
J9	11	18	28	12	20	18	10	27	17	-1

Fig. 1. Distance matrix

The chromosome comprises three important genes: due date assignment, scheduling, and delivery. The former consists of four rules while the latter two contain ten and nine rules respectively. Integration of these selected rules solves the problem at hand, with each set being formed through a combination of function values and route values to constitute a gene within the larger chromosome structure detailed in Fig. 2.

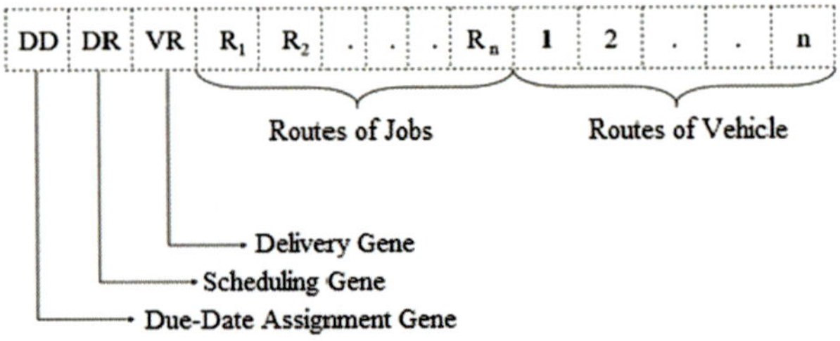

Fig. 2. Chromosome

For each rule, there is a gene value in the relevant gene. The values of the rules according to the functions are shown in Table 2.

Table 2. The Values of the Rules in the Gene

Gene Number	Due Date Assignment	Scheduling	Delivery
0	WSLK	WSPT	Single delivery
1	WPPW	WSOT	Batch delivery
2	WNOP	WLOT	Nearest neighbor
3	WTWK	WLPT	Savings algorithm
4		WATC	Sweep algorithm
5		ATC	Random delivery
6		MS	Hybrid delivery
7		WMS	Importance algorithm 1
8		EDD	Importance algorithm 2
9		WEDD	

4.2 Performance Criterion

The performance criterion is a function based on delivery time. This function penalizes the promised due date, tardiness, if any, and earliness, if any. The tardiness is calculated by Eq. (3) and earliness is calculated using Eq. (4).

$$T_j = \max(c_j - d_j, 0) \tag{3}$$

$$E_j = \max(d_j - c_j, 0) \tag{4}$$

The penalty values are determined using Eq. (5) and (6), where w_j is the customer's weight, c_j is the completion time and d_j is the delivery date, respectively.

$$P_E = w_j * (5 + 4 * (\frac{E}{480})) \tag{5}$$

$$P_T = w_j * (10 + 8 * (\frac{T}{480})) \tag{6}$$

In contrast to prevailing models in the literature, this study includes a promised due date because of its alignment with the just-in-time production philosophy's emphasis on timely job completion [9]. The philosophy prioritizes meeting deadlines precisely and considers early or late completion unfavorable. Thus, the performance function incorporates the promised due date, where higher values correspond to longer promises. However, given the problem nature at hand, lower performance function values are preferable. Therefore, assigning jobs closer to their time minimizes penalties calculated through Eq. (7) for non-adherence to due dates. The total penalty for a job (unit cp) is the sum of the P_D, P_E, P_T as formulated in Eq. (8). The performance criterion (PC) of the study is to minimize the sum of the penalties calculated for all jobs as shown in Eq. (9).

$$P_D = w_j * (8 * (\frac{D}{480})) \tag{7}$$

$$P_j = P_D + P_E + P_T \tag{8}$$

$$PC = \sum_{j=1}^{n} P_j \tag{9}$$

5 Results

For the problem, a software program is developed in Python programming language using PyCharm IDE on a computer with Intel(R) Core(TM) i7-4700HQ processor with 2.40 GHz and 16 GB RAM. NumPy, Matplotlib, random and math libraries are utilized.

First, ten different rule-free solutions are tried and averaged, and the results are recorded. Then, the results are obtained using the solution methods used in the study. For a more accurate evaluation and comparison, the iteration numbers are kept equal (100), and all random numbers used in the software are fixed. The results of SF 2 are shown in Table 3.

Table 3. The results of SF 2

Methods	ES	SA
SIRO-RDM	801806,4 cp	
Best (cp)	464487,2	504247,2
Improvement Rate	%42,07	%37,11

The results show that the highest rate of improvement is achieved with the ES. Each job represents a different customer. Each SF is solved by two different solution methods and the results of their performance are shown in Table 4.

Table 4. Results of the SF's

Methods	SF 1	SF 2	SF 3	SF 4
ES	**54222.7 cp**	**464487.2 cp**	**1144144.2 cp**	**1603461.2 cp**
SA	55937.4 cp	504247.2 cp	1203021.8 cp	1688062.2 cp

According to the results, the ES outperformed the SA in all four SF. In each SF, the performance of due date assignment, scheduling and delivery rules are analyzed for the chromosomes that performed best with ES. The analysis of the due date assignment rules is shown in Table 5.

The WSLK rule clearly outperforms other due date assignment rules. In all SFs, the WSLK rule had the best performance for the due date assignment. In SF 1, the

Table 5. The Analysis of the Due Date Assignment Rules

Rules	SF 1	SF 2	SF 3	SF 4
WSLK [0]	54222.7 cp	464487.2 cp	1144144.2 cp	1603461.2 cp
WPPW [1]	329531.4 cp	966147.6 cp	1391836.5 cp	1980590.6 cp
WNOP [2]	317710.2 cp	868822.8 cp	1309218.7 cp	1841577.8 cp
WTWK [3]	281890.2 cp	812662.8 cp	1269397.2 cp	1789976.5 cp

global performance of WSLK performed almost five times better than the other rules. Similar circumstances exist in other SFs. Nevertheless, it is important to note that the ratio declines as the number of employments grows. Table 6 displays the analysis of the scheduling rules.

Table 6. The Analysis of the Scheduling Rules

Rules	SF 1	SF 2	SF 3	SF 4
WSPT [0]	54222.7 cp	469933.2 cp	1143556.2 cp	1615156.6 cp
WSOT [1]	53575.5 cp	464487.2 cp	1147084.2 cp	1603461.2 cp
WLOT [2]	55403.5 cp	470309.2 cp	1147084.2 cp	1609669.2 cp
WLPT [3]	53575.5 cp	464487.2 cp	1147084.2 cp	1603461.2 cp
WATC [4]	54581.1 cp	464487.2 cp	1147084.2 cp	1603461.2 cp
ATC [5]	54581.1 cp	464487.2 cp	1147084.2 cp	**1603461.2 cp**
MS [6]	54222.7 cp	467365.8 cp	**1144144.2 cp**	1605789.2 cp
WMS [7]	54222.7 cp	467365.8 cp	1144144.2 cp	1605789.2 cp
EDD [8]	**54222.7 cp**	**464487.2 cp**	1147084.2 cp	1603461.2 cp
WEDD [9]	54222.7 cp	464487.2 cp	1147084.2 cp	1603461.2 cp

The scheduling rules don't show any clear dominance of one rule. The scheduling rules on the chromosome that produce the greatest results vary from SF to SF. The EDD (early due date) rule predominates in SF 1 and 2, the MS (minimum slack) rule in SF 3, and the ATC (apparent tardiness cost) rule in SF 4.

In SF 2, it is found that the scheduling rule WLOT for the chromosome with the best result leads to the greatest divergence from the best result. In this analysis, only the effect of the rule on a single chromosome is evaluated since the results obtained by changing only one gene (the scheduling gene) are analyzed. Therefore, at different iteration numbers and randomness, the results of the rules may differ. Changing a rule on the best chromosome can lead to a better result. However, it should be noted that a better solution is sometimes not achievable due to the small number of iterations, staying at the local optimum or the structure of the problem/rule. The results of the delivery rules for the best chromosome are shown in Table 7.

Table 7. The Analysis of the Delivery Rules

Rules	SF 1	SF 2	SF 3	SF 4
Single delivery [0]	98674.0 cp	930892.0 cp	2340006.0 cp	3511654.4 cp
Batch delivery [1]	64758.8 cp	632697.2 cp	1537654.4 cp	2428633.8 cp
Nearest neighbor [2]	**54222.7 cp**	490915.0 cp	1283879.0 cp	1724700.0 cp
Savings algorithm [3]	53864.2 cp	**464487.2 cp**	**1144144.2 cp**	**1603461.2 cp**
Sweep algorithm [4]	75726.7 cp	553585.2 cp	1335392.6 cp	1883753.2 cp
Random delivery [5]	75618.0 cp	617485.8 cp	1528558.4 cp	2233588.6 cp
Hybrid delivery [6]	55831.5 cp	524447.2 cp	1231944.6 cp	1827778.8 cp
Importance algorithm 1 [7]	58499.3 cp	522287.2 cp	1348608.6 cp	1789018.8 cp
Importance algorithm 2 [8]	58499.3 cp	525119.2 cp	1366528.6 cp	1812458.8 cp

In the analysis of the delivery rules, the savings algorithm is in the chromosomes that performed best in three of the four shop floors. The savings algorithm is, in fact, the top-performing rule in each of the SFs, but due to SF 1's constrained solution space, it wasn't in the best chromosome. However, the single delivery rule without batches performed 45% worse than the savings algorithm in SF 1 by 45%, SF 2 by 50%, SF 3 by 51%, and SF 4 by 54%. These percentages show that delivering in batches results in a more effective overall solution to the problem. It is obvious that single delivery [0], batch delivery [1], and random delivery [5] rules typically achieve the worst performances.

The findings indicate that altering the delivery rule on a given chromosome can have a notable impact on its performance, more so than modifying scheduling rules. Therefore, it can be concluded that delivery rules are more effective in changing the global solution than scheduling rules. On SF 1, the saving algorithm outperformed the closest alternative rule by 0.7%, while on SF 2, 3, and 4, the difference is even greater, at 5.4%, 7.1%, and 7.0%, respectively.

6 Discussions

This study aims to integrate the four basic manufacturing functions, namely process planning, scheduling, due date assignment, and delivery, for the first time in the literature. Moreover, customer priority levels are considered throughout the due date assignment, scheduling, and delivery stages, which is not commonly addressed in prior studies. However, in accordance with the Just-In-Time (JIT) philosophy, the objective function heavily weights the concepts of earliness, tardiness, and due date.

The problem involves four different shop floors, each containing 25, 50, 75, and 100 jobs respectively, belonging to different customers. Each shop floor has two machines and two different routes are considered for scheduling each job. The complexity and size of the problem necessitate the use of ES and SA algorithms to obtain a solution, where each solution is represented by a single chromosome. The performance of the shop floor, the two algorithms used for the solution, and each manufacturing function under different rules have been thoroughly evaluated.

The results indicate that the heuristic algorithms employed perform 42% better than SIRO-RDM, underscoring the importance of integration. ES outperforms SA in all shop floors among the solution methods. In the due date assignment rules, WSLK exhibits superior results on all shop floors. In scheduling rules, EDD (early due date) rule on the first and second shop floors, MS (minimum slack) rule on the third shop floor, and ATC (apparent tardiness cost) rule on the fourth shop floor stand out. Delivery-focused rules such as EDD and MS show more effective performance in scheduling. The savings algorithm performs better than other rules in delivery rules.

The study concludes that the due date assignment rules have a more prominent effect on the global performance of the functions than the scheduling and delivery rules. Thus, determining a good due date is crucial, and it is more effective for the company to determine the due date internally rather than from outside the factory without considering the conditions.

References

1. Demir, H.I., Canpolat, O., Erden, C., Şimşir, F.: Process planning and scheduling with WNOPPT weighted due-date assignment where earliness, tardiness and due-dates are penalized. J. Intell. Syst.: Theor. Appl. **1**(1), 16–25 (2018)
2. Yuan, J.: A note on the complexity of single-machine scheduling with a common due date, earliness-tardiness, and batch delivery costs. Eur. J. Oper. Res. **94**(1), 203–205 (1996). https://doi.org/10.1016/0377-2217(95)00168-9
3. Hutchison, J., Leong, K., Snyder, D., Ward, P.: Scheduling approaches for random job shop flexible manufacturing systems. Int. J. Prod. Res. **29**(5), 1053–1067 (1991). https://doi.org/10.1080/00207549108930119
4. Jiang, J., Hsiao, W.-C.: Mathematical programming for the scheduling problem with alternate process plans in FMS. In: Selected Papers from the 16th Annual Conference on Computers and Industrial Engineering, pp. 15–18. Pergamon Press, Inc., Elmsford (1994). http://dl.acm.org/citation.cfm?id=200436.200445. Accessed 10 Mar 2015
5. Zhang, H.C., Mallur, S.: An integrated model of process planning and production scheduling. Int. J. Comput. Integr. Manuf. **7**(6), 356–364 (1994). https://doi.org/10.1080/09511929408944623
6. Brandimarte, P.: Exploiting process plan flexibility in production scheduling: a multi-objective approach. Eur. J. Oper. Res. **114**(1), 59–71 (1999). https://doi.org/10.1016/S0377-2217(98)00029-0
7. Kim, K.-H., Egbelu, P.J.: Scheduling in a production environment with multiple process plans per job. Int. J. Prod. Res. **37**(12), 2725–2753 (1999). https://doi.org/10.1080/002075499190491
8. Yang, F., Davari, M., Wei, W., Hermans, B., Leus, R.: Scheduling a single parallel-batching machine with non-identical job sizes and incompatible job families. Eur. J. Oper. Res. **303**(2), 602–615 (2022). https://doi.org/10.1016/j.ejor.2022.03.027
9. Gordon, V., Proth, J.M., Chu, C.: A survey of the state-of-the-art of common due date assignment and scheduling research. Eur. J. Oper. Res. **139**(1), 1–25 (2002). https://doi.org/10.1016/S0377-2217(01)00181-3
10. Cheng, T.C.E., Kovalyov, M.Y., Lin, B.M.T.: Single machine scheduling to minimize batch delivery and job earliness penalties. SIAM J. Optim. **7**(2), 547–559 (1997). https://doi.org/10.1137/S1052623494269540

11. Zhao, C., Hsu, C.J., Lin, W.C., Liu, S.C., Yu, P.W.: Due date assignment and scheduling with time and positional dependent effects. J. Inf. Optim. Sci. 1–14 (2018). https://doi.org/10.1080/02522667.2017.1367515
12. Xiong, X., Wang, D., Cheng, T.C.E., Wu, C.-C., Yin, Y.: Single-machine scheduling and common due date assignment with potential machine disruption. Int. J. Prod. Res. **56**(3), 1345–1360 (2018). https://doi.org/10.1080/00207543.2017.1346317
13. Demir, H.I., Taskin, H.: Integrated process planning, scheduling and due-date assignment. Ph.D. thesis, Sakarya University (2005)
14. Çeven, E., Demir, H.I.: Benefits of integrating due-date assignment with process planning and scheduling. Master of Science Thesis, Sakarya University, Institute of Science (2007)
15. Demir, H.I., Cakar, T., Cil, I., Dugenci, M., Erden, C.: Integrating process planning WMS dispatching and WPPW weighted due date assignment using a genetic algorithm. Int. J. Comput. Inf. Eng. **10**(7), 1324–1332 (2016)
16. Demir, H.I., Cakar, T., Ipek, M., Erkayman, B., Canpolat, K.: Process planning and scheduling with PPW due-date assignment using hybrid search. MATTER: Int. J. Sci. Technol. **2**(1) (2016). https://www.grdspublishing.org/index.php/matter/article/view/152. Accessed 10 Sept 2018
17. Demir, H.I., Erden, C., Demiriz, A., Dugenci, M., Uygun, O.: Integrating process planning, WATC weighted scheduling, and WPPW weighted due-date assignment using pure and hybrid metaheuristics for weighted jobs. Int. J. Comput. Exp. Sci. Eng. **3**(1), 11–20 (2017). https://doi.org/10.22399/ijcesen.323860
18. Erden, C.: Dinamik bütünleşik süreç planlama, çizelgeleme ve teslim tarihi belirleme. Doktora Tezi, Sakarya Üniversitesi Fen Bilimleri Enstitüsü (2019)
19. Demir, H.I., Erden, C.: Dynamic integrated process planning, scheduling and due-date assignment using ant colony optimization. Comput. Ind. Eng. **149**, 106799 (2020). https://doi.org/10.1016/j.cie.2020.106799
20. Erden, C., Demir, H.I., Kökçam, A.H.: Solving integrated process planning, dynamic scheduling, and due date assignment using metaheuristic algorithms. Math. Probl. Eng. **2019**, 1572614 (2019). https://doi.org/10.1155/2019/1572614
21. Demir, H.I., Cakar, T., Ipek, M., Uygun, O., Sari, M.: Process planning and due-date assignment with ATC dispatching where earliness, tardiness and due-dates are punished. J. Ind. Intell. Inf. **3**(3), 197–204 (2015). https://doi.org/10.12720/jiii.3.3.197-204
22. Demir, H.I., Phanden, R.K.: Due-Date Agreement in Integrated Process Planning and Scheduling Environment Using Common Meta-Heuristics, pp. 161–184. CRC Press (2019). https://doi.org/10.1201/9780429021305-8
23. Demir, H.I., Erden, C., Kökçam, A., Göksu, A.: A tabu search and hybrid evolutionary strategies algorithms for the integrated process planning and scheduling with due-date agreement. J. Intell. Syst. Theory Appl. **4**(1), Article no. 1 (2021). https://doi.org/10.38016/jista.767154
24. Demir, H.I., Phanden, R., Kökçam, A., Erkayman, B., Erden, C.: Hybrid evolutionary strategy and simulated annealing algorithms for integrated process planning, scheduling and due-date assignment problem. Acad. Platform J. Eng. Sci. **9**(1), 86–91 (2021). https://doi.org/10.21541/apjes.764150
25. Tonizza Pereira, M., Seido Nagano, M.: Hybrid metaheuristics for the integrated and detailed scheduling of production and delivery operations in no-wait flow shop systems. Comput. Ind. Eng. **170**, 108255 (2022). https://doi.org/10.1016/j.cie.2022.108255
26. Garcia, J.M., Lozano, S.: Production and vehicle scheduling for ready-mix operations. Comput. Ind. Eng. **46**(4), 803–816 (2004). https://doi.org/10.1016/j.cie.2004.05.011
27. Li, F., Chen, Z.-L., Tang, L.: Integrated production, inventory and delivery problems: complexity and algorithms. INFORMS J. Comput. **29**(2), 232–250 (2017). https://doi.org/10.1287/ijoc.2016.0726

28. Han, D., Yang, Y., Wang, D., Cheng, T.C.E., Yin, Y.: Integrated production, inventory, and outbound distribution operations with fixed departure times in a three-stage supply chain. Transp. Res. Part E: Logist. Transp. Rev. **125**, 334–347 (2019). https://doi.org/10.1016/j.tre.2019.03.014
29. Zografos, K.G., Androutsopoulos, K.N.: A heuristic algorithm for solving hazardous materials distribution problems. Eur. J. Oper. Res. **152**(2), 507–519 (2004). https://doi.org/10.1016/S0377-2217(03)00041-9
30. Chen, H.-K., Hsueh, C.-F., Chang, M.-S.: Production scheduling and vehicle routing with time windows for perishable food products. Comput. Oper. Res. **36**(7), 2311–2319 (2009). https://doi.org/10.1016/j.cor.2008.09.010
31. Fu, L.-L., Aloulou, M.A., Triki, C.: Integrated production scheduling and vehicle routing problem with job splitting and delivery time windows. Int. J. Prod. Res. **55**(20), 5942–5957 (2017). https://doi.org/10.1080/00207543.2017.1308572
32. Uysal, M., Özcan, U.: Süpermarket Yerleşim Problemi Için Tavlama Benzetimi Algoritması Yaklaşımı. Karadeniz Fen Bilimleri Dergisi **9**(1), Article no. 1 (2019). https://doi.org/10.31466/kfbd.512098
33. Kuo, Y.: Using simulated annealing to minimize fuel consumption for the time-dependent vehicle routing problem. Comput. Ind. Eng. **59**(1), 157–165 (2010). https://doi.org/10.1016/j.cie.2010.03.012
34. Cura, T.: Modern Sezgisel Teknikler ve Uygulamaları. Papatya Yayincilik, Istanbul (2008). https://www.bkmkitap.com/modern-sezgisel-teknikler-ve-uygulamalari. Accessed 07 Jan 2023
35. Ayhan, E.: Kan Tedarik Zinciri Dağıtım Ağı Modellemesi ve Hibrit Genetik Algoritma ile Çözümü. Doktora Tezi, Yıldız Teknik Üniversitesi Fen Bilimleri Enstitüsü (2021)

Check for updates

ROS Compatible Local Planner and Controller Based on Reinforcement Learning

Muharrem Küçükyılmaz and Erkan Uslu(✉)

Computer Engineering, Yıldız Technical University, İstanbul, Turkey
muharrem.kucukyilmaz@std.yildiz.edu.tr, euslu@yildiz.edu.tr

Abstract. The study's main objective is to develop a ROS compatible local planner and controller for autonomous mobile robots based on reinforcement learning. Reinforcement learning based local planner and controller differs from classical linear or nonlinear deterministic control approaches using flexibility on newly encountered conditions and model free learning process. Two different reinforcement learning approaches are utilized in the study, namely Q-Learning and DQN, which are then compared with deterministic local planners such as TEB and DWA. Q-Learning agent is trained by positive reward on reaching goal point and negative reward on colliding obstacles or reaching the outer limits of the restricted movable area. The Q-Learning approach can reach an acceptable behaviour at around 70000 episodes, where the long training times are related to large state space that Q-Learning cannot handle well. The second employed DQN method can handle this large state space more easily, as an acceptable behaviour is reached around 7000 episodes, enabling the model to include the global path as a secondary measure for reward. Both models assume the map is fully or partially known and both models are supplied with a global plan that does not aware of the obstacle ahead. Both methods are expected to learn the required speed controls to be able to reach the goal point as soon as possible, avoiding the obstacles. Promising results from the study reflect the possibility of a more generic local planner that can consume in-between waypoints on the global path, even in dynamic environments, based on reinforcement learning.

Keywords: q-learning · deep reinforcement learning · DQN (deep q-learning) · ROS (robot operating system) · navigation · local planner

1 Introduction

In recent years, Autonomous Mobile Robots (AMR) have started to be preferred to improve operational efficiency, increase speed, provide precision, and increase safety in many intralogistics operations, like manufacturing, warehousing, and hospitals [1]. AMRs are a type of robot that can understand its environment and act independently. In order for AMR to understand the environment and act independently, it must know where it is and how to navigate. This topic has attracted the attention of many researchers and various methods have been proposed. One of the most widely accepted and used methods today is to first map the environment and use this map for localization and navigation.

Z. Şen et al. (Eds.): IMSS 2023, LNME, pp. 402–414, 2024.
https://doi.org/10.1007/978-981-99-6062-0_37

Localization and mapping play a key role in autonomous mobile robots. Simultaneous localization and mapping (SLAM) is a process by which a mobile robot can build a map of an unknown environment using sensors which perceive the environment and at the same time use this map to deduce its location [2]. For this task, sensors such as cameras, range finders using sonar, laser, and GPS are widely used. Localization can be defined as the information of where the robot is on the map. Localization information can be obtained from the wheel odometer. But the error in the wheel odometer is incremental over time [3]. Therefore, robot position can be found with triangulation formula by using beacon [4] or reflector [3]. In addition, it is possible to improve the position with object detection using a depth camera [5] or by matching the measurements taken from the real environment with the map using lidar sensors. Navigation can be defined as the process of generating the speed commands necessary for a robot to reach a destination successfully. The navigation methods for mobile robots can roughly be divided into two categories: global planner and local planner. Global planner methods such as A*, Dijkstra and Rapidly exploring Random Tree (RRT) aim to find a path consisting of free space between start and goal. The second submodule in navigation is a local planner. Local planner methods such as Artificial Potential Field, Dynamic Window Approach (DWA), Time Elastic Band (TEB) and Model Predictive Control (MPC) aim to create the necessary velocity commands to follow the collision free trajectory that respects the kinematic and dynamic motion constraints.

In the last decade, Reinforcement Learning (RL) methods have been used for navigation [6, 7] RL is an area of machine learning concerned with how agents take actions in an environment by trial and error to maximize cumulative rewards [8]. Many methods such as Value Function, Monte Carlo Methods, Temporal Difference Methods, and Model-based algorithms have been proposed to maximize the cumulative rewards for agents. Q-learning learning proceeds similarly to method of temporal differences (TD) is a form of model-free reinforcement learning [9]. However, since these models are insufficient in the complexity of the real world, more results are obtained with deep learning-based reinforcement learning methods such as Deep Q-Network (DQN) [1, 10]. In this study, a new local planner method called dqn_local_planner is proposed using the DQN algorithm. For DQN algorithm, a fully connected deep-network is used to generate velocity data for autonomous mobile robots using laser field scanner data, robot position and global path after a series of preprocessing. ROS framework, gazebo simulation and gym environment are used for model training and testing. To compare the proposed model with teb_local_planner and dwa_local_planner, static and dynamic environments are created in the gazebo. The paper is structured as follows: in Sect. 2, algorithms used for proposed method and comparison are mentioned, the explanation of proposed method is mentioned in Sect. 3, the experimental environment set up to test the model is mentioned in Sect. 4, the results obtained and the evaluation of the results are mentioned in Sect. 5 and a general evaluation is made in Sect. 6.

2 Methodology

2.1 Time Elastic Band

Time Elastic Band (TEB) is a motion planning method for local navigation. The classic 'elastic band' is described as a sequence of n intermediate robot poses. TEB is augmented by the time that the robot requires to transit from one configuration to the next configuration in sequence. Its purpose is to produce the most ideal velocity command for mobile robots by optimizing both the configuration and the time interval with weighted multi-objective optimization in real-time considering the kino-dynamic constraints such as velocity and acceleration limits. The objective function is defined as in Eq. (1), in which γ denotes weights and f_k denotes the constraints and objectives with respect to trajectory such as obstacle and fastest path.

$$f(B) = \sum\nolimits_k \gamma_k f_k(B) \tag{1}$$

The f_k are generalized in TEB as in Eq. (2), in which x_τ denotes the bound, S denotes the scaling and n denotes the polynomial order and ϵ denotes the small translation of the approximation.

$$e_\tau(x, x_\tau, \epsilon, S, n) \cong \begin{cases} \left(\frac{x-(x_\tau-\epsilon)}{S}\right)^n & if\ x > x_\tau - \epsilon \\ 0 & otherwise \end{cases} \tag{2}$$

2.2 Dynamic Window Approach

Dynamic Window Approach (DWA) aims to find the ideal velocity (v, w) search space containing the obstacles-safe areas under the dynamic constraints of the robot and find the maximum velocity in this space using objective function. To get the velocity search space, circular trajectories (curvatures) consisting of pairs (v, w) of translational and rotational velocities are determined. Then, this space is constrained by the admissible velocity at which it can stop without reaching the closed obstacle on the corresponding curvature. Finally, this space is constrained by the dynamic window which restricts the admissible velocities to those that can be reached within a brief time interval given the limited accelerations of the robot. The objective function as shown in Eq. (3) is maximized.

$$\mathrm{G(v, w)} = \sigma(\alpha * \mathrm{heading(v, w)} + \beta * \mathrm{dist(v, w)} + \gamma * \mathrm{vel(v, w)}) \tag{3}$$

The target heading, $heading(v, w)$, measures the alignment of the robot with the target direction. It is maximal if the robot moves directly towards the target. The clearance, $dist(v, w)$, is the distance to the closest obstacle that intersects with curvature. $vel(v, w)$ Function represents the forward velocity of robot.

2.3 Deep Q-Network

DQN is one of the methods of learning optimum behavior by interacting with environment by taking action, observing the environment, and rewarding. In general, observation at time t does not summarize the entire process, so previous observations and actions also should be included in the process. In DQN method, sequences of actions and observations, $s_t = x_1, a_1, x_2, : : : a_{t-1}, x_t$ are included as inputs. The agent's goal is to choose actions that will maximize future rewards as shown in Eq. (4), in which T is the time-step and γ is the discounted factor.

$$R_t = \sum_{t'=t}^{T} \gamma^{t'-t} r_t \tag{4}$$

The optimal action-value function $Q^*(s, a)$ allows us to determine the maximum reward we get when we take action a while in state s. The optimal action-value function can be explained as Eq. (5) with the Bellman equation.

$$Q^*(s, a) = \left[r + \gamma \max_{s'} Q^*(s', a') \mid s, a \right] \tag{5}$$

This can be obtained by iterative methods such as the value iteration algorithms, but it is impractical. It is more common to estimate the action-value function $Q(s, a; \theta) \approx Q^*(s, a)$ usually using linear function approximators and sometimes nonlinear approximators instead, such as neural network. Neural network function approximator with weights θ is used as Q-network. However, nonlinear approximators may cause reinforcement learning to be unstable or even to diverge. As a solution to this, replay memory is used that updates iteratively the target values periodically towards action-values Q. Deep Q-learning with experience replay algorithm can be given as follows:

Deep Q-learning with experience replay

1. Initialize replay memory D to capacity N
2. Initialize action-value function Q with random weights θ
3. Initialize target action-value function $\hat{Q}$ with weights $\theta^{-} = \theta$
4. For episode = 1, M do
   a. Initialize sequence $s_1 = \{x_1\}$ and pre-processed sequence $\phi_1 = \phi(s_1)$
   b. For $t = 1, T$ do
      (1) With probability ε select a random action a_t
      (2) Otherwise select $a_t = argmax_a Q(\phi(s_t), a; \theta)$
      (3) Execute action a_t in emulator and observe reward r_t and image x_{t+1}
      (4) Set $s_{t+1} = s_t, a_t x_{t+1}$ and pre-process $\phi_{t+1} = \phi(s_{t+1})$
      (5) Store transition $(\phi_j, a_j, r, \phi_{j+1})$ in D
      (6) Sample random minibatch of transitions $(\phi_j, a_j, r, \phi_{j+1})$ from D
      (7) Set $y_j = \begin{cases} r_j, & if\ episode\ term.\ at\ step\ j+1 \\ r_j + \gamma max_{a'} \hat{Q}(\phi_{j+1}, a'; \theta^{-}) & otherwise \end{cases}$
      (8) Perform a gradient descent step on $(y_j - Q(\phi_j, a_j; \theta))^2$ with respect to the network parameters θ
      (9) Every C steps reset $\hat{Q} = Q$
   c. End for
5. End for

3 Proposed Method

This study proposes a new local planner for Mobile Robots using the DQN method. The state space consists of 14 values. The continuous state space that consists of sensor measurements and robot-path measures is sampled at a certain resolution to obtain a discrete state space. One of the state space parameters is the angle difference as shown in Fig. 1a. The vector passing from the intermediate point as far as the lookahead distance of global path and robot's position is obtained. The angle difference (θ) is the difference between this vector and robot angle. The resolution of the angle difference is 0.35 rad. The second is the distance (L) of the robot to its closest point on the global path as shown in Fig. 1b the resolution of the distance is 0.2 m.

The third is the sensor data obtained from the preprocessed laser scanner. The 360-degree sensor data is divided into 12 sectors as shown in Fig. 2a. The smallest distance data in each sector is a state in the state space as shown in Fig. 2b. The resolution of this distance is determined according to Table 1. Thus, the model is able to review its decisions more precisely as it gets closer to the obstacle.

3 actions have been determined for our model. Go forward at 1 m/s linear velocity, slowly turn right at −0.5 rad/s angular velocity and 0.3 m/s linear velocity, slowly turn

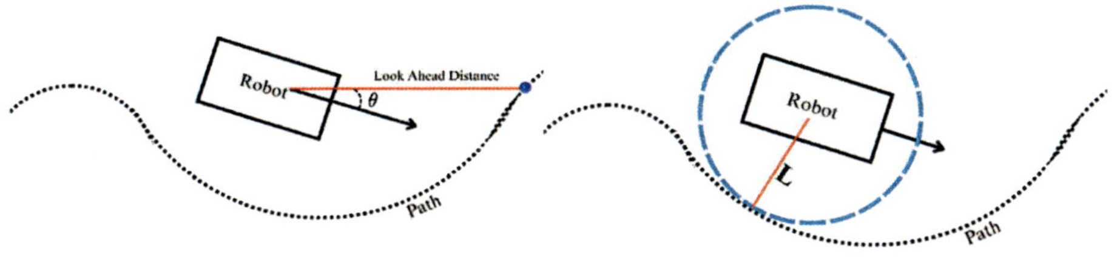

(a) (b)

Fig. 1. Two of the state space parameters a) angle difference between lookahead distance and robot heading, b) the distance (L) of the robot to its closest point on the global path.

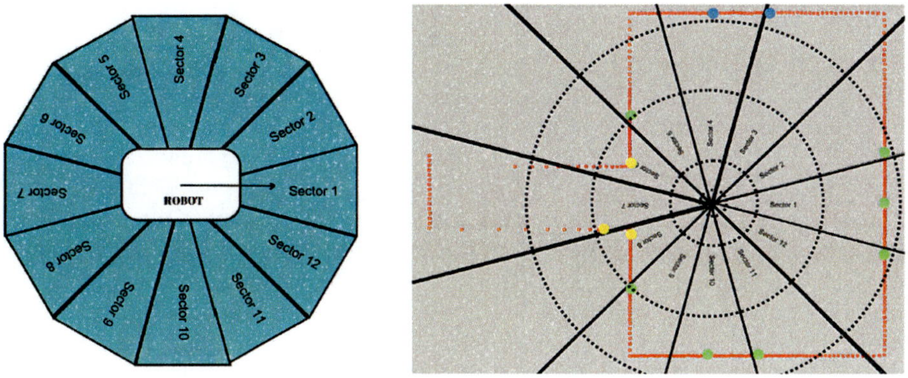

(a) (b)

Fig. 2. a) sector tiling around the robot, b) smallest distances of laser data on each sector

Table 1. The resolution table by distance

Min Distance (m)	Max Distance (m)	Resolution (m)
0.0	0.2	0.05
0.2	1.0	0.1
1.0	5.0	0.5
5.0	15.0	1.0
15.0	max range	5.0

left at 0.5 rad/s angular velocity and 0.3 m/s linear velocity. Our agent is rewarded with an inverse ratio to the closest distance to the global path, an inverse ratio to the angular difference and a reward if the robot reaches the goal. However, a negative reward is given if it is 3m away from the global path, if it got too close to the obstacle or if the angle difference was more than 2.44 rad. Expect for the first two rewards, the episode ends in other positive/negative reward cases. There are 3 fully connected layers with a rectified linear activation function in our network used for DQN agent. These layers consist of 64, 128, 32 and 3 output layers neurons as shown in Fig. 3.

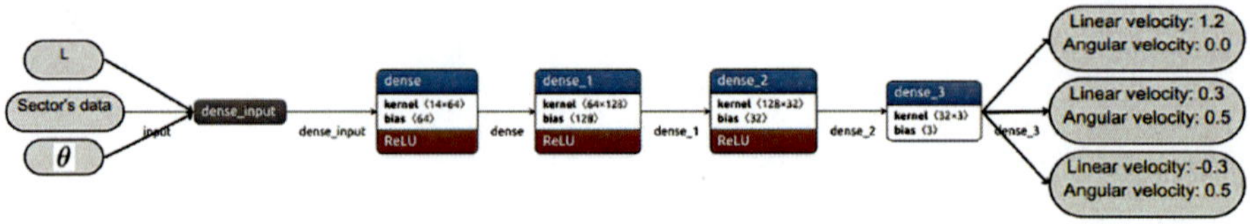

Fig. 3. Network architecture of DQN agent

4 Experimental Setup

The experiment is carried out in a simulation environment. Gazebo, an open-source robotic simulator, is chosen to simulate the environment. Turtlebot3 is chosen for mobile robot. The Turtlebot3 robot is both physically available and has a model in the gazebo environment. It is also compatible with the Robot Operating System (ROS). ROS, which we also use as our experimental setup, is a widely used open-source software for mobile robots. It allows complex algorithms developed for topics that form the basis of robotics such as mapping, localization, navigation used in robots to communicate with each other. In our study, we used 'gmapping' for mapping, 'amcl' for localization and 'move_base' for navigation. We used 'gym', the standard API for reinforcement learning, to observe the environment and determine the next action. We used the 'keras' library for model training. DQN was used as the reinforcement learning model.

In gazebo environment, a $(1, 3, 1)$ size box is placed in the center of the empty space as an obstacle. In each episode, the robot starts from the point $(-3.9, 0)$ and is expected to navigate to the goal at $(3, 0)$. In this way, DQN model is trained for about 5000 episodes. At the end of 5000 episodes, it is observed that the robot mostly navigates to the goal point. As the second stage, the location of the obstacle and goal point is changed in each episode. The obstacle is randomly generated between $(-0.5, 0.5)$ on the x-axis and $(-2, 2)$ on the y-axis. Path given to system as a global path is the straight line between the start point and the target point with 0.2 m waypoints. In this way, the model is trained for about 5000 more episodes. The trained model is integrated into ROS as local planner with the name dqn_local_planner. Scenarios in Table 2 are created to test our model and to compare it with teb_local_planner and dwa_local_planner.

To compare the local planners, the number of times the goal is successfully achieved, the total execution time and the sum of absolute areas between the planned path and the executed path metrics are used. Experiments are repeated 5 times. Reaching a 0.2 m radius circle with the goal in the center is classified as an achieved goal. Final heading angle is not considered.

Table 2. Scenarios

Scenario 1 (Fig. 4a): • Place the robot at point (−3.9, 0.0) • Give a point a. (2.0,0.0) as goal b. (4.0, 0.0) as goal c. (11.0, 0.0) as goal • Trigger global path planning using 'navfn' Place a (1.0, 3.0, 1.0) sized box at (0.0, 0.0) as a fixed obstacle	Scenario 2 (Fig. 4b): • Place the robot at point (−3.9, 0.0) • Give point (4.0, 0.0) as goal • Trigger global path planning using 'navfn' Place a (1.0, 3.0, 1.0) sized box at random $y \in [-2.0, 2.0]$ and fixed x = 0.0 and let the obstacle move repeatedly between −2.0 and 2.0 in y-axis with a speed of 0.2 m/s
Scenario 3 (Fig. 4c): • Place the robot to point (−3.9, 0.0) • Give point (10.0, 0.0) as goal • Trigger global path planning using 'navfn' Place a (1.0, 3.0, 1.0) sized box at random $y \in [-2.0, 8.0]$ and fixed y = 0.0 and let the obstacle first move away from the robot and then repeatedly move between −2.0 and 8.0 in x-axis with a speed of 0.1 m/s	Scenario 4 (Fig. 4d): • Place the robot at point (−3.9, 0.0) • Give point (10, 0.0) as goal • Trigger global path planning using 'navfn' Place a (1.0, 3.0, 1.0) sized box at random $x \in [-2.0, 8.0]$ and fixed y = 0.0 and let the obstacle first move towards the robot and then repeatedly move between 8.0 and 0.0 in x-axis with a speed of 0.1 m/s

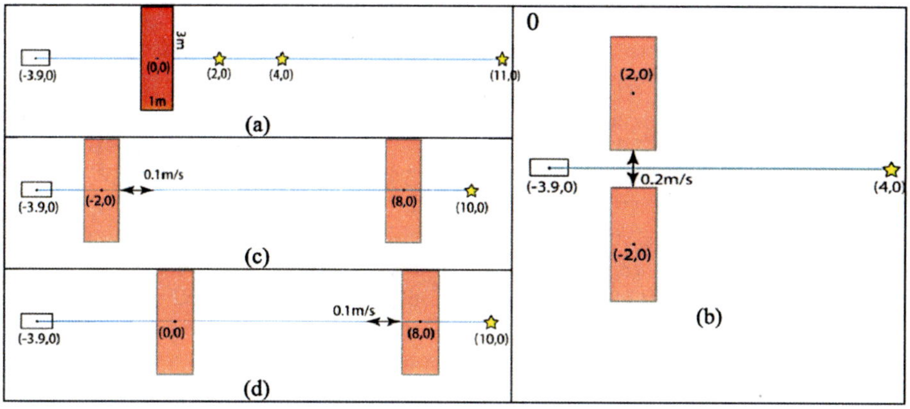

Fig. 4. 2D representation of a) scenario 1, b) scenario 2, c) scenario 3, d) scenario 4

5 Results and Discussion

The graph in Fig. 5 depicts the reward during the training of our model. The model is trained in 10050 episodes. At around episode 4000, our model has learned to go to a fixed goal in a static environment. Then the obstacles and the goal point are randomly generated in a large area and the model is trained for about 2000 more episodes. The drop on reward observed around episode 5000 is due to change in policy, but the robot quickly adapts to the unfamiliar environment. During the training, it is also observed that there was usually no obstacle between the robot and its goal. So, this area is narrowed.

Therefore, the model training decreased again after around episode 6000, but the robot quickly adapts again. It is observed that the model success does not increase after around episode10050, so the model training is terminated here.

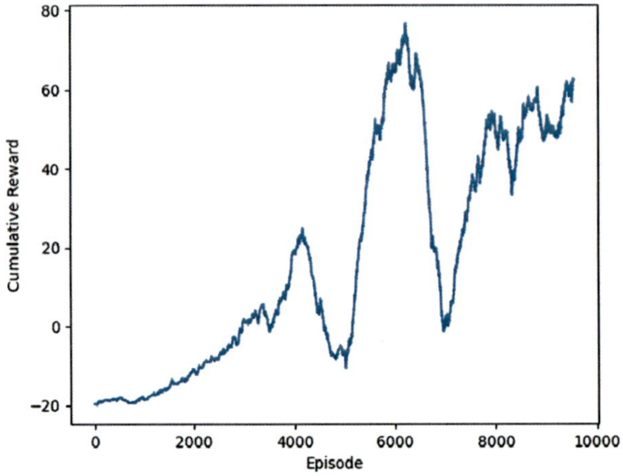

Fig. 5. Cumulative reward graph of proposed model

In the figures below, the blue line represents the global path, green arrows represent the robot's movement and red dots represent the laser sensor readings from the obstacle. Within the scope of scenario 1, the robot is given (4, 0) as goal point. The scenario is repeated 5 times. As shown in Fig. 6a and Fig. 6b, teb_local_planner and dwa_local_planner respectively, fails to pass the obstacle in any of the trials.teb_local_palnner and dwa_local_planner can trigger the move_base to recalculate the path in the case of suddenly encountering an obstacle, however DQN does not have this feature. Normally, there is a parameter in the ROS navigation stack for the global planner to detect dynamic obstacles as well. But here we have turned off that parameter to observe the performance of the pure local planner. As shown in Fig. 6c, the dqn_local_planner reaches the goal in an average of 33.59 ∓ 3.468 s. The robot has moved in an average of 12.032 ∓ 1.121 m away from its path during its entire movement. After passing the obstacle in each attempt, it does not immediately enter the path and goes directly to the goal point.

However, as shown in Fig. 7a, when the goal point is given as (2, 0), the robot could not reach the goal in all trials. If the projection of the robot's position to the path at any time becomes the goal point, it cannot reach the goal. When the goal point is given as (11, 0) as shown in Fig. 7b, the dqn_local_planner reaches the goal in an average of 57.395 ∓ 1.492 s. The robot has moved an average of 14.18 ∓ 1.876 m away from its path during its entire movement. The robot enters the path after passing the obstacle and continues to follow the path. In one of the experiments, the robot passed the goal without reaching the goal, so the experiment was terminated.

In scenario 2, the target is given when the obstacle is in various positions. The local path differs because the obstacle was a moving object as shown in Fig. 8a and Fig. 8b.

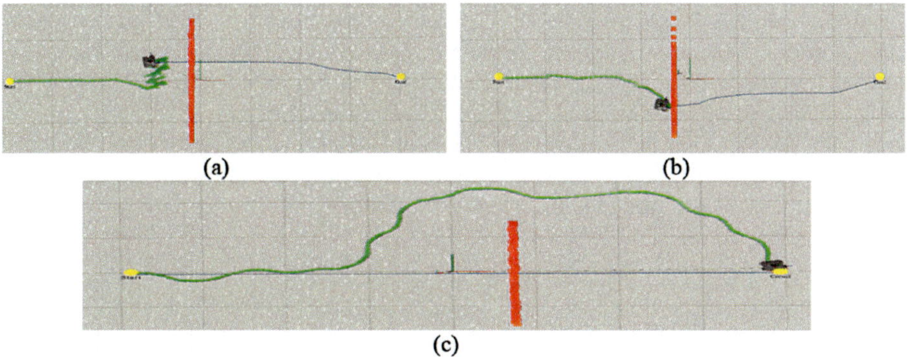

Fig. 6. Scenario 1b results for a) teb_local_planner, b) dwa_local_planner, c) dqn_local_planner

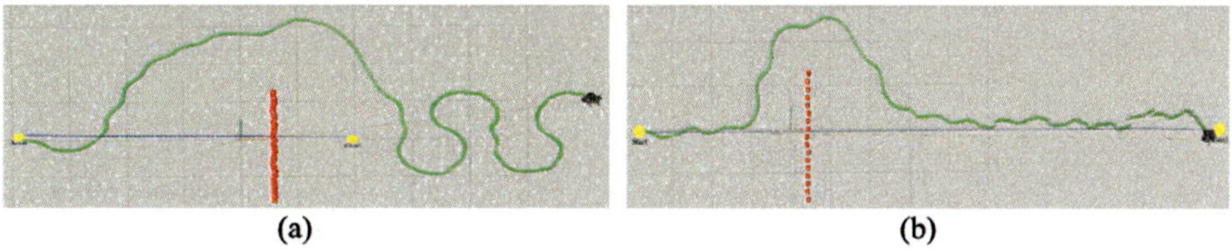

Fig. 7. dqn_local_planner behaviours for a) scenario 1a, b) scenario 1c

The dqn_loca_planner reached the target with an average of 38.88 ∓ 11.040 s. The robot has moved an average of 13.07 ∓ 16.86 m away from its path during its entire movement. It encounters an obstacle once and the alternative path cannot be reproduced. After the obstacle moves away from the robot, the robot continues its path as shown in Fig. 8c.

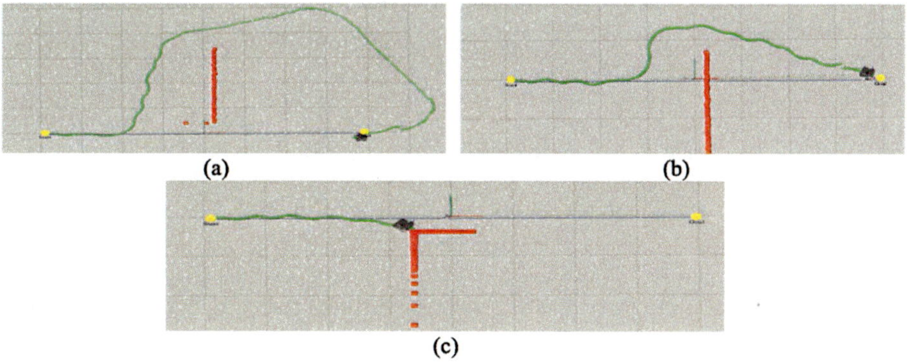

Fig. 8. dqn_local_planner behaviours from different runs for scenario 2 when obstacle is encountered at, a) $y > 0$, b) $y \cong 0$, c) $y < 0$

As shown in Fig. 9a, the teb_local_planner reaches the goal in an average of 25.9 ∓ 1.1 s. The robot has moved an average of 1.71 ∓ 0.244 m away from its path during its entire movement. When an obstacle is in front of the Robot, teb_local_planner waits until it gets out of the way. After the obstacle is removed, it continues its movement. As shown in Fig. 9b, the dwa_local_planner reaches the goal in an average of 25.13∓4.995 s. The robot has moved an average of 1.41 ∓ 1.374 m away from its path during its entire movement. DWA waits when an obstacle is in front of it and continues after it is removed. However, while the obstacle is moving towards the robot from the side of the robot, it cannot react adequately, and the obstacle hits the robot.

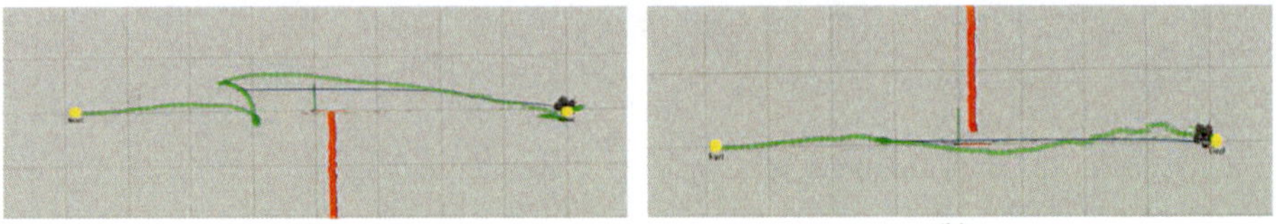

(a) (b)

Fig. 9. Scenario 2 behaviours for, a) teb_local_planner, b) dwa_local_planner

In scenario 3, the teb_local_planner and dwa_local_planner could not pass the obstacle as in the first scenario. As shown in Fig. 10, the dqn_local_planner reached the goal in an average of 49.23 ∓ 0.959 s. The robot has moved an average of 22.02 ∓ 5.17 m away from its path during its entire movement. Dqn_local_planner can easily pass slower objects in the same direction.

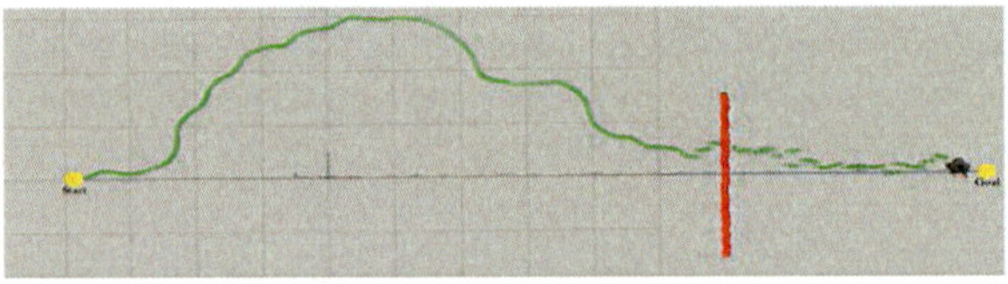

Fig. 10. Scenario 3 behaviour of dqn_local_planner.

Scenario 4 gave equivalent results to scenario 3. Thanks to the robot dqn_local_planner, it easily passed an obstacle coming towards it with a speed of 0.1 m/s.

To summarize the outputs of all scenarios, although teb_local_planner and dwa_local_planner give much better results when there are no obstacles or small obstacles on the path, dqn_local_planner gives better results if there are large obstacles on the robot's path. However, in the current model, if the robot accidentally passes the goal point while following the path, it is exceedingly difficult to return and reach the goal point again. It is usually stuck at the local minimum point as shown in Fig. 7a. The robot may also not take the shortest path while avoiding obstacles. If the robot is too close to the obstacle, it waits until the obstacle disappears and continues its movement. If the obstacle moves to the right in the y-axis forever and the robot encounters the obstacle, if it turns to pass to the right of the obstacle in the first step, they will go together forever in this way. The robot prefers not to go left at any given moment. Table 3 summarizes the result of DQN based local planner, when five tests are carried out for each scenario, by means of successful navigations, path execution times and error measure about the area between planned and executed paths.

Table 3. Comparison of all scenarios

Metrics	Scn 1			Scn 2	Scn 3	Scn 4
	a	b	c			
Global plan length (m)	5.9	7.9	14.9	7.9	13.9	13.9
The number of times the goal is successfully achieved	0	4	5	4	5	5
Total execution time (s)	–	33.59 ± 3.47	57.39 ± 1.49	38.88 ± 11.04	49.23 ± 0.96	49.34 ± 0.93
Sum of areas between planned and executed paths (m)	–	12.03 ± 1.12	14.18 ± 1.88	13.07 ± 16.86	22.02 ± 5.17	20.58 ± 3.59

6 Conclusion

In conclusion, a new method developed with DQN, which is a reinforcement learning method, is proposed for the local planner. In this method, the environment is first taught to the robot by using sensor data, robot location and global path. For testing the model, fixed, and moving (toward the robot, moving away from the robot, moving vertically) objects that are not in the environment map are added and it is observed whether the robot could reach the goal by avoiding obstacles. The same experiments are also performed with TEB and DWA. To test the performance of the pure local planner, the obstacles that are subsequently introduced into the environment are not added to the global path. While TEB and DWA planners are not successful in avoiding obstacles, our model is able to pass the obstacle easily in all scenarios. If the model does not learn the entire space, local minimum points may occur. In addition, if the robot and the obstacle start parallel movement in the same direction and at a similar speed, they will move together forever. However, our current model is trained in a specific environment and so works in a small world. More complex environments can be selected for training in later models. The action space of our model is also exceedingly small. It includes forward, turn right, and turn left actions. A subspace of each action can be created with a certain resolution.

Moreover, only a fully connected network is used in our model. While teaching more complex spaces, training can be continued with networks such as Convolutional Neural Network (CNN), Long Short-Term Memory (LSTM) network. As another future work it is planned also to include the dynamic objects' direction and speed values in the state space, in order to enable the robot to learn not to move in a crossing path with the dynamic object and let the robot move away from the obstacle in an efficient way.

References

1. Fragapane, G., De Koster, R., Sgarbossa, F., Strandhagen, J.O.: Planning and control of autonomous mobile robots for intralogistics: literature review and research agenda. Eur. J. Oper. Res. **294**, 405–426 (2021)
2. Durrant-Whyte, H., Bailey, T.: Simultaneous localization and mapping: part I. IEEE Robot. Autom. Mag. **13**, 99–110 (2006)
3. Betke, M., Gurvits, L.: Mobile robot localization using landmarks. IEEE Trans. Robot. Autom. **13**, 251–263 (1997)
4. Leonard, J.J., Durrant-Whyte, H.F.: Mobile robot localization by tracking geometric beacons. IEEE Trans. Robot. Autom. **7**, 376–382 (1991)
5. Biswas, J., Veloso, M.: Depth camera based indoor mobile robot localization and navigation. In: 2012 IEEE International Conference on Robotics and Automation (2012)
6. Ruan, X., Ren, D., Zhu, X., Huang, J.: Mobile robot navigation based on deep reinforcement learning. In: 2019 Chinese Control and Decision Conference (CCDC) (2019)
7. Heimann, D., Hohenfeld, H., Wiebe, F., Kirchner, F., Quantum deep reinforcement learning for robot navigation tasks, arXiv preprint arXiv:2202.12180 (2022)
8. François-Lavet, V., Henderson, P., Islam, R., Bellemare, M.G., Pineau, J., et al.: An introduction to deep reinforcement learning. Found. Trends® Mach. Learn. **11**, 219–354 (2018)
9. Watkins, C.J.C.H., Dayan, P.: Q-learning. Mach. Learn. **8**, 279–292 (1992)
10. Mnih, V., et al.: Human-level control through deep reinforcement learning. Nature **518**, 529–533 (2015)

Analyzing the Operations at a Textile Manufacturer's Logistics Center Using Lean Tools

Ahmet Can Günay[1], Onur Özbek[1], Filiz Mutlu[2], and Tülin Aktin[2](✉)

[1] Department of Electrical-Electronics Engineering and Department of Industrial Engineering, Istanbul Kültür University, Istanbul, Turkey
ahmetcangunay@outlook.com, onurozbek2000@gmail.com
[2] Department of Industrial Engineering, Istanbul Kültür University, Istanbul, Turkey
mutlu.filiz@outlook.com, t.aktin@iku.edu.tr

Abstract. Compliance with delivery times is crucial for businesses in the logistics sector. Numerous research has been conducted to improve distribution performance. Many of these studies touch on lean production as well. The strategies used in lean manufacturing are often employed by businesses and have a positive impact on performance.

This study focuses on the overseas shipping department of a textile company's logistics center. Workflow starts with product acceptance from manufacturers and ends with shipment to customers abroad. After a thorough examination, some bottlenecks that increase delivery times are observed.

Value Stream Mapping (VSM), which is a lean manufacturing technique, is chosen as the main method to be used. It aims to determine value added and non-value-added activities, resulting in minimizing or eliminating the non-value-added ones.

Initially, necessary data are gathered through workshops and interviews, and observations on Current State VSM are made. During these workshops, various improvements are proposed and evaluated together with the company's engineers. After takt time and cycle time calculations, label change station is identified as the bottleneck.

In the next step, Kaizens are suggested for the stations, and some lean techniques are employed to solve different workflow problems. Finally, short-term applicability of proposed improvements is discussed, and Future State VSM is drawn.

It can be concluded that significant improvements are achieved especially in lead time, changeover time, productivity rate and production speed. By reducing or eliminating non-value-added activities and identifying deficiencies that slow process flow, a standard, sustainable and developable process is proposed to the company.

Keywords: Lean Manufacturing Tools · Optimization · Lean Logistics · Operation Analysis

Z. Şen et al. (Eds.): IMSS 2023, LNME, pp. 415–426, 2024.
https://doi.org/10.1007/978-981-99-6062-0_38

1 Introduction

The aim of this study is to operate the workflow in the optimum way by scrutinizing the overseas product shipment process of a textile manufacturer's logistics center with various analysis techniques and lean tools.

The current process at the overseas shipment department, beginning from product acceptance and ending at shipment to customers abroad, is considered in detail. Lean tools and analysis techniques are used to reduce waiting times and wastes in the current situation. Consequently, the future state map is obtained as a result of these improvements.

The process starts with the arrival of the products taken from the Expa and continues with the affixing of the labels of the products and desi control. It ends with the pre-loading of the products and sending them to the countries where they are requested.

In the overseas shipment process, the following problems are observed:

- Wrong product is sent,
- The price tag has not been changed,
- Required products are missing,
- The automation does not read the barcode,
- It is unable to read the weight of the product in the master data.

These problems interrupt the process. At the same time, they cause loss of money, time, and work power.

In the first phase of the study, Value Stream Mapping, one of the lean tools, is found suitable for observing the current state. This method is used to identify value-added, non-value-added steps and wastes in the system [1] and is commonly implemented both in manufacturing and service industries. Setiawan et al. [2] provide a literature review of 45 VSM studies conducted in different service sectors and summarize the benefits of employing the method in companies operating in this industry.

The current state map of the overseas shipment department is drawn according to the data collected from the company. The data used in the study are obtained during the workshops, observations made in the operating departments of the company and interviews with company employees. At this stage, the value added, non-value-added steps and wastes of the process are identified and then non-value-added steps and wastes are removed with the Kaizen Methodology [3].

The study is organized as follows: Sect. 2 explains the steps of the applied methodology. Section 3 summarizes the implementation and the results obtained. Finally, the study is concluded in Sect. 4, together with some recommendations.

2 Methodology

The bottleneck process is analyzed with different lean techniques and some other techniques. This section represents the stages of the work. First, data and observations are obtained to determine the scope of this study. Then, the current state of the value stream map is created, and the relevant improvements are analyzed. Kaizen studies are carried out within the problem-solution relationship and necessary studies are performed together with the other lean production tools employed. As a result of these studies, the future state map is drawn. Finally, the improvements and results are discussed. Figure 1 below illustrates the methodology's flowchart in more detail.

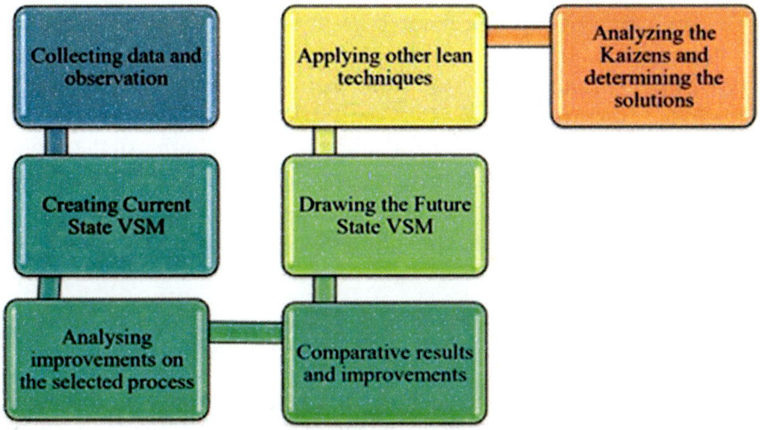

Fig. 1. Flowchart of the methodology.

2.1 Collecting Data and Observations

The first step starts with a visit to the logistics center and observe the processes. Before going to the company, the parts that need to be examined were planned and this plan was acted upon. During this time, there are crucial steps in defining the problem properly: identifying the necessary information to gather, choosing the process to be considered, and finally, setting the scope for a remediation study. In value stream mapping, material and information flows need to be mapped within the process flow. It is very important to observe the steps of all processes from the arrival of the order to the delivery of the product, to identify value-added and non-value-added activities. For this reason, answers to some predetermined questions are needed in order to collect the correct information. These questions are as follows:

- What is the number of workers working in each process?
- How many parcels are ordered daily?
- How many products are there in each box?
- How does the process flow at the general packaging line?
- How many shifts are there?
- What is the delivery time?
- What is the cycle time, usage rate, transport time of each line?
- What are the durations of value-added and non-value-added activities?
- How does the information flow between the internal departments?

The company can perform VSM analysis on the lines it chooses. By utilizing the method at the stages of planning, editing, and development, it helps the business run more efficiently. Therefore, the line selection may differ for each study. In such cases, Pareto Analysis can be used for decision making [4, 5]. According to their proportional relevance or worth in a particular situation, activities, processes, and items are categorized using this analysis. This approach to problem-solving is founded on the idea of separating the problem's primary causes from its less significant ones. From that point on, it becomes increasingly easier to decide which value streams to map for a specific case. In line with these data, the current state map is drawn. Kaizens discussed in the face of these problems are reported. Subsequently, by evaluating Kaizens in terms of the applicability of real-life conditions, decisions are made about their solutions, and then the future state map is drawn. In addition, improvements are proposed using other lean techniques such as Fishbone Diagram [6], which is a widely employed tool to perform a root cause analysis [7, 8], and 5 Whys [9].

2.2 Creating the Value Stream Map

The VSM method basically aims to identify the material flow from raw material to finished product, and the information flow that creates value-added and non-value-added activities. It is a tool for visualizing numerous processes and for identifying the value, waste, and waste sources in the value stream. To extract the VSM map, some basic calculations need to be done. These are listed as follows:

1. **Process Time (P/T):** The mean value of the processing times in a specific process, which considers the time taken to complete that activity.
2. **Standard Deviation:** It calculates the deviation around the mean value.
3. **Net Production Time:** Total amount of time during which a manufacturing process is actively producing goods, minus any scheduled downtime for maintenance, equipment changeovers, or other non-production activities.
4. **Workable Production Hours**: The number of hours during which a production process can be carried out effectively without compromising quality, safety, or productivity.
5. **Cycle Time (C/T):** It refers to the average time taken to complete a single unit.

$$C/T = \textit{Net Production Time}/\textit{Number of Units Produced} \tag{1}$$

6. **Takt Time (T/T):** The pace that exactly meets customer demand given workable production time.

$$T/T = \textit{Workable Production Hours}/\textit{Units Required}\,(\textit{Customer Demand}) \tag{2}$$

7. **Utilization:** It calculates utilization of processes based on customer demand.

$$\textit{Utilization} = C/T \big/ T/T \tag{3}$$

3 Implementation and Results

This section covers the application of the developed methodology at the logistics center. First, observations and data collection are made to determine the necessary information, to select the overseas shipping conveyor lines for the study, and to determine the scope of the study. Then, a Current State Value Stream Map (CVSM) is created with the data obtained from the previous section, and improvements and time losses are analyzed over the current state. Finally, comparative results and recommendations are given.

3.1 Current State Value Stream Map of the Process

A CVSM is drawn for all selected stations by collecting all the necessary data from the previous section. In the value stream map, information flow, process flow and timeline are created, respectively. Communication between the customer, suppliers and internal departments is indicated by moving arrows in the information flow. Data boxes are filled by measuring the process time, cycle time, transit time for each selected station. Finally, the value added and non-value-added activities to be included in the timeline are determined. This is important for understanding the system as a whole. By doing this, it provides ample time to analyze the waste and suggest improvements. The CVSM of the selected line is provided in Fig. 2.

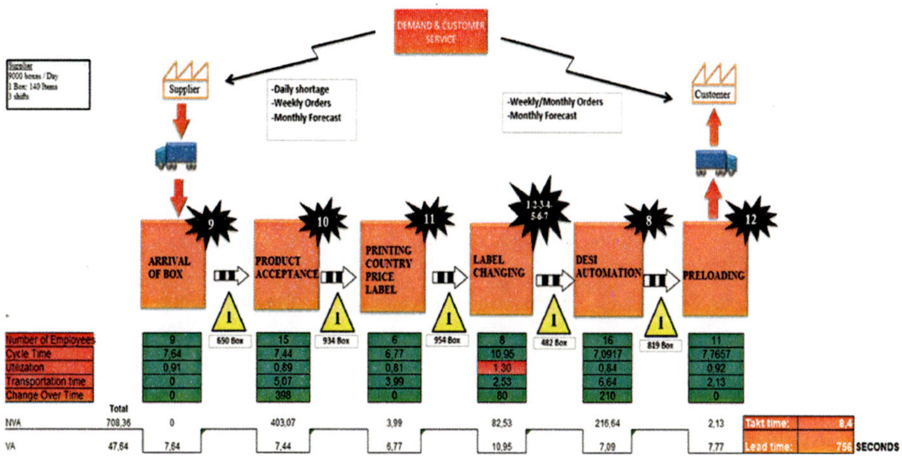

Fig. 2. CVSM of abroad shipment.

3.2 Proposed Kaizens

The Kaizens, which are proposed for the process, are identified after long brainstorming at the two-day workshop with engineers and managers from all departments. In the light of these data, they are evaluated one by one and determined the appropriate actions to produce a solution. Kaizens determined for the processes are shown in Table 1. In addition, the problems are scored according to their priorities and frequency of occurrence, and presented in Tables 2 and 3, respectively.

Table 1. Improvements achieved with kaizen suggestions.

Kaizen Burst	Frequency	Priority	Problem	Action
1	70%	A	It is difficult to attach cardboard labels on the products	Using sticker labels instead of cardboard labels increases staff productivity and facilitates their work. Thus, it saves time
2	80%	B	The environment in the work areas is messy and the work desks are out of adjustment	To create a new ergonomic working environment according to the wishes of the employees
3	15%	A	Having personnel with health problems	Compulsory check-ups must be made by the company at the specified time intervals for each employee
4	40%	C	Label printing machines are old	Label printing machines suitable for new technology should be purchased
5	20%	B	Having inexperienced staff	For inexperienced employees, a simulation of "How to print labels correctly" should be played on a computer or television in their working environment
6	30%	B	Belt jamming between the rollers in the conveyor system	It is necessary to have someone who constantly controls the flow of the system to prevent any substance from getting stuck between the rollers in the conveyor system and hindering the progress of the system
7	60%	A	Conveyor system maintenance stopping the system too much	The maintenance of the conveyor system should be done between the lunch breaks of the employees
8	40%	B	Excessive downtime of the process in the Desi machine	With the improvement in the Desi machine, the process time and downtime of the process have been reduced
9	60%	B	Slow operation of the elevator in the Expa	Time was saved by speeding up the elevator
10	50%	C	Loss of time due to slow hand pallet trucks	Using electric forklifts instead of hand pallet trucks
11	70%	B	Manually opening the box	Making with a box opening machine
12	60%	B	Carrying boxes with hand pallet truck	Carrying boxes with an electric forklift

Table 2. Kaizens' priority.

Kaizens' Frequency	Importance Coefficient
A	3
B	2
C	1

Table 3. Kaizens' frequency.

Kaizens' Frequency	Importance Coefficient
0–20	1
20–40	2
40–60	3
60–80	4
80–100	5

3.3 Pareto Chart

A Pareto chart is a useful tool for identifying the most important factors contributing to a particular problem. In the context of logistics, a Pareto chart can help identify the most significant sources of delays or inefficiencies in the supply chain.

After deciding on the stations in the conveyor line that are going to be examined, the production lines are decided, the last year's stoppage detail report is obtained to gain insights for each station, and these reports are then converted into a Pareto chart to categorize interruption definitions.

As a result of the calculations, the chart is drawn by arranging data according to the frequency of the problems obtained from the label change station, which is selected as the bottleneck. The Pareto chart of the station is shown in Fig. 3. According to this chart, it is decided to solve the 2nd, 1st, 7th, and 4th problems with priority.

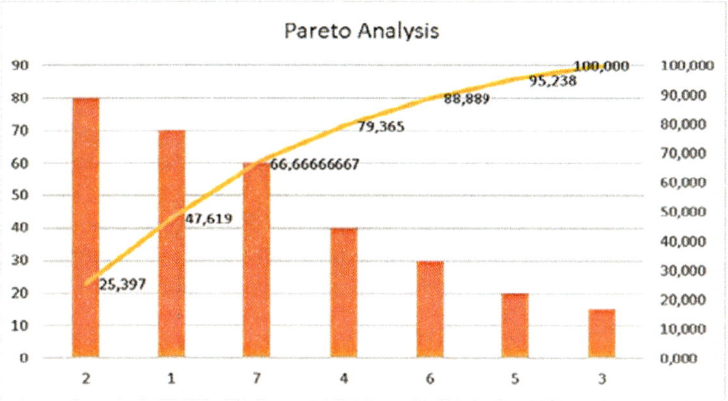

Fig. 3. Pareto chart of the responses.

3.4 Fishbone (Cause and Effect) Diagram

Cause-Effect Diagram is applied in the label changing station, which is determined as the bottleneck after the calculations. The bottleneck in this analysis refers to the problems at the station. The Fishbone Diagram presented in Fig. 4 includes the following subheadings: 1. Management problems, 2. Material problems, 3. Measurement problems, 4. Problems caused by environmental factors, 5. Method problems, 6. Machine-related problems, and 7. Human-induced problems.

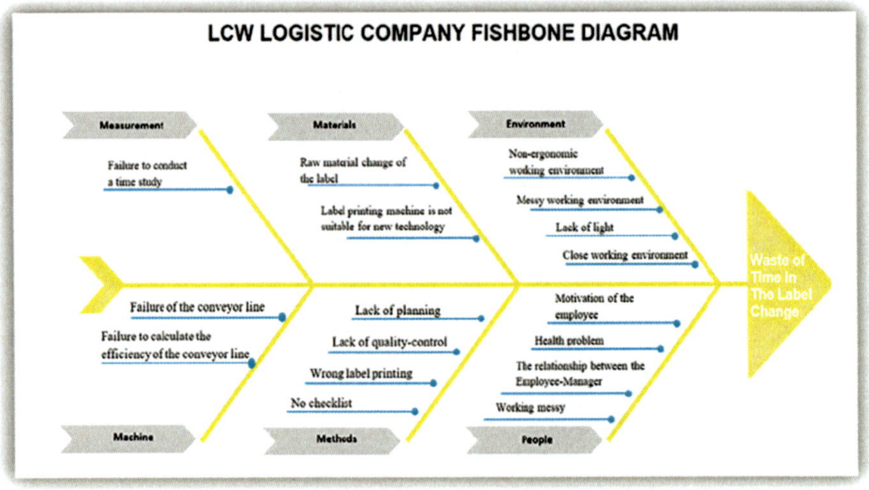

Fig. 4. Cause-effect chart of time waste in label change process.

After applying the Cause-Effect Diagram, the main and deeper causes are determined. To solve these problems, the causes of this problem must be eliminated.

3.5 5 Whys Analysis

5 Whys analysis is used to determine the root cause of the problems in the process. After 4 Whys questions, the root cause is detected.

As can be seen from Fig. 5, because of repeated "Why" questions, it is understood that the root cause of the interruption of the conveyor system is the long shift hours and fatigue of the working personnel. Once this root cause is resolved, the process will be cured.

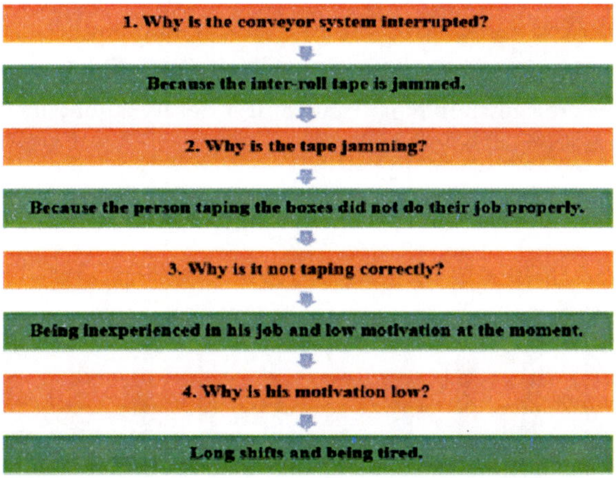

Fig. 5. 5 Whys analysis of conveyor system disruption.

3.6 Future State Value Stream Map

A Future State Value Stream Map (FVSM) is usually drawn after the baseline VSM is analyzed and opportunities for improvement are identified. FVSM is a visual representation of the desired state of the value stream after improvements are applied.

The FVSM of abroad shipment is drawn to represent the state of the value stream after the proposed improvements. This includes changes in process flow, layout, material flow, inventory levels, and other identified improvements. The FVSM of the selected line is given in Fig. 6.

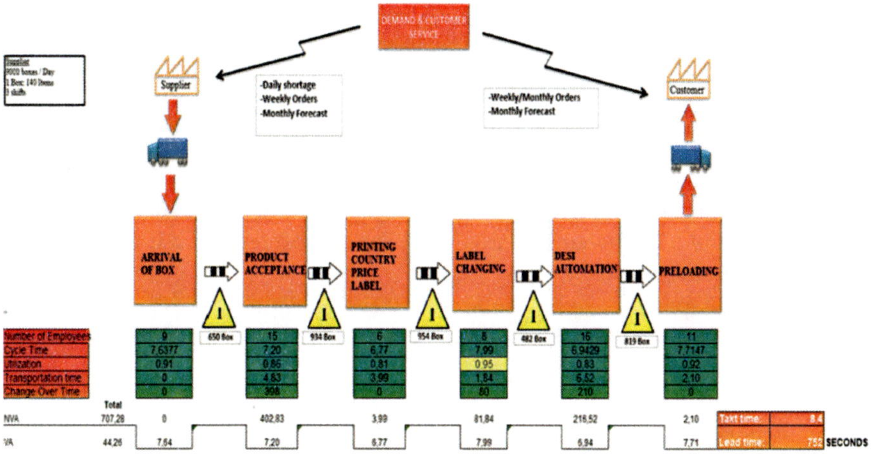

Fig. 6. FVSM of abroad shipment.

3.7 Comparative Results and Improvements

The current and future state maps are compared in terms of cycle time and utilization rate. These changes are shown in Figs. 7 and 8. It is observed that the usage rate of the label change station dropped below one.

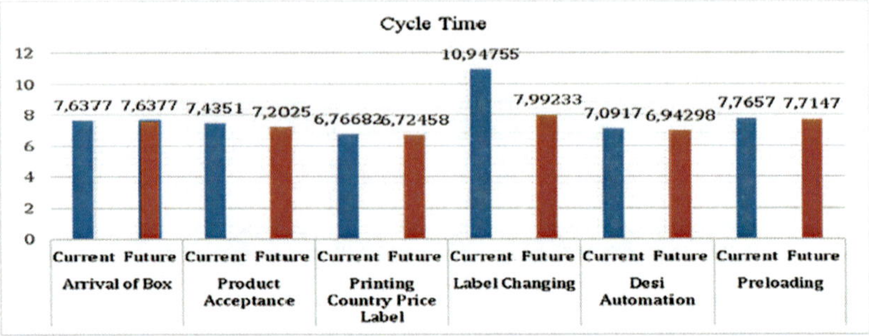

Fig. 7. The rate of change of cycle time between the CVSM and the FVSM.

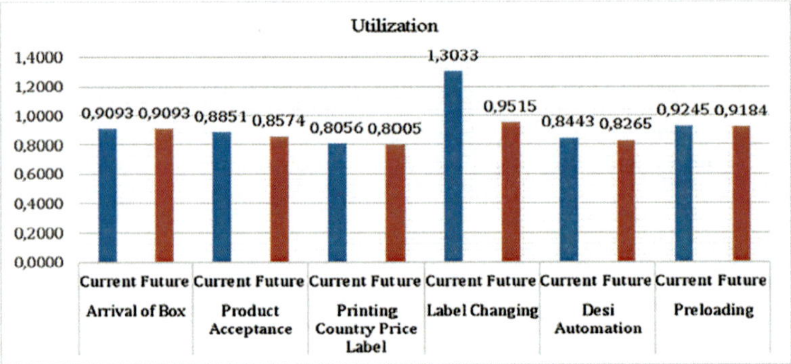

Fig. 8. The rate of change of utilization between the CVSM and the FVSM.

Table 4 summarizes the current and future usage rates, and the reduction percentages in these rates after the improvements are made. The highest reduction is obtained at the label changing station with 26.99%.

Table 4. Percentage changes in usage rates in current and future states.

Processes	States	Utilization	% Change (Reduction)
Arrival of Box	Current	0.9093	0.00%
	Future	0.9093	
Product Acceptance	Current	0.8851	3.13%
	Future	0.8574	
Printing Country Price Label	Current	0.8056	0.62%
	Future	0.8005	
Label Changing	Current	1.3033	**26.99%**
	Future	0.9515	
Desi Automation	Current	0.8443	2.10%
	Future	0.8265	
Preloading	Current	0.9245	0.66%
	Future	0.9184	

4 Conclusion and Recommendations

Businesses should strive to boost their productivity by making greater use of their production resources if they want to survive and develop in the global competition. Companies are increasingly implementing lean management techniques to achieve this aim and outperform other businesses. For this reason, lean brings the necessity of restructuring a management philosophy centered on excellence in order to provide a continuous flow to the customer by eliminating non-value-added activities and ensuring continuous improvement. This study exemplifies the implementation of a lean production system design at a textile manufacturer's logistics center operating in Turkey. It aims to support the design of conveyor lines to meet rapidly increasing demands and to increase efficiency by making improvements at the stations.

Work begins with observing processes and collecting data to determine the scope of the study. Measurements are made meticulously by taking samples from different shifts and days. The bottleneck process that causes prolongation of production times is identified. During the workshop organized, various improvements are proposed and evaluated with company engineers.

With the improvements made, significant improvements are achieved especially in lead time, changeover time, productivity rate and production speed. Slowdowns and time losses are analyzed with Fishbone Diagram. Kaizens are stated for each station in the overseas shipping department. Root cause analysis is performed with the 5 Whys method. The label change station is determined as the bottleneck, resulting from the calculated takt time and cycle time. Apart from the bottleneck, it is also understood that almost every activity works with great differences under takt time.

It is recommended to perform a cost analysis of the proposed improvements to assess the viability of this VSM study. Future research should further develop and validate these initial findings by performing a detailed simulation study, including planned stops, losses, and sensors. Also, a capacity analysis caused by defects can be explored to gain advantage in the challenging market.

References

1. Manos, T.: Value Stream Mapping-an introduction. Qual. Prog. **39**, 64 (2006)
2. Setiawan, I., Tumanggor, O.S.P., Purba, H.H.: Value Stream Mapping: literature review and implications for service industry. Jurnal Sistem Teknik Industri **23**, 155–166 (2021)
3. Imai, M.: Kaizen: The Key to Japan's Competitive Success. McGraw-Hill, New York (1986)
4. Karuppusami, G., Gandhinathan, R.: Pareto analysis of critical success factors of total quality management: a literature review and analysis. TQM Mag. **18**, 372–385 (2006)
5. Kumar, R., Singh, K., Jain, S.K.: Agile manufacturing: a literature review and Pareto analysis. Int. J. Qual. Reliab. Manag. **37**, 207–222 (2020)
6. Watson, G.: The legacy of Ishikawa. Qual. Prog. **37**, 47–54 (2004)
7. Luca, L., Luca, T.O.: Ishikawa diagram applied to identify causes which determines bearings defects from car wheels. In: IOP Conference Series: Materials Science and Engineering, vol. 564, pp. 1–6 (2019)
8. Sakdiyah, S.H., Eltivia, N., Afandi, A.: Root cause analysis using fishbone diagram: company management decision making. J. Appl. Bus. Taxation Econ. Res. **1**, 566–576 (2022)
9. Murugaiah, U., Benjamin, S., Marathamuthu, M.S., Muthaiyah, S.: Scrap loss reduction using the 5-whys analysis. Int. J. Qual. Reliab. Manag. **27**, 527–540 (2010)

Developing an RPA for Augmenting Sheet-Metal Die Design Process

Gul Cicek Zengin Bintas[1](✉), Harun Ozturk[1], and Koray Altun[2]

[1] Mubitek Tasarım Bilisim Mak. San. Ve Tic. Ltd. Şti., Bursa, Turkey
{gul.zengin,harun.ozturk}@mubitek.com
[2] Department of Industrial Engineering, Bursa Technical University, Bursa, Turkey
koray.altun@btu.edu.tr

Abstract. The mass production of identical products with high precision and accuracy heavily relies on dies and molds. In particular, the project-based die manufacturing process is crucial for introducing new products to the market. To produce a car body, an average of 1000–1250 unique dies are required, with no backups available. Simultaneously commissioning these dies is one of the most expensive and critical processes in automotive and other forming industries. However, with increasing technology and customer expectations, designers must design products quickly and efficiently to beat competitors to the market. Traditional production planning software is not well-suited for project-based work, such as die manufacturing, and often results in increased project duration due to a lack of integration and variability. To address this issue, this study proposes the use of Robotic Process Automation (RPA) with artificial intelligence methods to enable automatic data transfer from design to production. The findings of this study provide recommendations for efficient project management in the single manufacturing industry.

Keywords: Augmented Design · RPA-Robotic Process Automation · Project-based Manufacturing

1 Introduction

With Industry 4.0, the amount of data produced in the industrial sector is increasing rapidly, while design and production times are being reduced day by day. The excess of data and limited time can lead to increased stress levels for individuals, resulting in human errors and losses. Especially for companies engaged in project-based manufacturing, timely and error-free delivery of products is crucial to ensure uninterrupted supply chains. This study aims to prevent errors and losses in the design and production processes in the sheet metal industry, which operates on a project-based manufacturing model. By reducing human dependence in the industrial sector, increasing productivity and competitiveness can be achieved.

In recent years, the significance and function of metal forming processes within the manufacturing industry have been on a steady rise, mainly due to their efficient

Z. Şen et al. (Eds.): IMSS 2023, LNME, pp. 427–438, 2024.
https://doi.org/10.1007/978-981-99-6062-0_39

use of materials and cost-effectiveness [1]. In automotive industry, the most important process is the die design and manufacturing [2]. In project-based manufacturing sectors, production times are usually long. During the phase of tracking these projects with enterprise resource planning applications, it is necessary to spend time close to the project duration to create routes and recipes. However, these routes and recipes can be used in other projects as well. Therefore, the time spent is considered as wasted time. Moreover, extending the project duration reduces competitiveness, which is why enterprise resource planning applications are not always used. The necessity of using the data created in all systems has made it mandatory to generate production data from design data. At this point, an "augmented design" concept comes into play. By adding missing data parameters to the existing design tree to create the production tree and creating a collaborative platform, the product's routes and recipes can be easily created once the design process is completed.

An automobile consists of 350–400 sheet metal parts and about 1,250 molds are needed for their production. The total size of the publication data for a sheet metal die design can be 4 GB. During the production process, at least 5 revisions of this data are created and each revision is stored. When the product design data of the automobile, 3D models of plastic and other parts besides sheet metal are added, a significantly large amount of data is created that needs to be tracked and managed. Storing the data within the organization, ensuring its security, and conducting version control pose significant challenges. Due to deficiencies in the company's defined systems, data is transferred using methods such as USB, FTP, and WeTransfer, which create information security vulnerabilities.

While the use of digital systems that manage workflow from design to production is increasing in serial production, the situation is almost non-existent for project-based manufacturing. This is because the digital systems used in serial production are not suitable for the structure of project-based manufacturing. Therefore, in single production processes such as die manufacturing, which is the basis of manufacturing in the industrial sector, workflow management cannot be provided. The traditional workflow process from design to production is as follows (Fig. 1):

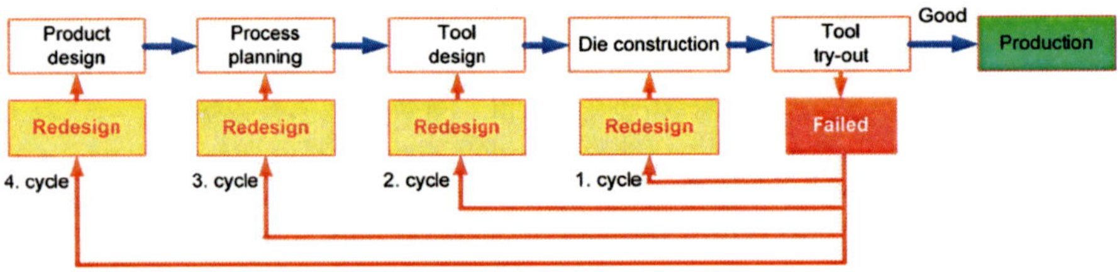

Fig. 1. From design to production flow chart [1].

The problems that led to this study can be summarized as follows:

- In project-based manufacturing, digital transformation has not been sufficiently implemented, and digital systems are attempted to be used with programs such as Excel or companies' own solutions and applications.

- When companies do not use workflow management systems, there is uncertainty about the people who will be assigned at any stage of the project. In addition to the person who will perform the task, other relevant individuals should also be informed via email, phone, etc. In such a situation, the people who need to be informed may be forgotten, which can create problems within the project.
- In project-based manufacturing, digital transformation is not adequately provided, and digital systems are attempted to be used by using programs such as Excel or companies' own solutions and applications. This results in spending too much time on non-value-added tasks such as manually entering data or collecting relevant materials from suppliers and business partners. This causes engineers to focus on inefficient tasks instead of optimizing designs and advancing innovation. In the industry, designers prepare a material list in Excel by individually identifying all the parts in their 3D design. The same applies to a 500-part design. This creates unnecessary workload for the designer, resulting in significant time and effort loss. At the same time, many problems are encountered, particularly in procurement and production, due to errors in the preparation of material lists.
- CAD models or engineering change orders are directed to team members or suppliers using traditional methods such as email. Due to the volume of daily emails, critical engineering data is lost or forgotten. Information buried deep in the inbox can lead to production delays or even worse, quality issues in the field, by causing a critical design flaw to remain unresolved.
- The CAM office staff takes screenshots of specific parts from the 3D design data, creates technical drawings, and communicates the necessary information to the operator through the CAM program. The CAM office staff plays a critical role in the design-to-production process (usually as a sole contributor). Any absence or inability to come to work by the staff member could lead to production delays, which in turn could result in cost, time, and even prestige loss. Therefore, it is necessary to eliminate the dependence of this task on a single staff member.
- The workshop supervisor needs certain critical information to decide which machine to use for which part in production. This information includes the expected surface area to be processed, hole tolerances and quantities in the design, expected surface roughness ratios, and so on. Without these data, the only thing the workshop supervisor can do is prioritize and prepare machine schedules using experiential or intuitive approaches. However, incorrect planning and/or the inability to estimate processing times can lead to time pressures on the operators. As a result, inefficiencies arise, such as having to remove an unfinished part from a machine and attaching another one.
- The personnel working on the assembly line follows the 3D design on the computer and separate the parts to perform the assembly. This process is susceptible to risks and causes time loss. If the mold is not monitored by a fixed team on the assembly line, the assembly and training steps become mixed up. People act without knowing what they are doing on which mold, causing inefficient operation in the workshop. Similar to the processing workshop, the assembly workshop also requires an organizer and controller in the form of an assembly supervisor. Without the necessary planning and data, the operations cannot proceed.

In the past, ensuring the feasibility of sheet metal parts and designing the corresponding dies was a task that relied heavily on experienced die designers. This involved a lot of calculations and as a result, was time-consuming. To address this issue, there is a need to create an intelligent system that combines suitable AI techniques and CAD systems to enable manufacturability assessments, concurrent planning, and rapid design of multi-operation dies [3]. Furthermore, it is essential to develop general design methods that are based on metal forming analysis and systematic experimental investigation [1].

The application of various methods of Computer Aided Engineering has become one of the most important topics in manufacturing industries and particularly in the automotive industry [1]. The aim of the study is to eliminate the problems identified above by utilizing Robotic Process Automation Methodologies and Artificial Intelligence Algorithms with a software that provides data flow from design to production. This will lead to a true digital transformation. In the designs carried out within the scope of the study, the Catia V5 CAD program, which is most preferred in the sheet metal die sector due to its capabilities, was used.

2 Method

The methodologies to be applied in the scope of the study can be summarized under 3 main headings: "Augmented Design", "Die Design Data Management" and "Die Production Management".

With "augmented design":

- Primary automotive manufacturers and customer standards will guide the design and the designer according to drawing information,
- Standard part libraries, solid models, selection processes and their management will be done quickly from the screen,
- The material list will be prepared by the program when the design is completed, sent to the purchasing unit and the processes will continue quickly, ensuring error-free material list, error-free purchasing and complete production,
- Thanks to intelligent coloring/parameters, quick data flow can be provided to those involved in predictive planning and creating weight information for 2D, 3D machining surfaces and raw materials necessary for cost calculation,
- Production information and processes will be defined at the design stage and transferred to the design through methods such as coloring and explanatory footnotes.

The die design data management includes:

- Management operations of interacting users within the company will be carried out,
- Data management will be secured through document management in the company's data center, preventing data repetition and losses,
- Project workflows and all related processes (notification mechanism, permission mechanism, etc.) will be carried out through the application,
- Form tools will be created to collect more data from users,
- Reporting formats will be prepared with the information obtained from the above modules, and these reports will provide information to the company.

The production management of dies:

- Old production parameter information from previous projects will be used for similar projects and new project parameters will be filled in,
- Image processing methods will be used to compare the surface model from the design with surface models from previous projects to identify similarities,
- Errors will be detected in the design data based on color and parameter information, and users will be alerted.

The desired outputs have been achieved through "Augmented Design" and "Design Data Management" methods. Work is still ongoing on the planned methods for "Die Production Management."

2.1 Augmented Design

With Augmented Design, designs will be made intelligent and standardized. In order to transfer data from design to production, the design must first be optimized and standardized. This allows access to the data and enables Robotic Process Automation to be performed. The target autonomy level within the scope of the project is "Level 4" as shown in Table 1. Level 5 represents full autonomy.

Table 1. Autonomy levels in "die design".

Autonomy level	Context	Base pattern/Template design	Environmental geometries/ Coloring	Production parameters/Hole	Documentation	System capacity	Level
Level 0	No autonomy	Designer	Designer	Designer	Designer	No	Just designer
Level 1	Design support	Designer/Macros	Designer	Designer	Designer	Some design abilities	With Support
Level 2	Partial autonomy	Designer/Macros	Designer	System	Designer	Some design abilities	Partial Autonomy
Level 3	Conditional autonomy	Designer/Macros	System	System	Designer	Some design abilities	Conditional autonomy
Level 4	High autonomy	System	System	System	System	Some design abilities	High autonomy
Level 5	Full autonomy	System	System	System	System	Full design	Full autonomy

In the Augmented Design method, 3D parametric die sets containing technical details specific to the customer are modeled by designers using Catia and the necessary parameters and rules are defined for the die sets. The Catia Knowledge Advisor module is used for these processes. Modern parametric 3D CAD systems, such as Catia V5 or SolidWorks from Dassault Systems, Unigraphics NX from Unigraphics Solutions, or

Pro/Engineer from Parametric Technology Corporation, allow the simultaneous use of the three basic elements in parametric geometry modeling [4]. The parametric information within the 3D data opened in the Catia CAD program is read and documentation data is prepared. If a standard part is to be used during sheet metal die design, standard parts are found from various company catalogs according to the need. These parts are added to the design either as ready-made or modeled. Within the scope of the study, a parametric standard part library has been created according to the customers' needs, and each part has been identified with a unique code (order code, company, material, size, etc.) (Fig. 2).

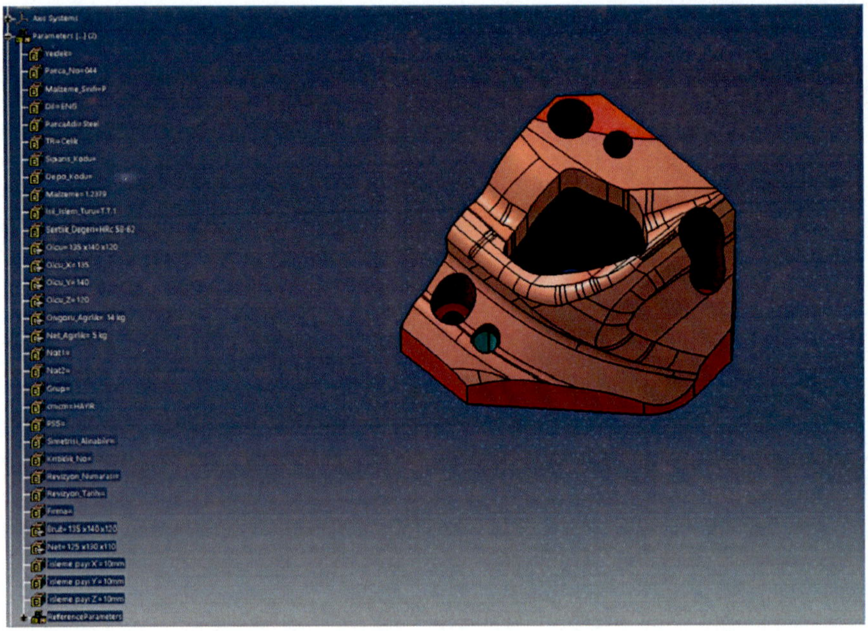

Fig. 2. Augmented design with design tree.

If a non-standard product will be used in sheet metal die design, the template file of the part is opened through a program. The template file contains the identity information of the part and the ready product tree parametrically. The designer can make necessary modeling through this file. The die sets, which are prepared parametrically and relationally linked in the application, also include production-related parameters. By defining user needs with the parameters in the mold sets, the design is brought to a certain level.

After creating the design library, classification and/or clustering operations will be performed using machine learning methods. Pattern recognition methods will be used to identify similarities between new designs and designs in the library. Predictions will be made for new designs using appropriate methods based on the data in the library. Inference, decision-making and/or optimization methods will also be applied.

When looking at the autonomous level of the base die and template design, inputs that can change the design, such as the size of the design company, the size of the design, and the standard parts to be used, are apparent. Classification will be made according to these inputs, and the most suitable base die and template design prediction will be made by regression method using decision trees, and the designer will be enabled to start from this design. These parameters are planned to provide data that can create routes and recipes in enterprise resource planning software.

Data from parameters such as material type, material class, order code, heat treatment type, hardness value, gross dimensions, net dimensions, predicted weight, net weight, warehouse code, company information, and processing allowances will be transferred directly to the enterprise resource planning software. Thus, the route and recipe can be automatically created, and all modules required for production will be triggered. Purchasing and planning processes can be carried out without errors and quickly.

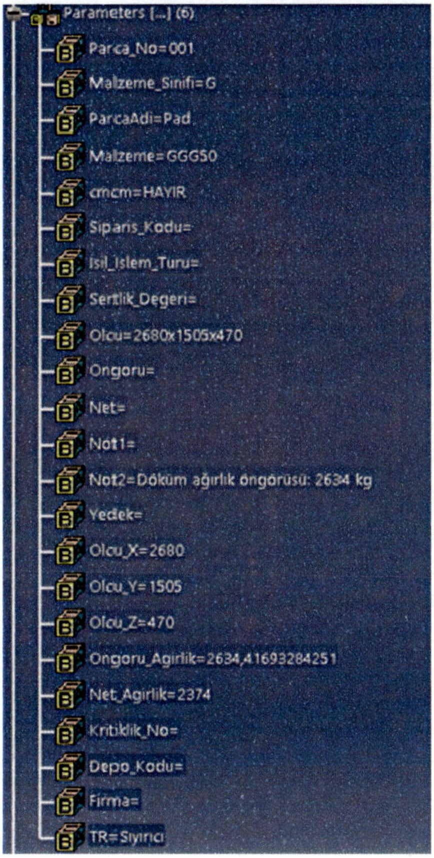

Fig. 3. Design tree with production parameters.

Another method used to improve the intelligence of the design is coloring. Coloring standards serve as a common language to facilitate communication between designers and manufacturers. These standards are based on assigning colors to parts during the design stage in a 3D CAD application. This helps determine which machine, methods, and tolerances will be used to manufacture the part, as well as its surface smoothness [5] (Fig. 4).

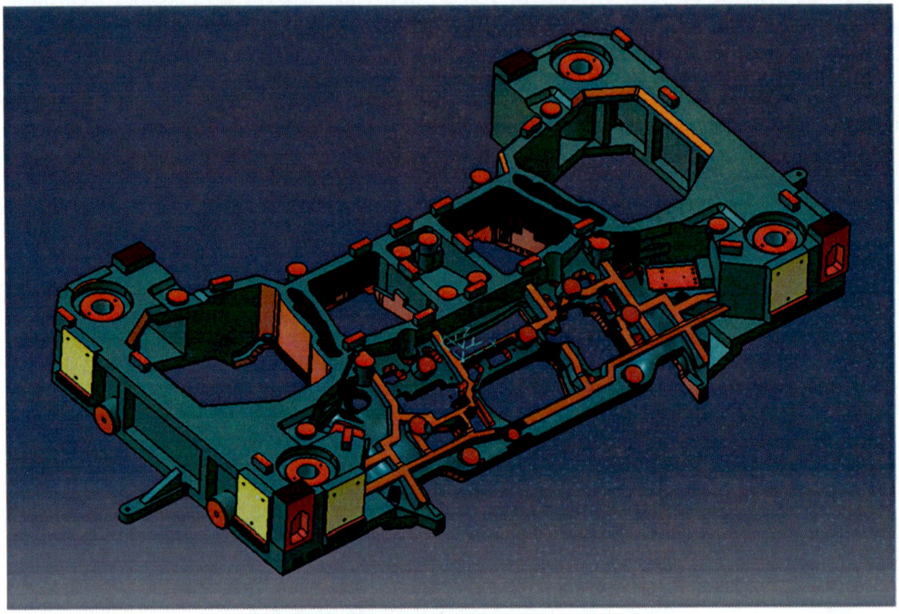

Fig. 4. Colored die design.

The red color in the die shown in Fig. 3 indicates 2D, dark red for rough machining, orange for controlled machining, purple for hole tapping, yellow for friction, and light green for precision surface machining. Inputs such as the type and dimensions of standard parts at the autonomous level, safety areas to check whether there is any crash and clearance with other standard parts, surface types, and processing areas are used for geometry and coloring. Cluster algorithms will be used based on these inputs. Surface coloring will be estimated, safety areas will be calculated, and errors made by designers will be prevented. Classification will be done based on the inputs, and production parameters will be predicted through regression using decision trees to determine where holes need to be created. Inputs such as the company where the design is prepared at the autonomous level, template type, and document type are included in the documentation. A clustering process will be performed based on these inputs, and the documents will be prepared and presented to the designer in a ready-to-use manner by scanning the product trees.

Die design must be inspected for compliance with customer specifications after it is completed. Designers are expected to create models that comply with these specifications. To ensure that the design meets the specifications, design control forms must be completed by the designer [6]. These forms contain a list of questions that are answered to verify the accuracy of the design. Typically, these forms are manually completed using an Excel template, which can lead to human error, data loss, incorrect data entry, and missed questions. To eliminate these problems, an application will be used to answer the questions and transfer the data directly to cloud systems for management. Automating this process will eliminate these risks and provide opportunities for improvement by keeping statistics on past quality errors.

Likewise, most of the metal stamping die design automation prototypes that have been reviewed are limited to specific application domains, or still require significant input from experienced designers to develop strip layouts, design die components, and model dies [3].

2.2 Design Data Management

The management system consists of two main panels. The first panel is where basic design activities that can be used by general users are carried out. The second panel is a place where only administrators can set limits and boundary conditions. In this center, called the Management Panel, users can add necessary rules and inputs to the interface where design activities can be carried out. For example, users can add specification documents required for design, add the company's standard template file for creating a material list, add parts to the standard parts library, add or remove design control items, and distribute all of these to users from a central location. To use the application on their computers, users will need to perform licensing procedures through the licensing interface. Information is collected from users through this panel for licensing procedures, and this information is stored in the Web Service database via the Web Service. Administrators will be able to view users from the management panel licensing interface. They can change users' usage times, cancel or reactivate licenses. The latest version of the application will be distributed to all users via the cloud system.

After the completion of documentation and modeling, the foundations of transitioning from design to production are established. In the program, data inputs are made through combo boxes with automatic lists for selection. For example, documents required for production such as material lists and model casting proposal forms are automatically filled based on design parameters. In addition, colorization and other processes are automatically performed to create enhanced designs. For instance, surfaces that require precise processing are colorized accordingly.

The bill of materials is created and the necessary parameters for production are transferred to the bill of materials through the production integration application module. The data organized here will be shaped according to the ERP program used by the customer, and the data collected from the Catia product tree will be processed into the ERP system. Then, the route and recipe of the product can be created through ERP programs.

In Document and Process Management, it has been identified that the most common problems that customers face are human errors and the inability to maintain document accuracy. All created documents and die data have been distributed with an authorization and versioning mechanism. The material list, revised die data, inspection checklists, operation cards, and CAM processing cards that will be created can be processed by authorized personnel who have the right to access and download them after the process is completed. The desired documents can be accessed through search functions, and reading, editing, and other operations can be carried out within the scope of authorization. By ensuring that the system always contains up-to-date data, errors that may arise from human errors have been prevented.

After the completion of the process, all data is transferred to the archive section for archiving. In the archive section, designs and processes can be filtered with various filters (such as part name, operation, etc.) and access to old data can be easily provided.

2.3 Design Production Management

In project-based single productions such as dies/fixtures/special machines, the process starts with design. The information required for production in MES-ERP within the scope of the project will be included in the product structure in the design and can be transferred to MES-ERP during the production phase through automated systems. In this sense, the software developed within the scope of the project can be named as Design Execution System (DES).

Integrated systems will be developed that consolidate decision support mechanisms containing algorithms that can optimize processes in real-time or provide support for related optimization. In die design, product trees will be production-compatible, starting from the smallest part and following the logic of assembly and sub-assembly relationships. With transfer software, they can be directly converted into ERP recipes and routes. Thus, a design-ERP-production-MES integrated system can be achieved.

When a new die request is received, part or process data is provided by the customer. As part of the project, similarity search will be performed using image processing methods in three-axis projection images. Thus, if there are similar projects that have been previously carried out, they can be predicted. In case this method fails, all die data and surface models will be kept in step format. The step format is a text-based open format, and many programs can import/export in step format. In this method, the incoming part or process data will be compared with the point clouds of the previous projects, and similar projects previously carried out can be predicted. Then, the production data, operations, and standard parts used in the previously carried out projects will be used as predicted data in the initial data of the new project.

The naming module we will develop will provide custom names for each part of the die in a project, and modules containing integration layers with automatic storage, WMS, and warehouse control will be developed. This way, the use of the correct parts without confusion will be ensured in die production.

3 Conclusion

Although there are some programs (such as Proleis, TeamCenter, etc.) whose contents are compatible, there is no software that fully meets the needs of project-based manufacturing worldwide, shortens the design phase, and integrates modules such as automatic storage, WMS, and warehouse control for customized part naming. Production management and design are addressed separately in these programs. However, there is no program that guides the designer to enter the necessary data for production and management and transfers the data to production. During the production process, applications such as surface creation and development using point clouds, generating CAM data and sending them to the machining workshop, standardizing delivery procedures, referring to measurement software, and supporting binding and parametric designs are implemented. However, planning and tracking are still done with traditional methods (such as excel, USB, manual forms, etc.). Programs used for production management are designed to collect data from production, analyze it, make decisions, and take action. As identifying and improving the deficiencies takes time, these programs fall behind the pace of the sector and cannot intervene in needs on time. The sector has not yet transitioned to intelligent systems in die design. There are no effective and efficient designs that ensure standardization in design.

In this work, we aim to ensure the correct, easy, and fast transfer of data to production, resolve integration problems of workflow, and execute data transfer with specific permissions for project-based manufacturing processes. Innovation processes that are not integrated lead to time losses, excess costs, and errors.

In order for digitalization to be fully achieved in the industry, digitalization efforts should start from the design phase. Design is a plan made for production, and design data needs to be prepared in a way that can be used as input for ERP. This can be achieved by creating parameters, identity cards, product trees, and RPAs in the design phase. Standardization of the design will reduce variability and ensure commonalities. For example, although the hood and handle sheet die of a car may differ significantly in size and shape, commonalities and similarities can be achieved in the product trees, recipes, and parameters that digital systems will use. With this commonality, end-to-end digital transformation can be achieved from project inception to equipment production completion. This study illustrates with examples how these objectives can be achieved in automotive die design, which is widely used in the industry.

References

1. Tisza, M., Lukacks, Z., Gal, G.G.: Integrated process simulation and die-design in sheet metal forming. In: International Deep-drawing Research Group IDDRG 2007 International Conference, pp. 1–10. Györ, Hungry (2007)
2. Bintas, M., Bintas. G., Kayitken, T., Kara, Y., Oz, C., Sari, U.: Development of intelligent computer aided design software for die design process. In: International Conference and Exhibition on Design and Production Machines and Dies/Molds, pp. 307–312, Atilim University, Antalya, Turkey (2013)
3. Naranje, V., Kumar, S.: AI applications to metal stamping die design–a review. Int. J. Mech. Ind. Aerosp. Sci. **3**(8), 721–727 (2010)
4. Marchenko, M., Behrens, B.A., Wrobel, G., Scheffler, R., Pleßow, M.: A new method of visualization and documentation of parametric information of 3D CAD models. Comput.-Aided Des. Appl. **8**(3), 435–448 (2011)

5. Bintas, M., Oz, C., Sari, U.: Die design coloring standard for manufacturing. In: International Conference and Exhibition on Design and Production Machines and Dies/Molds, pp. 313–316, Atilim University: Antalya, Turkey (2013)
6. Bintas, M.: Development of a computer aided die design software and die design process modelling. In: International Conference and Exhibition on Design and Production Machines and Dies/Molds, pp. 285–290, Atilim University (2011)

Detection of Cyber Attacks Targeting Autonomous Vehicles Using Machine Learning

Furkan Onur[1], Mehmet Ali Barışkan[1](✉), Serkan Gönen[1], Cemallettin Kubat[2], Mustafa Tunay[1], and Ercan Nurcan Yılmaz[3]

[1] Computer Engineering Department, Istanbul Gelisim University, İstanbul, Turkey
mabariskan@gelisim.edu.tr

[2] Aeronautical Engineering Department, Istanbul Gelisim University, İstanbul, Turkey

[3] Electrics and Electronics Engineering Department, Gazi University, İstanbul, Turkey

Abstract. The advent of Industry 4.0, characterized by the integration of digital technology into mechanical and electronic sectors, has led to the development of autonomous vehicles as a notable innovation. Despite their advanced driver assistance systems, these vehicles present potential security vulnerabilities, rendering them susceptible to cyberattacks. To address this, the study emphasized investigating these attack methodologies, underlining the need for robust safeguarding strategies for autonomous vehicles. Existing preventive or detection mechanisms encompass intrusion detection systems for Controller Area Networks and Vehicle-to-Vehicle communication, coupled with AI-driven attack identification. The critical role of artificial intelligence, specifically machine learning and deep learning subdomains, was emphasized, given their ability to dissect vehicular communications for attack detection. In this study, a mini autonomous vehicle served as the test environment, where the network was initially scanned, followed by the execution of Man-in-the-Middle, Deauthentication, DDoS, and Replay attacks. Network traffic was logged across all stages, enabling a comprehensive analysis of the attack impacts. Utilizing these recorded network packets, an AI system was trained to develop an attack detection mechanism. The resultant AI model was tested by transmitting new network packets, and its detection efficiency was subsequently evaluated. The study confirmed successful identification of the attacks, signifying the effectiveness of the AI-based model. Though the focus remained on autonomous vehicles, the study proposes that the derived methodology can be extended to other IoT systems, adhering to the steps delineated herein.

Keywords: IoT · Cyber Security · Machine Learning · IIoT

1 Introduction

The proliferation of Internet of Things (IoT) and Industrial Internet of Things (IIoT) has catalyzed substantial advancements in various sectors, including the automotive industry, exemplified by the emergence of autonomous vehicles. However, the cybernetic integration of IoT and IIoT systems within these vehicles elicits significant cybersecurity

Z. Şen et al. (Eds.): IMSS 2023, LNME, pp. 439–449, 2024.
https://doi.org/10.1007/978-981-99-6062-0_40

apprehensions. This study aims to scrutinize and counteract cyber threats—Man-in-the-Middle (MitM), Distributed Denial of Service (DDoS), Deauthentication (Deauth), and Replay attacks—impacting autonomous car networks, proposing a Gradient Boosting-oriented detection mechanism as a viable solution. With autonomous vehicles increasingly prone to cyberattacks due to their dependency on IoT and IIoT for data communication, processing, and decision-making, establishing rigorous security countermeasures is indispensable. The study underscores the utility of machine learning algorithms, namely Support Vector Machines (SVM), Random Forests (RF), and Neural Networks (NN), in the detection and mitigation of such threats. In particular, the superior efficacy of the Gradient Boosting algorithm in addressing these cyber threats within the IoT and IIoT landscape is demonstrated. This paper navigates through relevant literature, provides an overview of the targeted cyber-attacks and proposed detection mechanism, evaluates the performance of each algorithm, and concludes by encapsulating the findings, acknowledging study limitations, and suggesting future research directions.

2 Related Works

Within the framework of Industry 4.0, the domain of autonomous vehicles has sparked substantial interest. Despite the transformative potential of these vehicles, they remain susceptible to myriad cyber threats, necessitating focused research on their security. Recent studies have underscored the value of artificial intelligence (AI) in detecting and mitigating these threats. For instance, Kim et al. [1] analyzed the potential vulnerabilities and corresponding countermeasures in autonomous vehicles, advocating for enhanced anomaly detection via AI and machine learning. Nie et al. [2] demonstrated a successful remote attack on a Tesla Model S, exploiting its wireless connection to control the autonomous system.

Lee and Woo [3] proposed an insidious attack method, dubbed the CEDA, which stealthily attenuates CAN signals, thereby causing the targeted Electronic Control Unit (ECU) to ignore the received signals. Fowler et al. [4] deployed fuzz testing to identify security vulnerabilities in CAN prototypes, revealing software errors in ECU. Other studies, like those by Lim et al. [5] and Jakobsen et al. [6], focused on the vulnerabilities of the obstacle detection ultrasonic sensors and Lidar-camera sensor fusion, respectively.

Eriksson et al. [7] conducted an examination of in-vehicle Android Automotive application security utilizing static code analysis, while Cai et al. [8] spotlighted vulnerabilities in BMW's NBT Head Unit and Telematics Communication Box, emphasizing the necessity for all-encompassing security precautions. Zoppelt and Kolagari [9] investigated the prospect of cloud-based remote attacks on autonomous vehicles, deploying a Security Abstraction Model. Simultaneously, Maple et al. [10] introduced a hybrid model for attack surface analysis in connected autonomous vehicles, demonstrating its practical application through two use cases.

Other researchers like Miller and Valasek [11, 12] exploited a known Jeep Cherokee vulnerability to gain control of the vehicle, while Woo et al. [13] identified CAN as a security vulnerability and suggested a network address scrambling solution. Shrestha and Nam [14] proposed a regional block cipher for maintaining blockchain stability in VANETs, and Nasser and Ma [15] examined the Code Reuse security flaw, suggesting an HSM-based monitoring system. Zhang and Ma [16] put forth a hybrid IDS for high detection rates and minimal computational expense.

Subsequent research, including those by Zhou et al. [17], Olufowobi et al. [18], and Hamad et al. [19], concentrated on methods for ECU identification, message forgery attack detection, and intrusion response systems for autonomous vehicle networks. Song et al. [20] developed a deep convolutional neural network-based attack system for detecting malicious CAN traffic, while Tang et al. [21] reviewed machine learning techniques for future 6G vehicle networks.

Ahmad et al. [22] leveraged LSTM networks to mitigate relay attacks and verify driver identity, whereas Gundu and Maleki [23] improved Random Forest accuracy by incorporating time intervals. Kumar et al. [24] proposed BDEdge, a Blockchain and deep learning-based system for MEC server security. Alsulami et al. [25] developed a 99.95% accurate LSTM-based early detection system for False Data Injection, and Özgür [26] achieved a similar accuracy rate using Decision Analysis and Resolution.

While these studies have primarily centered around the communication systems of cars or core computer components like CAN and ECU, our research primarily focuses on the control systems such as gas, brake, and steering wheel, thereby offering a unique perspective on the security of autonomous vehicles.

3 Testing Infrastructure

3.1 Designed Autonomous System

The autonomous miniature vehicle, engineered utilizing Arduino, incorporates the ESP8266 NodeMCU module. This essential module enables bidirectional communication between the vehicle and peripheral computing devices such as desktops or mobile systems. Upon receiving distinct commands from these external interfaces, the vehicle, by leveraging the capabilities of the integrated communication module, initiates and executes the corresponding actions seamlessly (Fig. 1).

Fig. 1. Designed mini autonomous test vehicle.

The ESP8266 NodeMCU module establishes a localized network, wherein an external system can participate as a client, facilitating the transmission of HTTP GET commands. These command requests prompt interactions with the system, enabling the manual manipulation of the miniature autonomous test vehicle. For instance, such functionality can be deployed to remotely operate a stationary vehicle, initiating maneuvers such as exiting a parking space through a mobile device interface. This communication system's structural framework is depicted in Fig. 2.

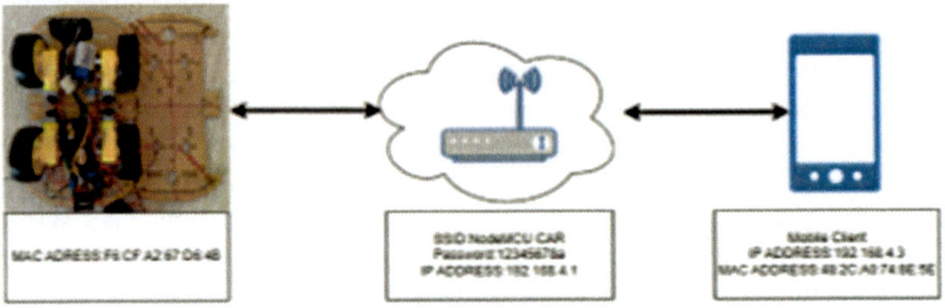

Fig. 2. Communication System

3.2 Preparation of Attack System

Deauthentication (Deauth), Denial-of-Service (DoS), Man-in-the-Middle (MitM), and Replay attacks were executed on the miniaturized autonomous test vehicle system utilized in this research. The assault methodologies were facilitated using network analysis and penetration tools including Nmap, hping3, airodump-ng, aireplay-ng, Ettercap, Wireshark, and Burp Suite. These tools were operated within the Kali Linux environment on the network topology described in Fig. 3. The precise attack workflow was depicted in Fig. 4.

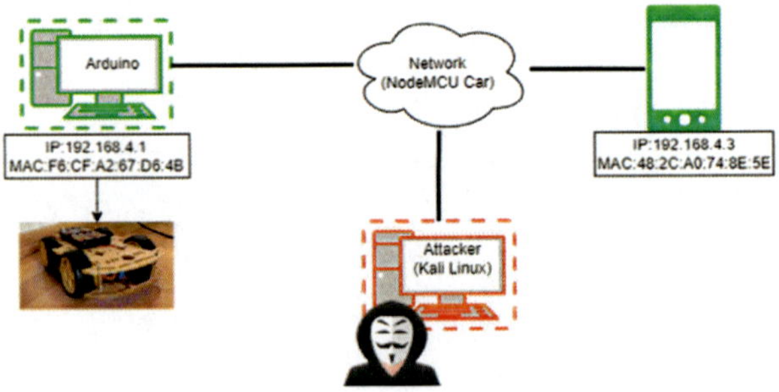

Fig. 3. Network Topology

When examining the general attack flowchart, it can be seen that the process was divided into five main stages. Specifically, the MitM and Replay attack procedures encompassed the first four phases: Discovery, Attack, Observation, and Repeated Attack. The final stage, Attack Detection via Artificial Intelligence, was a unique addition to this study, focusing on the automatic identification of both passive attacks (like MitM) and active attacks (like Replay attacks). This phase allows for early attack notification and immediate system responsiveness.

In the Discovery phase, an initial network scan was conducted to identify the target network, which was then subject to a specific scan. Subsequently, in the Attack phase, three distinct assaults were launched against the identified system: Deauth Attack, DoS, and MitM, as illustrated in the flowchart. The Observation phase followed, monitoring the impacted system to discern the consequences of the executed attacks. The Deauth Attack resulted in re-authentication, the DoS assault increased packet time intervals, and the MitM attack was discerned by observing duplicate packets via Wireshark. The successful modification of the victim device's ARP table by the attacker was also confirmed.

Subsequently, in the Repeated Attack phase, a Replay attack was enacted on the target system using the data gathered during the Observation stage. Finally, during the Detection phase, packet data obtained during the attack period was introduced to the machine learning system. This data encompassed packets from pre-attack, during attack, and post-attack stages, aiding in the detection and mitigation of future assaults.

4 Attack Analyses

4.1 Deauth Attack

A Deauthentication (Deauth) attack is a form of cyber-attack that disrupts network connectivity temporarily, thereby severing ongoing communications. The implications of this attack were evaluated in the context of a mini autonomous test vehicle system. The procedure encompassed several stages.

In the initial stage, the 'airodump-ng' command was utilized to detect all active networks in the proximity, subsequently providing comprehensive information regarding each one. The following stage involved identifying the specific target network for the attack - in this instance, the 'NodeMCU Car' network.

In the penultimate step, the 'aireplay-ng' command was employed to consistently transmit packets to the device tethered to the network until a threshold of 10,000 packets was reached, resulting in the device being ejected from the network.

This attack manifested in the disruption of communication between the smartphone and the autonomous vehicle, hindering real-time data transfer—a critical concern due to the halted flow of pertinent information regarding the autonomous vehicle. In both of the conducted tests, this disruption in communication was observed, thereby affirming the successful execution of the Deauth attack.

4.2 Denial of Service (DoS) Attack

A Denial of Service (DoS) attack is a form of cyber offensive aimed at overloading a computer system's resources, which consequently results in the denial of access services.

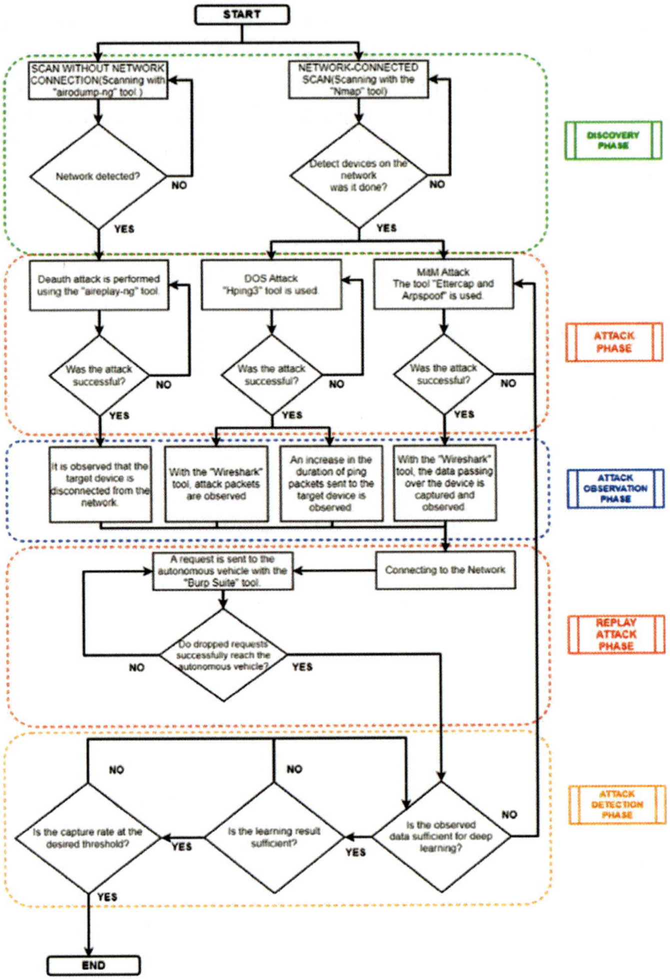

Fig. 4. Attack Flow Diagram

The ramifications of this attack form are assessed in the context of the designed mini autonomous test vehicle system.

4.3 Man-in-the-Middle (MitM) Attack

A Man-in-the-Middle (MitM) attack constitutes the interception, alteration, or manipulation of communication between two entities by an unauthorized third party. In wireless networks, packets are broadcast, enabling an attacker to capture all packets without necessitating preprocessing. A MitM attack on the developed mini autonomous car system is scrutinized herein. Initially, information concerning the target device was gleaned using the Nmap scan. As Wireshark initiated network listening, the network packets between the autonomous car and the phone weren't fully perceptible.

To gain access to all transmitted network packets, the Ettercap tool was utilized. Connected devices were identified via the Ettercap tool, and subsequently, an ARP poisoning attack was initiated by designating the phone as a target. Concurrently, all network packets became visible with the onset of the attack, facilitated by the Wireshark tool's listening function.

Upon manipulating one of the intercepted packets, a request was uncovered. Examination of the disclosed information ascertained that the data transmitted to the autonomous vehicle was both instantaneous and accurate.

4.4 Replay Attack

A Replay attack is a type of cyber offensive which involves obtaining unauthorized access or circumventing the authentication process by repetitively transmitting the same data employed in a preceding successful communication. This segment examines a Replay attack on the mini autonomous test vehicle system.

In order to modify the packets sent to the autonomous vehicle and transmit packets in the desired quantity and format, the Burp Suite tool was employed.

5 Detecting Attacks Through Artificial Intelligence Algorithms

In this section, the network traffic associated with the mini-autonomous test vehicle is subjected to various artificial intelligence algorithms to facilitate attack detection, as depicted in Fig. 14. The attack detection model employed in this study encompasses four stages. Initially, data amassed over the network is processed, and the dataset file, subsequent to the preprocessing stage, is integrated into the model. Subsequently, the prepared dataset is partitioned into 70% training and 30% validation data, and subsequently subjected to analysis using Neural Network (NN)-ReLU, kNN, Random Forest, Gradient Boosting, SVM, and Stochastic Gradient Descent artificial intelligence algorithms. The third stage involves visualization of the data procured from the artificial intelligence algorithms to enhance analysis. Finally, after evaluation, Gradient Boosting is selected as the artificial intelligence algorithm for attack detection, on account of its superior accuracy, F1, recall, and time values across all attacks, and preserved for application to real-time data.

5.1 Gradient Boosting

Gradient Boosting is a renowned machine-learning algorithm, which is utilized to tackle classification and regression problems. This algorithm represents an ensemble learning method, which amalgamates multiple weak learners into a singular strong learner. The fundamental concept of gradient boosting entails the construction of an ensemble of decision trees, where each subsequent decision tree aligns with the residual errors of the preceding tree.

The algorithm initiates with a simplistic decision tree tailored to the data. Subsequently, the model's residuals are computed, and a new tree aligns with these residuals. This process is iterated multiple times, with each new tree aligning with the residuals of the preceding trees. The final prediction is garnered by aggregating the predictions of all trees within the ensemble.

Gradient Boosting provides several advantages over alternative machine learning algorithms. It exhibits particular efficacy when handling high-dimensional data and can accommodate both numerical and categorical data. It also displays relative resistance to overfitting, which may pose an issue for other algorithms. Furthermore, Gradient Boosting is highly adaptable and compatible with numerous loss functions, rendering it a versatile algorithm for a myriad of problem types.

5.2 Model Creation and Training

Prior to the application of artificial intelligence algorithms, network packets underwent scrutiny. The dataset was divided, with 70% allocated to training and 30% to testing. Following the training of the model, results were evaluated based on various performance metrics such as training time, testing time, AUC, CA, F1, Precision, and Recall, as presented in Table 1.

Table 1. Model Comparison

Model	Train Time[s]	Test Time [s]	AUC	CA	F1	Precision	Recall
Gradient Boosting	332.383	1.404	0.997	0.987	0.987	0.987	0.987
Random Forest	18.112	0.979	0.986	0.981	0.981	0.981	0.981
Neural Network	382.897	0.9696	0.989	0.976	0.975	0.975	0.976
SGD	12.917	8.857	0.917	0.970	0.969	0.969	0.970
kNN	6.863	89.698	0.702	0.824	0.788	0.762	0.824
SVM	255.419	4.923	0.217	0.217	0.177	0.881	0.217

According to the time interval specified in Fig. 5 (X-axis), the network traffic of source hosts (Y-axis) was appraised. The red network packets, indicative of the packets dispatched by the attacker models, were tracked, leading to the successful visual detection of the attacker models in the reference model devoid of an attacker. The red packets marked as attack packets were attributed to the impact of the defined feature on the model.

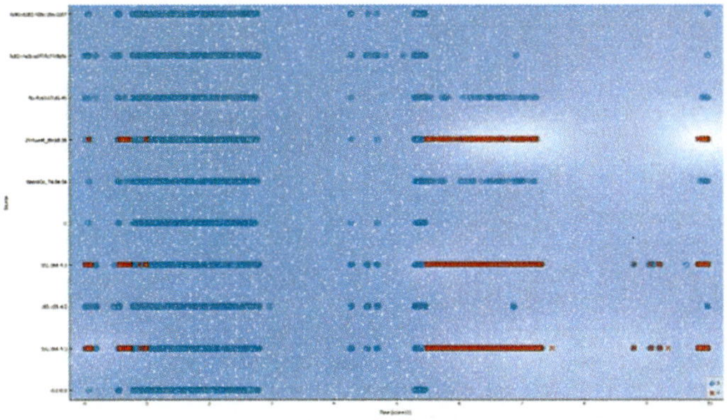

Fig. 5. Detection of attackers using the artificial intelligence model.

6 Discussion

In this research, we underscored autonomous vehicles' susceptibility to cyber-attacks, stressing the need for potent security measures. Utilizing machine learning algorithms, we presented a Gradient Boosting-based detection procedure, exhibiting remarkable performance in mitigating threats, with an accuracy of 99.7% and a precision score of 0.987. The study's outcomes highlight Gradient Boosting's potential in tackling IoT and IIoT cyber threats. As autonomous vehicles assume critical tasks, the importance of their security amplifies. This research contributes valuable insights towards developing robust countermeasures and underscores the necessity for continued exploration in autonomous vehicle cybersecurity to safeguard these systems and the interconnected environment.

7 Conclusion

To conclude, this investigation illustrates the efficacious application of a Gradient Boosting methodology for tackling cyber threats in autonomous vehicle networks. We exhibited the method's superiority in counteracting attacks such as MitM, DDoS, Deauth, and Replay through the utilization of diverse machine learning algorithms. Our results underline the essentiality of machine learning for autonomous vehicle security within the IoT and IIoT infrastructure. Continued research is imperative for the advancement of countermeasures and the adaptation to the progressively complex cyber threat landscape in autonomous vehicle technology.

References

1. Kim, K., Kim, J.S., Jeong, S., Park, J.H., Kim, H.K.: Cybersecurity for autonomous vehicles: review of attacks and defense. Comput. Secur. **103**, 102150 (2021)
2. Nie, S., Liu, L., Du, Y.: Free-fall: hacking tesla from wireless to can bus. Briefing, Black Hat USA **25**, 1–16 (2017)

3. Lee, Y., Woo, S.: CAN Signal Extinction-based DoS Attack on In-Vehicle Network. Secur. Commun. Netw. 2022 (2022)
4. Fowler, D.S., Bryans, J., Cheah, M., Wooderson, P., Shaikh, S.A.: A method for constructing automotive cybersecurity tests, a CAN fuzz testing example. In: 2019 IEEE 19th International Conference on Software Quality, Reliability and Security Companion (QRS-C), pp. 1–8. IEEE (2019)
5. Lim, B.S., Keoh, S.L., Thing, V.L.: Autonomous vehicle ultrasonic sensor vulnerability and impact assessment. In: 2018 IEEE 4th World Forum on Internet of Things (WF-IoT) (pp. 231–236). IEEE (2018)
6. Jakobsen, S.B., Knudsen, K.S., Andersen, B.: Analysis of sensor attacks against autonomous vehicles. In: 25th International Symposium on Wireless Personal Multimedia Communications. IEEE (2022)
7. Eriksson, B., Groth, J., Sabelfeld, A.: On the road with third-party apps: security analysis of an in-vehicle app platform. In: VEHITS, pp. 64–75 (2019)
8. Cai, Z., Wang, A., Zhang, W., Gruffke, M., Schweppe, H.: 0-days & mitigations: roadways to exploit and secure connected BMW cars. Black Hat USA **2019**, 39 (2019)
9. Zoppelt, M., Kolagari, R.T.: UnCle SAM: modeling cloud attacks with the automotive security abstraction model. Cloud Comput. 67–72 (2019)
10. Maple, C., Bradbury, M., Le, A.T., Ghirardello, K.: A connected and autonomous vehicle reference architecture for attack surface analysis. Appl. Sci. **9**(23), 5101 (2019)
11. Miller, C., Valasek, C.:. Remote exploitation of an unaltered passenger vehicle. Black Hat USA, **2015**(S 91), 1–91 (2015)
12. Miller, C.: Lessons learned from hacking a car. IEEE Design & Test **36**(6), 7–9 (2019)
13. Woo, S., Moon, D., Youn, T.Y., Lee, Y., Kim, Y.: Can id shuffling technique (cist): moving target defense strategy for protecting in-vehicle can. IEEE Access **7**, 15521–15536 (2019)
14. Shrestha, R., Nam, S.Y.: Regional blockchain for vehicular networks to prevent 51% attacks. IEEE Access **7**, 95033–95045 (2019)
15. Nasser, A., Ma, D.: Defending AUTOSAR safety critical systems against code reuse attacks. In: Proceedings of the ACM Workshop on Automotive Cybersecurity, pp. 15–18 (2019)
16. Zhang, L., Ma, D.: A hybrid approach toward efficient and accurate intrusion detection for in-vehicle networks. IEEE Access **10**, 10852–10866 (2022)
17. Zhou, J., Joshi, P., Zeng, H., Li, R.: Btmonitor: bit-time-based intrusion detection and attacker identification in controller area network. ACM Trans. Embed. Comput. Syst. (TECS) **18**(6), 1–23 (2019)
18. Olufowobi, H., Hounsinou, S., Bloom, G.: Controller area network intrusion prevention system leveraging fault recovery. In: Proceedings of the ACM Workshop on Cyber-Physical Systems Security & Privacy, pp. 63–73 (2019)
19. Hamad, M., Tsantekidis, M., Prevelakis, V.: Red-Zone: Towards an intrusion response framework for intra-vehicle system. In: VEHITS, pp. 148–158 (2019)
20. Song, H.M., Woo, J., Kim, H.K.: In-vehicle network intrusion detection using deep convolutional neural network. Veh. Commun. **21**, 100198 (2020)
21. Tang, F., Kawamoto, Y., Kato, N., Liu, J.: Future intelligent and secure vehicular network toward 6G: Machine-learning approaches. Proc. IEEE **108**(2), 292–307 (2019)
22. Ahmad, U., Song, H., Bilal, A., Alazab, M., Jolfaei, A.: Securing smart vehicles from relay attacks using machine learning. J. Supercomput. **76**, 2665–2682 (2020)
23. Gundu, R., Maleki, M.: Securing CAN bus in connected and autonomous vehicles using supervised machine learning approaches. In: 2022 IEEE International Conference on Electro Information Technology (eIT), pp. 042–046. IEEE (2022)
24. Kumar, P., Kumar, R., Gupta, G.P., Tripathi, R.: BDEdge: blockchain and deep-learning for secure edge-envisioned green CAVs. IEEE Trans. Green Commun. Netw. **6**(3), 1330–1339 (2022)

25. Alsulami, A.A., Abu Al-Haija, Q., Alqahtani, A., Alsini, R.: Symmetrical simulation scheme for anomaly detection in autonomous vehicles based on LSTM model. Symmetry **14**(7), 1450 (2022)
26. Özgür, A.: Classifier selection in resource limited hardware: decision analysis and resolution approach. J. Intell. Syst. Theory Appl. **4**(1), 37–42 (2021). https://doi.org/10.38016/jista.755419

Detection of Man-in-the-Middle Attack Through Artificial Intelligence Algorithm

Ahmet Nail Taştan[1], Serkan Gönen[2], Mehmet Ali Barışkan[1](✉), Cemallettin Kubat[3], Derya Yıltaş Kaplan[4], and Elham Pashaei[2]

[1] Computer Engineering Department, Istanbul Gelisim University, İstanbul, Turkey
mabariskan@gelisim.edu.tr
[2] Software Engineering Department, Istanbul Gelisim University, İstanbul, Turkey
[3] Aeronautical Engineering Department, Istanbul Gelisim University, İstanbul, Turkey
[4] Computer Engineering Department, Istanbul University-Cerrahpaşa, İstanbul, Turkey

Abstract. The amalgamation of information technologies and progressive wireless communication systems has profoundly impacted various facets of everyday life, encompassing communication mediums, occupational procedures, and living standards. This evolution, combined with enhanced wireless communication quality, has culminated in an exponential rise in interconnected devices, including domestic appliances, thereby birthing the Internet of Things (IoT) era. This proliferation, facilitated by cloud computing enabling remote device control, concurrently intensifies cybersecurity threats. Traditional Information and Communication Technology (ICT) architectures, characterized by a hub-and-spoke model, are inherently vulnerable to illicit access and Man-in-the-Middle (MITM) intrusions, thereby endangering information confidentiality. Leveraging Artificial Intelligence (AI) can ameliorate this scenario, enhancing threat training and detection capabilities, enabling precise and preemptive attack countermeasures. This research underscores the criticality of addressing the security implications accompanying technological advancements and implementing protective measures. Deploying AI algorithms facilitates efficient passive attack identification and alleviates network device burdens. Specifically, this study scrutinized the ramifications of an MITM attack on the system, emphasizing the detection of this elusive threat using AI. Our findings attest to AI's efficacy in detecting MITM attacks, promising significant contributions to future cybersecurity research.

Keywords: Man-in-the-Middle Attack · Cyber Security · Attack Detection · Artificial Intelligence

1 Introduction

The escalating proliferation and assimilation of the Internet of Things (IoT) and the Industrial Internet of Things (IIoT) technologies have engendered a radical transformation in contemporary industrial domains. These integrated systems, gadgets, and

Z. Şen et al. (Eds.): IMSS 2023, LNME, pp. 450–458, 2024.
https://doi.org/10.1007/978-981-99-6062-0_41

sensors proffer an array of advantages including amplified productivity, reduced operational expenditure, and enhanced decision-making processes. These innovative technologies have engendered novel business paradigms, revolutionized manufacturing procedures, and redefined human-machine interfaces. The increasing ubiquity of IoT and IIoT devices in critical sectors such as healthcare, transportation, energy, and manufacturing underscores their significance in our progressively interconnected world.

This extensive connectivity, however, escalates concerns regarding cybersecurity, particularly in relation to Man-in-the-Middle (MitM) attacks, posing substantial threats to individuals, organizations, and critical infrastructures. MitM attacks on IoT and IIoT systems present a broad spectrum of risks, encompassing data theft, unauthorized access to sensitive information, disruption of critical services, intellectual property loss, and potential physical damage in scenarios where compromised devices control safety-critical systems.

With IoT and IIoT environments becoming increasingly prevalent, it is imperative to formulate robust and efficacious methodologies to detect and mitigate such attacks, thereby ensuring these networks' security and reliability. In this context, we propose a comprehensive detection mechanism employing a multitude of machine learning algorithms to identify and counter MitM attacks within IoT and IIoT ecosystems.

Our evaluation of diverse machine learning algorithms, including logistic regression, support vector machines, k-nearest neighbors, and decision trees, revealed the superior efficacy of the Random Forest algorithm in detecting MitM attacks. The algorithm achieved an accuracy rate of 99.2%, an F1 score of 0.976, and a precision score of 0.977.

This research emphasizes the Random Forest algorithm's superior performance, thereby contributing to the broader effort of enhancing cybersecurity within IoT and IIoT environments. Additionally, we proffer insightful recommendations for both practitioners and researchers on employing this algorithm as a crucial facet of a robust and effective cybersecurity strategy amidst evolving threats and vulnerabilities. As IoT and IIoT technologies continue to advance, prioritizing the development and deployment of innovative solutions to safeguard these systems against the omnipresent dangers of MitM attacks is of paramount importance.

2 Literature Review

The critical triad of network security systems encompasses confidentiality, integrity, and availability (CIA) [1]. Other fundamental security principles include Authentication, Authorization, and Accounting (Triple-A) [2]. Confidentiality dictates that only authorized individuals can gain access to the system or information; integrity guarantees that the information remains intact and unaltered, and availability ensures data can be accessed by users when needed. As Solms V. et al. [3] suggests, cyber security has expanded beyond traditional information security to include protection of data resources and other associated assets.

In the digital realm where computer networks communicate, known as cyberspace [4], studies predominantly focus on detecting passive and active attacks through artificial intelligence algorithms on existing data sets. For instance, Otily Toutsop et al. [5]

analyzed the increased usage and interconnectedness of IoT devices, which inherently invites cyber threats. By applying Random Forest, Logistic Regression, and Decision Tree algorithms to attack data from the Hacking and Countermeasure Research Lab, they developed an intrusion detection system.

Pascal Moniriho et al. [6] devised an anomaly-based approach for IoT networks, which utilizes the Random Forest algorithm to categorize network traffic as normal or abnormal. The approach was evaluated using the IoTID20 dataset. Similarly, Robert A. Sowah et al. [7] used Artificial Neural Network (ANN) methods to identify intrusions in mobile ad hoc networks (MANETs).

Jayapandian Natarajan [8] proposed a reinforced learning-trained virtual agent to forecast cyber attacks in network transmissions, thereby enhancing the network communication security system. Also employing machine learning models, K. V. V. N. L Sai Kiran et al. [9] applied classifiers such as Naïve Bayes, SVM, decision tree, and Adaboost to recognize attacks in IoT networks.

Several other researchers have also proposed different models to tackle the pervasive issue of Man-in-the-Middle (MitM) attacks. JJ Kang et al. [10] devised a scheme for detecting such attacks with minimal interference and resources, utilizing a trusted time server and a learning-based algorithm to identify real anomalies in transmission time. Hitesh Mohapatra et al. [11] proposed a hybrid system employing machine learning for the detection and prevention of MitM intrusions, thereby creating a robust, attack-resistant intrusion detection system. In a similar vein, Anass Sebbar et al. [12] proposed a machine-learning technique to thwart MitM attempts.

On the other hand, researchers like Shikha Malik et al. [13], who addressed the dual aspects of IoT positives and cyber security threats for IoT devices, proposed a supervised machine learning model for IoT Security systems. Abebe Diro et al. [14] suggested a Long Short-Term Memory (LSTM)-based machine learning model to detect and counteract distributed cyber intrusion in fog-to-object communication in IoT.

From a comparative standpoint, Iqbal H. Sarker et al. [15] surveyed various machine and deep learning algorithms for IoT security, aiming to intelligently safeguard IoT devices against cyber attacks. In an innovative approach, Zhuo Ma et al. [16] proposed an artificial intelligence system that compares and selects successful protocols from a pool of over 2000, leveraging multi-criteria decision-making to increase success rates by 76%.

Despite such progress, Yang Li et al. [17] pointed out that existing research on MitM attacks fails to consider the characteristics of Communications-Based Train Control (CBTC) systems, particularly the limited computing capacity of onboard computers. Echoing this need for specialized solutions, Muhanna Saed et al. [18] used machine learning algorithms to detect and prevent MitM attacks, while Anarelli et al. [19] emphasized machine learning as a vital component for enhancing the resilience of cyber-physical systems.

Blockchain technology has also been utilized to tackle this issue, with Jinchun Choi et al. [20] proposing a blockchain-based MitM detection system to scan networks for anomalies, specifically within energy production and transfer systems. Patrick Wlazlo et al. [21] carried out focused research on smart grids, studying contingencies of attacks such as MitM. In contrast, Lv et al. [22] utilized a Deep Convolutional Generative

Adversarial Network (DCGAN) based fuzzy logic algorithm to generate fake data to boost the efficiency of their machine learning algorithm.

This study differs from the existing literature in its data collection methodology. The dataset is collected in real-time from a live system, simultaneously with real-time monitoring. This allows the system to detect attacks instantaneously.

3 Experimental Test Environment and Attack Analysis

3.1 Man in the Middle Attack

A Man-in-the-Middle (MitM) attack represents a cyber threat, where an adversary intercepts or alters communications within a public network. Succinctly, this attack involves interception and manipulation of data packets within the network. The network topology of the studied test environment is depicted below (Fig. 1).

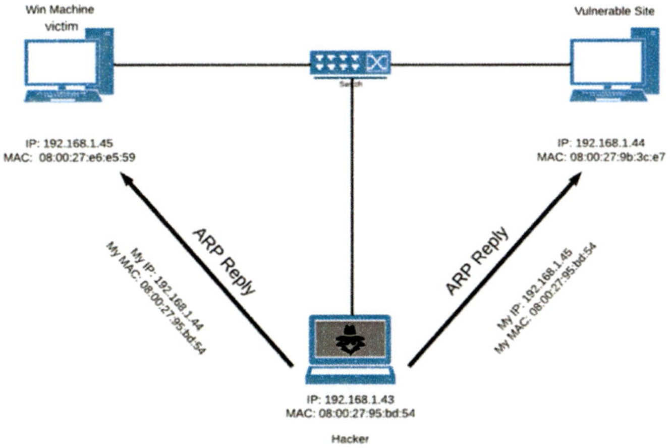

Fig. 1. Network topology

3.2 Experiment

Our study targets detection of Man-in-the-Middle (MitM) attacks through a deep learning algorithm, emphasizing feature selection on training and test packets. Attributes, including time, IP addresses, port addresses, protocol type (UDP, TCP, or Ping), and packet size are crucial, given their relevance in new attack methodologies. ARP poisoning, packet flags, segmentations, duplicate packets, and increased Round Trip Time (RTT) due to MitM attacks are also vital factors considered in the algorithm.

4 Attack Detection with Artificial Intelligence

This section presents a comparative evaluation of various artificial intelligence (AI) algorithms excluding the widely-employed Long Short-Term Memory (LSTM) algorithm, in the context of intruder detection. The proposed AI-centric intruder detection model comprises four stages. Initially, collected network data is preprocessed to form a suitable dataset for training AI algorithms. Subsequently, this dataset, divided into 85% training and 15% validation data, undergoes analysis via AI algorithms such as Neural Network (NN)-ReLU, Logistic Regression, Random Forest, Gradient Boosting, and Naive Bayes. Thirdly, graphical tools like Scatter Plot and Violin Plot facilitate enhanced comprehension of insights from the AI algorithms. The final stage involves performance evaluation of these algorithms, with the Random Forest algorithm emerging as the most efficacious in terms of accuracy, F1 score, Recall, and computational efficiency. A confusion matrix is constructed to further assess the model's performance, by juxtaposing predicted and actual target attribute values (Fig. 2).

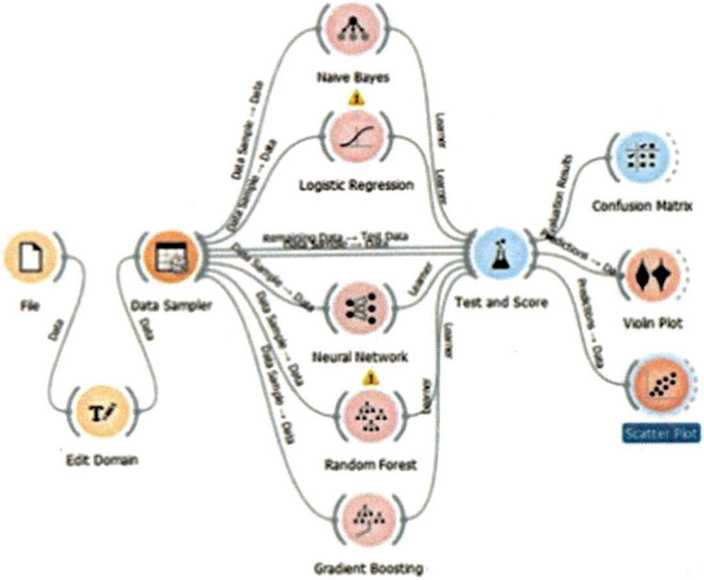

Fig. 2. Intrusion Detection with Artificial Intelligence Algorithms

4.1 Random Forest

Our goal in this section is to provide a concise but mathematically precise representation of the random forest generation algorithm. The general framework is non-parametric regression estimation, in which the input random vector is observed and the goal is $m(\mathbf{x}) = \mathbb{E}[Y \mid \mathbf{X} = \mathbf{x}]$ to estimate the square integratable random response by estimating the regression function $Y \in \mathbb{R}$. For this purpose, we assume that we are given a training example of $\mathcal{D}_n = ((\mathbf{X}_1, Y_1), \ldots, (\mathbf{X}_n, Y_n))$ independent random variables distributed as a pair of independent prototypes (X, Y). The goal is to use the dataset to generate the estimation $\mathcal{D}_n$ of the function m $m_n : \mathcal{X} \to \mathbb{R}$. In $\mathbb{E}[m_n(\mathbf{X}) - m(\mathbf{X})]^2 \to 0$ this context, we say that the regression function estimate mn (mean squared error) is consistent (evaluated by expectation X and its sample) $\mathcal{D}_n$.

The random forest is an estimator consisting of a collection of M random regression trees. J in the family For the tree, x is represented by the predicted value at the query point $m_n(\mathbf{x}; \Theta_j, \mathcal{D}_n)$, where $\Theta_1, \ldots, \Theta_M$ they are independent random variables, Θ distributed in the same way as a global random variable, and $\mathcal{D}_n$ independent of. In practice Θ the variable is used to resample the training set prior to the growth of individual trees and to select successive directions for splitting - more precise definitions will be given later. In mathematical terms, j. tree prediction takes the form

$$m_n(\mathbf{x}; \Theta_j, \mathcal{D}_n) = \sum_{i \in \mathcal{D}_n^*(\Theta_j)} \frac{\mathbb{1}_{\mathbf{X}_i \in A_n(\mathbf{x}; \Theta_j, \mathcal{D}_n)} Y_i}{N_n(\mathbf{x}; \Theta_j, \mathcal{D}_n)} \tag{1}$$

Here is $\mathcal{D}_n^*(\Theta_j)$ the set of data points selected before tree construction, $A_n(\mathbf{x}; \Theta_j, \mathcal{D}_n)$ the cell containing $N_n(\mathbf{x}; \Theta_j, \mathcal{D}_n)$ x and $A_n(\mathbf{x}; \Theta_j, \mathcal{D}_n)$ is the number of (pre-selected) points falling into it. We note that at this stage the trees are combined to form the (finite) forest estimate.

$$m_{M,n}(\mathbf{x}; \Theta_1, \ldots, \Theta_M, \mathcal{D}_n) = \frac{1}{M} \sum\nolimits_{j=1}^{M} m_n(\mathbf{x}; \Theta_j, \mathcal{D}_n). \tag{2}$$

The default value of M (number of trees in the forest) in the R package randomForest is ntree = 500. Since M can be chosen arbitrarily large (limited only by available computing resources), from a modeling point of view it makes sense to let M go to infinity and consider the (infinite) forest estimate instead of (2).

In this definition $\mathbb{E}_\Theta$, $\mathcal{D}_n$ it expresses the expectation according to the random parameter depending on the condition. In fact, the operation "$M \to \infty$" is justified by the law of large numbers, which asserts that it is almost certain depending on its condition.

Examining the attack detection success rates in Table 1, it can be seen that the highest accuracy rate was achieved with the Random Forest deep learning model, with a 99.2% success rate. Furthermore, it is observed that these high accuracy rates are supported by F1, precision, and recall values. Due to the high accuracy rate detected with the random forest deep learning model, the study continued with the Random Forest deep learning model as the chosen model. The results of the obtained values were successfully verified through continuous monitoring of network traffic using the Foren6 forensic analysis application.

The performance of the Random Forest model over time is illustrated in Fig. 3. Upon analyzing the results, it was observed that attackers were intercepting legitimate packets and duplicating them on themselves. In Fig. 3, the network traffic of source hosts (y-axis) is evaluated according to the specified time range (x-axis). The red-marked network packets are duplicated packets sent by the attacker. By tracking the red packets, the attacker attackers have been visually detected successfully.

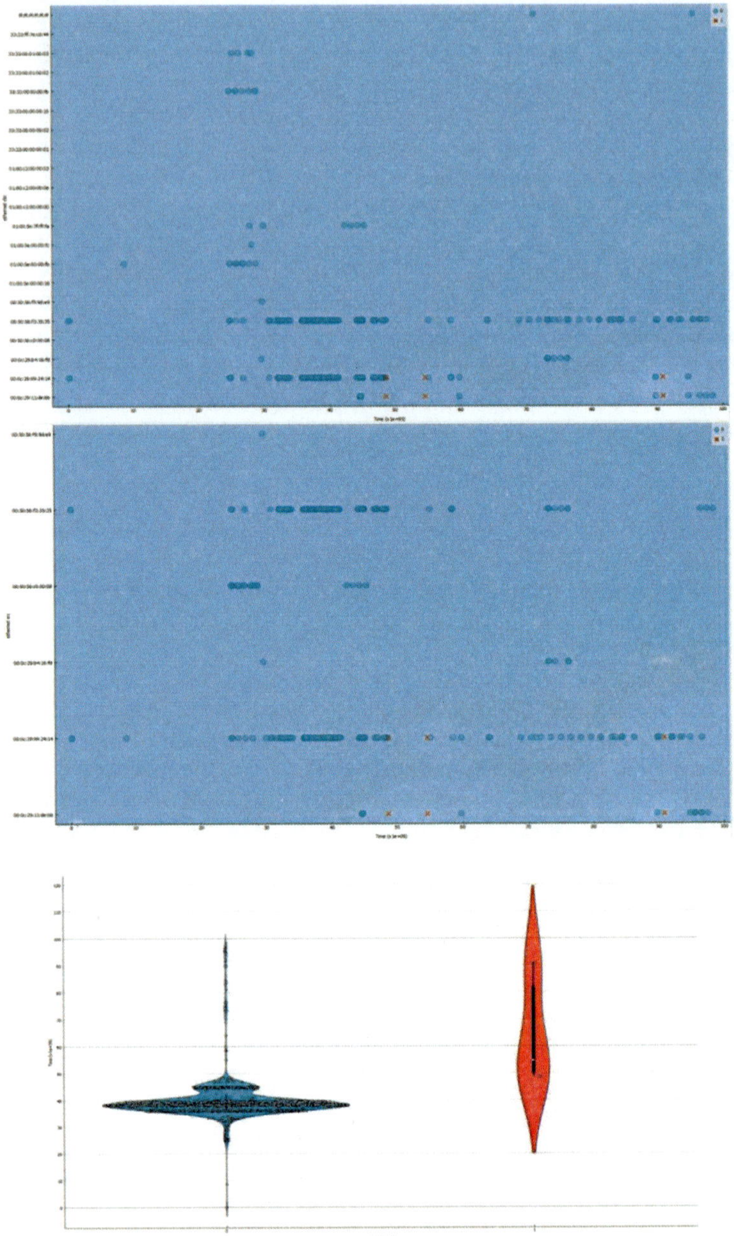

Fig. 3. Detection of attackers with artificial intelligence model

5 Discussion

The primary intent of this study was to devise a comprehensive detection procedure utilizing machine learning algorithms for the identification and mitigation of Man-in-the-Middle (MitM) attacks within IoT and IIoT environments. The evaluation of various machine learning algorithms concluded with the Random Forest algorithm demonstrating superior performance, achieving an accuracy of 99.2%. This can be attributed to its robustness against overfitting, ability to handle large datasets, and capacity for feature selection. Despite its promise, it is imperative to acknowledge that no single machine-learning technique offers total security against evolving MitM attacks. Consequently, future research should aim to corroborate these findings across varied datasets and conditions, while exploring a multi-pronged approach to enhance IoT and IIoT system security.

6 Conclusion and Recommendations

In conclusion, this research validates the preeminence of the Random Forest algorithm in detecting Man-in-the-Middle (MitM) attacks within IoT and IIoT systems, exhibiting an accuracy of 99.2%, an F1 score of 0.976, and a precision score of 0.977. The study recommends prioritizing its implementation within IoT and IIoT security mechanisms, necessitating continuous updates to confront emerging threats. Furthermore, the promotion of inter-disciplinary collaborations to devise innovative solutions for safeguarding against diverse cyber attacks is suggested. Moreover, organizations should accentuate employee security training and regular network infrastructure assessments. As IoT and IIoT technology progresses, prioritizing the development of effective security solutions becomes indispensable in protecting against MitM attacks and other cyber threats.

References

1. Simmonds, A., Sandilands, P., van Ekert, L.: An ontology for network security attacks. In: Manandhar, S., Austin, J., Desai, U., Oyanagi, Y., Talukder, A.K. (eds.) AACC 2004. LNCS, vol. 3285, pp. 317–323. Springer, Heidelberg (2004). https://doi.org/10.1007/978-3-540-30176-9_41
2. Stallings, W., Brown, L.: Computer Security Principles and Practice. Second penyunt (2012)
3. Von Solms, R., Van Niekerk, J.: From information security to cyber security. Comput. Secur. **38**, 97–102 (2013)
4. Boyd, B.L.: Cyber warfare: armageddon in a teacup? Army Command and General Staff College, Fort Leavenworth, KS (2009)
5. Toutsop, O., Harvey, P., Kornegay, K.: Monitoring and detection time optimization of man in the middle attacks using machine learning. In: 2020 IEEE Applied Imagery Pattern Recognition Workshop (AIPR). IEEE (2020)
6. Maniriho, P., et al.: Anomaly-based intrusion detection approach for IoT networks using machine learning. In: 2020 International Conference on Computer Engineering, Network, and Intelligent Multimedia (CENIM). IEEE (2020)
7. Sowah, R.A., et al.: Detection and prevention of man-in-the-middle spoofing attacks in MANETs using predictive techniques in Artificial Neural Networks (ANN). J. Comput. Netw. Commun. **2019**, 4683982 (2019)

8. Natarajan, J.: Cyber secure man-in-the-middle attack intrusion detection using machine learning algorithms. In: AI and Big Data's Potential for Disruptive Innovation, pp. 291–316. IGI Global (2020)
9. Kiran, K.S., et al.: Building an intrusion detection system for IoT environment using machine learning techniques. Procedia Comput. Sci. **171**, 2372–2379 (2020)
10. Kang, J.J., Fahd, K., Venkatraman, S.: Trusted time-based verification model for automatic man-in-the-middle attack detection in cybersecurity. Cryptography **2**(4), 38 (2018)
11. Mohapatra, H., et al.: Handling of a man-in-the-middle attack in WSN through intrusion detection system. Int. J. **8**(5), 1503–1510 (2020)
12. Sebbar, A., Karim, Z.K.I.K., Baddi, Y., Boulmalf, M., Kettani, M.-C.: MitM detection and defense mechanism CBNA-RF based on machine learning for large-scale SDN context. J. Ambient Intell. Human. Comput. **11**(12), 5875–5894 (2020). https://doi.org/10.1007/s12652-020-02099-4
13. Malik, S., Chauhan, R.: Securing the Internet of Things using machine learning: a review. In: 2020 International Conference on Convergence to Digital World-Quo Vadis (ICCDW). IEEE (2020)
14. Diro, A., Chilamkurti, N.: Leveraging LSTM networks for attack detection in fog-to-things communications. IEEE Commun. Mag. **56**(9), 124–130 (2018)
15. Sarker, I.H., et al.: Internet of Things (IoT) security intelligence: a comprehensive overview, machine learning solutions, and research directions. Mob. Netw. Appl. **28**, 296–312 (2023). https://doi.org/10.1007/s11036-022-01937-3
16. Ma, Z., Liu, Y., Wang, Z., Ge, H., Zhao, M.: A machine learning-based scheme for the security analysis of authentication and key agreement protocols. Neural Comput. Appl. **32**(22), 16819–16831 (2020). https://doi.org/10.1007/s00521-018-3929-8
17. Li, Y., et al.: A cross-layer defense scheme for edge intelligence-enabled CBTC systems against MitM attacks. IEEE Trans. Intell. Transp. Syst. **22**(4), 2286–2298 (2020)
18. Saed, M., Aljuhani, A.: Detection of man in the middle attack using machine learning. In: 2022 2nd International Conference on Computing and Information Technology (ICCIT). IEEE (2022)
19. Annarelli, A., Nonino, F., Palombi, G.: Understanding the management of cyber resilient systems. Comput. Ind. Eng. **149** (2020). https://doi.org/10.1016/j.cie.2020.106829
20. Choi, J., et al.: Blockchain-based man-in-the-middle (MITM) attack detection for photovoltaic systems. In: 2021 IEEE Design Methodologies Conference (DMC). IEEE (2021)
21. Wlazlo, P., et al.: Man-in-the-middle attacks and defense in a power system cyber-physical testbed. arXiv preprint arXiv:2102.11455 (2021)
22. Lv, W., Xiong, J., Shi, J., et al.: A deep convolution generative adversarial networks based fuzzing framework for industry control protocols. J. Intell. Manuf. **32**, 441–457 (2021). https://doi.org/10.1007/s10845-020-01584-z

A Novel Approach for RPL Based One and Multi-attacker Flood Attack Analysis

Serkan Gonen(✉)

Software Engineering Department, Istanbul Gelisim University, Istanbul, Turkey
sgonen@gelisim.edu.tr

Abstract. The Internet of Things (IoT) encompasses a vast network of interconnected devices, vehicles, appliances, and other items with embedded electronics, software, sensors, and connectivity, allowing them to collect and exchange data. However, the growing number of connected devices raises concerns about IoT cybersecurity. Ensuring the security of sensitive information transmitted by IoT devices is crucial to prevent data breaches and cyberattacks. IoT cybersecurity involves employing various technologies, standards, and best practices, including encryption, firewalls, and multi-factor authentication. Although IoT offers numerous benefits, addressing its security challenges is essential. In this study, a flood attack, a significant threat to IoT devices, was executed to assess the system's impact. A reference model without the attack was used to analyze network traffic involving single or multiple attackers. To prevent additional load on the operational system, network packets were mirrored via the cloud and transferred to artificial intelligence (AI) and forensic analysis tools in real-time. The study aimed to ensure continuity, a vital aspect of IoT system cybersecurity, by detecting the attacker using AI and analyzing real-time data with forensic analysis tools for continuous network monitoring. Various AI algorithms were evaluated for attacker detection, and the detection process proved successful.

Keywords: Internet of Things · Wireless Sensor Network · Flood Attaks

1 Introduction

The Internet of Things (IoT) and the Industrial Internet of Things (IIoT) refer to the interconnected networks of physical devices, machines, and sensors capable of collecting and exchanging data. The development of IoT and IIoT is transforming industries and the way we live, work, and interact with the world. As a key aspect of Industry 4.0, the fourth industrial revolution, the IoT and IIoT, are helping to improve efficiency, productivity, and convenience while creating new business opportunities and solutions. These devices can automate many tasks, reducing the need for manual intervention and freeing up time for more critical tasks. They can monitor and track health and safety conditions in real-time, alerting individuals and organizations to potential dangers and allowing quick action. They simplify and streamline many tasks, such as turning on and off lights, adjusting the temperature, or controlling entertainment systems. Thus, they help reduce

Z. Şen et al. (Eds.): IMSS 2023, LNME, pp. 459–468, 2024.
https://doi.org/10.1007/978-981-99-6062-0_42

waste, improve energy efficiency, and reduce the industry's environmental impact. As the number of connected devices continues to grow, IoT and IIoT technologies offer significant benefits and opportunities for various industries. However, the increasing interconnectedness also raises substantial concerns regarding cybersecurity and privacy, particularly in the IIoT, where sensitive industrial and personal information is collected and stored. Addressing these security concerns is essential to harness the benefits of IoT and IIoT without compromising the safety and privacy of individuals and organizations.

The widespread adoption of these technologies has left interconnected devices and systems vulnerable to major cybersecurity threats, such as unauthorized access, data breaches, distributed denial-of-service (DDoS) attacks, malware, ransomware, weak encryption, and physical tampering. Cybercriminals exploit these vulnerabilities to take control of devices, steal sensitive information, disrupt critical infrastructure, and jeopardize human safety. The challenges posed by the absence of standardized security protocols and the diverse nature of devices within IoT and IIoT networks amplify the difficulty of defending against these sophisticated threats. This highlights the urgent need for a comprehensive assessment of IoT and IIoT security risks and a collaborative approach among industry stakeholders, researchers, and policymakers to develop and implement robust security measures that protect the privacy and safety of individuals and organizations while maximizing the potential of these transformative technologies.

Therefore, in the study, Flood attacks, which are one of the significant attacks targeting IoT systems, were performed to examine the impact of the attack on the system. Subsequently, network traffic packets were mirrored to the expert system through the cloud to detect the attack. In this way, network traffic was continuously monitored through forensic analysis and analyzed using artificial intelligence algorithms.

In the continuing sections of the study, in the second section, similar studies are presented. In the third section, the testbed environment in which the attacks analyses have been carried out is described, and the attacks and their results are specified. In the fourth section, an artificial intelligence-based expert system is introduced for attack detection, and then studies are focused on applications for detecting flood attack carried out by one or multiple attackers. In the fifth section, the discussion section, general values, and solution proposals are stated. The study concluded in the conclusion section.

2 Literature

The integration of Information Technology, including the Internet of Things (IoT), big data, and artificial intelligence, with traditional manufacturing processes has transformed the industry from a closed structure to an open one, with increased digitalization and intricate IT networks both within and between factories. This has also expanded the number and intelligence of entry points for potential attacks on open-structure smart factories. The Industrial Internet of Things (IIoT) enhances connectivity for all industries, leading to the generation of valuable data and insights about operations. The IIoT is used across various industries to connect information, services, and people for intelligent operations in various management areas, such as smart energy, smart cities, healthcare, automation, agriculture, logistics, and transportation [1]. Malicious actors exploit security weaknesses in the Industrial Internet of Things (IIoT) to steal confidential information, interrupt operations, and potentially harm the image and interests of affected enterprises. Critical Infrastructure (CI) plays a vital role in businesses and industries. For instance, in 2019, a CI was subjected to an attack every 14 s on average. The study conducted by Morgan predicted that the financial losses resulting from cyberattacks targeting enterprises could reach as high as $11.5 billion by the start of 2022. [2] In research from Wu et al., they focused on industrial espionage, where information is stolen/chanced from 3D printers. They used kNN, Random Forest, and Anomaly Detection with a success rate of 96.1% [3]. There have been extensive research efforts focused on identifying unusual behavior and detecting intrusions in Industrial Control Systems, which is a part of the larger category of Cyber-Physical Systems [4–11]. Boateng et al. used a neural network with one class objective function on the SECURE WATER TREATMENT(SWAT) testbed, and they detected 15 of 36 attacks on this dataset [12]. Deep learning techniques have also been explored for detecting anomalies in crucial infrastructure systems. Muna et al. utilized deep neural networks and an autoencoder to identify anomalies in network traffic. In comparison to eight other anomaly detection systems, their experiments showed an improved detection rate and a reduced rate of false positives. However, the anomaly detection system requires fine-tuning various parameters optimally [13]. Kim et al., on the hand, proposed edge-computing with a lightweight DNN solution that has 98.93% accuracy on 25 different malware [14]. Latif, Shahid, et al. proposed an Intrusion Detection System (IDS) that leverages the benefits of deep random neural networks to secure and protect Industrial Internet of Things (IIoT) systems [1]. The system was evaluated using the UNSW-NB15 dataset to showcase its practicality and suitability for the IIoT. The results showed a high detection accuracy of 99.54% with a low rate of false alarms [1]. Yang, Kai, et al. employed deterministic finite automata to create unique profiles of each Industrial Control System (ICS) controller and carried out both active and passive intrusion detection in real-world experiments. The results showed a recall rate of 98% with a detection rate within 2 s [15]. On the other hand, Wu, Di et al. utilized a Long Short-Term Memory (LSTM) network and Naive Bayes deep learning algorithm on real-life time-series datasets to achieve 100% accuracy in detecting anomalies [16]. Leyi, Shi, et al. propose a method that uses CCN, Bİ-LSTM, and correlation information entropy on the gas pipeline dataset with an accuracy of 99.21% and a detection rate of 11.73s [17]. In contrast, Chu, Ankang, et al. employed GoogLeNet to extract features for the inception module and a Long Short-Term Memory (LSTM) network on a gas pipeline dataset, resulting in an accuracy rate of 97.56% [18]. Rachmadi et al. [19] conducted a study that introduces an Intrusion Detection System (IDS) based on the AdaBoost model for detecting Denial of Service (DoS) attacks. The study utilized the public IoT dataset MQTTset for evaluation and achieved an overall accuracy of up to 95.84%. On the other hand, Wahla et al. [20] developed a framework with AdaBoost for detecting Low-Rate One-Class Denial-of-Service (LRDoS) attacks, which are more challenging to identify than traditional DoS attacks. The proposed approach also addresses imbalanced dataset issues by adjusting the sample weights. The method was tested in a simulation environment based on NS2 and achieved a 97.06% detection rate for LRDoS attacks. On the contrary, Mohammed et al., using the XGBoost algorithm, identified a DDOS attack with 99% accuracy [21]. Nedeljkovic et al., on the other hand, used CNN based method on the SwaT testbed with an accuracy of 88% [22]. Shafiq et al. [23] conducted a study that demonstrates the effectiveness of a Decision Tree (DT)-based detection framework in identifying Bot-IoT attacks. The study used the public Bot-IoT dataset and achieved an accuracy rate of up to 99.99%.

3 Test Environment

This study investigates the effects of IoT Flood attacks, primarily Denial-of-Service, with variable attacker counts. The analysis proceeds through a rigorous, four-stage process (Fig. 1). Initially, the impact of increasing attacker count on system integrity is measured.

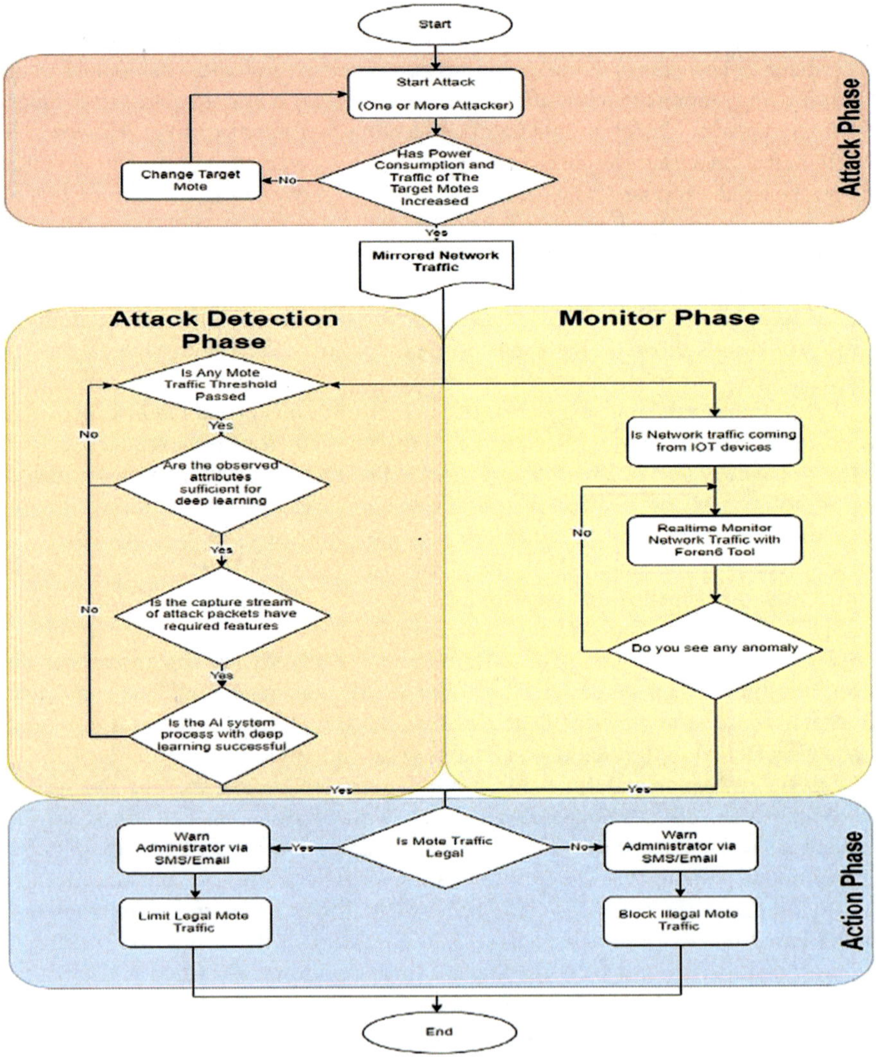

Fig. 1. Flow Diagram

Subsequently, cloud-mirrored network traffic is continuously monitored, emphasizing remote IoT devices. The intent is to identify anomalies through threshold deviation. Next, the efficacy of AI algorithms for attacker identification is evaluated within an expert system, using the network traffic data. Lastly, traffic source legitimacy is scrutinized and mitigating actions are undertaken for unauthorized access, maintaining system stability. Real-time communication with the administrator facilitates prompt response. This methodology promotes swift attack detection and network sustainability through constant monitoring and AI usage.

Within the study's purview, the testbed's topology is constructed to monitor an exclusively IoT-based network traffic via cloud mirroring. To safeguard the continuity of IoT systems, a crucial facet of cybersecurity, traffic directed to these devices is mirrored. This mirrored traffic is then transferred to both forensic tools and AI algorithms via a continuous monitoring expert system. This approach ensures that IoT system operations remain unaffected by security protocols, thereby enhancing the resilience and efficiency of these systems amidst potential cybersecurity threats.

4 Attack Analysis and Continuous Monitoring

This research investigated the impacts of prominent Denial-of-Service (DoS) threats, specifically flood attacks, on Internet of Things (IoT) infrastructures, utilizing simulations with one, three, and five adversaries. Notably, the study concentrated on the attacks' implications on power consumption. Foren6, a forensic analysis tool, was employed for continuous, real-time network traffic monitoring and anomaly identification.

Supplementing an AI model, the open-source Foren6 application was harnessed to validate the detected attacker motes. IoT systems, susceptible to flood attacks due to their typically constrained computational resources, warrant robust protective measures.

In assessing flood attack consequences, baseline network traffic sans attacker was scrutinized to establish thresholds. Figure 2 delineates packet volumes and power consumption in the IoT model's network traffic. Under no-attack conditions (Fig. 3), power consumption remains relatively uniform across devices. Analyzing further, it was found that Low Power Mode (LPM) values were uniform due to inter-mote communication, CPU usage varied with data processing tasks, and transmission conditions were contingent on proximity to the sink mote and communication status.

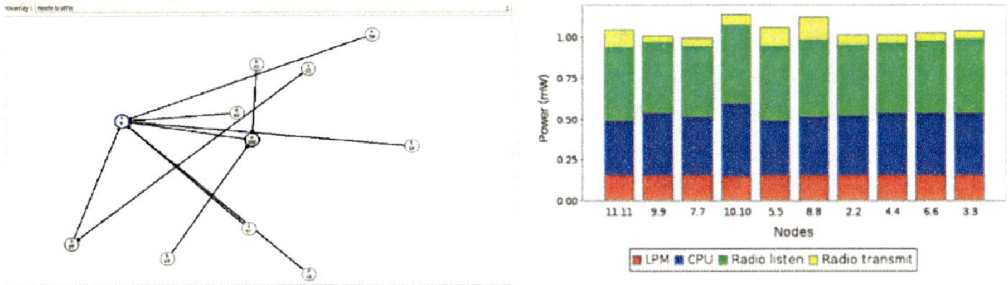

Fig. 2. IoT Reference Network Traffic and Power Consumption

Subsequent to baseline analysis, an attacker mote was introduced into the reference model, triggering a new network traffic flow. Figure 4 depicts real-time network traffic and packet dynamics during a single-attacker IoT flood attack. The attack's effect is evident in the escalated packet count and disrupted sink mote communication. As seen in Fig. 3, the attacker mote exhibits greater transmission traffic (highlighted in yellow) than other motes, including the sink mote. The CPU consumption, represented in blue, is notably higher in the attacker mote. Moreover, the highest listener count corresponds to the attacker (mote 12). In comparison to the reference model, all motes, barring the attacker, generate primarily response traffic due to the flood attack, thereby impeding the transmission of sensor data (e.g., temperature, humidity, pressure) to the system experts via the sink mote, undermining their intended functions.

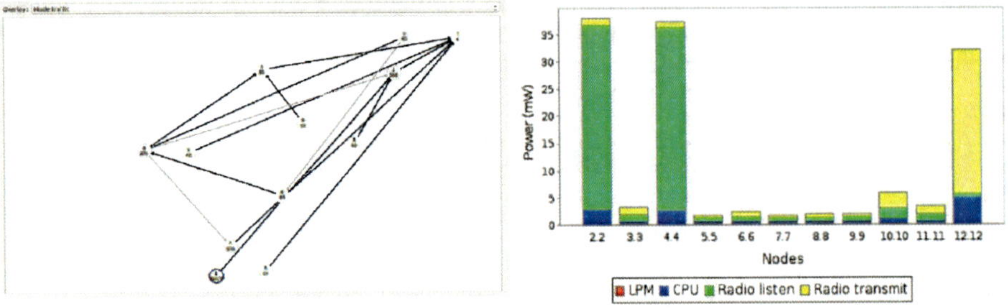

Fig. 3. IoT Network Traffic and Power Consumption with (1) Attacker

The terminal phase of the attack analysis introduced three, then five, attacker motes into the reference network for a flood attack. An upsurge in the number of transmitted and received packets and system communication disruptions were observed, relative to both the reference and the single-attacker models. As the number of attackers escalated, the attacked devices displayed heightened listening traffic, while attackers generated increased transmission traffic and consumed more CPU power. With five attackers, inter-mote communication was severely impeded, restricting data collection for certain modules. Yet, Foren6 forensic analysis application effectively identified the attacking motes. Test results revealed a dichotomy in roles, with attacking motes predominantly transmitting and targeted motes primarily listening.

5 Attack Detection with Artificial Intelligence

The study employed artificial intelligence (AI) algorithms to analyze network traffic mirrored to an expert system (Fig. 4) for attack detection. An Artificial Neural Network (ANN) expert system, incorporating a Rectified Linear Unit (ReLU) algorithm, was utilized to identify malicious packets during flood attack filtration. The collected data, following cleaning, was transformed into a whitelist dataset to facilitate attack analysis and detection.

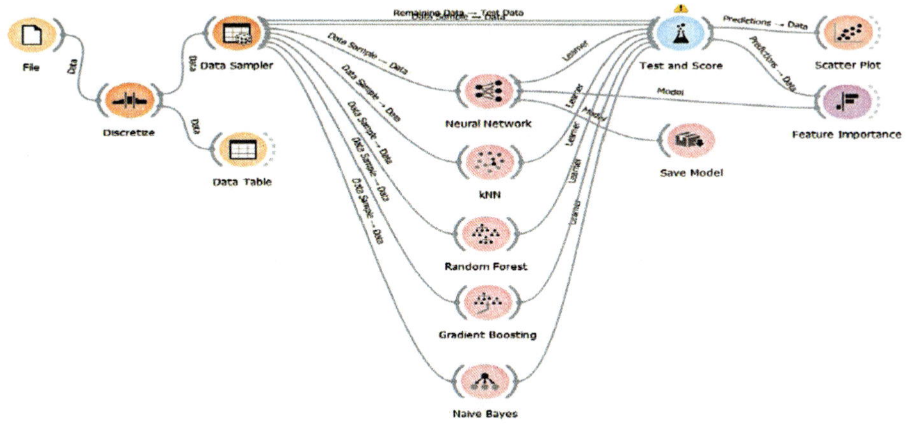

Fig. 4. Attacker Detection with Artificial Intelligence Algorithms

During the pre-processing stage, "6LoWPAN.Pattern" data, discovered in network packets, was identified as a significant feature for detecting flood attacks, thereby eliminating the need for additional attacker labelling. This feature enabled precise attacker detection in single and multiple attacker scenarios.

The dataset was split into training (70%) and testing (30%) segments. The highest-performing model, designated NN-ReLU, was selected as the expert system model. This model's specific algorithm is outlined in Table 1.

Precision, recall, F1 score, and classification accuracy (CA) were used as metrics to assess model performance. Upon analysis of Table-1, the NN model with the ReLU activation function achieved the highest accuracy rates, obtaining 99.99% for one attacker, 99.7% for three attackers, and 99.6% for five attackers. These high accuracy rates were substantiated by their corresponding F1, precision, and recall values. The study thus proceeded with the NN deep learning model due to its exceptional accuracy, and the validity of these results was confirmed through continuous network traffic monitoring with the Foren6 forensic analysis application.

Table 1. Artificial Network Results.

Model	Test Set	Train Time [s]	Test Time [s]	CA	F1	Precision	Recall
kNN	Beningn	0.407	1.111	0.987	0.986	0.985	0.987
	1 Attacker	13.809	477.502	0.979	0.974	0.973	0.979
	3 Attackers	30.242	1.776.882	0.971	0.969	0.968	0.971
	5 Attacers	26.232	1.204.322	0.963	0.954	0.955	0.963
Random Forest	Benign	1.247	0.162	0.981	0.991	0.981	1.0

(continued)

Table 1. (*continued*)

Model	Test Set	Train Time [s]	Test Time [s]	CA	F1	Precision	Recall
	1 Attacker	23.321	1.545	0.880	0.931	0.871	1.0
	3 Attackers	51.340	3.638	0.910	0.949	0.930	1.0
	5 Attackers	38.992	2.815	0.917	0.953	0.911	1.0
Naïve Bayes	Benign	0.155	0.017	0.981	0.972	0.963	0.981
	1 Attacker	5.473	0.291	0.196	0.064	0.038	0.196
	3 Attackers	10.113	0.512	0.162	0.045	0.026	0.162
	5 Attackers	8.278	0.382	0.158	0.043	0.025	0.158
NN Relu	Benign	8.418	0.202	**1.0**	**1.0**	**1.0**	**1.0**
	1 Attacker	168.926	1.603	**0.999**	**0.999**	**0.999**	**0.999**
	3 Attackers	349.466	3.659	**0.997**	**0.991**	**0.992**	**0.997**
	5 Attackers	285.656	3.396	**0.996**	**0.990**	**0.992**	**0.996**
	Benign	3.036	0.141	0.981	0.991	0.981	1.0
	1 Attacker	204.279	2.274	0.880	0.931	0.871	1.0
	3 Attackers	447.885	4.935	0.910	0.949	0.930	1.0
	5 Attackers	409.052	4.621	0.917	0.953	0.911	1.0

6 Discussion

This research scrutinized a denial-of-service flood attack on an IoT system, escalating from a single attacker to multiple attackers. As the number of attackers proliferated, system strain intensified, incapacitating several critical IoT devices. Power consumption data escalated alongside the number of attackers, as did network packet count. The investigation further assessed artificial intelligence algorithms, with the NN-ReLU model discerning attackers with 99.9% accuracy. IoT networks' pivotal cybersecurity component, system continuity, necessitates continuous network monitoring coupled with artificial intelligence to identify abnormal traffic and thwart attacks. Initial stages entailed observing IoT devices' real-time communication to establish reference network model threshold values. Subsequently, the influence of attackers on power consumption and artificial intelligence's capability to identify attacker motes was scrutinized. The study underscored the significance of real-time data monitoring via the open-source Foren6 Forensic Analysis Tool, along with artificial intelligence algorithms, for continuous monitoring and prevention of system vulnerability.

7 Conclusion

In the era of Industry 4.0, the pervasive integration of information technologies, notably the Internet of Things (IoT), significantly influences critical infrastructures, smart cities, and key societal sectors. However, this digital transformation has inadvertently resulted in notable vulnerabilities due to cybersecurity oversights. This research scrutinizes a Denial of Service (DoS)-specifically a Flood attack-within an IoT-focused test environment, evaluating its systemic impact and exploring mitigation strategies through continuous monitoring and artificial intelligence (AI) algorithms. These mechanisms collectively bolster the system's resilience against the often formidable DoS attacks. The analyses revealed that, when compared using AI algorithms, the NN deep learning model deploying a ReLU activator yielded the highest accuracy in attacker mote detection amidst the limited traffic collected.

References

1. Latif, S., Idrees, Z., Zou, Z., Ahmad, J.: DRaNN: a deep random neural network model for intrusion detection in industrial IoT. In: 2020 International Conference on UK-China Emerging Technologies (UCET), pp. 1–4. IEEE (2020)
2. Morgan, S.: Global ransomware damage costs predicted to hit $11.5 billion by 2019. Cybercrime Magazine (2018). https://cybersecurityventures.com/ransomware-damage-report-2017-part-2/. Accessed 11 Feb 2023
3. Wu, M., Song, Z., Moon, Y.B.: Detecting cyber-physical attacks in CyberManufacturing systems with machine learning methods. J. Intell. Manuf. **30**, 1111–1123 (2019). https://doi.org/10.1007/s10845-017-1315-5
4. Narasimhan, S., Biswas, G.: Model-based diagnosis of hybrid systems. IEEE Trans. Syst. Man, Cybern.-Part A: Syst. Hum. **37**(3), 348–361 (2007)
5. Pasqualetti, F., Dörfler, F., Bullo, F.: Cyber-physical attacks in power networks: models, fundamental limitations and monitor design. In: 2011 50th IEEE Conference on Decision and Control and European Control Conference, pp. 2195–2201. IEEE (2011)
6. Teixeira, A., Pérez, D., Sandberg, H., Johansson, K.H.: Attack models and scenarios for networked control systems. In: Proceedings of 1st International Conference on High Confidence Networked System, pp. 55–64 (2012)
7. Boateng, E.A.: Anomaly detection for industrial control systems based on neural networks with one-class objective function. In: Proceedings of Student Research Creative Inquiry Day, vol. 5 (2021). https://publish.tntech.edu/index.php/PSRCI/article/view/810/321. Accessed 11 Feb 2023
8. Zhao, F., Koutsoukos, X., Haussecker, H., Reich, J., Cheung, P.: Monitoring and fault diagnosis of hybrid systems. IEEE Transactions on Systems, Man, and Cybernetics. Part B, Cybernetics : a publication of the IEEE Systems, Man, and Cybernetics Society. **35**, 1225-1240 (2006). https://doi.org/10.1109/TSMCB.2005.850178
9. Abidi, M.H., Alkhalefah, H., Umer, U.: Fuzzy harmony search based optimal control strategy for wireless cyber physical system with industry 4.0. J. Intell. Manuf. **33**, 1795–1812 (2022). https://doi.org/10.1007/s10845-021-01757-4

10. Colabianchi, S., Costantino, F., Di Gravio, G., Nonino, F., Patriarca, R.: Discussing resilience in the context of cyber physical systems. Comput. Ind. Eng. **160**, 107534 (2021). https://doi.org/10.1016/j.cie.2021.107534
11. Lambán, M.P., Morella, P., Royo, J., Sánchez, J.C.: Using industry 4.0 to face the challenges of predictive maintenance: a key performance indicators development in a cyber physical system. Comput. Ind. Eng. **171**, 108400 (2022). https://doi.org/10.1016/j.cie.2022.108400
12. Boateng, E.A., Bruce, J.W., Talbert, D.A.: Anomaly detection for a water treatment system based on one-class neural network. IEEE Access **10**, 115179–115191 (2022). https://doi.org/10.1109/ACCESS.2022.3218624
13. Muna, A.H., Moustafa, N., Sitnikova, E.: Identification of malicious activities in industrial internet of things based on deep learning models. J. Inf. Secur. Appl. **41**, 1–11 (2018)
14. Kim, H., Lee, K.: IIoT malware detection using edge computing and deep learning for cybersecurity in smart factories. Appl. Sci. **12**(15), 7679 (2022). https://doi.org/10.3390/app12157679
15. Yang, K., Li, Q., Lin, X., Chen, X., Sun, L.: iFinger: intrusion detection in industrial control systems via register-based fingerprinting. IEEE J. Sel. Areas Commun. **38**(5), 955–967 (2020)
16. Di, W., Jiang, Z., Xie, X., Wei, X., Weiren, Y., Li, R.: LSTM learning with Bayesian and Gaussian processing for anomaly detection in industrial IoT. IEEE Trans. Industr. Inf. **16**(8), 5244–5253 (2019)
17. Leyi, S., Hongqiang, Z., Yihao, L., Jia, L.: Intrusion detection of industrial control system based on correlation information entropy and CNN-BiLSTM. J. Comput. Res. Dev. **56**(11), 2330–2338 (2019). https://doi.org/10.7544/issn1000-1239.2019.20190376
18. Chu, A., Lai, Y., Liu, J.: Industrial control intrusion detection approach based on multiclassification GoogLeNet-LSTM model. Secur. Commun. Netw. **2019**, 1–11 (2019)
19. Rachmadi, S., Mandala, S., Oktaria, D.: Detection of DoS attack using AdaBoost algorithm on IoT system. In: Proceedings of the 2021 International Conference on Data Science and Its Applications (ICoDSA'21). IEEE, pp. 28–33. Los Alamitos, CA (2021)
20. Wahla, A.H., Chen, L., Wang, Y., Chen, R., Fan, W.: Automatic wireless signal classification in multimedia Internet of Things: an adaptive boosting enabled approach. IEEE Access **7**(2019), 160334–160344 (2019)
21. Mohammed, A.S., Anthi, E., Rana, O., Saxena, N., Burnap, P.: Detection and mitigation of field flooding attacks on oil and gas critical infrastructure communication. Comput. Secur. **124**, 103007 (2023)
22. Nedeljkovic, D., Jakovljevic, Z.: CNN based method for the development of cyber-attacks detection algorithms in industrial control systems. Comput. Secur. **114**, 102585 (2022)
23. Shafiq, M., Tian, Z., Sun, Y., Du, X., Guizani, M.: Selection of effective machine learning algorithm and Bot-IoT attacks traffic identification for Internet of Things in smart city. Future Gener. Comput. Syst. **107**, 433–442 (2020). https://doi.org/10.1016/j.future.2020.02.017

Investigation of DataViz as a Big Data Visualization Tool

Fehmi Skender[1](✉), Violeta Manevska[2], Ilija Hristoski[3], and Nikola Rendevski[2]

[1] Faculty of Engineering and Architecture, International Vision University Gostivar, Gostivar, Republic of North Macedonia
fehmi.skender@vizyon.edu.mk

[2] Faculty of Information and Communication Technologies Bitola, University "St. Kliment Ohridski", Bitola, Republic of North Macedonia
{violeta.manevska,nikola.rendevski}@uklo.edu.mk

[3] Faculty of Economics-Prilep, University "St. Kliment Ohridski" Bitola, Bitola, Republic of North Macedonia
ilija.hristoski@uklo.edu.mk

Abstract. Big Data hype is increasing extremely fast. A few quadrillion bytes of data in various formats are generated almost daily.These data are the subject of extensive data analysis, especially visual data analytics. New applications and technologies for processing and visualizing Big Data are constantly developing and evolving daily. Therefore, being well-informed and up to date with Big Data processing and visualization software is of very high importance, yet inevitable in contemporary data analysis, providing better decisions and solutions for both businesses and managers. This research is based on a thorough analysis of reviews of current Big Data visualization tools made by the world's most popular authorities. The study's main objective is to describe a new visualization tool based on algorithms, primarily aiming to improve existing Big Data visualization software capabilities. Considering the visualization aspects of Big Data and the previously noted criteria, we present a custom-made application named DataViz, developed using Python. The DataViz application is simple and easy to use since it has an intuitive user interface to serve various users, including those without enhanced computer skills. Regarding the analysis of current visualization tools, the DataViz application considers and implements several important criteria, including Accuracy, Empowering, Releasing, and Succinct. The development of such an application fills the gap among different commercially available Big Data visualization tools delivering enhanced visualization capabilities and optimization. As such, it can provide a solid basis for further improvement and transformation into a fully functional software tool.

Keywords: Big data · Data visualization · Python

Z. Şen et al. (Eds.): IMSS 2023, LNME, pp. 469–478, 2024.
https://doi.org/10.1007/978-981-99-6062-0_43

1 Introduction

In their everyday life, communication, and work, people usually prefer visual cognition. There are several reasons for this behavior. First, the human brain processes visual information more easily as compared to text information. When looking at charts or diagrams, the brain interprets the values and relationships between them very easily and simultaneously, which is not the case with text information. Visualizations can also be very effective in presenting various types of information and in understanding complex concepts and relationships. For instance, complex concepts and trends can be inferred based on a few important points on a chart or diagram. At the same time, one of the greatest reasons for visual cognition is that visual presentations can be used to present spatial or multidimensional data. For instance, maps can be useful for presenting data on the spread of diseases in a particular region, which is difficult to be achieved by textually displaying data [1]. In short, visuals can effectively communicate information with a wide audience, because everyone can understand them regardless of their level of education and knowledge of the field. The visual representation of data is important for communicating information with users, meaning that a graphical representation can be a better way to share information than a text data view. Data visualization has many advantages, such as the ability to help in detecting trends and relationships within data [2]. There are various ways to visualize data, including charts, diagrams, maps, etc. Choosing the right visualization type depends on the data type and the purpose of visualization. Graphs are the most commonly used ways to visualize data [3]. They can be linear, columnar, shaped like circles, pillars, etc. Charts can be useful in presenting time series data, matching values, or displaying percentage parts. Diagrams can be just as effective in visualizing data. They can be used for displaying relationships among the values, identifying categories, and displaying percentage parts [4]. Maps can be useful for visualizing spatial and/or temporal data, such as data for spreading diseases, cadastre needs, etc. Today, there are several powerful Big Data visualization tools on the market, offering various features, including support for a variety of visual styles, simplicity and easiness to use, and the capability to handle a large volume of data. Choosing the right one has to take into account multiple criteria. Still, all of them can be sublimed into four basic criteria: Accuracy, Empowering, Release, and Succinct [5]. The aim of the paper is twofold: (a) To present some of the most compelling visualization tools that are often used in Big Data; and (b) To elaborate on a brand new, custom-made software tool named DataViz, which was created as a relatively optimal solution for Big Data processing and visualization. The paper is organized as follows. Section 2 briefly elaborates on the notion of data visualization and its importance, especially in the context of Big Data. In Sect. 3, the data and methodology used to fulfill the goals of the research are described. The mean features of the custom-made DataViz application are described in Sect. 4, whilst in Sect. 5 a comparative analysis between the DataViz app and a set of selected commercial visualization tools has been carried out.

2 Data Visualization

Data visualization is the process of displaying information or data in charts, diagrams, images, or other visual formats to facilitate the understanding and interpretation of data. This visualization may include various techniques for a visual representation of data, such as line charts, pillar diagrams, pies, histograms, maps, etc. [6]. It is the process of converting information into a visual context, such as a graph or map, to make it easier for the human brain to absorb and extract insights. In addition, it is one of the main phases in the data science process that asserts that after data has been collected, processed, and modeled, it must be visualized before conclusions can be drawn. The primary purpose of data visualization is to make it simpler to discover trends, patterns, and outliers in massive data sets [7].

For instance, photography can transmit much more information than a text or speech, i.e., it can send a powerful message, convey emotions, illustrate context, display details, and activate viewers' minds. Length, width, and height are the basis of human spatial perception and the ability to understand the three-dimensional nature of the world around it. Although our senses may receive information about other dimensions, such as time and temperature, the human perception system is unable to process that information identically as it does with three-dimensional data. The human brain processes visual information 60,000 times faster than text [8]. The growing popularity of Big Data and data analysis initiatives has fostered the need for proper visualization more than ever. Machine learning (ML) and predictive analytics are rapidly being used by businesses to process vast volumes of data that can be difficult and time-consuming to sort through, interpret, and explain [9]. Visualization may help to accelerate this process and deliver facts to business executives and other interested parties in ways that they can understand. Since visualizations of complex algorithms' results are generally easier to interpret than their numerical outputs, it becomes of utmost importance to visualize the outputs to monitor results and ensure that models are performing as intended. Big Data visualization frequently goes beyond the approaches employed in traditional visualization. It instead employs more sophisticated visualizations, such as heat maps [10]. However, Big Data visualization often necessitates the use of sophisticated computer systems based on powerful computer hardware, efficient storage solutions, software, and even a transfer to the cloud to collect raw data, interpret it, and convert it into graphical representations that people can utilize to efficiently conclude. It is worth mentioning that the insights produced by Big Data visualization are only as accurate as the data being employed and analyzed [11]. As a result, it is critical to have both people and procedures in place to manage and control the quality of data, metadata, and data sources, as well as to identify the best data sets and visualization styles to guarantee organizations are optimizing the use of their data. The importance of data visualization is based on facilitating the detection of trends and regularities in data (patterns), identifying abnormalities and inaccurate information, generating new ideas, and the power to deliver specific messages [12]. Today's data visualization applications concentrate on data collection, storage, and use of data in communication, i.e. extracting information from the user's point of interest. Data visualization provides a quick and efficient means to present information in a universal fashion using visual information. Other advantages of visualization techniques include the potential to absorb information faster, improve insights, and make faster decisions;

increased awareness of the next actions to be taken to improve the organization; an improved capacity to sustain the interest of the audience with information they can understand; easy dissemination of information that increases the opportunity to share insights with everyone involved; the elimination of the need to use data science, since data is more accessible and understandable; and an increased ability to achieve success with greater speed and fewer mistakes. Because of all these, data visualization tools and techniques are nowadays extensively used in business, sales, marketing, finance, logistics, politics, healthcare, science, and research [13–16]. Leading companies like Amazon, Twitter, Apple, Facebook, and Google extensively use data visualization to make appropriate business decisions. There are many data visualization tools available in the market today, but according to Haan & Watts, the best data visualization software of 2023 includes Microsoft Power BI, Tableau, Qlik Sense, Klipfolio, Looker, Zoho Analytics etc. [17].

3 Data and Research Methodology

The research is first based on a review of literature related to authoritarian websites and portals for ranking visualization tools. The experimental method further confirms the ranking of the tools. The best-ranked tool using the method to collect primary data through an experiment promotes the new DataViz tool. As an example study, data with earthquakes recorded by the Kaggle.com center [18] for the period from 01.01.2020 to 31.12.2020. In addition to the magnitude, each event is represented by its hypocentral depth, corresponding geographical coordinates, hypocentral time, and magnitude scale type. These are data on earthquakes given with the basic parameters–magnitude of the earthquake ML $\geq$ 5.0, hypocentral depth and spatial distribution of the earthquakes defined by the corresponding coordinates.

Data analysis was done with the DataViz tool. Then a comparison of the data was made with the different tools - Tableau, Microsoft Power BI, Qlik Sense, Klipfolio, Looker, and Zoho Analytics (the six best-ranked visualization tools according to Capterra, Forbes, Peerspot, and Crozdesk), taken as a tool with high accuracy in the data processing. The purpose of this comparison is to show that the DataViz tool when processing the data, gives results almost the same as the other tools.

4 Data Visualization with the DataViz Application

DataViz is a heavily modified custom-made application based on DataExplore, designed to meet the requirements of Python 3.9. Additionally, it has been adapted to be converted into an.exe file, enabling its availability on any machine for local use. Its primary purpose is to analyze and visually present data (Fig. 1). On the left of the user interface, the data is shown as a spreadsheet table. On the bottom right part of the user interface, there are specific parameters to adjust the visualization graph on the top right part of the interface.

DataViz's internal structure and functionalities are developed and programmed in Python using its standard library and Tkinter.Tkinter was used for the construction of the interface. The PandasTable library is a Python module that provides a table widget for Tkinter, which is the standard GUI toolkit for Python. The library is designed for use in applications where a table is needed to store and manipulate large amounts of data. It uses the powerful data manipulation capabilities of the Pandas library, which is a popular open-source Python library for data analysis and provides high-performance data structures and tools. PandasTable is intended to be used by Python/Tkinter GUI developers who need to incorporate a table into their applications for handling large data sets. It also caters to non-programmers who may not be familiar with Python or the Pandas API but want to use the included DataExplore application for data manipulation and visualization. Additionally, data analysts and programmers who want to quickly and interactively explore tabular data without writing code can find it useful. By leveraging the capabilities of Pandas and Tkinter, PandasTable provides a convenient and efficient way to display, manipulate, and visualize tabular data in a GUI environment. It can be a valuable tool for a wide range of users, from GUI developers to data analysts, for efficiently working with large data sets and performing data exploration tasks in a user-friendly manner [19].

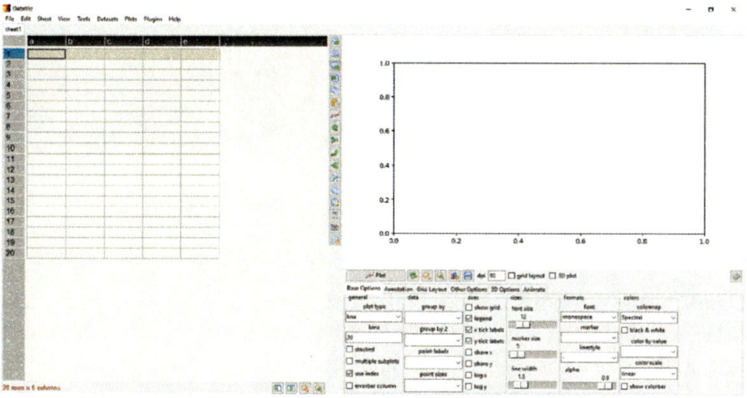

Fig. 1. The DataViz application user interface

The DataViz app consists of multiple files, images, and scripts that are interconnected and make the tool functional. The tool is also designed to work as a dependent program on the program side for operating and testing the application. At the same time, the DataViz app works independently. The design of an independent application is based on the auto-py-to-exe library, an external library of the Python programming language. The Pandas table library guarantees the overall value and functionality of the application in the visual display part of the data, their communication with the options that come with the charts, and the changes that each option implements the data as a source and way to provide. The application's graphical user interface is divided into three separate areas: (1) In the first, the table-shaped data is displayed; (2) In the second area, the data chart is displayed; and (3) The third area consists of various options and functions intended to apply modifications to the chart. For demonstration purposes, the developed DataViz application processes and visualizes data from the earthquake database for the period 01.01.2020 – 31.12.2020, as depicted in Fig. 2.

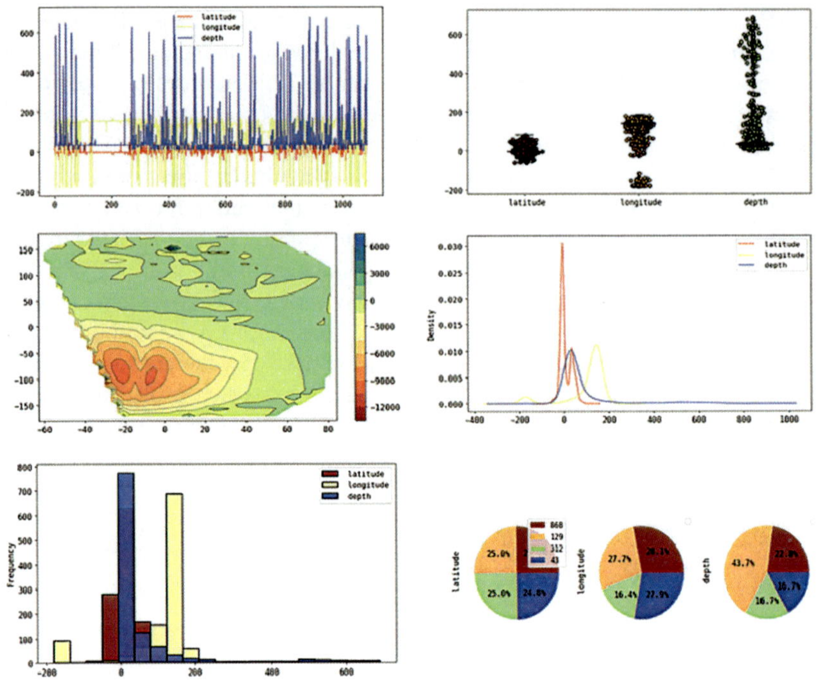

Fig. 2. Some visualization types offered by the DataViz application.

The DataViz app supports all popular chart types, making it quite functional. In addition to the usual plot types, the DataViz tool has other settings about certain attributes such as data, axes, sizes, formats, colors, and other settings, which make the tool quite useful.

5 Raking of the Current Visualization Tools

Ranking current visualization tools is important as it helps users make informed decisions, allocate resources effectively, evaluate performance, foster innovation, and prioritize user experience, all of which contribute to better data visualization practices and outcomes. Figure 3 shows the results of rankings made by Forbes, Capterra, G2, Peerspot, and Crozdesk. The table contains the rankings of Forbes, Capterra, G2, Peerspot, and Crozdesk, in 5 rows, and PowerBI, Tableau, QlikSense, Klipfolio, Looker and Zoho Analytics in 6 columns, with an index of 1 to 5 depending on whether the tool is ranked by ranking authorities, Only Crozdesk, which makes the ranking on a scale of 0 to 100 but the ranking is adjusted to the rankings of other authority.

After viewing the results of the literature review, i.e. the rankings, it is obvious that Microsoft Power BI is one of the best tools for visualizing data. Figure 4 portrays the comparison between Microsoft Power BI and the DataViz app in visualizing specimen data for three different visualizations: Line plot (a and b), Stacked bar chart (c and d),

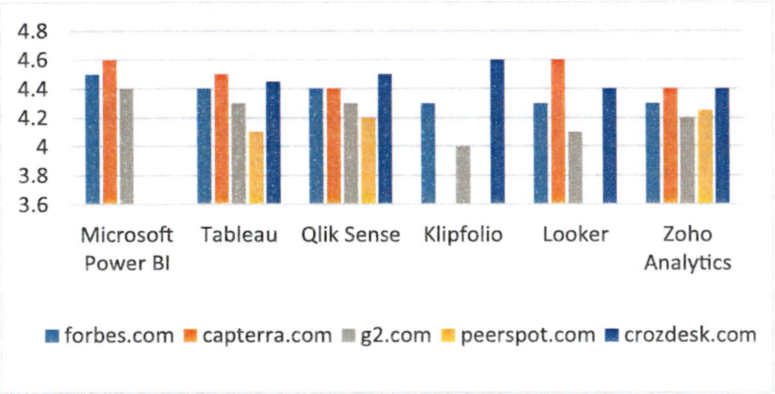

Fig. 3. Rankings of the visualization tools according to world web authority software.

Pie chart (e and f). When observing the different graphical views of data, care is taken of the following aspects: Identical tip of schedule, axes and scales, colors and styles, proportions and scales, data and information. Hypothetically, any well-ranked visualization tool cannot always be perfect. The MS PowerBI tool, like any other commercial software, is already developed. Looking at the positives of the DataViz app, it is a relative solution when it comes to visual processing, interactivity, and simplicity regarding its functionality. The four criteria, stating that each tool should be open to reward and strain, can be briefly explained as follows:

Accuracy: The accuracy of visualization refers to the extent to which the information presented accurately represents the underlying data. An accurate visualization allows viewers to make informed decisions based on the information shown. Power BI enables users to obtain accurate, reliable, and visually appealing results from their data, making it an excellent choice for data analysis and visualization. Whereas, DataViz enables the transfer of accurate information, allowing a basis for analysis.

Empowering: Empowering visualizations are those that allow viewers to interact with the data and gain a deeper understanding of the information presented. Such visualizations should provide tools for exploring, manipulating, and customizing the data. Power BI allows viewers to interact with the data and gain a deeper understanding of the information presented. Whereas DataViz provides tools for exploring, manipulating, and customizing the data.

Releasing: A releasing visualization conveys information clearly and effectively without requiring extensive prior knowledge or interpretation. Such visualizations should be easy to understand and interpret, even for those who are not experts in the subject. Power BI and DataViz convey information clearly and effectively without requiring extensive prior knowledge or interpretation.

Succinct: Succinct visualizations convey information concisely and efficiently. They should present only the most important information and avoid clutter and unnecessary details. Power BI and DataViz convey information concisely and efficiently. They present the most important information while avoiding unnecessary details.

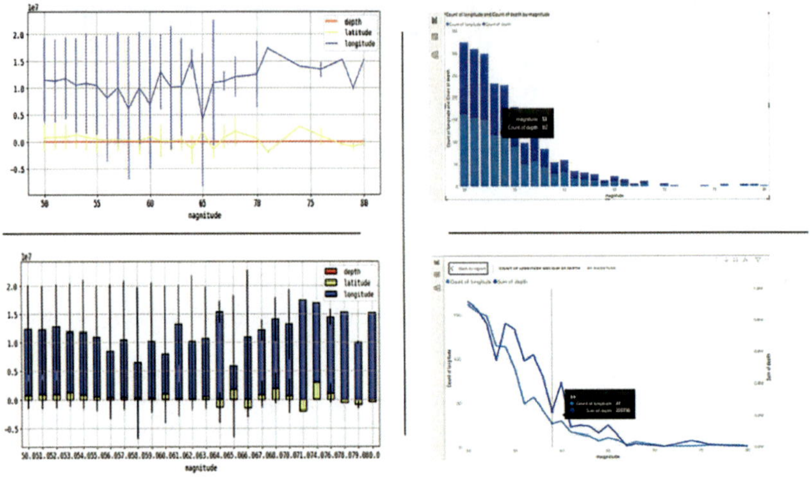

Fig.4. Comparative visualization between Microsoft Power BI and DataViz

It is important to note that these criteria are not mutually exclusive and can often overlap. A high-scoring visualization should balance all these criteria, effectively conveying information accurately, empowering viewers to interact with the data, releasing information clearly and intuitively, and doing so concisely and efficiently. As a common conclusion in terms of visualization tools:

Chart-Based Tools: These tools provide a wide range of pre-defined chart types, such as bar charts, line charts, pie charts, and more. Users can select a chart type, import data, and customize various chart elements such as labels, colors, and legends. Some tools may also offer interactive features, such as tooltips and drill-down options, allowing users to explore data in more detail. Many chart-based tools have options for data updates, allowing users to refresh the data source or connect to live data feeds to keep visualizations up to date.

Dashboard Tools: Dashboard tools allow users to create interactive dashboards with multiple visualizations, such as charts, tables, and maps, arranged in a visually appealing and meaningful way. Users can customize the layout, size, and appearance of each visualization, and may also have options for interactivity, filtering, and data drilling. Dashboards can often be connected to live data sources, enabling automatic updates as new data becomes available.

Data Storytelling Tools: These tools combine data visualizations with narrative elements to tell a compelling story using data. Users can create visualizations and add annotations, explanations, and context to guide readers through the data story. Data storytelling tools may also allow for updates, as users can update the underlying data and refresh the visualizations and narratives to reflect the latest information.

Custom Coding Tools: Some data visualization tools provide options for users to write custom code to create visualizations using programming languages such as JavaScript, Python, or R. This allows for greater flexibility and customization but may require more technical expertise. Custom coding tools can also handle updates by allowing users to modify the code to accommodate changes in data sources or data structure.

Cloud-Based Tools: Cloud-based data visualization tools store data and visualizations in the cloud, allowing for easy collaboration, sharing, and updates. Users can typically update the data source in the cloud, and the visualizations will automatically reflect the changes. Cloud-based tools also offer scalability and accessibility, as visualizations can be accessed from any device with an internet connection.

In terms of possibilities for updates, data visualization tools may offer various options, such as manual updates where users can upload new data or connect to different data sources, automatic updates where visualizations are refreshed periodically based on predefined schedules, or real-time updates where visualizations are connected to live data feeds and reflect changes in real-time. It is important to note that the availability and capabilities of updates may vary depending on the specific data visualization tool and its features. Users should carefully review the documentation and features of the tool they are using to understand its update capabilities and ensure that it meets their specific needs.

6 Conclusion

Big Data visualization tools are software applications that enable users to create visual representations of big volumes of data, making it easier to interpret and understand complex information. Such tools typically have various structures and features that allow users to create and customize visualizations, and they may also offer additional options for data updates to keep the visualizations current and relevant. The number of visualization tools available on the market may be the reason for making a good choice and an opportunity to see not only their strengths but also their shortcomings and limitations. The hypothesis that any visualization tool can be optimized has been proven by the newly developed DataViz tool, which appears to be a relatively light solution, synthesized based on the shortcomings of current commercial tools. With the DataViz tool, the hypothesis that each tool can be optimized to improve the quality of visualization has been fully proven and acts encouraging for future research. Accuracy, Empowering, Release, and Succinct criteria are also important to enhance the visual changes in colors, formats, styles, chart types, or any other aspect of visualization that will improve the understanding and interpretation of the data. According to the table, Qlik Sense has the highest estimates of accuracy, enabling and publishing the visualizations of Kaggle earthquake data. Tableau and Looker also have good ratings for these categories. Power BI and Zoho Analytics have medium accuracy, enabling, and publishing ratings, while Klipfolio has smaller ratings in all categories. Regarding the shortness of visualization, Qlik Sense and Tableau receive high ratings, while Power BI and Klipfolio receive lower ratings. Looker and Zoho Analytics have average ratings for this category. The development of the new tool DataViz is based on the conclusions drawn from observing and analyzing the construction and functionality of current data visualization tools. With the help of Python libraries and the tools themselves, many significant optimizations can be made. However, in this specific case, the intervention of Python code in the tools has been replaced by creating a brand new tool as a relatively light solution based on the four pre-established criteria.

References

1. Wilke, C.O.: Fundamentals of Data Visualization: A Primer on Making Informative and Compelling Figures, 1st edn. O'Reilly Media Inc, Sebastopol (2019)
2. Healy, K.: Data Visualization: A Practical Introduction, 1st edn. Princeton University Press, New Jersey/Oxfordshire (2019)
3. Yi, M., Restori, M.: How to Choose the Right Data Visualization, Chartio.com (2021). https://chartio.com/learn/charts/how-to-choose-data-visualization/. Accessed 1 Apr 2023
4. Gutman, A.J., Goldmeier, J.: Becoming a Data Head: How to Think, Speak, and Understand Data Science, Statistics, and Machine Learning, 1st edn. John Wiley & Sons Inc, Indianapolis (2021)
5. www.capterra.com, "Data Visualization Software," Capterra.com (2023). Available: https://www.capterra.com/data-visualization-software/. Accessed 28 Mar 2023
6. Schwabish, J.: Better Data Visualizations: A Guide for Scholars, Researchers, and Wonks. Columbia University Press, West Sussex (2021)
7. Brush, K., Burns, E.: Data Visualization. TechTarget.com (2023). https://www.techtarget.com/searchbusinessanalytics/definition/data-visualization. Accessed 28 Mar 2023
8. oit.williams.edu, "Using Images Effectively in Media," The Media Education Center. https://oit.williams.edu/files/2010/02/using-images-effectively.pdf. Accessed 29 Mar 2023
9. Lantz, B.: Machine Learning with R: Expert Techniques for Predictive Modeling, 3rd edn. Packt Publishing, Birmingham (2019)
10. Campbell, A.: Data Visualization: Ultimate Guide to Data Mining and Visualization, Independently published by Alex Campbell (2020)
11. Rockswold, G., Krieger, T.A., Rockswold, J.C.: College Algebra with Modeling and Visualization, 6th edn. Pearson Education Inc., New York (2018)
12. Burns, K.B.: Definition of Data Visualization. TechTarget.com (2022). https://www.techtarget.com/searchbusinessanalytics/definition/data-visualization. Accessed 19 Mar 2023
13. Sayilgan, E., Yuce, Y.K., Isler, Y.: Frequency recognition from temporal and frequency depth of the brain-computer interface based on steady-state visual evoked potentials. J. Intell. Syst. Appl. **4**(1), 68–73 (2021). https://doi.org/10.54856/jiswa.202105160
14. Skender, F., Ali, I.: Big data in health and the importance of data visualization tools. J. Intell. Syst. Appl. **5**(1), 33–37 (2022). https://doi.org/10.54856/jiswa.202205198
15. Degirmenci, M., Yuce, Y.K., Isler, Y.: Motor imaginary task classification using statistically significant time domain and frequency domain EEG features. J. Intell. Syst. Appl. **5**(1), 49–54 (2022). https://doi.org/10.54856/jiswa.202205203
16. Degirmenci, M., Yuce, Y.K., Isler, Y.: Classification of multi-class motor imaginary tasks using poincare measurements extracted from EEG signals. J. Intell. Syst. Appl. **5**(2), 74–78 (2022). https://doi.org/10.54856/jiswa.202212204
17. Haan, K., Watts, R.: The best data visualization tools of 2023. Forbes.com. https://www.forbes.com/advisor/business/software/best-data-visualization-tools/. Accessed 18 Mar 2023
18. Fine Open Datasets and Machine Learning Projects. https://www.kaggle.com
19. Farrell, D.: DataExplore: an application for general data analysis in research and education. J. Open Res. Softw. **4**(1), e9 (2016). https://doi.org/10.5334/jors.94

A Development of Electrified Monorail System (EMS) for an Automobile Production Line

İlyas Hüseyin Güvenç[1](✉) and H. Metin Ertunç[2]

[1] Robo Automation and Engineering Company, Gölcük, Kocaeli, Turkey
ilyashguvenc@gmail.com
[2] Mechatronics Engineering Department, Kocaeli University, Kocaeli, Turkey
hmertunc@kocaeli.edu.tr

Abstract. The production of several materials on the same production line is very important for efficiency. Electrified Monorail System (EMS) is a product transport system with rails and this system can reach every production plant in the manufactory. If EMS is designed and used effectively, the production will be fast and efficient. In this study, an existing EMS was improved on a vehicle production line.

EMS carries the parts of vehicles to different areas in the factory environment. When the production type varies, then the transportation of the parts will be complicated. In other words, the increasing of the production variant makes the operation of the system be difficult. EMS lines are needed to develop solving transportation problems that are mismatch buffer, new lines design, flexible using of the carriers of the body parts and smart flexible automation. In this project, an enhanced EMS study related to flow algorithm of the production line was carried out to increase the efficiency of the manufacturing. Thus, the motion algorithms and coordination algorithms of the system were developed. Carriers that adapt flexible manufacturing system were designed to carry different car parts.

Keywords: Electrified Monorail System · Flexible Production · Carrier · Transportation

1 Introduction

Nowadays competitive market requires a production system that quickly responds to ever-changing demands and customer needs. Although flexible manufacturing and customization has long been an issue in modern manufacturing society, improving the responsiveness of a manufacturing system remains a major challenge in practice, especially for high investment and highly complex manufacturing industries such as the automotive industry [1].

The electric monorail system is an overhead conveyor system for in-plant material flow. It consists of independent vehicles that run freely on the track system. The running track is suspended from the ceiling by brackets. This frees up floor space for business and production purposes. The track system allows any number of vehicles to be run

Z. Şen et al. (Eds.): IMSS 2023, LNME, pp. 479–486, 2024.
https://doi.org/10.1007/978-981-99-6062-0_44

continuously or in cycles. The modular design provides short assembly times as well as a variety of load configurations. Excellent stopping accuracy and high speeds, thanks to frequency-controlled steering with integrated distance measurement, distinguish this system from others [2].

In this study, improvements were made on a production line with an EMS system. Control, PLC, Robotic automation developments have been made in automobile production lines. Cycle time is very important in automation lines, especially in automobile production lines. As a result of the research and developments, the cycle time has been improved by applying new algorithms integrated into the PLC software system and making the EMS carrier system suitable for multiple production. In addition, a new electric vehicle automation line was added to the production line. The complexity that occurred when the new model added to the existing EMS line was integrated into the system was solved with PLC software [1, 3, 4].

2 Material and Method

2.1 Mechanical Development

Carriers are used to transport car-parts in the production lines with EMS. There may be many carriers in line at the same time. If carriers transport one part at a time, carriers cost too much resources. Thus, a carrier is designed that can carry multiple car parts at the same time in Fig. 1.

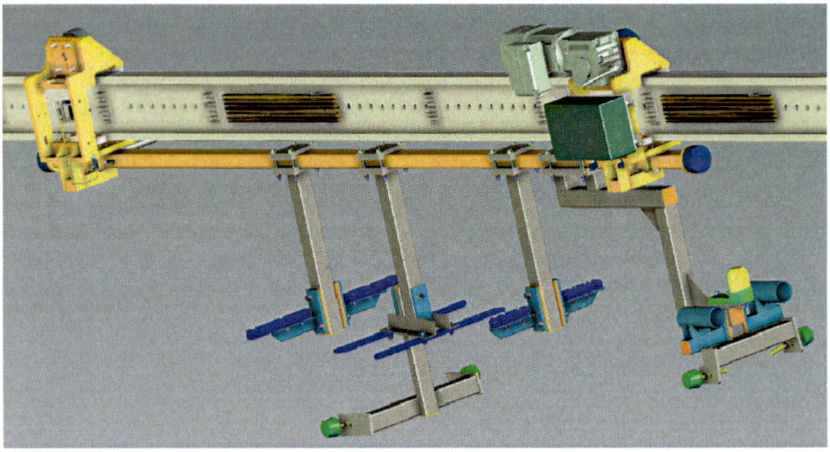

Fig. 1. Flexible EMS Carrier

A new add-on was designed to adapt the existing carrier (carrier) to the new model. The designed add-on was mounted on the carrier and was made to carry the new model. With the new model, the carrier algorithm was created and added to the system (Fig. 2).

The efficiency of the software system has been increased by adding additional paths to the EMS line. The switch points of the roads, accumulation, loading and unloading operations have been improved in Fig. 3.

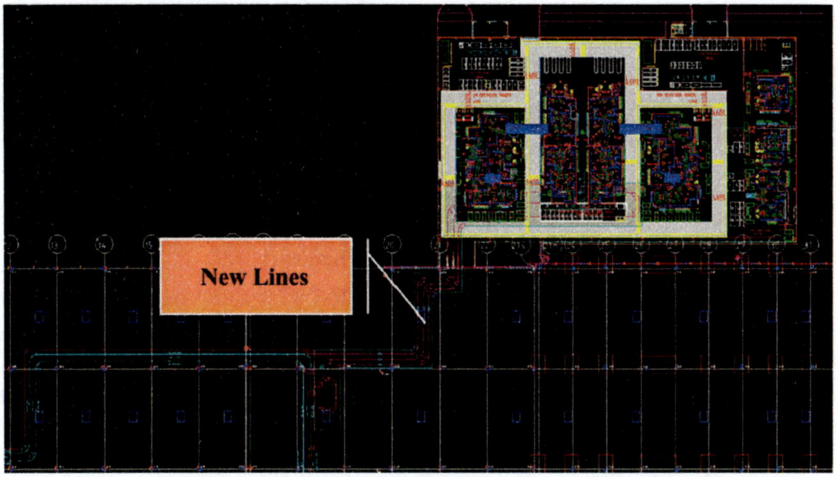

Fig. 2. EMS Additional line into the existing lines

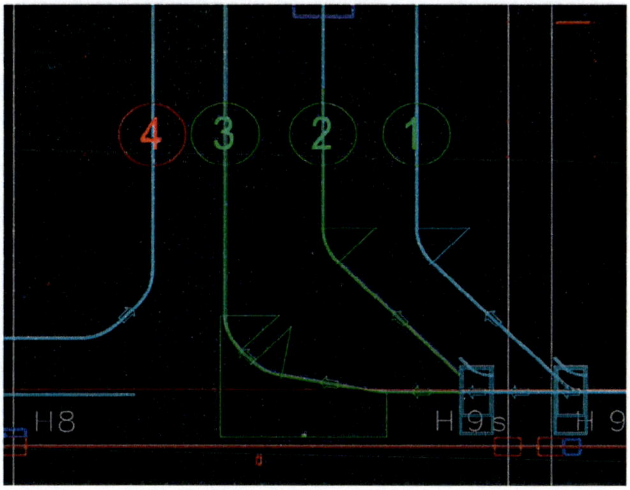

Fig. 3. Line rearrangements

Increasing the capacity of the existing buffer area with the new model in Fig. 4. During this project, the safety of the transport area was also improved. The software system has been improved with the growth of the Buffer line and the inclusion of the new model vehicle in the line.

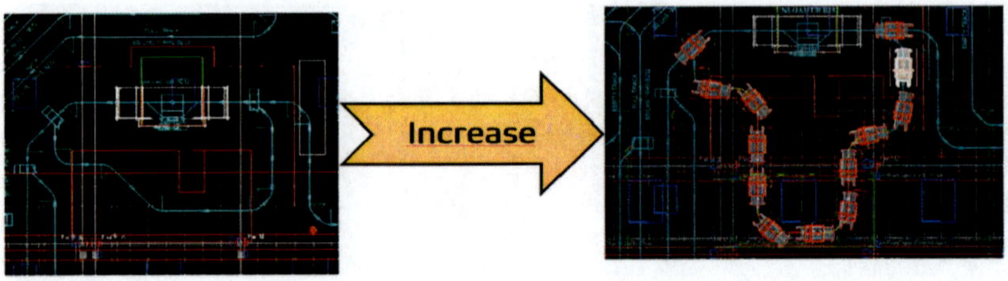

Fig. 4. Buffer Capacity Increase

2.2 PLC and Control Concept

As is known, PLC systems are quite common in the field of automation. PLC, good electrical system means good control of the system. One of the big handicaps that we have not studied is the size and complexity of the PLC system. There are dozens of carrier systems and PLCs with split sector control. Each sector works among itself and communicates with other sectors. Figure 5 shows the relationship between the communication system and the carriers.

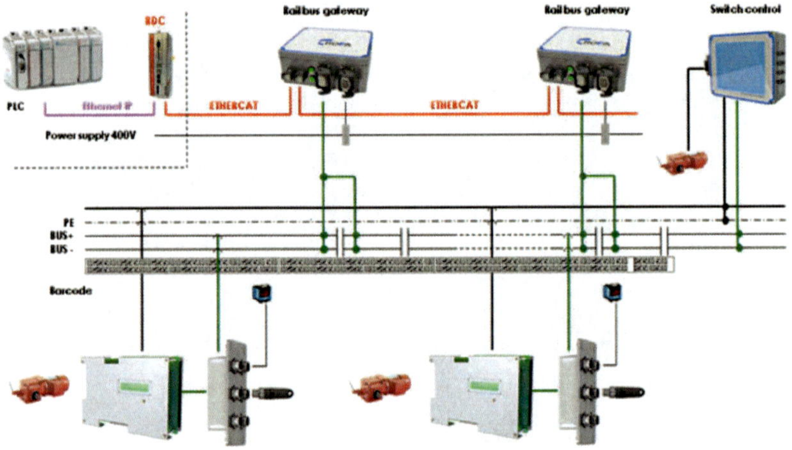

Fig. 5. PLC system schema

Master-Slave method was used to control the system. As seen in Fig. 6, the carriers are controlled by a Master PLC. In this system, which belongs to the Framing station, the routes of the parts are controlled. It is ensured that any part goes before or after according to the production order. Here Buffer Fields are used to swap parts in order. For example, if the third part will be taken forward in a row where there are five parts, a new order is created by entering the others into the Buffer Area.

One of the important parts of the PLC system is the switches. Switches are located at junction points on different routes and provide routing between monorails. In Fig. 7, the switches can be observed.

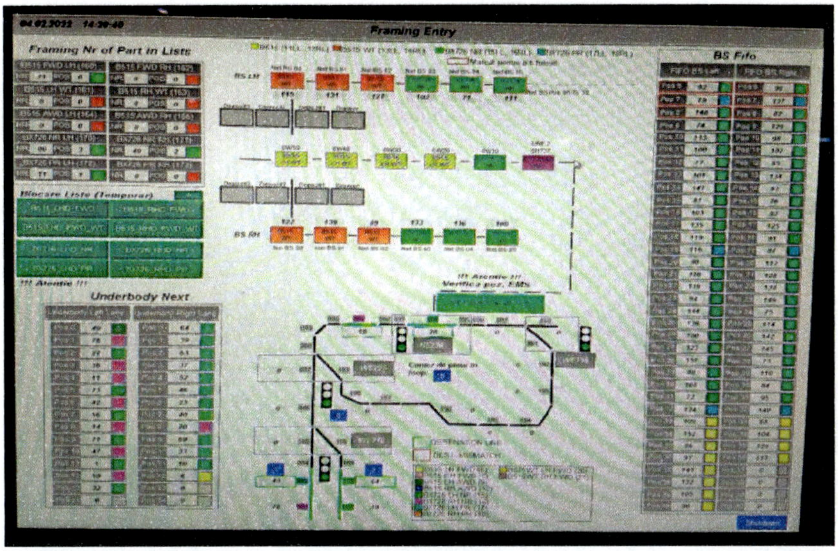

Fig. 6. PLC control panel

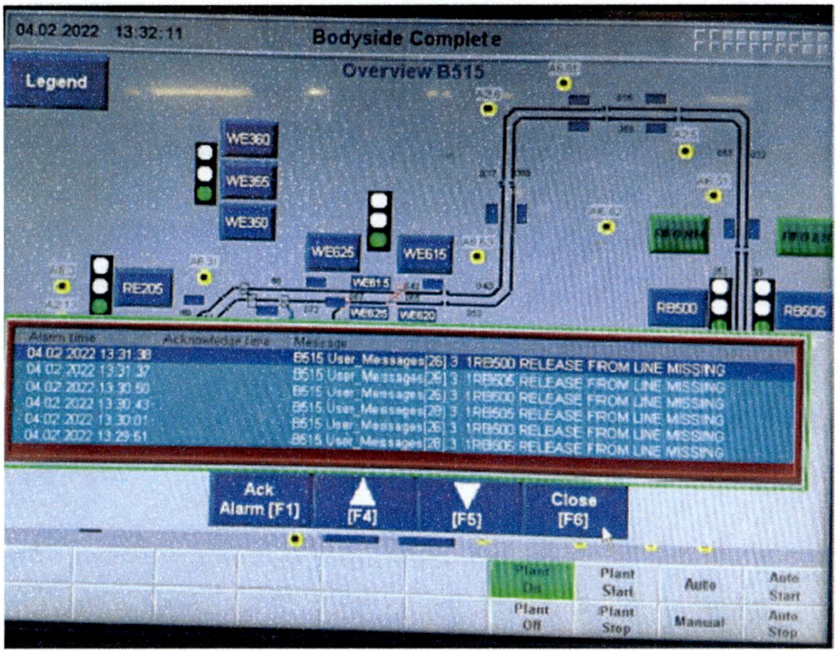

Fig. 7. PLC Carrier Control System

3 Installation and Test

One of the challenges in this study is system setup and testing. The biggest challenge in the installation phases is to install a new system while production is in progress. The main reason for this is that the operating cost of the factory is very high. A one-minute stop can cost thousands of dollars. So much so that with a well-designed installation plan and competent employees, it can be completed for very little cost. In Fig. 8, the switches are added to the system and tested.

Fig. 8. Switch installation

It is the most important step in test automation systems to bring the system to the desired efficiency. Because the test is the place where changes are made for the system to work without errors. In addition, the process can be made more efficient by producing better solutions than expected while testing. In this study, an image from the testing moment of the PLC software of the monorail system in Fig. 9.

Fig. 9. PLC test

4 Conclusion and Discussion

In this study, the development of an electric monorail system in the automobile production line has been made. EMS is of great importance for automobile production lines. This system, which is located in all stations of the factory, carries out the transportation operations on the production lines. With the improvement of each stage of the EMS, the software development of the carriers, transport lines, buffer areas, PLC system has been made and tested.

The net results obtained in the study: increasing the capacity of production due to transportation thanks to the increased carrying capacity, providing flexible production by producing different model vehicles on the same production line, making the system efficient by improving the PLC algorithms, and producing the same production of 8 vehicle types in total, 2 models and 4 different versions. Production on the line.

Although the level of difficulty is very high since the work was done on an existing production line. The aim of the work was successfully completed.

References

1. Wang, J., Chang, Q., Xiao, G., Wang, N., Li, S.: Data driven production modeling and simulation of complex automobile general assembly plant. Comput. Ind. **62**(7), 765–775 (2011). https://doi.org/10.1016/j.compind.2011.05.004. ISSN 0166-3615
2. Kumar, P., Prasad, S.B., Patel, D., et al.: Production improvement on the assembly line through cycle time optimization. Int. J. Interact. Des. Manuf. (2022). https://doi.org/10.1007/s12008-022-01031-8

3. Grzechca, W.: Assembly line balancing problem with reduced number of workstations. IFAC Proc. Volumes **47**(3), 6180–6185 (2014). https://doi.org/10.3182/20140824-6-ZA-1003.02530. ISSN 1474-6670. ISBN 9783902823625
4. Wu, Y.: Automatic signal recognition method of hanging sorting experimental device based on PLC. In: Proceedings of the 3rd Asia-Pacific Conference on Image Processing, Electronics and Computers (IPEC 2022), pp. 132–136. Association for Computing Machinery, New York (2022). https://doi.org/10.1145/3544109.3544133
5. Bao, Y., Li, Y., Ding, J.: A case study of dynamic response analysis and safety assessment for a suspended monorail system. Int. J. Environ. Res. Public Health **13**(11), 1121 (2016)
6. Electric monorail system with magnetic levitations and linear induction motors for contactless delivery applications. In: 8th International Conference on Power Electronics-ECCE Asia, pp. 2462–2465. IEEE
7. Slip control of linear induction motor in electric monorail system to reduce normal force considering magnetic characteristics. In: Proceedings of the Korean Magnestics Society Conference, p. 223. The Korean Magnetics Society

A Modified Bacterial Foraging Algorithm for Three-Index Assignment Problem

Ayşe Hande Erol Bingüler[1(✉)], Alper Türkyılmaz[2], İrem Ünal[2], and Serol Bulkan[1]

[1] Marmara University, Istanbul, Turkey
{hande.erol,sbulkan}@marmara.edu.tr
[2] Marmara University Institute of Pure and Applied Sciences, Istanbul, Turkey
{alper.turkyilmaz,irem.unal}@marun.edu.tr

Abstract. The Three-Index Assignment Problem (3-AP) is well-known combinatorial optimization problem which has been shown to be NP-hard. Since it is very difficult to find the best result in polynomial time, many heuristic methods have been proposed to obtain near optimal solutions in reasonable time. In this paper, a modified Bacterial Foraging Optimization Algorithm (BFOA) is proposed to solve 3-AP. BFOA is inspired by the social foraging behaviour of Escherichia coli (Ecoli). Algorithm imitates the behaviour of the foraging bacteria Ecoli and aims to eliminate those bacteria that have weak foraging methods and maintaining those bacteria that have strong foraging methods. The Hungarian method (most known method for solving the classical linear two-dimensional assignment problem) is integrated to BFOA algorithm at repositioning phase to swim farther and faster to find the best solution. Proposed algorithm has been tested and benchmarked with other algorithms in literature and results show that the new algorithm outperforms other heuristics in literature in terms of solution quality.

Keywords: Three-index assignment problem · bacterial foraging optimization · Hungarian Method · combinatorial optimization

1 Introduction

In recent years, bio-inspired optimization algorithms have been widely used to solve complex optimization problems. The BFO algorithm, in particular, has been applied to various optimization tasks such as function optimization, image processing, and data mining [1–4]. However, it has not been widely used to solve the 3-dimensional assignment problem specifically.

The Hungarian method, on the other hand, is a well-known combinatorial optimization algorithm that is used to solve the assignment problem. It has been widely used for finding an optimal one-to-one assignment of tasks to agents and is known for its efficiency and accuracy. However, the literature about the Hungarian method on solving the 3-dimensional assignment problem is scarce.

There have been several studies that have proposed solutions for the 3-dimensional assignment problem (3-AP).

Z. Şen et al. (Eds.): IMSS 2023, LNME, pp. 487–498, 2024.
https://doi.org/10.1007/978-981-99-6062-0_45

In [5], the authors proposed a fast separation algorithm for the 3-dimensional assignment problem, which is based on linear programming and the Hungarian method. The authors propose a new algorithm for solving this problem that is faster than previous methods. They also demonstrate the effectiveness of their algorithm through computational experiments. Overall, the article presents a new approach for solving a specific type of combinatorial optimization problem and shows that it is more efficient than existing methods.

In [6], a hybrid genetic algorithm is proposed to solve the 3-dimensional assignment problem. It aims to improve the efficiency and accuracy of solving the problem by combining the strengths of two optimization techniques, Genetic Algorithm, and Simulated Annealing. The algorithm is tested on various instances of the problem and compared with other methods, and it showed to be more efficient and accurate.

In [7], an approximate muscle-guided global optimization algorithm is proposed to solve the 3-dimensional assignment problem. The algorithm uses a "Muscle" mechanism to guide the search for the optimal solution, which improves the efficiency and accuracy of the algorithm. The algorithm is tested on various instances of the problem and shown to be more efficient and accurate than other methods.

In [8], a new decomposition-based approach is proposed for the multi-dimensional assignment problem with decomposable costs. These types of problems involve assigning a set of items to a set of agents, where each item has multiple attributes or indices that must be considered in the assignment, and the costs associated with the assignment are decomposable, meaning that they can be broken down into simpler costs. The authors present an experimental study of the performance of these algorithms, comparing them with other known algorithms and showing that they perform well in practice.

In [9], the researchers discuss the complexities and challenges associated with multi-index assignment problems in combinatorial optimization. The author examines the difficulty of solving these problems and suggests methods for approximating solutions. The author argues that the Multi-Index Assignment Problem (MIAP) is NP-hard, meaning that the problem is computationally intractable. In other words, the time required to solve the problem increases exponentially with the size of the problem, making it impractical to solve for large instances of the problem. However, the author suggests that, by using approximation algorithms, it is possible to get a solution that is close to optimal. These algorithms trade off optimality for computational tractability.

The article also discusses the various application areas where MIAPs appear and the importance of solving such problems in these areas. These areas include scheduling, transportation, and resource allocation in various fields such as operations research, logistics, and computer science.

In [10], the researchers present a mathematical model and an algorithm for solving three-dimensional axial assignment problems with the coefficients of decomposable cost. The authors introduce the concept of decomposable cost coefficients and show how it can be used to solve these types of problems. They also analyse the computational complexity of the algorithm, providing insight into its performance and limitations.

To the best of our knowledge, there are no studies that have proposed hybridizing BFO with the Hungarian method to solve the 3-dimensional assignment problem. In this article, we propose a new algorithm that hybridizes the BFO algorithm with the Hungarian method to solve the 3-dimensional assignment problem. The Hungarian method is used during the chemotactic phase of the BFO algorithm to search for better food regions which allows the algorithm to converge faster and find better solutions.

The rest of this paper is organized as follows. Section II describes definition and the formulation of the three-index assignment problem are given. In Section III, we give brief information about the original Bacterial Foraging Optimization Algorithm (BFOA) that gives us inspiration to develop our algorithm. Section IV explains in detail our hybrid-modified BFOA heuristic algorithm. Experimental results and the comparison are presented in Section V. Finally, conclusion part is stated in the last section, in Section VI.

2 Problem Description and Formulation

The multidimensional assignment problem (MAP) is a generalization of the classical assignment problem where multiple agents are assigned to multiple tasks that have multiple attributes or criteria. The goal is to assign agents to tasks in such a way that the overall cost or objectives function is minimized. The most studied case of MAP is 3-AP, however the problems with larger values in dimension have also a number of applications [11].

In Three-Dimensional Assignment Problem (3-AP), there are three sets of tasks, each requiring three resources. The 3-AP can be formulated as a linear programming problem, where the objective is to maximize the total profit or utility of the assignment subject to the constraints of each agent's capacity and the allocation of resources to each task. The constraints of the 3-AP can be represented by a three-dimensional matrix, where each element represents the cost of assigning a resource of a specific type to a task of a specific set, and an agent's capacity is represented by the sum of the assigned costs.

The 3-AP has various practical applications, such as in production planning, scheduling, and logistics. For example, in production planning, the problem can be used to assign workers to machines for a manufacturing process, where each worker has different skills, and each machine requires different types of resources. Similarly, in logistics, the problem can be used to assign delivery routes to vehicles, where each route requires different types of resources, such as fuel, time, and capacity.

The 3-AP falls under the category of combinatorial optimization problems, where the goal is to find the optimal solution from a large set of possible assignments.

The formulation of the three-dimensional assignment problem is as follows [12]:

$$\min \sum_{i=1}^{m} \sum_{j=1}^{m} \sum_{k=1}^{m} c_{i,j,k} x_{i,j,k} \tag{1}$$

subject to;

$$\sum_{j=1}^{m} \sum_{k=1}^{m} x_{i,j,k} = 1, \quad \forall i = 1, 2, \ldots, m \tag{2}$$

$$\sum_{i=1}^{m}\sum_{k=1}^{m} x_{i,j,k} = 1, \quad \forall j = 1, 2, \ldots, m \tag{3}$$

$$\sum_{i=1}^{m}\sum_{j=1}^{m} x_{i,j,k} = 1, \quad \forall k = 1, 2, \ldots, m \tag{4}$$

$$x_{i,j,k} \in \{0, 1\},$$

$$\forall i = 1, 2, \ldots, m,$$

$$\forall j = 1, 2, \ldots, m,$$

$$\forall k = 1, 2, \ldots, m$$

Karp [13], in his study, showed that the 3-AP is NP-hard, which means that finding the optimal solution requires an exponentially growing amount of time as the problem size increases. There are different types of algorithms that can be used to solve the 3-AP, such as exact methods, heuristic methods, and metaheuristic methods. Exact methods guarantee to find the optimal solution, but are computationally expensive for large-sized problems. Heuristic methods provide fast solutions but do not guarantee optimality. Metaheuristic methods combine the advantages of both exact and heuristic methods, and can efficiently find near-optimal solutions for large-sized problems.

3 Bacterial Foraging Optimization Algorithm (BFOA)

Bacterial Foraging Optimization Algorithm (BFOA) is a population-based optimization algorithm inspired by the foraging behavior of bacteria in their natural habitats. It was first proposed by Kevin M. Passino [14]. The BFOA is a method of solving global optimization problems using the behavior of E. coli bacteria as inspiration. This algorithm focuses on the search and avoidance strategy of the bacteria, with the goal of selecting the bacteria with the most effective foraging methods to maximize energy obtained per unit time while eliminating those with weaker foraging methods. During the decision-making process, bacteria communicate with each other through signals and move to the next step to collect nutrients if certain conditions have been met. The BFOA consists of four main components: chemotactic movement, swarming behavior, reproduction, and elimination and dispersal of bacteria.

3.1 Chemotactic

E. coli bacteria have two distinct modes of movement, which are simulated in the BFOA algorithm. These modes are accomplished through the use of flagella, which enable the bacterium to swim and tumble. Swimming involves the bacterium moving in a straight line in a particular direction to search for nutrients, while tumbling involves the bacterium changing its direction by making random, abrupt movements. E. coli bacteria switch between these two modes of movement throughout their lifetime in order to effectively forage for nutrients. The flagella of E. coli bacteria facilitate these two basic movement operations, allowing them to navigate their environment in search of food.

3.2 Swarm

The BFOA algorithm is inspired by the collective behavior of E. coli bacteria, which have been observed to form complex and stable swarms in semi-solid nutrient environments. This group behavior is thought to be an attempt by healthy bacteria to stimulate and attract other bacteria to a central location, in order to find solutions to environmental challenges more quickly.

This behavior is triggered by the release of a chemical substance called succinate, which causes the bacteria to excrete amino acids like serine and aspartate. These excretions attract other bacteria to aggregate into groups, which move together in concentric patterns of swarms with high bacterial density. This phenomenon allows the bacteria to efficiently explore and exploit their environment, in order to find the most beneficial sources of nutrients.

3.3 Reproduction

The reproduction phase of the BFOA algorithm occurs following the chemotactic and swarming phases. After the bacteria have moved and grouped together, some individuals will have collected sufficient nutrients, while others may have been less successful in their search. The least healthy bacteria will die off, while the healthier bacteria - those with lower objective function values - will reproduce asexually by splitting into two new bacteria at the same location. This process helps to maintain a constant population size within the swarm.

By reproducing only the healthier bacteria, the algorithm ensures that subsequent generations will have a higher proportion of individuals with favorable characteristics, as determined by the objective function. This allows the swarm to continue searching for the global optimum by selecting for increasingly fit individuals over time.

3.4 Elimination and Dispersal

The population of bacteria may decline due to various environmental factors such as overcrowding or sudden changes in temperature. Similarly, in natural settings, bacteria may die due to certain events or get dispersed to new locations. To mimic this behavior, the BFOA algorithm randomly eliminates a few bacteria from the population with a predetermined probability when such events occur, and replaces them with new individuals generated randomly across the search space.

4 Modified BFOA With Parallel Hungarian Method Execution (MoBFOA-PHM)

The Bacterial Foraging Optimization Algorithm (BFOA) has been acknowledged as a potent algorithm for searching solutions, not only due to its robustness, but also because it accurately models the behaviour of E. coli bacterium and the evolutionary designed control system within it. Researchers have attempted to combine BFOA with other algorithms to investigate its local and global search properties in isolation. Furthermore, the BFOA has been employed in various real-world problems and has outperformed many versions of the Genetic Algorithm (GA) and Particle Swarm Optimization (PSO).

The proposed algorithm (MoBFOA-PHM) hybridizes the BFOA with Hungarian Algorithm. It has three main steps derived from the BFO algorithm: Chemotactic, Reproduction, Eliminate & Disperse. Hungarian Algorithm which is developed by Harold Khun [15] in 1955 to solve 2-dimensional assignment problems (2-AP) in polynomial time. Proposed algorithm integrates the Hungarian method in its chemotactic phase to represent the swimming behaviour of the bacteria and to swim farther and faster to find the best solution within the nearby solution space. Moreover, the chemotactic step is executed in parallel over the colony to reduce the overhead of the time-consuming swim operation.

The flow of the algorithm is given in Fig. 1, the algorithm parameters are given in Table 1 and the pseudo-code is as follows:

1) Initialization parameters $S, N_c, N_w, N_{re}, N_{ed}, P_{ed}$.
2) Initialize the colony of size S by randomly creating the bacteria and calculate the fitness function J_i.
3) Get the minimum cost through the fitness function and setting it as the best result achieved thus far.
4) Increment the counter for Elimination-dispersal loop: $1 = 1 + 1$
5) Increment the counter for Reproduction loop: $k = k + 1$
6) Increment the counter for Chemotactic loop: $j = j + 1$

   Divide the colony into N_w workers to parallelly process the chemotactic phase. Assign each worker ($W_{size} = S/N_w$) bacteria to process.

   Figure 2 shows the flowchart of the chemotactic step. In this step, at each worker and for each bacterium assigned to it:

   i) Apply tumble behavior.
   ii) Compute the fitness function, if the new value (J_{comp}) smaller than the best value (J_{best}), copy and set the bacterium as the best solution
   iii) Apply swim behavior.
   iv) Compute the fitness function, if the new value (J_{hun}) smaller than the best value (J_{best}), copy and set the bacterium as the best solution.
   v) If $j < N_c$ continue with the next chemo-tactic step till number of chemotactic steps are reached.
7) Sort the colony in ascending order based on fitness function.
8) Copy the first half of the colony over the second half of the colony. If $k < N_{re}$ continue with the next reproduction step till number of reproduction steps are reached.
9) For the second half of the colony, with the probability P_{ed}, delete the bacteria from the colony and replace it with a randomly created bacterium. If $l < N_{ed}$ continue with the next eliminate & disperse step till number of eliminate & disperse steps are reached.
10) End of the algorithm, output the best solution (bacterium).

4.1 Initialization

For the initial colony of size S, each bacterium is set to be a pair of randomly generated permutations of length P. In 3-AP problems, it is common way to represent a solution

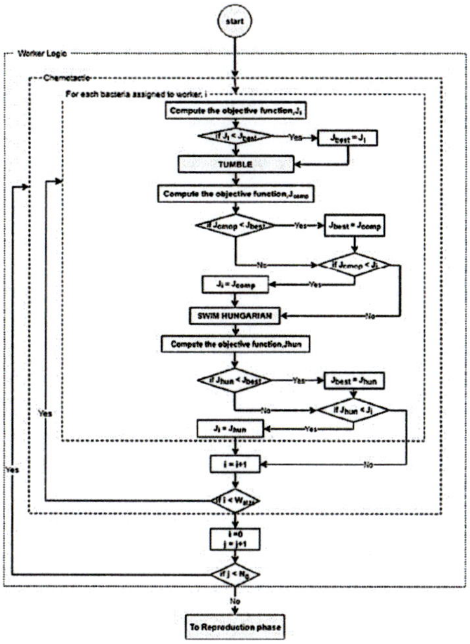

Fig. 1. MoBFOA-PHM Flow Diagram

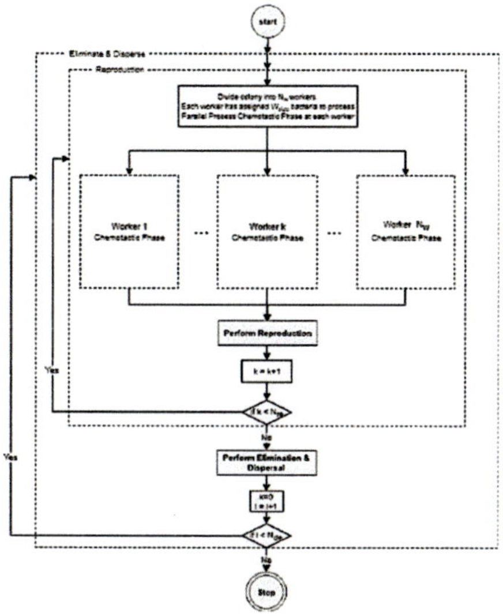

Fig. 2. Chemotactic step Flow Diagram

Table 1. BFO Algorithm parameters

Parameter	Definition	Parameter	Definition
S	Colony size (population size)	N_{re}	Reproduction step size
N_c	Chemotactic step size	N_{ed}	Elimination-dispersal step size
N_w	Worker size	P_{ed}	The probability of a bacteria being eliminated or dispersed

with two permutations (the 2 dimensions) which is always a feasible solution. The third dimension can be thought as virtual array of size P whose values are fix and sorted from 1 to P. In Fig. 3, two sample bacterium (solution) are presented for a problem size of $P = 5$. A new solution is generated and added to the initial colony by permutating the dimension j and k, while keeping the dimension i fixed.

The 3-dimensional cost matrix $C(dim_1, dim_2, dim_3)$ is given and cost (fitness function) of the bacterium can be calculated using the Eq. 5:

$$cost\ of\ bacterium = \sum_{m=1}^{P} C(i[m], j[m], k[m]),\ where,\ P = problem\ size \quad (5)$$

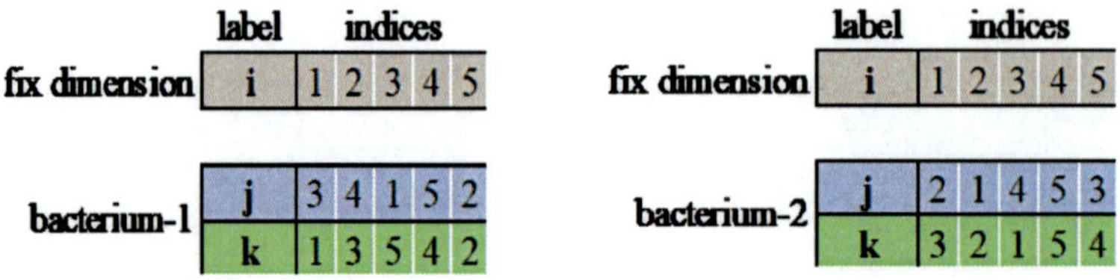

$$cost_{bacterium1} = C(1,3,1) + C(2,4,3) + C(3,1,5) + C(4,5,4) + C(5,2,2)$$

$$cost_{bacterium2} = C(1,2,3) + C(2,1,2) + C(3,4,1) + C(4,5,5) + C(5,3,4)$$

Fig. 3. Two sample bacterium representation with cost calculation

The cost calculation is applied to all the bacteria in the colony and the best solution is stored as best found so far.

4.2 Chemotactic Step

This process simulates the movement of a bacterium through swimming and tumbling to find a favorable medium for foraging. The proposed algorithm uses following strategies for the two behaviors.

4.2.1 Tumbling

Bacterium changes its direction to another direction with tumbling behavior. By applying a perturbation to the one of the permutations of the bacterium, this behavior is simulated. This is achieved by randomly selecting one of the swap mutation or insertion mutation and then applying it to the randomly selected permutation (one of the dimension j, k) of the bacterium.

Algorithm-1 tumble():

```
random_value ← Rand(2)
tumble_index = Ø
if (random_value == 1)
      then tumble_index ← j
else tumble_index ← k
if random_value == 1
     then swapMutation(tumble_index)
   else insertionMutation(tumble_index)
```

4.2.2 Swimming

The swimming behavior is for moving in same direction to search food. To simulate movement in the same direction to find better medium for foraging, we optimize one permutation while keeping the other dimensions fixed. The optimized permutation is the dimension selected in the tumbling phase. In Fig. 4, the cost-matrix table and a random solution with an initial cost of 283 of a sample 3-AP problem of size 4 are presented. The corresponding cost values of the assignments are highlighted in the cost matrix table. For this particular example, suppose the *tumble_index* is selected as or *k* in the tumbling phase. Our proposed algorithm makes the bacterium moves in the same direction as the *tumble_index* by means of optimizing the permutation *k* by applying the Hungarian Method.

C	K1				K2				K3				K4			
I/J	J1	J2	J3	J4	J1	J2	J3	J4	J1	J2	J3	J4	J1	J2	J3	J4
I1	76	66	69	72	14	33	10	57	65	58	78	84	28	29	92	94
I2	71	6	77	68	74	56	27	94	48	41	65	96	83	69	54	79
I3	24	83	75	7	66	11	1	92	63	90	68	76	79	92	97	43
I4	85	59	55	41	97	81	41	54	22	68	66	26	65	88	51	53

	label	indices			
	i	1	2	3	4
bacterium	j	4	1	3	2
	k	2	4	1	3

initial, cost = 283

Fig. 4. Cost matrix (*C*) and a sample solution for a 3-AP problem of size 4

In Fig. 5, the cost-matrix (HCM) that will undergo to Hungarian method has been presented. The HCM (the matrix on the right in Fig. 5) is constructed by taking all the cost values of *tumble_index* (which is *k* for this example) while keeping the other two indices (*i* & *j*) fix. The formula for constructing the HCM is shown in Eq. 6:

$$HCM\,(n,m) = \begin{cases} C(i(n), j(n), m), & \text{if } tumbleindex = k \\ C(i(n), m, k(n)), & \text{if } tumbleindex = j \end{cases} \quad n, m = 1, 2, .., \mathrm{P} \tag{6}$$

where, $i(n), j(n), k(n)$ are the nth value in the corresponding permutation and *P* is the problem size.

C	K1				K2				K3				K4			
I/J	J1	J2	J3	J4	J1	J2	J3	J4	J1	J2	J3	J4	J1	J2	J3	J4
I1	76	66	69	72	14	33	10	57	65	58	78	84	28	29	92	94
I2	71	6	77	68	74	56	27	94	48	41	65	96	83	69	54	79
I3	24	83	75	7	66	11	1	92	63	90	68	76	79	92	97	43
I4	85	59	55	41	97	81	41	54	22	68	66	26	65	88	51	53

HCM	K1	K2	K3	K4
I1	C(1,4,1)	C(1,4,2)	C(1,4,3)	C(1,4,4)
I2	C(2,1,1)	C(2,1,2)	C(2,2,3)	C(2,2,4)
I3	C(3,3,1)	C(3,3,2)	C(3,3,3)	C(3,3,4)
I4	C(4,2,1)	C(4,2,2)	C(4,2,3)	C(4,2,4)

HCM	K1	K2	K3	K4
I1	72	57	84	94
I2	71	74	48	83
I3	75	1	68	97
I4	59	81	68	88

Fig. 5. Hungarian cost matrix (*HCM*)

In Fig. 6, the result of the Hungarian algorithm that is applied to the HCM and the new bacterium are presented. The cost is reduced from 283 to 202.

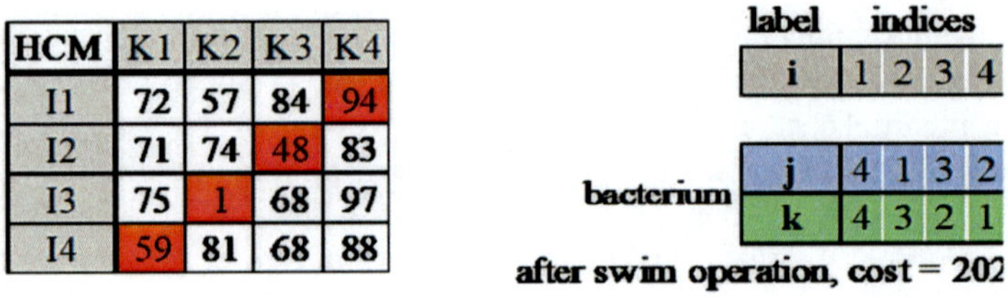

Fig. 6. The result of the Hungarian algorithm and the new bacteria with reduced cost.

5 Experimental Results

This section describes the evaluation of the MoBFOA-PHM algorithm's efficacy using experimental results over the 3-AP benchmark dataset. The algorithm was developed using Java SDK version 8 and tested on a computer with Windows 10 Pro (64 bit) operating system and an Intel i7-4710HQ processor (2.50 GHz) having 8 cores and 16 GB memory. To assess the proposed algorithm's effectiveness and efficiency, problem instances from the literature were simulated, and the algorithm's performance was compared with that of other algorithms from the literature. The parameters of the MoBFOA-PHM algorithm were chosen based on the results of the experiments and empirically determined to achieve optimal performance.

Balas and Saltzman [16] created their dataset, which consists of 60 test cases for the three-dimensional assignment problem. The problem size ranges from n = 4 to n = 26, with increments of 2. For each problem size, five instances are randomly generated, and the cost coefficients c(i,j,k) are integers uniformly distributed between 0 and 100. The costs found by the proposed algorithm for each instance of each problem size are presented in Table 2. In addition, Table 3 displays the outcomes of the experiments conducted on this dataset, where each row presents the average score of the five instances with the same size as the benchmarked algorithms.

The presented table contains the comparison results of different algorithms on the three-dimensional assignment problem dataset. The "Optimal" column displays the optimal solutions obtained from the dataset creators, while the "B-S" column represents the solutions obtained by using their Variable Depth Interchange heuristic. The "AMGO" column displays the outcomes from Jiang's research [7], and the "LSGA" column shows the results of Huang's paper [6]. The last column, "MoBFOA-PHM," represents the average costs achieved by the proposed algorithm in this study. The best results are highlighted in bold.

According to Table 3, our proposed algorithm outperforms other benchmarked algorithms, except for problem sizes 16 and 22. For problem sizes up to 14, both AMGO, LSGA, and our proposed algorithm are able to find the optimal solutions. Additionally, our proposed MoBFOA-PHM algorithm is able to find the optimal solutions for problem sizes 18 and 20.

Table 2. Costs found by the proposed algorithm for each instance

Problem Size	Test Instances				
	1	2	3	4	5
4	46	47	48	53	17
6	46	42	42	43	28
8	19	32	23	37	8
10	25	11	21	21	17
12	17	22	11	14	14
14	9	9	13	10	9
16	12	7	13	11	10
18	6	5	7	8	6
20	5	5	3	7	4
22	5	3	6	6	4
24	4	3	2	2	2
26	1	2	2	2	2

Table 3. Balas and Saltzman Dataset (12 * 5 instances)

Problem Size	Avg. Cost Optimal	Avg. Cost MoBFOA-PHM	Avg. Cost B-S	Avg. Cost AMGO	Avg. Cost LSGA
4	42,2	**42,2**	43,2	**42,2**	**42,2**
6	40,2	**40,2**	45,4	**40,2**	**40,2**
8	23,8	**23,8**	33,6	**23,8**	**23,8**
10	19	**19**	40,8	**19**	**19**
12	15,6	**15,6**	24	**15,6**	**15,6**
14	10	**10**	22,4	**10**	**10**
16	10	10,6	25	**10**	**10**
18	6,4	**6,4**	17,6	6,8	7,2
20	4,8	**4,8**	27,4	5	5,2
22	4	4,8	18,8	**4,4**	5,6
24	1,8	**2,6**	14	2,8	3,2
26	1,3	**1,8**	15,7	2,4	3,6

**** Best solutions are written in bold***

6 Conclusion

The Three-Index Assignment Problem (3-AP) is studied in this paper. We hybridized BFOA algorithm with Hungarian Method and parallelly applied the chemotactic step to the members of the colony. The Hungarian method increases the algorithm's ability to find the best solution within the nearby solution space.

The experimental outcomes revealed that our hybrid method (MoBFOA-PHM) outperformed other available heuristics. The proposed algorithm produced better solutions in similar execution times for the tested benchmark samples. Furthermore, it was noted that parallel execution of the chemotactic step was efficient in reducing the computational time, particularly for large-sized problems. In future research, other data sets will be tested for comparison of algorithm performance.

References

1. Sanyal, N., Chatterjee, A., Munshi, S.: Bacterial foraging optimization algorithm with varying population for entropy maximization based image segmentation. In: Proceedings of the 2014 International Conference on Control, Instrumentation, Energy and Communication (CIEC), Calcutta, India, pp. 641–645 (2014)
2. Bhaladhare, P.R., Jinwala, D.C.: A clustering approach using fractional calculus-bacterial foraging optimization algorithm for k-anonymization in privacy preserving data mining. Int. J. Inf. Secur. Priv. (IJISP) **10**(1), 45–65 (2016)
3. Bakhshali, M.A., Shamsi, M.: Facial skin segmentation using bacterial foraging optimization algorithm. J. Med. Signals Sens. **2**(4), 203–210 (2012)
4. Kim, D.H., Abraham, A., Cho, J.H.: A hybrid genetic algorithm and bacterial foraging approach for global optimization. Inf. Sci. **177**(18), 3918–3937 (2007)
5. Dokka, T., Mourtos, I., Spieksma, F.C.R.: Fast separation for the three-index assignment problem. Math. Prog. Comput. **9**, 39–59 (2017)

6. Huang, G., Lim, A.: A hybrid genetic algorithm for the Three-Index Assignment Problem. Eur. J. Oper. Res. **172**(1), 249–257 (2006)
7. Jiang, H., Xuan, J., Zhang, X.: An approximate muscle guided global optimization algorithm for the Three-Index Assignment Problem. In: 2008 IEEE Congress on Evolutionary Computation (IEEE World Congress on Computational Intelligence), Hong Kong, China, pp. 2404–2410 (2008)
8. Bandelt, H.-J., Crama, Y., Spieksma, F.C.R.: Approximation algorithms for multi-dimensional assignment problems with decomposable costs. Discret. Appl. Math. **49**(1–3), 25–50 (1994)
9. Spieksma, F.C.R.: Multi index assignment problems: complexity, approximation, applications. In: Pardalos, P.M., Pitsoulis, L.S. (eds.) Nonlinear Assignment Problems, pp. 1–12. Springer, Boston (2000). https://doi.org/10.1007/978-1-4757-3155-2_1
10. Burkard, R.E., Rudolf, R., Woeginger, G.J.: Three dimensional axial assignment problems with decomposable cost coefficients. Discret. Appl. Math. **65**, 123–169 (1996)
11. Gutin, G., Karapetyan, D.: A memetic algorithm for the multidimensional assignment problem. In: Stützle, T., Birattari, M., Hoos, H.H. (eds.) Engineering Stochastic Local Search Algorithms. Designing, Implementing and Analyzing Effective Heuristics, pp. 125–129. Springer, Heidelberg (2009). https://doi.org/10.1007/978-3-642-03751-1_12
12. Krokhmal, P.A., Grundel, D.A., Pardalos, P.M.: Asymptotic behavior of the expected optimal value of the multidimensional assignment problem. Math. Program. **109**, 525–551 (2007)
13. Karp, R.: Reducibility among combinatorial problems. In: Miller, R., Thatcher, J. (eds.) Complexity of Computer Computations, pp. 85–103. Plenum Press (1972)
14. Passino, K.M.: Biomimicry of bacterial foraging for distributed optimization and control. IEEE Control Syst. Mag. **22**(3), 52–67 (2002)
15. Kuhn, H.W.: The Hungarian Method for the assignment problem. Nav. Res. Logist. Q. **2**, 83–97 (1955)
16. Balas, E., Saltzman, M.J.: An algorithm for the three-index assignment problem. Oper. Res. **39**, 150–161 (1991)

EFQM Based Supplier Selection

Ozlem Senvar(✉) and Mustafa Ozan Nesanir

Department of Industrial Engineering, Marmara University, Maltepe, 34854 Istanbul, Turkey
ozlem.senvar@marmara.edu.tr, ozlemsenvarresearch@gmail.com

Abstract. In this study, the importance of quality management practices in achievement of operational results and customer satisfaction in logistics is handled. The objective of this study is to propose a quality management framework through multi criteria decision making (MCDM) based on the criteria utilized in the European Foundation for Quality Management (EFQM) for assessment of suppliers. In this regard, this study proposes EFQM based integrated methodology using Analytic Hierarchy Process (AHP) and Technique for Order Preference by Similarity to an Ideal Solution (TOPSIS) method for supplier selection problem.

Keywords: EFQM · Supplier Selection · Multi Criteria Decision Making

1 Introduction

Provision of the required product and quantity at the right place, at the right time, at an affordable cost, in an efficient manner and in right condition is inevitable for logistics management [26]. Since importance given to quality in the field of logistics is increasing [4], it is critical to choose a competent supplier in order to provide the necessary quality in logistics management.

Provision of the required product and quantity at the right place, at the right time, at an affordable cost, in an efficient manner and in right condition is inevitable for logistics management [26]. Since importance given to quality in the field of logistics is increasing [4], it is critical to choose a competent supplier in order to provide the necessary quality in logistics management.

For evaluating alternatives, MCDM can be used instead of RADAR which is EFQM Models' own evaluation system. With RADAR, companies are required to share all relevant information so that they can be fully examined. However, this is not possible among companies in the same industry. Also, independent authority is required for RADAR evaluation system. For these reasons, alternatives can be examined with multi criteria decision making methods instead of the RADAR system.

Bozorgmehr and Tavakoli [3] emphasises that agile decision-making tool is necessary for accurate supplier evaluation considering supplier performance verification to enhance long term partnership leading to reliable supplier selection process. Metaxas et al. [12] developed integrated methodology via fuzzy AHP to determine the weights of the criteria by decision makers and TOPSIS to rank alternatives. The proposed instrument is used

Z. Şen et al. (Eds.): IMSS 2023, LNME, pp. 499–509, 2024.
https://doi.org/10.1007/978-981-99-6062-0_46

to evaluate Sustainable Business Excellence Index (SBEI) and its potential impact on the formulation of firm strategy.

The weights of the criteria are determined through AHP method. This study aims to propose an EFQM based integrated methodology using AHP and TOPSIS methods for supplier selection problem. Integrated AHP and TOPSIS approach is used considering the criteria of EFQM Model 2020. The weights of the criteria are determined through AHP method.

This study presents a case study considering Ekol Logistics, which is Turkey's leading logistics exporter and increasingly growing company in Europe. Since Ekol Logistics provides services in Europe, it aims to evaluate suppliers on the basis of EFQM criteria in cases of using outsourcing. Three suppliers, which are logistics service providers, are considered as alternatives. Three companies, which provide services via road, water and railway, have been in the logistics industry for more than 30 years. Supplier names are coded as A, B and C for confidentiality purposes.

This study presents a case study considering Ekol Logistics, which is Turkey's leading logistics exporter and increasingly growing company in Europe. Since Ekol Logistics provides services in Europe, it aims to evaluate suppliers on the basis of EFQM criteria in cases of using outsourcing. Three suppliers, which are logistics service providers, are considered as alternatives. Three companies, which provide services via road, water and railway, have been in the logistics industry for more than 30 years. Supplier names are coded as A, B and C for confidentiality purposes. The rest of the study is organized as follows: Sect. 2 briefly presents literature review on EFQM and MCDM methods with EFQM. Section 3 explains methodology. Section 4 explains EFQM 2020 Model. Section 5 presents application and results. Section 6 provides conclusion.

2 Literature Review

The literature review section consists of two parts, which are overview to EFQM and MCMD methods with EFQM.

2.1 Overview to EFQM

EFQM is a non-profit organization created in 1988 by fourteen leading European businesses with a mission of sustainable excellence in Europe and a vision of a world. Recognizing the challenge, the EFQM was established to promote world-class approaches that will achieve sustainable excellence in the management of European organisations Its members are industry and service companies, as well as universities, research agencies and institutions in academia [5, 28].

With the EFQM Excellence Model, organizations see where they are on the road to corporate excellence and areas for improvement. It is ensured that a common way of thinking is formed so that the organization can live its ideas inside and outside. It allows repetitive jobs to be seen in organization. The basic EFQM Excellence Model was originally developed jointly by practitioners and scientists and published in 1991. Since 1991, it has been constantly reviewed and improved. A major revision of the

Excellence Model took place in 1999. Innovation and learning, as well as collaboration with external partners, was identified as critical to corporate performance [28].

The EFQM Excellence Model 2013 came after the EFQM Excellence Model 2010. The EFQM Excellence Model 2013 was revised to align with business needs and trends. There is a more balanced scoring system in EFQM Excellence Model 2013 compared to the 2010 model. The EFQM Excellence Model 2013 aims to achieve standards of excellence through leadership, to establish policies and strategies to achieve quality, and to use resources and personnel appropriately by directing all company processes on a customer-based basis. These processes provide a positive impact on society in addition to customer and staff satisfaction [9].

EFQM Model 2020 came after the EFQM Excellence Model 2013. The United Nations' 17 Sustainable Development Goals were an important factor in shaping the EFQM Model 2020. The concept of "excellence" has been removed from the model's name. While EFQM Model 2020 consists of seven criteria, EFQM Excellence Model 2013 consists of nine criteria. Sustainability is given more place in the EFQM Model 2020 [6, 7, 15].

2.2 MCDM Methods with EFQM Model

Related studies in the literature can be seen in Table 1.

Table 1. Summary of EFQM related studies with MCDM

Authors (year)	Title	Methodology
Jalaliyoon et al. (2012) [10]	Implementation of Balanced Score Card and European Foundation for Quality Management Using TOPSIS Method	TOPSIS
Serkani et al. (2013) [22]	Using AHP and ANP Approaches for Selecting Improvement Projects of Iranian Excellence Model in Healthcare Sector	Analytic Network Process (ANP) and AHP
Sajedi et al. (2013) [21]	An Improved TOPSIS/EFQM Methodology for Evaluating the Performance of Establishments	TOPSIS

(continued)

Table 1. (*continued*)

Authors (year)	Title	Methodology
Bozorgmehr and Tavakoli (2015) [3]	An EFQM-Based Supplier Evaluation and Verification Methodology Based on Product's Manufacturing Policy and Technology Intensity: Industrial Gas Turbine Industry as a Case Study	AHP, Weighted Sum Method
Najafi and Naji (2018) [14]	Hybrid Multiple Criteria Decision Making and Balanced Scorecard Approach	Hybrid Fuzzy ANP, Fuzzy ELECTRE and Fuzzy TOPSIS
Raziei (2019) [19]	EFQM Excellence Model Based on Multi Criteria Processes Fuzzy AHP, Fuzzy DEMATEL, Fuzzy TOPSIS, and Fuzzy VIKOR; A Comparative Survey	Hybrid Fuzzy AHP, Fuzzy DEMATEL, Fuzzy TOPSIS, and Fuzzy VIKOR
Uygun et al. (2020) [25]	A Novel Assessment Approach to EFQM Driven Institutionalization Using Integrated Fuzzy Multi Criteria Decision-Making Methods	Integrated fuzzy DEMATEL, fuzzy ANP, and VIKOR

3 Integrated AHP and TOPSIS Approach

AHP method is used for criteria weighting. TOPSIS method is used to rank the alternatives. AHP method was developed by Thomas Saaty for the solution of complex problems involving multiple criteria [27]. The AHP method gives significant contributions to the activity of structuring and analyzing decision problems, which they deal with on very different issues, and is a method that is heavily used [1].

The TOPSIS method was presented in 1992 by Chen and Hwang, with reference to the work of Hwang and Yoon in 1981. The basic principle is that the chosen alternative is the shortest distance from the ideal solution and the farthest distance from the negative ideal solution [17]. Integrated AHP and TOPSIS are executed by combining the following steps [8, 16, 18].

Step 1: Hierarchical structure: The first step in AHP is to create a hierarchical structure with criteria and sub criteria related to the problem. It is necessary to start with this step, in order to better understand and evaluate the problem of interest [13].
Step 2: Pairwise comparison matrix: Pairwise comparison allows to focus on the answers to some simple questions, thus allowing the weighting to be done correctly (13). In order

to make comparisons, it is necessary to use Saaty's scale that shows how many times more important or dominant one criterion is than the other criterion [20].

Step 3: Calculation of relative priority values of criteria: Relative priority values of each criterion are calculated and normalization is performed. After the relative priority matrix is formed, the consistency test is performed with the ratio CR = CI/RI [11]. As a result of the calculation, if the Consistency Ratio (CR) is less than 0.1, it is consistent [23].

Step 4: Determining the decision matrix: In the decision matrix, the columns are the criteria to be used for comparison, and the rows are the alternatives of the decision maker.

Step 5: Performing the standardization process: The standard decision matrix is created by performing the standardization process.

Step 6: Making the weighted standard decision matrix: The weight values of criteria are determined so that the weight totals are 1. Then, the standard decision matrix and the determined weight values are multiplied to obtain the weighted standard decision matrix.

Step 7: Identifying ideal and negative ideal solutions: In the 4th step, ideal solution and negative ideal solution are calculated [2]. Ideal solution consists of alternatives that get the best values in each criterion. Negative ideal solution consists of alternatives with worst values in each criterion.

Step 8: Developing negative ideal and positive ideal distance measures: The distances from the negative and positive ideal solution set are measured by the Euclidean distance approximation. The deviation values of the obtained alternatives regarding the criteria are called "Negative Ideal Discrimination (Si−) "and "Positive Ideal Discrimination (Si+)" measure.

Step 9: Calculating the relative closeness to the ideal solution: By using positive and negative discrimination measures, the closeness coefficient (Ci*) is calculated according to the ideal solution for each alternative of the decision maker. The closeness coefficient is represented as an Eq. (1).

$$c_i^* = \frac{S_i^-}{S_i^+ + S_i^-} \tag{1}$$

Ci* value is equal to or between 0 and 1. Ci* value close to 0 indicates that it is far from the ideal solution, and a Ci* value close to 1 indicates that it is close to the ideal solution.

4 EFQM Model 2020

EFQM Model 2020 have been updated in relation to global trends and global changes. 17 sustainable development goals of the United Nations played an important role in shaping the EFQM Model 2020 [7]. 17 sustainable development goals guide humanity in lifting humanity out of poverty, ensuring a healthy planet for future generations, eliminating inequalities and ensuring dignified lives for all [24].

4.1 Criteria of EFQM Model 2020

The EFQM Model 2020 was launched in 2019. The latest version of this model, the EFQM Model 2020, was designed using years of experience in changing markets to understand the benefits of organizational analysis, future forecasting and predictive intelligence. There are seven criteria which are "Purpose, Vision and Strategy", "Organisational Culture and Leadership", "Engaging Stakeholders", "Creating Sustainable Value", "Driving Performance and Transformation", "Stakeholder Perception", "Strategic and Operational Performance" for the EFQM Model 2020. And There are totally 23 sub criteria [7, 15].

5 Application

In this part, it was aimed to select a logistics supplier for Ekol Logistics company for the European market. Three logistics supplier which can serve as road, water or railway were examined. Three suppliers have been in the logistics industry for more than 30 years. Supplier names are shown as A, B and C for confidentiality purposes.

AHP method was used for evaluating criteria weights. TOPSIS was used for logistics supplier selection. Three experts, who have experiences in Central Planning & Freight and Planning, Business Intelligence & Freight Reporting and Sustainability Strategies, evaluates the alternatives with respect to the criteria. Judgements of three experts are taken considering criteria evaluation for both operational and sustainability perspectives.

For creation of decision matrix which consists of alternatives and criteria, three experts compared the alternatives with respect to the criterion. These subjective judgments were made via numerical value using Saaty's scale. Creating of single judgement in group decision, can be made by taking geometric mean. After determining the decision matrix, experts constructed the pairwise comparison matrices by using scale given in Saaty (20). Single pairwise comparison matrix was created by taking geometric mean of three experts' judgements. Weight coefficient for each criterion was determined over the single pairwise comparison.

CR value was computed below 0.1 (consistent) for both of the pairwise comparison matrix of all three experts and pairwise comparison matrix calculated using geometric mean. After the weights of the criteria were determined, ranking of alternatives was assessed with TOPSIS method.

5.1 Implementation of Steps

Figure 1 shows steps of methodology applied for implementation of this study.

As the first step, a hierarchical structure including the goal, criteria and alternatives was created (see Fig. 2).

As the second and third step, relative priority values of the criteria were determined with group decision and the values were shown as the W matrix in Table 2.

After calculating the relative priority values of the criteria, the Consistency Ratio (CR) value was calculated. For the CR value, first, the C and W matrices were multiplied. Then, the CR value was obtained by calculating the λ value [7.14] and CI value (0.02), by taking the RI value as 1.32 (EFQM Model 2020 – 7 criteria). CR value (0.02) was below 0.1 (consistent) both for the pairwise comparison matrices of three experts and for the single pairwise comparison matrix calculated with the geometric mean.

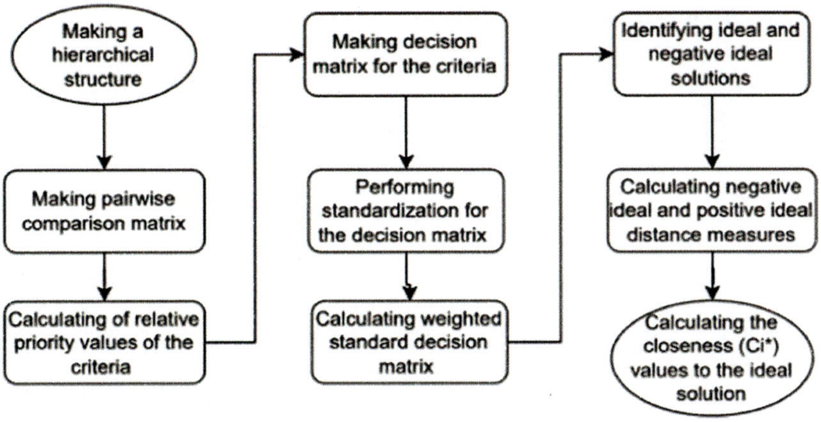

Fig. 1. Flow chart for steps of methodology

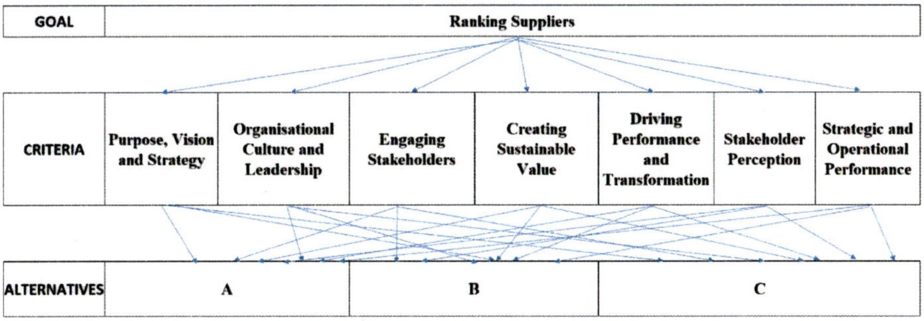

Fig. 2. Hierarchical Structure for supplier selection based on EFQM Model 2020

As the fourth step, with Saaty's Scale, three experts (Central Planning & Freight Manager; Planning, Business Intelligence & Freight Reporting Expert; Sustainability Strategies Expert) evaluated alternatives respectively to the criteria. Then, Table 3 was obtained by calculating the geometric mean of three experts' judgements.

As the fifth step, the standardization process was performed (Table 4) for the decision matrix created in the fourth step. Standardization was made by dividing each value in each column by the square root of the sum of the squares of the values in the column.

As the sixth step, calculated criteria weights (Table 2) were written above in Table 5. The weight of the relevant criterion was multiplied with the relevant values in the matrix.

As the seventh step, ideal and negative ideal solutions are determined. All criteria are benefit criteria. And since the criteria are benefit criteria, the maximum values in the columns of the weighted standardized decision matrix are the ideal solution values. The minimum values in the columns are the negative ideal solution values. In Table 6, ideal values were written as V+ and negative ideal values as V−.

As the eight step, negative ideal and positive ideal distance measures were calculated (Table 7). Si+ shows the distance of the relevant alternative from the positive ideal

Table 2. Relative Priority Values of Criteria for EFQM Model 2020

	Purpose, Vision and Strategy	Organisational Culture and Leadership	Engaging Stakeholders	Creating Sustainable Value	Driving Performance and Transformation	Stakeholder Perception	Strategic and Operational Performance	W
Purpose, Vision and Strategy	0.12	0.18	0.12	0.13	0.09	0.17	0.10	0.13
Organisational Culture and Leadership	0.06	0.09	0.10	0.11	0.09	0.09	0.08	0.09
Engaging Stakeholders	0.13	0.12	0.13	0.15	0.13	0.15	0.12	0.13
Creating Sustainable Value	0.13	0.12	0.13	0.15	0.15	0.13	0.20	0.15
Driving Performance and Transformation	0.29	0.22	0.23	0.22	0.22	0.25	0.17	0.23
Stakeholder Perception	0.07	0.10	0.09	0.11	0.09	0.10	0.17	0.10
Strategic and Operational Performance	0.21	0.18	0.19	0.13	0.22	0.10	0.17	0.17

Table 3. Decision Matrix for the Criteria of the EFQM Excellence Model 2020

Alternatives	Purpose, Vision and Strategy	Organisational Culture and Leadership	Engaging Stakeholders	Creating Sustainable Value	Driving Performance and Transformation	Stakeholder Perception	Strategic and Operational Performance
A	7.00	7.00	8.32	7.32	7.61	7.56	7.65
B	6.32	7.00	6.65	7.32	6.95	7.27	6.95
C	7.00	6.65	7.32	6.65	6.65	6.21	7.32

Table 4. Standardization for EFQM Model 2020

Alternatives	Purpose, Vision and Strategy	Organisational Culture and Leadership	Engaging Stakeholders	Creating Sustainable Value	Driving Performance and Transformation	Stakeholder Perception	Strategic and Operational Performance
A	0.596	0.587	0.644	0.595	0.620	0.620	0.604
B	0.538	0.587	0.515	0.595	0.567	0.596	0.549
C	0.596	0.558	0.566	0.541	0.542	0.510	0.578

distance, while Si− shows the distance of the relevant alternative from the negative ideal distance.

As the nineth and last step, the closeness (Ci*) values to the ideal solution were calculated (Table 8). The Ci* value was obtained by dividing the negative ideal distance

Table 5. Weighted Standard Decision Matrix for EFQM Model 2020

W	0.13	0.09	0.13	0.15	0.23	0.10	0.17
Alternatives	Purpose, Vision and Strategy	Organisational Culture and Leadership	Engaging Stakeholders	Creating Sustainable Value	Driving Performance and Transformation	Stakeholder Perception	Strategic and Operational Performance
A	0.596	0.587	0.644	0.595	0.620	0.620	0.604
B	0.538	0.587	0.515	0.595	0.567	0.596	0.549
C	0.596	0.558	0.566	0.541	0.542	0.510	0.578

Table 6. Ideal and Negative Ideal Solutions for EFQM Model 2020

W	0.13	0.09	0.13	0.15	0.23	0.10	0.17
Alternatives	Purpose, Vision and Strategy	Organisational Culture and Leadership	Engaging Stakeholders	Creating Sustainable Value	Driving Performance and Transformation	Stakeholder Perception	Strategic and Operational Performance
A	0.078	0.052	0.085	0.087	0.141	0.065	0.104
B	0.070	0.052	0.068	0.087	0.129	0.062	0.094
C	0.078	0.049	0.075	0.079	0.123	0.053	0.099
V+	0.078	0.052	0.085	0.087	0.141	0.065	0.104
V−	0.0701	0.0493	0.0679	0.0788	0.1233	0.0534	0.0941

Table 7. Negative Ideal and Positive Ideal Distance Measure for EFQM Model 2020

Alternatives	Si+	Si−
A	0.000	0.031
B	0.024	0.014
C	0.025	0.011

for the relevant alternative by the sum of the positive ideal distance and the negative ideal distance. The alternative with a Ci* value close to 1 is better. Alternative A was ahead of B and C.

Table 8. Ci* Value for EFQM Model 2020

Alternatives	Si+
A	1.000
B	0.357
C	0.309

6 Conclusion

In this study, three experts, which were "Central Planning & Freight Manager", "Planning, Business Intelligence & Freight Reporting Expert" and "Sustainability Strategies Expert", judged the criteria and alternatives. In this way, both operational and sustainability perspective were obtained for evaluating. In addition, the "Sustainability Strategies Expert" knows the strategy of the Ekol Logistics. This brought also strategy perspective for evaluation.

Supplier A is ahead of supplier B and supplier C based on EFQM Model 2020. As a result of the evaluation, supplier B and supplier C received almost the same score. Supplier B is slightly ahead of Supplier C. Supplier A can be selected for transportation in Europe. For further direction, other multi criteria decision making methodology can be adapted to solve the considered problem and compared with the results of this study.

Acknowledgments. The authors would like to thank Ekol Logistics for financial support to this research project.

References

1. Aktaş, R., Doğanay, M.M., Gökmen, Y., Gazibey, Y., Türen, U.: Sayisal Karar Verme Yöntemleri. Beta Basım A. Ş., İstanbul (2015)
2. Lezki, Ş., Sönmez, H., Işıklar, E., Özdemir, A., Alptekin, N.: İşletmelerde Karar Verme Teknikleri. In: Durucasu, H. (ed.) Anadolu Üniversitesi Yayınları, Eskişehir, pp. 102–137 (2016)
3. Bozorgmehr, M., Tavakoli, L.: An EFQM-based supplier evaluation and verification methodology based on product's manufacturing policy and technology intensity: industrial gas turbine industry as a case study. Int. J. Integr. Supply Manag. **9**(4), 276–306 (2015)
4. Czajkowska, A., Stasiak-Betlejewska, R.: Quality management tools applying in the strategy of logistics services quality improvement. Serbian J. Manag. **10**(2), 225–234 (2015)
5. The Fundamental Concepts of Excellence. https://www.onecaribbean.org/wp-content/uploads/Fundamental-Concepts-of-EFQM.pdf. Accessed 11 Jan 2023
6. EFQM: EFQM Excellence Model 2013. Brussels, Belgium (2012)
7. EFQM: The EFQM Model. 2nd edn (2021)
8. Ertuğrul, İ, Özçil, A.: Çok Kriterli Karar Vermede TOPSIS Ve VIKOR Yöntemleriyle Klima Seçimi. Çankırı Karatekin Üniversitesi İktisadi ve İdari Bilimler Fakültesi Dergisi **4**(1), 267–282 (2014)
9. Rivera, D.E., Terradellas Piferrer, M.R., Benito Mundet, M.H.: Measuring territorial social responsibility and sustainability using the EFQM excellence model. Sustainability **13**(4), 1–25 (2021)
10. Jalaliyoon, N., Bakar, N.A., Taherdoost, H.: Implementation of balanced score card and European foundation for quality management using TOPSIS method. Manag. Int. J. Acad. Res. Manag. (IJARM) **1**(1), 26–35 (2012)
11. Kuru, A., Akın, B.: Entegre Yönetim Sistemlerinde Çok Kriterli Karar Verme Tekniklerinin Kullanımına Yönelik Yaklaşımlar ve Uygulamaları. Öneri Dergisi **10**(38), 129–144 (2012)
12. Metaxas, I.N., Koulouriotis, D.E., Spartalis, S.H.: A multicriteria model on calculating the sustainable business excellence index of a firm with fuzzy AHP and TOPSIS. Benchmarking Int. J. **23**(6), 1522–1557 (2016). https://doi.org/10.1108/BIJ-07-2015-0072

13. Millet, I.: Ethical decision making using the analytic hierarchy process. J. Bus. Ethics **17**(11), 1197–1204 (1998)
14. Najafi, A., Naji, E.: Improving projects of the EFQM model using fuzzy hybrid multiple criteria decision making and balanced scorecard approach. Indian J. Sci. Technol. **9**(44), 1–7 (2016)
15. Nenadál, J.: The new EFQM model: what is really new and could be considered as a suitable tool with respect to quality 4.0 concept? Qual. Innov. Prosperity **24**(1), 17 (2020). https://doi.org/10.12776/qip.v24i1.1415
16. Olson, D.L.: Comparison of weights in TOPSIS models. Math. Comput. Model. **40**(7–8), 721–727 (2004)
17. Opricovic, S., Tzeng, G.H.: Compromise solution by MCDM methods: a comparative analysis of VIKOR and TOPSIS. Eur. J. Oper. Res. **156**(2), 445–455 (2004)
18. Özkan, H., Kocaoğlu, B.: Özkan, M: Bir Eğitim Kurumunun Yemek Hizmeti Alımında Analitik Hiyerarşi Sürecine Göre Tedarikçi Seçimi. Uluslararası Sosyal Araştırmalar Dergisi **11**(59), 1048–1062 (2018)
19. Raziei, S.: EFQM excellence model based on multi-criteria processes fuzzy AHP, fuzzy dematel, fuzzy TOPSIS, and fuzzy VIKOR; a comparative survey. Int. J. Sci. Technol. Res. **8**(04), 248–260 (2019)
20. Saaty, T.L.: Relative measurement and its generalization in decision making why pairwise comparisons are central in mathematics for the measurement of intangible factors the analytic hierarchy/network process. RACSAM-Revista De La Real Academia De Ciencias Exactas, Fisicas Y Naturales. Serie A. Matematicas **102**, 251–318 (2008)
21. Gholamzadeh, M., Mojahed, M.: An improved TOPSIS/EFQM methodology for evaluating the performance of organizations. Life Sci. J. **10**(1), 4315–4322 (2013)
22. Serkani, E.S., Mardi, M., Najafi, E., Jahanian, K., Herat, A.T.: Using AHP and ANP approaches for selecting improvement projects of iranian excellence model in healthcare sector. Afr. J. Bus. Manag. **7**(23), 2271–2283 (2013)
23. Taha, H.A.: Yöneylem Araştırması. Baray, Ş., Esnaf, Ş. (translation). Literatür Yayıncılık, İstanbul (original press date 1997) (2014)
24. United Nations. The Sustainable Development Goals Report 2017. United Nations Publications, New York (2017)
25. Uygun, O., Yalcin, S., Kiraz, A., Erkan, E.F.: A novel assessment approach to EFQM driven institutionalization using integrated fuzzy multi-criteria decision-making methods. Scientia Iranica **27**(2), 880–892 (2020)
26. Wang, K.: 4.0 solution-new challenges and opportunities. In: 6th International Workshop of Advanced Manufacturing and Automation, pp. 68–74. Atlantis Press (2016)
27. Winston, L.W.: Operations Research Applications and Algorithms, 4th edn. Thomson Learning, Belmont (2004)
28. Zink, K., Vob, W.: The new EFQM excellence model and its impact on higher education institutions. Sinergie Rapporti di Ricerca **9**(2000), 241–255 (1999)

Classification of Rice Varieties Using a Deep Neural Network Model

Nuran Peker(✉)

Sakarya University, 54050 Serdivan/Sakarya, Turkey
nuranpeker41@gmail.com

Abstract. Deep learning is a machine learning approach that has been widely used in many different fields in recent years. It is used in agriculture for various purposes, such as product classification and diagnosis of agricultural diseases. In this study, we propose a deep-learning model for the classification of rice species. Rice is an agricultural product that is widely consumed in Turkey as well as in the world. In our study, a rice data set that contains 7 morphological features obtained by using 3810 rice grains belonging to two species is used. Our model consists of three hidden layers and two dropouts (3H2D) added to these layers to prevent overfitting in classification. The success of the model is compared with Logistic Regression (LR), Multilayer Perceptron (MLP), Support Vector Machine (SVM), Decision Tree (DT), Random Forest (RF), Naïve Bayes (NB), K-Nearest Neighbors (KNN), Ada Boost (AB), Bagging (BG), and Voting (VT) classifiers. The success rates of these methods are as follows: 93.02%, 92.86%, 92.83%, 92.49%, 92.39%, 91.71%, 88.58%, 92.34%, 91.68%, and 90.52% respectively. The success rate of the proposed method is 94.09%. According to the results obtained, the proposed method is more successful than all of these machine learning methods.

Keywords: Deep Neural Networks · Machine Learning · Rice Data · Classification

1 Introduction

Deep learning is a machine learning technique with sub-branches such as deep neural networks, convolutional neural networks, deep belief networks, and recurrent neural networks. In this study, a deep neural network is used. Deep neural networks refer to a type of learning model in which multiple layers of nodes are used to derive high-level functions from the model input information. The more layers it needs to process to achieve the result, the greater the depth of the network. The use of methods based on deep learning has increased in the field of agriculture, as well as in many other fields, in recent years. In one study [1], deep learning-based methods are applied for weed detection in agricultural crops including carrots, sunflowers, sugar beets, soybeans, and maize. In the study [2], which presents a constructed model using computer vision and a deep convolutional neural network to aid in the prediction of crop diseases, 14 crop types and 26 diseases can be identified using a dataset of both diseased and healthy plant leaves.

Z. Şen et al. (Eds.): IMSS 2023, LNME, pp. 510–521, 2024.
https://doi.org/10.1007/978-981-99-6062-0_47

Another study [3] presents a deep learning-based model for early prediction of crop frost to help farmers save crops. In the study [4] on products including different fruits and vegetables, it is observed that the application of deep learning methods provides better accuracy for precise agriculture.

In this study, a deep learning-based classification model is proposed for two rice types patented in Turkey. Rice is an important food source for the majority of the world's population and has a large share of the global agricultural industry [5]. Rice has a big significance in human nutrition in Turkey, as well as in the world, in terms of being economical and nutritious. In recent years, there have been many studies on rice using via deep-learning approach. In the study [6], a pixel-level classification model is presented that is used convolutional neural networks to detect rice in complex landscape areas, where rice is easily mixed up with its surroundings. The study [7] shows that to improve the accuracy of rice counting and detection in the field, a system is developed for the detection and counting of panicles based on advanced region-based convolutional neural networks. For collecting images, an unmanned aerial vehicle that has a high-definition RGB camera is used. In the study [8], an end-to-end prediction method for rice yield estimation is proposed, combining two backpropagation neural networks with an independently recurrent neural network. The study [9] introduces a computational model based on convolution neural networks to detect N6-methyladenine sites in the rice genome. In the study [10], a model with a custom memory-efficient convolutional neural network (namely RiceNet) is presented for the detection of rice grain diseases automatically. In their study, Yang et al. [11] propose an approach to identify the main growth stages of rice from RGB images using convolutional neural network architecture. In their study, Shi et al. [12] propose a collaborative method that combines deep learning and machine learning theory to monitor rice quality variance. In the study [13], a faster region-based convolutional neural network is utilized for the real-time determination of three rice leaf diseases including hispa, brown spot, and, rice blast. The study [14] proposes a method that combines the learning capability of deep learning with the simplicity of phenological methods to map the rice paddies with high accuracy without the requirement of field samples. In the literature, there are also many different studies using deep learning-based classification models, the main subject of which is rice. In the study [15], the convolutional neural network is used to classify the broken rice and the whole rice. In the study [16], a hyperspectral imaging model is presented for the classification of 10 different high-quality rice varieties using a deep learning network. In the study [17], a label-free classification model is proposed for different rice varieties using a deep neural network-assisted imaging technique. In the study [18], a deep learning-based method is used to classify both grain-shaped and milled images of five varieties of Spanish rice. In the study [19], a classification model for rice seedling images is proposed using a convolutional neural network algorithm. In the study [20], a classification model is proposed using a convolutional neural network with 18 layers for seven different rice varieties mostly grown in Pakistan. In the study [21], a model is proposed to classify five different types of rice (semi, sd, integ, blanco, vapo) in flour or grain format with a convolutional neural network using thermographic images of rice.

As it can be understood from the examples mentioned above, the studies on the classification of rice in the literature generally consist of convolutional neural network models in which rice images are used as the training data. In our study, differently, we present a deep neural network model for the classification of rice data. For this purpose, we use the Rice dataset, which consists of numerical equivalents of rice images, details of which are described in Sect. 2.1. In other words, we classify a data set that includes not the images of rice, but the numerical equivalents obtained from these images. These numerical data consist of the morphological data of the rice grains through the images obtained by computer vision.

2 Material and Methods

2.1 Dataset

In this study, the Rice dataset created by Cinar and Koklu [22], via computer vision of rice grains of Osmancik and Cammeo species which are native to Turkey is used. In their work, the images of rice samples are obtained and the images are processed with the image processing techniques. The images are converted to grayscale images and then the binary ones. In the next step, the noise is removed from the images. In the last step, morphological features are extracted from the acquired images. The Rice dataset consists of numerical values obtained from these morphological characteristics of rice grains. Table 1 shows the morphological features and explanations of the Rice data. Table 2 shows the statistics of the Rice data.

Table 1. The morphological features and explanations of the Rice data

Feature	Explanation
Area	States the number of pixels therein the boundaries of the grain of rice
Perimeter	Computes the perimeter by reckoning the distance among pixels around the boundaries of the rice grain
MajorAxisLength	States the longest line that can be drawn on the grain of rice, that is, the big axis distance
MinorAxisLength	States the shortest line that can be drawn on the grain of rice, that is, the small axis distance
Eccentricity	Measures how round the ellipse is, which has identical moments as the grain of rice
ConvexArea	States the pixel count of the smallest convex shell of the region created by the rice grain
Extent	States the ratio of the region created by the rice grain to the bounding box pixels
Class	Class distribution: 2180 Osmancik, 1630 Cammeo

Table 2. The statistics of the Rice data

Features	Min	Max	Mean	Std. Dev.	Skewness	Kurtosis
Area	7551	18913	12667.73	1732.37	0.3252	−0.4311
Perimeter	359.1	548.446	454.2392	35.5971	0.2214	−0.8402
MajorAxisLength	145.2645	239.0105	188.7762	17.4487	0.2602	−0.9518
MinorAxisLength	59.5324	107.5424	86.3138	5.7298	−0.1349	0.5621
Eccentricity	0.7772	0.948	0.8869	0.0208	−0.4492	0.0711
ConvexArea	7723	19099	12952.50	1776.97	0.3198	−0.4658
Extent	0.4974	0.861	0.6619	0.0772	0.3438	−1.0301

2.2 Methods

Logistic Regression. Logistic regression-LR [24] is a statistical approach that helps to define the relationship between dependent and independent variables and to establish an acceptable model, using the least variable to have the best fit. In its basic form, the dependent variable is estimated from one or more variables using a logistic function. In the binary LR model, a dependent variable has two possible values, such as Osmancik/Cammeo. These values are encoded as 0 and 1. Each of the independent variables can be a binary variable or a continuous one.

Multilayer Perceptron. Multilayer perceptron-MLP [25] is a type of deep neural network that consists of at least three layers. These layers are the input layer, hidden layer, and output layer. The number of hidden layers varies according to the problem, at least one, and is adjusted according to the need. The output of each layer becomes the input of the next layer. Each node is connected to all nodes in the next layer. The output layer processes data from the previous layers and determines the output of the network. The output number of the system is equal to the number of nodes in the output layer.

Cinar and Koklu use an MLP architecture with the most general form in their study [22]. The method proposed in this study is also an MLP. However, it has a much more complex structure than the architecture used in their work. Figure 1 shows the overall architecture of the MLP used in their study.

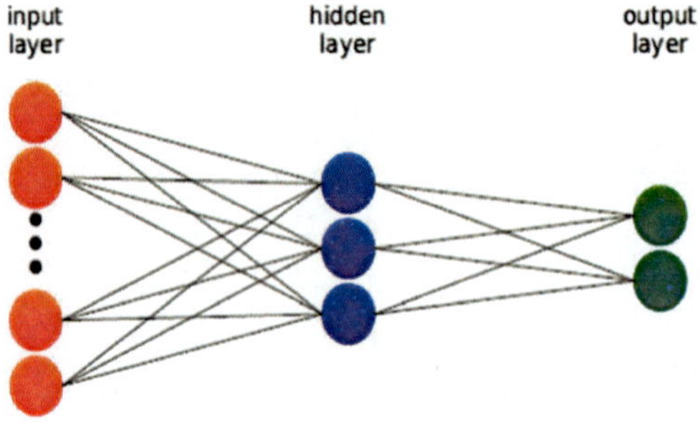

Fig. 1. The overall architecture of the MLP

Support Vector Machine. Support Vector Machine-SVM [26] is a vector space-based machine learning method that finds a decision boundary between the two classes that are furthest from any point in the training data. SVM classifies data linearly in two-dimensional space. In cases where it is not possible to classify the data set linearly, it maps each data to the upper feature space using the kernel function and performs the classification with the help of a hyperplane in this new space.

Decision Tree. A Decision Tree-DT [27] is a machine learning approach used to classify a dataset containing a large number of records by applying a set of decision rules. In other words, it is a structure that helps to separate large amounts of records into small groups of records by applying simple decision-making steps. DT is started at the root of the tree to predict the class label of an observed data point. The value of the root attribute is compared with the attribute of the record. On the basis of this comparison, the branch corresponding to that value is followed and jumped to the next node. The last node on the branch is the decision node, which is the class label.

Random Forest. Random Forest-RF [28] is an ensemble classifier consisting of multiple DTs. One of the biggest problems of decision trees is overfitting. To solve this problem, RF selects different subsets from DTs and trains them. When making a new classification, RF evaluates each DT's individual classification for the entries and selects the prediction with the most votes. RF can work efficiently with datasets containing large amounts of data, but its results are more difficult to understand and interpret than the results of decision trees.

Naive Bayes. Naive Bayes is a probabilistic and supervised learning algorithm based on the application of Bayes' [29] with the assumption of conditional independence of the class variable between each pair of features whose value is given [30]. In Naïve Bayes classification, a certain amount of training data known as the class label is presented to the system. A model is created with the probability operations performed on the training data. The test data newly given to the system is run according to the previously obtained probability values and it is tried to determine which class the data belongs to. The greater the number of training data, the greater the ability to identify the true class of the test data. However, NB can also get successful results with a small number of data. It is a fast, easy-to-implement, and scalable algorithm.

K-Nearest Neighbors. The k-nearest neighbour-KNN [31] is a non-parametric, lazy learning-based method used for classification and regression. Being non-parametric means that it does not make any assumptions about the data; being lazy means that it does not use the training data points to make any generalizations. In KNN classification, the output is a class membership. Where k is a positive number; an observation point is classified according to its distance from its k-nearest neighbors. In other words, it is classified on the basis of simple majority votes of the k-nearest neighbors or weighted votes calculated by multiplying the k-nearest neighbors by a weight proportional to the inverse of their distance from it. In the study, k is selected as 1.

Ada Boost. Ada Boost-AB [32] is a well-known ensemble method that improves the simple boosting algorithm via an iterative process. The main idea behind this algorithm is to create a strong learner using many weak learners. The algorithm aims to give more focus to the patterns that are harder to classify. Initially, all patterns are allocated the same weight. The weights of all misclassified samples are improved in each iteration while the weights of properly classified samples are reduced. As a result, by performing extra iterations and generating more classifiers, the weak learner is compelled to concentrate on the hard samples of the training set. In addition, each individual classifier is allocated a weight. This weight measures the general accuracy of the classifier. Also, it is a function of the sum weight of the correctly classified patterns. The higher weights are therefore provided to more accurate classifiers. These weights are used for the classification of novel patterns.

Bagging. The bagging algorithm-BG [33] is a method that retrains the base learner by deriving new training sets from an existing training set. The technique seeks to improve the accuracy of a classifier by generating an enhanced composite classifier and combining the different outputs of learned classifiers into a single prediction. Each classifier is applied to a sample of instances taken with a substitution from the training set and the decisions are combined by voting.

Voting. The basic idea behind this method is to unite conceptually several machine learning classifiers into a meta-classifier that has a preferable generalization performance than each individual classifier alone [34]. Voting has two types called soft voting and hard voting. The voting in which the class with the highest probability average is the output class is called soft voting. The voting in which the class with the highest number of votes is the output class is called hard voting.

2.3 Proposed Method

Our model is a deep neural network consisting of 3 hidden layers and 2 dropout layers (called 3H2D). Deep neural networks are quite robust learning systems. However, overfitting is an important problem for these networks. Because the large networks are slow to process, this makes it difficult to deal with overfitting by uniting the estimations of different big neural networks at the test time. Dropout [35] is a method to figure out this problem. The main idea behind the method is to randomly drop nodes (with their connections) from the neural network in the training process. This prevents the nodes from fitting too much. In 3H2D, we use ReLU (Rectified Linear Unit) activation function in the first hidden layer and the sigmoid activation function in the other hidden layers. The dropout probability of the model is set to 0.5. This means that half of the available nodes which are randomly selected are disabled where the dropout is applied. The initial random weights of the layers are determined uniformly (it generates values with a uniform distribution). Adam is selected as an optimizer with a 0.001 learning rate. The activation function of the output layer is set to softmax. We train the model with 50 epochs, and 5 different batch_size values between 32–512. The best accuracy value is obtained when the batch_size is 32. All these hyper-parameters are chosen intuitively at the beginning and after different trials, the parameters that show the best performance are preferred. Python's Keras library [36] is used for the implementation of the model.

The Mean Square Error (MSE) in Eq. 1 is preferred as the loss function. We normalize the data in the range [0–1]. We divide the dataset into 80% training and 20% testing. To reduce variance and overfitting, we shuffle the dataset before dividing. A more general model is created as shuffling the data will ensure the training, test, and validation sets represent the overall distribution of the data. We use 10% of the training data for validation during the training phase. The seed is taken as 1 -seed(1)- for the reproducibility of the results.

$$MSE = \frac{1}{n}\sum_{i=1}^{n}\left(y_i - \widehat{y_i}\right)^2 \tag{1}$$

where y is the vector of the actual values of the variable being predicted, $\hat{y}$ is the predicted values of the variable. Figure 2 shows the architecture of the proposed model.

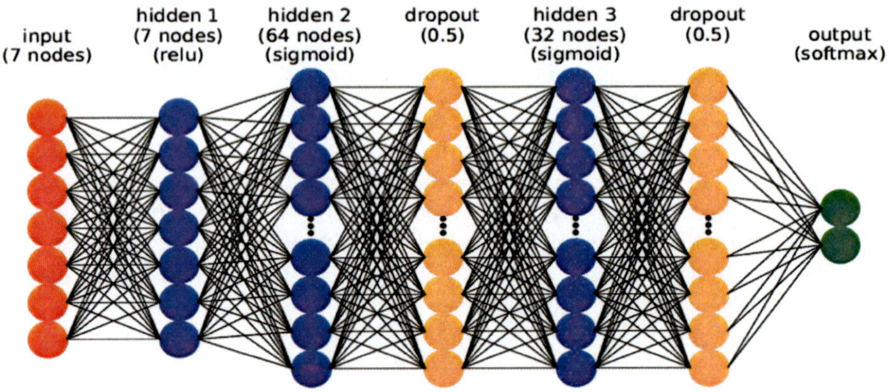

Fig. 2. The architecture of the 3H2D

3 Results and Discussion

In this study, we classify with a deep neural network on the Rice dataset, which includes the Osmancik type grown in Turkey since 1997 and the Cammeo type grown since 2014. We prefer two ways of classification. First, we run our model with different batch_size values and save the results. Secondly, we compare the highest score of our model with 10 different machine learning methods. Batch_size specifies the number of samples to be used to train the neural network at a time. In other words, it is the number of samples used to update the parameters of the network at each step. The larger the batch_size value, the shorter the time required for classification. However, it is quite important to choose the most appropriate value as it directly affects the classification success. In our model, the highest classification accuracy is obtained with 94.09%, when the batch_size value is 32. The time spent by the model for this is 8.57 s. When the batch_size value is taken as 512, the classification success decreases to 90.16%, but the time shortens to 2.14 s. From this point of view, it can be concluded that it is possible to shorten the time by increasing the batch_size value for large data sets whose training process will take a lot of time, in a way that will minimally affect the classification success. Table 3 shows the performances of the 3H2D for different batch_size values. Figure 3 shows the learning and loss curves of the 3H2D.

Table 3. The performances of the 3H2D for different batch_size values

Batch_size	Accuracy %	Train_loss	Test_loss	Time(second)
32	94.09	0.0613	0.0484	8.57
64	93.44	0.0616	0.0495	5.16
128	93.44	0.0646	0.0501	3.49
256	92.39	0.0715	0.0554	2.74
512	90.16	0.8494	0.0879	2.14

The testing (or validation) loss is smaller than the training loss as we use the dropout regularization layer in our network. Because the dropout mechanism is enabled during the training process but disabled during the testing process. Also, the training error (or loss) of the model is the average of the errors (or losses) for each batch of the training data, throughout the current epoch. On the other hand, the testing error (or loss) for an epoch is calculated at the end of the epoch, resulting in a lower error. Table 4 shows the performance of the proposed method (3H2D) in comparison with LR, SVM, DT, MLP, RF, NB, KNN, AB, BG, and VT. The performance of the models is evaluated in terms of accuracy, precision, sensitivity, and f1-score values. Considering the results in Table 4, it is seen that 3H2D has more successful results than all the classification algorithms being compared. In Fig. 4, it is also possible to see the performance of the models visually. It is clear from the graphs in the figures that the proposed method performs better.

Table 4. The comparative performances of the methods

Method	Accuracy	Precision	Sensitivity	F_1-score
LR	93.02	91.35	92.26	91.80
MLP	92.86	91.04	92.17	91.60
SVM	92.83	91.53	91.70	91.62
DT	92.49	91.29	91.18	91.23
RF	92.39	90.80	91.36	91.08
NB	91.71	89.63	90.86	90.24
KNN	88.58	87.06	86.37	86.71
AB	92.34	92.18	92.16	92.17
BG	91.68	91.47	91.55	91.51
VT	90.52	90.40	90.22	90.30
3H2D	94.09	93.71	92.87	93.29

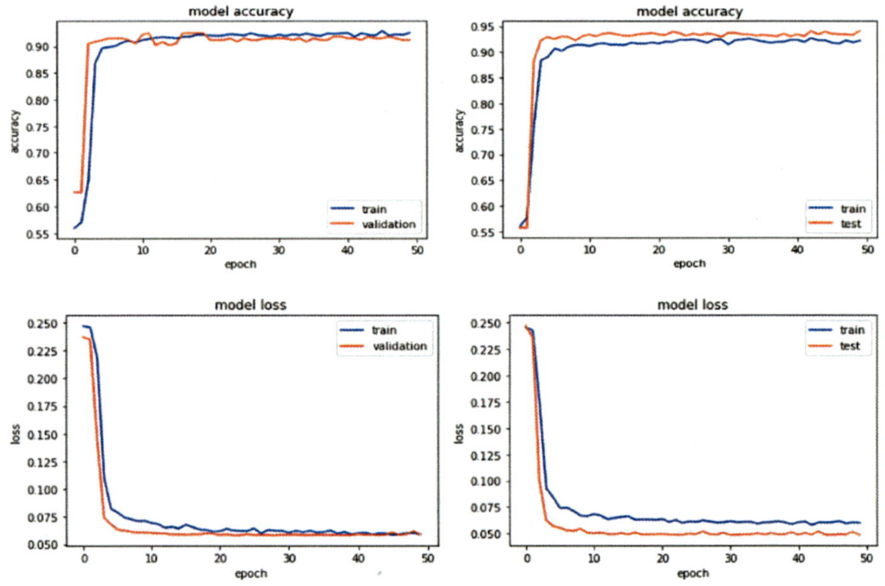

Fig. 3. The learning and loss curves of the 3H2D (batch_size = 32)

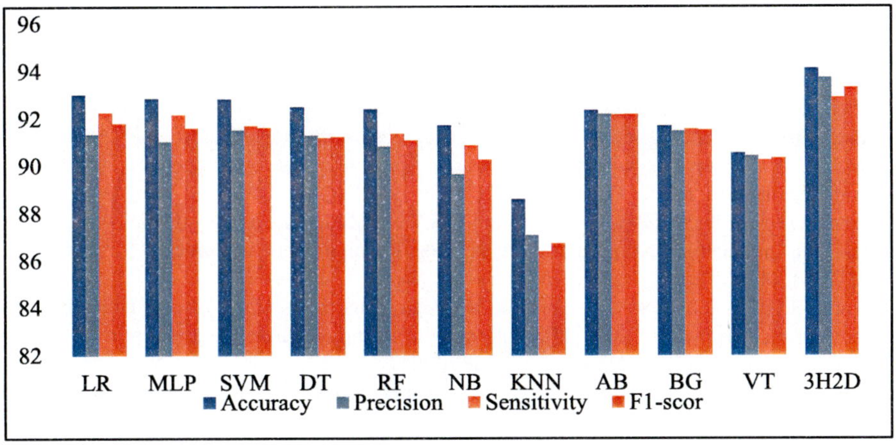

Fig. 4. The performances of the models

4 Conclusion

In this study, we classify two rice varieties that have been patented in Turkey, with a model we created using deep neural networks. The network model we use consists of 3 hidden layers and 2 dropout layers (3H2D). The purpose of adding the dropout layers is to avoid overfitting problem in classification. We compare our model (3H2D) with 10 different machine learning algorithms using accuracy, precision, recall, and f1-score metrics. Among the other machine learning algorithms, LR achieves the best classification success rate of 93.02%. On the other hand, the classification success rate of the proposed method is 94.09%. The obtained results reveal that our model outperforms all machine learning algorithms it is compared to, for all metric values. In addition, the proposed method has a very flexible and useful model. It is also functional to use it for datasets with multi-class or from different fields by changing its hyper-parameters.

References

1. Moazzam, S.I., et al.: A review of application of deep learning for weeds and crops classification in agriculture. In: 2019 International Conference on Robotics and Automation in Industry (ICRAI), pp. 1–6. IEEE (2019)
2. Pallagani, V., et al.: DCrop: a deep-learning-based framework for accurate prediction of diseases of crops in smart agriculture. In: 2019 IEEE International Symposium on Smart Electronic Systems (ISES), pp. 29–33 (2019)
3. Guillén-Navarro, M.A., et al.: A deep learning model to predict lower temperatures in agriculture. J. Ambient Intell. Smart Environ. **12**(1), 21–34 (2020)
4. Darwin, B., et al.: Recognition of bloom/yield in crop images using deep learning models for smart agriculture: a review. Agronomy **11**(4), 646 (2021)
5. Kelly, S., Tolvanen, J.P.: Domain-Specific Modeling: Enabling Full Code Generation. Wiley IEEE Computer Society Press, Hoboken (2008)

6. Jiang, T., Liu, X., Wu, L.: Method for mapping rice fields in complex landscape areas based on pre-trained convolutional neural network from HJ-1 A/B data. ISPRS Int. J. Geo Inf. **7**(11), 418 (2018)
7. Zhou, C., et al.: Automated counting of rice panicle by applying deep learning model to images from unmanned aerial vehicle platform. Sensors **19**(14), 3106 (2019)
8. Chu, Z., Yu, J.: An end-to-end model for rice yield prediction using deep learning fusion. Comput. Electron. Agric. **174**, 105471 (2020)
9. Park, S., et al.: i6mA DNC: prediction of DNA N6-Methyladenosine sites in rice genome based on dinucleotide representation using deep learning. Chemom. Intell. Lab. Syst. **204**, 104102 (2020)
10. Emon, S.H., Mridha, M.A.H., Shovon, M.: Automated recognition of rice grain diseases using deep learning. In: 2020 11th International Conference on Electrical and Computer Engineering (ICECE), pp. 230–233. IEEE (2020)
11. Yang, Q., et al.: A near real-time deep learning approach for detecting rice phenology based on UAV images. Agric. For. Meteorol. **287**, 107938 (2020)
12. Shi, Y., et al.: Improving performance: a collaborative strategy for the multi-data fusion of electronic nose and hyperspectral to track the quality difference of rice. Sens. Actuators B Chem. **333**, 129546 (2021)
13. Bari, B.S., et al.: A real-time approach of diagnosing rice leaf disease using deep learning-based faster R-CNN framework. PeerJ Comput. Sci. **7**, e432 (2021)
14. Zhu, A., et al.: Mapping rice paddy distribution using remote sensing by coupling deep learning with phenological characteristics. Remote Sens. **13**(7), 1360 (2021)
15. Son, N.H., Thai-Nghe, N.: Deep learning for rice quality classification. In: 2019 International Conference on Advanced Computing and Applications, pp. 92–96. IEEE (2019)
16. Weng, S., et al.: Hyperspectral imaging for accurate determination of rice variety using a deep learning network with multi-feature fusion. Spectrochim. Acta Part A Mol. Biomol. Spectrosc. **234**, 118237 (2020)
17. Joshi, D., et al.: Label-free non-invasive classification of rice seeds using optical coherence tomography assisted with deep neural network. Opt. Laser Technol. **137**, 106861 (2021)
18. Pradana-López, S., et al.: Low requirement imaging enables sensitive and robust rice adulteration quantification via transfer learning. Food Control **127**, 108122 (2021). https://doi.org/10.1016/j.foodcont.2021.108122
19. Yang, M.D., et al.: A UAV open dataset of rice paddies for deep learning practice. Remote Sens. **13**(7), 1358 (2021)
20. Gilanie, G., et al.: RiceNet: convolutional neural networks-based model to classify Pakistani grown rice seed types. Multimed. Syst. **27**(5), 867–875 (2021). https://doi.org/10.1007/s00530-021-00760-2
21. Estrada-Pérez, L.V., et al.: Thermal imaging of rice grains and flours to design convolutional systems to ensure quality and safety. Food Control **121**, 107572 (2021). https://doi.org/10.1016/j.foodcont.2020.107572
22. Cinar, I., Koklu, M.: Classification of rice varieties using artificial intelligence methods. Int. J. Intell. Syst. Appl. Eng. **7**(3), 188–194 (2019)
23. Peker, N., Kubat, C.: Application of Chi-square discretization algorithms to ensemble classification methods. Expert Syst. Appl. **185**, 115540 (2021). https://doi.org/10.1016/j.eswa.2021.115540
24. Berkson, J.: Application of the logistic function to bio assay. J. Am. Stat. Assoc. **39**(227), 357–365 (1944)
25. Rumelhart, D.E., Hinton, G.E., Williams, R.J.: Learning representations by back-propagating errors. Nature **323**(6088), 533–536 (1986)

26. Boser, B.E., Guyon, I.M., Vapnik, V.N.: A training algorithm for optimal margin classifiers. In: Proceedings of the Fifth Annual Workshop on Computational Learning Theory, pp. 144–152 (1986)
27. Breiman, L., et al.: Classification and Regression Trees. CRC Press (1986)
28. Breiman, L.: Random forests. Mach. Learn. **45**(1), 5–32 (2001)
29. Bayes, T.: LII. An essay towards solving a problem in the doctrine of chances. By the late Rev. Mr. Bayes, FRS communicated by Mr. Price, in a letter to John Canton, AMFR S. Philos. Trans. R. Soc. Lond. **53**, 370–418 (1763)
30. Narasimha Murty, M., Susheela Devi, V.: Pattern Recognition. Springer, London (2011)
31. Cover, T., Hart, P.: Nearest neighbor pattern classification. IEEE Trans. Inf. Theory **13**(1), 21–27 (1967)
32. Freund, Y., Schapire, R.E.: Experiments with a new boosting algorithm. In: Proceedings of the 13th International Conference on Machine Learning (ICML), vol. 96, pp. 148–156 (1996)
33. Breiman, L.: Bagging predictors. Mach. Learn. **24**(2), 123–140 (1996)
34. Raschka, S.: Python Machine Learning. Packt Publishing Ltd. (2015)
35. Srivastava, N., et al.: Dropout: a simple way to prevent neural networks from overfitting. J. Mach. Learn. Res. **15**(1), 1929–1958 (2014)
36. Chollet, F.: Keras: The Python deep learning library. Astrophysics Source Code Library, ascl-1806 (2018)

Elevation Based Outdoor Navigation with Coordinated Heterogeneous Robot Team

Ömer Faruk Kaya and Erkan Uslu(✉)

Department of Computer Engineering, Yildiz Technical University, Istanbul, Turkey
euslu@yildiz.edu.tr

Abstract. Today, the use of heterogeneous robot teams is increasing in military operations and monitoring the environment. However, single 2D autonomous navigation systems fail in rough terrains. In this study, a solution to this problem is proposed using 2.5D navigation that takes into account the slopes on the elevation map. Using ROS and Gazebo, we coordinate drones and ground vehicles to process terrain elevations. The simulation world used in the study reflects a real-world rough terrain and also some urban artifacts are added to the simulation world. Husky simulation model is used as the ground vehicle utilizing 3D LIDAR, GPS and 9 DOF IMU sensors, which can output 3D map and 3D localization using 3D SLAM. Using the SLAM localization, a 2.5D map is created on the ground vehicle. Drone simulation model, similarly equipped, follows the ground vehicle with a GPS-based waypoint navigation and can create a 2.5D map using its sensors. A global plan is created for the ground vehicle by cooperative effort of both robots, using the map information from ground vehicle where available and using the map information from the drone where ground vehicle's map is insufficient. 2.5D navigation of the ground vehicle is carried out by the local planner taking into account the calculated cooperative global path. Proposed method results in shorter routes and fewer path planning issues. This is shown by the comparative analysis where the ground vehicle or the drone is used alone.

Keywords: elevation mapping · localization · 3D mapping · 2.5D outdoor navigation · drone · ground vehicle · heterogeneous robot teams

1 Introduction

The use of unmanned vehicle systems is increasing day by day in military, commercial, and academic environments. These systems, which operate in areas where humans cannot or do not want to enter, make people's lives easier. In unmanned vehicle systems, air and ground vehicles are generally used together. Although fixed-wing air vehicles can perform tasks for longer periods, they usually require large areas for takeoff, so multi-rotor vehicles are preferred. Thanks to the air vehicles transmitting information from high altitudes to ground vehicles using their onboard sensors, ground vehicles can complete their tasks more quickly and safely, providing them with a broader view of the area.

Z. Şen et al. (Eds.): IMSS 2023, LNME, pp. 522–533, 2024.
https://doi.org/10.1007/978-981-99-6062-0_48

Three-Dimensional (3D) Simultaneous Localization and Mapping (SLAM) is very important in the field of robotics due to its ability to generate map and odometry information. Various 3D SLAM algorithms have been proposed in the literature that demonstrate significant improvements in accuracy and robustness. Among these algorithms, the LeGO-LOAM [2], LIO-SAM [3], and ORB-SLAM3 [1] algorithms have gained popularity due to their efficient and effective approaches. LeGO-LOAM is a real-time algorithm that uses a low-cost 3D lidar and a stereo camera to achieve high accuracy and robustness in dynamic environments. LIO-SAM is another real-time SLAM algorithm that uses a LiDAR-Inertial Odometry for 3D mapping and localization. ORB-SLAM3, on the other hand, is a feature-based algorithm that utilizes visual and inertial measurements for accurate and robust 3D mapping and localization. These algorithms have shown promising results in various scenarios, including outdoor, indoor, and dynamic environments.

Global planning is a fundamental task in autonomous navigation systems that aims to find a collision-free path from the current robot's pose to the desired goal position. One of the most popular global planning algorithms is A* [4] which is a heuristic search algorithm that finds the shortest path in a graph. Another widely used algorithm is Dijkstra's algorithm [5] which also finds the shortest path but with a higher computational cost. Other global planning algorithms such as D* Lite [6], RRT [7], and RRT* [8] have been proposed and are receiving significant interest in the robotics community. These algorithms address various challenges in global planning such as dynamic obstacle avoidance, uncertainty, and high-dimensional state spaces. For example, D* Lite is an incremental version of D* that updates the path only on modified areas, while RRT and RRT* use sampling-based techniques to explore the state space and create collision-free paths.

Global path following is also an essential part of autonomous vehicle navigation. A variety of path tracking algorithms have been proposed in the literature to ensure that the vehicle follows the desired path accurately and safely. Model predictive control (MPC) [9] has become a popular approach in recent years due to its ability to handle non-linear systems and constraints. Studies have demonstrated the effectiveness of MPC-based path tracking algorithms in various applications, such as autonomous cars, unmanned aerial vehicles, and mobile robots. Another popular approach is the proportional-integral-derivative (PID) control [10], which is widely used in industrial control applications due to its simplicity and effectiveness. Another approach is sliding mode control (SMC) [11], which proposes a method that considers model-independent adaptive feedback actions.

In the literature, the cooperative operation of drones and ground vehicles contributes to completing tasks more quickly and accurately. The movements of the ground vehicle have been optimized in an environment where the map is neuron topologically modelled with the Dragonfly Algorithm [15]. The map of the drone and the ground vehicle, which create the map of the environment using their built-in cameras, has been combined with the proposed method [16].

This aimed to create a more accurate map of the environment. In the simulation study [17] with three drones and a ground vehicle, finite state machines were used to ensure that the coordinated robot team avoids obstacles and follows the defined sequence.

Ground vehicles complete the missions in a longer time or unsuccessfully when only use their own sensors in rough terrains. Due to sensor constraints, they can choose the longer routes, or they end up unsuccessfully by oscillating while trying to reach their destination. In the proposed method, the performance of the tasks has been increased in terms of distance and time by using the map created with the sensors of the drone following the ground vehicle with the GPS-Waypoint method. The method was tested in a world created from a height map in the Gazebo simulation environment, with vehicles and trees added to it.

The paper is structured as follows: in Sect. 2, system design and used methods are explained, the test environment and tests used are explained in Sect. 3, test results and evaluations are given in Sect. 4, in Sect. 5 potential future studies and conclusion of this study are given.

2 System Design

In our study, three-dimensional simultaneous localization and mapping (SLAM) is used for obtaining location information which is performed in a terrain environment that includes bumps and holes where it is hard to be traversed with two-dimensional navigation approaches. Navigation is performed by generating elevation-based costmaps. The system design for the proposed method is provided in Fig. 1.

2.1 Simultaneous Localization and Mapping (SLAM)

In the study, the LIO-SAM package, which is based on the LOAM package that generates odometry with LIDAR and can combine IMU and LIDAR data, is used. LIO-SAM is a graph-based 3D mapping method that performs two-step position optimization. It produces odometry data by adding IMU bias to the odometry data produced by combining LIDAR odometry generated by LOAM [12], which extracts the edge and surface points from consecutive scans in the LIDAR data and compares them with the next time data. The optimization process is performed using the general-purpose factor graph library GTSAM [13]. In the LIO-SAM system diagram given in Fig. 2, the inputs and outputs of the system are provided. There are 4 main modules implemented within the system. The first of these is the imuPreintegration part, which uses the generated odometry and IMU data to produce a higher frequency odometry. The imageProjection part uses IMU and lidar data to deskew the point cloud and generate an initial guess. The featureExtraction part, which receives the deskewed point cloud data, extracts edge and planar features. In the mapOptimization part, these edge and planar features are optimized by the factor graph using either GPS (if available) or only the lidar odometry factor, producing odometry.

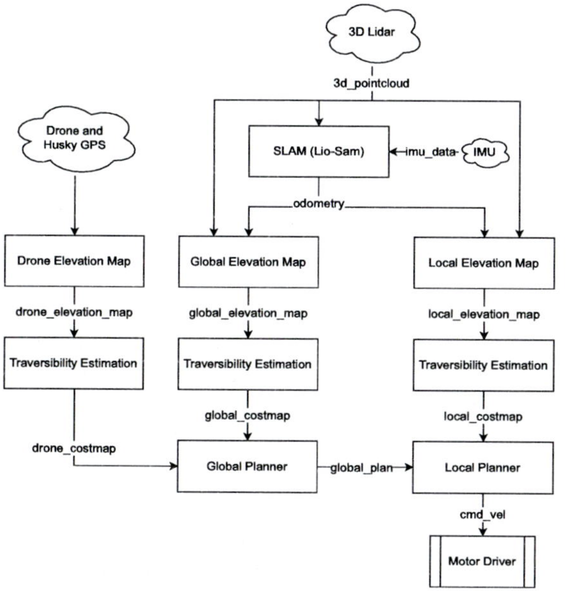

Fig. 1. System Design

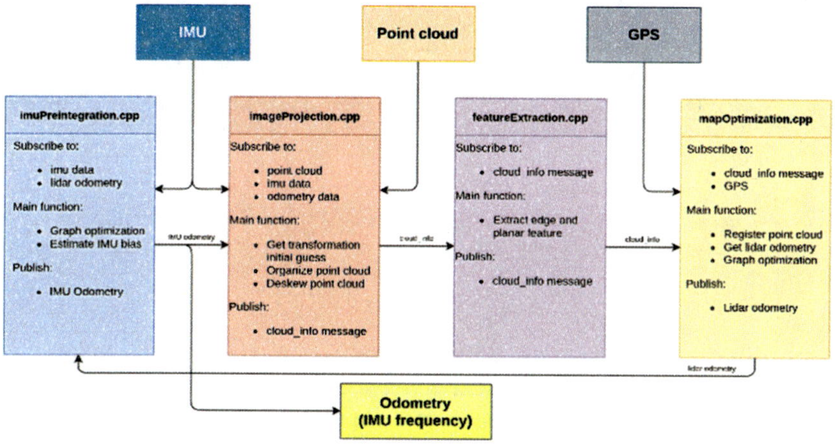

Fig. 2. Lio-Sam System Architecture [3]

2.2 Navigation

In the study, elevation maps are created for both aerial and ground vehicles to enable navigation. Navigable areas are analysed based on these elevation maps, and cost maps are generated as a result. These cost maps are then used to plan a global path, with the ground vehicle's map is prioritized for planning within its boundaries, and the aerial vehicle's map is utilized to plan the most optimal path outside of those boundaries. The path is then tracked using a PID-based local planner.

Elevation Map. In the study, a modified version of the costmap_2d structure, which is readily available in ROS, is used to represents each cost map cell as an "8-bit unsigned char" where integer values range between 0 and 255 can be stored. The original structure makes it difficult to directly represent the environment with elevations and pits as required in the study. To solve these problems, the elevation_mapping package is used, which utilizes the grid_map structure mentioned in article [14]. The elevation_mapping package, which is an elevation-based mapping method is used in the study, that continuously updates a probabilistic map using the real heights measured from the 3D LIDAR in the terrain. The elevation map output of the Gazebo world shown in Fig. 3a is provided in Fig. 3b.

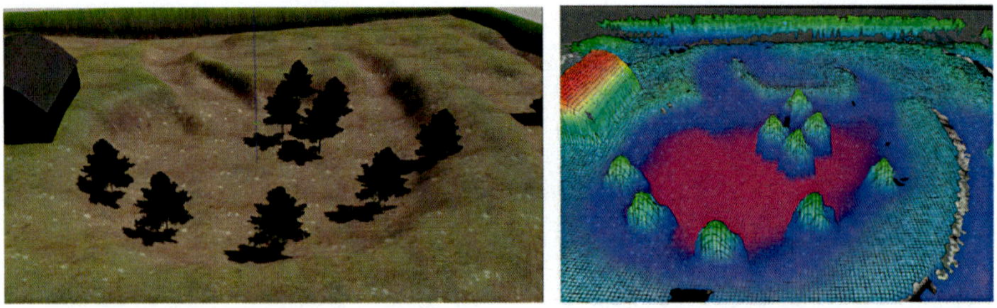

(a) (b)

Fig. 3. a) Gazebo World, b) Elevation Map of Gazebo World

Traversibility Map. The traversability map used in the study is generated by processing the elevation map with the filter chain given in Fig. 4. The two most important factors in this filter chain are slope and roughness. For roughness determination, the height of each cell in the elevation map is compared with the height of surrounding cells, and an estimation is made and published as a grid_map layer. For slope determination, the surface normals in the upward direction of the cells on the map grid are extracted, and these surface normals are combined to form a curve. The slope of this curve follows the slope of the map, and this resulting slope map is also published as a grid_map layer. The published slope and roughness layers are then combined to form a traversability map. This traversability map is reduced to a costmap_2d format for path planning purposes.

In Fig. 5, the traversability map is shown, which is generated by sequentially applying the normal filter, slope filter, step filter, and roughness filter to the elevation map output given in Fig. 3b, and then taking the weighted average of all of them. The pink areas represent non-traversable regions such as steep bumps, trees, and buildings, the yellow areas represent high-cost regions, and the red areas represent low-cost regions.

Fig. 4. Filter Chain

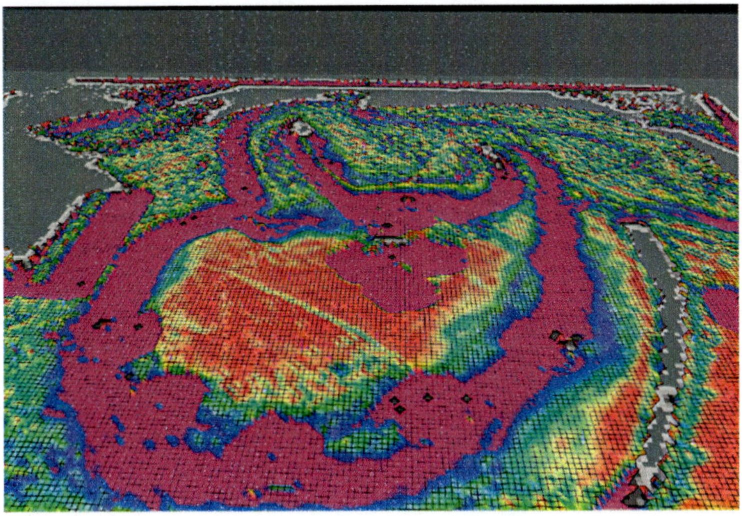

Fig. 5. Traversibility Map

Global Planning. A* and Dijkstra algorithms are two commonly used pathfinding algorithms. Dijkstra's algorithm is a classic algorithm that finds the shortest path between two nodes in a graph by exploring all possible paths from the starting node to the destination node. It is guaranteed to find the shortest path, but it can be slow if the graph is large or complex. A* algorithm is a more efficient version of Dijkstra's algorithm that uses a heuristic function to guide the search towards the destination node. This heuristic function estimates the distance between the current node and the destination node, allowing A* to prioritize the nodes that are most likely to lead to the shortest path. A* is often faster than Dijkstra's algorithm, especially in large or complex graphs. However, the quality of the path found by A* may not always be optimal if the heuristic function is not well-designed.

In this study, if the target point is on the ground vehicle's cost map, path planning is performed using only the ground vehicle's cost map with these methods. However, if the target point is outside the ground vehicle's map, a path from the vehicle's location to the target location is planned on the wider but less detailed map created by the aerial vehicle. A point at the determined look-ahead distance on the planned path is selected, and a path is planned from the ground vehicle's detailed map to this selected point.

Local Planning. In the study, a PID (Proportional-Integral-Derivative) based controller is used to track the path generated by the global planner, considering the obstacles in the environment. The PID controller is a feedback control system that generates an output signal by using the current error (difference), the time integral of the error, and the derivative of the error.

$$v(t) = K_p \cdot e(t) + K_i \cdot \int (e(t)dt + K_d \cdot \frac{(de(t))}{dt}) \tag{1}$$

The PID equation is shown in (1), u(t) represents the output signal, e(t) represents the current error, and Kp, Ki, and Kd represent proportional, integral, and derivative coefficients, respectively.

Applying a PID controller for the global plan follower requires comparing the vehicle position with the target path and calculating the difference between them. This difference is used as the error and is processed by the PID controller to steer the vehicle towards the target path. If the path generated by the global planner is no longer valid, new paths are drawn.

It does not perform any extra path planning internally and only generates speed commands suitable for the given global path. As a result, it provides more accurate and stable path following compared to other global path tracking algorithms.

3 Experiment Setup

In this study, the Gazebo simulation environment, which can simulate the real-world indoor and outdoor environments successfully, is used. A world with trees, buildings, and elevations is created and used as a benchmark for comparisons in the study. Two vehicles, one ground and one aerial, are used in the simulation environment and necessary sensors are added to them for the system.

For the test environment, the Gazebo world shown in Fig. 6a is created from the height map given in Fig. 6b, and then trees and vehicles are added to it.

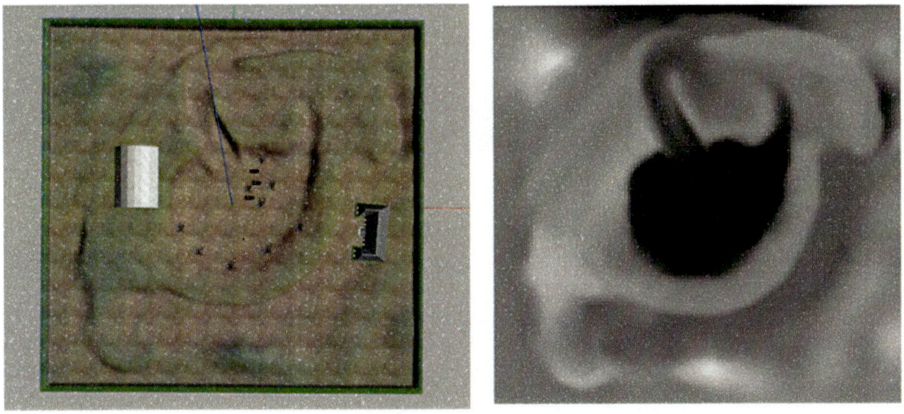

(a) (b)

Fig. 6. a) Gazebo World, b) Height Map which used for creation Gazebo World

3.1 Husky

Husky, which is represented in Fig. 7a, is a medium-sized four-wheeled and differential-driven ground vehicle. The simulated ground vehicle is equipped with GPS having a 0.0001 m/s drift, 3D LiDAR having a 0.0008 m gaussian noise, and 9 DOF IMU sensor having a 0.005 m/s drift, that enables it to create 3D and 2.5D maps, navigate, and determine its position.

3.2 Hector Quadrotor

Hector Quadrotor, which is represented in Fig. 7b, is a popular open-source platform for aerial robotics research and experimentation. The simulated aerial vehicle is equipped with GPS having a 0.0001 m/s drift, 3D LiDAR having a 0.0008 m gaussian noise, and 9 DOF IMU sensor having a 0.005 m/s drift that enables it to create maps and follow ground vehicles using GPS-based waypoints.

In the study, three target points represented by different colours in Fig. 8 with varying difficulty levels are selected for benchmarking. Among these, the one shown in green is chosen in a location without any obstacles, elevations, or pits. The target point represented in yellow is selected to be on top of a hill and reachable by two paths, one longer than the other. The target point represented in purple is selected to be on top of a hill and far from the dimensions of the global cost map when approached.

(a)

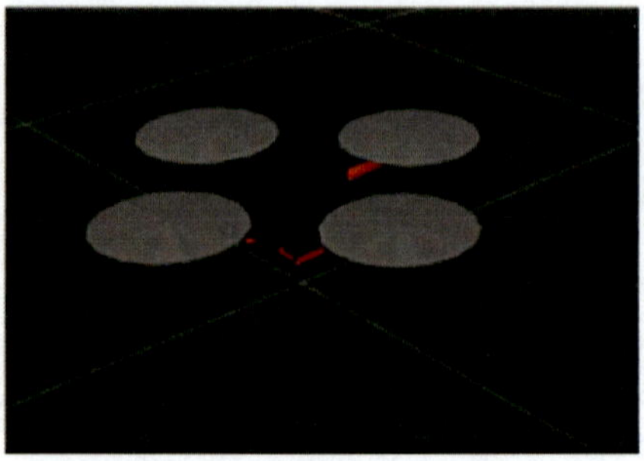

(b)

Fig. 7. Robots: a) Husky Robot, b) Hector Quadrotor

In the experiments, drone and ground vehicle are started from the point represented by red point each time, and after the drone reached an altitude of 30 m, the target points were tested with only local map, 2D map and navigation, using both 2.5D global and local maps, and with the proposed method.

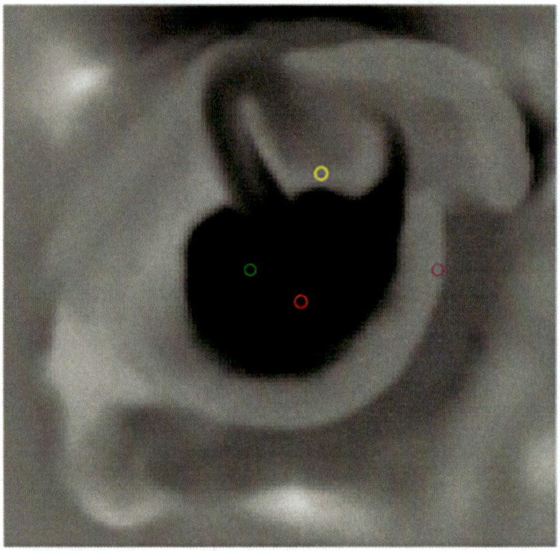

Fig. 8. Test Setup

4 Results and Discussion

In the experiment environment shown in Fig. 8, vehicles are started from the red point. The points marked with green, yellow and purple are given as targets to the vehicles in sequence. Paths to the targets are calculates using Dijkstra and A* algorithms. In order to compare with the proposed method three other approaches, firstly, a local map created only by the vehicle's own sensors is provided, secondly a 2D map created with 2D LiDAR, thirdly a 2.5D global and local map are tested. Arrival times and distances travelled are used as comparison metrics. If the travelled distance reaches to a value that is 10 times longer than the distance between the vehicle and the target, but the vehicle still could not reach the target, the test is stopped and considered unsuccessful. The test results are provided in Table 1. The paths calculated to the targets using the proposed method are given in Fig. 9.

As can be seen in the table, all methods have exhibited similar performances on obstacle-free and flat surfaces. However, when there is an obstacle between the target and the robot, a local map alone is not sufficient, and the need for a global map has been observed. Also, it can be noticed that, if the global map

(a) (b) (c)

Fig. 9. Calculated paths with proposed method to points indicated in a) Green, b) Yellow, c) Purple

Table 1. Experimental results

Goal		Only Local Map	2D Mapping	2.5D Local and Global Map	Proposed Method
Green	Time (s)	29.04	29.25	29.19	29.08
	Trajectory (m)	30.01	30.05	30.1	30.02
Yellow	Time (s)	Oscillation	No Path	170.45	150.27
	Trajectory (m)	Oscillation	No Path	155.37	138.84
Purple	Time (s)	Oscillation	No Path	Oscillation	180.39
	Trajectory (m)	Oscillation	No Path	Oscillation	147.43

is 2D, the path could not be calculated due to the bumps in the terrain. If a 2.5D map is used and traversibility calculations are made according to the slopes in the environment, it was possible to reach the goals higher than the starting point. The proposed method has reached the target in a shorter time and with a shorter path compared to the method using both global and local maps. In cases where the length of the obstacle between the robot and the target is larger than the global map of the vehicle, all methods except the proposed one have entered oscillation and failed to reach their targets.

5 Conclusion

The study deals with cooperative navigation problem of a UGV and a UAV, where UGV utilizes an elevation-based 2.5D map and UAV utilizes GPS based leader following. The method is developed and implemented in a simulation environment that allows the ground vehicle to use the map generated by the drone when the given goal is outside its own map. In situations where multiple paths exist or optimal and accurate paths cannot be reached using only local sensors or 2D map, the A* method has been used to calculate the paths using drone and ground vehicles maps. The calculated global path has been followed using a PID-based path-following algorithm. The proposed cooperative navigation method enabled the UGV to plan and execute outdoor paths that cannot be achieved by 2D map-based navigation, 3D sensor-based navigation or 2.5D map-based navigation approaches.

In future studies, this system will be made usable in GPS-denied environments by using map merging-based methods instead of GPS.

References

1. Campos, C., Elvira, R., Rodríguez, J., Montiel, J.M., Tardós, J.D.: ORB-SLAM3: an accurate open-source library for visual, visual-inertial, and multimap SLAM. IEEE Trans. Robot. **37**, 1874–1890 (2021)
2. Shan, T., Englot, B.: LeGO-LOAM: lightweight and ground-optimized lidar odometry and mapping on variable terrain. In: 2018 IEEE/RSJ International Conference on Intelligent Robots and Systems (IROS), pp. 4758–4765 (2018). https://doi.org/10.1109/IROS.2018.8594299
3. Shan, T., Englot, B., Meyers, D., Wang, W., Ratti, C., Rus, D.: LIO-SAM: tightly-coupled lidar inertial odometry via smoothing and mapping. In: 2020 IEEE/RSJ International Conference on Intelligent Robots and Systems (IROS), pp. 5135–5142 (2020). https://doi.org/10.1109/IROS45743.2020.9341176
4. Hart, P., Nilsson, N., Raphael, B.: A formal basis for the heuristic determination of minimum cost paths. IEEE Trans. Syst. Sci. Cybern. **4**, 100–107 (1968). https://doi.org/10.1109/TSSC.1968.300136
5. Dijkstra, E.: A note on two problems in connexion with graphs. Numer. Math. **1**, 269–271 (1959). https://doi.org/10.1007/BF01386390
6. Koenig, S., Likhachev, M.: D*Lite. In: Proceedings of the National Conference on Artificial Intelligence, pp. 476–483 (2002)

7. LaValle, S.M.: Rapidly-exploring random trees: a new tool for path planning. Research Report 9811 (1998)
8. Karaman, S., Frazzoli, E.: Sampling-based algorithms for optimal motion planning. Int. J. Robot. Res. **30**, 846–894 (2011). https://doi.org/10.1177/0278364911406761
9. Ming, T., Deng, W., Zhang, S., Zhu, B.: MPC-Based Trajectory Tracking Control for Intelligent Vehicles (2016)
10. Åström, K., Hägglund, T.: The future of PID control. Control Eng. Pract. **9**, 1163–1175 (2001)
11. Oh, K., Seo, J.: Development of a sliding-mode-control-based path-tracking algorithm with model-free adaptive feedback action for autonomous vehicles. Sensors **23**, 405 (2023)
12. Zhang, J., Singh, S.: LOAM: lidar odometry and mapping in real-time In: Robotics: Science and Systems Conference (RSS), pp. 109–111 (2014)
13. Dellaert, F., Kaess, M.: Factor Graphs for Robot Perception (2017)
14. Fankhauser, P., Hutter, M.: A universal grid map library: implementation and use case for rough terrain navigation. In: Koubaa, A. (ed.) Robot Operating System (ROS). SCI, vol. 625, pp. 99–120. Springer, Cham (2016). https://doi.org/10.1007/978-3-319-26054-9_5
15. Ni, J., Wang, X., Tang, M., Cao, W., Shi, P., Yang, S.: An improved real-time path planning method based on dragonfly algorithm for heterogeneous multi-robot system. IEEE Access **8**, 140558–140568 (2020). https://doi.org/10.1109/ACCESS.2020.3012886
16. Potena, C., Khanna, R., Nieto, J., Siegwart, R., Nardi, D., Pretto, A.: AgriColMap: aerial-ground collaborative 3D mapping for precision farming. IEEE Robot. Autom. Lett. **4**, 1085–1092 (2019). https://doi.org/10.1109/LRA.2019.2894468
17. Ju, C., Son, H.: Hybrid systems based modeling and control of heterogeneous agricultural robots for field operations (2019)

Investigation of the Potentials of the Agrivoltaic Systems in Turkey

Sena Dere[1], Elif Elçin Günay[1](✉), and Ufuk Kula[2]

[1] Department of Industrial Engineering, Sakarya University, Serdivan, Sakarya, Turkey
ekabeloglu@sakarya.edu.tr

[2] College of Engineering and Technology, American University of the Middle East, Al Ahmadi, Kuwait

Abstract. The demand for renewable energy to mitigate CO2 emissions is continuously growing due to the risk of environmental degradation. In this regard, agrivoltaic systems emerged to provide a potential solution for environmental deterioration. Agrivoltaic systems are the common use of land for crop and energy production. These systems are designed to efficiently share the same land between photovoltaics and agriculture so that food and energy can be produced conjointly. In this study, we investigate the potential of agrivoltaic systems in Turkey. With this aim, three provinces with the highest agricultural production are selected, and their capabilities are investigated by using GIS (Geographic Information System) with a fuzzy analytical hierarchy process (FAHP) method. First, the literature is reviewed to determine the criteria for agrivoltaics installation. Then, criteria weights are determined using FAHP in order to capture the associated uncertainty in decision-maker's (DMs) judgments and preferences. Finally, suitability analysis is conducted using the criteria weights and the GIS data to reveal potential sites for agrivoltaics development. The study results contribute to practitioners deciding the target locations to sustainably generate solar energy while allowing crop production.

Keywords: Renewable Energy · Solar Energy · Agrivoltaics · Fuzzy AHP

1 Introduction

The widespread use of fossil fuels is a major contributor to climate change, air pollution, and other serious environmental issues. Combustion of these fuels emits dangerous gases such as CO2 and methane and causes the Earth's temperature to increase. In contrast, renewable energy sources such as biomass, wind, or solar are environmentally friendly and do not release any harmful gases to our society. Among these renewable resources, photovoltaic (PV) solar energy production has remarkably grown in which daylight is absorbed and converted into electrical energy. The International Energy Agency (IEA) predicts that by 2050, 16% of the total energy produced will be based on PV power [1]. In order to realize this prediction and meet a larger share of the world's total energy demand, massive areas are needed for the construction of PV systems. However, the

Z. Şen et al. (Eds.): IMSS 2023, LNME, pp. 534–545, 2024.
https://doi.org/10.1007/978-981-99-6062-0_49

need for a large surface area brings a land occupation conflict between solar farms and cultivation. Agrivoltaics have been introduced to provide a viable solution to our energy needs and a pathway to more sustainable agriculture [2].

Agrivoltaics allow the dual use of land for PV panel installation and cultivation so that the vital asset land can be used for food and energy production [3]. Agrivoltaic systems involve installing ground-mounted solar PV panels built at a suitable height for cultivation. This placement of the elevated solar panels parallel throughout the field allows land to be used for planting between - and under-mounted solar panels. The strategy of a common use of the land for crop and energy production leads to notable growth for the Agrivoltaics.

This study aims to investigate the potential of agrivoltaic systems in Turkey. In this manner, a GIS-based multi-criteria decision-making (MCDM) model is used to identify suitable fields for installing agrivoltaic systems in three provinces: Konya, Ankara, and Şanlıurfa, which have the highest cultivated area in Turkey. First, the evaluation criteria for decision-making were determined through a literature review. Then, GIS data was collected for each evaluation criterion for three provinces. Based on expert opinion and the findings from the literature, the weights of the criteria were determined by the FAHP method. Using the weights of each criterion, a suitability map was created in ArcGIS ESRI 10.8 software, and suitable locations were determined for the three provinces.

2 Literature Review

Researchers have attempted to use GIS with MCDM techniques to evaluate the suitability of the regions for renewable energy sources installations. GIS is a spatial system that creates, manages, analyzes, and maps data. By organizing layers of information together and analyzing the spatial locations, GIS visualizes maps and creates 3D scenes [4]. In the integrated use of GIS and MCDM, GIS provides spatial data while MCDM methods are deployed for criteria weight determination.

Among many MCDM techniques, AHP is dominant in many studies due to its ease of use, straightforward calculation capability, convenience for experts from different backgrounds, and integration of DMs consistency in evaluation [5]. Some research papers that use GIS with AHP are as follows. Uyan [6] studied solar farm selection in the Karapinar region of Konya, Turkey considering environmental and economic factors. Colak et al. [7] investigated the optimum locations for solar power plants in Malatya Province. Saraswat et al. [8] examined the suitability of solar farm sites in India, considering technical, economic, and socio-environmental factors. Wang et al. [9] developed a participatory least-conflicting sitting framework to evaluate suitable solar development sites for Southwestern Taiwan, considering environmental, economic, and social factors.

On the other hand, due to fully capturing the subjective evaluation and judgments of the DM based on linguistic terms, fuzzy MCDM approaches are also utilized with GIS. For instance, Noorollahi et al. [10] studied land suitability analysis for solar farms by integrating GIS and FAHP in Iran. Again, a similar study was conducted by Asakereh et al. [11] for a different region in Iran. Studies mainly consider solar farms or solar plant construction, as the literature shows. However, by employing FAHP to model vague and uncertain expert opinions and subsequently incorporating GIS data, our research enhances the existing literature on agrivoltaic systems in Turkey by exploring their potential.

3 Methodology

Our methodology includes four main stages, as shown in Fig. 1. In the first stage, results of a literature review and expert judgments are utilized to decide on the evaluation criteria for agrivoltaics system construction. The second stage comprises the GIS data collection concerning the criteria selected in the first stage. In the third stage, the expert's judgment supported by literature is used to assign weights for each criterion using FAHP. Additionally, layers for each criterion are created. Then, total scores representing the region's suitability are calculated using the layer information and the weights. Finally, a suitability map is created to reveal potential areas for agrivoltaics construction. Each stage is presented in Sects. 3.1–3.4, respectively.

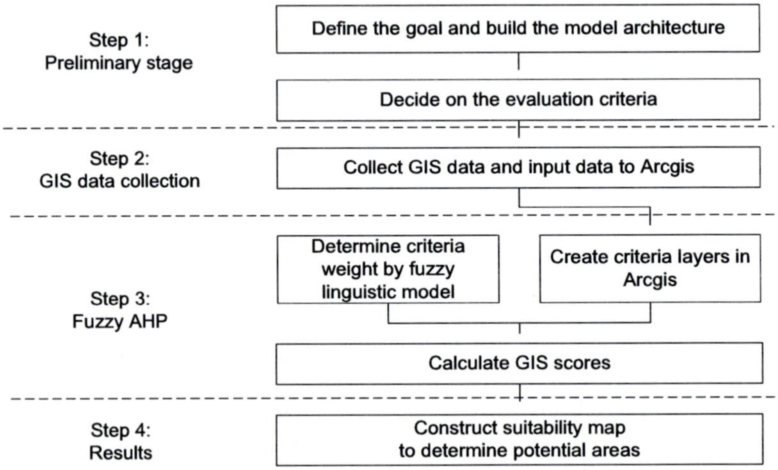

Fig. 1. Stages of solution methodology

3.1 Preliminary Stage

Considering the scope of the paper, "GIS-based MCDM", "FAHP", in the field of "agrivoltaics", and "solar farm" are selected as our keywords to determine the evaluation criteria. Google Scholar search engine was used for the literature search. The search was limited to journal articles in English and master theses in Turkish. Considering the relevance of the study with the objective, we selected two of them as the main resources.

Selected evaluation criteria used in decision-making are as follows: C1. Solar Energy Potential: We measured the solar energy potential of a location by the amount of Global Horizontal Irradiation (GHI) it received. GHI refers to the total amount of solar radiation a horizontal surface receives at a given location, including direct and diffuse radiation. Because solar radiation provides energy converted into electricity by solar panels, higher GHI values are desired; C2. Aspect: Aspect refers to the angle of arrival of solar rays, which is crucial for the efficient use of solar systems. Since Turkey is located in the Northern Hemisphere, it receives the sun mostly from the South. Therefore, the maps were reclassified by scoring the highest value to the South and the lowest to the North; C3. Slope: Sloped terrains are important to efficiently convert sunlight to electrical energy. Even on flat floors, solar panels are tilted to have the highest exposure to the sun. However, since our study is motivated by the dual use of the land for agriculture and solar energy, regions with a slope of 20% are eliminated due to difficulties in planting. C4. Land cover: Living spaces, protected areas, or restricted areas where agriculture cannot take place are removed, and only regions available for agriculture are considered in the study; C5. Proximity to roads: This criterion shows the distance of the region to the roads. Regions close to the roads are desired for lower transportation costs; C6. Proximity to waterways: Proximity to water resources is important since agriculture is also involved in addition to the solar panel system. The potential area should not be far from water sources to create irrigation channels, but it should also not be too close to minimize possible flood and flood risks; C7. Proximity to energy lines: The energy produced in power plants is stored in power distribution units and distributed by energy transmission lines. Therefore, regions close to energy transmission lines are desired; C8. Proximity to natural gas lines: The potential area should be away from the natural gas line or any pipeline to eliminate the risk of explosion or leak in natural gas pipelines; C9. Proximity to fault line: Turkey has many active faults due to its geological location. Regions with low earthquake risk should be considered.

3.2 Collection of GIS Data

GIS data collected from multiple resources are deployed for spatial analysis by ESRI ArcGIS 10.8 [12]. Table 1 presents the data types, sources, and analysis for all evaluation criteria.

Table 1. Data resources used in GIS analysis

Criteria		Source	Data Type	Scale/Cell Size
C1	Solar Power Energy	Global Solar Atlas (2018)	Raster	100 m
C2	Aspect	EarthExplorer USGS	Raster	25 m
C3	Slope	EarthExplorer USGS	Raster	25 m
C4	Land cover	CORINE Land Cover (2018)	Raster	100 m
C5	Proximity to roads	OpenStreetMap (2023)	Vector - Line layer	1/25.000
C6	Proximity to waterways	OpenStreetMap (2023)	Vector - Line layer	1/25.000
C7	Proximity to energy lines	OpenStreetMap (2023)	Vector - Line layer	1/25.000
C8	Proximity to natural gas lines	OpenStreetMap (2023)	Vector - Line layer	1/25.000
C9	Proximity to fault lines	MTA (2001)	Vector - Line layer	1/25.000

3.3 Fuzzy Analytical Hierarchy Process

AHP, introduced by Saaty [5], considers MCDM problems by performing a pairwise comparison of criteria and alternatives for decision-making. However, AHP feels short of presenting the uncertain judgments and opinions of the DMs due to the incorporation of crisp number-based calculations. Therefore, FAHP was developed to fully cover experts' linguistic judgments and preferences, which are uncertain and imprecise. In FAHP, fuzzy numbers better represent the scaling scheme in judgment matrices [13].

In this paper, we considered triangular fuzzy numbers (TFN) to define the uncertainty associated with DM. A triangular fuzzy number can be represented as M = (l, m, u), where l, m, and u refer to the smallest possible value, the most promising value, and the largest possible value, respectively. A fuzzy number is defined by its membership function μ_M (x). Let X is a set of items known as the universe, and a fuzzy subset M in X is represented by a membership function μ_M (x) as in Eq. (1).

$$\mu_M(x) = \left\{ \begin{array}{ll} 0 & x < u, x > l \\ \frac{x-l}{m-l} & l \le x \le m \\ \frac{u-x}{u-m} & l \le x \le m \end{array} \right\} \quad (1)$$

FAHP based on the geometric mean method is applied according to [14] as follows:

Step 1: Perform a pairwise comparison among performance criteria to integrate the DM's judgments using the linguistic terms for the relative importance [15] and then create a fuzzy judgment matrix $\tilde{\mathrm{M}}$.

$$\tilde{M} = \begin{bmatrix} 1 & \tilde{m}_{12} & \cdots & \tilde{m}_{1n} \\ \tilde{m}_{21} & 1 & \cdots & \tilde{m}_{2n} \\ \vdots & \vdots & \ddots & \vdots \\ \tilde{m}_{n1} & \tilde{m}_{n2} & \cdots & 1 \end{bmatrix} = \begin{bmatrix} 1 & \tilde{m}_{12} & \cdots & \tilde{m}_{1n} \\ 1/\tilde{m}_{12} & 1 & \cdots & \tilde{m}_{2n} \\ \vdots & \vdots & \ddots & \vdots \\ 1/\tilde{m}_{1n} & 1/\tilde{m}_{2n} & \cdots & 1 \end{bmatrix}$$

where

$$\tilde{m}_{ij} = \begin{cases} \tilde{9}^{-1}, \tilde{8}^{-1}, \tilde{7}^{-1}, \tilde{6}^{-1}, \tilde{5}^{-1}, \tilde{4}^{-1}, \tilde{3}^{-1}, \tilde{2}^{-1}, \tilde{1}^{-1}, \tilde{1}^{1}, \tilde{2}^{1}, \tilde{3}^{1}, \tilde{4}^{1}, \tilde{5}^{1}, \tilde{6}^{1}, \tilde{7}^{1}, \tilde{8}^{1}, \tilde{9}^{1}, & 1, i \neq j \\ 1, & i = j \end{cases}$$

Step 2: Define the fuzzy geometric mean and the fuzzy weights of each criterion as follows as in Eqs. (2–3), respectively:

$$\tilde{r}_i = \left(\tilde{m}_{i1} \otimes \ldots \otimes \tilde{m}_{ij} \otimes \ldots \tilde{m}_{in}\right)^{1/n} \quad (2)$$

$$\tilde{w}_i = \left(\tilde{r}_i \otimes \left[\tilde{r}_1 \oplus \ldots \oplus \tilde{r}_i \oplus \ldots \tilde{r}_n\right]\right)^{-1} \quad (3)$$

where $\tilde{m}_{ij}$ denotes the fuzzy comparison value of ith criterion with respect to j^{th} criterion. The notation $\tilde{r}_i$ denotes the geometric mean of fuzzy comparison value of criterion i to each criterion, and $\tilde{w}_i$ shows the fuzzy weight of the ith criterion, which is defined by a TFN.

3.4 Construct Suitability Map

After fuzzy weights for each criterion are determined by the FAHP method, the center of area method [14] is applied to calculate de-fuzzified weights. Using de-fuzzified weights, a multiplication operation is performed between each cell value within the map layers and the corresponding criteria weights. Subsequently, multiplications are aggregated to generate the suitability map for all potential areas.

4 Results

In order to apply the proposed GIS-based FAHP model, three provinces where agriculture is intensively cultivated based on 2020 statistics were selected; Konya, Ankara, and Şanlıurfa [16]. First, the GIS data collected from resources shown in Table 1 for all evaluation criteria are input to ArcGIS ESRI 10.8 software. Then, spatial analysis such as Euclidean distance, slope, reclassify, and weighted overlay analysis are used. In Euclidean distance, the closest distance between each point and the closest source is calculated. The Euclidean analysis is performed for criteria 5-9. In aspect analysis, ArcToolBox – Spatial Analysis – Aspect tool is used to determine the aspect of the region. Slope analysis is deployed to identify the slope of the region. In the reclassification analysis, values in the input rasters are rescaled to ensure a common evaluation scale, where "1" refers to the lowest score and "10" refers to the highest score depending on the characteristic of the criteria, such as minimum better or maximum better. Reclassification analysis is conducted for all evaluation criteria based on [7]. Figures 2, 3 and 4 show the map layers of each criterion based on reclassification analysis. In Figs. 2, 3 and 4, lighter colors represent lower scores, while dark red represents higher scores. After getting the map layers for each criterion in Figs. 2, 3 and 4, a weighted overlay analysis is conducted. In weighted overlay analysis, each cell value in the map layers is multiplied by the weight of importance of the criteria, and then all the multiplications are aggregated to produce the suitability map. However, prior to weighted overlay analysis, criteria weights should be determined.

In order to calculate the criteria weights, the FAHP method is applied. First, using the literature studies and an expert's assistance, a pairwise comparison matrix among criteria is constructed using the linguistic scale in [15]. The results of the comparison matrix are presented in Table 2. It is important to mention that the consistency ratio of the pairwise matrix is below 0.1. We used SpiceLogic AHP software for criteria weight determination. The pairwise comparison matrix in Table 2 is input to SpiceLogic AHP, and criteria weights are calculated by the FAHP based on geometric mean method presented in Sect. 3.3. Table 3 shows the fuzzy and the crisp (defuzzified) values of the weights. Defuzzification is conducted by the center of area method that calculates the average of the smallest possible, the most promising, and the largest possible value of each criterion.

Finally, using the defuzzified weights in Table 3, a suitability map for each province is generated using weighted overlay analysis in Figs. 5, 6 and 7. In Fig. 5, the potential areas for agrivoltaics are mainly located in the southern part of Konya; specifically, districts Çumra, Karapınar, and Ereğli locate the most suitable locations. In Fig. 6, districts Polatlı, Balat, and Şereflikoçhisar in southern Ankara have the highest potential

Table 2. Pairwise comparison among criteria

	C1			C2			C3			C4			C5			C6			C7			C8			C9		
	l	*m*	*u*	*l*	*m*	*u*	*l*	*m*	*u*	*l*	*m*	*u*	*l*	*m*	*u*	*l*	*m*	*u*	*l*	*m*	*u*	*l*	*m*	*u*	*l*	*m*	*u*
C1	1	1	1	2	3	4	3	4	5	2	3	4	6	7	8	6	7	8	2	3	4	9	9	9	9	9	9
C2	0.25	0.33	0.5	1	1	1	1	2	3	0.33	0.50	1	4	5	6	4	5	6	2	3	4	6	7	8	6	7	8
C3	0.20	0.25	0.33	0.33	0.50	1	1	1	1	0.33	0.50	1	5	6	7	4	5	6	1	2	3	9	9	9	9	9	9
C4	0.25	0.33	0.50	1	2	3	1	2	3	1	1	1	3	4	5	3	4	5	1	1	1	7	8	9	7	8	9
C5	0.13	0.14	0.17	0.17	0.20	0.25	0.14	0.17	0.20	0.20	0.25	0.33	1	1	1	1	2	3	0.17	0.20	0.25	2	3	4	2	3	4
C6	0.13	0.14	0.17	0.17	0.2	0.25	0.17	0.20	0.25	0.20	0.25	0.33	0.33	0.50	1	1	1	1	0.17	0.20	0.25	1	2	3	1	2	3
C7	0.25	0.33	0.50	0.25	0.33	0.50	0.33	0.50	1	1	1	1	4	5	6	4	5	6	1	1	1	7	8	9	9	9	9
C8	0.11	0.11	0.11	0.13	0.14	0.17	0.11	0.11	0.11	0.11	0.13	0.14	0.25	0.33	0.50	0.25	0.33	0.50	0.33	0.50	1	1	1	1	1	2	3
C9	0.11	0.11	0.11	0.13	0.14	0.17	0.11	0.11	0.11	0.11	0.13	0.14	0.25	0.33	0.50	0.25	0.33	0.50	0.33	0.50	1	0.33	0.50	1	1	1	1

locations for agrivoltaics. For Şanlıurfa, districts in northern counties, mainly Siverek and Viranşehir, locate the most suitable fields for agrivoltaics in Fig. 7.

Table 3. Criteria weights

Criteria	Weights			
	l	*m*	*u*	Crisp value
C1. Solar Energy Potential	0.20	0.31	0.45	0.30
C2. Aspect	0.10	0.16	0.26	0.16
C3. Slope	0.09	0.14	0.23	0.14
C4. Land cover	0.10	0.16	0.25	0.16
C5. Proximity to roads	0.02	0.04	0.06	0.04
C6. Proximity to waterways	0.02	0.03	0.05	0.03
C7. Proximity to energy lines	0.08	0.12	0.19	0.12
C8. Proximity to natural gas lines	0.01	0.02	0.04	0.02
C9. Proximity to fault lines	0.01	0.02	0.03	0.02

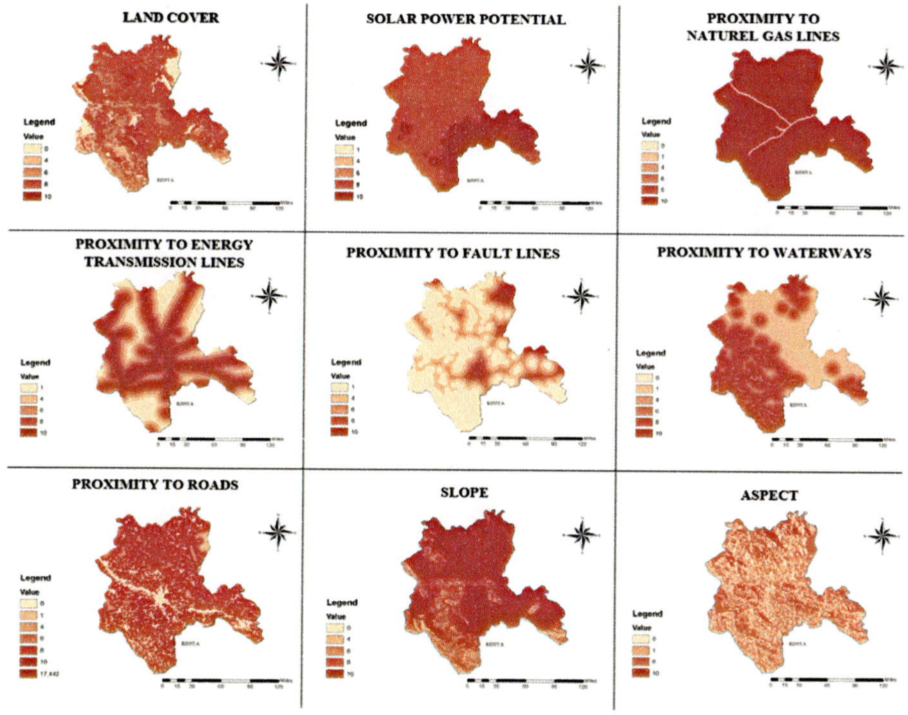

Fig. 2. Map layers of each criterion with reclassified values for Konya

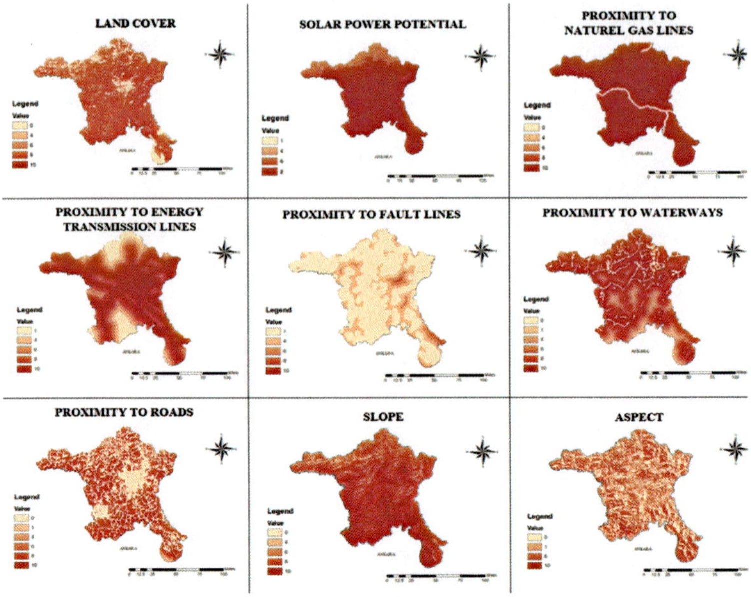

Fig. 3. Map layers of each criterion with reclassified values for Ankara

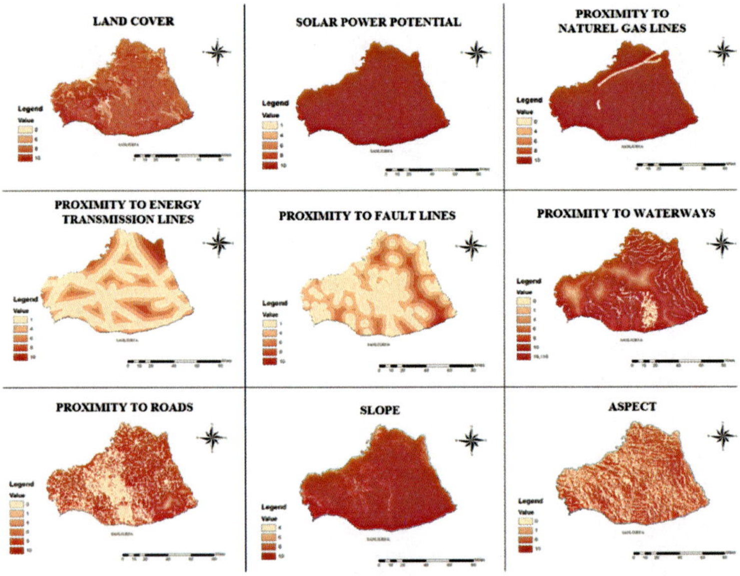

Fig. 4. Map layers of each criterion with reclassified values for Şanlıurfa

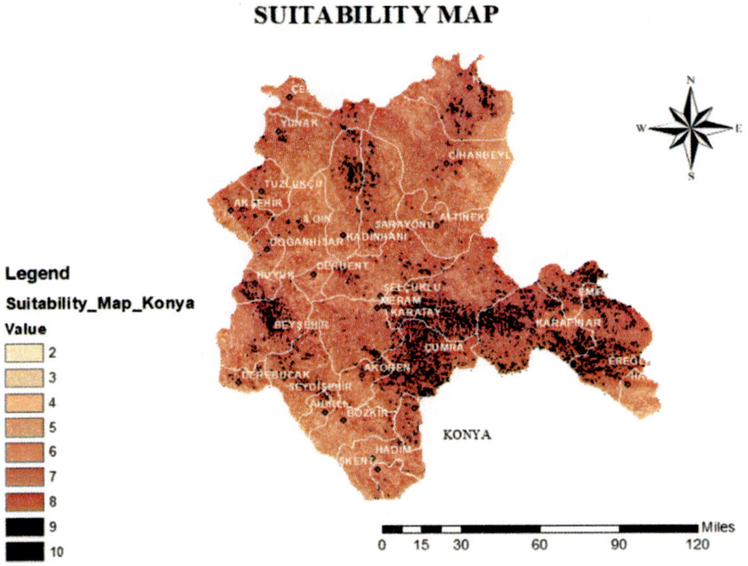

Fig. 5. Suitability map for Konya

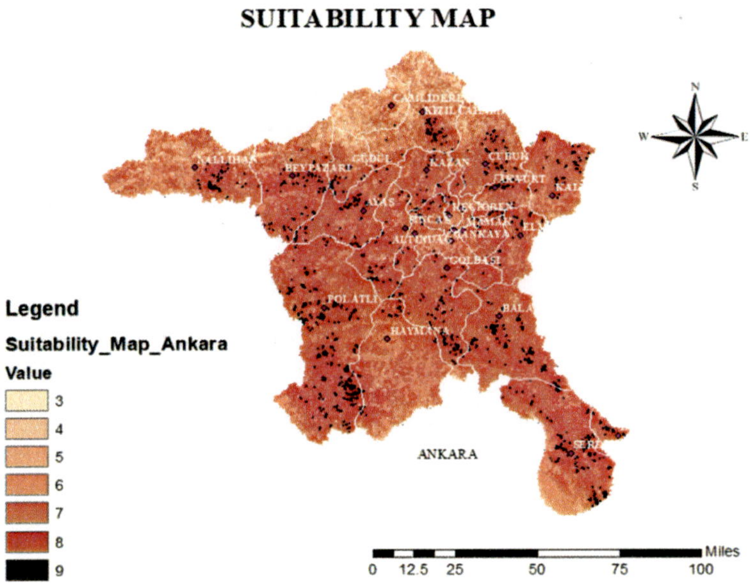

Fig. 6. Suitability map for Ankara

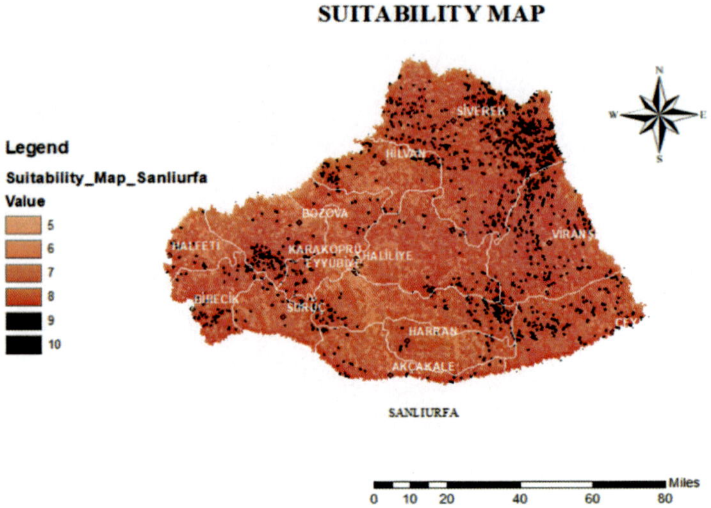

Fig. 7. Suitability map for Şanlıurfa

5 Conclusion

This study uses GIS-based FAHP to analyze the potential of agrivoltaic systems for three provinces in Turkey. First, evaluation criteria were identified through a literature review. Then, GIS data was collected for each criterion, and the criteria weights were determined by FAHP. Finally, a suitability map for the three provinces was created using GIS data and criteria weights. The results obtained in the study have potential benefits for efficient energy and crop production. Suitability maps created help practitioners in deciding the appropriate fields for agrivoltaics to gain the most benefit.

References

1. S. P. Energy: Technology roadmap. Technical report, IEA (2014). https://www.iea.org/reports/technology-roadmap-energy-storage. Accessed 10 Apr 2023
2. Goetzberger, A., Zastrow, A.: On the coexistence of solar-energy conversion and plant cultivation. Int. J. Sol. Energy **1**(1), 55–69 (1982)
3. Dinesh, H., Pearce, J.M.: The potential of agrivoltaic systems. Renew. Sustain. Energy Rev. **54**, 299–308 (2016)
4. ESRI, Environmental Systems Research Institute. https://www.esri.com/en-us/what-is-gis/overview. Accessed 10 Apr 2023
5. Saaty, T.L.: The Analytic Hierarchy Process. McGraw-Hill, New York (1980)
6. Uyan, M.: GIS-based solar farms site selection using analytic hierarchy process (AHP) in Karapinar region, Konya/Turkey. Renew. Sustain. Energy Rev. **28**, 11–17 (2013)
7. Colak, H.E., Memisoglu, T., Gercek, Y.: Optimal site selection for solar photovoltaic (PV) power plants using GIS and AHP: a case study of Malatya Province, Turkey. Renew. Energy **149**, 565–576 (2020)

8. Saraswat, S.K., Digalwar, A.K., Yadav, S.S., Kumar, G.: MCDM and GIS based modelling technique for assessment of solar and wind farm locations in India. Renew. Energy **169**, 865–884 (2021)
9. Wang, H.W., Dodd, A., Ko, Y.: Resolving the conflict of greens: a GIS-based and participatory least-conflict siting framework for solar energy development in southwest Taiwan. Renew. Energy **197**, 879–892 (2022)
10. Noorollahi, E., Fadai, D., Akbarpour Shirazi, M., Ghodsipour, S.: Land suitability analysis for solar farms exploitation using GIS and fuzzy analytic hierarchy process (FAHP)—a case study of Iran. Energies **9**, 643 (2016)
11. Asakereh, M., Soleymani, M., Sheikhdavoodi, M.J.: A GIS-based Fuzzy-AHP method for the evaluation of solar farms locations: case study in Khuzestan province Iran. Sol. Energy **155**, 342–353 (2017)
12. ESRI 2023. ArcGIS 10.8. Environmental Systems Research Institute, Redlands
13. Cheng, C.H., Mon, D.L.: Evaluating weapon system by analytical hierarchy process based on fuzzy scales. Fuzzy Sets Syst. **63**(1), 1–10 (1994)
14. Sun, C.C.: A performance evaluation model by integrating FAHP and fuzzy TOPSIS methods. Expert Syst. Appl. **37**(12), 7745–7754 (2010)
15. Kannan, D., Khodaverdi, R., Olfat, L., Jafarian, A., Diabat, A.: Integrated fuzzy multi criteria decision making method and multi-objective programming approach for supplier selection and order allocation in a green supply chain. J. Clean. Prod. **47**, 355–367 (2013)
16. STATAGRI. https://www.statagri.com/tarim-alanlari/. Accessed 10 Apr 2023

Analyzing Replenishment Policies for Automated Teller Machines

Deniz Orhan[1(✉)] and Müjde Erol Genevois[2]

[1] Institute of Science, Galatasaray University, 34349 Istanbul, Turkey
deniz.orhan@ogr.gsu.edu.tr
[2] Department of Industrial Engineering, Galatasaray University, 34349 Istanbul, Turkey
merol@gsu.edu.tr

Abstract. Logistics has become one of the most significant parts of the process in many business areas. For an efficient logistics system, each stage of the operation needs to be designed carefully. Logistics approaches are also applied in the financial sector such as Automatic Teller Machines (ATM) cash management. ATMs provide efficient service for a financial institution to its customers through a self-service, time-independent, and simple-to-use mechanism. For daily financial transactions, ATM is the fastest, safest, and most practical banking tool. Many challenges have risen in the network design of the cash and these problems can be solved by using the optimized solution. This solution aims to satisfy the customer at the ATM and at the same time, minimize loss for banks. This paper states the solution for the replenishment of ATMs. Firstly, data is analyzed by using different policies from several approaches after then an efficient metric system is applied to compare the results of it. In the end, the method selected has the appropriate results according to metric. Furthermore, to avoid bottlenecks and become quicker in the procedures, inventory control connects the supply of cash and customer demand in the ATMs. The replenishment policy starts with forecasting cash withdrawals by applying various methods such as statical methodologies (ARIMA and SARIMA) and machine learning methods (Prophet and DNN). By creating a decision support system, several methods are applied in order to visit ATMs by using different inventory control methodologies.

Keywords: Cash replenishment · ATM · Inventory control · Decision support system

1 Introduction

In logistics activities, meeting customer expectations for a wide range of high-quality goods is a significant obstacle. To provide high-quality service, various areas need to be arranged effectively. In our research, the most important problem starts with finding an optimal cash replenishment policy. Firstly, determining the cash replenishment strategy will be a huge effect on the outcome. As crucial as the accuracy of the cash prediction is, the forecast must be made in such a way that it can meet consumer demands while also not loading too much money that is not withdrawn from the ATM for a long time. An

Z. Şen et al. (Eds.): IMSS 2023, LNME, pp. 546–554, 2024.
https://doi.org/10.1007/978-981-99-6062-0_50

efficient cash inventory management system should generate a balance between holding costs and customer service levels; a huge amount of inventory entails significant financial expenses, yet an adequate inventory level is crucial to meet consumer cash demand [1]. As a result, a suitable estimated cash management strategy should be devised. The first solution may be refilling the ATM with a low amount of money, but this solution has some negative sides such as a lack of demanding cash. This causes dissatisfaction among customers that is the most undesirable situation. To solve this problem, loading the ATM with a lot of cash can cause idle cash. Idle cash means inactive money in the ATM over a long period. This idle cash may classify as a loss for banks due to not being traded at the exchange. Moreover, this unprofitable strategy of banks is the second most undesirable situation. As a result, the goal is to prevent the replenishment of more money than is required. The most efficient way to optimize ATM cash management is to maintain these constraints in balance.

This study, it is aimed to construct effective forecasting for ATMs. In the literature, most of the studies applied traditional methods but the accuracy of the predictions directly affects customer satisfaction with ATM service. First of all, various forecasting methodologies are applied to the NN5 data. These are statical methodologies which are ARIMA and SARIMA, and machine learning methods which are Prophet, and DNN. From the literature review, Prophet is never used before for ATM cash replenishment policies. To analyze the result of the methodologies, MAPE (Mean absolute percentage error) method is applied.

Similar methods have always been used in most studies in the literature. However, how adequate these methods are to meet the requirements of today's world is discussable. These methods show great results, but more effective outcomes can be found by using new-generation methods such as Prophet. From this point of view, it is possible to say that there is a deficiency in the literature. In this study, it is explained how the use of the most recent methods gives optimal results in today's demanding world.

There are four more sections in the paper. The second section provides a research summary of previous literature, and the third section explains the methodologies and mathematical formulation of the suggested method. Results are shown in the fourth section. Conclusions and future works are stated in the last section.

2 Literature Review

Several industries, including finance, entertainment, manufacturing, etc., have service point issues. The primary point of contact for the financial industry is the ATM. In terms of, increasing client loyalty, a quick and competent system is crucial. This chapter categorizes articles that are concerned with service point challenges which are forecasting problems cost-minimizing.

2.1 Forecasting Problem

The difficulty of forecasting the right quantity of cash for each day's customer needs which means that the minimal amount of cash is always accessible until the next replenishment is problematic. A machine learning technique to forecast ATM replenishment amounts utilizing a data-driven approach for estimating the correct amount for ATM problems [2]. Machine learning technique is the most common method to solve forecasting problems in recent years. Another machine learning code for problem represented to solve forecasting of daily cash withdraws from ATMs. By using Artificial Neural Network (ANN), daily cash withdrawals for certain ATMs are claimed as predictable. In addition, the study provides a cost function for determining the most efficient money and replenishment days. Cash withdrawals adopt an estimated seasonal pattern depending on date and time parameters [3]. Machine learning methods, which give efficient results in forecasting problems, have developed in recent years.

2.2 Cost Minimization

Today's service sector faces many issues while providing quality services and keeping service costs low. One of the significant types of research which used real data from banks was generated by Van Anholt et al. [4] who introduced, designed, and solved a complex multiperiod inventory-routing problem involving pickups and deliveries inspired by ATM replenishment in the Netherlands. This paper anticipates this development by providing cash supply chain parties based on improving cash management operations and reducing costs. Bolduc et al. [5] introduce the new model to minimize overall transportation and inventory costs by defining which customers will be served by a specific distributor, as well as the delivery routes for those supplied by the private fleet. Outcomes showed an improvement in overall solution quality. In today's modern age, the decreasing cost is the main concern of many businesses.

3 Methodologies

3.1 Forecasting Methods

Prophet Method

Prophet is a Facebook-developed open-source library based on decomposable (trend + seasonality + holidays) models. It is one of the most useful machine-learning methods in the literature. Prophet is a time series prediction model released in 2017. Taylor & Letham claim that a basic, modular regression model was used that frequently works well with default settings and allows analysts to pick and choose which components are important to their forecasting challenge and make changes as needed [6]. Prophet has various aspects from other machine learning methodologies. Seasonality and the holiday's approach of the Prophet are the core differences when we compare with forecasting strategies. Trend term, seasonal period term, and holidays are the three essential aspects of the model. Aside from excellent prediction abilities, Prophet can also efficiently cope with data having periodic and cyclic changes, as well as significant anomalous values

[7]. Because of the many features, ATM data have seasonal changes and also include noise data. When used to estimate ATM withdrawal, the Prophet method may transcend the constraints of standard forecasting models and provide optimal forecasting results. In the literature, Prophet is not used to estimate the cash demand of ATMs before. It is generally used for production planning or sales forecasting. But it can be also suitable for our problem.

Mathematical Formulation of Prophet

Forecasting is a curial accomplishment for business in terms of production planning schedules, workload prediction, or inspecting weak points. Prophet employs a decomposable time series model that consists of three primary components: trend, seasonality, and holidays. The following equation combines them. The mathematical formulation of Prophet is taken from Taylor, S. J. et al. [6].

$$y(t) = g(t) + s(t) + h(t) + \varepsilon \tag{1}$$

g(t) is a trend item, *s(t)* is a seasonal change, *h(t)* is the holiday factor, ε is an error term.

Trend parameters discuss in two ways. The first one is nonlinear growth which reaches a saturation point.

$$g(t) = \frac{C(t)}{1 + \exp(-(k + a(t)^T\delta(t - (m + a(t)^T\gamma)))} \tag{2}$$

C(t) is the time-varying capacity, *k* is the base rate, *m* is offset, $k + a(t)^T\delta$ is a growth rate of time varying, to connect the endpoints of segments $m + a(t)^T\gamma$ is adjusted as an offset parameter and δ is the change in the growth rate.

For linear growth, a piece-wise constant rate of growth ensures an efficient model most of the time.

$$g(t) = (k + a(t)^T\delta)t + (m + a(t)^T\gamma) \tag{3}$$

where k is the growth rate, δ is rate adjustments, m is offset and γ is to make the function continuous.

Seasonality models must be expressed as periodic functions of t. In order to imitate seasonality, fourier terms are used in regression models by using sine and cosine terms.

$$s(t) = \sum_{n=1}^{N}\left(a_n \cos\left(\frac{2\pi nt}{P}\right) + b_n \sin\left(\frac{2\pi nt}{P}\right)\right) \tag{4}$$

P is the regular period that is expected, 365 for annual data and 7 for weekly data. $N = 10$ and $N = 3$ for yearly and weekly seasonality work well for most cases.

Some holidays are on certain dates in the year but some of them can be varying every year. Therefore, the model needs to fit this change. Resuming that the impacts of vacations are independent. Assign each holiday a parameter κ_i, which is the corresponding change in the prediction, and an indicator function showing whether time t is during holiday i.

$$h(t) = Z(t)\kappa \tag{5}$$

ARIMA and SARIMA Methods

ARIMA (autoregressive integrated moving average) is a statistical analysis model which employs time series data to optimize the analysis of data or forecast future trends. An autoregressive integrated moving average model is a type of regression analysis that determines how steady one dependent variable is in comparison to other changing variables. An ARIMA model has three constants in terms of; p for a number of autoregressive terms, d for the order of differencing, and q for the number of moving-average terms. Generally, ARIMA duration would be expressed as ARIMA (p, d, q). To determine this forecast properly by ARIMA, several steps are required according to Nahmias. There are four core steps to implement this forecasting method ARIMA also known as Box-Jenkins. These are respectively, data transformation, model identification, parameter estimation, forecasting, and evaluation [8]. To forecast the cash in the ATM, ARIMA is the most common method for this problem. Khanarsa, P. et al. state that a set number of subsequence patterns are produced and fitted using ARIMA models, and the best-fitted model is identified using the symmetric mean absolute percentage error (SMAPE) [9].

Moreover, to overcome seasonality in the data SARIMA can be a useful method. SARIMA similarly not also used past data like ARIMA but also take into account the seasonality of the pattern as a parameter. SARIMA become more suitable for forecasting complex data spaces containing trends. SARIMA duration would be expressed $(P, D, Q)_m$. m is the seasonality term which is the number of observations per year. SARIMA become the most suitable method where seasonality is known. SARIMA models applied to certain subseries may be valuable tools in characterizing the structure of daily withdrawals regardless of ATM location, according to the empirical findings [10].

Neural Network Method

Machine learning also has a huge share in the literature to forecast optimal solutions. Especially neural networks have become the most popular trends in machine learning. Artificial neural networks (ANNs) and simulated neural networks (SNNs) are a subset of machine learning that is at the heart of deep learning methods. This diagram depicts a multilayer feed-forward network, in which each layer of nodes gets input from the former layers. A weighted linear combination is used to combine the inputs to each node. Before being output, the result is changed by a nonlinear function [11] (Figs. 1, 2 and 3).

Neural networks may be used for cash demand forecasting. Venkatesh et al. state that four neural networks are used which are general regression neural network (GRNN), multi-layer feed-forward neural network (MLFF), group method of data handling (GMDH), and wavelet neural network (WNN) built to predict an ATM center's cash demand and GRNN has shown the best result according to the SMAPE [12]. The unknown future values of a collection of 100-time series of daily ATM cash withdrawals were modeled and forecasted using a multi-layer feed-forward neural network model. The outputs of the forecasting model depict the complicated dynamics of the time series data that have been properly described [13]. One of the best ways to find an efficient solution is to compare the different methods and classify their results.

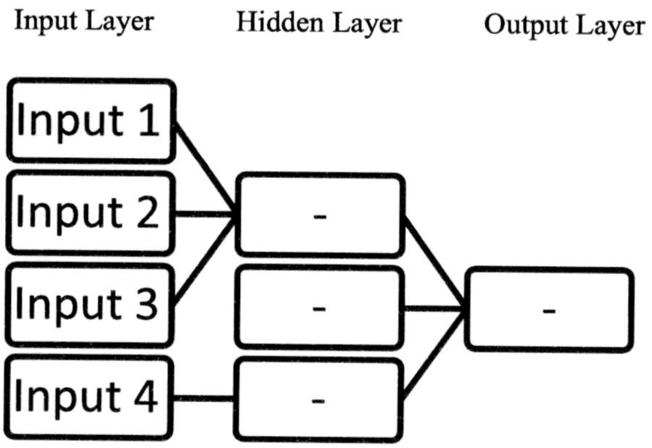

Fig. 1. Multilayer feed-forward network

3.2 Comparison of Methodologies

Firstly, machine learning methods may be compared with each other. Nonlinear prediction methods, such as artificial neural networks (ANNs) and support vector regression (SVR), have been devised and widely employed in artificial intelligence domains to overcome the flaws of linear models. Moreover, the constraints of SVR and LSTM (Long Short Term Memory) models impair the prediction ability of the models when the data is increasingly complicated or changes dramatically [7]. Therefore, these methods are not suitable for noisy data.

In the Prophet, unlike ARIMA models, the measurements do not need to be evenly spaced, and missing data are not interpolated. In most of the studies, ARIMA compares with the Prophet, and better results are getting from Prophet.

4 Results

The data is comprised of two years' worth of daily cash demand from several automated teller machines located in England. These statistics are the results of forecasted NN5 competitiveness. This data includes the 111 randomly chosen cash machines at various places. We transform the cash withdrawal data to 13 ATMs for 56 days thus analyzing data becomes easier. Finding the original data's average is the first step in the process.

For ATM cash replenishment problems, statistical approaches, machine learning methods, and hybrid methods can be used. After a review of the literature, the seasonality parameter which is significant for our data led to the selection of the SARIMA approach over static alternatives. The machine learning category includes a large number of neural networks. The most used one is ANN, which is inappropriate for our data since it requires a larger dataset for an effective outcome. Compared to other neural network techniques, DNN showed greater results. Moreover, Prophet, the most recent machine learning technique used in this work, is also applied. The ARIMA-SARIMA-DNN-Prophet with realized the quantity of money is included in our compression.

We decided to run the model in terms of forecast the last 8 days by using the first 48 days' data. Moreover, the first 48 days are recalculated using the predicted 8-day standard deviation. Because of it, the entire collection of data is now compatible with one another. The initial quantity of cash for one randomly chosen ATM 103 is used to compare the procedures in the graph below.

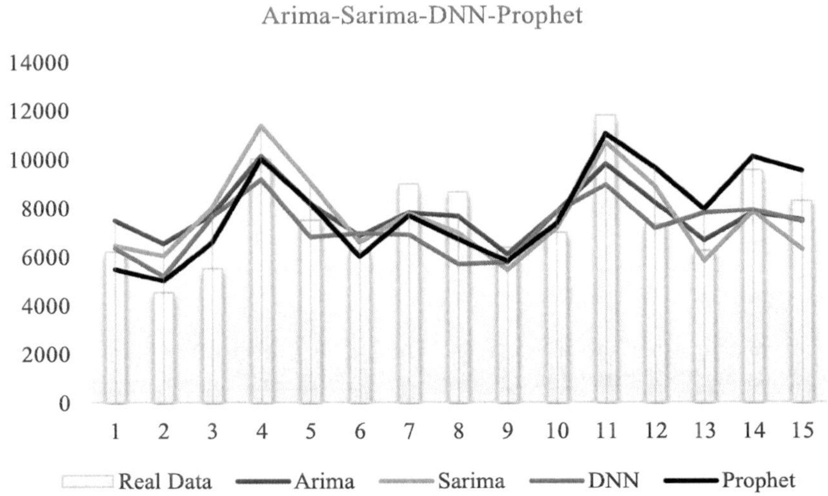

Fig. 2. Comparison of methods for ATM 103 which is randomly selected.

In addition, the mean absolute percentage error (MAPE) is calculated for all methods. MAPE of ARIMA, SARIMA, DNN, and Prophet is 19,60%, 19,72%, 19,05%, and 17,89% respectively. Prophet is the best methodology by the smallest error rate in this group for ATM 103.

After then, MAPE is calculated for the whole data to identify the data in the aggregate. The results of the methods are shown below.

When we examine the averages of MAPE percentages for each method, the results are almost similar to the randomly selected ATM result. 23,33%, 22,82%, 21,67%, and 18,52% for ARIMA, SARIMA, DNN, and Prophet respectively. Prophet showed the best results.

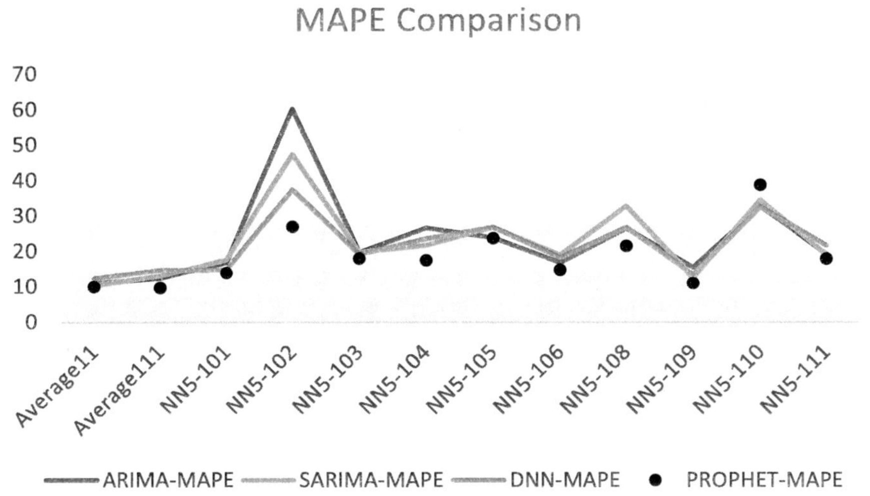

Fig. 3. MAPE Comparison of methods for all ATMs

5 Conclusion and Future Study

The decision support system is constructed for ATM cash replenishment. Methods from several perspectives have analyzed each other. After examining the outcomes of the metric system, the one with the lowest error value was selected. The Prophet forecasting method is used in this study. On many benchmark situations from the literature, we demonstrated the importance of the problem and why it should be done optimally. In addition, as a future work, VRP can be applied according to Prophet's forecast resulting in minimizing cost. In this manner, realistic solutions can be found in the banking industry.

Acknowledgment. The authors acknowledge that this research was financially supported by Galatasaray University Research Fund (Project Number: FOA-2022–1128).

References

1. Baker, T., Jayaraman, V., Ashley, N.: A data-driven inventory control policy for cash logistics operations: An exploratory case study application at a financial institution: data-driven inventory control policy for cash logistics operations. Decis. Sci. **44**(1), 205–226 (2013)
2. Asad, M., Shahzaib, M., Abbasi, Y., Rafi, M.: A long-short-term-memory based model for predicting ATM replenishment amount (2020)
3. Serengil, S.I., Özpınar, A.: ATM cash flow prediction and replenishment optimization with ANN. Int. J. Eng. Res. Dev. **11**(1), 402–408 (2019)
4. Anholt, R.G., Coelho, L.C., Laporte, G., Vis, I.F.A.: An inventory-routing problem with pickups and deliveries arising in the replenishment of automated teller machines, pp. 1077-1091 (2016)
5. Bolduc, M.-C., Laporte, G., Renaud, J., Boctor, F.F.: A tabu search heuristic for the split delivery vehicle routing problem with production and demand calendars. In: European Journal of Operational Research, pp. 122–130 (2010)

6. Taylor, S.J., Letham, B.: Prophet: Forecasting at scale, pp. 37–45 (2017)
7. Wang, D., Meng, Y., Chen, S., Xie, C., and Liu, Z.: A hybrid model for vessel traffic flow prediction based on wavelet and prophet, 1–16 (2021)
8. Nahmias, S.: Production and Operation Analysis, 6th edn. McGraw-Hill Eduction (Asia) Co. and Tsinghua University Press (2013)
9. Khanarsa, P., Sinapiromsaran, K.: Multiple ARIMA subsequences aggregate time series model to forecast cash in ATM. In: 9th International Conference on Knowledge and Smart Technology: Crunching Information of Everything, pp. 83–88. (2017). https://doi.org/10.1109/KST.2017.7886096
10. Gurgul, H., Suder, M.: Modeling of withdrawals from selected ATMs of the "Euronet" network, pp. 65–82 (2013)
11. Hyndman, R.J., Athanasopoulos, G.: Forecasting: Principles and Practice, 2nd edn. OTexts, Melbourne, Australia (2018)
12. Venkatesh, K., Ravi, V., Prinzie, A., Poel, D.V.D.: Cash demand forecasting in ATMs by clustering and neural networks. Eur. J. Oper. Res. **232**(2), 383–392 (2014). https://doi.org/10.1016/j.ejor.2013.07.027
13. Atsalaki, I.G., Atsalakis, G.S., Zopounidis, C.D.: Cash withdrawals forecasting by neural networks. J. Comput. Optim. Econ. Finance **3**(2), 133–142 (2011)

The Significance of Human Performance in Production Processes: An Extensive Review of Simulation-Integrated Techniques for Assessing Fatigue and Workload

Halil İbrahim Koruca[1](✉), Kemal Burak Urgancı[1], and Samia Chehbi Gamoura[2]

[1] Industrial Engineering Department, Süleyman Demirel University, Isparta, Turkey
{halilkoruca,kemalurganci}@sdu.edu.tr
[2] HuManis Laboratory of EM Strasbourg Business School, Strasbourg, France
samia.gamoura@em-strasbourg.eu

Abstract. Human involvement in manufacturing is indispensable. The performance and reliability of the human element directly impacts the success of a system. The complex interplay between fatigue, workload, and human performance necessitates recognition of the importance of human factors in manufacturing to achieve optimal outcomes. The efficiency of employees in production systems can be influenced by a number of factors, including exhaustion, cognitive and physical abilities, and the amount of time available for tasks. In general, these factors can have an impact on human reliability, which in turn can affect the reliability of production systems. A key objective of workload research is to identify and analyze potential performance declines, along with their associated factors. The pursuit of successful outcomes in an operation is not enough, as the negative impacts of such outcomes on the system or personnel must also be minimized. Once a possible issue or decrease has been determined, corrective measures are put in place to address such instances. Such foresight serves to prevent the costs that may arise from making process modifications, and to optimize resource utilization. This study conducts a comprehensive review of related literature, focusing on simulation integration. A taxonomy of fatigue and workload is used to contextualize and assess these factors. Methodologies and techniques are described, and evaluations of specific assessment techniques are provided, along with discussions of available information. This research makes a persuasive case for the importance of evaluating fatigue and workload among operators.

Keywords: Human performance · simulation · fatigue · workload assessment

1 Introduction

Ergonomics is a scientific discipline that aims to optimize the interaction among humans and other elements of a system to enhance human well-being and overall system performance. Initially, ergonomics focused on physical work environments to maximize

Z. Şen et al. (Eds.): IMSS 2023, LNME, pp. 555–566, 2024.
https://doi.org/10.1007/978-981-99-6062-0_51

productivity [1]. However, traditional ergonomic evaluations only considered peripheral outcomes and disregarded the contributions of the brain during work. The primary objective of ergonomics is to ensure that work demands are within the capability of the operator [2].

Ergonomics is responsible for adapting tasks to suit the needs and abilities of humans, ensuring their safety, health, and well-being in the workplace. Incorporating human factors and ergonomics principles into the manufacturing process is essential for promoting safe and effective work practices [3]. With the advancement of technology, there has been a significant shift in the level of human control over the manufacturing process. As machines become more sophisticated and complex, it is essential to consider the impact on human resources and their performance [4].

Various factors, such as mental and physical fatigue, skill level, and available time, can influence the performance of human resources and subsequently impact the reliability and efficiency of production systems [5]. Neglecting the significance of human factors and ergonomics principles can lead to adverse effects on both human well-being and the success of the business.

The review highlights the importance of considering the impact of machine capabilities on human performance and the significance of analyzing operator fatigue and recovery in designing and managing production systems by providing an overview of the different methods used to measure workload and fatigue and evaluate their strengths and limitations. The concept of workload and fatigue is discussed along with some definitions. The literature on fatigue and workload is classified based on multiple dimensions, such as the methods for identification and measurement, analytic models applied for fatigue and workload assessment, and their applications within manufacturing.

2 Introduction

Physical ergonomics focuses on human anatomic, anthropometric, physiological, and biomechanical characteristics, while cognitive human factors focus on mental processes that affect interactions among humans and other elements of a system [6]. As most work systems involve both physical and cognitive processing, it is necessary to consider them together when examining human behavior at work [2]. High cognitive demands can affect physical capabilities, and physical demands can affect cognitive processing. Therefore, it is essential to integrate physical and cognitive subsystems when evaluating and redesigning work situations. The conventional evaluation of work demands involves assessing biological and physiological effects, such as joint force, muscle function, and heart rate, in both laboratory and real-world environments [7].

This research to contribute to the interdisciplinary Human Factors and Ergonomics field by covering topics such as quantifying human behavior, predictive tools for assessing multidimensional work demands, new methodologies, and evidence on the interplay between physical and cognitive processes.

2.1 Workload Techniques and Approaches

Workload, defined as the amount of work assigned to or expected from a worker in a specified time period, is a multidimensional concept that is commonly used in the evaluation of human performance in complex systems. A clear understanding of workload and workload analysis is crucial for identifying potential performance issues, predicting workload, and preventing mission failure and damage to operators or machines. In this subsection, we discuss analytical techniques for workload analysis, the factors that must be taken into account in workload management, and taxonomies for workload assessment techniques.

Analytical techniques for workload analysis are used to predict workload during system concept development and preliminary system design. The focus of these techniques is on workload analysis that may be applied without operators-in-the-loop. These techniques are used to predict workload early in system development, where the greatest design flexibility is available with the least impact on system cost. The analytical techniques can be classified into five categories: comparison, expert opinion, mathematical models, task analysis methods, and simulation models [8].

To effectively manage workload, it is necessary to establish a valid and quantifiable system for measuring it. The concept of workload is used as a tool in this effort. To handle workload effectively, various factors such as the particular tasks involved, the context in which they are performed, the grouping of tasks within a work period, the prioritization of tasks, the unique qualities of each operator, and the level of cognitive or physical exertion required by the tasks must be taken into account. A comprehensive approach to workload management or evaluation should encompass all of these considerations [9].

Workload due to human activities in systems is comprised of four broad categories: perceptual tasks, cognitive tasks, communication tasks, and motor processes [10]. The workload technique taxonomy can be stated as two main categories: analytical techniques and empirical techniques [11]. Analytical techniques produce results that are used to predict performance and estimate workload without actually having human operators exercise the system during system concept development and preliminary system design. In contrast, empirical techniques require a human operator to interact with the system in question. The identification and development of useful analytical procedures for estimating workload and predicting performance continue to be an actively pursued goal, especially in the applied sector, where system developers need to assess workload early in the design process while conceptual designs are easily modified [12].

Workload management involves considering various factors such as the particular tasks involved, the context in which they are performed, and the level of cognitive or physical exertion required by the tasks. Taxonomies have been developed to classify workload methods and techniques into descriptive categories, including analytical and empirical techniques. The development of analytical procedures for estimating workload and predicting performance continues to be an actively pursued goal [13].

Workload refers to the total demand placed on an individual when performing a task. The concept of workload is important in assessing task performance and predicting how individuals will respond to different workloads. According to the theory developed by [14], workload is not the result of one central processing resource but rather the use of several processing resources. This theory explains why humans are unable to talk and listen simultaneously, yet they can effortlessly chew gum while walking. This subsection explores the Multiple Resource Theory (MRT) and its use of visual, auditory, cognitive, and psychomotor components (VACP) to assess workload. It also looks at VACP rating scales that have been developed for each component and offers examples of how MRT has been applied in different scenarios.

Information processing is reliant on multiple resources, which can be characterized by four distinct components: visual, auditory, cognitive, and psychomotor, often referred to as VACP. These components can be employed to deconstruct any task executed by an individual. Specifically, the visual and auditory components pertain to the external stimuli that are attended to, while the cognitive component relates to the level of information processing required. Lastly, the psychomotor component encompasses the physical actions involved in performing the task.

For each VACP component, rating scales have been developed to offer a comparative evaluation of the extent to which each element of the resource is utilized [15]. These scales are particularly useful in determining the level of workload that a task demands from an individual. VACP rating scales are provided in Table 1, and as the scale value increases, the usage of the resource component also increases.

MRT has been applied in various scenarios to evaluate workload demands on individuals. For instance, Mao [16] used MRT to examine the amount of work that tank crew members are required to perform under conditions of informatization using DES. The analysis was conducted using IMPRINT, a tool developed based on MRT by Wickens [17] in 2002. IMPRINT utilizes VACP scales and divides human physiological resources. Another simulation-based framework proposed by Seo [18] estimated physical demands and corresponding muscle fatigue to evaluate the impact of muscle fatigue during construction operations. The framework estimates physical demands from a planned operation modeled using discrete event simulation (DES) through biomechanical analyses and uses dynamic fatigue models. It also models workers' strategies to mitigate muscle fatigue to understand how muscle fatigue affects time and cost performance of the planned operation. MRT provides a useful framework for assessing workload demands on individuals. Its VACP components and rating scales can be employed to deconstruct tasks and determine the level of workload that a task demands from an individual. This information can be valuable in predicting task performance and designing interventions to improve performance.

2.2 Workplace Fatigue

Fatigue is a common phenomenon experienced in daily life and in workplaces, resulting from various factors such as work type and duration, workload, and changes in working posture. It is a multidimensional concept that affects employees in both physical and mental ways [19]. Fatigue assessment in the workplace is critical to prevent worker injuries, reduce expenses associated with workers' compensation, and ensure optimal work schedules. Fatigue is a complex and multi-dimensional construct that can affect workers in various ways, including decreased productivity, increased errors, and increased risk of accidents and injuries [20]. Therefore, assessing fatigue in the workplace is essential for maintaining worker safety and productivity.

Workplace fatigue can be categorized as either localized muscle fatigue or whole-body fatigue, depending on the type of task [22]. Localized muscle fatigue affects specific muscle groups, such as during assembly line or precision work, while whole-body fatigue is more cardiorespiratory in nature and occurs during manual materials handling. A decrease in force generating capacity is a common ergonomic sign of localized muscle fatigue [20].

Table 1. VACP Values and Descriptors [21]

Scale	Descriptor	Scale	Descriptor	Scale	Descriptor	Scale	Descriptor
Visual		Auditory		Cognitive		Psychomotor	
0.0	No visual activity	0.0	No auditory activity	0.0	No cognitive activity	0.0	No Psychomotor activity
1.0	Visually detect	1.0	Detect sound	1.0	Automatic	1.0	Speech
3.7	Visually discriminate	2.0	Orient to sound(general)	1.2	Alternative selection	2.2	Discrete actuation
4.0	Visually check	4.2	Orient to sound(selective)	3.7	Signal recognition	2.6	Continuous adjustive
5.0	Visually locate	4.3	Verify feedback	4.6	Evaluation (single aspect)	4.6	Manipulative
5.4	Visually track	4.9	Interpret speech	5.3	Encoding/Decoding	5.8	Discrete adjustive
5.9	Visually read	6.6	Discriminate sound	6.8	Evaluation (several aspect)	6.5	Symbolic production
7.0	Visually search	7.0	Interpret sound patterns	7.0	Estimation, calculation, conversion	7.0	Serial Discrete Manipulation

Understanding the level of fatigue is imperative to assess recovery subsequent to task completion. Employees can suffer from mental and physical fatigue depending on the cognitive and physical demands of their jobs. Mental fatigue is the result of energy imbalance in the brain in exactly the same way that physical fatigue eventually derives from energy deprivation in the muscle [23]. If an operator's mental workload is either too high or too low, human-system performance in the workplace may decline, potentially endangering safety [24]. Therefore, it is important to measure the level of fatigue to manage its impact on employees' performance and well-being.

Two types of techniques and instruments are used to evaluate fatigue: objective and subjective methods [2]. Objective measures are physical measurements that include heart rate, respiratory rate, skin temperature, and percent of maximum heart rate. In contrast, subjective measures of fatigue are based on explicit factors, which are consciously perceived, and can be evaluated through self-reports, physical measurement, level of tiredness surveys, and fatigue questionnaires. Explicit factors are different from implicit components of fatigue, which are unconsciously perceived by the worker. To evaluate implicit fatigue, objective response measures such as psychophysiology or behavioral responses are necessary.

Techniques and Instruments for Assessing Fatigue. Actigraphy, EEG, physiological measures, and subjective measures are some methods used to assess fatigue in the workplace. Actigraphy is a non-invasive method that measures and records movement and light levels to estimate sleep duration, quality, and timing. It has been shown to be reliable in assessing fatigue in various populations [25]. EEG records the electrical activity of the brain and can be used to measure sleep stages and fatigue-related changes in brain activity [26]. However, EEG requires specialized equipment and trained personnel, and may not be feasible in some workplace settings. Physiological measures, such as heart rate variability, cortisol levels, and pupillometry, provide objective data on physiological changes associated with fatigue [27]. For example, heart rate variability has been shown to be a reliable and valid measure of fatigue in airline pilots. Subjective measures rely on self-report, such as questionnaires or visual analogue scales, and are easy to administer but prone to bias. It is recommended to use subjective measures in conjunction with objective measures when assessing fatigue in the workplace [28, 29].

2.3 Rest Allowance and Fatigue Models

Assessing fatigue in the workplace is essential for maintaining worker safety and productivity. Work-rest scheduling is one of the most crucial components of fatigue management. Rest breaks help workers recover from fatigue and improve their well-being. The management of rest intervals during a work shift, including their quantity, timing, and duration, is the subject of work-rest scheduling.

Several models are used to estimate rest allowances and fatigue. For instance, Rohmert's model [30] uses holding time as a fraction of maximum holding time (fMHT), while Rose et al.'s model uses maximum holding time (MHT) directly. Relative intensity of the exertion is described as a fraction of maximum voluntary contraction (fMVC). To compute the rest allowance, fMVC and fMHT are input variables required for the calculation. MHT models published in the literature that can be either used independent of the muscle group involved in the static exertion or can be muscle specific e.g., one for upper limbs and another one for the back and hips. Several rest allowances and fatigue models are proposed in Table 2.

Maximum endurance time (MET) in static force exertions is utilized to assess working postures. MET is used to evaluate the maximum duration of tolerable static muscular contraction [31]. MET models are used to calculate the necessary recovery time after static muscle work. The estimation of MET is essential to determine appropriate work-rest regimens for static muscular work [32, 33]. MET pertains to static muscle work and signifies the maximum duration for which a static muscular load can be sustained. This measurement is primarily employed in assessing the duration of maintaining a static muscular contraction that is deemed tolerable. Furthermore, the MET is utilized in models for rest allocation, serving as a basis for computing the necessary recovery time following static muscle work [15]. MET models are most commonly used when a worker performs repetitive tasks with consistent muscular force [16].

Static and intermittent static muscular work is a known risk factor for musculoskeletal disorders (MSD) [35]. Estimating MET to accommodate a percentage of the worker population is necessary to determine appropriate work-rest regimens for static muscular work in industrial settings. El Ahrache [34], discusses the effects of static muscular work on workers and recommends sufficient and well-spaced rest periods to prevent fatigue, discomfort, and musculoskeletal disorders. An approach to determining rest periods for static muscular work in manufacturing workstations is presented and compared with four rest allowances models. The approach was applied to seven workstations in three

Table 2. Rest allowance and Fatigue models

Rohmert (1973)	$RA = 18 \times (fMHT)^{1.4} \times fMVC - 0.15)^{0.5} \times 100$	*RA* state for rest allowances *f*MVC is the fraction of maximum voluntary contraction of a muscle to perform any task and *f*MHT is fraction of maximum holding time
Rose et al. (1992)	$RA = 0.164 \times \left[4.61 + ln\left(\frac{1}{100-(fMHT)^{-1}}\right)\right]^{-1} \times 100$ $MET = (7.96)e^{-4.16(\%MVC)}$	*f*MHT required
Ma et al. (2009)	$\frac{F_{cem}(t)}{MVC} = e^{\int_0^t -[F_{load}(u)/MVC]du}$	$F_{cem}(t)$ is current exertable maximum force (current muscle strength); $F_{load}(t) =$ forces required for the task (e.g., workloads); and t = current time (seconds)
Jaber and Neumann (2010)	$L^i_{max} = MLC * f_i * MET_i$	L^i_{max} is the maximum fatigue index for task i; *MLC* is maximum load capability; f_i is an individual's maximum capability
Perez et al. (2014)	$fF_i = \frac{t_i}{MET_i}$	fF_i is the fraction of fatigue contribution per task *i*; MET_i is the maximum endurance time for task *i*

(*continued*)

Table 2. (*continued*)

Dode et al. (2016)	$MET = (7.96)e^{-4.16(\%MVC/100)}$ $RA = 3MET^{-1.52}$	*MVC* is the maximum voluntary contraction of a muscle to perform any task
Glock et al. (2019)	$F_s(t_a) = 1 - e^{-\theta_a t_a}$ $F_r(t_w) = \beta F_s(t_w)$	$F_s(t_a)$ is the fatigue accumulated by time t_a; θ_a is a fatigue growth parameter when a worker completes an action a during time t_a; $F_r(t_w)$ is the dynamic (real) fatigue

industrial sectors. The results showed substantial differences in the recommendations from the different rest allowances models.

To estimate fatigue rates and potential ergonomic impacts of proposed system design alternatives, a multi-pronged approach utilizing Discrete Event Simulation (DES), biomechanical analysis, and static fatigue models was employed by Perez [39]. This approach provides a comprehensive understanding of the potential ergonomic impacts of proposed alternatives in system design. Perez's study demonstrated the efficacy of using DES as an ergonomic tool in predicting and preventing negative outcomes of human exposure to work-related practices in the early stages of system design. The duration and load of each task obtained from the simulation model were recorded in a preset spreadsheet to calculate fatigue accumulation, which could be applied to other aspects of operator exposure and performance, ensuring sustainable working conditions for operators in the early design stages when the potential impact on health and performance is greatest, and costs are lowest.

The learning-forgetting-fatigue-recovery model, which addresses worker capabilities and restrictions in manufacturing environments, taking into account the physical loading required to perform a task, leading to worker fatigue and the need for rest breaks, was presented by Jaber and Neumann [36]. Incorporating learning into the production process was found to reduce fatigue and improve performance, while recovery breaks of sufficient length were necessary to alleviate some fatigue, but longer breaks could lead to forgetting and reduced performance.

The general MET model for operator fatigue prediction was proposed by Dode [38], and it was integrated with learning and human fatigue-recovery patterns into DES models of production systems. Their approach demonstrated the efficacy of early-stage fatigue management, resulting in cost savings and increased efficiency. The study developed a new approach to predict productivity and quality based on system design configurations, predicting the accumulation of operator fatigue, fatigue-related quality effects, and productivity changes. In the demonstration comparison, the proposed system with human factors taken into consideration at the engineering design stage resulted in 7–33% lower fatigue dosage.

To address the dynamic aspects of the packaging process, Glock et al. [35] proposed a method of calculating dynamic real fatigue by determining an empirical coefficient that is multiplied by the estimated total static fatigue level.

To address the dynamic aspects of the packaging process, Glock et al. [35] proposed a method of calculating dynamic real fatigue by determining an empirical coefficient that is multiplied by the estimated total static fatigue level. Their model includes a biomechanical model to estimate fatigue-recovery parameters, an optimization model to minimize the total relevant cost, and an integration of the fatigue-recovery approach to determine the optimal box size and work schedule that minimize cost while satisfying the upper bound on accumulated fatigue level. However, estimating total static fatigue through a model is not sufficient to replicate the dynamic aspects of the packaging process, and concerns were raised about the validity of existing fatigue and recovery models, particularly in more dynamic circumstances.

3 Discussion

The literature review conducted in this study highlights those certain models, policies, and shifts regarding work-rest schedules are effective in reducing fatigue levels among employees involved in manufacturing and construction activities, thereby enhancing job performance. Glock et al. [35] and Jaber et al. [36] provide evidence in support of this. However, further computational validation is needed to determine the effectiveness of work-rest strategies in reducing accumulated fatigue and enhancing operational performance.

Incorporating simulations, such as DES, along with objective measures, is necessary for predicting workload and fatigue and taking appropriate precautions in various industries [37]. DES enables researchers to predict the impact of various factors on workload and fatigue levels and evaluate the effectiveness of different interventions before their implementation. This approach provides a powerful tool for improving the design of manufacturing systems, reducing the likelihood of fatigue-related accidents, and enhancing employee well-being [38, 39]. Organizations can adopt proactive strategies to minimize the occurrence and severity of fatigue. The use of DES and human factors modeling can help system designers comprehend the fatigue-related effects of proposed alternatives at the system design stage. Early warning systems are also recommended for collecting timely information and conducting fatigue risk assessments. Despite limitations in muscle modeling, it can be utilized to understand the mechanical exposure and fatigue-related effects of proposed alternatives in system design stages, and can contribute to 'Virtual Human Factors' approaches for proactive ergonomics capability.

Moreover, objective measures such as actigraphy and EEG provide quantitative data on fatigue levels and are less susceptible to subjective biases than self-reported measures, making them a valuable tool for fatigue assessment and risk management [40].

The use of simulations and objective measures in predicting and managing workload and fatigue is essential for improving workplace safety, productivity, and employee well-being. By incorporating these techniques into system design and evaluation, researchers can develop more effective interventions and improve the overall performance of manufacturing systems. As such, continued research in this area is critical to advancing our understanding of the complex relationship between workload, fatigue, and human performance.

4 Conclusion

This paper provides a literature review on workload and fatigue, their measurement, and the utilization of simulations for prediction. The integration of human factors in manufacturing system design is essential for the development of efficient and advanced manufacturing systems. Appropriate design and management of production systems necessitate analyzing operator fatigue and recovery as part of conventional decision-making support models. Understanding and modelling human behaviour and actions in manufacturing processes are essential for the design and implementation of effective and efficient manufacturing systems that can adjust to the dynamic nature of human involvement. Research on task-specific workload estimation methods is vital to improve our comprehension of the intricate relationship between workload and human performance. This understanding will aid in predicting, preventing, managing, and reducing negative factors impacting human performance.

References

1. Kiran, D.R.: Ergonomics and work study. Work Organ. Methods Eng. Product. 219–232 (2020)
2. Mehta, R.K.: Integrating physical and cognitive ergonomics. IIE Trans. Occup. Ergon. Hum. Factors **4**(2–3), 83–87 (2016)
3. Papetti, A., Gregori, F., Pandolfi, M., Peruzzini, M., Germani, M.: A method to improve workers' well-being toward human-centered connected factories. J. Comput. Des. Eng. **7**(5), 630–643 (2020)
4. Pacaux-Lemoine, M.P., Trentesaux, D., Rey, G.Z., Millot, P.: Designing intelligent manufacturing systems through Human-Machine Cooperation principles: a human-centered approach. Comput. Ind. Eng. **111**, 581–595 (2017)
5. Baines, T.S., Kay, J.M.: Human performance modelling as an aid in the process of manufacturing system design: a pilot study. Int. J. Prod. Res. **40**(10), 2321–2334 (2002)
6. Reiman, A., Kaivo-oja, J., Parviainen, E., Takala, E.P., Lauraeus, T.: Human factors and ergonomics in manufacturing in the industry 4.0 context – a scoping review. Technol. Soc. **65**, 101572 (2021)
7. Karwowski, W., Siemionow, W., Gielo-Perczak, K.: Physical neuroergonomics: the human brain in control of physical work activities. Theor. Issues Ergon. Sci. **4**(1–2), 175–199 (2003)
8. Dick, A.O., Hill, S.G., Plamondon, B.D., Wierwille, W.W., Lysaght, R.J.: Analytic techniques for the assessment of operator workload. In: Proceedings of the Human Factors Society Annual Meeting, vol. 31, no. 3, pp. 368–372 (1987)
9. National Aeronautics and Space Administration: Human Integration Design Handbook (Hidh) (2014)
10. Knisely, B.M., Joyner, J.S., Vaughn-Cooke, M.: Cognitive task analysis and workload classification. MethodsX **8**, 101235 (2021)

11. Hancock, P.A., Eggemeier, F.T.: Human Mental Workload Properties of Workload Assessment Techniques (1988)
12. Wierwille, W.W., Williges, R.C.: Survey and Analysis of Operator Workload Assessment Techniques. Springer, Heidelberg (1978)
13. Ma, Q.G., Sun, X.L., Fu, H.J., Zhao, D.C., Guo, J.F.: Manufacturing process design based on mental and physical workload analysis. Appl. Mech. Mater. **345**, 482–485 (2013)
14. Wickens, C.D.: Multiple resources and mental workload. Hum. Factors **50**(3), 449–455 (2008)
15. Schuck, M.M.: Development of equal-interval task rating scales and task conflict matrices as predictors of attentional demand. Ergonomics **39**(3), 345–357 (1996)
16. Mao, M., Xie, F., Hu, J., Su, B.: Analysis of workload of tank crew under the conditions of informatization. Def. Technol. **10**(1), 17–21 (2014)
17. Wickens, C.D., Hollands, J.G., Banbury, S., Parasuraman, R.: Engineering psychology and human performance (2000)
18. Seo, J., Lee, S., Seo, J.: Simulation-based assessment of workers' muscle fatigue and its impact on construction operations. J. Constr. Eng. Manag. **142**(11), 04016063 (2016)
19. Matthews, G., Desmond, P.A., Neubauer, C., Hancock, P.A.: The Handbook of Operator Fatigue, 1st edn. CRC Press, Boca Raton (2012)
20. Caldwell, J.A., Caldwell, J.L., Thompson, L.A., Lieberman, H.R.: Fatigue and its management in the workplace. Neurosci. Biobehav. Rev. **96**, 272–289 (2018)
21. Keller, J.: Human performance modeling for discrete-event simulation: workload. In: Winter Simulation Conference Proceedings, vol. 1, pp. 157–162 (2002)
22. Lu, L., Megahed, F.M., Cavuoto, L.A.: Interventions to mitigate fatigue induced by physical work: a systematic review of research quality and levels of evidence for intervention efficacy. Hum. Factors **63**(1), 151–191 (2021)
23. Noakes, T.D.: Fatigue is a brain-derived emotion that regulates the exercise behavior to ensure the protection of whole body homeostasis. Front. Physiol. **3**, 82 (2012)
24. Hockey, G.R.J.: Applied attention theory. Ergonomics **52**(2), 270–271 (2009)
25. Ibáñez, V., Silva, J., Cauli, O.: A survey on sleep assessment methods. PeerJ **6**, 1–26 (2018)
26. Cummings, M.A.: American society of electroneurodiagnostic technologists, Inc. Guidelines on intraoperative electroencephalography for technologists. Am. J. Electroneurodiagnostic Technol. **38**(3), 204–225 (1998)
27. Chen, M.L., Lu, S.Y., Mao, I.F.: Subjective symptoms and physiological measures of fatigue in air traffic controllers. Int. J. Ind. Ergon. **70**(110), 1–8 (2019)
28. Mazur, L.M., Mosaly, P.R., Hoyle, L.M., Jones, E.L., Marks, L.B.: Subjective and objective quantification of physician's workload and performance during radiation therapy planning tasks. Pract. Radiat. Oncol. **3**(4), e171–e177 (2013)
29. Brookings, J.B., Wilson, G.F., Swain, C.R.: Psychophysiological responses to changes in workload during simulated air traffic control. Biol. Psychol. **42**(3), 361–377 (1996)
30. Rohmert, W.: Problems of determination of rest allowances part 2: determining rest allowances in different human tasks. Appl. Ergon. **4**(3), 158–162 (1973)
31. Rose, L.M., Neumann, W.P., Hägg, G.M., Kenttä, G.: Fatigue and recovery during and after static loading. Ergonomics **57**(11), 1696–1710 (2014)
32. Delleman, N., Boocock, M., Kapitaniak, B., Schaefer, P., Schaub, K.: ISO/FDIS 11226: evaluation of static working postures. In: Proceedings of the Human Factors and Ergonomics Society Annual Meeting, pp. 6–442 (2000)
33. Rohmert, W., Wangenheim, M., Mainzer, J., Zipp, P., Lesser, W.: A study stressing the need for a static postural force model for work analysis. Ergonomics **29**(10), 1235–1249 (1986)
34. El Ahrache, K., Imbeau, D., Farbos, B.: Percentile values for determining maximum endurance times for static muscular work. Int. J. Ind. Ergon. **36**(2), 99–108 (2006)

35. Glock, C.H., Grosse, E.H., Kim, T., Neumann, W.P., Sobhani, A.: An integrated cost and worker fatigue evaluation model of a packaging process. Int. J. Prod. Econ. **207**, 107–124 (2018)
36. Jaber, M.Y., Neumann, W.P.: Modelling worker fatigue and recovery in dual-resource constrained systems. Comput. Ind. Eng. **59**(1), 75–84 (2010)
37. Völker, I., Kirchner, C., Bock, O.L.: On the relationship between subjective and objective measures of fatigue. Ergonomics **59**(9), 1259–1263 (2016)
38. Dode, P., Greig, M., Zolfaghari, S., Neumann, W.P.: Integrating human factors into discrete event simulation: a proactive approach to simultaneously design for system performance and employees well being. Int. J. Prod. Res. **54**(10), 3105–3117 (2016)
39. Perez, J., de Looze, M.P., Bosch, T., Neumann, W.P.: Discrete event simulation as an ergonomic tool to predict workload exposures during systems design. Int. J. Ind. Ergon. **44**(2), 298–306 (2014)
40. Pilcher, J.J., Huffcutt, A.I.: Effects of sleep deprivation on performance: a meta-analysis. Sleep **19**(4), 318–326 (1996)

Sentiment Analysis of Twitter Data of Hepsiburada E-commerce Site Customers with Natural Language Processing

İsmail Şimşek[1], Abdullah Hulusi Kökçam[1](✉), Halil Ibrahim Demir[1], and Caner Erden[2]

[1] Industrial Engineering Department, Sakarya University, Sakarya, Turkey
ismail.simsek3@ogr.sakarya.edu.tr, {akokcam, hidemir}@sakarya.edu.tr
[2] Faculty of Applied Sciences, Sakarya University of Applied Sciences, Sakarya, Turkey
cerden@subu.edu.tr

Abstract. Social media usage has significantly increased, allowing people to freely express their wishes and complaints through these platforms. Consequently, this has caused a significant increase in data, providing information about users, companies, services, and products. However, making sense of this data with human effort is impossible, necessitating various methods. Sentiment analysis is one such method that helps us understand customers' thoughts on products and companies. In this study, the emotions of e-commerce site users were analyzed using Turkish Twitter data. Text mining techniques were applied to Twitter data, which were analyzed and classified as positive and negative. It was found that 66.9% of the tweets were negative, and 33.1% were positive. The classification results were evaluated using precision, Recall, and F1 criteria. As a result of the evaluation, the sensitivity criterion for negative comments was 94%, and the precision criterion for positive comments showed the highest value with 86%. When looking at the F1 score, 85% for negative comments and 69% for positive comments were calculated. The accuracy rate of the model was found to be 79%.

Keywords: Classification · Sentiment Analysis · Social Media · Text Mining

1 Introduction

With the rapid development of technology and the internet, social media platforms have become increasingly popular, significantly increasing people's interactions through social media tools. Consequently, massive data is generated, which needs to be analyzed. However, with the development of data mining techniques, it has become possible to draw meaningful results from this data stack. As a result, many companies have started to store and analyze the data of their users. One of the most commonly used data analysis techniques is sentiment analysis.

Z. Şen et al. (Eds.): IMSS 2023, LNME, pp. 567–578, 2024.
https://doi.org/10.1007/978-981-99-6062-0_52

In personal social media accounts, people tend to share their thoughts, which can serve as an accurate reflection of their opinions about a product or a company. However, the content shared on social media is often unstructured data, which requires text mining methods to transform it into a structured form. Text mining involves using natural language processing (NLP) tools and data mining techniques to analyze the data. NLP is a sub-branch of artificial intelligence that focuses on processing and understanding human language.

In this study, sentiment analysis was performed by applying text mining to Twitter data, one of the most popular social media platforms. The objective was to determine whether customers' comments about the company were positive or negative. By using text mining techniques, the data was transformed into a structured form that could be analyzed to obtain valuable insights. The following studies in the literature were examined to provide a better understanding of the subject.

Akgül et al. [1] aimed to classify tweets containing a specific keyword into emotional categories of positive, negative, and neutral. The researchers utilized both dictionary and n-gram models for tagging the tweets, achieving a success rate of approximately 70% using the dictionary method and 69% using the character-based n-gram method.

Albayrak et al. [2] conducted sentiment analysis on social media to aid decision-makers in evaluating the conscription debate. The authors employed NLP for data cleaning and utilized the SentiTurkNet dictionary and n-gram method for sentiment analysis. Their analysis showed that 16% of the tweets were positive, 5% were negative, and 79% were neutral.

Onan [3] performed sentiment analysis on Turkish tweets and opinions expressed by Twitter users using a machine learning approach. The author applied the Naive Bayes algorithm, support vector machines, logistic regression, and n-gram techniques. The experimental studies revealed that the best performance (77.78%) was achieved by combining 1-g and 2-g feature sets of the dataset and using the Naive Bayes algorithm.

Aydın et al. [4] utilized a multi-PSO-based classification method to apply text mining on Twitter datasets and perform sentiment analysis. The goal was to extract significant information from the collected Twitter data and classify the emotions in the text as positive, negative, or ambiguous. The PSO method yielded accuracy rates of 89.6%, 83.19%, and 69.95% for three different Twitter datasets. The researchers compared this classification method with other machine learning techniques and found that it produced better results.

Ayan et al. [5] conducted a study to detect "Islamophobia" in tweets through sentiment analysis. Linear ridge regression and Naive Bayes classifiers were trained, and precision, Recall, and f1 criteria were calculated. The Naive Bayes classifier outperformed the Ridge model in positive tweets. The study achieved 96.3% accuracy with Ridge Regression and 95.3% with the Naive Bayes Classifier.

Demir et al. [6] aimed to perform sentiment analysis on Turkish texts using two methods. The first method involved using three different dictionaries: Afinn, Bing, and NRC. The second method involved classifying the texts according to their polarity (positive or negative) and category (weak or strong). Both methods were applied to three different datasets, and the first method achieved accuracy rates of 77.14%, 72.78%, and 74.17%, while the second method achieved rates of 82.85%, 74.92%, and 77.50%. However, the study found that the system gave erroneous results due to allusive interpretations, metaphorical meanings, misspellings, and idioms with more than one meaning, which could have positive or negative connotations depending on the individual.

To perform sentiment analysis on Turkish texts, Toçoğlu et al. [7] conducted a study to test various machine learning methods and compared the results. The algorithms were evaluated based on their accuracy rates. Among the tested methods, ANN achieved the highest accuracy of approximately 86%, followed by SVM with 85.67%, RF with 81.17%, and KEYS with 70.33%.

Büyükeke et al. [8] aimed to provide hotel businesses with strategic development suggestions by utilizing current text mining methods on social media notifications regarding travel experiences in tourism and travel destination preferences. In order to achieve this, the study employed Logistic Regression, Support Vector Machine, and Naive Bayes methods. The results showed that 80% of the comments analyzed were positive, while the remaining 20% were negative. The comments were categorized into three groups: "Experience" accounted for 26.70%, "Value and entertainment" for 24.68%, and "Complaint" for 20.41%.

Çelik et al. [9] conducted sentiment analysis on comments received on Facebook accounts of various companies using different machine learning methods, including Random Forest, Logistic Regression, Naive Bayes, Support Vector Machines, KNN, CountVectorizer, and TF-IDF. The study resulted in accuracy rates of 65% for Naive Bayes, 65% for Random Forest, 70% for Logistic Regression, and 67% for Support Vector Machines.

Kırcı and Gülbak [10] conducted a study to compare the use of emojis by Instagram users for two smartphone brands. The analysis was conducted using Python libraries and the VADER dictionary. The study found that both brands generally received positive comments, but Brand 2 received more positive comments than Brand 1. Additionally, while there were neutral comments for Brand 1, no neutral comments were observed for Brand 2.

The study conducted by Tuzcu [11] aimed to perform a sentiment analysis of reader comments on a book sales website. To achieve this goal, four different machine learning algorithms, including Multilayer Perceptron (MLP), Naive Bayes (NB), Support Vector Machines (SVM), and Logistic Regression (LR), were utilized. The MLP algorithm achieved the highest classification success rate of 89%, while NB, SVM, and LR achieved success rates of 77.57%, 80.93%, and 84.07%, respectively.

In a study conducted by Küçükkartal [12], sentiment analysis was performed on 10,000 English tweets that contained the word "vaccine." The most frequently mentioned words in these tweets were "vaccine," "flu," "hpv," "tumor," "ebola," "dementia," and "polio." The study found that the majority of people who wrote about vaccines discussed diseases. The analysis revealed that approximately 42% of the tweets were positive, 14% were negative, and 43% were neutral.

Uma and Prabha [13] aimed to explore the use of scenario review methods in breaking down the complex information present in a series of markers that focus on the changing patterns of tweet dialects and tweet sizes on Twitter. Specifically, they focused on classifying stages in sentiment analysis and developing a process to identify and classify positive, negative, and neutral sentences.

Atılgan and Yoğurtçu [14] aimed to determine customers' sentiment towards a cargo company using text mining and TF-IDF techniques. They classified emotions as positive, negative, or neutral. The analysis revealed that 47.6% of the notifications were negative, 31.4% were neutral, and 21.0% were positive. The researchers identified that the words "obstacle," "far," and "stay" were frequently used in the negative notifications.

Behera et al. [15] aimed to compare the performance of deep learning and traditional machine learning models in sentiment analysis. The study found that deep learning models outperformed traditional machine learning algorithms. Among the models tested, Co-LSTM, SVM, and CNN achieved the highest accuracy rates of 83.13%, 83.11%, and 82%, respectively.

Beşkirli et al. [16] conducted a sentiment analysis on tweets made before and after the Covid-19 vaccine announcements to observe the change in emotions. Text mining, TF-IDF, and classification methods were used in the study. The analysis revealed that 31% of the tweets before the vaccine announcement had a positive sentiment, 29% had a negative sentiment, and 41% were neutral. After the vaccine announcement, 53% of the tweets had a positive sentiment, 14% had a negative sentiment, and 33% were neutral.

Bostancı and Albayrak [17] conducted a study to identify the emotional aspects of users who are in the process of selecting a university and to create advertisements suitable for their user profile. The study involved categorizing data collected from social media into 25 keywords and using the TF-IDF technique to determine the frequency of these keywords. Emotional states were identified as "optimistic, pessimistic, humorous, productive, and extroverted" using this technique. Results of the analysis revealed that the rate of extroversion was higher than the other categories, at a rate of 77.82%. Based on these findings, five advertising posters were created for Düzce University.

Göçgün and Onan [18] conducted a sentiment analysis on product reviews from Amazon. They classified the reviews as "positive, negative, or neutral" and used machine learning methods and the TF-IDF method to convert the text to numerical data. The classification methods they used were Decision Tree, Naive Bayes, Support Vector Machines, and Logistic Regression. The results showed that the decision tree method achieved high performance, with an accuracy rate of 94% and an F1 score of 75%. On the other hand, Naive Bayes showed the lowest performance, with an accuracy of 90%.

Indulkar and Patil [19] developed three algorithms to perform sentiment analysis on Twitter data and compared their performance on two different datasets. The algorithms used were Logistic Regression, Multinomial Naïve Bayes, Random Forest, Google Word2Vec, and Mean Cross-Validation Accuracy (MCVA). The study found that the Random Forest algorithm gave the best results out of the three algorithms. The accuracy rate was 96.3% for the Uber dataset and 83.4% for the Ola dataset.

The work of Pathak et al. [20] proposes an online sentiment analysis model based on deep learning. The study involves newly collected datasets from Twitter under the hashtags "#ethereum", "#bitcoin," and "#facebook". The proposed model aims to support online response and scale with the flow of short text data and achieve significant performance for subject-level sentiment analysis. The analysis resulted in an accuracy rate of 88.9%.

Park et al. [21] focused on how to analyze comments that are not made in standard words. The case study shows which parts of automobiles cause customer dissatisfaction and, therefore, where investment and investigation are needed.

Kumaş [22] conducted sentiment analysis on Twitter data using five different classification algorithms: Naive Bayes, KNN, Support Vector Machine, Logistic Regression, and Decision Tree. The aim was to classify the sentiment of tweets as either positive or negative. The classifiers were ranked according to their F1 scores, which were determined as 70%, 65%, 73%, 71%, and 69% for Naive Bayes, KNN, Support Vector Machines, Logistic Regression, and Decision Tree, respectively.

With the rapid development of technology and the increasing number of users, the amount of data stored on the internet has significantly increased. Companies are striving to gain valuable insights from this data to develop their strategies, with a key focus on their customers. Understanding customer opinions and preferences is crucial for building customer loyalty, gaining new customers, and ultimately growing businesses. To achieve this, companies are adopting various methods to learn about their customers' ideas and feelings. Social media platforms provide an excellent opportunity for users to express their opinions and emotions about a company or product. To analyze customer sentiment, companies are utilizing various algorithms, as manual analysis of customer feedback is not feasible. This study will examine the real feelings of e-commerce site users about a company or product and explore how customer relationship management and company strategies can be developed based on this sentiment analysis.

2 Methodology

In this section, we will describe the methodology used to perform sentiment analysis on the Twitter data of Hepsiburada e-commerce site customers. The methodology involved several steps, including data collection, pre-processing of tweets, application of natural language processing (NLP) techniques, sentiment analysis using the BERT (Bidirectional Encoder Representations from Transformers) model, and data visualization.

The study was conducted using the Python programming language on the Jupyter Notebook environment, with the Pandas, NLTK, Matplotlib, Sklearn, Wordcloud, and TurkishStemmer libraries being utilized. The flow chart of the applied sentiment analysis methodology is depicted in Fig. 1.

2.1 Data Collection

The data was acquired from Twitter using the name of the e-commerce site (Hepsiburada) as a keyword. Tweets posted with this keyword were collected in Microsoft Excel, and the data was extracted and processed for further analysis.

2.2 Pre-processing of Tweets

The pre-processing steps were applied to the raw Twitter data before performing sentiment analysis. Firstly, all uppercase letters in the text were converted to lowercase to standardize the text and avoid confusion between words with different capitalization.

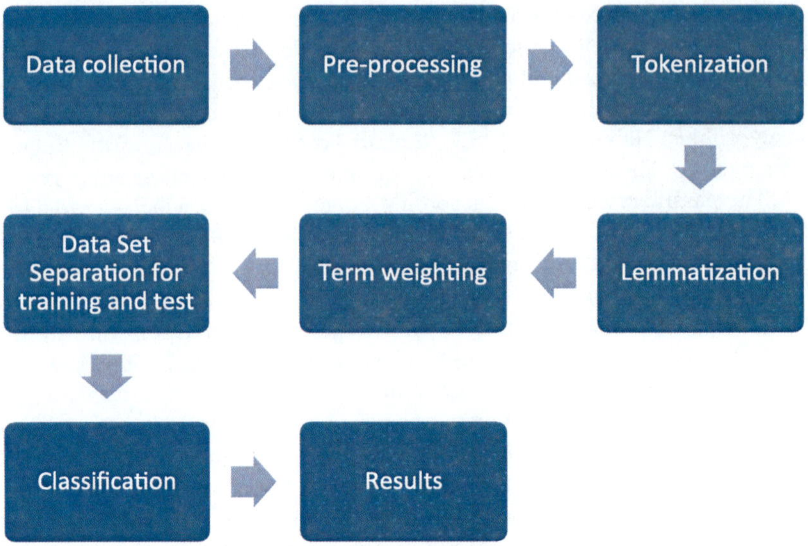

Fig. 1. Flow chart of the applied sentiment analysis methodology

Next, words with less than three characters in length were omitted as they often do not carry much meaning and can add noise to the data. Additionally, any extra spaces in the sentences were removed to make the text consistent and easier to work with.

2.3 Tokenization

Tokenization is defined as the process of breaking up texts into parts according to desired features. In this study, tokenization was performed to remove frequently used conjunctions or words such as "or" from the text so that they do not affect the accuracy of the model.

Moreover, tweets often contain symbolic or descriptive elements such as hashtags, usernames, URL addresses, "rt," emojis, punctuation, and numbers that do not make sense on their own. Cleaning these elements is crucial for the accuracy of the analysis, as emphasized by Kumaş [22].

2.4 Lemmatization

The second step in the cleaning process is the concept of lemmatization. Lemmatization involves the process of separating words in a text by their root form. To carry out this operation, tweets were divided according to their words [22]. By applying lemmatization, the words were transformed into their base or root form, which helps to reduce the complexity of the data and eliminate redundancy.

2.5 Sentiment Analysis Using BERT

The trained BERT (Bidirectional Encoder Representations from Transformers) Model is used to perform sentiment analysis. This model, unlike other models, evaluates the sentence from both left to right and from right to left [23].

BERT Algorithm is an algorithm developed to provide a better understanding of questions, such as updating other algorithms of Google and serving users better. BERT algorithm is a language processing technique consisting of the initials of the words "Bidirectional Encoder Representations from Transformers" and in which artificial intelligence and machine learning technologies are used together. This technique was created to enable Google users' search queries to encounter faster and more accurate results. This algorithm aims to increase accuracy in the end by allowing the words in the query to be evaluated with the words in front and at the end, or with synonyms and similar words, instead of dealing with them alone. The algorithm will help to better understand the queries with complex or long expressions [24].

In this study, Bert-base Turkish sentiment model is used for sentiment analysis which is based on BERTurk for the Turkish language. BERTurk is a developed BERT model for Turkish, focused on the Turkish NLP community [25].

The first five rows of the data set are given as an example, which is formed by combining the "Label" and "Score" columns found as a result of the classification, and the "No" and "Tweet" columns in the data set, are shown in Table 1. Negative comments are labeled with 0, and positive comments are labeled with 1.

Table 1. The first 5 rows of the combined data set as a result of the classification

No	Tweet	Label	Score
0	ürünü satın almak ıstediğim zaman içeriğini ve...	0	0.996846
1	reklamlarınızda oynatacağınız başka kimse yokm...	1	0.833912
2	türkiye deki birçok alışveriş sitesinden sırf ...	0	0.999340
3	tohum sayfasında soğan satıyorsunuz yorumları ...	0	0.994700
4	bugüne kadar opetten çalıştığım firmanın talim...	1	0.933309

Positive words in the tweets, according to the scores found, are shown in Fig. 2, and negative words are shown in Fig. 3.

Fig. 2. Positive words in Tweets

The results of the sentiment analysis performed as a result of the model and the positive and negative classification results of the comments made by Twitter users using the "hepsiburada" tag are shown in Fig. 4.

After examining the words used in the tweets and the results of the sentiment analysis, it becomes apparent that people's sentiments are influenced by the current economic conditions. Many users tweeted about various transactions, such as site usage, product returns, and shipping, and expressed their dissatisfaction with the issues they experienced. Additionally, users also mentioned other companies and advertisements in their tweets.

3 Results

The pre-processed dataset was divided into 90% training and 10% testing data. The training data was used to train the model, with TF-IDF (Term Frequency-Inverse Document Frequency) used as the vectorizer and the Stochastic Gradient Descent algorithm used as the classifier [26].

TF-IDF is a statistical value that determines the significance of a word in a document. It is computed by multiplying the term frequency (TF) of a word with its inverse document frequency (IDF). The pre-processing techniques involved in text classification

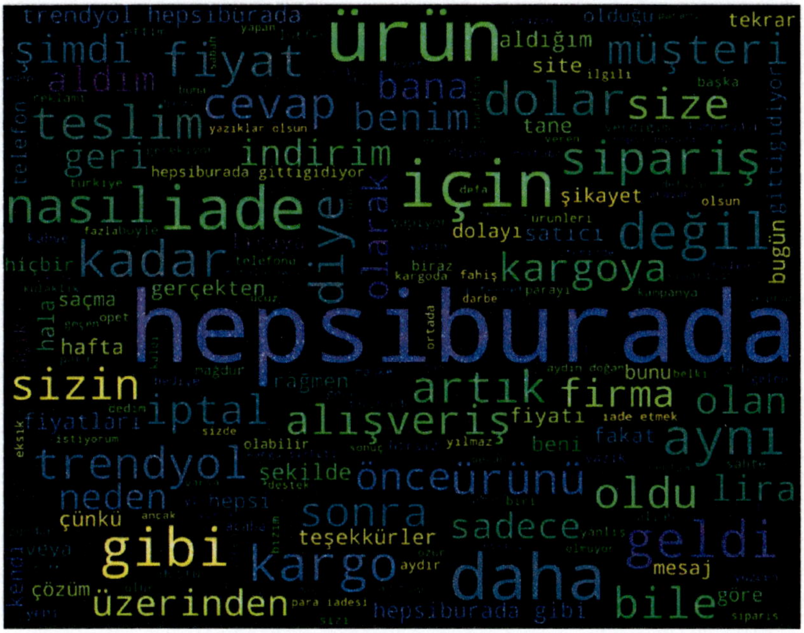

Fig. 3. Negative words in Tweets

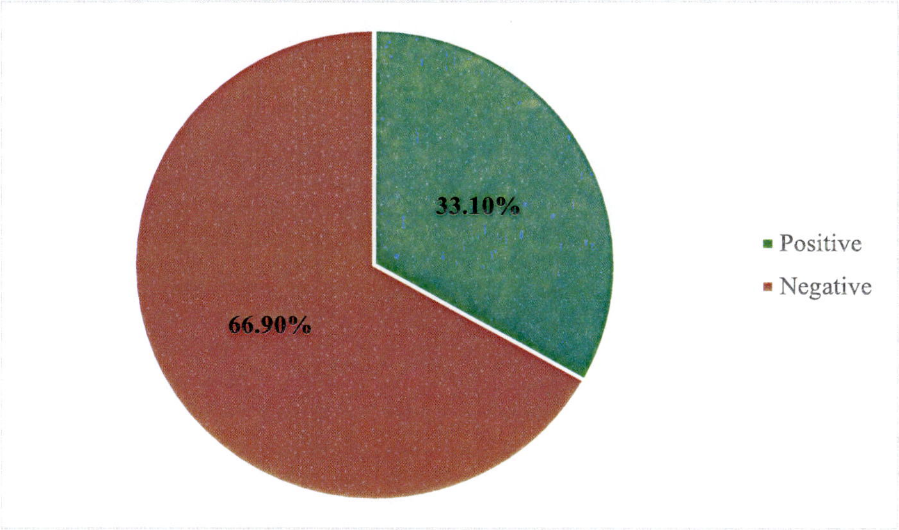

Fig. 4. Results of classifying comments as positive and negative

and topic modeling include the removal of stop words, stemming, and tokenization. TF is calculated by dividing the number of occurrences of the word in the document by the total number of words in the document, while DF is calculated by dividing the number of documents containing the word by the total number of documents in the corpus. IDF is obtained by taking the logarithm of the DF value [27].

After applying term weighting and classification techniques, the results were evaluated using precision, Recall, and F1 scores. True Positive (TP) indicates the correct prediction of positive values, while False Positive (FP) represents the incorrect prediction of positive values. True Negative (TN) indicates the correct prediction of negative values, while False Negative (FN) represents the incorrect prediction of negative values. Specificity measures the number of correct negative predictions made. Accuracy is the ratio of correctly predicted observations to the total number of observations (Eq. (1)) [28]. Precision measures the number of positive predictions that are correct (Eq. (2)), while Recall measures the number of positive cases that are correctly predicted overall positive cases in the data (Eq. (3)) [29]. F1-Score is the weighted average of precision and recall values, providing a comprehensive evaluation metric that captures many aspects at once (Eq. (4)) [28].

$$accuracy = \frac{TP + TN}{TP + FP + FN + TN} \tag{1}$$

$$Precision = \frac{TP}{TP + FP} \tag{2}$$

$$Recall = \frac{TP}{TP + FN} \tag{3}$$

$$F1 = \frac{2 * (recall * precision)}{recall + precision} \tag{4}$$

The accuracy and precision, sensitivity, and F1 scores of the model realized in this study are shown in Table 2.

Table 2. Model's accuracy and precision, recall and F1 score values

	Precision	Recall	F1-score	Support
Negative	0.77	0.94	0.85	32
Positive	0.86	0.57	0.69	21
Accuracy			0.79	53
Macro average	0.81	0.75	0.77	53
Weighted average	0.80	0.79	0.78	53

Based on the analysis results, it can be seen that the model performs well in identifying negative comments, achieving an accuracy of 94% with the best sensitivity score. On the other hand, for positive comments, the model achieves an accuracy of 86% with a precision score. Looking at the F1 score, the model correctly predicts negative comments 85% of the time, while positive comments can be identified with a rate of 69%. The overall accuracy rate of the model is 79%. However, it should be noted that the model is more successful in identifying negative comments than positive comments.

4 Conclusion

This study conducted sentiment analysis on tweets about Hepsiburada e-commerce site users using natural language processing and the Bert Model. After pre-processing the data and transforming them using TF-IDF vectorization, the data were classified into 90% training and 10% test set. The results showed that 66.9% of the tweets were negative, while 33.1% were positive. The precision metric showed the highest accuracy in detecting positive comments with 86%, while the sensitivity metric showed the best results in detecting negative comments with 94%. The accuracy rate of the model was calculated as 79%, which is considered sufficient for the Turkish language compared to other machine learning methods due to the difficulty of machines understanding the natural structure of the language and the scarcity of models developed for the Turkish language.

In future studies, sentiment analysis in Turkish texts can be expanded by increasing the size of the data set and using different machine learning and deep learning methods to increase the accuracy of the models in perceiving the Turkish language. Overall, this study demonstrates the potential of natural language processing and machine learning techniques for sentiment analysis and highlights the importance of developing models specific to the natural structure of languages.

References

1. Akgül, E.S., Ertano, C., Banu, D.: Sentiment analysis with Twitter. Pamukkale Üniversitesi Mühendislik Bilimleri Dergisi **22**(2), 106–110 (2016)
2. Albayrak, M., Topal, K., Altıntaş, V.: "Sosyal medya üzerinde veri analizi: Twitter". Süleyman Demirel Üniversitesi İktisadi ve İdari Bilimler Fakültesi Dergisi, vol. 22, no. Kayfor 15 Özel Sayısı, pp. 1991–1998 (2017)
3. Onan, A.: Twitter mesajlari üzerinde makine öğrenmesi yöntemlerine dayali duygu analizi. Yönetim Bilişim Sistemleri Dergisi **3**(2), 1–14 (2017)
4. Aydın, İ, Salur, M.U., Başkaya, F.: Duygu analizi için çoklu populasyon tabanlı parçacık sürü optimizasyonu. Türkiye Bilişim Vakfı Bilgisayar Bilimleri ve Mühendisliği Dergisi **11**(1), 52–64 (2018)
5. Ayan, B., Kuyumcu, B., Ciylan, B.: Twitter Üzerindeki İslamofobik twitlerin duygu analizi ile tespiti. Gazi Univ. J. Sci. Part C Des. Technol. **7**(2), 495–502 (2019)
6. Demir, Ö., Chawai, A.I.B., Doğan, B.: Türkçe metinlerde sözlük tabanli yaklaşimla duygu analizi ve görselleştirme. Int. Periodical Recent Technol. Appl. Eng. **1**(2), 58–66 (2019)
7. Toçoğlu, M.A., Çelikten, A., Aygün, İ, Alpkoçak, A.: Türkçe metinlerde duygu analizi için farklı makine öğrenmesi yöntemlerinin karşılaştırılması. Dokuz Eylül Üniversitesi Mühendislik Fakültesi Fen ve Mühendislik Dergisi **21**(63), 719–725 (2019)
8. Büyükeke, A., Sökmen, A., Gencer, C.: Metin madenciliği ve duygu analizi yöntemleri ile sosyal medya verilerinden rekabetçi avantaj elde etme: Turizm sektöründe bir araştırma. J. Tourism Gastronomy Studies **8**(1), 322–335 (2020)
9. Çelik, Ö., Osmanoğlu, U.Ö., Çanakcı, B.: Sentiment analysis from social media comments. Mühendislik Bilimleri ve Tasarım Dergisi **8**(2), 366–374 (2020)
10. Kırcı, P., Gülbak, E.: Instagram verileri ile duygu analizi. Avrupa Bilim ve Teknoloji Dergisi, pp. 360–364 (2020)
11. Tuzcu, S.: Çevrimiçi kullanici yorumlarinin duygu analizi ile siniflandirilmasi. Eskişehir Türk Dünyası Uygulama ve Araştırma Merkezi Bilişim Dergisi **1**(2), 1–5 (2020)

12. Küçükkartal, H.K.: Twitter'daki verilere metin madenciliği yöntemlerinin uygulanmasi. Eskişehir Türk Dünyası Uygulama ve Araştırma Merkezi Bilişim Dergisi **1**(2), 10–13 (2020)
13. Uma, J., Prabha, K.: Sentiment analysis in machine learning using twitter data analysis in Python. **11**(12), 3042–3053 (2020). https://doi.org/10.34218/IJARET.11.12.2020.286
14. Atılgan, K.Ö., Yoğrurtçu, H.: Kargo firmasi müşterilerinin twitter gönderilerinin duygu analizi. Çağ Üniversitesi Sosyal Bilimler Dergisi **18**(1), 31–39 (2021)
15. Behera, R.K., Jena, M., Rath, S.K., Misra, S.: Co-LSTM: convolutional LSTM model for sentiment analysis in social big data. Inf. Process. Manage. **58**(1), 102435 (2021)
16. Beşkirli, A., Gülbandılar, E., Dağ, İ: Metin madenciliği yöntemleri ile twitter verilerinden bilgi keşfi. Eskişehir Türk Dünyası Uygulama ve Araştırma Merkezi Bilişim Dergisi **2**(1), 21–25 (2021)
17. Bostancı, B., Albayrak, A.: Duygu analizi ile kişiye özel içerik önermek. Veri Bilimi **4**(1), 53–60 (2021)
18. Göçgün, Ö.F., Onan, A.: Amazon ürün değerlendirmeleri üzerinde derin öğrenme/makine öğrenmesi tabanlı duygu analizi yapılması. Avrupa Bilim ve Teknoloji Dergisi **24**, 445–448 (2021)
19. Indulkar, Y., Patil, A.: Comparative study of machine learning algorithms for twitter sentiment analysis. In: 2021 International Conference on Emerging Smart Computing and Informatics (ESCI). IEEE, pp. 295–299 (2021)
20. Pathak, A.R., Pandey, M., Rautaray, S.: Topic-level sentiment analysis of social media data using deep learning. Appl. Soft Comput. **108**, 107440 (2021)
21. Park, S., Cho, J., Park, K., Shin, H.: Customer sentiment analysis with more sensibility. Eng. Appl. Artif. Intell. **104**, 104356 (2021)
22. Kumaş, E.: Türkçe twitter verilerinden duygu analizi yapilirken siniflandiricilarin karşilaştirilmasi. Eskişehir Türk Dünyası Uygulama ve Araştırma Merkezi Bilişim Dergisi **2**(2), 1–5 (2021)
23. Uçar, K.T.: BERT Modeli ile Türkçe Metinlerde Sınıflandırma Yapmak. Medium (2020). https://medium.com/@ktoprakucar/bert-modeli-ile-türkçe-metinlerde-sınıflandırma-yapmak-260f15a65611. Accessed 10 Apr 2023
24. Urhan, M.: BERT Algoritması Nedir? Örneklerle BERT Algoritması. (2020). https://zeo.org/tr/kaynaklar/blog/bert-algoritmasi-nedir. Accessed 10 Apr 2023
25. Yıldırım, S.: savasy/bert-base-turkish-sentiment-cased · Hugging Face, Hugging Face (2021). https://huggingface.co/savasy/bert-base-turkish-sentiment-cased. Accessed 10 Apr 2023
26. Mete, E.: Doğal Dil İşleme (NLP) ile Sentiment (Duygu) Analizi Tespiti, WordPress. (2018). https://emrahmete.wordpress.com/2018/11/25/dogal-dil-isleme-nlp-ile-sentiment-duygu-analizi-tespiti/. Accessed 10 Apr 2023
27. Scott, W.: TF-IDF for Document Ranking from scratch in python on real world dataset. Medium (2019). https://towardsdatascience.com/tf-idf-for-document-ranking-from-scratch-in-python-on-real-world-dataset-796d339a4089. Accessed 10 Apr 2023
28. Tiwari, A.: Supervised learning: From theory to applications. In: Pandey, R., Khatri, S.K., Kumar, N.S., Verma, P., (eds.) Artificial Intelligence and Machine Learning for EDGE Computing. Academic Press, pp. 23–32 (2022). https://doi.org/10.1016/B978-0-12-824054-0.00026-5
29. Kanstrén, T.: A Look at Precision, Recall, and F1-Score, Medium (2020). https://towardsdatascience.com/a-look-at-precision-recall-and-f1-score-36b5fd0dd3ec. Accessed 10 Apr 2023

Chaotic Perspective on a Novel Supply Chain Model and Its Synchronization

Neslihan Açıkgöz[1(✉)], Gültekin Çağıl[2], and Yılmaz Uyaroğlu[3]

[1] Department of Public Relations and Publicity, Vocational School of Mersin University, Mersin, Turkey
neslihan@mersin.edu.tr

[2] Department of Industrial Engineering, Faculty of Engineering, Sakarya University, Sakarya, Turkey
cagil@sakarya.edu.tr

[3] Department of Electrical Electronics Engineering, Faculty of Engineering, Sakarya University, Sakarya, Turkey
uyaroglu@sakarya.edu.tr

Abstract. Today, scarce resources or many unpredictable factors such as demand have increased the importance of the supply chain. The motivation of this study is the need for the design and analysis of the dynamic supply chain model that will shed light on the companies. In the study, four different dynamic nonlinear supply chain models, which will be an example for companies to reveal their structures, are summarized and a new chaotic supply chain dynamic model developed for citrus production from perishable products, which has not yet been studied in the literature, is presented. The chaotic structure of this new model is demonstrated with time series, phase portraits, bifurcation diagrams, and Lyapunov exponents. In addition, with the active control technique, in which control parameters are added to all the equations of the supply chain system, the chaotic structure of the system was brought under control and synchronous operation was ensured with a different system. Thus, the production amount, demand, and stock data, which are the supply chain status variables of a company's factories in a different area, can have similar values with an error close to zero.

Keywords: Perishable Products · Supply Chain · Dynamical Analysis · Synchronization

1 Introduction

Supply chain management aims to integrate the main business processes between each element in the chain, taking into account customer satisfaction, cycle time, and costs.

The main purpose of supply chain management is to increase customer satisfaction, reduce cycle time, and reduce inventory and operating costs. To achieve these goals, system optimization and control are required, which are primarily based on modeling the system. Various alternative methods have been proposed for modeling the supply chain [1]. Economic game-theoretical model [2], deterministic dynamic model [3–10],

Z. Şen et al. (Eds.): IMSS 2023, LNME, pp. 579–591, 2024.
https://doi.org/10.1007/978-981-99-6062-0_53

stochastic model [11], nonlinear dynamic model [12–15], simulation model [16, 17] can be given as examples. In the literature, the decision variables of supply chain dynamic models had been taken as "demand quantity and production quantity" [14], "stock, price and demand" [17], or "demand, stock, and production quantity" [12, 13, 15, 18]. These system dynamics, which are considered in supply chain modeling, vary according to the product type, the problem, and the size of the chain.

In this study, perishable products were taken as the focus. Perishable products lose their quality and value after a certain time period, even when used correctly throughout the supply chain. Therefore, especially food products often require specialized, more advanced transportation and storage solutions [19–21]. For products that are defined as perishable products, prescription drugs, pharmaceuticals (vitamins and cosmetics), chemicals (household cleaning products), batteries, photographic films, frozen foods, fresh products, dairy products, fruit and vegetables, cut flowers, blood cells can be counted. Special handling, storage techniques, and equipment are required to prevent damage, deterioration, and contamination. This includes handling, washing, rinsing, grading, storage, packaging, temperature control, and daily or hourly shelf life quality testing, and also incurs a separate cost for all these activities. Disrupting the integrity of the cold chain can destroy the entire season's gains. For this reason, careful supply chain management of perishable products is important in terms of customer satisfaction, costs, and operating profit. Önal et al. [22] developed a mixed-integer nonlinear programming model to maximize retailer profit. Gerbecks [23] quantitatively modeled an e-retailer's perishable supply chain, taking into account lost sales, expenditure, and operating costs. In another study, a traditional three-level supply chain consisting of the manufacturer, retailer, and customer is modeled to manage orders and stocks of perishable products [24]. In the literature, it is seen that the optimization of the activities (such as pricing, and stock) in the supply chain of perishable products has been studied. But, any study has not been found in which demand, production quantity, and stock variables are considered together, and the supply chain of perishable products is modeled with the differential equation system. In the light of this information, the supply chain of citrus products, which are considered perishable products, is modeled with a differential equation system with three decision variables [18]. The dynamic behavior of this model, which is revealed by phase portraits and bifurcation graphs, can be explained by chaos theory.

In this study, synchronization which will allow two factories connected to an enterprise to work synchronously with each other is also emphasized. The two chaotic systems discussed are synchronized using the active control technique and the system outputs are shown with time-series graphics, where the system outputs take the same values with zero error after a certain period of time.

This study is organized as follows: In Sect. 1, a new supply chain model proposed for perishable products is introduced with its notations. In Sect. 2, the system is synchronized using the active control technique. Finally, the differences between the developed new model with other models are presented in the conclusion section.

2 A Novel Supply Chain Model for Perishable Products

This section describes the main characteristics and assumptions of the model and then examines its dynamic properties using bifurcation diagram, Lyapunov exponents, and phase portraits.

2.1 Model Development and Assumptions

Food supply chains are global networks encompassing production, processing, distribution, and even sieving [25], and supply chains of perishable products are more complex and unstable than others [26]. In addition, each link of the supply chain manages its inventory and the lack of communication between each other causes the supplier's deterioration and demand information to be delayed, thus not meeting consumer needs quickly and accurately [26].

In this study, the focus is on the supply chain model of perishable products, based on the assumption that system dynamics vary depending on the relevant processes related to each product type, such as food, oil, and consumer products [13].

The notations used in the model are given in Table 1.

Table 1. Notations used to describe the model

Symbol	Denotation
x(t)	Quantity demanded at time t
y(t)	Stock amount at time t (finished packaged product stock)
z(t)	Amount of production at time t
k	Amount of distortion
m	The rate of deterioration of the product in stock
n	The rate of deterioration of the product in production
s	Customer satisfaction rate
d	Customer satisfaction ($0 \leq d \leq 1$). (Even if all of the customer's demands are met, meeting the customer's demand in the required quality will also affect customer satisfaction.)

x, y, and z represent state variables; m, n, and d represent parameters.

How the equations that make up the model are obtained are given in detail below, respectively. The demand for the t + 1 period depends on our production amount in the previous period and the customer satisfaction rate multiplied.

$$x_{t+1} = (z_t + y_t) * s \tag{1}$$

The customer satisfaction rate depends on the rate at which the customer's demand is met by the company. In other words, customer satisfaction is achieved depending on how much of the company's production capacity meet the demand. Even if all of the customer's demands are met, meeting the demands of the customers in the required quality will also affect customer satisfaction. Accordingly, customer satisfaction was taken as the combined effect of demand fulfillment rate and customer satisfaction.

$$s = \left(\frac{z_t + y_t}{x_t}\right) * d \tag{2}$$

According to Eq. (1) and (2), Eq. (1) will take the following form:

$$x_{t+1} = (z_t + y_t) * s = (z_t + y_t) * \left(\frac{z_t + y_t}{x_t}\right) * d \tag{3}$$

In many studies on inventory management in the literature, it is assumed that products have unlimited lifetimes and that the demand is independent of product age. However, although this assumption is not valid for short-lived, perishable products, it is seen that the demand is directly related to the age of the product [27]. Especially in the production and processing of fresh vegetables and fruits, there may be deterioration in the product during the process steps such as collecting, transmitting, processing or keeping the product. These deterioration rates are especially important for the inventory level.

In Eq. (4), the current inventory level is obtained by subtracting the amount of demand from the amount of production and the amount of deterioration.

$$y_{t+1} = z_t + y_t - x_t - k_t \tag{4}$$

The amount of deterioration (k) is equal to the sum of the amount of deterioration in production and the amount of deterioration in stock, as given below. The rate of deterioration in different perishable foods varies considerably. For example, canned food gradually declines in quality, while fresh fish and delicatessen products can deteriorate within a few hours. Therefore, the rate of deterioration can greatly affect the replacement policy and pricing strategy [28].

$$k_t = m.y_t + n.z_t \tag{5}$$

According to Eq. (4) and (5), the difference equation giving the inventory level is:

$$y_{t+1} = (1 - n)z_t + (1 - m)y_t - x_t \tag{6}$$

The production amount is obtained from the difference between the demand level and the stock amount.

$$z_{t+1} = x_{t+1} - y_{t+1} + m * y_t \tag{7}$$

The supply chain system consisting of the continuous form of the difference equations that determine the demand, stock, and production amount explained in detail above, is given by Eq. (8).

$$\begin{aligned} \dot{x} &= -x + (z + y) * \left[\left(\frac{z + y}{x}\right) * d\right] \\ \dot{y} &= -x - m * y + (1 - n) * z \\ \dot{z} &= x - (1 - m) * y - z \end{aligned} \tag{8}$$

2.2 Dynamical Properties of the Proposed Model

The novel perishable product supply chain model given above was simulated in Matlab R2021 with the initial values given in Table 2 and their dynamical behavior is summarized in Table 3.

Table 2. Initial values of the new model variables

Symbol	Value
x(0)	0.013
y(0)	−0.01
z(0)	0.01

Table 3. The dynamical behaviours summaries of the new suggested novel perishable product supply chain system according to different parameters

Parameter values	Dynamics	Phase portraits
$m = -1.04$, $n = -10.4005$, $d = 0.302$	Periodic 2	Figure 1.c
$m = -1.04$, $n = -10.4005$, $d = 0.323$	Periodic 4	Figure 1.b
$m = -1.04$, $n = -10.4005$, $d = 0.33202$	Chaos	Figure 1.a

In the simulation results; the phase portraits in Fig. 1 show that the system exhibits a chaotic state when it moves away from the chaotic structure as the customer satisfaction value increases.

The bifurcation graph in Fig. 2 a gives the change in the equilibrium points calculated by using 0.01 step number in the [−2;0.5] interval of parameter m for $d = 0.302$. In this case, it is clearly seen that the m parameter, which expresses the stock deterioration rate to which the system is sensitive, will put the system in a chaotic state as of −1.05. At the same time, in Fig. 2a, bifurcation diagram gives the change in the equilibrium points calculated by using 0.01 step number in the [−0.5;2] interval of parameter m for $d = 0.33202$. That is, for values of stock deterioration ratio greater than 1.05, it will put the system in order.

The path that a dynamic model follows when its behavior is studied in phase space is called a trajectory. The Lyapunov exponent is used as a measure of how far an orbit is away from the nearest orbit. If the Lyapunov exponent is positive, it is understood that the orbits are very close to each other at first and get further and further apart. As this positive value increases quantitatively, the rate of divergence of the orbits also increases. The negative value of the Lyapunov exponent indicates that the distant orbits approach each other over time. The increase in exponential value indicates the degree of complexity, that is, unpredictability. A system with at least one positive Lyapunov exponent in three-dimensional phase space is said to exhibit chaotic behavior [29]. As seen in Fig. 3, it is seen that a Lyapunov exponent is positive with $\lambda_1 = 4{,}9084$, and it

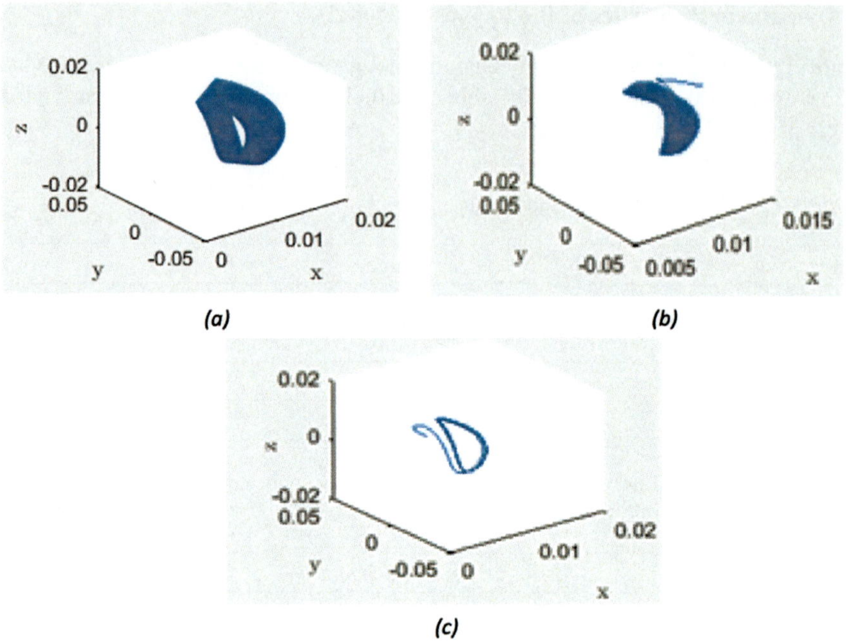

Fig. 1. 3D phase portraits of the new supply chain model for (a) d = 0.302; (b) d = 0.323; (c) d = 0.33202

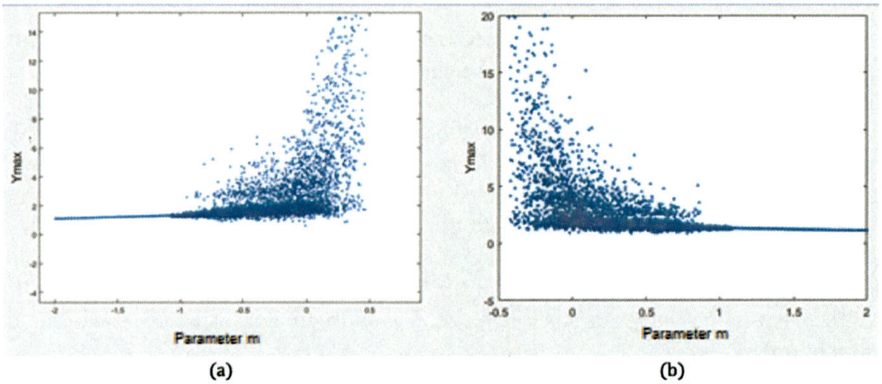

Fig. 2. (a)The bifurcation graph of the stock disruption rate parameter, m, in the range [−2;0.5] for **d = 0.302,** (b) The bifurcation graph of the stock disruption rate parameter, m, in the range [−0.5;2] for **d = 0.33202.**

proves that the new supply chain system is in a chaotic structure with the given initial conditions and parameters values. Because a system with at least one positive Lyapunov exponent is chaotic [30, 31].

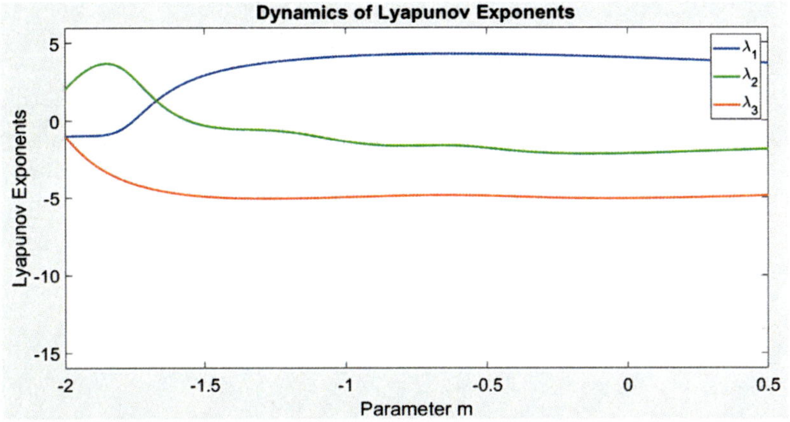

Fig. 3. Graph of Lyapunov Exponents in the range of [−5:5] of *m* stock decay rate parameter

3 Synchronization of the New Chaotic Supply Chain Model with Active Control Method

In this study, two chaotic systems discussed are synchronized using active control. According to the active control technique, the manager and implementer systems are introduced below:

Manager System

$$\dot{x_1} = -x_1 + (z_1 + y_1) * \left[\left(\frac{z_1 + y_1}{x_1}\right) * 0,302\right]$$

$$\dot{y_1} = -x_1 + 1,04 * y_1 + 11,4005 * z_1 \tag{9}$$

$$\dot{z_1} = x_1 - 2,04 * y_1 - z_1$$

Implementer System

$$\dot{x_2} = -x_2 + (z_2 + y_2) * \left[\left(\frac{z_2 + y_2}{x_2}\right) * 0,302\right] + u_1$$

$$\dot{y_2} = -x_2 + 2,04 * y_2 + 11,4005 * z_2 + u_2 \tag{10}$$

$$\dot{z_2} = x_2 - 2,04 * y_2 - z_2 + u_3$$

u_1, u_2, u_3 are active control functions in the implementer system. The primary purpose of control signals is to enable the implementing system to follow the master system necessary to achieve synchronization.

For state variables, the error is defined as follows:

$$e_1 = x_2 - x_1$$

$$e_2 = y_2 - y_1$$
$$e_3 = z_2 - z_1 \tag{11}$$

Following active control design procedures, fault dynamics are obtained using manager and implementer system equations and fault definitions.

$$\dot{e_1} = -e_1 + (e_3 + e_2) * \left[\left(\frac{e_3 + e_2}{e_1}\right) * 0,302\right] + u_1$$
$$\dot{e_2} = -e_1 + 2,04 * e_2 + 11,4005 * e_3 + u_2 \tag{12}$$
$$\dot{e_3} = e_1 - 2,04 * e_2 - e_3 + u_3$$

According to the fault dynamics, the control functions are redefined as follows:

$$u_1 = -\left((e_3 + e_2) * \left[\left(\frac{e_3 + e_2}{e_1}\right) * 0,302\right]\right) + v_1$$
$$u_2 = v_2 \tag{13}$$
$$u_3 = v_3$$

So the error dynamics Eq. (12) becomes:

$$\dot{e_1} = -e_1 + v_1$$
$$\dot{e_2} = -e_1 + 2,04 * e_2 + 11,4005 * z + v_2 \tag{14}$$
$$\dot{e_3} = e_1 - 2,04 * e_2 - e_3 + v_3$$

In the active control method, a fixed matrix A is chosen to control the error dynamics (14).

$$\begin{bmatrix} v_1 \\ v_2 \\ v_3 \end{bmatrix} = A * \begin{bmatrix} e_1 \\ e_2 \\ e_3 \end{bmatrix} \tag{15}$$

There are several options for obtaining controller coefficients A_{ij}s to obtain a stable closed-loop system. Here, the following matrix is processed that meets the Routh-Hurwitz criteria calculated for the stability of the synchronous state:

$$A = \begin{bmatrix} 0 & 0 & 0 \\ 1 & -3,04 & -11,4005 \\ -1 & 2,04 & 0 \end{bmatrix} \tag{16}$$

Provided that the eigenvalues (λ_1, λ_2, λ_3) are negative, (λ_1, λ_2, λ_3) = (−1, − 1, − 1) is chosen for ease of calculation.

The error dynamical equations and control functions are given below:

$$\dot{e}_1 = -e_1$$

$$\dot{e}_2 = -e_2 \tag{17}$$

$$\dot{e}_3 = -e_3$$

$$u_1 = -\left((e_3 + e_2) * \left[\left(\frac{e_3 + e_2}{e_1}\right) * 0,302\right]\right)$$

$$u_2 = e_1 - 3,04 * e_2 - 11,4005 * e_3 \tag{18}$$

$$u_3 = -e_1 + 2,04 * e_2$$

Numerical experiments were made using the simulation program, and the initial conditions of the manager and implementer systems were taken as $x_1(0) = 0,12$, $y_1(0) = -0,01$, $z_1(0) = -0,01$ and $x_2(0) = -5$, $y_2(0) = -0,3$ $z_2(0) = -1$, respectively. Numerical results are given graphically to validate the proposed method.

In Fig. 4, the change of state variables of manager and implementer systems over time is shown. Again, Fig. 4 gives the time-dependent variation of error vectors with x_1 and x_2, y_1 and y_2, z_1 and z_2. As seen in more detail in Fig. 5, control signals are activated at t = 0. After the control signals are activated, the error vectors rapidly approach zero from the moment t = 50. Accordingly, it is seen that the active controller synchronizes the manager and implementer systems.

This study shows that synchronization of chaotic systems occurs through active control. The numerical results confirm the validity and effectiveness of the generalized active control method.

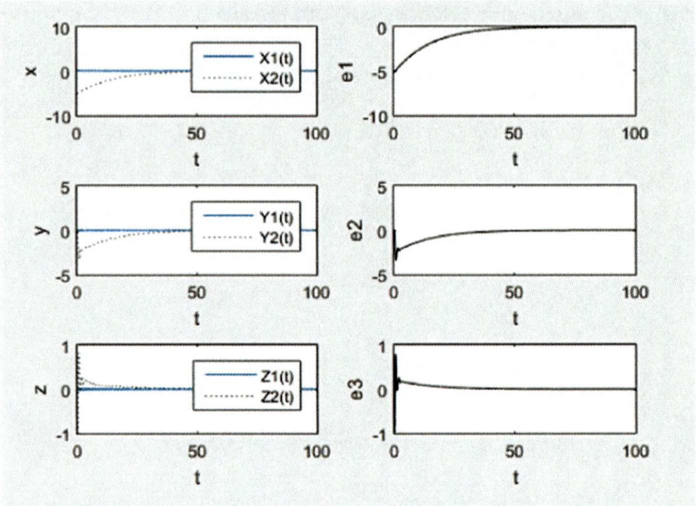

Fig. 4. Time-dependent variation graphs of the state variables of manager and implementing systems and the differences (errors) between these variables

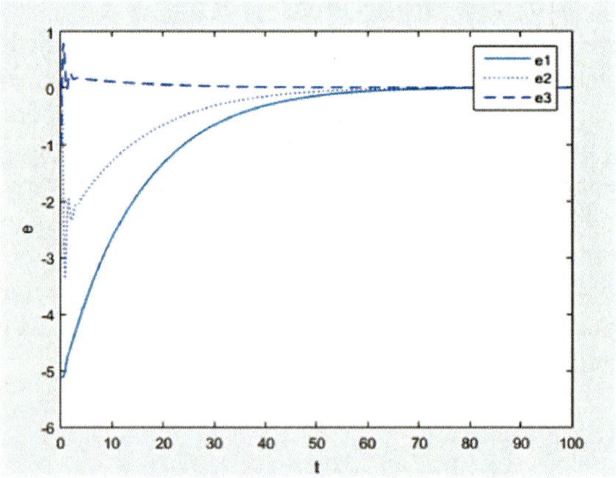

Fig. 5. Time dependent variation of error vectors

4 Conclusion and Suggestions for Future Work

We modeled a three-stage nonlinear supply chain consisting of the grower, the citrus processing plant and the customer. For perishable products, product deterioration may occur during processing or keeping the citrus. In this study, the most important difference from the supply chain dynamic models in the literature mentioned is that these deterioration rates affecting the amount of stock, production and demand are added to the model.

When customer satisfaction *"d"* is *0.302* in the proposed new supply chain model, the system exhibits chaotic behavior as seen in Fig. 1. The system was examined for customer satisfaction ratio is *0.302, 0.323,* and *0.33202*. Thus, it was revealed that a small increase in *d* ratio approached the system to a steady state. Moreover, it has been observed that a dynamic system of perishable products is sensitively dependent on customer satisfaction. In Fig. 3, it is seen that at least one Lyapunov exponent with $\lambda_1 = 4.9084$ is positive and proves that the new supply chain system is in a chaotic structure with the given initial conditions and parameter values.

The active control method is used for chaos synchronization in the proposed model. Control signals are activated at the start time, and the error approaches zero beyond the moment $t = 50$. So that, the manager and implementer systems synchronize with the active control parameters. This analysis has shown us that two possible supply chain systems (for example, two factories of the firm in different cities) can operate in sync.

Consequently, obtained results are relevant for managing complex supply chain systems. In future studies, different control techniques can be used in the synchronization of the new supply chain system modeled for perishable products, as well as dynamic supply chain models for different sectors or different product types can be developed.

References

1. Sarimveis, H., Patrinos, P., Tarantilis, C.D., Kiranoudis, C.T.: Dynamic modeling and control of supply chain systems: A review. Comput. Oper. Res. **35**, 3530–3561 (2008)
2. Agiza, H.N., Elsadany, A.A.: Chaotic dynamics in nonlinear duopoly game with heterogeneous players. Appl. Math. Comput. **149**(3), 843–860 (2004). https://doi.org/10.1016/S0096-3003(03)00190-5
3. Simon, H.A.: On the application of servomechanism theory to the study of production control. Econometrica **20**, 247–268 (1952)
4. Vassian, H.J.: Application of discrete variable servo theory to inventory control. Oper. Res. **3**, 272–282 (1955)
5. Towill, D.R.: Optimization of an inventory system and order based control system. Int. J. Prod. Res. **20**, 671–687 (1982)
6. Blanchini, F., Rinaldi, F., Ukovich, W.: A network design problem for a distribution system with uncertain demands. SIAM J. Optim. **7**, 560–578 (1997)
7. Blanchini, F., Miani, S., Pesenti, R., Rinaldi, F., Ukovich, W.: Robust control of production-distribution systems. In: Moheimani, S.O.R., (ed.), Perspectives in Robust Control, Lecture Notes in Control and Information Sciences, n. 268. Springer, Berlin, pp. 13–28 (2001). https://doi.org/10.1007/BFb0110611
8. Das, S.K., Abdel-Malek, L.: Modeling the flexibility of order quantities and lead-times in supply chains. Int. J. Prod. Econ. **85**, 171–181 (2003)
9. Kumara, S.R.T., Ranjan, P., Surana, A., Narayanan, V.: Decision making in logistics: A chaos theory based approach. CIRP Ann. **52**(1), 381–384 (2003). https://doi.org/10.1016/S0007-8506(07)60606-4
10. Lin, P.-H., Wong, D.S.-H., Jang, S.-S., Shieh, S.-S., Chu, J.-Z.: Controller design and reduction of bullwhip for a model supply chain system using z-transform analysis. J. Process Control **14**, 487–499 (2004)
11. Gallego, G., van Ryzin, G.: Optimal dynamic pricing of inventories with stochastic demand over finite horizons. Manage. Sci. **40**(8), 999–1020 (1994)

12. Zhang, L., Li, Y.J., Xu, Y.Q.: Chaos synchronization of bullwhip effect in a supply chain. In: ICMSE 2006, International Conference on Management Science and Engineering, pp.557–560 (2006)
13. Anne, K.R., Chedjou, J.C., Kyamakya, K.: Bifurcation analysis and synchronization issues in a three-echelon supply chain. Int. J. Log. Res. Appl. **12**(5), 347–362 (2009). https://doi.org/10.1080/13675560903181527
14. Dong, M.A.: Research on supply chain models and its dynamical character based on complex system view. J. Appl. Sci. **14**(9), 932–937 (2014)
15. Mondal, S.: A new supply chain model and its synchronization behaviour. Chaos Solitons Fractals **123**, 140–148 (2019)
16. Larsen, E.R., Morecroft, J.D.V., Thomsen, J.S.: Complex behaviour in a production-distribution model. Eur. J. Oper. Res. **119**, 61–74 (1999)
17. Wu, Y., Zhang, D.Z.: Demand fluctuation and chaotic behaviour by interaction between customers and suppliers. Int. J. Prod. Econ. **107**, 250–259 (2007)
18. Açıkgöz, N.: Chaotic Structure and Control of Supply Chain Management: A Model Proposed for Perishable Products. Sakarya University, Doctoral Thesis (2021)
19. Zhang, G., Habenicht, W., Spieß, W.: Improving the structure of deep frozen and chilled food chain with tabu search procedure. J. Food Eng. **60**(1), 67–79 (2003). https://doi.org/10.1016/S0260-8774(03)00019-0
20. Lowe, T.J., Preckel, P.V.: Decision technologies for agribusiness problems: a brief review of selected literature and a call for research. Manuf. Serv. Oper. Manag. **6**(3), 201–208 (2004). https://doi.org/10.1287/msom.1040.0051
21. Rong, A., Akkerman, R., Grunow, M.: An optimization approach for managing fresh food quality throughout the supply chain. Int. J. Prod. Econ. **131**(1), 421–429 (2011). https://doi.org/10.1016/j.ijpe.2009.11.026
22. Önal, M., Yenipazarli, A., Kundakcioglu, O.E.: A mathematical model for perishable products with price-and displayed-stock-dependent demand. Comput. Ind. Eng. **102**, 246–258 (2016). https://doi.org/10.1016/j.cie.2016.11.002
23. Gerbecks, W.T.M.: A model for deciding on the supply chain structure of the perishable products assortment of an online supermarket with unmanned automated pick-up points. Master Thesis, BSc Industrial Engineering - Eindhoven University of Technology (2012)
24. Campuzano-Bolarín, F., Mula, J., Díaz-Madroñero, M.: A supply chain dynamics model for managing perishable products under different e-business scenarios. In: 2015 International Conference on Industrial Engineering and Systems Management (IESM). Seville, Spain, pp. 329–337 (2015). https://doi.org/10.1109/IESM.2015.7380179
25. Yu, M., Nagurney, A.: Competitive food supply chain networks with application to fresh produce. Eur. J. Oper. Res. **224**(2), 273–282 (2013). https://doi.org/10.1016/j.ejor.2012.07.033
26. Wang, W.: Analysis of bullwhip effects in perishable product supply chain-based on system dynamics model. In: 2011 Fourth International Conference on Intelligent Computation Technology and Automation, pp. 1018-1021 (2011). https://doi.org/10.1109/ICICTA.2011.255
27. Kaya, O.: Kısa ömürlü ürünler için koordineli bir stok ve fiyat yönetimi model. Anadolu Univ. J. Sci. Technol. A- Appl. Sci. Eng. **17**(2), 423–437 (2016). https://doi.org/10.18038/btda.00617
28. Yang, S., Xiao, Y., Kuo, Y.-H.: The supply chain design for perishable food with stochastic demand. Sustainability **9**, 1195 (2017). https://doi.org/10.3390/su9071195
29. Shin, K., Hammond, J.K.: The instantaneous Lyapunov exponent and its application to chaotic dynamical systems. J. Sound Vib. **218**(3), 389–403 (1998)

30. Wolf, A., Swift, J.B., Swinney, H.L., Vastano, J.A.: Determining lyapunov exponents from a time series. Physica D **16**, 285–317 (1985). https://doi.org/10.1016/0167-2789(85)90011-9
31. Van Opstall, M.: Quantifying chaos in dynamical systems with Lyapunov exponents. Furman Univ. Electron. J. Undergraduate Math. **4**(1), 1–8 (1998)

Maximizing Efficiency in Digital Twin Generation Through Hyperparameter Optimization

Elif Cesur(✉), Muhammet Raşit Cesur, and Elif Alptekin

Industrial Engineering Department, Istanbul Medeniyet University, Istanbul, Turkey
19120606025@ismu.edu.tr, {rasit.cesur, elif.karakaya}@medeniyet.edu.tr

Abstract. In recent years, digitalization has become widespread in many fields, including manufacturing systems. Following the deployment of Industry 4.0 technologies in manufacturing, significant improvements have been noted in many production areas, such as predictive maintenance, production planning, dynamic scheduling, and more. The digital twin is one of the technologies used in this area. The objective of this paper is to develop an automated and executable digital twin process within the concept of industrial artificial intelligence. For this purpose, hyperparameter optimization is used, a method to find the best external parameters that should be set by machine learning algorithm developers. It is used to automate the process and increase its accuracy by selecting the optimal parameters that are appropriate for the dataset. In this study, a digital twin was created using the random forest method on a dataset obtained from a CNC machine for a predictive maintenance application. Hyperparameter optimization is integrated into the machine learning process and the random forest hyperparameters such as bootstrap, maximum depth, maximum features, minimum sample leaf, minimum sample split, and the number of estimators is optimized. This study is a digital twin application under the concept of industrial artificial intelligence. A machine learning algorithm with integrated hyperparameter optimization is proposed, which is more efficient and faster in the field of predictive maintenance in production. For this purpose, the Hyperopt library in the Python programming language is used.

Keywords: Hyperparameter Optimization · Random Forest · Industrial Artificial Intelligence · Automated Digital Twin Generation

1 Introduction

Industry 4.0 and digitization technologies have affected the manufacturing area hugely. These technologies started to be used in many areas of manufacturing from product design to quality control. By using the Internet of things, cloud technologies, cyber-physical systems, human-robot collaboration, and many other technologies in manufacturing, more efficient and flexible production is aimed. The digital twin is one of these technologies. The digital twin is a connection between the physical and digital environment. The twin in the digital environment works with real-time data. Thus, a dynamic and bidirectional model of the physical object is created.

Z. Şen et al. (Eds.): IMSS 2023, LNME, pp. 592–599, 2024.
https://doi.org/10.1007/978-981-99-6062-0_54

Machine learning algorithms can be used for modeling the working of machines for digital twin applications. Some parameters in these algorithms must be chosen by the users and have a significant impact on the efficiency of the algorithm. These parameters are called hyperparameters. Hyperparameters can be decided with the trial-and-error method or expert opinion approach. But these methods are not analytic and can be caused high computational costs. Hyperparameter optimization uses heuristic, meta-heuristic algorithms, and mathematical and statistical methods to select the best hyperparameter set.

Industrial artificial intelligence is described as a systematic discipline that aimed to develop and deploy sustainable and repeatable processes with consistent successes [1]. Hyperparameters can be changed for different datasets. For that reason, algorithms cannot repeat themselves without the user deciding on the hyperparameters. In hyperparameter optimization integrated machine learning algorithm, the decision-making task moves from user to computer. Thus, a repeatable and automated process can be achieved.

This paper comprises five sections. After the introduction part as in Sect. 1, the literature review is discussed in Sect. 2. Section 3 explains the methodology of the paper. The case study has been applied and implications are made in Sect. 4. The conclusions are described in Sect. 5 as the last part of the study.

2 Literature Review

Digital twin technology has first applied in 2010 by NASA to mirror the life of its flying physical twin. In the year digital twin literature has expanded with applications in multiple areas such as aviation, smart city applications, health, agriculture, product design, risk analysis, and manufacturing. Liu et al. created a digital twin of an aero-engine gear production workshop to purpose intelligent and dynamic scheduling. By optimizing the scheduling process in the digital twin, the efficiency of the production line increased [2]. Luo et al. have purposed a predictive maintenance algorithm based on digital twins for the CNC bench [3].

Machine learning algorithms are one of the tools that are used for creating digital twins. Hyperparameter tuning is an important step to develop a machine-learning process. In recent years, hyperparameter optimization studies have increased in the literature. Heuristic and meta-heuristic algorithms, statistical methods, and software such as TUPAQ, Auto-Weka, and Hyperopt are the methods used for hyperparameter optimization. Ma et al. studied a self-learning error control mechanism and they used long-short term memory and neural network for that purpose. To optimize the hyperparameters of the neural network, they applied Bayesian optimization. They showed the hybrid Bayesian-LTSM neural network model works better than LTSM neural network with random hyperparameters [4]. Gülcü and Kuş studied a review of genetic, particle swarm, differential improvement algorithms, and Bayesian optimization for convolutional neural network hyperparameter optimization [5]. Lermer and Reich used the trial-and-error method for choosing hyperparameters of neural network algorithms and suggested more advanced techniques like biogeography-based optimization [6].

Hyperparameter optimization integrated machine learning algorithm model is important to create the automated, repeatable, executable process in the context of industrial artificial intelligence. Lee et al. characterized industrial AI's key elements as analytics technology, big data technology, cloud or cyber technology, domain know-how, and evidence [7]. Peres et al. showed most industrial AI projects are about process optimization, quality control, predictive maintenance, human-robot collaboration, and ergonomics [8]. In the industrial AI concept, automated digital twin projects are searched in the literature. Lugaresi and Matta purposed an automated digital twin model that can create digital twins in a short time for variable processes and requires minimum external intervention. They tried the model in two test cases and a production line and the purposed model successfully worked [9].

In literature, predictions are made by using classic artificial intelligence or machine learning algorithms. As a difference from the other studies, this paper integrates hyperparameter optimization into the prediction process in a digital twin context and shows that performance metrics can be improved. Making the digital twin concept applicable in the industry with less coding background is the main contribution of this study. Thus, digital twin technology will be available for every part of the industrial area, not only for researchers.

3 Methods

In this paper, a random forest is used to generate a digital twin model. Hyperopt-Sklearn libraries are used for hyperparameter optimization applications to increase the efficiency of the digital model.

3.1 Random Forest

The random forest algorithm was first mentioned by Leo Breiman in 2001. To solve regression and classification problems, the random forest used many decision trees that are created with random samples from the original data set. Each decision tree is estimated with different samples and result of that each decision tree creates different results. If the problem type is regression, the random forest takes the average of the different results (Eq. 1). In classification problems, the mode of the different results will be the result (Eq. 2).

$$\hat{f}_{ro}(x) = \frac{1}{n_{tree}} \sum\nolimits_{k=1}^{n_{tree}} \hat{f}_k(x). \tag{1}$$

$$\hat{f}_{ro}(x) = mode(\hat{f}_1(x), \hat{f}_2(x), \hat{f}_3(x), \ldots., \hat{f}_{n_{tree}}(x)). \tag{2}$$

Figure 1 shows the model of the random forest algorithm. As seen in Fig. 1, the random forest uses 'n' different decision trees. Random forest algorithm reduces the overfitting problem by using the estimation of n number of decision trees with low correlation and parallelly trained.

In Table 1, there are the hyperparameters of the random forest method and their explanation [11].

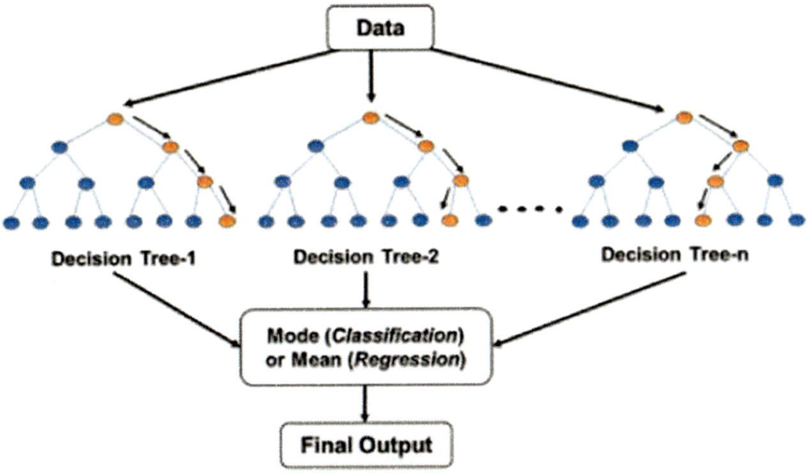

Fig. 1. Random Forest Model [10]

Table 1. Hyperparameters of Random Forest

Hyperparameter	Explanation
Mtry	Number of drawn candidate variables in each split
Sample size	Number of observations in each tree
Replacement	Draw observations with or without replacement
Node size	Minimum number of observations in a terminal node
Number of trees	Number of trees in the forest
Splitting rule	Splitting rule for nodes

3.2 Hyperopt-Sklearn

Hyperopt is a library software in Python programming language. It was first introduced in the ICML workshop on automated machine learning (AutoML) in 2014. This library is used for hyperparameter optimization of machine learning algorithms by using the Sklearn library for machine learning algorithms. Three things need to be defined to optimize hyperparameters by using the Hyperopt library.

A Search Domain. It's needed to define the functions/statistical distributions to indicate the hyperparameters' possible values. In the Hyperopt library, there is a function called 'hp' to define the search domain. It contains several statistical distributions like the normal distribution, uniform distribution for numerical hyperparameters, and a function called 'choice' for categorical hyperparameters.

An Objective Function. The machine learning algorithm, hyperparameters, and performance metric are defined in the objective function.

An Optimization Algorithm. There are three optimization algorithms defined in the Hyperopt library: random search, annealing, and tree-structured parzen estimators. Random search searches the domain randomly and purposes to find the best solution. Annealing is a search algorithm designed based on slow-cooling metals. Tree-structured parzen estimators is a search algorithm based on a sequential model optimization (SBMO) approach. The 'Suggest' function is used for defining the optimization algorithm. It can be selected from those three algorithms or created as a mixed algorithm with these algorithms.

By using the "fmin" function of the Hyperopt library, hyperparameter optimization can be applied [12].

4 Case Study

The data used for this application is obtained from the sensors on the CNC workbench. First, the data set that contains duration, speed, and motion is obtained and the data set is expanded with a time column that is calculated as motion/speed. Table 2 shows some parts of the data set used. The purpose is to predict the duration by using speed, motion, and time data that is in 4 columns and a 115-row dataset.

Table 2. Data Set Used

Duration	Speed	Motion	Time
7712.5	125	10	0.08
11490	125	15	0.12
19041.38	125	25	0.2
26593.5	125	35	0.28
37921.63	125	50	0.4
49247.88	125	65	0.52
56801.13	125	75	0.6
64352	125	85	0.68
75682.13	125	100	0.8
87014.13	125	115	0.92
97742.83	125	125	1
102121.4	125	135	1.08
113451.1	125	150	1.2
124777.4	125	165	1.32
132329.4	125	175	1.4

Firstly, prediction work was performed using the "RandomForestRegressor" library in the R program. As a performance metric, the mean absolute percentage error (MAPE) was used and the accuracy value was found to be 78.50963%.

The goal is to increase the accuracy value by optimizing the hyperparameters in the random forest algorithm. In line with this goal, Hyperopt and Sklearn libraries in Python programming language are used. For hyperparameter optimization, firstly hyperparameter space is determined. In determining the search space, the "hp" function in the Hyperopt library is utilized. For categorical parameters "choice" option of "hp", for numerical parameters uniform distribution is used. The objective function is defined and the MAPE is calculated as a performance metric. The accuracy score is defined as 1-MAPE. After defining hyperparameter space and objective function, the "fmin" function is used for optimization. "fmin" function optimizes hyperparameters with 3 different algorithms. The accuracy values obtained with the optimized hyperparameters for 3 different methods are shown in Table 3.

Table 3. Method-Accuracy Score

Method	Accuracy Score
Hyperopt-rand	95.11069
Hyperopt-anneal	95.15747
Hyperopt-tpe	95.18777

When the Hyperopt library is run with random search, the hyperparameter set that gives 95.11069% accuracy is shown in Table 4.

Table 4. Optimum Hyperparameter Set

Hyperparameter	Value of Hyperparameter
bootstrap	True
Max depth	40.0
Max features	Auto
Min samples leaf	1.8620689595597015
Min samples split	2.773157165416073
Number of estimators	1100.0

In a random search application, 100 different hyperparameter sets were scanned. No slope was observed among the accuracy values for 100 different hyperparameter sets. Accuracy scores are obtained between %70 and %95.11069.

Figure 2 shows the accuracy values obtained in 100 searches and the histogram graph of these values.

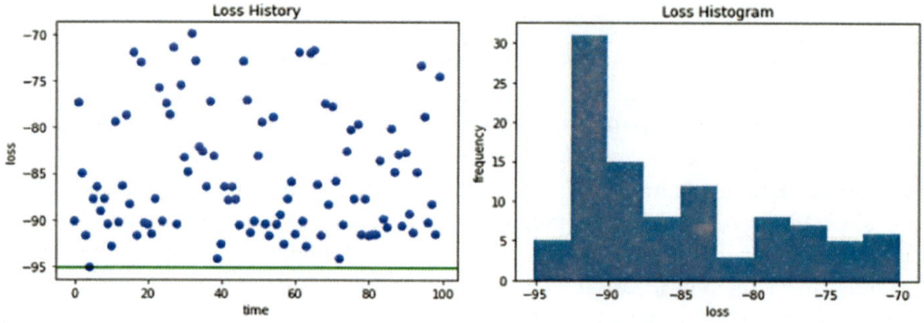

Fig. 2. Random Search Accuracy Score Graphics

5 Conclusion

This paper aimed to create an automated and executable digital twin process in the industrial artificial intelligence concept. Successfully repeatable digital twin generation will provide digital twin applications for everyone with less coding background. This process also aimed to increase the efficiency of the process by using hyperparameter optimization techniques instead of classic methods to decide on hyperparameters.

For these aims, random forests are used as a machine-learning algorithm for prediction. To decide random forest's hyperparameters, Hyperopt-Sklearn libraries were used. After applications, results show the difference and importance of hyperparameter optimization. Firstly, the random forest algorithm ran in R programming language without hyperparameter optimization and the accuracy score (1-MAPE) turned out 78.50963%. Then in Python language, random forest is applied with Hyperopt-Sklearn libraries. Random search, annealing, and "tree-structured parzen" estimators were used as optimization techniques, and the results were observed. Accuracy scores were found around %95 for these 3 methods. The efficiency of the digital twin generation has increased through hyperparameter optimization. For future works, hyperparameter optimization can be done by meta-heuristic algorithms such as evolutionary algorithms.

References

1. Lee, J., Singh, J., Azamfar, M.: Industrial Artificial Intelligence (2019)
2. Liu, Z., Chen, W., Zhang, C., Yang, C., Cheng, Q.: Intelligent scheduling of a feature-process-machine tool supernetwork based on digital twin workshop. J. Manuf. Syst. **58**, 157–167 (2021). https://doi.org/10.1016/j.jmsy.2020.07.016
3. Luo, W., Hu, T., Ye, Y., Zhang, C., Wei, Y.: A hybrid predictive maintenance approach for CNC machine tool driven by Digital Twin. Robot. Comput.-Integr. Manuf. **65**, 101974 (2020). https://doi.org/10.1016/j.rcim.2020.101974
4. Ma, C., Gui, H., Liu, J.: Self learning-empowered thermal error control method of precision machine tools based on digital twin. J. Intell. Manuf. **34**(2), 695–717 (2021). https://doi.org/10.1007/s10845-021-01821-z
5. Gülcü, A., Kuş, Z.: Konvolüsyonel Sinir Ağlarında Hiper-Parametre Optimizasyonu Yöntemlerinin İncelenmesi. Gazi Üniversitesi Fen Bilimleri Dergisi Part C: Tasarım ve Teknoloji **7**(2), 503–522 (2019). https://doi.org/10.29109/gujsc.514483

6. Lermer, M., Reich, C.: Creation of digital twins by combining fuzzy rules with artificial neural networks. In: IECON Proceedings (Industrial Electronics Conference), vol. 2019-Octob, pp. 5849–5854 (2019). https://doi.org/10.1109/IECON.2019.8926914
7. Lee, J., Davari, H., Singh, J., Pandhare, V.: Industrial Artificial Intelligence for industry 4.0-based manufacturing systems. Manuf. Lett. **18**, 20–23 (2018). https://doi.org/10.1016/j.mfglet.2018.09.002
8. Peres, R.S., Jia, X., Lee, J., Sun, K., Colombo, A.W., Barata, J.: Industrial artificial intelligence in industry 4.0 – systematic review, challenges and outlook. IEEE Access **8**, 220121–220139 (2020). https://doi.org/10.1109/ACCESS.2020.3042874
9. Lugaresi, G., Matta, A.: Automated manufacturing system discovery and digital twin generation. J. Manuf. Syst. **59**(February), 51–66 (2021). https://doi.org/10.1016/j.jmsy.2021.01.005
10. Rani, A., Kumar, N., Kumar, J., Sinha, N.K.: Machine learning for soil moisture assessment. Deep Learn. Sustain. Agric. **2022**, 143–168 (2022). https://doi.org/10.1016/B978-0-323-85214-2.00001-X
11. Probst, P., Wright, M.N., Boulesteix, A.L.: Hyperparameters and tuning strategies for random forest. Wiley Interdisc. Rev.: Data Min. Knowl. Discovery **9**(3), 1–15 (2019). https://doi.org/10.1002/widm.1301
12. Hutter, F., Kotthoff, L., Vanschoren, J.:Automated Machine Learning (2019). https://doi.org/10.1007/978-981-16-2233-5_11

The Effect of Parameters on the Success of Heuristic Algorithms in Personalized Personnel Scheduling

Esra Gülmez[1], Kemal Burak Urgancı[2], Halil İbrahim Koruca[2](✉), and Mehmet Emin Aydin[3]

[1] Momentum BT, Technopolis Antalya R&D-2 Building, Antalya, Turkey
[2] Department of Industrial Engineering, Suleyman Demirel University, Isparta, Turkey
{kemalurganci,halilkoruca}@sdu.edu.tr
[3] University of the West of England, Bristol, UK
mehmet.aydin@uwe.ac.uk

Abstract. Work-life balance is an approach that aims to enable employees to balance their work, family, and private lives. It is seen that the factors in the work-life balance are not relevant to work and family, considering the activities that one wishes for oneself, friendships and social life. For this reason, it has become mandatory for working people to devote enough time to their business life, family life and private life to protect their physical and mental health. This is particularly important in health-care sector, where the peace of mind of workers influences the outcoming service significantly. Within the scope of this study, a framework (also implemented as a software) for health-care workers has been developed in order to make weekly and monthly scheduling suitable for work-life balance. Three population-based heuristic algorithms for scheduling, namely genetic algorithm, ant colony and particle swarm optimization algorithms, are integrated into the system to optimize the schedules. The system aims to show optimal schedules for both which personnel is to work in which time zones of the hospital administrators and in which time zones the individual personnel is planned to work. The proposed approach allows doctors, as the most flexible workers, to opt the working hours and periods in a hospital flexibly. This study provides comparative results to demonstrate the performance of the three algorithms integrated to optimize the generated personal schedules optimized with respect to one's own preferences. Furthermore, it identifies the boundaries of the parameters affecting the success with Taguchi Method.

Keywords: Work Life Balance · Scheduling · Meta-Heuristic Algorithms · Taguchi Method

1 Introduction

To adapt to the technological advances of business life, employees reduce time for themselves and their families due to the expansion of technology learning, increasing responsibilities and the ability to work from home. Work-life balance refers to the ability

Z. Şen et al. (Eds.): IMSS 2023, LNME, pp. 600–611, 2024.
https://doi.org/10.1007/978-981-99-6062-0_55

of an employee to avoid conflicts between work and family responsibilities, and to have enough time for oneself, one's family and one's work, and to be satisfied in all areas [1].

"Work", as we know it, refers to labor for which a wage is paid and creates a value. Since the beginning of history, people have been trying to fulfill their desires and needs with the wages they receive in return for their knowledge, skills, and abilities. "Private life", or in other words "non-work life", refers to the part of individuals' lives in which they do not work. Family life is only a part of private life [2]. The word "balance" means the equal distribution of a quantity. Working people do not always allocate equal amounts of time to work and private life. Depending on the time and individuals, it is seen that sometimes work life and sometimes private life dominate. When discussing the concept of balance, it would be more accurate to use the word "balance" in the sense of "stability of the body or mind" in physical and psychological terms. Balance may differ from person to person [2]. Work Life Balance (WLB), in its shortest definition, is the state in which a person's work-related demands and personal life demands are in balance [3]. Work-life balance can also be defined as the ability of employees to fulfill both family and work responsibilities without any problems. Accordingly, work-life balance can also be defined as the ability of employees to determine their own working time and periods to minimize the conflict between the tasks they undertake in their work and non-work (private) lives [4].

In the healthcare sector, work-life balance is of paramount importance due to long shifts and the fact that employees face serious challenges during their working hours. According to research, 44% of physicians report working more than 60 h per week and more than one-third of their work time is consumed by administrative tasks (e.g., billing, staff scheduling, scheduling) [5].

This study aims to further increase the self-selection rates of working time and starting hours to contribute to the work-life balance of healthcare workers. For this purpose, two new algorithm modules (Ant colony and Particle swarm algorithms) were developed and added to the Work-Life Balance Tool (WLB-Tool) [6, 7]. The assignment performances of three meta-heuristic algorithms (Genetic algorithm, Ant colony and Particle swarm algorithms) in the scheduling software WLB-Tool are compared in this study.

In the first stage, it is necessary to enter the time periods determined by the hospital manager in accordance with the work-life balance and the number of personnel to work in these time periods with the help of the interface of the developed software. In the second stage, employees need to select the time period they want to work per week and the time slots determined by the manager that are appropriate for this period. In the third stage, meta-heuristic algorithms (Ant colony algorithm and particle swarm algorithm) are used to select assignments for the entered requests and the algorithm is run and personnel assignments/scheduling are performed by the staff scheduling software. At the last stage, the results of the assignments are shown, and the list of which personnel are assigned for which time periods for the enterprise, the time periods in which each personnel will work and the rate at which their demands are met, and the weekly personnel costs can be obtained.

With the help of the software, one of the algorithms is run when the demand entry processes for hospital management (determining the time slot interval and the number of personnel required in this interval) and employees (selecting the weekly working time and the time slot to work) are completed.

The software assignment process aims to meet 100% of the manager's shift demands and 100% of the scheduling demands of each employee. After the software completes the assignment of employees.

a) a list of which personnel are assigned to which shift (for the hospital, a chart showing who is assigned to which shifts),
b) It is possible to obtain a list showing the time periods assigned to each staff member, in other words, a list showing the personalized working times for the week or month in which they will work, and a list showing the rate at which the requests of each staff member are met (Personalized schedules can be obtained for staff members).
c) A report calculating personnel costs for the relevant week can be obtained. Scheduling results for different work-life balance scenarios for a hospital are presented and interpreted in the experimental results section of the paper.

The rest of this paper is organized as follows: Sect. 2 presents the related background and literature review, Sect. 3 presents the methods and materials; Genetic algorithm, Ant colony algorithm, Particle swarm algorithm, Constraints of algorithms. In Sect. 3, methods, and materials; Genetic algorithm, Ant colony algorithm, Particle swarm algorithm, Constraints of algorithms. In Sect. 4 experimental results, Taguchi method and Sect. 5 provide evaluation the results of different algorithms are presented.

2 Background and Literature Review

In a questionnaire study on the regulation of working hours in a university hospital, employees' satisfaction, and the effects of flexible working hours and/or work-life balance, working time and starting hours were investigated. 41.3% of the employees stated that their current position did not cause any problems between them and their families and 29.7% of the employees stated that they had problems with their family and work life due to their position. 66% of the employees stated that their opinions and suggestions were not taken into consideration in determining working hours. When asked about the appropriateness of working hours, 30% of the employees find it appropriate, while 38.4% do not find it appropriate. The remaining 40.6% of the employees answered "partially", indicating that they are not fully satisfied with the current working hours. From the survey results, it is understood that a large proportion of employees are not satisfied with the current working time and working schedule and would be satisfied with a flexible working arrangement where they can determine their own working time and starting time [8].

With the Industrial Revolution in the early eighteenth century, especially in Western Europe, issues such as the role of child laborers in business life and the working hours of all employees became more on the agenda and because of these problems, legal regulations were introduced by states in the nineteenth century. In Turkey, legal regulations were introduced at the beginning of the twentieth century for working life, which was continued until the end of the nineteenth century with the tradition of ahi community, and working hours, employer and employee rights were defined. The International Labor Organization (ILO) decided that working hours should be 8 h per day and 48 h per week. Working hours in Turkey were determined as 48 h, with the Law No. 3008, which entered into force in 1936, In addition, special arrangements have been made regarding the working conditions of women and child workers. Legal regulations have been made on many important issues such as obtaining the consent of the worker for working overtime and increased wage payment in case of overtime work [9].

Working time, responsibilities and duties of personnel are determined by laws and regulations. According to the law, personnel must complete 45 h of working time per week (4857 Labor Law, Article 63). For doctors and nurses, they must complete their weekly working time in different time intervals, such as night shifts, day shifts, etc., according to the number of personnel required by the hospital within a 7-day, 24-h period. The assignment of doctors, nurses, and technical staff to shifts is usually done by the hospital management. This can lead to unfair distribution and start times between doctors and other staff. Flexicurity in working life, including temporary employment, and telecommuting, entered into force on 06.06.2016 after Article 7 of the 4857 Labor Law was amended by Law No. 6715. The law is based on the harmony of work life and family life, provides real job security for part-time workers on maternity leave, and transforms unregistered work at home and seasonal work into jobs with employment security and insurance. Regulations such as compensatory work, intensified work week, etc. have also been included in the labor law. Considering the basic provisions mentioned above, there are small differences in the weekly working hours and duration in the European Union countries for 2015, ranging from 38.4 h in Norway to 45.1 h in Iceland [10]. Work balance is defined as an individual's subjective experience of having the right amount and the right variation between different activities in daily life, namely work, home and family activities, leisure, rest and sleep [11].

Work-life balance is an approach that aims to enable employees to balance their work, family, and private lives. As a social being, human beings are not limited to family life. Considering their personal activities, friendships, and social life, it can be seen that the factors in work-life balance are not limited to work and families. From this perspective, in recent years, researchers have defined the term work-life balance to represent activities outside of work that are not limited to family matters, but also include personal matters, friends and community [12–14].

Planning doctors' work schedules is a complex task. Legal requirements, different qualification levels and preferences for different working hours increase the difficulty of finding a solution that fulfills all requirements at the same time. Unplanned absences, e.g., due to illness, are also reported to further increase the complexity [13].

The achievements in the use of meta-heuristic-based approaches to tackle optimization problems with different combinations and constraints have been increasing tremendously over the last decade. These achievements have led researchers in the field to apply numerous meta-heuristic-based approaches for such high-dimensional problems [14]. The study of determining personalized working hours in hospitals was determined to be appropriate to be solved using meta-heuristic algorithms since the constraints are not specific. A population-based algorithms genetic algorithm was applied for the study conducted in a hospital. In the previous study in this field; a) According to priority rules (age, seniority, etc.), b) Balancing algorithm (considering the wishes of the employees with justice and balance), c) Optimization of employee requests with genetic algorithm [15].

3 Materials and Methods

To demonstrate the practicality of our proposed method, we generated potential work-life balance scenarios for doctors in the internal medicine clinic of Suleyman Demirel University Hospital in Turkey. These scenarios were designed based on the preferences of the doctors working in the unit. Next, we utilized the WLB-Tool to schedule their work hours, considering their weekly working hours and preferred starting times. The outcomes and effectiveness of each scheduling method were also included in our study.

Metaheuristics are a variety of approximate optimization algorithms that provide exit strategies from local optimal points. In the population-based optimization technique, the optimization process starts with a set of random solutions and evolves by repetition. A set of solutions are considered as the desired solutions [16]. The genetic algorithm, ant colony algorithm and particle swarm algorithm used in the solution of the problem were preferred because they provide the appropriate modeling for assigning employees to the shifts they requested.

3.1 Compared Algorithms and Their Parameters

Metaheuristic algorithms seek solutions to problems with probabilities from an infinite space universe. For this reason, it offers infinite variations on the way to the solution. In the case of a problem, there is a search for solutions to many situations such as 100% fulfillment of the manager's shift assignment requests, balanced shift assignments among personnel, and maximum contribution to the work-life balance of personnel. This requires many tests to evaluate each variation on the way to the optimum solution and makes it difficult to make the right decision on the parameters. Taguchi method was used to find a solution to this situation.

Genetic Algorithm (GA)
The Genetic Algorithm (GA) method is a mathematical model that is based on the relationship between genes and chromosomes. Genetic algorithms are widely used for optimization problems such as scheduling and routing.

Genetic algorithms are a heuristic method that aims to find the optimum result by random search technique. GA is one of the optimization methods used in complex and multidimensional problems with stochastic structure [15].

Parameters,

- Mutation rate: It is the parameter that shows the mutation rate at which the genetic algorithm will work.
- Number of iterations: It is the parameter that shows the number of steps of the genetic algorithm.
- Number of populations: It is the parameter that shows the target group size [15].

In the operation of the fitness function, the effect of everyone in the population is measured and optimized. The mutation rate is effective to produce the best result in the algorithm. The mutation rate is expected to be a value between 0 and 1. The optimal range should be determined according to the case problem.

Ant Colony Algorithm (ACA)

The basic idea of ACA is to mimic the cooperative behavior of ant colonies. As soon as the ant finds a food source, it evaluates it and carries some food back to the nest. During its foraging journey, the ant leaves a pheromone trail on the ground. The pheromone accumulated on the foraging trail, the amount of which depends on the quantity and quality of the food, directs other ants to the food source [17]. The amount of pheromone on the arc decreases over time due to evaporation. Each ant decides on a path according to the amount of pheromone. The more pheromone trails, the shorter the path instead of the longer one. The technique is based on updating the pheromone trail of good solutions and there are many such ant models [18]. Upon locating a sustenance reservoir, an ant promptly engages in an assessment of the resource's attributes and subsequently transports a portion of it to its abode. In the course of this home ward journey, the ant meticulously dispenses a trail of pheromones onto the terrain. The extent of pheromone deposition, contingent upon the abundance and caliber of the sustenance, serves as a navigational cue for fellow ants, directing them towards the identified food source [19].

If there are different paths for the same source, then the pheromones released on the shortest path will last longer than the pheromones released on the longest path, so the shortest path will be more attractive for new ants leaving the nest, gradually they will all start to take the shortest path and the pheromone concentration will increase on this path, while the pheromone concentration on the longest path will gradually decrease [20].

Parameters,

- Number of ants: It is the parameter that determines the target group to be included in the problem.
- Number of iterations: It is the parameter that shows the number of iterations (steps) the algorithm will run.
- Pheromone reinforcement ratio (α): It is the parameter that shows the degree of importance of pheromones between the nodes.
- Heuristic strengthening ratio (β): A parameter indicating the importance of the distance between the identified nodes.
- Pheromone evaporation rate (ρ): A parameter indicating the rate at which pheromones between nodes will evaporate when each step is completed.

To implement the ant colony algorithm, the parameters α (alpha), β (beta) and ρ (RHO) must be entered and should be between 0 and 1. In the considered problem, it was observed that as the value entered for alpha and beta approaches 1, the percentage of fulfillment of personnel requests in assignments decreases. The RHO value, which is the pheromone evaporation rate, should be between $0 < \rho < 1$. The parameter RHO (ρ)

is used to prevent unlimited accumulation of pheromone trails and allows the algorithm to forget bad decisions made previously.

Particle Swarm Algorithm (PSA)

Particle Swarm Optimization (PSO) is one of the newly proposed metaheuristic techniques by Kennedy and Eberhart that is based on the natural flocking and swarming behavior of birds and insects. It starts with a population of random solutions and updates generations to search for the optimum. In PSO, potential solutions, called particles, move in the problem space by following the current optimal particles [20].

In PSO, individuals in the solution population move in the search space trying to find the optimal solution. This process is like Genetic Algorithm (GA) as each movement can be viewed as a new generation of solutions generated after a series of rules are followed, like GA. However, the way particles calculate their new positions is structurally different from GA. Additionally, PSO's memory is not based on weights that need to be recalculated in each generation, but rather on individuals [21].

In PSO, individuals have a position and a velocity. The PSO algorithm works by attracting particles to high fitness positions in the search space. Each particle has a memory function and adjusts its trajectory based on two pieces of information: the best position it has visited so far, and the best position reached by the entire swarm [20]. Each particle maintains a memory of its previous best position (p_{best}), and a particle considers the entire population as its topological neighbors, where the best value is the global best (g_{best}). The PSO algorithm runs through iterations, and each solution generated in each iteration is compared to both its own local best and the global best of the swarm [22].

PSO initiates a swarm of particles traversing the search space to find an optimal global best. In fact, each particle represents a potential solution [24].

Parameters:

- C1 and C2 (Cognitive and Social Parameters): Also known as acceleration constants, they play a significant role in determining the trajectory of particles.
- C1 (Cognitive Parameter): Represents the confidence of the particle in finding the best solution.
- C2 (Social Parameter): Represents the confidence of the swarm.
- ϑmax (Maximum Velocity Vector): Maximum velocity allowed for particles during each iteration on a given path.
- ω (Inertia Weight): In PSO, the inertia weight enables particles to move towards the optimum point. [23].

In the shift scheduling using Particle Swarm Optimization (PSO), the population size is considered equal to the total number of employees. The management requests are positioned as the target locations of the particles within the swarm. The number of employees is matched with the number of particles in the algorithm. The constants used in PSO are in the interface for the manager to modify before the scheduling.

C1, C2, and ω parameters are required for PSO. Usually, C1 and C2 are set to 2, but different values can be considered based on $0 < C1 + C2 < 4$. ω is typically set to be less than 1 ($\omega < 1$) but can be updated during iterations.

Proposed Problem-Solving Approach: Main and Specific Constraints.

The developed software is planned to provide the maximum possible customization options according to the demands of employees and management. The algorithms are designed based on the necessary constraints specific to the problem at hand. The required constraints were identified before the adaptations were made.

Main constraints:

- 100% fulfilment of the shift requests of the hospital management is required.
- When assigning personnel, it is checked whether there are different assignments on the same day in the same shift.
- If there are not enough personnel request entries for the assignment requests entered by the management, personnel assignments are made according to the contract and shift criteria.

Specific constraints:

- When an employee logs in, their working hours specified in the contract are checked. It is not allowed to make request entries exceeding the contract working hours.

The parameter values that need to be used according to the main and specific constraints defined in the developed software were determined using the Taguchi method. The analysis results conducted in Minitab application are presented in Table 1 of Sect. 4. In addition, the software run results based on the parameters determined using the Taguchi method are provided in Table 2 of Sect. 4.

4 Experimental Results

In determining the parameter values of the software prepared using Taguchi method, a hybrid structure was established. First, 10 experiments were conducted in each model using Minitab application to determine the method's largest best, smallest best, and nominal best values in the mathematical model created with the Taguchi method. Each algorithm used in the software was evaluated separately in the experiments. The values generated because of the study are presented in Table 1. According to these values, the genetic algorithm and particle swarm algorithm produced the largest best optimal result, while the ant colony algorithm produced the nominal best optimal result. In addition, after the Taguchi method results were obtained, an "ANOVA Analysis" was performed in the Minitab application, and the accuracy of the method was determined to be 99.20%.

Furthermore, to determine the default algorithm to be used in the result report, the results were reported by comparing all three algorithms. The optimal bests generated for the three algorithms are highlighted in yellow in Table 1.

The abbreviations used in Table 1 are shown as IT = Iteration Count, MO = Mutation Rate, PS = Population Count, AVE = Average Assignment to Desired Shift, AS = Number of Assignments to Desired Shift.

Table 1. Results of Taguchi Method Application

# of Exp.	Genetic Algorithm (GA)					Ant Colony Algorithm (ACA)						Particle Swarm Algorithm (PSA)					
	Control Factors			Exp. Results	S/N	Control Factors				Exp. Results	S/N	Control Factors				Exp. Results	S/N
	MO	IT	PS	AVE (%)	AS	α	β	ρ	IT	AVE (%)	AS	C_1	C_2	ω	IT	AVE (%)	AS
1	0	150	12	52%	12	0.4	0.3	2	150	79%	21	2,21	1,2	0,45	150	43%	9
2	0.12	100	20	31%	7	0.01	0.6	1	100	36%	7	1,23	2,1	0,56	100	54%	11
3	0.29	75	210	76%	19	0.32	0.76	4	75	54%	12	0,89	0,98	0,76	870	21%	3
4	0.35	200	100	63%	16	0.12	0.9	2	200	32%	5	0	3	0,9	200	38%	7
5	0.42	800	5	21%	4	0.9	0.9	1	800	41%	8	1	1	1	800	24%	6
6	0.54	210	90	42%	9	0.87	1	1	210	64%	17	2	2	0	210	46%	8
7	0.58	200	300	43%	9	0.65	0	3	200	53%	11	2,7	0,8	0,63	200	45%	8
8	0.7	870	100	54%	13	0.99	0	2	870	21%	4	0,89	0,72	0,78	75	52%	10
9	0.89	210	250	36%	8	1	0.23	4	210	59%	15	0,92	2,32	0,91	210	33%	6
10	1	100	320	60%	14	0	0.43	3	100	53%	13	1,65	1,65	1	100	28%	5
	Average			48%	11.1	Average				49%	11.3	Average				38%	7.3

For evaluating the algorithm results, runs were conducted for each scenario according to the parameters determined by the Taguchi method. A team of 12 people was selected as a sample set in the runs. Random data entry was made for the 12 people according to the shift requests entered by the manager. The results obtained are presented comparatively in Table 2.

Table 2 Algorithm results run for different scenarios

	A0-1 (Scenario with only shift selection)			**A0-2** (Scenario where both shift selection and weekly working hours selection are made)			**S1** (4 Shifts of 6 hours)			**S2** (5 shifts with 8-hour, 5-hour, 4-hour, 3-hour, and 4-hour shifts)			**S3** (6 shifts, each of them 4 hours)		
	GA	**KKA**	**PSA**	**GA**	**KKA**	**PSA**	**GA**	**KKA**	**PSA**	**GA**	**KKA**	**PSA**	**GA**	**KKA**	**PSA**
PID	**%**	**%**	**%**	**%**	**%**	**%**	**%**	**%**	**%**	**%**	**%**	**%**	**%**	**%**	**%**
P1	32	65	20	62	59	18	16	83	68	35	10	56	92	57	13
P2	56	58	31	80	79	32	75	66	52	35	36	65	68	93	94
P3	45	54	48	14	83	39	60	65	44	67	23	58	40	50	31
P4	21	32	65	10	65	82	64	70	10	25	14	28	92	88	96
P5	76	43	41	28	20	96	84	29	36	29	87	19	39	97	64
P6	43	63	48	19	91	34	97	75	10	28	78	25	40	14	94
P7	65	54	47	32	45	59	10	62	27	97	79	90	34	53	22
P8	67	65	25	83	77	15	73	35	80	88	32	21	59	96	65
P9	43	76	75	94	68	88	86	84	52	95	19	52	52	88	88
P10	24	32	65	53	73	60	75	83	10	21	68	93	98	54	87
P11	65	57	60	75	54	53	84	94	81	83	36	12	77	69	38
P12	87	54	45	56	51	28	68	84	43	91	45	75	45	24	52
Ave. ($\bar{x}$)	**52.00**	**54.42**	**47.50**	**50.50**	**63.75**	**50.33**	**66.00**	**69.17**	**42.75**	**57.83**	**43.92**	**49.50**	**61.33**	**65.25**	**62.00**
S.D. (σ)	**20.75**	**13.21**	**16.81**	**29.23**	**19.48**	**27.31**	**26.80**	**19.85**	**25.53**	**31.39**	**27.30**	**28.21**	**23.41**	**28.09**	**30.39**

As can be seen in Table 1, the parameters that produced the best results were mutation rate = 0.29, iteration number = 75, population number = 210 for genetic algorithm assignments, alpha = 0.4, beta = 0.3, RHO = 2, iteration number = 150 for ant colony algorithm assignments, and c1 = 0.89, c2 = 0.72, omega = 0.78, and iteration number = 75 for particle swarm algorithm assignments. The algorithm results that achieved the best average and standard deviation among the three algorithms for different scenarios were highlighted in yellow.

5 Conclusions

A personalised personnel scheduling framework has been proposed and developed to solve the shift scheduling problem using three different population-based meta-heuristic algorithms in order to improve personal work-life balance. Unlike traditional and ongoing approaches, the employees' requests for scheduling have been considered in the system dynamically. To ensure the optimal outcome achieved, a thorough parametric study has been conducted for each algorithm using the Taguchi method. The optimized values in this regard were transferred and integrated into the developed software to enable the management panel to produce faster and more accurate results. Statistical analysis with ANOVA was performed to evaluate the results of the Taguchi method with a confidence level of 99.20%.

The proposed approach aims to ensure work-life balance for employees and increase satisfaction for both employees and managers by presenting a problem-solving perspective to shift scheduling, such as determining the shift periods and shift lengths, which are determined by employees and managers. To achieve this, a dynamic structure has been established in the algorithm. The addition of primary and specific constraints to the algorithm, based on the results of the analysis, has contributed to achieving optimal results.

References

1. Mudd, E.H.: Women's conflicting values. J. Marriage Family Living **8**(3), 50–65 (2002)
2. Efeoglu, I., Ozgen, H.: The effects of work, family life conflict on job stress, job satisfaction and organizational commitment: a study in the pharmaceutical industry. Cukurova Univ. J. Soc. Sci. Instit. **16**(2), 237–254 (2007)
3. Lockwood, N.R.: Work-life balance: challanges and solutions. HR Magazine (2003)
4. Friedmann, O, Christensen, P., Degroot, J.: Work and Life. Harvard Business Review: Work and Life Balance. Mess Publications 14 (2001).
5. Shanafelt, T.D., Hasan, O., Dyrbye, L.N., et al.: Changes in burnout and satisfaction with work-life balance in physicians and the general US working population between 2011 and 2014. Mayo Clin. Proc. **90**, 1600–1613 (2015)
6. Gulmez, E.: Personalized Staff Scheduling with Ant Colony and Particle Swarm Algorithms: Hospital Application Example. Suleyman Demirel University Institute of Science, Isparta (2021)
7. Koruca, H.İ, Emek, M.S., Gulmez, E.: Development of a new personalized staff-scheduling method with a work-life balance perspective: case of a hospital. Ann. Oper. Res. **328**(1), 793–820 (2023)
8. Koruca, H.I., Bosgelmez, G.: Evaluation of the effect of work-life balance and flexible workingsystem on employee satisfaction. SDU J. Health Sci **9**(4), 32–36 (2018)
9. Durmaz, O.: An Essay on the Birth of Labor Law in Turkey (1909–1947) 1–3 (2017)
10. Kapiz Ozen, S.: Work family life balance and a new approach to balance theory of boundary. Dokuz Eylul Univ. J. Soc. Sci. Instit. **4**(3), (2002)
11. Wagman, P., Håkansson, C., Björklund, A.: Occupational balance as used in occupational therapy: a concept analysis. Scand. J. Occup. Ther. **19**(4), 322–327 (2012)
12. Benito-Osorio, D., Muñoz-Aguado, L., Villar, C.: The impact of family and work-life balance policies on the performance of Spanish listed companies. Manag. (17), 214–236, (2014)
13. Hughes, J., Bozionelos, N.: Work-life balance as source of job dissatisfaction and withdrawal attitudes: an exploratory study on the views of male workers. Personnel Rev. **36**(1), 145–154 (2007)
14. Meller, C., Aronsson, G., Kecklund, G.: Boundary management preferences, boundary control, and work-life balance among full-time employed professionals in knowledge-intensive. Flexible Work, Nordic J. Work. Life Stud. **4**(4), 8–23 (2014)
15. Gross, C.N., Fugener, A., Brunner, J.O.: Online rescheduling of physicians in hospitals. Flex. Serv. Manuf. J. **30**, 296–328 (2018)

16. Bolaji, L.A., Bamigbola, F.A., Shola, P.B.: Late acceptance hill climbing algorithm for solving patient admission scheduling problem. Know. Based Syst. **145**, 197–206 (2017)
17. Emek, S.: Scheduling of hospital staff with the genetic algorithm method from the perspective of work-life balance. Suleyman Demirel University, Institute of Science and Technology, Isparta (2016)
18. Loscos, D., Martí-Oliet, N., Rodríguez, I.: Generalization and completeness of stochastic local search algorithms. Swarm Evol. Comput. **68**, 100982 (2022). https://doi.org/10.1016/j.swevo.2021.100982
19. Socha, K., Dorigo, M.: Ant colony optimization for continuous domains. Eur. J. Oper. Res. **185**, 1155–1173 (2008)
20. Kiran, M.S., Ozceylan, E., Gunduz, M., Paksoy, T.: A novel hybrid approach based on Particle Swarm Optimization and Ant Colony Algorithm to forecast energy demand of Turkey. Energy Convers. Manag. **53**, 75–83 (2011)
21. Deng, W., Xu, J., Zhao, H.: An improved ant colony optimization algorithm based on hybrid strategies for scheduling problem. IEEE Access **7**, 20281–20292 (2019)
22. Kuo, R.J., Hong, S.Y., Huang, Y.C.: Integration of particle swarm optimization-based fuzzy neural network and artificial neural network for supplier selection. Appl. Math. Model **34**, 3976–3990 (2010)
23. Loscos, D., Martí-Oliet, N., Rodríguez, I.: Generalization and completeness of stochastic local search algorithms. Swarm Evolutionary Comput. **68**, 100982 (2021)
24. Qawqzeh, Y., Alharbi, M.T., Jaradat, A., Sattar, K.N.A.: A review of swarm intelligence algorithms deployment for scheduling and optimization in cloud computing environments. PeerJ. Comput. Sci. **7**, e696 (2021)

A Decision Support System Design Proposal for Agricultural Planning

Fatmanur Varlik(✉), Zeynep Özçelik, Eda Börü, and Zehra Kamişli Öztürk

Industrial Engineering Department, Eskisehir Technical University, Eskisehir, Turkey
{fatmanurvarlik,zeynepozcelik,edaboru}@ogr.eskisehir.edu.tr,
zkamisli@eskisehir.edu.tr

Abstract. According to estimates, the world population is expected to be around 9 billion by 2050. Related to this, the number of people suffering from hunger is increasing day by day. Unconscious use of agricultural lands, climate changes, and the effect of increasing population, the problem of food needs increase the pressure on agriculture. In order to meet the need for food effectively, studies in areas such as sustainable land/forest management, improvement of cultivation areas, and agricultural policies should be carried out urgently. Increasing productivity by improving agricultural activities is important in terms of realizing the potential of agricultural lands.

In this study, a plant production plan is created according to the relevant data on the agricultural lands in a city. The aim is to create a sustainable production plan and to present a model that will increase the economic return that maximizes the return that was established. In the study, a mixed-integer model that maximizes the return is developed and it is aimed to determine which plants will grow in which region and to make a production plan accordingly. The distinctive value of the study is to propose a new decision support system for planning studies in Turkey in the field of agricultural production. As a result of the success of the project in the desired way, a system that will help the decision-makers in the relevant field will be gained, as well as contribute to the literature.

Keywords: Mixed integer programming · mathematical modeling · agricultural revenue · agricultural planning · decision support system

1 Introduction

According to estimates, the world population is expected to be around 9 billion by 2050. According to reports published by the United Nations, the number of people suffering from hunger is increasing every day. With the unconscious use of agricultural land, climate changes and the increasing population, the problem of food demand increases the pressure on agriculture. In order to effectively meet the food need, studies in areas such as sustainable land/forest management, improvement of cultivation areas and agricultural policies should be carried out urgently.

The most basic way to meet the food needs of the increasing population in a sustainable way is the effective planning of natural resources. The agricultural sector has a

Z. Şen et al. (Eds.): IMSS 2023, LNME, pp. 612–621, 2024.
https://doi.org/10.1007/978-981-99-6062-0_56

multifunctional structure for food security, economic income and environmental targets. Agricultural planning has very uncertain elements in terms of its components and there are issues to be considered while planning.

The aim of the study is to develop a model that will ensure the most efficient use of agricultural areas due to the increasing population, climate crisis and food security reasons. It is aimed to examine the productivity of agricultural lands in Eskişehir and to make improvements to increase productivity. In order to achieve this goal, it is aimed to establish a mathematical model to determine the suitable product group to be grown in an agricultural land and to draw up a production plan accordingly. With this model to be established, since the harvest will be made in line with the characteristics of the agricultural land and the plant to be grown, the structure of the soil will not deteriorate and the amount of product obtained and income will increase.

1.1 Literature Review

In this research conducted, Eren [7] and Ersoy [12] studies were mostly referred. In Eren's [7] project, mixed integer linear programming was used to find the 3-year cropping pattern and maximize the revenue in determined agricultural areas. On the other hand, Ersoy [12] conducted a short-, medium- and long-term agricultural production planning for profit maximization in Antalya using mixed integer linear programming. The articles reviewed were tabulated, summarized and presented in Table 1.

Table 1. Literature Research Summary Table

Article Author	Method	Goals
Ahmed et al. (2011) [1]	Linear Programming	Maximum profit
Alabdulkader et al. (2012) [2]	Linear Programming	Maximum profit + efficient use of water resources and arable land
Andreea ve Adrian (2012) [3]	Linear Programming	Minimum cost & Maximum profit
Lu et al. (2013) [4]	(IPAPSOM=PPM+IPM) (PPM = Probabilistic Programming Model) (IPM = Inexact Interval Programming Model)	Maximum profit, Ensuring food security, Increasing the welfare of farmers, Increasing the variety of food more utilization of the workforce
Jeyavanan et al. (2017) [5]	Goal Programming	Increasing agricultural yield
Li ve Ma (2017) [6]	Fuzzy Programming & Genetic Algorithm	Maximum profit & Maximum ecological benefit
Eren (2017) [7]	Mixed Integer Linear Programming	Maximum profit
R. Poldaru et al. (2018) [8]	Linear Programming	Minimum arable area + Meeting food demand
Martina Grubišić et al. (2019) [9]	Linear Programming	Maximum profit + Determining the optimum production
Zaród (2020) [10]	Multi-criteria Linear Programming	Maximum profit & Maximum production & Minimum amount of organic matter in the soil
Al-Omari ve Mohtar (2020) [11]	Multi Objective Linear Programming	Maximum profit
Ersoy (2020) [12]	Mixed Integer Programming	Maximum profit
El Sayed (2021) [13]	Fuzzy Goal Programming	Maximum profit, Maximum production, Minimum investment, Selection of low fertilizer crops & Water saving
Gündüz (2021) [14]	Multi Criteria ABC Analysis & GIS Multi-Criteria Evaluation	Maximum profit

With the project made, it enables the producers dealing with herbal agriculture, especially the provincial directorates of agriculture, to have an effective production plan.

During the literature review, not many studies on agricultural production planning were found in Turkey. In a master's thesis, a 3-year rotation plan was created and a crop pattern was established. In this study, the decision whether to plant a crop or not was evaluated over a period of years. This situation allows only a maximum of one crop to be planted on agricultural land within a year. For this reason, the time period was determined as a month in this study and it allowed more than one crop to be planted on a parcel within a year if the necessary constraints were met.

Another unique element of the study is the creation of a decision support system design. No decision support system has been found in the scanned studies and the decision support system planned to be established is aimed to be integrated into the model created to make production planning by making the project more unique.

2 Data Preparation

This study aims to contribute to a more effective agricultural planning in Turkey. A production plan that will be created by taking into account the soil structures and climatic characteristics of Turkey will increase the yield to be obtained from agricultural lands and move Turkey to the top of the agricultural product export ranking.

In order to create the mathematical model of the problem under investigation, the plants to be used in this model and the agricultural areas where the evaluation will be carried out were determined. Eskişehir province was selected as the pilot region for the study. The annual reports for 2021 published by Eskişehir Provincial Directorate of Agriculture and Forestry were examined and according to this report, crop species suitable for cultivation in Eskişehir were determined. Data such as the area and land distribution of agricultural lands for plant production of Eskişehir province on district basis are also given in the annual reports.

2022 Producer Price Index table from TUIK (TurkStat) page for the price parameter to be used in the model and the prices determined for September 2022 were added to the model as a parameter. Various sources and website of Eskişehir Provincial Directorate of Agriculture and Forestry were examined for generate parameter tables.

Ecological requirements of the selected plants were compiled from various sources and given Table 2.

Table 2. Ecological Request Table

Crop	Average Vegetation Duration (day)	Texture	pH	EC (mms)	Organic Material (%)	Yield (watery) (kg/da)	Yield (arid) (kg/da)	Min Temperature (°C)	Max Temperature (°C)	Relative Humidity (%)	Water Requirement (mm)	Price (Turkish Lira)
Onion	160	loamy, sandy	6.5-7.5	1,2	%2<	5000-6000		7	27	50-70	450-500	4,49
Cucumber	55	loamy	5,8-6,5	2,5	%2<	5000-8000		15	30	50-70	300-425	4,55
Watermelon	95	sandy-loamy, clayed-loamy	6-6,5	2	%2<	3000-7000	1500-3000	18	35	50-70	400-600	2
Melon	100	sandy	6-6,7	1	%2<	1000-2000	700	15	32	50-70	400-600	3,92
Tomato	100	loamy	5,5-7	25	%2<	6000-12000		16	32	50-70	800-900	5,33
Pepper	75	loamy, sandy-loamy	6-6,5	1,5	%2<	2000-3000		16	32	50-70	600-900	8,4
Potato	100	sandy-loamy	5,5-7	1,7	%2<	3000-4500		18	30	50-70	300-400	5,21
Brocoli	110	clayed-loamy	5,5-6,5	2,8	%2<	800-2000		5	20	50-70	450-600	4,6
Lettuce	80	clayed,sandy,loamy	6.Tem	1,3	%2<	1000-1500		7	24	50-70	450-600	5,42
Sunflower	125	sandy,clayed	6-7,5	4,8	%2<	300-500	250-300	5	40	50-70	700-800	14,35
Parsley	80	sandy	5,8-7,3	1,8	%2<	100-500		7	24	50-70	500-700	3

According to these ecological requirements, a parcel-month suitability matrix showing whether the crops can be grown on a month basis in a 12-month period was derived. The information in the table was compiled from different sources on the internet. The matrix is given Table 3.

Table 3. Crop-Month Suitability Table

Crop	January	February	March	April	May	June	July	August	September	October	November	December
onion	1	1	1	0	0	0	0	0	0	0	0	0
cucumber	0	0	0	1	1	1	1	1	1	1	0	0
watermelon	0	0	0	1	1	0	0	0	0	0	0	0
melon	0	0	0	1	1	0	0	0	0	0	0	0
tomato	0	0	1	1	0	0	0	0	0	0	0	0
pepper	0	0	1	1	0	0	0	0	0	0	0	0
potato	0	1	0	1	0	0	1	1	0	0	0	0
brocoli	0	0	1	1	1	0	0	0	1	1	1	0
lettuce	0	0	0	1	1	1	1	1	1	0	0	0
sunflower	0	1	1	1	0	0	0	0	0	0	0	0
parsley	0	0	0	1	1	1	1	1	0	0	0	0

Based on the data obtained, a crop-parcel suitability matrix showing whether a plant is suitable for growing in a parcel was created. While preparing this table, the plant family matrix table was utilized. The matrix is given Table 4.

Table 4. Crop-Parcel Suitability Table

District	onion	cucumber	watermelon	melon	tomato	pepper	potato	brocoli	lettuce	sunflower	parsley
Alpu	1	1	1	1	1	1	1	0	1	1	1
Beylikova	1	1	1	1	1	1	1	0	1	1	1
Çifteler	0	1	1	1	1	1	1	1	0	1	1
Günyüzü	1	1	1	1	1	1	1	1	1	0	1
Han	1	1	1	1	1	1	1	0	1	1	1
İnönü	1	1	1	1	1	1	1	1	1	1	1
Mahmudiye	1	1	1	1	1	1	1	1	1	1	1
Mihalgazi	1	1	1	1	1	1	1	1	1	1	1
Mihalıççık	1	1	1	1	1	1	1	1	0	1	1
Odunpazarı	1	1	1	1	1	1	1	1	1	1	1
Sarıcakaya	1	1	1	1	1	1	1	0	1	1	0
Seyitgazi	1	1	1	1	1	1	1	1	0	1	1
Sivrihisar	1	1	1	1	1	1	1	1	1	1	1
Tepebaşı	1	1	1	1	1	1	1	1	1	1	1

Crop family matrix is given Table 5.

Table 5. Crop Family Table

Crop	onion	cucumber	watermelon	melon	tomato	pepper	potato	brocoli	lettuce	sunflower	parsley
onion	1	0	0	0	0	0	0	0	0	0	0
cucumber	0	1	1	1	0	0	0	0	0	0	0
watermelon	0	1	1	1	0	0	0	0	0	0	0
melon	0	1	1	1	0	0	0	0	0	0	0
tomato	0	0	0	0	1	1	1	0	0	0	0
pepper	0	0	0	0	1	1	1	0	0	0	0
potato	0	0	0	0	1	1	1	0	0	0	0
brocoli	0	0	0	0	0	0	0	1	0	0	0
lettuce	0	0	0	0	0	0	0	0	1	0	0
sunflower	0	0	0	0	0	0	0	0	0	1	0
parsley	0	0	0	0	0	0	0	0	0	0	1

Parcel sizes were used to determine the amount of plants to be planted in the parcel. Since there was no clear information about the parcel size in the data obtained, the values written in the table were determined randomly. The parcel size table for the parcels considered is given Table 6.

Table 6. Parcel Size

Parcel	Land Size (decare)	Parcel	Land Size (decare)
Alpu	23	Mihalgazi	5
Beylikova	12	Mihalıççık	13
Çifteler	19	Odunpazarı	18
Günyüzü	18	Sarıcakaya	9
Han	10	Seyitgazi	14
İnönü	17	Sivrihisar	15
Mahmudiye	24	Tepebaşı	20

3 Methods and Techniques

The main problem of this study is to identify the appropriate crop pattern to be cultivated on an agricultural land and to prepare a production plan accordingly. Many variables need to be considered to come up with this production plan, which is a real-life problem. Some of the variables to be considered are real numbers and some are integers.

Mixed integer programming is an extension of linear programming. In this method, while the objective function and constraints are linear as in linear programming, what makes the method different is the addition of a constraint that ensures that some of the decision variables in the model take integer values greater than or equal to zero. By this means, it prevents undesirable and unrealistic results in linear programming.

The decision variables in the model are written in such a way that one variable determines whether a crop is suitable for cultivating in a given area and the other variable represents the amount of production. In this context, one of the decision variables is binary, taking the value 1 if the crop is suitable for the land and 0 if it is not; the decision variable that gives the amount of production is integer.

The objective function of the mathematical model to be established is to make crop planning that will provide maximum revenue to Eskişehir province on an annual basis. To create this objective function, the producer price information of the products in September 2022 and their yields on dry/irrigated land were taken into consideration.

The assumptions used in this study are given as follows.

- Soil properties are assumed to be homogeneous.
- To preserve diversity and soil mineral values, one crop is allowed to be planted once in a parcel.
- It is assumed that the lands considered in the model are not neighbors.

3.1 Mathematical Model

Sets and Indices

I: $\{i|i = 1, ..., 11\}$ index of crop
P: $\{p|p = 1, ..., 14\}$ index of parcels
T: $\{t|t = 1, ..., 12\}$ index of months

Parameters

n : total number of crops
m: total number of parcels
Bp: land size of parcel p
Fi: market prices of crop i
Hi: average harvest time of crop i
VSi: the average yield of crop i in irrigated land
VKi: the average yield of crop i in dry land
PSp: if parcel is irrigated PSp is 1, otherwise 0
Eip: if the crop i is suitable for the soil texture of the parcel p Eip is 1, otherwise 0
Mit: if the crop i is suitable for growing in month Mit is 1, otherwise 0

Decision Variables

$yipt$ = If crop I is planted in parcel p at month t,1 otherwise 0
$xipt$ = Production quantity of crop I is planted in parcel p at month t

Objective Function

$$maxZ = \sum_p \sum_i \sum_t F_i * x_{ipt} * (PS_p * VS_i + (1 - PS_p) * VK_i) \tag{1}$$

Constraints

$$\sum_i y_{ipt} \leq 1 \forall (p, t) \tag{2}$$

$$x_{ipt} \leq B_p * y_{ipt} \forall (i, p, t) \tag{3}$$

$$\sum_p \sum_y y_{ipt} \geq 1 \forall (i) \tag{4}$$

$$\sum_{t=1}^{12} y_{ipt} \leq 1 \forall (i, p) \tag{5}$$

$$E_{ip} * M_{it} \geq y_{ipt} \forall (i, p, t) \tag{6}$$

$$y_{ipt} + \sum_j^n y_{ipt} \leq 1 \forall (i, p, t, l) l \in (t, t + H\_i) \tag{7}$$

$$y_{ipt}, y_{ipl} \in \{0, 1\} \forall (i, p, t, l) \tag{8}$$

$$x_{ipt} \geq 0 \forall (i, p, t) \tag{9}$$

Constraint (1) provides that at most one crop can be planted in a parcel at month t. The amount of crops planted in a parcel cannot exceed the size of that parcel, constraint (2) forces that *xipt* can not be exceed *Bp*. .. Then, each crop is guaranteed to be cultivated at least once by the constraint (3). In order to prevent negative effects on soil structure characteristics, constraint (4) is written for the situation that the same crop can be cultivated at most once in a parcel. *yipt* can only take the value 1 if it is suitable for cultivation in the parcel p (if *Eip* is equal to 1) and if it is suitable for cultivation in month t (if *Mit* is equal to 1) with constraint (5). Constraint (6) prevent the cultivation of another crop in a parcel while a crop is already cultivated. With constraint (7) and constraint (8), *yipt* takes the values 0 or 1, *xipt* takes positive integer values.

As a result of the mathematical model created in the study, a decision support system interface design has been created to ensure the sustainability of the model. The designed decision support system is given in Fig. 1, databases of the decision support system design are given Fig. 2.

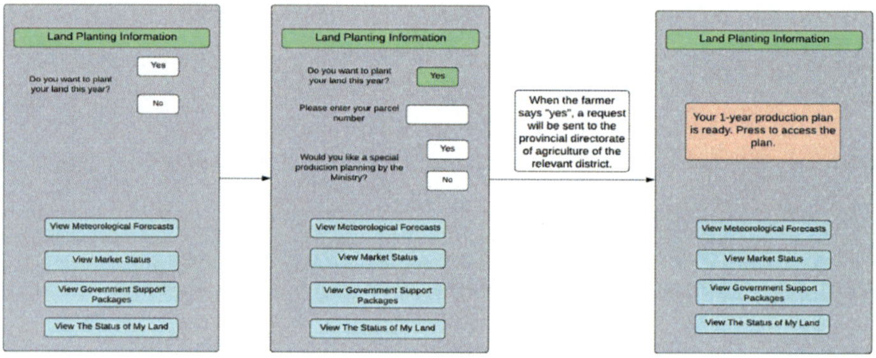

Fig. 1. Interface of DSS

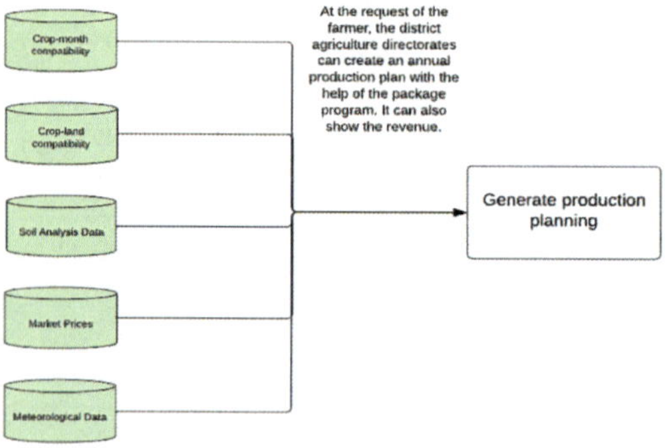

Fig. 2. Databases of DSS

4 Numerical Results

The mathematical model established with 11 crops and 14 parcels was coded by "GAMS Neos Solver".

The results of the model are shown Fig. 3.

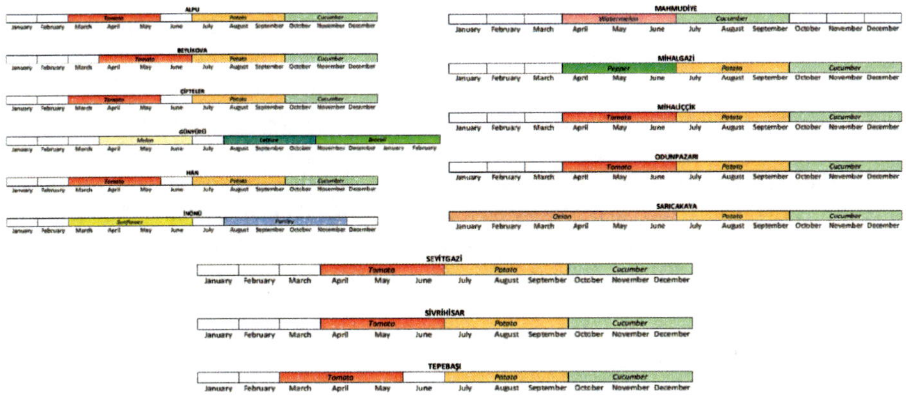

Fig. 3. Production Planning Results of Each Parcel

According to the assignment results, it was ensured that each plant was assigned to a parcel at least once. Further, the sowing of other plants other than the plant sowed in a parcel was allowed to increase the yield to be obtained from that parcel. So that, the values in the soil were preserved.

In line with the results obtained, the revenues obtained from the parcels are shown in Table 7.

Table 7. Revenues from Each Parcel

Parcel	Revenue	Parcel	Revenue	Parcel	Revenue	Parcel	Revenue
Alpu	₺ 2.110.308,00	Han	₺ 917.525,00	Mihaliccik	₺ 1.192.782,00	Sivrihisar	₺ 1.376.288,00
Beylikova	₺ 1.101.030,00	İnonu	₺ 67.086,50	Odunpazari	₺ 1.651.545,00	Tepebasi	₺ 1.835.050,00
Cifteler	₺ 1.743.298,00	Mahmudiye	₺ 132.000,00	Saricakaya	₺ 664.267,50		
Gunyuzu	₺ 49.392,00	Mihalgazi	₺ 350.562,50	Seyitgazi	₺ 1.284.535,00		
Total				₺14.475.669,50			

In order to make the study more realistic, it is important to plan crop production on a monthly rather than a yearly basis. Considering that in real life farmers can cultivate agricultural land more than once a year, the time interval t was determined on a monthly basis and a production plan was made accordingly.

In the related project, after the mathematical model was solved, a decision support system interface design was created who can be used by Provincial Directorates of Agriculture and farmers. The designed decision support system will provide the user with the set of crops that can be cultivated on a parcel, the production plan of the land and the expected return value. Hence, it is thought that this study will set an example for other cities of our country and contribute to the improvement of crop production planning. Through this system, which offers a production plan specialized for the farmer's land, a sustainable development goal in crop agriculture will be realized.

With the decision support system design worked on, it is learned whether the farmers will plant the parcels they own in the current year and then this information is transferred to the Provincial Directorates of Agriculture. In the interface to be designed for the Provincial Directorates of Agriculture, the decision support system to be operated in line with the soil data of the related land and the information received from the farmer will provide the production plan in the related parcel. With the decision support system design worked on, it is learned whether the farmers will plant the parcels they own in the current year and then this information is transferred to the Provincial Directorates of Agriculture. In the interface to be designed for the Provincial Directorates of Agriculture, the decision support system to be operated in line with the soil data of the related land and the information received from the farmer will provide the production plan in the related parcel.

5 Conclusion and Discussion

With the obtained crop production planning, the plants were assigned to the parcels suitable for growing and the revenue to be obtained was maximized. At the same time, provides a production plan to guide users. In addition, by ensuring that appropriate crops are planted on the parcels, it has contributed to the protection of the physical qualities of the soil and to more conscious agriculture.

When the mathematical model was established, crop-month adaptation and crop-parcel matrices were determined and planning was made according to these matrices. The matrices were prepared as a result of various research, and it is important to prepare more accurate matrices with a team of experts in the field in order to come to the optimal solutions suitable for real life.

Agriculture brings with it many uncertainties such as climate and price fluctuations. In the scope of this study, these uncertainties are not considered. By developing constraints that take these uncertainties into account, it is thought that results including various scenarios can be obtained.

The model is a pilot study for Eskisehir province and was created with a small number of parcels and plants. Therefore, optimal results could be obtained through GAMS application. However, GAMS will not be able to provide an optimal solution for a large number of parcels and plants. In this case, metaheuristic algorithms can be used to obtain feasible cultivation plans. During the literature search, it was predicted that Genetic Algorithm, which is one of the metaheuristic algorithms, can provide suitable results for this problem.

Methods such as sensors, irrigation systems, drones, automatic harvesters, etc., especially artificial intelligence, can determine the inputs needed by the soil, provide control and reduce labour. With the mathematical model we have established, a very high level of efficiency can be achieved in agriculture through the use of these intelligent agricultural systems.

References

1. Ahmed, E., Sulaiman, J., Mohd, S.: Agricultural resources allocation and field crops competition. Am. J. Agric. Biol. Sci. **6**(3), 384–392 (2011)
2. Alabdulkader, A.M., Al-Amoud, A.I., Awad, F.S.: Optimization of the cropping pattern in Saudi Arabia using a mathematical programming sector model. Agric. Econ. **58**(2), 56–60 (2012)
3. Andreea, I.R., Adrian, T.R.: Linear programming in Agriculture: case study in region of development South-Mountenia. Int. J. Sustain. Economies Manag. (IJSEM) **1**(1), 51–60 (2012)
4. Lu, S.S., Liu, Y.S., Long, H.L., Xing-Hang, G.U.A.N.: Agricultural production structure optimization: a case study of major grain producing areas, China. J. Integr. Agric. **12**(1), 184–197 (2013)
5. Jeyavanan, S., Siyambalapitiya, S.B., Jeyavanan, K.: Application of goal programming on yield optimization of selected agricultural crops. Int. J. Innovat. Res. Sci. Eng. Technol. **6**(7), 12472–12477 (2017)
6. Li, X., Ma, X.: An uncertain programming model for land use structure optimization to promote effectiveness of land use planning. Chin. Geogra. Sci. **27**(6), 974–988 (2017)
7. Eren, H.: Determination of cropping pattern with mathematical modeling technique in special agricultural lands. Master's thesis. Erciyes University, Institute of Educational Sciences, Kayseri (2019)
8. Põldaru, R., Viira, A. H., & Roots, J. (2018). Optimization of arable land use to guarantee food security in Estonia
9. Grubišić, M., Kolarec, B., Mamić, M.: A linear programming approach to land allocation in vegetable production: a case study from Croatia. Int. J. Model. Optim. **9**(3), 160–165 (2019)
10. Zarod, J.: Agricultural production planning using a multicriteria optimization model. Eur. Res. Stud. J. **XXIII**(Issue 4), 481–490 (2020)
11. Al-Omari, A., Mohtar, R.: Optimum water allocation for the agricultural sector in South Central Texas. Fresenius Environ. Bull. **29**(9A), 8168–8176 (2020)
12. Ersoy, M.C.: Modeling study on agricultural investment planning-Gazipaşa case. Master's thesis, Alanya Alaaddin Keykubat University (2020)
13. El Sayed, L.M.: Determining an optimum cropping pattern for Egypt (2021)
14. Gündüz, H.: Multi-criteria evaluation of product and production region selection problem in good agricultural practices: van province example. Master's thesis. Kocaeli University, Institute of Educational Sciences (2021)

Cyber Attack Detection with Encrypted Network Connection Analysis

Serkan Gonen[1](✉), Gokce Karacayilmaz[2], Harun Artuner[3], Mehmet Ali Bariskan[4], and Ercan Nurcan Yilmaz[5]

[1] Software Engineering Department, Istanbul Gelisim University, Istanbul, Turkey
sgonen@gelisim.edu.tr
[2] Hacettepe University Institute of Science, Ankara, Turkey
[3] Computer Engineering Department, Hacattepe University, Ankara, Turkey
[4] Computer Engineering Department, Istanbul Gelisim University, Istanbul, Turkey
[5] Electrical – Electronic Engineering Department, Gazi University, Ankara, Turkey

Abstract. The evolution of science and technology has led to increasingly complex cyber security threats, with advanced evasion techniques and encrypted communication channels making attacks harder to detect. While encryption has improved privacy and confidentiality for users, it has also provided a new avenue for attackers to exploit. Traditional intrusion detection systems, which transitioned from signature-based to behavior-based approaches, have struggled to keep up with these challenges. To address this issue, researchers have turned to continuous system monitoring and network traffic packet analysis. However, this method can be resource-intensive and time-consuming, particularly when analyzing encrypted packets. In this study, the JA3 fingerprint infrastructure was examined as a potential solution for quickly detecting attacks conducted over encrypted sessions while minimizing system downtime and damage. The results demonstrated that the JA3 infrastructure effectively detected attacks carried out via encrypted channels. Although Windows 10 and Kali 2020.4 operating systems were used as the victim and attacker systems respectively, the methodology can be applied to other operating systems and network hardware by following the outlined steps. This research is expected to make a significant contribution to the field of encryption-based attack prevention.

Keywords: JA3 · Continious Monitoring · TLS · Cyber Security · Network · Cryptography · Finger Print

1 Introduction

The JA3 fingerprinting method was developed by John Althouse, Jeff Atkinson, and Josh Atkins at Salesforce in 2017 [1]. The method is based on Lee Brotherson's work in this field and the FingerprintTLS program [2]. JA3 is currently the most widely adopted TLS fingerprinting algorithm [3], and it has broad industry support within the context of network monitoring systems [4–6]. JA3 is particularly preferred in networks where high-speed traffic is desired due to its predictable fingerprint format and fixed fingerprint size.

Z. Şen et al. (Eds.): IMSS 2023, LNME, pp. 622–629, 2024.
https://doi.org/10.1007/978-981-99-6062-0_57

JA3 generates a fingerprint using the fields of network traffic packets indicated in Fig. 1:

- TLS version
- Cipher
- Extensions
- Elliptic curve
- Elliptic curve format

TLS Alanı	TLS Versiyonu	Şifre	Uzantılar	Eliptik Egri	Eliptik Egri Formatı
Onluk Deger (Deger tanımlı degilse boş bırakılmalı)	769	47-53-5-10-49161-49162-49171-49172-50-56-19-4	0-10-11	23-24-25	0
MD5 Özeti (JA3 Parmak İzi)	ada70206e40642a3e4461f35503241d5				

Fig. 1. Creating the JA3 fingerprint for a TLS connection

If any of these fields are missing, they are replaced with an empty string. For example, the following is a valid JA3 fingerprint string with missing Elliptic Curve and Elliptic Curve Format fields, where they are replaced with a dash:

$$769, 4-5-47-51-50-10-22-19-9-21-18-3-8-20-17-255, , ,$$

The JA3 fingerprinting standard also hashes the strings with the MD5 hashing algorithm. This makes the strings portable and easily shareable across different software applications. The hash value for this particular JA3 fingerprint string would be b677934e592ece9e09805bf36cd68d8a.

Furthermore, when a similar algorithm is used to generate fingerprints on the server-side, it can be observed in Fig. 2 that the JA3 fingerprints are largely dependent on the operating system family and version [7, 8].

The advantages of the JA3 and JA3s fingerprinting methods are as follows:

1.1 Lightweight

Although the cost of analyzing encrypted packets is very high, the JA3 fingerprinting method can be used for security analysis in a lighter and cheaper way than packet analysis because it does not concern itself with the content of the packets. The security performance of this method is particularly higher in static networks with IoT/IIoT devices [9]. Client–server relationships for IoT/IIoT devices are unique and often unchanging. Because the less the encrypted session information on the system changes, the easier it will be to track and detect new fingerprints entering the system.

1.2 Prevention Against DNS Over HTTPS (DoH):

Due to security concerns arising from open DNS traffic, HTTPS over DNS (DoH), a new DNS encryption format where DNS requests are sent over HTTPS (port 443), has become widespread [10]. By using JA3 and JA3s results together, web pages that serve from the same server for DNS traffic sent using TLS can be identified, and thus, attacks on DNS (such as DNS spoofing, DNS poisoning) can be prevented.

1.3 Fast Response Time

Thanks to the JA3 fingerprinting method in current communication over HTTPS, the fingerprint that is independent of content reveals the encryption algorithm, operating system, and hardware information (firmware), and thus, early intervention can be provided quickly as suspicious devices are alerted unless the newly added system or hardware is added to the table of relevant units.

2 Literature

Research paper by Karaçay, L and et al., discusses a protocol for privately evaluating detection models on system data while protecting the privacy of both the system data and detection models. The main approach is to provide end-to-end encryption for the system data and detection models utilizing lattice-based cryptography that allows homomorphic operations over cipher text [11]. Fu, Z. et al., on the other hand introduces a dataset that contains encrypted synthetic attacks and benign traffic. This dataset conforms to two requirements: the content requirements, which focus on the produced dataset, and the process requirements, which focus on how the dataset is built [12].Similar to Karaçay M. Shen et al., discusses traffic analysis and how it can be used to monitor network activities, discover specific patterns, and glean valuable information from network traffic. The article provides a comprehensive survey of machine learning-powered encrypted network traffic analysis [13].

3 Analysis

In this study, the JA3 fingerprinting method was used to passively detect anomalies in encrypted network traffic of a continuously monitored network. The stages of encrypted connection attack detection and prevention analysis, as illustrated in Fig. 2, were followed in this study. Firstly, a network scan was performed to detect the operating systems (Windows, Linux, IoS, etc.) and/or devices (IoT, IIoT, PLC, etc.) on the network. Windows 10 operating system was defined as the target victim device in this study. In the second stage, a malicious implant that establishes encrypted reverse connection from the attacker device to the target operating system was placed, and the attacker was granted access to the encrypted connection. After successfully observing the flow of target system information to the attacker through the connection, the final stage of attack detection and prevention was initiated. In this stage, the JA3 and JA3s fingerprint values of the TLS packets obtained over encrypted network traffic were analysed.

As a result of the analysis, it was determined that the encrypted connection was a "Meterpreter" session. Finally, the current attack was prevented and the system functionality and continuity were ensured.

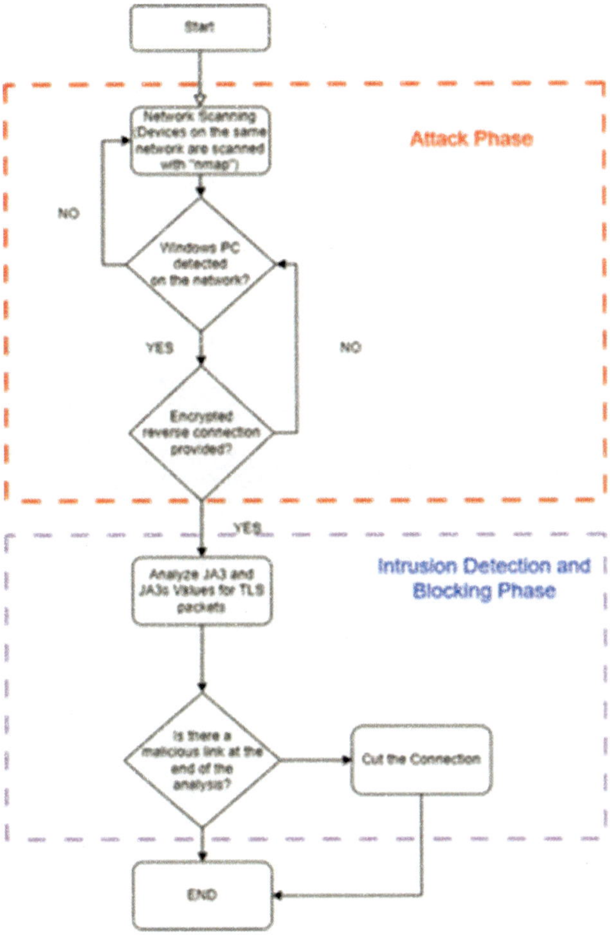

Fig. 2. Stages of Attack Scanning, Detection, and Prevention Flow Diagram

The proposed method is based on identifying anomalous SSL/TLS connections that use different certificates from those used in the continuously monitored network. The method was evaluated on a dataset of both malicious and normal network traffic. The JA3 fingerprinting method was preferred for the analysis due to its high accuracy in detecting malicious samples developed by the most dangerous attacker groups and its very low false-positive rate on legal encrypted traffic.

In this study, the Arkime graphic interface was used for continuous monitoring of encrypted network session (TLS) traffic. Arkime, also known as Moloch, is an open-source web-based network analysis tool that uses Elasticsearch database to examine, search, analyze, and export network traffic data. As seen in Fig. 3, the Arkime interface allows security experts to visually monitor traffic for various characteristics in real-time and perform various queries (e.g. tls.ja3 = = EXISTS!) on the data.

In the scenario created for the purpose of detecting attacks carried out over encrypted sessions, the attacker injected malicious software into the victim Windows 10 operating system via Kali 2020.4 operating system. The malicious software used to establish the connection via Meterpreter was prepared as a TLS encrypted Reverse Shell. As revealed by the examination conducted through the Arkime application, the client (192.168.163.131/Windows 10) was connected to the attacker (192.168.163.130/Kali 2020.4) via TLS v1.2 over port 4444, as shown in Fig. 3

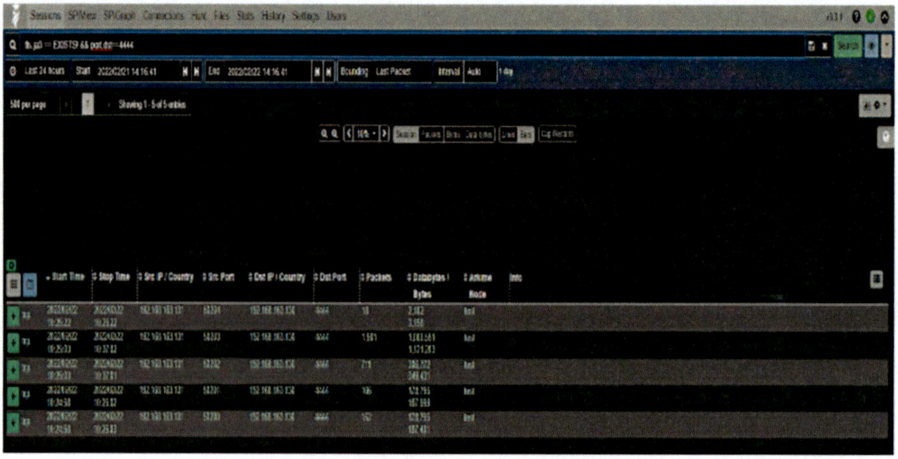

Fig. 3. Analysis of TLS network traffic

As a reference, the unique combination of Kali 2020.4 Meterpreter and Windows 10 TLS domain connection used in the study is provided below (Fig. 4):

JA3=> 72a589da586844d7f0818ce684948eea (Windows 10)

JA3s=> ec74a5c51106f0419184d0dd08fb05bc (Kali 2020.4)

Fig. 4. Meterpreter and Windows 10 TLS domain connection used in the study

During the attack analysis phase of the study, an attack was carried out on the Windows 10 system with the Kali 2020.4 attacker machine as seen in Fig. 5, and the resulting JA3 and JA3s fingerprint values were formed as indicated below. By combining

JA3 and JA3s into a fingerprint, false positive values were significantly reduced, thereby reducing the likelihood of security managers receiving false alarms.

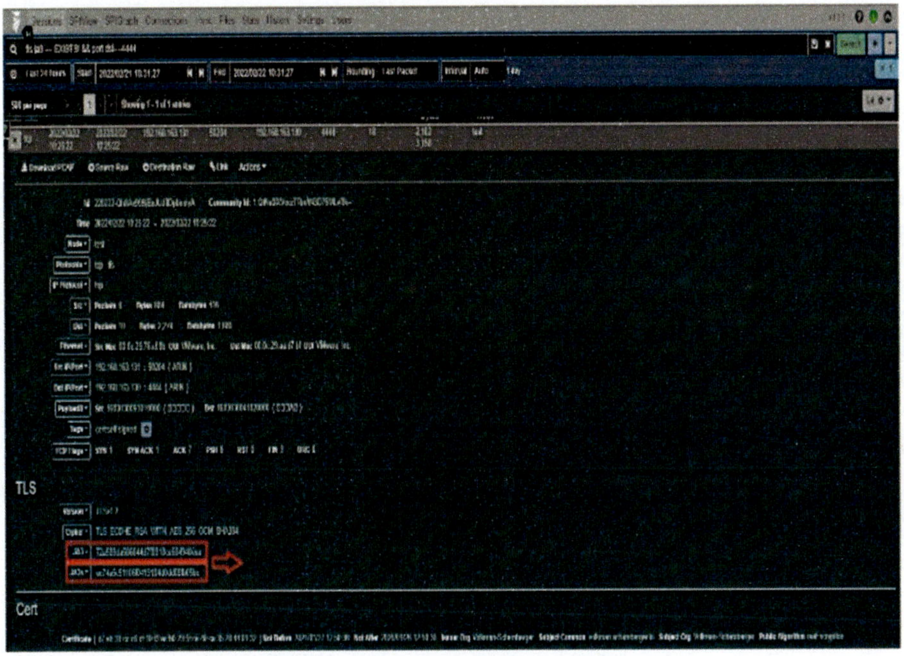

Fig. 5. JA3 + JA3s Fingerprint Analysis

During the attack analysis phase, the JA3 fingerprint value for the attack was obtained using TLS values. This fingerprint was quickly detected and prevented.

4 Discussion and Implementations

When the results obtained during the analysis phase of the study are evaluated in general, it is seen that the use of JA3 fingerprinting makes a significant contribution to the detection and prevention of illegal connections and attacks performed based on the encryption infrastructure of the existing HTTPS/TLS infrastructure without causing any damage to the systems. In the testbed created within the scope of the study, a reverse shell was obtained on the victim computer as a result of the attack carried out from the attacker computer, and an encrypted transmission channel was created from the victim computer to the attacker computer by adding the TLS infrastructure. Thus, even if the network traffic was examined, the flowing traffic could not be instantaneously interpreted due to being encrypted. In order to detect such attacks, the JA3 fingerprinting infrastructure was used in the testbed and thus, the detection of attacks carried out in an encrypted manner was provided. In the detection phase of the attacks, as stated within the analyses, the detection and prevention of all similar attacks that can be performed on components with known hardware and system properties on the network using the same method will also be provided.

5 Conclusion

With the advancements in science and technology, our daily lives, including education, healthcare, shopping, and banking, are rapidly transitioning to the digital world. The increased importance of privacy and confidentiality in this process has led to the widespread use of encryption. However, this has also led to the emergence of new methods for cyber-attacks, where encryption algorithms are used to conceal attacks and make them more complex to detect and mitigate. Moreover, the lack of consensus among countries in the field of cyber law and the relatively low penalties for cyber-crimes compared to physical crimes have facilitated the transition of attackers to the virtual world.

In this context, this study aimed to detect and prevent attacks carried out using encryption algorithms, using entirely open-source tools and a test environment. The JA3 fingerprint method was used for the detection phase of the analysis. The results showed that attacks carried out using malware that utilizes encryption algorithms could be detected and prevented successfully and in a short time frame. While the Windows 10 operating system was chosen as the target system in this study, the steps outlined in the analysis phase can be applied to detect and prevent encrypted malicious connections on other operating systems and network devices.

References

1. Althouse, J.: Open Sourcing JA3 SSL/TLS Client Fingerprinting for Malware Detection. Şubat, Erişim (2022). https://engineering.salesforce.com/open-sourcing-ja3-92c9e53c3c41
2. Althouse, J., Atkinson, J. ve Atkins, J.: "Salesforce/ja3", Şubat, Erişim (2022). https://github.com/salesforce/ja3
3. Eric, R.: RFC 8446 – The Transport Layer Security (TLS) Protocol Version 1.3. Şubat, Erişim (2022). https://datatracker.ietf.org/doc/rfc8446/.
4. Kane, W.A., Vlach, T., Luks, R.: Encrypted traffic analysis. flowmon, 2019D. In: Sarunyagate (ed.), Lasers. McGraw-Hill, New York (1996)
5. Suricata. 6.17. "JA3 Keywords — Suricata 6.0.4 documentation.", Şubat 2022, Erişim: https://suricata.readthedocs.io/en/suricata-6.0.4/rules/ja3-keywords.html.K. Schwalbe, Information Technology Project Management, 3rd ed. Course Technology, Boston (2004)
6. Telesoft. JA3 Fingerprinting: Encrypted Thread Detection. (2020)
7. Matoušek, P., Burgetová, I., Ryšavý, O., Victor, M.: On reliability of JA3 hashes for fingerprinting mobile applications. In: Goel, S., Gladyshev, P., Johnson, D., Pourzandi, M., Majumdar, S. (eds.) Digital Forensics and Cyber Crime: 11th EAI International Conference, ICDF2C 2020, Boston, MA, USA, 15–16 Oct 2020, Proceedings, pp. 1–22. Springer International Publishing, Cham (2021). https://doi.org/10.1007/978-3-030-68734-2_1
8. Hejcman, L.: Fingerprinting and Identification of TLS Connections.
9. Deri, L., Fusco, F.: Using deep packet inspection in CyberTraffic analysis. In: 2021 IEEE International Conference on Cyber Security and Resilience (CSR), pp. 89–94. IEEE (2021)
10. Choi, J.S.: DNS over HTTPS (DoH) (2021)
11. Karaçay, L., Savaş, E., Alptekin, H.: Intrusion detection over encrypted network data. The Comput. J. **63**(4), 604–619 (2019). https://doi.org/10.1093/comjnl/bxz111

12. Fu, Z., et al.: Encrypted malware traffic detection via graph-based network analysis. In: Proceedings of the 25th International Symposium on Research in Attacks, Intrusions and Defenses, pp. 495–509. Presented at the Limassol, Cyprus (2022). https://doi.org/10.1145/3545948.3545983
13. Shen, M., et al.: Machine learning-powered encrypted network traffic analysis: a comprehensive survey. IEEE Commun. Surv. Tutorials **25**(1), 791–824 (2023). https://doi.org/10.1109/COMST.2022.3208196

Blockchain Enabled Lateral Transshipment System for the Redistribution of Unsold Textile Products in a Circular Economy

Hatice Büşra Gökbunar[1] and Banu Soylu[2(✉)]

[1] Graduate School of Natural and Applied Sciences, Erciyes University, Kayseri, Turkey
[2] Industrial Engineering Department, Erciyes University, Kayseri, Turkey
bsoylu@erciyes.edu.tr

Abstract. With the rise of carbon footprint awareness and the circular economy, the need to improve supply chain traceability and visibility has increased. Traditionally, companies have used database-driven platforms to improve traceability, reuse, and reduce waste. In practice, lateral transshipment among stakeholders also aims to improve reuse and reduce waste by redistributing unsold products. Although the optimization issues are well-studied in this field, there are many challenges in sharing data about ingredients, processes, delivery stages, and the lifecycle of a product. Manufacturers, retailers, and consumers have only limited access to this information. Blockchain technology can help to manage the complexities of managing a circular supply chain, especially in the context of lateral transshipment. This study presents a framework for blockchain and smart contract-enabled lateral transshipment among fast fashion retailers. There are three main components of the framework: inventory management system by the organizer, introducing retailers' excess inventory and orders into the inventory sharing pool, delivery with execution of smart contracts. The proposed framework not only contributes to the visibility and traceability of the system but also increases responsiveness, flexibility and efficiency.

Keywords: Blockchain · inventory sharing · smart contracts · closed-loop supply chain · sustainability

1 Introduction

The textile industry is one of the most polluting manufacturing industries [1], which can be reduced by transitioning to a circular economy. All textile manufacturing processes are highly water-intensive, using an average of 93 billion cubic meters of water per year, corresponding 4% of global freshwater withdrawals. The dyeing, bleaching, washing and treatment of textiles has a huge environmental impact, including the oil used to produce synthetic fibers, the fertilizers used to grow cotton, the chemicals used to manufacture, dye and finish fibers and textiles, and the huge number of accessories and packaging used to cover them [2]. The textile industry also pollutes the oceans with microfibers found in textile waste. The amount of textile waste is increasing on a global scale, and it has

Z. Şen et al. (Eds.): IMSS 2023, LNME, pp. 630–640, 2024.
https://doi.org/10.1007/978-981-99-6062-0_58

been suggested that recycling or reusing textile products could reduce new waste from virgin materials [3]. In addition, the annual carbon footprint of the fashion industry's product lifecycle (3.3 billion tons of CO2 emissions) is almost equal to the carbon footprint of the 28 countries/regions of the EU (3.5 billion tons) [4]. This highlights the importance of exploring solutions to improve the textile and apparel supply chain for greater sustainability and circularity.

Transitioning to circular supply chains in the textile industry means developing an entire complex supply chain, including design, materials, production, distribution, reverse logistics, collection, sorting and recycling to achieve "closed loop" supply chain. To this end, the 9Rs (refuse, rethink, reduce, reuse, repair, refurbish, remanufacture, buy back, recycle, recover) have recently been proposed [5]. Figure 1 illustrates these elements from a holistic perspective. The textile industry faces challenges related to the efficient use of resources and the management of waste. The concept of the circular economy has emerged as a solution to these challenges, aiming to minimize the consumption of natural resources, reduce waste and promote the reuse of materials. In addition, all stakeholders have a significant impact on circular economy activities in the textile industry, such as fiber producers, yarn manufacturers, fabric manufacturers, garment manufacturers, designers, retailers, end users, collectors and recyclers. In a typical circular textile supply chain, collectors and recyclers collect used or unused textile waste for reuse or recycling, sort it by color and material type, and remove accessories such as staples and zippers.

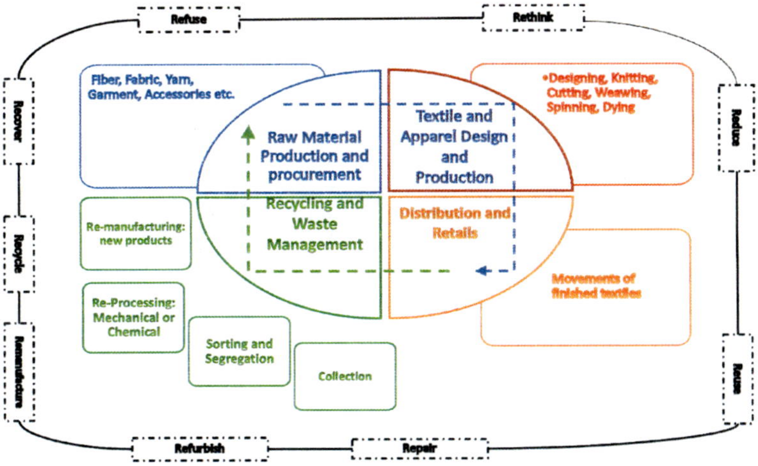

Fig. 1. The elements of typical circular textile supply chain

Although recycling is one of the functional elements of circularity, the complexity of these processes during collection, sorting and recycling makes it the main barrier in the transition to a circular textile supply chain [6]. Before this complex stage, we can promote sustainability by intervening in the distribution channel. By focusing on the movement of unsold products between retailers, we can reduce unnecessary waste and loss of resources. In line with this theme, this study proposes to strengthen the effectiveness of the redistribution channel of unsold products with lateral transshipment through the cooperation of retailers to achieve more circularity. In this way, a supply chain system can become more agile, responsive, and profitable by increasing the ability to transfer excess inventory from one member to the scarce member.

A well-designed lateral transshipment system promotes the textile and apparel supply chain circularity [7]. However, the effective implementation of lateral transshipment practices in the textile industry is challenged by the lack of cooperation and communication between retailers. To cover these deficiencies, the blockchain technology can provide transparency and traceability in lateral transshipment by creating a secure and immutable database that records all inventory events and transactions. Therefore, we propose a blockchain-enabled lateral transshipment to improve the sustainability and efficiency of the circular economy. As highlighted by researchers, blockchain applications have recently begun to revolutionize various aspects of supply chain and operations management for the development of real-time capabilities [8]. The proposed framework contributes to the literature on blockchain applications in the circular textile supply chain, specifically intervening in the distribution and retailer stage through the design of smart contracts.

The rest of the paper is organized as follows. In Sect. 2, we review the lateral transshipment topic and the related literature. In Sect. 3, we review the blockchain and smart contract applications in circular supply. In Sect. 4, we present the blockchain-enabled lateral transshipment (BELT) framework and implementation details. Finally, we give the conclusions and future work in Sect. 6.

2 Lateral Transshipment in Supply Chain

Lateral transshipment within an inventory system can be conducted periodically at predetermined points in time to proactively redistribute stock, or reactively to meet demand. Transfers are allowed between stock points of the same echelon and can be either actual products or on demand [7]. Two main branches of literature on lateral transshipment can be identified: *proactive transshipment* and *reactive transshipment*. Proactive transshipment is a strategic way of redistributing inventory among points at the same echelon. This branch typically focuses on demand forecasting and inventory optimization models. In particular, the focus on timing ensures that everything is in the right place at the right time. On the other hand, considering more dynamic, uncertain and response-oriented systems like textile and apparel supply chain, reactive transshipment is used when one stock point lacks stock while another has sufficient stock [9]. The implementation of lateral transshipment not only minimizes the need for new production, but also reduces the risk of stock-outs, balances inventory levels, and increases responsiveness [7].

Several studies have investigated the potential of lateral transshipments in contributing to the circular economy. Nakandala et al. [10] proposed a model for perishable products to improve the deterioration and other related costs. The results show that lateral transshipment is a well-performing strategy to minimize the total cost of decision rules in cost scenarios. Ekren et al. [11] used a simulation technique to compare the performance of the food supply chain where the lateral inventory sharing strategy is applied. Momenitabar et al. [12] demonstrated that the concept of lateral resupply and backup supplier could significantly reduce the total cost and reduce shortages under demand uncertainties. The study proposes a fuzzy multi-objective mixed-integer linear programming model to optimize total and environmental costs, including energy consumption and pollutant emissions, while increasing job opportunities for designing an efficient sustainable closed-loop supply chain network toward a circular economy.

The textile industry faces challenges in collaboration and communication among stakeholders, which hinders the effective implementation of circular economy practices [6]. The implementation of lateral transshipment requires the collection, sharing, and analysis of data from various sources. However, data-related challenges, such as data access, ownership, sharing, and privacy, can hinder the effective implementation. Blockchain technology and smart contracts are seen as a solution to these problems.

3 Blockchain Technology in Circular Supply Chain

Blockchain technology, a decentralized and distributed ledger system that enables secure and transparent data sharing, has the potential to address some of the aforementioned challenges. By using blockchain technology, stakeholders can have access to a secure and immutable database that records all transactions and events in the system. This can provide transparency and traceability, which are essential for building trust among stakeholders [13]. Blockchain technology solutions for the implementation of circular supply chain and also in the context of lateral transshipment are as follows.

Reliability. Blockchain technology offers reliability and security by creating a decentralized network that ensures no single entity has complete control over the network. This means that the data on the blockchain is immutable and cannot be changed without consensus from the network participants. This makes data immutable and tamper-proof, eliminating the risk of fraud or human error. Blockchain records all relevant information about the supply chain, such as origin and stakeholders involved, and makes it visible to all stakeholders with individual access rights. This level of transparency, verifiability, and authenticity allows companies to track imports and exports with a new level of peer-to-peer interaction [14].

Transparency. Transparency is becoming one of the fundamental keys to improving supply chain performance. Blockchain has the potential to increase the transparency of the system, and greater transparency can increase productivity, provide better service to customers, and reduce costs [15].

Blockchain provides transparency by creating a decentralized ledger that allows all parties to view and track the transfer of goods in real time. This transparency fosters trust between stakeholders, reduces the likelihood of disputes, and ensures that all parties are working toward the same goal. It also ensures that the transfer of goods is visible to all parties involved, reducing the risk of lost or stolen items. The use of blockchain technology as an infrastructure provides a robust and integrated system for recording, storing, and disseminating relevant information. Blockchain technology enables comprehensive, transparent, and reliable material traceability and information tracking not only of a material's source but also of its current state [16].

Traceability. Blockchain allows goods to be tracked throughout the supply chain. Every transaction on the blockchain is recorded in a secure and transparent manner, allowing stakeholders to trace the origin and movement of goods from one supply chain to another. Blockchain enables secure information sharing, monitoring, real-time data collection, accountability and traceability across the supply chain. This technology can also be used to monitor and maintain the performance of the built environment and manage the lifecycle of a product from production to end-of-life. It can also be used to track the environmental impact of a product, ensuring that companies meet their sustainability goals [16].

In a circular economy, stakeholders may have invested in inventory or other assets that do not generate revenue or income. This capital is tied up and cannot be used for other business purposes until the inventory or asset is sold. Blockchain allows companies to share and track the flow of products in a circular economy network, and hence can address the challenge of capital tied up. This also provides feedback for manufacturing strategies [17].

Authenticity. Blockchain provides authenticity by using digital signatures and cryptographic hashes to ensure that the data on the blockchain is authentic and cannot be altered [18].

Trusted Agreements. Blockchain can create smart contracts that automate the transfer of goods between stakeholders in supply chain. These contracts are self-executing and ensure that all parties involved agree to the terms of the transfer. This eliminates the need for intermediaries and ensures that the transfer of goods is secure and transparent. These contracts can also execute rules such as limiting the transfer amounts between retailers, or may return alerts to warn the management center.

3.1 Smart Contracts

A smart contract is software that can be executed by all consensus nodes on the blockchain. Basically, it contains program code, files and account balances. On the blockchain, any user can post a transaction to create a contract. Once a contract is created, the program code cannot be modified [19]. The program logic of a contract is executed by updating the blockchain accordingly. The contract's code is executed each time it receives a message, either from a user or from another contract. During the execution of its code, the contract can read from its memory file or write to its memory file. A contract can also receive money into its account balance, and it can send money to other contracts or to other users [20].

Reducing the risk of fraud, providing more secure and transparent transactions, and reducing the costs associated with manual verification and paperwork are some of the benefits of smart contracts. Confidence in documentation can lead to a reduction in waste, risk and insurance premiums [21].

Conversely, the disadvantages of smart contracts include the potential for errors and bugs in the code, the lack of a legal framework, and the potential for abuse by bad actors. However, despite these potential drawbacks, smart contracts have great potential to improve the efficiency, security, and reliability of transactions across a wide range of industries.

3.2 Advantages of Smart Contracts

Some advantages of smart contracts and implementation details are explained as follows.

Automation. Smart contracts automate the execution of contracts, eliminating the need for intermediaries and reducing the risk of errors or disputes [14].

Trust. Smart contracts operate on a blockchain network, providing a transparent and immutable record of all transactions between stakeholders. Smart contracts allow information to be shared, ensure responsibility is assigned, and reduce business costs. The verification that the process is direct and trustworthy between the participants, in turn, guarantees the automation of the agreement, the fight against fraud and the malfunctioning of traditional third parties, and avoids the uncertainties associated with system failures due to manipulation or fraudulent attacks [22].

Efficiency. Smart contracts automate many of the processes involved in transferring textile products, reducing the time and resources required to complete transactions [23].

To make efficient and effective business decisions, blockchain technology enables an almost unique real-time tracking system that allows status data to be updated in a timely and automated manner. Smart contract data on the private blockchain can provide companies with proof of what has been agreed, and this benefit could encourage companies to add their orders and contracts at an early stage, allowing other parties to have the data earlier and start forecasting earlier in the process. Stakeholders can use this data to improve their planning and reduce the bullwhip effect [24, 25].

Security. Smart contracts are stored on a distributed blockchain network, providing enhanced security and reducing the risk of data breaches and hacks [20].

3.3 Disadvantages of Smart Contract

Some disadvantages that are commonly encountered in applications are explained as follows.

Complexity. Smart contracts can be complex and require significant technical expertise to develop and maintain. On the other hand transactions take a certain amount of time in network bottlenecks. In addition, a number of confirmation blocks are typically recommended to ensure that the transaction is not accidentally or intentionally removed. So, delay is also an issue [26].

Furthermore, due to the uncertain environment and the complexity of operating a smart contract, make it difficult to create certainty through technological means. Smart contracts are more complex than traditional contracts [27].

Inflexibility. Smart contracts are self-executing and cannot be modified once deployed. This can be a disadvantage if the contract needs to be revised or updated [19].

Legal Challenges. Even in academic publications, smart contracts are seen as a technological solution to commercial and legal uncertainty, but they are still relatively new and there are legal challenges related to their enforceability and validity [19, 27].

3.4 Blockchain and Smart Contract in Lateral Transshipment

In the context of the intersection of blockchain and lateral transshipment, blockchain technology can provide transparency and traceability in lateral transshipment by creating a secure and immutable database that records all transactions and events in the distribution channel. This transparency and traceability can help build trust among stakeholders by providing a clear and reliable source of information about the product or shipment. This can be particularly important in lateral transshipment, where goods move between multiple parties [28].

Each transaction is stored in a block that is linked to the previous block, creating a chain of blocks. This allows stakeholders to see record of the entire history of a product or shipment, from its origin to its final destination, without relying on a trusted third party [29].

The role of smart contracts in blockchain-enabled lateral transshipment in a circular textile supply chain is to automate and enforce the rules and conditions for the transfer of textile products between retailers. For example, the contract can be automatically executed to propose the (some portion of) inventory to an inventory sharing pool when the circulation rate of the product falls below a pre-defined level. Or it can limit the amount of inventory pulled from the pool based on the circulation rate.

4 BELT Framework

The proposed framework, called the Blockchain Enabled Lateral Transshipment (BELT) framework, uses distributed ledger technology to securely share textile and apparel inventory information among retailers and distribution centers. We refer to retailers and distribution centers as nodes. The dual-ledger system allows nodes to keep their own records securely and also share some information regarding demand and supply among nodes. Figure 2 shows the main components of the BELT framework and the interactions between nodes. Here, the distribution center coordinates all retailers from product assortment to store and shelf design. Therefore, inventory sharing can be easily performed between retailers without needing money transfer since all inventories are actually owned by the distribution center. However, retailers are responsible for tracking their sales, transfers and turnover rates and may propose to submit inventory to the pool.

The key elements of the framework are as follows.

Private ledger. It allows individual nodes to evaluate their own protected data, current inventory, past usage and turnover ratio of each product. Accordingly, the node determines its own inventory needs or excess.

Public Consortium Ledger. It contains aggregate information publicly available to all nodes in the lateral transshipment network. The inventory information shared by the nodes in the pool can be tracked from this ledger. All nodes have secure access to this part of the system, and they are able to fulfill their demand from this pool, or they can submit their excess inventory into the pool. The ledger keeps track of inventory information as well as physical transactions between nodes.

Smart Contract. It is a self-executable script designed with the information gathered from ledgers to ensure rules, procedures and protocols. Accordingly, it can automatically

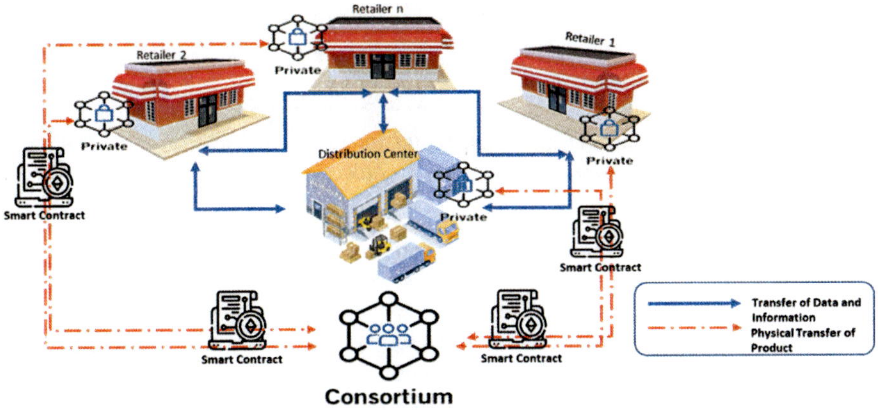

Fig. 2. Interactions between nodes of the proposed BELT framework

trigger all transfers in the workflow. There are two types of smart contracts in the BELT framework: When a retailer has excess inventory, it is submitted to the pool using an *excess smart contract* and according to its content. So, they are available to all retailers. On the other hand, when a retailer needs something from this inventory, he uses the *demand smart contract* to fulfill his need. The data recorded on the blockchain can be used to track the transfer route of a product and to decide whether it is outdated or not. In addition, the optimal distribution plan for physical transfers can be prepared based on the daily confirmed contracts.

A private ledger-based inventory management system can be used to analyze inventory levels, demand, and sales trends. This data can be used to optimize the supply chain by identifying opportunities to reduce costs, improve responsiveness, and increase customer satisfaction. In addition, all distribution nodes upload information to the public consortium ledger. The private ledger blockchain-based system provides a mechanism that is automatically executed when inventory levels in stores fall below a pre-defined threshold, or when the unsold period of a product exceeds a certain time threshold. Once the product has been demanded by the retailer, the public ledger is checked to see if the demand is fulfilled before the product is ordered from the manufacturer. So, this contributes to sustainability by encouraging reuse. The proposed BELT framework emphasizes the definition of an efficient textile and apparel product redistribution strategy for the distribution network, which is reviewed and approved by the nodes. Regarding the framework, a distribution node labels excess inventory and updates the public consortium ledger, and a demand node pulls the scarce products from the pool of available products, then the lateral transhipment is initiated. All these events are continuously monitored.

The BELT framework allows you to track and dynamically respond to discrepancies and evolving operating conditions, reducing inaccurate or incomplete deliveries while limiting excess inventory and ultimately reducing out-of-date inventory. The implementation details are further explained below.

Implementing Access Control. You can use blockchain-based access control mechanisms to ensure that only authorized parties have access to the framework. This is done by using smart contracts that define roles and permissions for each participant.

Defining the asset and its attributes: Define the asset being transshipped and its attributes, such as its product ID, quantity, expiration date, and location. This is done by using a smart contract that defines the asset's structure.

Creating an Automatic Trigger and Confirmation Mechanism for Excess Products. The smart contracts include an upper limit on the amount that each store can pull from the pool according to its own turnover ratio. This also reduces the number of times a product is labeled as excess in the pool, thereby reducing unnecessary reallocation costs. The public consortium ledger can also be used to track the history of each transfer, including the source and destination of the product. After the products are redistributed, their inventory records are updated in real-time on the blockchain. All the transaction details are recorded on the blockchain, creating a transparent and auditable information sharing pool. This information can be used to track product movements, monitor inventory levels, and optimize supply chain processes.

Implement Rules and Regulations. To ensure that the framework operates fairly and transparently, you can implement rules and regulations using smart contracts. For example, a feedback tool can define rules about the correctness of the quality and quantity of products transshipped, the adequacy of the response time, and the penalties for retailers who do not comply with the rules.

5 Conclusion

In this paper, we have discussed the benefits of blockchain technology applied to lateral transshipment in supply chain distribution operations. The proposed blockchain-based lateral transshipment (BELT) framework aims to improve the redistribution channel of unsold products through lateral transshipment, using blockchain technology and smart contracts. The framework includes three key components: the inventory management system based on private ledger, the lateral transshipment platform driven by public consortium ledger, and the smart contracts. The inventory management system on private ledger allows retailers to monitor their inventory levels and identify excess inventory or demand. The public ledger facilitates communication and coordination between retailers for the exchange of excess inventory. The smart contract system automates the exchange process and ensures that all parties abide by the agreed terms. By implementing the BELT framework, the unsold products are dynamically reallocated to reduce outdates and shortages by creating new solutions for inventory routing problem. It will significantly facilitate decision making in complex networks while maintaining privacy of information sharing and networking by using a dual ledger structure (private ledgers at each node and a single public consortium ledger). Thus, retailers can reduce unnecessary waste and loss of resources by transferring excess inventory to other retailers who need it. This promotes circularity in the textile supply chain by extending the lifespan of products and reducing the need for new production. Additionally, the use of blockchain technology and smart contracts provides transparency and traceability in the exchange process, which enhances trust and reduces the risk of fraud or error. In conclusion, the BELT framework has the potential to improve the sustainability and efficiency of the circular textile supply chain by promoting lateral transshipment and using blockchain technology and smart contracts to enhance transparency and traceability. Overall, this paper highlights the importance of innovative and collaborative approaches to address the environmental challenges facing the textile industry and to achieve a more sustainable future. Furthermore, the methodological developments of this work can be readily applied to similarly complex and demanding supply chain networks, such as vaccines, spare parts, construction, food pantries or stores, and other applications that depend on effective management of limited perishable resources in a disaster or crisis situation.

References

1. Jia, F., Yin, S., Chen, L., Chen, X.: The circular economy in the textile and apparel industry: a systematic literature review. J. Cleaner Product. **259**, 120728 (2020)
2. MacArthur, E.: Beyond plastic waste. Science **358**(6365), 843–843 (2017)
3. Dahlbo, H., Aalto, K., Eskelinen, H., Salmenperä, H.: Increasing textile circulation—Consequences and requirements. Sustain. Product. Consumption **9**, 44–57 (2017). https://doi.org/10.1016/j.spc.2016.06.005
4. House of Commons Environmental Audit Committee: Fixing Fashion: Clothing Consumption and Sustainability. House of Commons, London (2019)
5. Potting, J., Hekkert, M.P., Worrell, E., Hanemaaijer, A.: Circular economy: measuring innovation in the product chain. Planbureau voor de Leefomgeving (2544) (2017)
6. Kazancoglu, I., Kazancoglu, Y., Kahraman, A., Yarimoglu, E., Soni, G.: Investigating barriers to circular supply chain in the textile industry from Stakeholders' perspective. Int. J. Logistics Res. Appl. **25**(4–5), 521–548 (2022)
7. Paterson, C., Kiesmüller, G., Teunter, R., Glazebrook, K.: Inventory models with lateral transshipments: a review. Eur. J. Operat. Res. **210**(2), 125–136 (2011)
8. Dolgui, A., Ivanov, D., Potryasaev, S., Sokolov, B., Ivanova, M., Werner, F.: Blockchain-oriented dynamic modelling of smart contract design and execution in the supply chain. Int. J. Product. Res. **58**(7), 2184–2199 (2020)
9. Burton, J., Banerjee, A.: Cost-parametric analysis of lateral transshipment policies in two-echelon supply chains. Int. J. Product. Econ. **93**, 169–178 (2005)
10. Nakandala, D., Lau, H., Shum, P.K.: A lateral transshipment model for perishable inventory management. Int. J. Product. Res. **55**(18), 5341–5354 (2017)
11. Ekren, B.Y., Eroglu, E., Kazancoglu, Y., Kumar, V.: Lateral inventory share based business model for IoT enabled sustainable food supply chain network. In: Proceedings of the 5th NA International Conference on Industrial Engineering and Operations Management, IEOM Society, Detroit, MI, USA, pp. 9–11 (2020)
12. Momenitabar, M., Dehdari Ebrahimi, Z., Arani, M., Mattson, J., Ghasemi, P.: Designing a sustainable closed-loop supply chain network considering lateral resupply and backup suppliers using fuzzy inference system. Env, Dev. Sustain. 1–34 (2022)
13. Wang, B., Luo, W., Zhang, A., Tian, Z., Li, Z.: Blockchain-enabled circular supply chain management: a system architecture for fast fashion. Comput. Ind. **123**, 103324 (2020)
14. Kouhizadeh, M., Sarkis, J., Zhu, Q.: At the nexus of blockchain technology, the circular economy, and product deletion. Appl. Sci. **9**(8), 1712 (2019)
15. Zelbst, P.J., Green, K.W., Sower, V.E., Bond, P.L.: The impact of RFID, IIoT, and Blockchain technologies on supply chain transparency. J. Manuf. Technol. Manag. **31**(3), 441–457 (2020)

16. Shojaei, A., Ketabi, R., Razkenari, M., Hakim, H., Wang, J.: Enabling a circular economy in the built environment sector through blockchain technology. J. Cleaner Product. **294**, 126352 (2021)
17. Raj, P.V.R.P., Jauhar, S.K., Ramkumar, M., Pratap, S.: Procurement, traceability and advance cash credit payment transactions in supply chain using blockchain smart contracts. Comput. Indust. Eng. **167**, 108038 (2022)
18. Mougayar, W.: The Business Blockchain: Promise, Practice, and Application of the Next Internet Technology. John Wiley & Sons (2016)
19. Delmolino, K., Arnett, M., Kosba, A., Miller, A., Shi, E.: Step by step towards creating a safe smart contract: Lessons and insights from a cryptocurrency lab. In: Clark, J., Meiklejohn, S., Ryan, P.Y.A., Wallach, D., Brenner, M., Rohloff, K. (eds.) FC 2016. LNCS, vol. 9604, pp. 79–94. Springer, Heidelberg (2016). https://doi.org/10.1007/978-3-662-53357-4_6
20. Abeyratne, S.A., Monfared, R.P.: Blockchain ready manufacturing supply chain using distributed ledger. International Journal of Research in Engineering and Technology **5**(9), 1–10 (2016)
21. Saberi, S., Kouhizadeh, M., Sarkis, J., Shen, L.: Blockchain technology and its relationships to sustainable supply chain management. Int. J. Product. Res. **57**(7), 2117–2135 (2019)
22. Wang, Q., Lau, R.Y.K., Mao, X.: Blockchain-enabled smart contracts for enhancing distributor-to-consumer transactions. IEEE Consumer Electron. Mag. **8**(6), 22–28 (2019)
23. Casado-Vara, R., Prieto, J., De la Prieta, F., Corchado, J.M.: How blockchain improves the supply chain: case study alimentary supply chain. Procedia Comput. Sci. **134**, 393–398 (2018)
24. van Engelenburg, S., Janssen, M., Klievink, B.: A blockchain architecture for reducing the bullwhip effect. In: Shishkov, B. (ed.) BMSD 2018. LNBIP, vol. 319, pp. 69–82. Springer, Cham (2018). https://doi.org/10.1007/978-3-319-94214-8_5
25. Ivanov, D., Dolgui, A., Sokolov, B.: The impact of digital technology and Industry 4.0 on the ripple effect and supply chain risk analytics. Int. J. Product. Res. **57**(3), 829–846 (2019)
26. Mendling, J., et al.: Blockchains for business process management-challenges and opportunities. ACM Trans. Manag. Inform. Syst. **9**(1), 1–16 (2018)
27. Mik, E.: Smart contracts: terminology, technical limitations and real world complexity. Law Innov. Technol. **9**(2), 269–300 (2017)
28. Kumar, N.M., Chopra, S.S.: Leveraging blockchain and smart contract technologies to overcome circular economy implementation challenges. Sustainability **14**(15), 9492 (2022)
29. Ates, A., Ozsoy, A.: Blockchain based lateral transshipment. In: 2021 Third International Conference on Blockchain Computing and Applications (BCCA), pp. 49–56 (2021)

Automl-Based Predictive Maintenance Model for Accurate Failure Detection

Elif Cesur[1](✉), M. Raşit Cesur[1], and Şeyma Duymaz[2]

[1] Faculty of Engineering and Natural Sciences, Industrial Engineering, İstanbul Medeniyet University, İstanbul, Türkiye
rasit.cesur@medeniyet.edu.tr

[2] Faculty of Machinery, Industrial Engineering, Yildiz Technical University, İstanbul, Türkiye

Abstract. This study focuses on predictive maintenance, a critical maintenance policy that benefits from the development of the Digital Twin (DT) philosophy. To implement predictive maintenance, it is essential to predict potential failures. In this study, machine learning algorithms are used to detect failure conditions. Five different types of failures are classified by examining parameters such as air temperature, process temperature, rotation speed, torque, and tool wear. The study utilizes Automatic Machine Learning (AutoML), which runs machine learning algorithms and returns the best method, its hyperparameters, and many outputs, such as accuracy and performance metrics. The literature on machine learning algorithms in predictive maintenance has focused on finding the best algorithm by applying selected methods. However, this study aims to contribute to the literature by finding the algorithm that provides the best results among all methods using AutoML in predictive maintenance.

Keywords: AutoML · Predictive Maintenance · Digital Twin · Classification

1 Introduction

Failures that occur in unexpected situations are challenging for companies in many respects. Therefore, it is necessary to develop an effective maintenance policy for machines. With the development of the Digital Twin (DT) concept, which is the modeling of the behavior of physical products in real situations and transferring them to the virtual environment, predictive maintenance studies have been operating. Today, DT studies focus on predictive maintenance activities. DT applications can have enough processing power to detect errors, gain new insights, and even determine how to improve each part of the entire system by predicting its behavior under stressful conditions [1]. Companies must predict and prevent malfunctions in machines before they happen, to minimize disruptions in production, and reduce losses caused by malfunctions.

In maintenance planning, an estimation can be done to predict the failure and take precautions. Machine Learning (ML) is one of the most common applications used in industry for predicting failure. ML is a subset of artificial intelligence (AI) that can learn from and make predictions from data and focuses on algorithms. Algorithms such as

Z. Şen et al. (Eds.): IMSS 2023, LNME, pp. 641–650, 2024.
https://doi.org/10.1007/978-981-99-6062-0_59

Support Vector Machine, Random Forest, Naive Bayes, KNN, and Logistic Regression are machine learning algorithms.

Algorithms are used according to the characteristics of the data in the studies. It takes a large number of trials to find the algorithms and parameters that give the best results. As an alternative to this, there are some libraries in the literature to find the algorithm that gives the best result. Automated Machine Learning (AutoML) application has been developed as an important solution for this process. It is known as a system that automates the ML process. It tries ML algorithms with various parameters in the background and returns the best model and best parameter as output. Some of the open-source libraries are; Autokeras, TPOT, Autosklearn, and H2O are libraries. Automated feature selection, Model (Algorithm) Selection (Automatic Algorithm/Model selection), and Hyperparameter Optimization (Optimization of Hyperparameters) are performed with AutoML. In this way, it will be possible to achieve the best result and the acquisition time will be improved. This study aims to contribute to the literature by finding the algorithm that gives the best results among all the methods using AutoML in predictive maintenance.

The general structure of the article is as follows: Sect. 2 provides information about the available literature. Section 3 mentions the AutoML method used in the study. Section 4 makes predictions on machine failures on a dataset with 5 entries. The results are presented in Sect. 5.

2 Literature Review

Methods using predictive maintenance have been investigated in the literature. In general, studies have been carried out on fault detection with machine learning techniques. Machine learning techniques used in studies for predictive maintenance are 33% Random Forest (RF), 27% Artificial neural networks (ANN), 25% Support vector machine (SVM), and 13% K-means algorithms [2]. Panda et al. tackled the problem of reducing long downtimes of a large automotive system using a machine learning algorithm. ML models, three different algorithms such as C5.0, and C5.0 with boosting and classification and regression tree (CART) are used [3]. Santos et al. detected short-circuit faults in induction motors using RF. The results, which consist of two classes, indicate whether there is a short circuit or not [4]. Uhlmann et al. used the k-means algorithm to diagnose laser melting benches by paying attention to 3 factors (temperature, oxygen percentage, and pressure) [5]. Guo et al. propose a performance benchmarking method for residential systems using only smart thermostats as data sources. The proposed method analyzes the so-called steady-state behavior of the system and compares a critical property called the weighted average difference of cooling effort both between systems and within each system [6]. Dangut et al. present a new deep learning technique based on autoencoder and recurrent unit networks with bidirectional gates to process extremely rare failure predictions in aircraft predictive maintenance modeling [7]. Shaheen et al. used combined neural network architectures and cumulative neural network designs to predict the failure state of a mechanical component and estimate its remaining useful life [8]. Einabadi et al. used artificial neural networks to reduce machine downtime with predictive maintenance [9]. Dangut et al. analyzed the log records of error messages on airplanes by using pattern mining methods [10]. Calabrase et al. used a data-driven approach based on machine learning applied to woodworking industrial machinery for a large Italian woodworking company. The predicted failure probabilities were calculated using tree-based classification models (Gradient Boosting, Random Forest, and Extreme Gradient Boosting) [11]. Predictive maintenance applications have been widely used with machine learning techniques. AutoML applications have started to be used to obtain algorithms and parameters that achieve better results.

Vincent et al. examine Bayesian optimization in depth and suggest using the evolutionary strategy of genetic algorithm, differential evolution, and covariance matrix adaptation for acquisition function optimization. The performance of hyperparameter optimization techniques was compared with the help of AutoML models [12]. Škrlj et al. tried to automate simultaneous learning for both images and text. It offers an AutoML (automatic machine learning) approach to automatic machine learning model configuration definition for data consisting of two modalities [13]. Sahin et al. describe a new AutoML framework to predict soil liquefaction potential problems based on stacking community learning (SEL) combined with a greedy search algorithm. The overall concept of the AutoML framework consists of three main steps: data preparation, greedy feature selection, and greedy stacking aggregation [14]. Raj et al. propose the integration of Convolutional Neural Networks (CNN), Vision Transformers (ViT), and AutoML to obtain slice-level predictions as well as patient-by-patient prediction results [15]. AutoML usage area is increasing. There are limited studies on predictive maintenance in the literature.

Using recently collected data from a Portuguese software company client, Ferreira et al. conducted a benchmark comparison study with Supervised AutoML tools and the proposed AutoOneClass method to estimate the number of days until the next failure of a piece of equipment, as well as to determine whether the equipment will fail after a certain period [16]. Cinar et al. monitored autonomous transfer vehicle and electric motor status to detect equipment failures and operational anomalies and were automated with AutoML and workflow automation Technologies [17]. Rivas et al. developed several models with sampling techniques using the partial discharges measurement dataset to evaluate the health of insulation in a general power system on covered conductors from power lines. An AutoML model without a multi-algorithm resampling technique performed better and achieved the best results [18]. Kocbek et al. have shown that AutoML approaches result in better performance compared to competitive winning solutions and have excellent potential to build robust predictive models in the rail industry [19].

3 Theoretical Background

3.1 Maintenance Policies

A maintenance policy is divided into two planned and unplanned maintenance. Unplanned maintenance is a policy based on performing maintenance only in the event of a breakdown. In the event of a malfunction, it is performed by replacing either the part or the whole that caused the malfunction. It is used for equipment that is not a bottleneck, can be easily repaired, or has spare parts. It is a reactive approach. Planned maintenance is divided into 3 as periodic, preventive, and predictive maintenance. Types of Maintenance policies are given in Fig. 1.

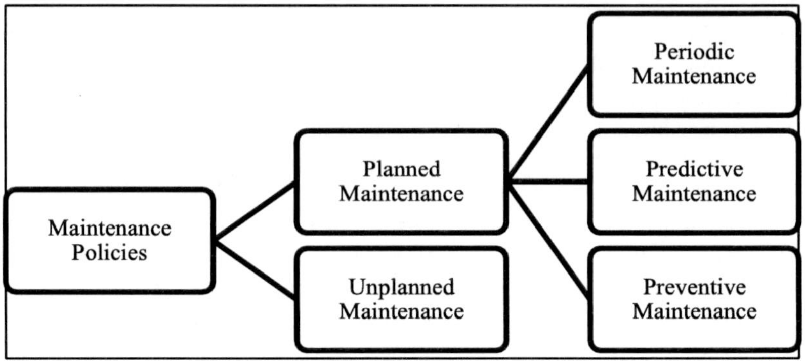

Fig. 1. Types of maintenance policies.

Periodic Maintenance. Periodic maintenance is a maintenance policy made at regular intervals. Equipment and parts are reviewed regularly. In this method, no malfunction is expected for maintenance. In all cases, maintenance is carried out at specified time intervals. The period in which maintenance will be carried out is determined and planned by the authorities of the company.

Preventive Maintenance. Preventive maintenance includes taking the necessary measures to prevent the malfunction from occurring rather than predicting or detecting it. It is to prevent the occurrence of malfunctions in the initial state. It is aimed to eliminate the fault together with its causes.

Predictive Maintenance. Predictive maintenance, on the other hand, is a policy where it is decided whether to perform maintenance or not by making various measurements. It detects the risk of failure. There are 3 basic approaches: fault and anomaly detection, estimation of remaining useful life, and fault detection by classification.

The aim is to foresee the time when normal working conditions will be exceeded. The remaining life approach; is divided into three risk-based, fault detection, and threshold value usage. Risk-based is based on planning maintenance activities based on probability distributions obtained from historical data. The probability distribution will depend on the failure rate depending on the amount of work, or it can be determined by considering various environmental factors (temperature, pressure, etc.) that are correlated with the failure rate. Fault analysis is the analysis of the behavior of a component with the failure data of the same or equivalent components to analyze whether there is a situation like the failures seen previously in the component. The threshold value is the control of whether certain characteristics of a system (vibration, current, etc.) are within normal operating limits. Normal operating limits can be determined based on risk, or they can be obtained because of various analyzes. By measuring the time-dependent change in the characteristics of the system, it can be calculated how long it takes for the system to go out of its normal operating limits.

For fault detection (machine learning algorithms, statistical techniques, etc.) estimation tools are used. Predictive maintenance provides a longer operating life than other maintenance strategies and allows for preventive maintenance activities. Predictive maintenance is the approach that emerged because of the digital twin philosophy. The digital twin is the modeling of the system in the physical environment and transferring it to the virtual environment.

3.2 Automated Machine Learning (AutoML)

In studies on data sets, it may not always be possible to find the algorithm that gives the best result among the results. There are many ML algorithms such as Random Forest, Support vector machines, and Artificial neural networks. AutoML is an application that automates the use of these ML algorithms. It is an emerging field where the process of creating machine learning models to model data is automated. There are libraries such as autosklearn, auto-keras, h2o, tpot. A comparison of the AutoML libraries used is given in Fig. 2. Tpot and lazypredict libraries were used in the study.

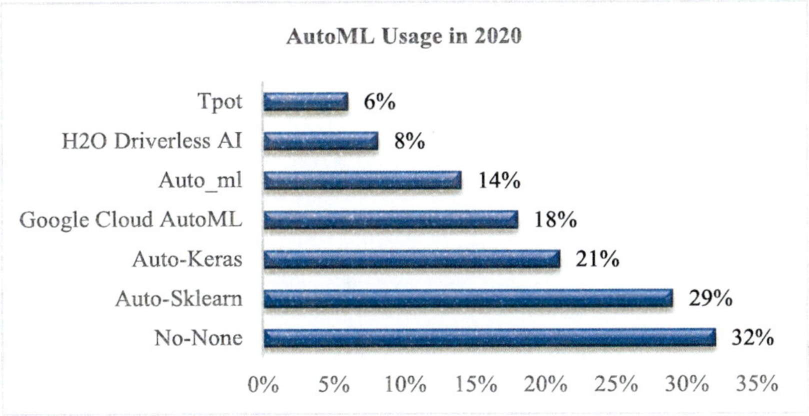

Fig. 2. Comparison of autoML libraries.

Tpot. The Tree-Based Pipeline Optimization Tool (TPOT) is one of the first AutoML methods and open-source software packages developed for the data science community. It is a Python-based tool. TPOT, Dr. Randal Olson at the University of Pennsylvania's Computational Genetics Laboratory. Developed with Jason H. Moore. TPOT automates the modeling pipeline with a greater emphasis on data preparation alongside modeling algorithms and model hyperparameters. The goal of TPOT is to automate the creation of machine learning pipelines by combining a flexible expression tree representation of pipelines with stochastic search algorithms such as genetic programming. Not suitable for natural language processing studies.

Lazypredict. It is a method used to select the best-performing machine learning algorithms. It is one of the best Python libraries for automating ML applications. Many basic models are created without using a lot of code and it helps to understand which models work better without any parameter adjustments.

4 Case Study

4.1 Data Description

A data set of 10000 rows, which contains malfunctions in the machines, was used. There are 5 different factors in this data set, namely air temperature, process temperature, rotation speed, torque, and tool wear. The distributions of the factors are given in Fig. 3.

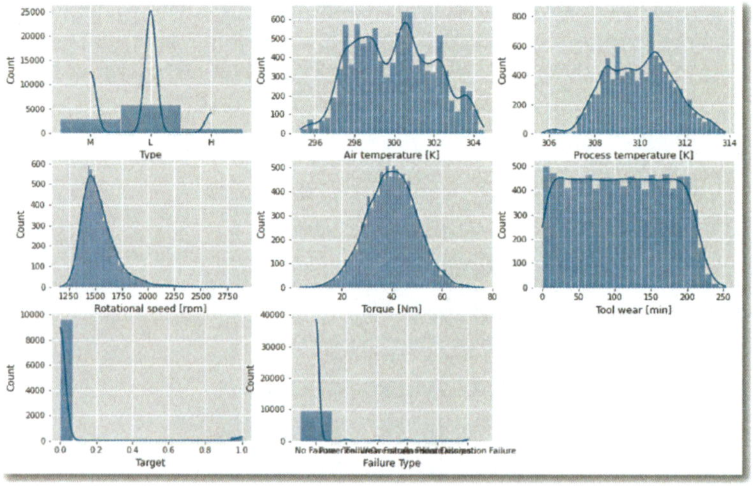

Fig. 3. Distribution of 5 different factors.

The relationships between the factors were examined. According to Fig. 4, the strongest relationship between air temperature and process temperature is seen.

As a result of these factors, 6 different failure types were determined. Fault condition number 6; represents the state of no fault. One; Heat Dissipation Failure, 2; Overstrain Failure, 3; Power failure, 4; Tool Wear Failure, 5; It is designated as Random Failure. The distribution of fault types is given in Fig. 5.

4.2 Method Comparison

Two AutoML methods, Tpot and LazyPredict, were utilized within the parameters of this study to estimate the machine defect rates. The findings are shown below, respectively (see Fig. 6).

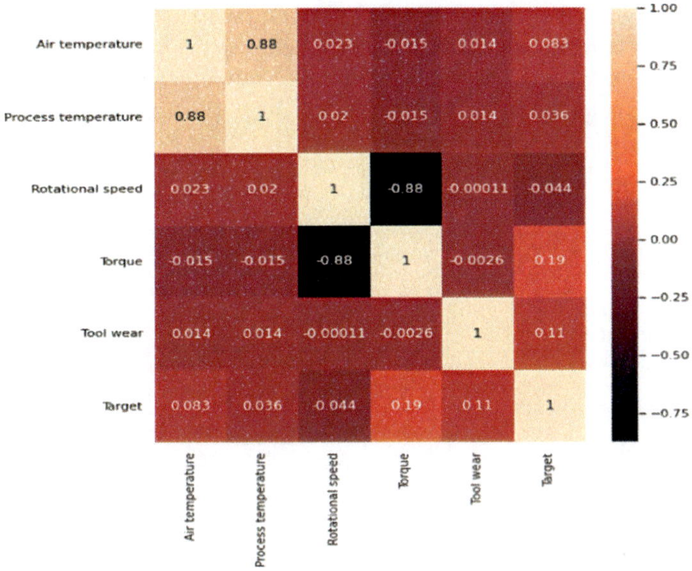

Fig. 4. Relationships between factors.

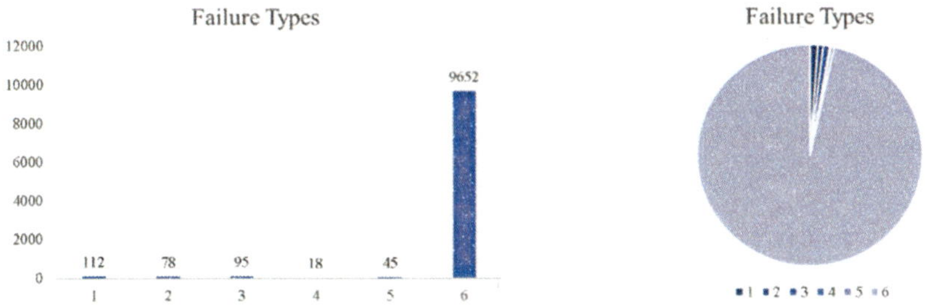

Fig. 5. Distribution of faults by type.

```
Generation 1 - Current best internal CV score: 0.968

Generation 2 - Current best internal CV score: 0.9763

Generation 3 - Current best internal CV score: 0.9763

Generation 4 - Current best internal CV score: 0.9763

Generation 5 - Current best internal CV score: 0.9763

Best pipeline: XGBClassifier(PCA(input_matrix, iterated_power=9, svd_solver=randomized), learning_rate=0.5, max_depth=
4, min_child_weight=13, n_estimators=100, n_jobs=1, subsample=0.9000000000000001, verbosity=0)
```

Fig. 6. Result of tpot autoML.

Figure 6 above displays the Tpot output. The XGBooting Classifier algorithm returns the best accuracy number and displays the parameters of the best algorithm as a consequence of this AutoML library. For instance, the XGBooting algorithm indicated that the learning rate should be 0.5 and the maximum depth should be 4 (Table 1).

Table 1. LazyPredict results of the best ML Algorithms.

Model	Accuracy	Balanced Accuracy	F1 Score
NearestCentroid	0.49	0.69	0.64
QuadraticDiscriminantAnalysis	0.98	0.60	0.98
LGBMClassifier	0.98	0.56	0.98
BaggingClassifier	0.98	0.56	0.98
DecisionTreeClassifier	0.98	0.54	0.98
RandomForestClassifier	0.98	0.54	0.98
GaussianNB	0.96	0.53	0.96
ExtraTreeClassifier	0.97	0.46	0.97
ExtraTreesClassifier	0.98	0.44	0.98
SGDClassifier	0.98	0.43	0.97
LabelPropagation	0.97	0.42	0.97
LabelSpreading	0.97	0.41	0.97
PassiveAggressiveClassifier	0.98	0.38	0.97
LogisticRegression	0.98	0.38	0.97
KNeighborsClassifier	0.97	0.36	0.97
CalibratedClassifierCV	0.97	0.33	0.97
Perceptron	0.97	0.32	0.97
SVC	0.97	0.31	0.97
LinearSVC	0.97	0.30	0.96
LinearDiscriminantAnalysis	0.97	0.28	0.96
RidgeClassifier	0.97	0.17	0.95
DummyClassifier	0.97	0.17	0.95

The algorithms with the best outcomes after using the LazyPredict AutoML classifier library are displayed in the table above in order of their balanced score. Although the NearestCentroid algorithm has a low accuracy number, it has the best-balanced score value. When F1 and Accuracy values are taken into account, BaggingClassifier, DecisionTreeClassifier, and RandomForestClassifier all produce better results.

5 Conclusion

Malfunctions in machines are very important for companies. They need to develop effective maintenance policies to deal with failures. This study has adopted the predictive maintenance policy that emerged as a result of the digital twin philosophy to predict the failures in the machines. In this study, 5 factors (air temperature, process temperature, rotation speed, torque, and tool wear) affecting the failure status were examined. In the data set consisting of 10000 rows, 6 different types of faults occurred. Fault condition number 6; represents the state of no fault. What's that; Heat Dissipation Failure, 2; Overstrain Failure, 3; Power failure, 4; Tool Wear Failure, 5; It is designated as Random Failure. In the literature, classical machine learning algorithms are generally used in predictive maintenance and fault prediction studies. In our study, AutoML libraries were used to predict failure. Thanks to the use of AutoML, the algorithm, and parameters that give the best results have been obtained. Among AutoML's frequently used libraries, Tpot is preferred. In addition to these, lazy predict, one of the Python open libraries that gives the best results by trying the algorithms, is used.

Contribution of the study to the literature; It is the development of AutoML studies on predictive maintenance and fault detection, which are limited in number, using lazyPredict and different libraries. AutoML is one of the trending applications. At the same time, this study provided diversification of the usage area. In the future, more complex and more factor-affected failure situations will be detected with AutoML in the study. AutoML has more than 10 popular libraries. The results will be enriched by using other libraries in the future.

References

1. Wang, Y.G., Chen, Y., Zhao, Y.W.: Chemical mechanical planarization of silicon wafers at natural pH for green manufacturing. Int. J. Precis. Eng. Manuf. **16**(9), 2049–2054 (2015)
2. Carvalho, T., Soares, F., Vita, R., Francisco, R., Basto, J., Alcalá, S.: A systematic literature review of machine learning methods applied to predictive maintenance. Comput. Ind. Eng. **137**, 106024 (2019)
3. Panda, C., Singh, T.R.: ML-based vehicle downtime reduction: a case of air compressor failure detection. Eng. Appl. Artif. Intell. **122**, 106031 (2023)
4. Dos Santos, T., Ferreira, F.J.T.E., Pires, J.M., Damasio, C.: Stator winding short-circuit fault diagnosis in induction motors using random forest. In: 2017 IEEE International Electric Machines and Drives Conference, IEEE, Miami, FL, USA (2017)
5. Uhlmann, E., Pontes, R.P., Geisert, C., Hohwieler, E.: Cluster identification of sensor data for predictive maintenance in a selective laser melting machine tool. Procedia Manuf. **24**, 60–65 (2018)
6. Guo, F., Rasmussen, B.: Performance benchmarking of residential air conditioning systems using smart thermostat data. Appl. Therm. Eng. **225**, 120195 (2023)
7. Dangut, M.D., Jennions, I.K., King, S., Skaf, Z.: A rare failure detection model for aircraft predictive maintenance using a deep hybrid learning approach. Neural Comput. Appl. **35**(4), 2991–3009 (2023)
8. Shaheen, B., Kocsis, Á., Németh, I.: Data-driven failure prediction and RUL estimation of mechanical components using accumulative artificial neural networks. Eng. Appl. Artif. Intell. **119**, 105749 (2023)

9. Einabadi, B., Baboli, A., Ebrahimi, M.: Dynamic predictive maintenance in industry 4.0 based on real time information: case study in automotive industries. IFAC **52**(13), 1069–1074 (2019)
10. Dangut, M.D., Skaf, Z., Jennions, I.K.: An integrated machine learning model for aircraft components rare failure prognostics with log-based dataset. ISA Trans. **113**, 127–139 (2021)
11. Calabrese, M., et al.: SOPHIA: an event-based IoT and machine learning architecture for predictive maintenance in industry 4.0. Information **11**(4), 202 (2020)
12. Vincent, A.M., Jidesh, P.: An improved hyperparameter optimization framework for AutoML systems using evolutionary algorithms. Sci. Rep. **13**(1), 4737 (2023)
13. Škrlj, B., Bevec, M., Lavrač, N.: Multimodal AutoML via representation evolution. Mach. Learn. Knowl. Extr. **5**(1), 1–13 (2022)
14. Sahin, E.K., Demir, S.: Greedy-AutoML: a novel greedy-based stacking ensemble learning framework for assessing soil liquefaction potential. Eng. Appl. Artif. Intell. **119**, 105732 (2023)
15. Raj, R., Mathew, J., Kannath, S.K., Rajan, J.: StrokeViT with AutoML for brain stroke classification. Eng. Appl. Artif. Intell. **119**, 105772 (2023)
16. Ferreira, L., Pilastri, A., Romano, F., Cortez, P.: Using supervised and one-class automated machine learning for predictive maintenance. Appl. Soft Comput. **131**, 109820 (2022)
17. Cinar, E., Kalay, S., Saricicek, I.: A predictive maintenance system design and implementation for intelligent manufacturing. Machines **10**(11), 1006 (2022)
18. Rivas, J., Boya-Lara, C., Poveda, H.: Partial discharge detection in power lines using automated machine learning. In: Proceedings - 2022 8th International Engineering, Sciences and Technology Conference, pp. 223–230, IESTEC (2022)
19. Kocbek, S., Gabrys, B.: Automated machine learning techniques in prognostics of railway track defects. In: IEEE International Conference on Data Mining Workshops, pp. 777–784, ICDMW (2019)

An Intelligent System Proposal for Providing Driving Data for Autonomous Drive Simulations

Muhammet Raşit Cesur(✉), Elif Cesur, and Abdülsamet Kara

Faculty of Engineering and Natural Sciences, Industrial Engineering, İstanbul Medeniyet University, İstanbul, Turkey
rasit.cesur@medeniyet.edu.tr

Abstract. Simulation technology is being used to reduce the costs of development and testing processes in autonomous driving studies. Autonomous systems gaining driving experience in simulation is faster and more cost-effective than real-world work. However, systems developed through simulation are becoming increasingly distant from reality, and it is uncertain how these systems will respond to situations that may occur in the real world. Therefore, having a simulation environment that is close to reality in which an autonomous driving system will be developed will contribute to the more efficient operation of the systems developed in the simulation environment in the real world. To create a more realistic driving simulation, a large amount of driving data from the real world is needed. In this study, a smart system has been developed that analyzes various driving videos and produces the necessary data for the simulation. The proposed system calculates the position change of the vehicles in each frame of the video. Depending on the position change, the vehicle's speed, acceleration, total displacement, and maneuvering style are revealed. This allows for data on different driving styles to be obtained, transferred to the simulation environment, and enables the simulated vehicles to behave like real drivers. In addition, the vehicles and stationary objects in the video are detected by an intelligent system, and the vehicles in consecutive frames are matched and scaled to made frames identical. By this way we eliminate the effects of extrinsic parameters such as camera transformation, angle change, and zoom in/zoom out to produce the required data.

Keywords: Object detection · image processing · autonomous drive

1 Introduction

Currently, numerous studies are being conducted on autonomous driving algorithms. The prevailing consensus among most of these studies is the requirement to incorporate human-like behaviors [3]. This is due to the rising number of autonomous vehicles on the roads, leading to a blend of autonomous and human-driven traffic, commonly referred as hybrid traffic. As decision-making in such a complex environment is a challenging process, researchers have turned to simulation-based solutions to mitigate the risks associated with potential mistakes, as the consequences could be catastrophic and result in loss of human life. Furthermore, the high costs associated with research and

Z. Şen et al. (Eds.): IMSS 2023, LNME, pp. 651–655, 2024.
https://doi.org/10.1007/978-981-99-6062-0_60

development, as well as testing, have prompted autonomous driving researchers to utilize simulation during the development phase.

To create a simulation model that is suitable for developing an autonomous driving algorithm, it must closely resemble the real world in terms of its level of detail. Driver behaviors, physical constraints and relations, various traffic scenarios and various goal functions must be implemented in the simulation study. As part of our study, we aim to generate a driving dataset from real traffic videos to incorporate driver behaviors and various traffic patterns into the simulation. This will enable us to create a simulation model that is more realistic and reflective of actual driving conditions.

2 Literature Review

Numerous studies on autonomous driving have highlighted the necessity of incorporating human-compatible behaviors into algorithms to effectively solve the autonomous driving problem [3, 4, 9]. Notably, humans are highly diverse due to factors such as personal taste and preference, randomness, aggressiveness, and driving experience [6]. Drivers' cognitive ability, skills, attention, fast processing, and interpretation of traffic situations and rules all affect their decision-making process [3], and autonomous driving algorithms must be designed to react accordingly to the decisions of all drivers on the road, whether autonomous or human. To achieve this, researchers in autonomous driving studies widely adopt sampling-based motion planning methods [11]. A simulation model that incorporates driver behavior must also include a physical model [13] or differential models such as Kalman Filters [8] to represent dynamic actions in the simulation over time. Modelling the dynamism and physical relations are very important, because autonomous drive systems and human have different constraints in driving such as following distance and maximum speed limit [9]. This difference affects the traffic and road capacity utilization, as well. To understand the effects of autonomous driving to traffic many simulation-based drive test scenarios are generated, and test generation mechanism are proposed [7, 10].

To satisfy data needs of the simulation model image processing studies can be utilized to analyze traffic videos. Such a system must be able to detect moving vehicles, speed and total displacement of vehicles. The FlowNet can be utilized on detection of moving vehicles [5]. Also foreground [1] and background detection [15] algorithms can be utilized when a moving camera such as police camera [2, 16] is used. In recent years, it is understood that supervised methods (deep learning) outperform traditional unsupervised methods [16], such as neighborhood search [12]. In this case it is also possible to use tensor or QR decomposition-based methods [14] to detect the scene for analysis of both standing and moving objects, to calculate velocity of moving objects captured by moving camera.

3 Method

The system depicted in Fig. 1 analyzes traffic videos captured by different cameras and under various environmental conditions to generate driving patterns related to displacement and velocity. Two consecutive frames in the traffic videos are matched to determine the speed and displacement amount of the tracked vehicle. By calculating the time elapsed between the two frames and the absolute displacement of the vehicle, its speed can be accurately determined. However, to calculate the absolute displacement correctly, both intrinsic and extrinsic camera parameters are required, including camera speed, shooting angle, field of view, and focal length. Unfortunately, it is not feasible to obtain these parameters for a large amount of traffic videos available on the internet. However, if the physical properties of the tracked vehicle are known, displacement amounts can be calculated without intrinsic parameters. Additionally, if a stationary object is detected in two consecutive images, the absolute displacement amount can be calculated without extrinsic parameters.

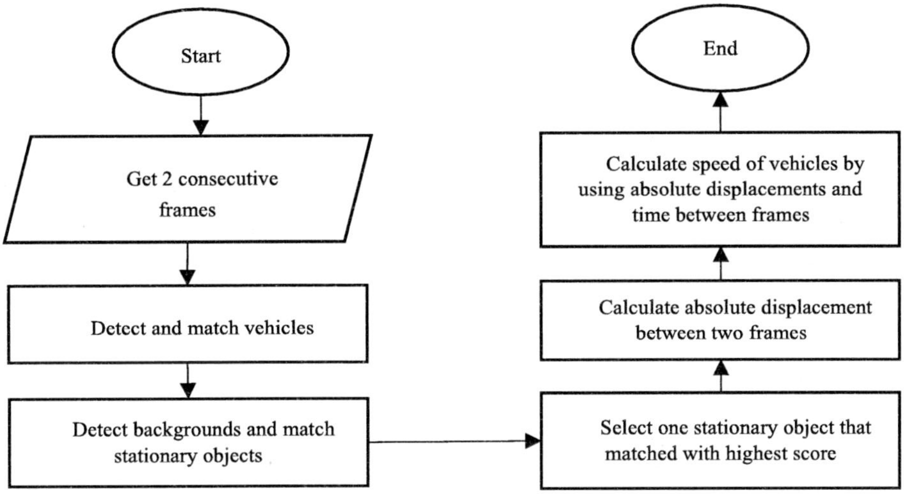

Fig. 1. Flow Diagram of the Proposed Algorithm

The YOLOv5 model is an accurate and fast deep learning model for vehicle detection, making it suitable for real-time applications and performing well on a single GPU. Figure 2 displays the algorithm's results on a frame extracted from a sample video. The algorithm output provides the coordinates of each vehicle, which are then used to obtain specific information about the driver being tracked, including the vehicle's identification and maneuvers. The system tracks each vehicle by determining the center point of the rectangle formed by the model's bounding boxes. Once this information is gathered to identify the correct driver profile, the system proceeds to the next stage.

To determine the absolute displacement between two frames, the SIFT (Scale-Invariant Feature Transform) algorithm is used to locate feature points in the background of both frames. By calculating the displacement of a stationary object in the field of view of the camera, the absolute displacement of the vehicles can be determined without the need for intrinsic camera parameters. This process involves locating a point in the first frame and finding its corresponding location in the second frame to determine the vehicle's total displacement. The system also calculates the vehicle's speed and acceleration

Fig. 2. Detection of a Stationary Object and a Car

using the video's frames per second feature. Binary frames are compared throughout the video to track changes in velocity, acceleration, and other relevant data.

4 Conclusion

We have proposed a semi-automatic algorithm to generate a driving pattern dataset by analyzing traffic videos. Our system focuses on background detection to identify stationary objects and object classification to detect vehicles, as we lack knowledge of the intrinsic and extrinsic parameters of cameras used to capture the traffic videos available on the internet. Therefore, we rely on features extracted from videos to eliminate the need for intrinsic and extrinsic parameters. Once the length of each vehicle is provided to the system, it can automatically generate the driving dataset.

References

1. Ben Khalifa, A., Alouani, I., Mahjoub, M.A., Ben Amara, N.E.: Pedestrian detection using a moving camera: a novel framework for foreground detection. Cogn. Syst. Res. **60**, 77–96 (2020)
2. Chapel, M.-N., Bouwmans, T.: Moving objects detection with a moving camera: a comprehensive review. Comput. Sci. Rev. **38**, 100310 (2020)
3. Cina, M., Rad, A.B.: Categorized review of drive simulators and driver behavior analysis focusing on ACT-R architecture in autonomous vehicles. Sustain. Energy Technol. Assessments. **56**, 103044 (2023)
4. Coskun, S.: Autonomous overtaking in highways: a receding horizon trajectory generator with embedded safety feature. Eng. Sci. Technol. Int. J. **24**, 1049–1058 (2021)
5. Delibasoglu, I., Kosesoy, I., Kotan, M., Selamet, F.: Motion detection in moving camera videos using background modeling and FlowNet. J. Vis. Commun. Image Represent. **88**, 103616 (2022)
6. Di, X., Shi, R.: A survey on autonomous vehicle control in the era of mixed-autonomy: from physics-based to AI-guided driving policy learning. Transp. Res. Part C Emerg. Technol. **125**, 103008 (2021)

7. Gao, L., Bai, W., Leary, R., Varadarajan, K., Brennan, S.: ROS integration of external vehicle motion simulations with an AIMSUN traffic simulator as a tool to assess CAV impacts on traffic. IFAC-PapersOnLine. **54**, 870–875 (2021)
8. Jung, J., Kim, H., Park, K., Lee, K.Y.: State estimation for autonomous guided vehicle using the extended Kalman filter. IFAC Proc. **35**, 487–492 (2002)
9. Li, T., Guo, F., Krishnan, R., Sivakumar, A., Polak, J.: Right-of-way reallocation for mixed flow of autonomous vehicles and human driven vehicles. Transp. Res. Part C Emerg. Technol. **115**, 102630 (2020)
10. Li, Y., Tao, J., Wotawa, F.: Ontology-based test generation for automated and autonomous driving functions. Inf. Softw. Technol. **117**, 106200 (2020)
11. Li, Z., Zhan, W., Sun, L., Chan, C.-Y., Tomizuka, M.: Adaptive sampling-based motion planning with a non-conservatively defensive strategy for autonomous driving. IFAC-PapersOnLine. **53**, 15632–15638 (2020)
12. Minematsu, T., Uchiyama, H., Shimada, A., Nagahara, H., Taniguchi, R.-I.: Adaptive background model registration for moving cameras. Pattern Recognit. Lett. **96**, 86–95 (2017)
13. Nan, J., Shang, B., Deng, W., Ren, B., Liu, Y.: MPC-based path tracking control with forward compensation for autonomous driving. IFAC-PapersOnLine. **54**, 443–448 (2021)
14. Sabat, N., Raj, S., George, S.N., Kumar, S.: A computationally efficient moving object detection technique using tensor QR decomposition based TRPCA framework. J. Vis. Commun. Image Represent. **92**, 103785 (2023)
15. Yun, K., Lim, J., Choi, J.Y.: Scene conditional background update for moving object detection in a moving camera. Pattern Recognit. Lett. **88**, 57–63 (2017)
16. Zhao, X., Wang, G., He, Z., Jiang, H.: A survey of moving object detection methods: a practical perspective. Neurocomputing **503**, 28–48 (2022)

A Stochastic Bilevel Programming Model for an Industrial Symbiosis Network

G. Sena Daş[1(✉)], Murat Yeşilkaya[2], Büşra Altinkaynak[3], and Burak Birgören[1]

[1] Department of Industrial Engineering, Kirikkale University, Kirikkale, Turkey
senadas@kku.edu.tr, birgoren@kku.edu.tr

[2] Niksar Vocational School of Technical Sciences, Tokat Gaziosmanpaşa University, Tokat, Turkey
murat.yesilkaya@gop.edu.tr

[3] Department of Industrial Engineering, Alparslan Turkeş Science and Technology University, Adana, Turkey
baltinkaynak@atu.edu.tr

Abstract. A stochastic bilevel model for an industrial symbiosis (IS) network in an eco-industrial park (EIP) is presented in this paper. Companies in IS networks produce final products using mostly by-products rather than raw materials. Companies in these networks are known to share a variety of synergies, including wastes and by-products. They mostly negotiate on the price and amount of the by-products. This negotiation is critical as by-products should be available on time at the right amount for smooth production. Moreover, the use of these by-products should provide savings compared to the use of raw materials so that companies could benefit from IS. However, the fluctuations in the final product demand affect both the EIP management (the leader) and companies (followers) in the network. Therefore, we proposed a stochastic bilevel model where the lower-level problem is a two-stage stochastic programming model. Our findings show that the demand uncertainty has an impact on both parties.

Keywords: industrial symbiosis networks · bilevel programming · stochastic programming

1 Introduction

In this paper, we present a stochastic bi-level model for a theoretical industrial symbiosis (IS) network. A sawmill, a paper mill, a particle board producer, a fiberboard producer, and a pellet producer are all part of this network that is situated in an eco-industrial park. In this network the production of final products such as lumber and paper results in the generation of by-products such as waste wood, wood chips, wood dust, and fuel wood. Plants can either utilize raw materials or by-products to manufacture final goods. The sawmill processes lumber and supplies by-products to the network. The pellet producer on the other hand only buys by-products from the network. The fiberboard plant and

Z. Şen et al. (Eds.): IMSS 2023, LNME, pp. 656–663, 2024.
https://doi.org/10.1007/978-981-99-6062-0_61

the particleboard producer may use by-products instead of raw materials. These plants also supply by-products to the network. Interested readers can obtain detailed information about the data regarding the companies in the IS network from [3].

In our case. the park authority/leader wants to minimize the environmental impact of production by asking companies to use fewer raw materials or more by-products. Companies, on the other hand, want to maximize the profit network by considering the leader's objective [2]. The final product demand of each plant is known to be uncertain. We employ a two-stage stochastic programming approach to address this uncertainty, which has an impact on how the amount of by-products generated in the network. The lower-level problem in the offered bi-level model is represented by this stochastic model. In the next section, we introduce this stochastic bilevel model.

2 A Stochastic Bilevel Programming Model for an Industrial Symbiosis Network

We present a bilevel model where park management is the upper-level decision maker/leader and the companies in the network are the lower-level decision makers/followers. The leaders' objective indicates the park authority's intention to encourage the network's use of by-products by reducing overall raw material consumption. On the other hand, the lower-level objective reflects the objective of the followers. The aim of the followers is to maximize their profit. Revenue $f_{REV,i}$ is computed by considering the amount obtained from the purchasing of final products and by-products as shown in Eq. 1. Cost is calculated for each plant taking into account all costs associated with production as well as taxes connected to carbon emissions. Total production-related costs $f_{ECO,i}$ of a plant i shown in Eq. 2 and consist of purchasing costs, transportation costs of raw material and by-products, energy costs, and inventory costs. If a plant produces more than the demand, then an inventory cost for the remaining amount applies. Finally, carbon tax $f_{CO_2,i}$ that should be paid by plant i due to emissions that occur during transportation and production is defined in Eq. 3.

$$f_{REV,i} = p_i^{fp} Q_i^{fp} + \sum_{j \in I} \sum_{k \in K} p_k Q_{ijk}^{bp,in} + \sum_{k \in K} p_k Q_{ik}^{bp,out} \tag{1}$$

$$\begin{aligned} f_{ECO,i} = & \sum_{m \in M} p_m^{rm} Q_{im}^{rm} + \sum_{j \in I} \sum_{k \in K} p_k Q_{ijk}^{bp,buy} + \\ & \sum_{m \in M} C_m^{trn,rm} d_i Q_{im}^{rm} + \sum_{j \in I} \sum_{k \in K} C_k^{trn,bp} d_{ij} Q_{ijk}^{bp,buy} + \\ & \sum_{m \in M} C^e a_m^{e,rm} Q_{im}^{rm} + \sum_{k \in K} C^e a_k^{e,bp} Q_{ik}^{bp,use} + C_i^{sto} Inv_i \end{aligned} \tag{2}$$

$$f_{CO_2,i} = \sum_{k \in K} C^{CO_2} w_k^{bp,CO_2} Q_{ik}^{bp,use} + \sum_{j \in I} \sum_{k \in K} C^{CO_2} w_k^{trn,CO_2} d_{ij} Q_{ijk}^{bp,buy} +$$

$$\sum_{m\in M} Q_{im}^{rm}(C^{CO_2}w_m^{rm,CO_2} + C^{CO_2}w_m^{trn,CO_2}d_i) \tag{3}$$

We can determine how much of the final product will be produced based on the amount of raw materials and by-products consumed, as stated in Eq. 4.

$$Q_i^{fp} = \sum_{m\in M} w_{im}^{rm}Q_{im}^{rm} + \sum_{k\in K} w_{ik}^{bp}Q_{ik}^{bp,use} \tag{4}$$

We offer a two-stage stochastic programming model for the lower-level problem as follows;

$$S1_L\colon \max\sum_{i\in I} f_{REV,i} - \Bigg(f_{CO_2,i} + \sum_{m\in M} p_m^{rm}Q_{im}^{rm} + \sum_{j\in I}\sum_{k\in K} p_k Q_{ijk}^{bp,buy} +$$
$$\sum_{m\in M} C_m^{trn,rm}d_iQ_{im}^{rm} + \sum_{j\in I}\sum_{k\in K} C_k^{trn,bp}d_{ij}Q_{ijk}^{bp,buy} +$$
$$\sum_{m\in M} C^e a_m^{e,rm}Q_{im}^{rm} + \sum_{k\in K} C^e a_k^{e,bp}Q_{ik}^{bp,use} + E_\xi\left[Q(Q_i^{fp},\xi)\right]\Bigg) \tag{5}$$

s.t.

$$\sum_{k\in K} Q_{ik}^{bp,use} \geq \theta_i^{low}\sum_{m\in M} Q_{im}^{rm}, \quad i\in I \tag{6}$$

$$\sum_{k\in K} Q_{ik}^{bp,use} \leq \theta_i^{upp}\sum_{m\in M} Q_{im}^{rm}, \quad i\in I \tag{7}$$

$$f_{REV,i} - (f_{ECO,i} + f_{CO_2,i}) \geq 0, \quad i\in I \tag{8}$$

$$Q_i^{fp} - Inv_i^+ + Inv_i^- = D_i, \quad i\in I \tag{9}$$

$$\sum_{m\in M}\sum_{k\in K} w_{imk}^{rm,bp}Q_{im}^{rm} = \sum_{j\in I}\sum_{k\in K} Q_{ijk}^{bp,in} + \sum_{k\in K} Q_{ik}^{bp,out}, \quad i\in I \tag{10}$$

$$Q_{ijk}^{bp,in} = Q_{jik}^{bp,buy}, \quad i,j\in I, k\in K \tag{11}$$

$$Q_{ik}^{bp,use} \leq \sum_{j\in I} Q_{ijk}^{bp,buy}, \quad i\in I, k\in K \tag{12}$$

$$\sum_{m\in M} w_{im}^{rm}Q_{im}^{rm} + \sum_{k\in K} w_{ik}^{bp}Q_{ik}^{bp,use} \leq Cap_i, \quad i\in I \tag{13}$$

$$\sum_{m\in M} w_{im}^{rm}Q_{im}^{rm} + \sum_{k\in K} w_{ik}^{bp}Q_{ik}^{bp,use} = Q_i^{fp}, \quad i\in I \tag{14}$$

$$Inv_i^+ \leq Cap_i^{Inv^+}, \quad i\in I \tag{15}$$

$$Inv_i^- \leq Cap_i^{Inv^-}, \quad i\in I \tag{16}$$

$$Q_{ijk}^{bp,buy}, Q_{ijk}^{bp,in}, Q_{ik}^{bp,out}, Q_{ik}^{bp,use}, Q_{im}^{rm}, Q_i^{fp}, Inv_i^+, Inv_i^- \geq 0,$$
$$i,j\in I, k\in K, m\in M \tag{17}$$

where $Q(Q_i^{fp}, \xi)$ is obtained by solving the second stage model which is presented below.

$$Q(Q_i^{fp}, \xi) = \min \left[C_i^{sto} Inv_i^+ + C_i^{loss} Inv_i^- \right] \tag{18}$$

$$\text{s.t.} \tag{19}$$

$$Q_i^{fp} - Inv_i^+ + Inv_i^- = D_i(\xi), \quad i \in I \tag{20}$$

$$Inv_i^+ \leq Cap_i^{Inv^+}, \quad i \in I \tag{21}$$

$$Inv_i^- \leq Cap_i^{Inv^-}, \quad i \in I \tag{22}$$

$$Q_i^{fp}, Inv_i^+, Inv_i^- \geq 0 \tag{23}$$

Then the deterministic equivalent of the $S1_L$ is presented below;

$$\min \sum_{i \in I} \sum_{m \in M} Q_{im}^{rm} \tag{24}$$

s.t.

$$\max \sum_{i \in I} f_{REV,i} - \Big(f_{CO_2,i} + \sum_{m \in M} p_m^{rm} Q_{im}^{rm} + \sum_{j \in I} \sum_{k \in K} p_k Q_{ijk}^{bp,buy} + \sum_{m \in M} C_m^{trn,rm} d_i Q_{im}^{rm} + \sum_{j \in I} \sum_{k \in K} C_k^{trn,bp} d_{ij} Q_{ijk}^{bp,buy} + \sum_{m \in M} C^e a_m^{e,rm} Q_{im}^{rm} + \sum_{k \in K} C^e a_k^{e,bp} Q_{ik}^{bp,use} + E_\xi \left[Q(Q_i^{fp}, \xi) \right] \Big) \tag{25}$$

s.t.

$$\sum_{k \in K} Q_{ik}^{bp,use} \geq \theta_i^{low} \sum_{m \in M} Q_{im}^{rm}, \quad i \in I \tag{26}$$

$$\sum_{k \in K} Q_{ik}^{bp,use} \leq \theta_i^{upp} \sum_{m \in M} Q_{im}^{rm}, \quad i \in I \tag{27}$$

$$f_{REV,i} - (f_{ECO,i} + f_{CO_2,i}) \geq 0, \quad i \in I \tag{28}$$

$$Q_i^{fp} - Inv_i^+ + Inv_i^- = D_i, \quad i \in I \tag{29}$$

$$\sum_{m \in M} \sum_{k \in K} w_{imk}^{rm,bp} Q_{im}^{rm} = \sum_{j \in I} \sum_{k \in K} Q_{ijk}^{bp,in} + \sum_{k \in K} Q_{ik}^{bp,out}, \quad i \in I \tag{30}$$

$$Q_{ijk}^{bp,in} = Q_{jik}^{bp,buy}, \quad i, j \in I, k \in K \tag{31}$$

$$Q_{ik}^{bp,use} \leq \sum_{j \in I} Q_{ijk}^{bp,buy}, \quad i \in I, k \in K \tag{32}$$

$$\sum_{m \in M} w_{im}^{rm} Q_{im}^{rm} + \sum_{k \in K} w_{ik}^{bp} Q_{ik}^{bp,use} \leq Cap_i, \quad i \in I \tag{33}$$

$$\sum_{m \in M} w_{im}^{rm} Q_{im}^{rm} + \sum_{k \in K} w_{ik}^{bp} Q_{ik}^{bp,use} = Q_i^{fp}, \quad i \in I \tag{34}$$

$$Inv_i^{+} \leq Cap_i^{Inv^{+}}, \quad i \in I \tag{35}$$

$$Inv_i^{-} \leq Cap_i^{Inv^{-}}, \quad i \in I \tag{36}$$

$$Q_{ijk}^{bp,buy}, Q_{ijk}^{bp,in}, Q_{ik}^{bp,out}, Q_{ik}^{bp,use}, Q_{im}^{rm}, Q_i^{fp}, Inv_i^{+}, Inv_i^{-} \geq 0,$$
$$i, j \in I, k \in K, m \in M \tag{37}$$

Constraints (26–27) representing the ratio of by-product to raw material. In our case, final products can not be produced only from raw materials but at least a minimum amount of by-products should be used. Constraint (28) guarantees that each company in the network is still profitable. The constraint (29) shows that if the demand can not be satisfied, the final product inventory or loss of sales can occur. Constraint (30) indicates that by-products will be obtained only as a result of raw material use. The obtained by-product can be sold to other plants in the network or outside the network. Constraint (31) is the balance constraint that ensures that the amount of by-product sold to plant j is equal to the amount of by-product bought by plant j. Constraint (32) ensures that the by-product used by a plant can not be more than the amount bought. Constraint (33) represents the production capacity of each plant. Constraint (34) indicates that the final product can be manufactured using both raw material and by-products. An upper bound for final product inventory and amount of loss sale is defined in constraints (35, 36). Finally, decision variables are presented in Eq. 37.

3 The Results of the Stochastic Bilevel Model

We assumed that the average (μ) of demand(D_{is}) for each plant is known and normally distributed. The standard deviation of a plant's demand is calculated as $\sigma = c_v * \mu$ where c_v is a coefficient of variation determined by the decision maker. This value indicates the magnitude of the fluctuation in demand. In this study, c_v value is set to 0.1. The solutions obtained for the bi-level stochastic model are compared with two single objective models, the follower's and the leader's model. The follower's model aims to maximize the profit of the network. The leader's model, on the other hand, aims to minimize the amount of raw material used. The results of these models are given in Table 1.

When the results are analyzed, the highest raw material usage is observed in the follower's model whereas the lowest raw material usage is seen in the leader's model. This leads to a low profit in the network. The highest profit for plants is achieved in the followers model as expected. In the stochastic bi-level model, a result that reflects a consensus between the leader and followers is obtained. Compared to the leader's model more raw material is used in the stochastic bilevel model. However, less use of raw material is observed in the stochastic bilevel model compared to the follower's model.

In terms of by-product use, more by-product is produced in the stochastic bilevel model. Unlike other models, the entire amount produced in the network is

Table 1. Results of the compared models

	Leader's model	Follower's model	Stochastic Bilevel model
Total raw material used (tonne)	**1652906**	**1954137**	**1859452**
Sawmill Plant	620075	691219	620075
Particleboard Plant	278667	308191	330533
Fiberboard Plant	682797	864617	823529
Pellet Plant	213	243	256
Pulp and Paper Plant	71154	89867	85059
Total by-product used (tonne)	**487791**	**324167**	**539347**
Sawmill Plant	0	0	0
Particleboard Plant	183920	203406	218152
Fiberboard Plant	213024	8646	212516
Pellet Plant	21116	24045	25321
Pulp and Paper Plant	69731	88070	83358
Total by-product produced (tonne)	**487791**	**574727**	**539347**
Sawmill Plant	232528	259207	232528
Particleboard Plant	47931	53009	56852
Fiberboard Plant	198011	250739	238823
Pellet Plant	0	0	0
Pulp and Paper Plant	9321	11772	11143
Total by-product sold the outside (tonne)	**0**	**250561**	**0**
Sawmill Plant	0	250560	0
Particleboard Plant	0	0	0
Fiberboard Plant	0	0	0
Pellet Plant	0	0	0
Pulp and Paper Plant	0	0	0
Total final product produced (m^3)	**1282883**	**1454774**	**1450133**
Sawmill Plant	387547	432012	387547
Particleboard Plant	302220	334239	358470
Fiberboard Plant	505060	578714	598810
Pellet Plant	11301	12868	13551
Pulp and Paper Plant	76755	96941	91755
Total production cost ($)	**302534850**	**331891680**	**335732056**
Sawmill Plant	107191000	117281000	107191000
Particleboard Plant	47800400	50796600	56309800
Fiberboard Plant	121646000	131275000	142888000
Pellet Plant	1015350	1129980	994156
Pulp and Paper Plant	24882100	31409100	28349100
Total CO_2 emission related cost ($)	**13750966**	**14340021**	**15924488**
Sawmill Plant	831335	926718	831335
Particleboard Plant	2707690	2994540	3211650
Fiberboard Plant	9356200	9342310	10858400
Pellet Plant	34865	39693	41806
Pulp and Paper Plant	820876	1036760	981297
Total profit ($)	**91141183**	**121330484**	**109983438**
Sawmill Plant	11186700	14679300	11186700
Particleboard Plant	16699200	20536700	20194700
Fiberboard Plant	34428300	49701000	42521400
Pellet Plant	305883	374484	590138
Pulp and Paper Plant	28521100	36039000	35490500

consumed. Compared to the deterministic version presented in [1] an increase is observed both in the amount of by-products and final products produced. Despite this increase, a loss of profit is observed in the total profit of the network.

4 Conclusion

In this study, we proposed a stochastic bilevel programming model for an IS network. Our aim is to examine how the demand uncertainties of the companies affect the leader and the followers in the network. When the results were examined, we observed an increase both in the amount of final products and by-products produced according to the deterministic scenario. However, inventory costs caused a decline in total profit. The findings suggested that further research is needed to fully comprehend the implications of demand uncertainty in IS networks.

Acknowledgements. This work is supported by TUBITAK under grant no: 120M223.

Appendix

Notations used in this paper are as follows;

Sets	
I	is the set of plants in EIP $(i, j \in I)$
K	is the set of bp types $(k \in K)$
M	is the set of rm types $(m \in M)$
Parameters	
D_i	demand of plant i for fp (m^3)
Cap_i	production capacity of plant i (m^3)
$w_{imk}^{rm,bp}$	conversion rate (cr) for obtaining bp type k by processing rm type m from plant i
w_{ik}^{bp}	cr of obtaining fp by using bp type k in plant i
w_{im}^{rm}	cr of obtaining fp by using rm type m in plant i
w_m^{rm,CO_2}	the amount of CO_2 released while processing rm type m (tonne)
w_k^{bp,CO_2}	the amount of CO_2 released while processing bp type k (tonne)
w^{trn,CO_2}	the amount of CO_2 emitted during transporting per km (tonne)

p_m^{rm}	unit purchasing price of rm type m (\$/tonne)
p_k^{bp}	unit sales price of bp type k (\$/tonne)
p_i^{fp}	unit sales price of fp produced by plant i (\$/m^3)
C^{CO_2}	tax paid for produced carbon dioxide (\$/tonne)
$C_m^{trn,rm}$	unit transportation cost per km for rm type m (\$/tonne)
C^e	unit cost of energy (\$/kWh)
C_{ik}^{sto}	inventory holding cost of bp type k of plant i (\$/tonne)
θ_i^{low}	lower bound of bp to rm
θ_i^{upp}	upper bound of bp to rm
$C_k^{trn,bp}$	unit transportation cost per km for bp type k (\$/tonne)
$a_k^{e,bp}$	the amount of energy required to process bp type k (kWh/tonne)
$a_m^{e,rm}$	the amount of energy required to process rm type m (kWh/tonne)
d_{ij}	distance between plant i and plant j (km)
d_i	distance between the supplier and plant i (km)

Decision Variables

Q_{im}^{rm}	amount of rm type m purchased by plant i (tonne)
$Q_{ijk}^{bp,in}$	amount of bp type k sold to plant j from plant i (tonne)
$Q_{ik}^{bp,out}$	amount of bp type k purchased by plant i of plant j (tonne)
$Q_{ik}^{bp,use}$	amount of bp type k used (tonne)
Inv_i^+	amount of final product inventory of plant i (tonne)
Inv_i^-	amount of final product back order of plant i (tonne)

References

1. Altinkaynak, B., Göçken, T., Yeşilkaya, M., Birgören, B., Daş, G.S.: Comparison of fuzzy solution approaches for a bilevel linear programming problem. In: Aliev, R.A., Kacprzyk, J., Pedrycz, W., Jamshidi, M., Babanli, M.B., Sadikoglu, F. (eds.) ICAFS 2022. LNCS, vol. 610, pp. 58–66. Springer, Cham (2022). https://doi.org/10.1007/978-3-031-25252-5_13
2. Das, G.S., Yesilkaya, M., Altinkaynak, B., Birgoren, B.: Modeling an industrial symbiosis network using bilevel programming. In: 2021 62nd International Scientific Conference on Information Technology and Management Science of Riga Technical University (ITMS), pp. 1-6. IEEE (2021)
3. Yeşilkaya, M., Daş, G.S., Türker, A.K.: A multi-objective multi-period mathematical model for an industrial symbiosis network based on the forest products industry. Comput. Ind. Eng. **150**, 106883 (2020)

Examining the Role of Industry 4.0 in Supply Chain Optimization Through Additive Manufacturing

Shubhendu Singh(✉), Subhas Chandra Misra, and Gaurvendra Singh

Indian Institute of Technology, Kanpur, India
shubhendusingh77@gmail.com

Abstract. The implementation of Industry 4.0 enabling technologies to optimize supply chain processes has become an essential aspect of modern manufacturing. Supply chain optimization (SCO) through additive manufacturing technology (AMT) is one area where Industry 4.0 is having a considerable impact. This study, therefore, explores the aspects of Industry 4.0 enabled by the incorporation of AMT into the production system that contribute to supply chain optimization. Ten such features have been identified and analyzed in this study using the novel Grey Influence analysis (GINA) technique. GINA methodology in this research helps to circumvent the restrictions of contemporary causal modeling methods such as the Decision-making trial and evaluation laboratory (DEMATEL) and Interpretative structural modeling (ISM) with the cumulative integration of responses, thus mitigating data loss. The results of the investigation demonstrate that cloud manufacturing, sustainable manufacturing, on-demand manufacturing, and distributed manufacturing are the predominant features of Industry 4.0 enabled by AMT. The findings of this study can be utilized by businesses to take informed decisions concerning incorporation, investment, and optimal application of AMT as well as to determine the significance of AMT to optimize their supply chains.

Keywords: Industry 4.0 · Additive manufacturing (AMT) · Supply chain optimization (SCO) · Grey influence analysis (GINA)

1 Introduction

Industry 4.0, also known as the Fourth Industrial Revolution, is transforming the manufacturing sector through the incorporation of cutting-edge technologies such as robotics, artificial intelligence, additive manufacturing technology (AMT), and the Internet of Things. (IoT) [1–3]. Supply chain (SC) optimization through AMT is one of the areas where Industry 4.0 is having an enormous effect. The optimization of the supply chain is a process that involves the design, planning, and realization of an SC in an effective and efficient manner, that can adapt rapidly and proficiently to fluctuations in demand, reduce delivery times, and enhance profitability [4, 5]. Examining methods of diminishing redundant expenses, refining quality, and upgrading customer service are some of the ways of optimizing SC.

Z. Şen et al. (Eds.): IMSS 2023, LNME, pp. 664–674, 2024.
https://doi.org/10.1007/978-981-99-6062-0_62

AMT also known as 3D printing is the process of layering materials to create a three-dimensional object from a digital model [6]. The application of additive manufacturing has revolutionized the manner in which products are conceptualized, created, and generated. For many years, AMT has been employed for prototyping, however, recent breakthroughs in the realm of materials and machines have made it feasible to utilize AMT for manufacturing parts of the highest quality [7]. The implementation of AMT is especially beneficial for supply chain management, as it facilitates on-demand production, thus reducing the need for substantial inventory stockpiles and the related costs [8]. Utilizing AMT provides companies the ability to fabricate products as dictated by demand, thereby eliminating the necessity of stockpiling large inventories [9]. Organizations are also able to swiftly react to fluctuations in demand, thereby minimizing the chances of their possessing superfluous stock. Furthermore, AMT facilitates a higher level of customization and personalization, enabling products to be tailored to meet specific customer requirements [10]. Additionally, AMT also has positive implications for supply chain logistics as it diminishes the requirement for extensive distance conveyance and related transfer expenses. The adoption of AMT thus has the potential to mitigate the carbon footprint of the supply chain, due to the elimination of the requirement for long-distance shipping of products.

Despite the perceived benefits, the capacity of AMT to optimize SC processes is yet to be explored. This study thus examines and prioritizes twelve critical attributes of Industry 4.0 facilitated by the adoption of AMT that aids in supply chain optimization. The three major objectives of this study are.

- Identification of Industry 4.0 features enabled by AMT adoption that aids in supply chain optimization.
- Prioritization of identified features based on their capability to enhance supply chain performance.
- Application of novel Grey Influence Analysis (GINA) techniques to examine and rank features.

The rest of the study is organized as follows. Section 2 discusses the previous literature of works on Industry 4.0, AMT, and supply chain optimization. Section 3 explains the methodology employed in this study while Sect. 4 highlights the results obtained from the analysis. Finally, Sect. 5 outlines the study's findings.

2 Literature Review

Numerous studies have been conducted on Industry 4.0, which are characterized by the incorporation of advanced technologies such as the Internet of Things (IoT), artificial intelligence (AI), AMT, and robotics to automate and optimize manufacturing processes [1, 11, 12]. Industry 4.0 allows for the regulation of production processes through the implementation of real-time synchronization of production flows, consequently allowing for the production of both individualized and customized products [13]. It centers significantly on AMT to realize its goal of smart, efficient, and sustainable manufacturing processes [14]. AMT is highly compatible with Industry 4.0's emphasis on digitalization and connectivity because it can be readily integrated into digital operations, enabling the construction of digital models and the automation of production processes. In the context of Industry 4.0, AMT can be used to manufacture customized components on demand, reducing the need for large inventories and minimizing waste. Moreover, AMT can be combined with other Industry 4.0 technologies, such as IoT sensors and artificial intelligence, to optimize production processes and facilitate predictive maintenance. Study by [1] offers a broad perspective on AMT in the framework of Industry 4.0. The paper presents an in-depth analysis of the fundamental principles and technological advances of Industry 4.0 and investigates the application of AMT in a variety of industrial sectors and evaluates how it can be incorporated with other Industry 4.0 advancements to attain superior productivity and resilience. [14] explores the influences on the incorporation of AMT in manufacturing sectors. The research reveals that elements such as technological readiness, resource accessibility, and organizational support are pivotal in the implementation of AMT. [15] analyzed the costs and benefits of implementing AMT and its implications on supply chain logistics, production procedures, and inventory supervision. Insightful information for designers and engineers seeking to optimize AMT is provided by [16]. [6] highlights the need for a strategic approach to implement AMT for sustainability, underlining the significance of factoring in product design, material choice, and supply chain management. The paper also emphasizes AMT's role in reducing manufacturing waste and emissions and facilitating circular economy models. Even though there have been several studies discussing the adoption and implementation of AMT and its manufacturing application, there are lack of studies exploring the supply chain optimization potential of Industry 4.0 enabled by AMT. Ten enabling features of AMT have been identified as can be seen in Table 1, which have been evaluated in this study by using the novel GINA technique.

Table 1. List of features of Industry 4.0 enabled by AMT adoption.

Flexibility	[10, 17]	Incorporation of AMT facilitates the production of items with greater design autonomy and enables businesses to effectively regulate their SC and swiftly adjust to alterations in the market
Agility	[18, 19]	Adoption of AMT may improve agility, thus allowing companies to regulate their SC effectively, improve responsiveness, and acquire a competitive advantage in the market
Customization	[10, 20]	Advancements in AMT have enabled businesses to produce tailored products at scale, diminishing the reliance on traditional mass production and storage. This capability addresses the specific requirements and inclinations of each customer, thereby enhancing customer gratification and dedication

(*continued*)

Table 1. *(continued)*

On-demand manufacturing	[8, 21]	Utilizing AMT, organizations are able to fabricate parts and items in accordance with demand, thereby hastening lead times and augmenting SC responsiveness
Risk management	[22, 23]	Implementation of AMT allows risk management, as well as increased optimization of the SC, by minimizing disruptions, mitigating quality issues, and enhancing risk assessment
Sustainable manufacturing	[10, 24]	AMTs provide a more sustainable approach to manufacturing by reducing waste, energy consumption, and carbon footprint, and use of recycled materials
Integration	[1, 25]	AMTs facilitate the incorporation of digital technologies, such as the IoT and cloud computing, into the manufacturing process, which enables real-time communication and data sharing throughout the SC, resulting in enhanced efficiency, transparency, and decision-making
Cloud manufacturing	[26, 27]	AMTs facilitate cloud manufacturing by allowing the creation of digital design files that can be stored in the form of digital inventory, and shared across the network It reduces the need for maintaining physical inventory and transportation of materials and products
Distributed Manufacturing	[9, 28]	Local production of components and products using AMT reduces the necessity for long-distance transportation and the environmental impact of the SC
Collaboration	[19, 29]	AMT can facilitate collaboration between various players of the SC, such as designers, manufacturers, and suppliers, thereby enhancing communication and improving efficiency

3 Research Methodology

This study implements an innovative and novel influence analysis technique, Grey Influence Analysis (GINA), that is mainly applicable to any survey-based or expert response-based study with numerous responses. GINA's purpose is to understand and measure the influence linkages among a set of factors or features utilizing the fundamentals of grey theory [30] and input-output models [31]. The application of causal modeling approaches such as ISM and DEMATEL, among others, is frequently employed by researchers in the supply chain management domain to assess the interrelations between various attributes. These methods, however, generally involve a small group of experts who offer opinions on either a crisp value or a grey/fuzzy spectrum [32]. The application of the GINA methodology in this research may help to circumvent the restrictions of contemporary causal modeling methods with the cumulative integration of results, thus mitigating data loss [33]. This approach is recommended to be adopted in order to incorporate a substantial quantity of responses into any influence analysis/causal modeling scenario to improve the accuracy of the results derived from the causal analysis. A thorough explanation of every step is provided below:

Step 1: Identification of features of Industry 4.0 enabled through AMT.

Step 2: Determine the influence relations (IR) among the features.

Step 3: Cleansing and sorting of the responses.

Step 4: Transformation of IR into grey values.

Suppose that $\otimes m_{xy}, x, y = 1, 2, 3, \ldots, k$ expresses the influence of the xth feature over yth feature in grey values, then, $\underline{\otimes} m_{xy}$ is the influence matrix lower bound while $\overline{\otimes} m_{xy}$ is the influence matrix upper bound. The influence matrix will now be presented as follows,

$$M(\otimes) = \left[\otimes m_{xy}\right]_{k \times k} = \begin{bmatrix} \otimes m_{11} & \cdots & \otimes m_{1k} \\ \vdots & \ddots & \vdots \\ \otimes m_{k1} & \cdots & \otimes m_{kk} \end{bmatrix} \quad (1)$$

Step 5: Aggregate the IR in grey scales.

The aggregated IR can then be compiled and represented in the following manner:

$$\sum\nolimits_{h=1}^{k} \otimes m_{xy}^{h} = \left(\sum\nolimits_{h=1}^{k} \underline{\otimes} m_{xy}^{h}, \sum\nolimits_{h=1}^{k} \overline{\otimes} m_{xy}^{h} \right) \quad (2)$$

where h is an index ranging from $1, 2, 3, \ldots, u$.

Step 6: Whitenization of grey scores into crisp values.

In this step, three whitenization models namely critical, ideal, and typical are used, where the lower limit, upper limit, and the average of the upper and lower limit values of the aggregate influences are utilized to turn grey values into crisp scores for the three models respectively.

Step 7: Calculate the direct influence coefficient.

The direct influence coefficient of the three models with respect to their corresponding crisp values (a = 1, 2, 3) can be conveyed as

$$d_{xy}^{a} = \frac{m_{xy}^{a}}{m_{y}^{a}}; x, y = 1, 2, 3, \ldots, k; a = 1, 2, 3 \quad (3)$$

Equation (4) can be utilized to characterize direct influence coefficient matrices that indicate the values of d_{xy}^{a}.

$$D^{a} = \left[d_{xy}^{a}\right]_{k \times k} = \begin{bmatrix} d_{11}^{a} & \cdots & d_{1k}^{a} \\ \vdots & \ddots & \vdots \\ d_{k1}^{a} & \cdots & d_{kk}^{a} \end{bmatrix}; a = 1, 2, 3 \quad (4)$$

Step 8: Ascertain the complete influence matrix.

Equation (5) can be employed in order to ascertain the overall influence coefficient matrix.

$$E^A = \left(\left(I - D^A\right)^{-1} - I\right) \tag{5}$$

Here, I suggest an identity matrix of the equivalent order as that of E^A which may be represented as,

$$E^A = \left[e_{xy}^a\right]_{k\times k} = \begin{bmatrix} e_{11}^a & \cdots & e_{1k}^a \\ \vdots & \ddots & \vdots \\ e_{k1}^a & \cdots & e_{kk}^a \end{bmatrix}; a = 1, 2, 3 \tag{6}$$

Step 9: Establish the grey responsibility coefficient and influence coefficients.

The grey responsibility and influence coefficient can be derived from Eqs. (7) and (8), respectively.

$$r_x^a = \frac{\sum_{y=1}^{k} e_{xy}^a}{\frac{1}{k}\sum_{x=1}^{k}\sum_{y=1}^{k} e_{xy}^a}; a = 1, 2, 3 \tag{7}$$

$$s_y^a = \frac{\sum_{x=1}^{k} e_{xy}^a}{\frac{1}{k}\sum_{x=1}^{k}\sum_{y=1}^{k} e_{xy}^a}; a = 1, 2, 3 \tag{8}$$

Step 10: Determine the total influence coefficient.

$$t_{xy}^a = r_x^a + s_y^a \forall x = y; a = 1, 2, 3 \tag{9}$$

Step 11: Determine and assess the most influential factors.

Using Eq. 10, the mean of the t_{xy}^a values are calculated which shows the most substantial feature

$$t_{xy} = \frac{(t_{xy}^1 + \left(4 \times t_{xy}^2\right) + t_{xy}^3)}{h} \tag{10}$$

where h is an arbitrary number (often 1,2….6).

In order to put into effect, the GINA approach, an exploratory survey-based assessment has been conducted to examine the features of Industry 4.0 enabled through the adoption of AMT that aids in optimizing the supply chain. The study has identified ten features through a literature review and the input of experts, as presented in Table 1. The influence of each feature is assessed via a Likert scale of 1 to 5, with 1 indicating no influence and 5 indicating maximum influence (as shown in Table 2). A group of experts with experience in Industry 4.0, AMT, and supply chain optimization from academia, businesses, and research were consulted to gather inputs. A questionnaire was sent to a randomly selected sample of 136 individuals with expertise in these domains, resulting in 87 responses. After inspecting the data, 63 valid and complete responses were used for analysis. Summary of the experts can be seen in Table 3. The subsequent section discusses and elaborates on the analysis findings.

Table 2. Linguistic terms and corresponding grey values

Ratings	Linguistic terms	Grey Values
1	Very Low	(0, 3)
2	Low	(2, 5)
3	Medium	(4, 7)
4	High	(6, 9)
5	Very High	(8, 10)

Table 3. Summary of the experts

Position		Experience	
Industry		1-5 years	38
Operation manager	12	6-10 years	17
Supply chain analyst	6	>10 years	8
Supply chain optimization analyst	9		
Industrial engineer	3		
Production planner	7		
Academia			
Supply chain management faculty	8		
Additive manufacturing faculty	7		
AMT researcher	11		

4 Results and Discussion

This study employs Grey influence analysis (GINA) technique to prioritize features of Industry 4.0 enabled by AMT that can optimize supply chain. Ten such features have been identified from the literature and analyzed. The results of the analysis can be seen in Table 4, which shows the grey responsibility, influence, and total influence coefficients of the features while Table 5 lists the features' rankings according to their total influence scores.

Investigation of our results reveals that *Cloud manufacturing (F8)*, which has the highest value of total influence score is the predominant feature of AMT that aids in optimizing SC. Cloud manufacturing enabled by AMT allows for the creation of a repository of digital designs in the cloud, which can be accessed whenever needed, thus minimizing the necessity of maintaining physical inventory.

Manufacturers can leverage digital inventory to produce products with greater proximity to the point of consumption, subsequently reducing lead times and decreasing transportation expenditures. Sustainable manufacturing (F6) is ranked second when it comes to optimizing SC through AMT. AMT enables sustainable manufacturing practices by reducing waste, energy consumption, and the need for transportation, allowing

Table 4. Grey responsibility, influence, and total influence coefficients

	1	2	3	4	5	6	7	8	9	10
Critical										
Grey inf. Coeff	0.1	0.099	0.1	0.099	0.099	0.1	0.099	0.1	0.1	0.099
Grey resp coeff	0.084	0.091	0.091	0.125	0.077	0.144	0.075	0.147	0.094	0.069
Sum (GIC, GRC)	0.184	0.191	0.192	0.225	0.177	0.244	0.174	0.247	0.194	0.169
Ideal										
Grey inf. Coeff	0.1	0.099	0.1	0.099	0.1	0.1	0.1	0.1	0.1	0.1
Grey resp coeff	0.09	0.095	0.095	0.117	0.086	0.125	0.082	0.128	0.098	0.079
Sum (GIC, GRC)	0.19	0.195	0.195	0.217	0.186	0.225	0.182	0.228	0.198	0.179
Typical										
Grey inf. Coeff	0.099	0.1	0.099	0.1	0.1	0.099	0.1	0.099	0.1	0.1
Grey resp coeff	0.088	0.094	0.094	0.12	0.082	0.132	0.079	0.134	0.096	0.076
Sum (GIC, GRC)	0.188	0.194	0.194	0.22	0.182	0.232	0.179	0.234	0.196	0.176
Aggregate										
Grey inf. Coeff	0.099	0.1	0.099	0.1	0.1	0.1	0.1	0.1	0.1	0.1
Grey resp coeff	0.088	0.0939	0.094	0.12	0.082	0.133	0.079	0.135	0.096	0.075
Total Influence	0.188	0.193	0.194	0.22	0.182	0.233	0.179	0.235	0.196	0.175

businesses to optimize their supply chain by decreasing lead times, inventory costs, and transportation emissions.

On-demand (F4) and distributed manufacturing (F9) are the third and fourth-ranked Industry 4.0 features attained through the implementation of AMT that can have a significant impact on supply chain optimization. Using on-demand production methods, businesses can make items only when they are ordered, cutting down on inventory and storage expenses. By decreasing wait times, improving adaptability, and reducing wastage, this feature can aid in supply chain optimization. On the other hand, by using distributed manufacturing features, businesses can create products closer to the final

Table 5. Rank of features of Industry 4.0 enabled by AMT based on Total influence score.

Features of Industry 4.0 enabled by AMT	Total Influence score	Rank
Flexibility (F1)	0.1882	7
Agility (F2)	0.1939	6
Customization (F3)	0.1943	5
On-demand manufacturing (F4)	0.2208	3
Risk management (F5)	0.1824	8
Sustainable manufacturing (F6)	0.2331	2
Integration (F7)	0.1796	9
Cloud manufacturing (F8)	0.2356	1
Distributed Manufacturing (F9)	0.1966	4
Collaboration (F10)	0.1755	10

consumers, cutting down on the need for long-distance shipping and the associated carbon impact. Supply chain optimization is optimized by this method as well, as transit costs are decreased, and response times are reduced. Fifth-ranked feature enabled by AMT is *customization (F3).* By enabling manufacturers to produce complex components and products with high precision and versatility, AMT facilitates customization which improves consumer satisfaction and loyalty, reduces waste and inventory costs, and enhances the resilience and versatility of the SC.

5 Conclusion

This study is exploratory research that aims to examine the features of Industry 4.0, which are made possible by AMT. This research utilizes a novel technique called GINA, which helps to evaluate the impact of various factors on supply chain optimization and performance by analyzing a large sample size of responses while minimizing data loss during aggregation. The study has identified ten distinct features of Industry 4.0 that are enabled by AMT and have been prioritized based on their total influence coefficient value, which determines the level of importance of a feature. Investigation of the results shows that *Cloud manufacturing (F8)* is the prominent feature enabled by AMT that aids in supply chain optimization. Other important features include *sustainable manufacturing (F6), on-demand (F4), and distributed manufacturing (F9).* The findings of this study will assist supply chain managers and practitioners in making knowledgeable choices regarding investments in AMT and other enabling technologies of Industry 4.0, considering the benefits identified for optimizing the supply chain. They can also enhance their workflows and streamline operations for maximum efficiency and minimal waste by comprehending how AMT can enhance the supply chain. Finally, by utilizing the analysis to adopt AMT and optimizing their supply chain, companies can potentially gain a competitive edge. This can lead to cost savings, improved efficiency, and higher-quality products, which can help companies differentiate themselves from competitors and entice more consumers.

References

1. Horst, D.J., Duvoisin, C.A., de Almeida Vieira, R.: Additive manufacturing at industry 4.0: a review. Int. J. Eng. Tech. Res. **0869**(8), 3 (2018). www.erpublication.org
2. Stentoft, J., Rajkumar, C.: The relevance of Industry 4.0 and its relationship with moving manufacturing out, back and staying at home. Int. J. Prod. Res. **58**(10), 2953–2973 (2020). https://doi.org/10.1080/00207543.2019.1660823
3. Garrido-hidalgo, C., Olivares, T., Ramirez, F.J., Roda-sanchez, L.: An end-to-end Internet of Things solution for reverse supply chain management in industry 4.0. Comput. Ind. **112**, 103127 (2019). https://doi.org/10.1016/j.compind.2019.103127
4. Garcia, D.J., You, F.: Supply chain design and optimization: challenges and opportunities. Comput. Chem. Eng. **81**, 153–170 (2015). https://doi.org/10.1016/j.compchemeng.2015.03.015
5. Beheshtinia, M.A., Feizollahy, P., Fathi, M.: Supply chain optimization considering sustainability aspects. Sustain. **13**(21), 11873 (2021). https://doi.org/10.3390/su132111873
6. Sonar, H., Ghosh, S., Singh, R.K., Khanzode, V., Akarte, M., Ghag, N.: Implementing additive manufacturing for sustainability in operations: analysis of enabling factors. IEEE Trans. Eng. Manag., pp. 1–15 (2022). https://doi.org/10.1109/TEM.2022.3206234
7. Niaki, M.K., Nonino, F., Palombi, G., Torabi, S.A.: Economic sustainability of additive manufacturing: contextual factors driving its performance in rapid prototyping. J. Manuf. Technol. Manag. **30**(2), 353–365 (2019). https://doi.org/10.1108/JMTM-05-2018-0131
8. Song, J.S., Zhang, Y.: Stock or print? Impact of 3-D printing on spare parts logistics. Manage. Sci. **66**(9), 3860–3878 (2020). https://doi.org/10.1287/mnsc.2019.3409
9. Liu, P., Huang, S.H., Mokasdar, A., Zhou, H., Hou, L.: The impact of additive manufacturing in the aircraft spare parts supply chain: Supply chain operation reference (scor) model based analysis. Prod. Plan. Control **25**, 1169–1181 (2014). https://doi.org/10.1080/09537287.2013.808835
10. Mehrpouya, M., Vosooghnia, A., Dehghanghadikolaei, A., Fotovvati, B.: The benefits of additive manufacturing for sustainable design and production (2021). https://doi.org/10.1016/B978-0-12-818115-7.00009-2
11. Ntinas, C., Gkortzis, D., Papadopoulos, A., Ioannidis, D.: Industry 4.0 sustainable supply chains: an application of an IoT enabled scrap metal management solution. J. Clean. Prod. **269**, 122377 (2020). https://doi.org/10.1016/j.jclepro.2020.122377
12. Kumar, V., Vrat, P., Shankar, R.: Prioritization of strategies to overcome the barriers in industry 4.0: a hybrid MCDM approach, **58**(3), 711–750. Springer India (2021). https://doi.org/10.1007/s12597-020-00505-1
13. Moeuf, A., Lamouri, S., Pellerin, R., Tamayo-Giraldo, S., Tobon-Valencia, E., Eburdy, R.: Identification of critical success factors, risks and opportunities of Industry 4.0 in SMEs. Int. J. Prod. Res. **58**(5), 1384–1400 (2020). https://doi.org/10.1080/00207543.2019.1636323
14. Sonar, H., Khanzode, V., Akarte, M.: Investigating additive manufacturing implementation factors using integrated ISM-MICMAC approach. Rapid Prototyp. J. **26**(10), 1837–1851 (2020). https://doi.org/10.1108/RPJ-02-2020-0038
15. Thomas, D.: Costs , benefits , and adoption of additive manufacturing : a supply chain perspective. Int. J. Adv. Manuf. Technol. **85**, 1857–1876 (2016). https://doi.org/10.1007/s00170-015-7973-6
16. Stavropoulos, P., Foteinopoulos, P., Papapacharalampopoulos, A.: On the impact of additive manufacturing processes complexity on modelling. Appl. Sci. 11(16), 7743 (2021). https://doi.org/10.3390/app11167743
17. Eyers, D.R., Potter, A.T., Gosling, J., Naim, M.M.: The flexibility of industrial additive manufacturing systems. Int. J. Oper. Prod. Manag. **38**(12), 2313–2343 (2018). https://doi.org/10.1108/IJOPM-04-2016-0200

18. Basu, J.R., Abdulrahman, M.D., Yuvaraj, M.: Improving agility and resilience of automotive spares supply chain: the additive manufacturing enabled truck model. Socioecon. Plann. Sci., **85**, 101401 (2022). https://doi.org/10.1016/j.seps.2022.101401
19. Belhadi, A., Kamble, S.S., Venkatesh, M., Jabbour, C.J.C., Benkhati, I.: Building supply chain resilience and efficiency through additive manufacturing: an ambidextrous perspective on the dynamic capability view. Int. J. Prod. Econ., **249**, 108516 (2022). https://doi.org/10.1016/j.ijpe.2022.108516
20. Berman, B.: 3-D printing: the new industrial revolution. Bus. Horiz. **55**(2), 155–162 (2012). https://doi.org/10.1016/j.bushor.2011.11.003
21. Huang, S.H., Liu, P., Mokasdar, A., Hou, L.: Additive manufacturing and its societal impact: a literature review. Int. J. Adv. Manuf. Technol. **67**(5–8), 1191–1203 (2013). https://doi.org/10.1007/s00170-012-4558-5
22. Moreno-Cabezali, B.M., Fernandez-Crehuet, J.M.: Application of a fuzzy-logic based model for risk assessment in additive manufacturing R&D projects. Comput. Ind. Eng., **145**, 106529 (2020). https://doi.org/10.1016/j.cie.2020.106529
23. Meyer, M.M., Glas, A.H., Eßig, M.: A Delphi study on the supply risk-mitigating effect of additive manufacturing during SARS-COV-2. J. Purch. Supply Manag. **28**(4), 100791. (2022). https://doi.org/10.1016/j.pursup.2022.100791
24. Javaid, M., Haleem, A., Singh, R.P., Suman, R., Rab, S.: Role of additive manufacturing applications towards environmental sustainability. Adv. Ind. Eng. Polym. Res. **4**, 312–322 (2021). https://doi.org/10.1016/j.aiepr.2021.07.005
25. Eyers, D., Managing 3D printing: Operations Management for Additive Manufacturing. Springer International Publishing, Berlin (2020).https://doi.org/10.1007/978-3-030-23323-5
26. Rauch, E., Unterhofer, M., Dallasega, P.: Industry sector analysis for the application of additive manufacturing in smart and distributed manufacturing systems. Manuf. Lett. **15**, 126–131 (2018). https://doi.org/10.1016/j.mfglet.2017.12.011
27. De Giovanni, P., Belvedere, V., Grando, A.: The selection of industry 4.0 technologies through Bayesian networks: an operational perspective. IEEE Trans. Eng. Manag., 1–16 (2022). https://doi.org/10.1109/TEM.2022.3200868
28. Khajavi, S.H., Holmström, J., Partanen, J.: Additive manufacturing in the spare parts supply chain: hub configuration and technology maturity. Rapid Prototyp. J. **24**(7), 1178–1192 (2018). https://doi.org/10.1108/RPJ-03-2017-0052
29. Akmal, J.S., Salmi, M., Björkstrand, R., Partanen, J., Holmström, J.: Switchover to industrial additive manufacturing: dynamic decision-making for problematic spare parts. Int. J. Oper. Prod. Manag. **42**(13), 358–384 (2022). https://doi.org/10.1108/IJOPM-01-2022-0054
30. Ju-Long, D.: Control problems of grey systems. Syst. Control Lett. **1**(5), 288–294 (1982). https://doi.org/10.1016/S0167-6911(82)80025-X
31. Liu, S., Forrest, J.Y.L.: Grey Systems: Theory and Applications. Springer, Berlin Heidelberg (2010). https://books.google.co.in/books?id=hJsRBwAAQBAJ
32. Vishwakarma, A., Dangayach, G.S., Meena, M.L., Gupta, S.: Analysing barriers of sustainable supply chain in apparel & textile sector: a hybrid ISM-MICMAC and DEMATEL approach. Clean. Logist. Supply Chain **5**, 100073 (2022). https://doi.org/10.1016/j.clscn.2022.100073
33. Rajesh, R.: An introduction to grey influence analysis (GINA): applications to causal modelling in marketing and supply chain research. Expert Syst. Appl. **212**, 0–2 (2023). https://doi.org/10.1016/j.eswa.2022.118816

Mathematical Models for the Reviewer Assignment Problem in Project Management and a Case Study

Zeynep Rabia Hosgor(✉), Elifnaz Ozbulak, Elif Melis Gecginci, and Zeynep Idil Erzurum Cicek

Eskisehir Technical University/Industrial Engineering, Eskisehir, Turkey
{zrhosgor,elifnazozbulak, elifmelisgecginci}@ogr.eskisehir.edu.tr, zierzurum@eskisehir.edu.tr

Abstract. Project management is a critical process for every institution and/or organization. This process should be managed as best as possible in order to manage resources and time well and at the same time achieve successful results. One of the reasons that make project management difficult is the increase in project proposals with the increase in reading rates and incentives. The evaluation process of the project proposals includes the assignment of an expert who will evaluate the project. This stage is solved with reviewer assignment problems in the literature. When the literature is examined, it is seen that the general aim of reviewer assignment problems is to maximize the degree of reviewer-project match. This study, in addition to the literature, it is aimed to minimize the evaluation time of the reviewers' project proposals and to ensure a balanced distribution among the reviewers. When we look at the results of the test problems for this study, which has two objectives, maximum matching degree and minimum evaluation time, it is seen that the objectives have been met. In this way, the assignment, which was made manually and caused a waste of time, was completed in a fair and reliable way.

Keywords: project management · reviewer assignment problem · mathematical model · optimization

1 Introduction

With the increasing population and developing technology, the efficient use of resources has become an important issue day by day. Time and workforce are some of the most important resources we have. Project management, which aims to use time and workforce efficiently, is a very critical process. Looking at the project management process, in the literature, there are reviewer assignment problems in determining the reviewers to be assigned to the project. Reviewer assignment problem processes generally include determining the subjects where the reviewers are experts, determining the degree of compatibility between the project and reviewer expertise, and assigning reviewers to

Z. Şen et al. (Eds.): IMSS 2023, LNME, pp. 675–682, 2024.
https://doi.org/10.1007/978-981-99-6062-0_63

the project in line with this compatibility. In this study, two mathematical models were developed with the aim of maximum matching degree and minimum evaluation time. The developed mathematical models were tested for data from a scientific research project unit. Studies show that it adheres to maximum matching degree and minimum evaluation time, and a balanced assignment is achieved while achieving these goals.

2 Literature

Reviewer assignment problems in the literature studies were examined. In one of these studies, Hartvigsen et al. (1999) created a pool of reviewers in their study and then assigned each article to several reviewers. The solution method consists of a two-stage optimization approach. For each article, it is intended to assign a certain number of reviewers and to have as great a level of expertise as possible, with the appointment also being above threshold. A network model was created to examine the constraints determined in the study. An important feature of the solution under review is that each article is assigned to at least one reviewer who is "as expert as possible" for that article. In this study, a technique for assigning an expert reviewer to articles is presented [1].

Li and Watanabe (2013) discussed the issue of assigning reviewers to articles. Combining the preference-based and subject-based approach found in the literature, they proposed a method to model the problem depending on the degree of match between the reviewers and the articles. When examining the degree of match between the reviewer and the article; The degree of expertise of the reviewer and the degree of compatibility between the reviewer and the article were discussed. These two criteria were prioritized with the Analytical Hierarchy Process (AHP) method. As a result, the assignment algorithm is given and the evaluation results are shown in comparison with the Hungarian algorithm [2].

Liu et al. (2016) integrated heuristics and operations research techniques developed in reviewer assignment. In this approach, decision models are used to realize the best reviewer assignment that maximizes the total expertise level of the reviewers assigned to the projects. In the assignment of the projects to the reviewers, four aspects have been determined that consider each proposal to be reviewed by a certain number of reviewers, avoid conflicts of interest, a balanced distribution of proposals at different levels, a fair opportunity for each project, and the workload of each reviewer to be as balanced as possible. In the study, traditional optimization (Simplex Method) is used to solve the assignment model. Liu et al. suggests that more efficient metaheuristic algorithms can be used to increase system efficiency, and information rules can also be developed to avoid conflicts of interest [3].

Pradhan et al. (2000) examined a similar problem in their study in terms of maximizing the rate of reviewer-article compatibility, minimizing conflict of interest and balancing the workload of reviewers. They first examined these three factors individually and then proposed a meta-heuristic method called matrix factorization based greedy algorithm to solve the general maximization type multi-objective reviewer assignment problem. The compatibility between the reviewer and the article was determined by a subject modelling technique called Latent Dirichlet Allocation. Conflict of interest was handled using the co-authorship graph instead of relying on the self-reports of the reviewers and authors. In this way, they argued, the time required to collect such preliminary information from reviewers could be avoided. The proposed method showed that the average quality of assignment was superior [4].

Kat (2021) developed an algorithm that lists potential panelist candidates in order to create the most suitable panelist cluster in panels where more than one project proposal is evaluated, and a decision support system called PaneLIST that uses this algorithm. In the developed decision support system, five criteria, namely similarity module, field of activity intersection module, project score module and panel score module, were determined in order to create the most appropriate panelist cluster [5].

The aim of reviewer assignment problems in the literature is to maximize the degree of reviewer-article matching. In addition to maximizing the degree of reviewer-article matching, a balanced assignment among the reviewers and minimizing the total evaluation time of the projects is a unique value of this project.

3 Problem Definition and Model

Scientific Research Projects, which are an important source of experience for academicians and students in universities, are the whole of projects that can contribute to science, economy and art at a social or universal level. Generally, higher education institutions attach importance to these research projects and allocate certain funds for this. Therefore, there is a project management unit that works meticulously to ensure that the evaluation of projects is done by experts in a fair way. The decision on the assignment of the projects to the reviewers is taken by this unit. The absence of a digital system here can lead to loss of time and the inability to make a fair and balanced assignment due to human error. In this project, two different models were established for the reviewer assignment problem and solved with GAMS. The data that will contribute to the development and implementation of the models were obtained from the Scientific Research Projects Unit.

Based on the mathematical model defined in the study of Kat (2021), two models were customized and developed.

In the first model developed, the maximum matching degree is aimed and is given below.

$$\mathrm{Maks} \sum_{i \in I} \sum_{j \in J} c_{ij} * x_{ij} \tag{1}$$

$$s.t.$$

$$\sum_{i \in I} x_{ij} = 1, \forall j \tag{2}$$

$$\sum_{j \in J} q_j * x_{ij} \le A, \forall i \tag{3}$$

$$x_{ij} \le c_{ij}, \forall (i, j) \tag{4}$$

$$x_{ij} \in \{0, 1\} \forall (i, j)$$

In the first model given above, the I represent the set of the reviewers, while the J represents the set of the project proposals, defense and final report. C_{ij} shows the degree of matching of the reviewer i with the project j. x_{ij} is a binary variable that takes the value 1 if the reviewer i is assigned to project j and 0 otherwise. q_j indicates the degree of difficulty according to the type of project. Here, the project proposal has a difficulty level of 3, the final report 2 and the defense petition 1. The maximum number of projects that can be assigned for each reviewer is indicated by the fixed number A. A value is calculated by the ratio of the sum of the difficulty levels of each project to the total number of reviewers. Since this value must be an integer, the decimal number is rounded to the next smallest integer. Equation 2 is written to ensure that the current situation is that each project is evaluated by a reviewer. This equality ensures that 1 reviewer is assigned to each j project. Equation 3 limits the number of projects each reviewer will evaluate. Finally, Eq. 4 provides that if the matching degree between the reviewer and the project is zero, the i reviewer will not be assigned to the j project.

In the second model developed, the aim is to minimize the evaluation time. The model created is given below.

$$\text{Min} \sum_{i \in I} \sum_{j \in J} t_i * x_{ij} \tag{5}$$

$$s.t.$$

$$\sum_{i \in I} x_{ij} = 1, \forall j \tag{6}$$

$$\sum_{j \in J} q_j * x_{ij} \leq A, \forall i \tag{7}$$

$$x_{ij} \leq c_{ij}, \forall (i, j) \tag{8}$$

$$x_{ij} \in \{0, 1\} \forall (i, j)$$

The difference of the second model given above from the first model is the objective function. With the objective function, the sum of the evaluation times of the reviewers' projects will be minimized. i. reviewer's average evaluation time is expressed as t_i.

4 Experimental Study

4.1 Test Problem

The test problem was created in line with the data received from the Scientific Research Projects Unit and the developed models were solved using the GAMS package program.

There are 266 projects and 6 reviewers in the created test problem. Each project was examined according to its type and the values of the project difficulty levels q_j were determined. In this problem, the constant number A was found to be 101 by dividing the total q_j value by the number of reviewers. The expert opinion method was applied in the determination of the matching degree c_{ij} which was examined as another parameter. Finally, t_i values were determined on a daily basis based on the average evaluation times of the reviewers.

4.2 Experimental Results

The results obtained by solving the the first mathematical model for which the maximum degree of matching is aimed are summarized in Fig. 1 and Table 1.

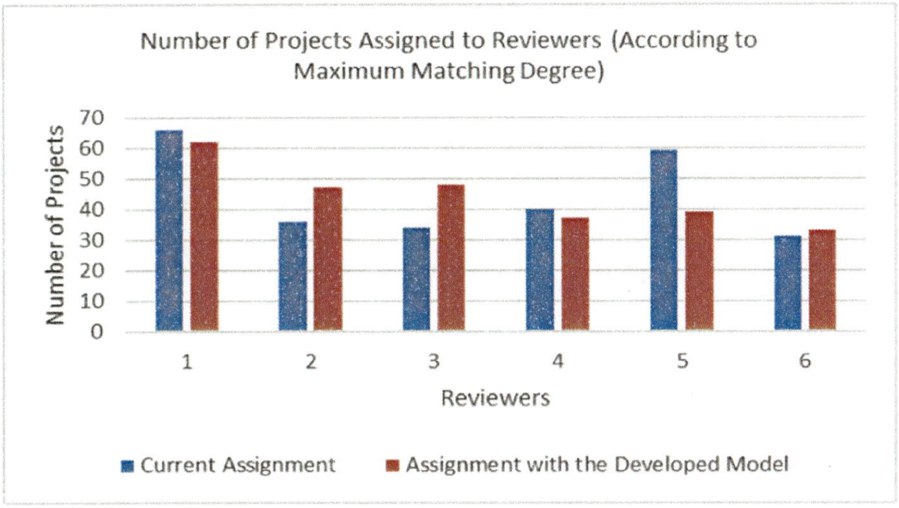

Fig. 1. Number of projects assigned to reviewers (according to maximum matching degree)

Table 1. Number of projects assigned to reviewers according to project type and maximum degree of compliance.

Type of Project						
Reviewer	1	2	3	4	5	6
Defense Petition	23	8	4	1	0	0
Final Report	39	24	35	10	16	0
Project Proposal	0	15	9	26	23	33
Upper Limit (A)	101	101	101	99	101	99

In the current system, assignments take hours manually. Assignments made with the developed model give results in an average of 2 s.

When Fig. 1 is examined, it is observed that the number of assignments made with the developed mathematical model and the number of existing assignments are close. In other words, it is possible to make new appointments by using the workforce used in the current system more efficiently. However, it is not possible to say that a balanced assignment has been made just by looking at Fig. 1. At this point, it would be more accurate to examine the assignments according to the type of project and therefore the difficulties, by looking at Table 1. In order to understand whether a balanced assignment has been made, the sum of the product of the assigned project and the related project difficulty should be looked at instead of the number of projects. In this direction, when the results in Table 1 are examined, it is observed that the value of 101, which is determined as the constant number of *A*, is not exceeded and this number is close to each other for each reviewer. As a result, it is possible to say that a balanced assignment has taken place.

The results obtained by running the program for another mathematical model in which it is aimed to minimize the sum of the evaluation times of the reviewers' projects are summarized in Fig. 2 and Table 2.

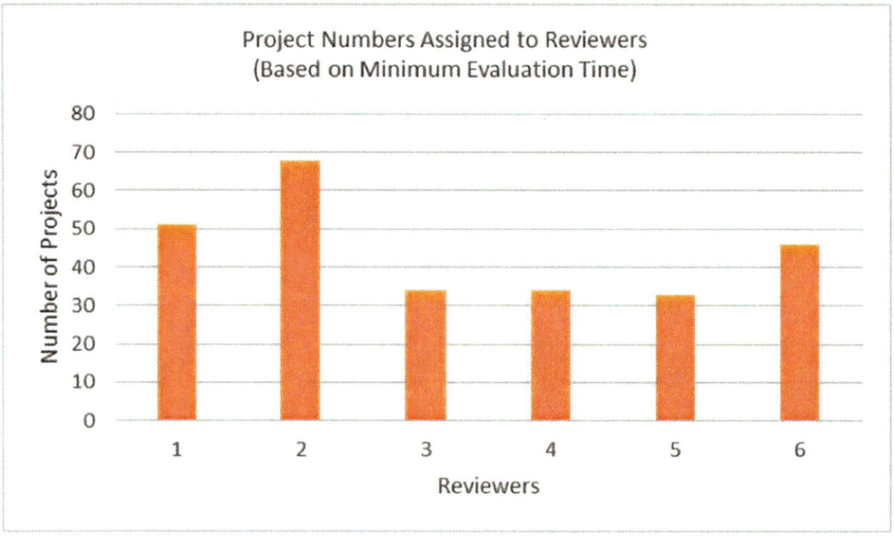

Fig. 2. Number of projects assigned to reviewers (according to the minimum evaluation period)

Table 2. Number of projects assigned to reviewers by project type and minimum evaluation time.

Type of Project						
Reviewer	1	2	3	4	5	6
Defense Petition	1	35	0	0	0	0
Final Report	50	33	1	1	1	38
Project Proposal	0	0	33	33	32	8
Upper Limit (A)	101	101	101	101	98	100

Similarly, when Table 2, which includes the results of the model aiming at the minimum evaluation time, is examined, it is observed that the value of 101 determined as the fixed number of A is not exceeded and this number is close to each other for each reviewer. As a result, it is possible to say that a balanced assignment has been achieved for this model as well.

The average evaluation period of the projects of the reviewers and the number of assigned projects with GAMS output are given in Table 3.

Table 3. The average evaluation period of the reviewers' projects and the number of assigned projects

Reviewer	Average Evaluation Times (days)	Number of Projects Assigned
1	8	51
2	3	68
3	17	34
4	18	34
5	29	33
6	15	46

When Table 3 is examined, more projects were assigned to reviewers with a minimum evaluation time. Therefore, it was ensured that the total evaluation time aimed in the model was minimized.

5 Conclusion

In this study, two mathematical models were developed with the aim of maximum matching degree and minimum evaluation time for reviewer assignment problems. The developed mathematical models were tested with the data received from the scientific research projects unit. Studies show that it meets the objectives of maximum compatibility and minimum evaluation time, and a balanced assignment is achieved while achieving these objectives. In addition, the models give results in an average of 2 s, avoiding hours of manual assignments.

After this point, the objective function of the problem will be considered as a multi-objective problem. The future work includes the development of a decision support system design that can prevent unfair and unbalanced assignments that may occur due to human error, save time, be reliable, user-friendly and include keywords belonging to projects.

Acknowledgments. This study is supported by TUBITAK 2209-A - Research Project Support Programme for Undergraduate Students and Eskisehir Technical University, Scientific Research Projects Committee (22LOP392).

References

1. Hartvigsen, D., Wei, J.C.: The conference paper-reviewer assignment problem. Decis. Sci. **30**(3), 865–876 (1999)

2. Xinlian, L., Watanabe, T.: Automatic paper-to-reviewer assignment, based on the matching degree of the reviewers. Procedia Comput. Sci. **22**, 633–642 (2013)
3. Liu, O., Wang, J., Ma, J., Sun, Y.: An intelligent decision support approach for reviewer assignment in R&D project selection. Comput. Ind. **76**, 1–10 (2015)
4. Pradhan, D.K., Chakrabot, J., Choudhary, P., Nandi, S.: An automated conflict of interest based greedy approach for conference paper assignment system. J. Informetrics **14**(2), 101022 (2020)
5. Kat, B.: An algorithm and a decision support system for the panelist assignment problem: the case of TUBITAK. J. Fac. Eng. Archit. Gazi Univ. **36**(1), 69–87 (2021)

Support Management System Model Proposal for the Student Affairs of Faculty

Ilknur Teke[1](✉) and Cigdem Tarhan[2]

[1] Management Information Systems, Dokuz Eylül University, Izmir, Turkey
ilknur.teke@cbu.edu.tr

[2] Management Information Systems - Regional Development and Business Sciences Research and Application Center (DEÜ-BİMER), Dokuz Eylül University, Izmir, Turkey
cigdem.tarhan@deu.edu.tr

Abstract. In today's technology age, digitalization, which aims to enable institutions to carry out their processes independently of time and space, can be briefly summarized as transferring physically carried out processes to electronic environment, monitoring and managing from there. In this study, a model has been proposed for the Student Affairs Unit, which operates within the Faculty of Economics and Administrative Sciences of Dokuz Eylül University, to move its communication with the students to the digital environment. The proposed model is web-based and responsive, which requires user authentication for the current students. The model includes two different functions. The first function is to create and categorize the frequently answered questions (FAQs) for the faculty and present them to the students with a user-friendly design. The other one includes the process where the student can submit a request as a ticket for the subject to consult for the situations other than the FAQs, and the request is answered by the relevant personnel in the student affairs unit. The proposed model provides business analytics information in the form of response times for requests, requests according to departments, and requests according to categories, in order to be beneficial for Managers as well.

Keywords: Student Affairs · Support Management System · Digitalization · Business Analytics

1 Introduction

Considering the rapidly changing and constantly developing technological developments in the current time, digitalization has become inevitable for most institutions. The 4th industrial revolution, the digitization of the industry, is driving the development of all institutions and ways of doing business, including higher education. [1–4]. When evaluated from the perspective of Higher Education Institutions, the digitalization process revealed institutional processes covering administrative tasks and academic processes covering educational activities. Examples of institutional processes include different software applications based on their respective subjects such as the personnel

Z. Şen et al. (Eds.): IMSS 2023, LNME, pp. 683–692, 2024.
https://doi.org/10.1007/978-981-99-6062-0_64

information system which stores and processes personnel information, tracks personnel affairs and the electronic document management system which includes official records and electronic signature processes. On the other hand, examples of academic processes include the student information system which stores student information and runs academic processes such as courses, exams and grading, and the importance of providing a digital education platform was highlighted during the Covid-19 crisis when education was carried out remotely [5, 6]. As digital and social technologies continue to advance, there is an opportunity to unlock new and innovative ways to interact with the student and university communities [7–9]. From a student's perspective, a platform that includes academic processes, online access to resources, and software support is seen as essential [10].

The aim of this study to propose a solution to digitalize and strengthen the communication process between the unit of student affairs and the student for the face-to-face educational conditions via using management information systems. The proposed model in this study is designed as ticket-based support management system including frequently answered questions also. Dokuz Eylül University (DEU), Faculty of Economics and Administrative Sciences (FEAS) is chosen as case area of this study. When the literature studies in the similar field are examined, it has been seen that the studies on this subject have been carried out by considering the open and distance education conditions, including the education processes. Preparing this study by considering face-to-face education on the physical campus conditions and presenting it as a student affairs support management system distinguishes it from its counterparts.

2 Existing Situation

It has been observed that the student affairs unit operating at DEU/FEAS does not use any information system to manage communication with student. Students applying to the unit in person or trying to reach them via phone cause an undetectable and unmanageable workload in the unit, preventing the unit from fulfilling its other duties and responsibilities. In addition, during recurring processes such as registration periods and graduations, students asking similar questions and staff providing repetitive answers does not add value to the work being done in the unit and leads to the constant repetition of the same tasks. Due to these situations, there are consequences such as not being able to reach the necessary audience with a small announcement when needed, and not being able to find answers to students' questions at other times.

Considering this; it is understood that the most of these problems arise from the lack of communication between the student affairs unit and the students. The solution model presented in the study is designed with the aim of minimizing face-to-face meetings by enabling students to find solutions to their questions about student affairs through the online platform.

3 Literature

Ayaz and Arakaya have mentioned that research on service quality for universities is limited in their studies and emphasized that such a study is especially necessary for newly established universities [11]. In the study, research was conducted on the service quality of the student affairs department, where the students receiving face-to-face education receive service. As a method, a questionnaire study was applied to the students selected from different departments and the demographic characteristics of these students were also evaluated. As a result, the order of priority in the quality of service expected by the students participating in the study was obtained as follows: Reliability, responsiveness, trust, physical characteristics and empathy.

Bozkurt has stated that universities with more than one hundred thousand students are considered as mega universities and presented a study investigating the support services in these universities [12]. In the study, which gave priority to literature research, support services were divided into categories and the part that includes student affairs such as registration procedures and course schedules was specified as administrative services.

Cabellon and Junco has emphasized why it is necessary to keep up with the digital age in their work and provided student affairs practitioners a foundation and framework for the use of digital and social technologies in working with and serving students effectively. However, they noted that the opportunity to use digital information provides university administrators with the ability to make more informed decisions and take timely action to support student success [7].

Genç Kumtepe et al. have made the definition of support services in their studies including course materials. They emphasized that students who receive education through open and distance learning methods need support services more than those who receive education face-to-face, and stated that the most important tool for communication between these students and the institution is web pages. For this reason, content analysis for the support services on the web pages of 60 different universities was carried out and a support services model was proposed according to the results obtained [13].

Johnson Jr. and Yen have associated student affairs with management information systems and mentioned the benefits of using management information systems in processes such as grading processes, lesson planning, transcript processes and fees [14].

Tait has divided the support services in open and distance learning into sections and made a detailed definition. This definition he made was used in most of the subsequent studies in the literature. He stated in general the structure and features of the structure required for the development of the student support service and emphasized that a universal method could not be applied in this field due to the dynamics such as student groups, working styles and educational cultures [15].

4 Method

Within the scope of the study, a focus interview was conducted with Dokuz Eylül University FEAS Student Affairs Unit as a qualitative data collection method. The problems identified after the interview were collected under 3 main headings: infrastructure problems arising from technical reasons, communication problems between students and the unit, and administrative problems. The following summary table is created (Table 1).

Infrastructure Problems: the problems that are related to software using currently in the University. The most important problem about this is the software that are running independently with each other. In detail, the use of different software with different

Table 1. Problems of Student Affairs Unit

Infrastructure Problems	Communication Problems	Administrative Problems
Incompatible technologies of existing systems	The absence of organized and categorized Frequently Asked Questions (FAQs)	Workload caused by more than 1,000 students per staff
The lack of integration between software systems	The announcements not reaching the students due to being shared only on the faculty's website	Inability to specialize in a field due to the assignment of tasks based on faculty departments rather than job nature
Weakness of the support and help sections of the software	Incomplete or incorrect contact information of students	The fact that the staff working in the info desk do not have enough information about the processes
The lack of user guidance for software	Inefficient use of social media accounts	
	Constantly asking similar questions in routine processes by students	
	Difficulties that occur when student participation is not known in advance	

databases that are not integrated with each other causes more than one different value for the same data.

Communication Problems: the problems that are related to communication between student affairs unit and students. There are two important problems in this regard: The unit is unable to provide necessary information flow because of the lack of communication information of students and students apply physically to unit for their administrative process needing.

Administrative Problems: These are the problems arising from the effort to maximize the man/hour productivity within the student affairs unit.

4.1 System Development Life Cycle

In order to solve the problems mentioned within the scope of the study, the support management system aimed to digitalize the communication between student affairs unit and students is designed by following the steps of system development life cycle (Fig. 1).

Defining of Problems, Opportunities and Goals. In this step, difficulties arising from the physical application of students due to the lack of software that provides communication between the student affairs unit and the student is defined as the problem of this study. And developing support management system in order to strengthen student-staff communication, monitor the workload caused by student support and track work is defined as the aim of this study.

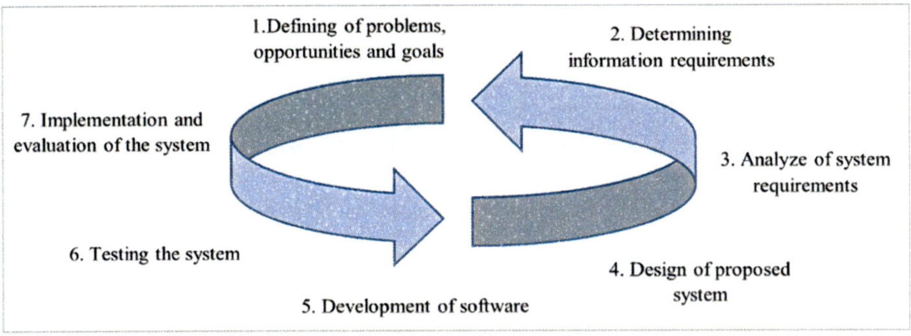

Fig. 1. System Development Life Cycle [16]

Determining Information Requirements. In order to determine the information requirements, a focus interview is held with DEU/FEAS student affairs unit, the problems experienced is discussed and the expectations of the unit regarding the system model is investigated.

Analyze of System Requirements. Considering the system as a whole with its functions, using the JavaScript-based MEAN Stack is deemed appropriate in terms of both efficiency and security. The MEAN Stack is a popular web development stack used to build full-stack JavaScript applications. It consists of four key technologies: MongoDB, Express JS, Angular, and Node JS [17, 18]. To develop a system with MEAN Stack, the following system requirements table is created (Table 2).

Design of Proposed System. In this study, a system model is designed to propose a solution to the problems that are determined after the focus interview with the student affairs unit of DEU/FEAS and compiled under the heading of communication problems in the summary table above. This system model is designed considering the duties, authorities and responsibilities of student affairs unit (Table 2).

The proposed model is designed to be have these features as a support management system:

- Web-based application that is sensitive for the different screen sizes (responsive structure)
- It requires to be log-in for both staff members and students
- It includes FAQs divided into categories that staff can also add questions to
- Tickets can be created by students about what he/she wants to consult and submit it
- Student affairs unit staff can reply for the ticket added by a student
- It includes a business analytics section to track the transactions made within the system, obtain summary data and view the workload.

A flowchart is created showing the operation of the designed model within the scope of this study. The flowchart includes the operations performed by two different types of users, one student and one staff member with user accounts in the system (Fig. 2).

Table 2. System Requirements

Operating System	The MEAN stack is cross-platform and can be developed on Windows, macOS, or Linux
Text Editor or IDE	A text editor or an Integrated Development Environment (IDE) is required for coding. Popular choices for MEAN stack development include Visual Studio Code, Atom, Sublime Text, WebStorm, and Eclipse
Node JS	Node.js is a JavaScript runtime environment that is required for running JavaScript on the server side
MongoDB	MongoDB is a NoSQL (Not Only Structured Query Language) database that stores data in a JSON-like format and is used in MEAN stack development
Angular	Angular is a front-end JavaScript framework that is used for building dynamic web applications
Express JS	Express JS is a back-end JavaScript framework that is used for building web applications and APIs
Additional Libraries	MEAN stack development requires additional libraries and dependencies, such as Mongoose for connection to database, Body-parser, Nodemon, and Cors
Other	• The latest versions of Node JS, MongoDB, Angular and Express JS should be installed on the system • It is also recommended to have a good internet connection and a minimum of 4 GB of RAM to ensure smooth development experience

There are two types of user groups that can login to the system with user authentication. One of them is the students and the other is the staff members of the student affairs unit.

Student who login to the system is presented with categorized FAQs according to her/his department and class information. It is designed to present FAQs under three main headings according to the time of entry to the system (current date), class information of the student and subject headings.

If the FAQs are not sufficient, the student creates a ticket within the system regarding the subject he/she wants to consult. The content of this ticket is expected to be within the scope of the duty and authority of the student affairs unit. The tracking of the ticket-related process is done within the system with the unique tracking number given by the system at the time of application.

While creating a ticket, the student's department information is saved to the database along with the other information about the request. The created ticket is sent to a staff's user account working in that department, according to the department information of the student who created it. The schema showing the creation and response process of a ticket is created as follows (Fig. 3).

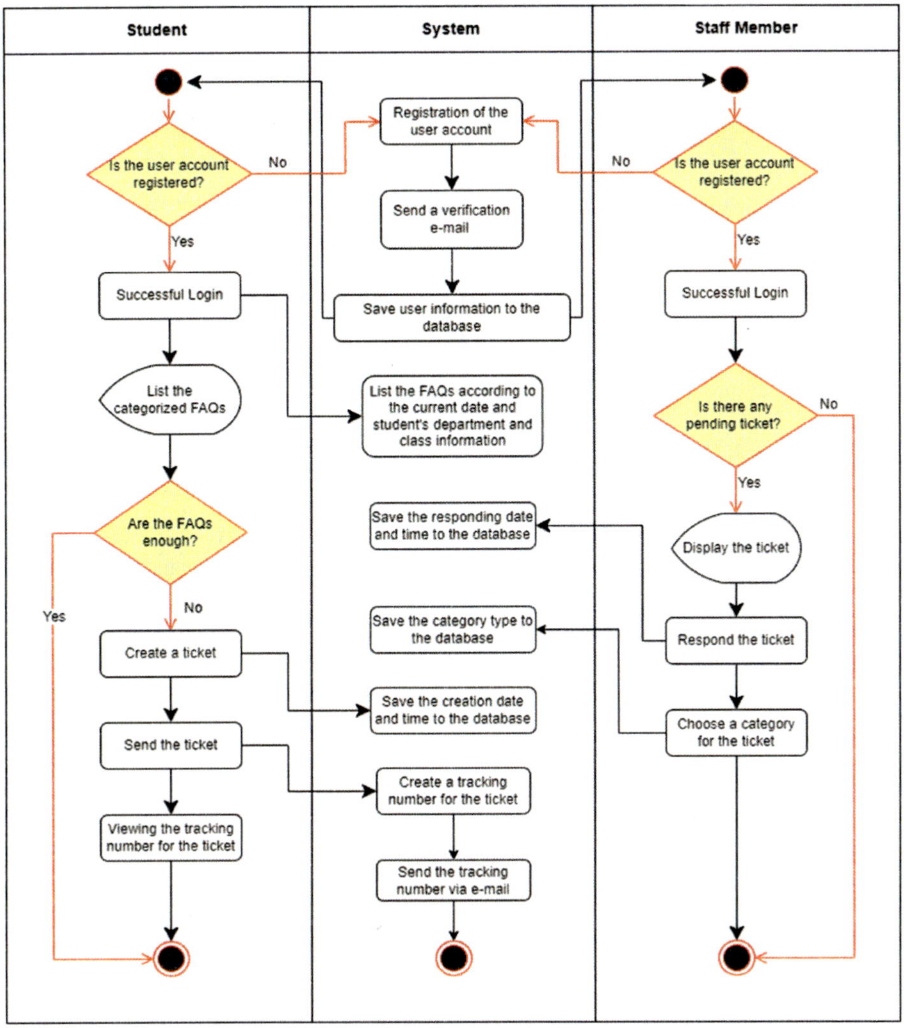

Fig. 2. The Flowchart of Design Model Operations

Student affairs unit personnel (if any) logging into the system displays and replies the pending tickets with the tracking number. In order to categorize and obtain statistical data in the next process, the personnel should choose the category related to the ticket after the answering process.

In this study, a web-based model, which is planned to be used as a support management system, is designed and presented in order to strengthen the communication of DEU/FEAS student affairs unit with students. Other steps of the system development life cycle, which continues with software development, will be carried out in future studies.

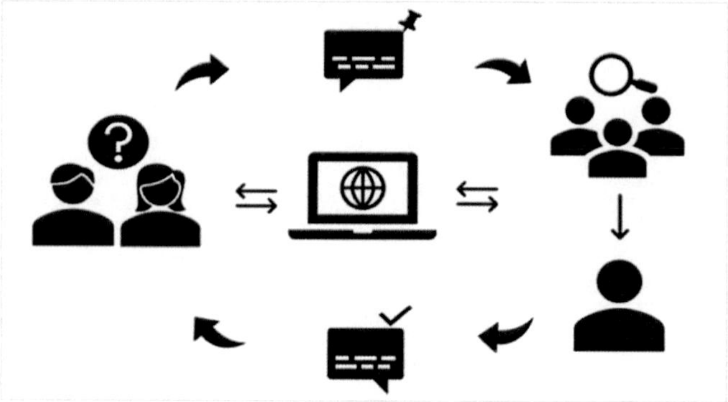

Fig. 3. Staff Selection Process by Department for A Ticket

5 Conclusion

In today's conditions, where digitalization is also important for universities, the lack of software to manage the communication relations between the student affairs units of faculties and the physical applications of students for consulting are identified as the problem of this study.

The model presented as a solution in the study is a responsive web-based application. In the proposed model, it is aimed to speed up the communication between the student and the staff of the faculty student affairs unit and to benefit both parties. Students of the faculty who are actively studying can access the system with user registration in the proposed model. Additionally, frequently asked questions are also categorized in the system. If the requested information is not found among the questions in the system, there is a process in which the student makes the application using the ticket creation method and the staff provides feedback in the form of responding to the ticket.

In the method section of the study, the work flow chart showing the way the model works is presented. The expected benefits of using the support management system that includes this flow are as follows:

- Minimizing students' physical applications to the unit,
- Providing communication by way of creating a ticket within the system instead of communicating by the students via telephone,
- Easily accessible by categorizing frequently asked questions concerning students in the system,
- Keeping the contact information of the students up-to-date by asking the faculty students to verify their e-mail information via OTP (One Time Password) during the creation of a user account,
- Obtaining statistical information over time by placing the created requests into categories by the personnel,
- With the business analytics section, student communication processes of the student affairs unit can be monitored and followed.

In future studies, it is planned to develop the student affairs support management system model, which is designed to strengthen and accelerate the communication between the student affairs unit and students, using MEAN JavaScript stack technology.

In addition to future studies, it is planned to add an artificial intelligence (AI) module to this proposed system. It's aimed to enhance system by providing faster and more accurate responses to common students' tickets. Here are some general steps can be followed to add an AI module to the system:

- Identification of ticket types that can be automated: The selection of the category related to the ticket by the staff in the answering process aims at this. The categorisation of tickets is part of the automation process and is planned to be used in the future to give suggestions to the student during the ticket creation phase.
- Choosing an AI tool or platform: There are many AI tools and platforms available that can help to add an AI module to a support system. Some popular options include Dialogflow, Watson Assistant, and Amazon Lex.
- Training the AI model: Once an AI tool or platform has been selected, it is necessary to train the AI model using relevant data. This can include past student tickets and responses. The more data provided, the better the AI model will be able to understand and respond to tickets.
- Integration of the AI module with the system: Once the AI model has been trained, it required to be integrated with the system. This will involve setting triggers to automatically route specific tickets to the AI module and ensuring that the AI module can generate relevant responses.
- Testing and optimising the AI module: It is important to continuously test and optimise the AI module to ensure that it provides accurate and useful responses to tickets. The AI model may need to be fine-tuned based on feedback from students and staff.

References

1. Telukdarie, A., Munsamy, M.: Digitization of higher education institutions. In: 2019 IEEE International Conference on Industrial Engineering and Engineering Management (IEEM), Macao, China, pp. 716–721 (2019). https://doi.org/10.1109/IEEM44572.2019.8978701
2. Lee, M., et al.: How to respond to the fourth industrial revolution, or the second information technology revolution? Dynamic new combinations between technology, market, and society through open innovation. J. Open Innov.: Technol. Mark. Complexity **4**(3), 21 (2018). https://doi.org/10.3390/joitmc4030021
3. Fomunyam, K.G.: Education and the fourth industrial revolution: challenges and possibilities for engineering education. Int. J. Mech. Eng. Technol. (IJMET) **10**(08), 271–284 (2019)
4. Tri, N.M., Hoang, P.D., Dung, N.T.: Impact of the industrial revolution 4.0 on higher education in Vietnam: challenges and opportunities. Linguist. Cult. Rev. **5**(3), 1–15 (2021). https://doi.org/10.21744/lingcure.v5ns3.1350
5. Bygstad, B., Øvrelid, E., Ludvigsen, S., Dæhlen, M.: From dual digitalization to digital learning space: Exploring the digital transformation of higher education. Comput. Educ. **182**, 104463 (2022). https://doi.org/10.1016/j.compedu.2022.104463
6. Coman, C., Țîru, L.G., Meseșan-Schmitz, L., Stanciu, C., Bularca, M.C.: Online teaching and learning in higher education during the coronavirus pandemic: Students' perspective. Sustainability **12**(24), 10367 (2020). https://doi.org/10.3390/su122410367

7. Cabellon, E.T., Junco, R.: The digital age of student affairs. New Dir. Stud. Serv. **2015**(151), 49–61 (2015). https://doi.org/10.1002/ss.20137
8. Mujalli, A., Khan, T., Almgrashi, A.: University accounting students and faculty members using the blackboard platform during COVID-19; proposed modification of the UTAUT model and an empirical study. Sustainability **14**(4), 2360 (2022). https://doi.org/10.3390/su14042360
9. Szyszka, M., Tomczyk, Ł, Kochanowicz, A.M.: Digitalisation of schools from the perspective of teachers' opinions and experiences: the frequency of ICT use in education, attitudes towards new media, and support from management. Sustainability **14**(14), 8339 (2022). https://doi.org/10.3390/su14148339
10. Thoring, A., Rudolph, D., Vogl, R.: Digitalization of higher education from a student's point of view. In: Book of Proceedings EUNIS 23rd Annual Congress - Shaping the Digital Future of Universities, pp. 279–288 (2017)
11. Ayaz, N., Arakaya, A.: Yükseköğretimde hizmet kalitesi ölçümü: öğrenci işleri daire başkanlığı örneği. Yükseköğretim ve Bilim Dergisi **1**, 123–133 (2019)
12. Bozkurt, A.: Mega üniversitelerde öğrenci destek hizmetleri. https://ab.org.tr/ab13/kitap/bozkurt_mega_AB13.pdf. Accessed 24 Jan 2023
13. Kumtepe, E.G., Toprak, E., Öztürk, A., Büyükköse, G.T., Kılınç, H., Menderis, İ.A.: Açık ve uzaktan öğrenmede destek hizmetleri: Yerelden küresele bir model önerisi. Açıköğretim Uygulamaları ve Araştırmaları Dergisi **5**(3), 41–80 (2019)
14. Johnson, D.E., Yen, D.: (Chi-Chung): management information systems and student affairs. J. Res. Comput. Educ. **23**(1), 127–139 (1990). https://doi.org/10.1080/08886504.1990.10781948
15. Tait, A.: Planning student support for open and distance learning. Open Learn.: J. Open Dist. e-Learn. **15**(3), 287–299 (2000). https://doi.org/10.1080/713688410
16. Sistem Geliştirme Yaşam Döngüsü. https://vahaptecim.com.tr/sistem-gelistirme-yasam-dongusu/. Accessed 08 Feb 2023
17. Mean Stack Tutorial | Mean.js Tutorial – javatpoint. www.javatpoint.com, https://www.javatpoint.com/mean-stack-tutorial. Accessed 10 Feb 2023
18. What is the MEAN Stack? Introduction & Examples, MongoDB. https://www.mongodb.com/mean-stack. Accessed 10 Feb 2023

A Hybrid Decision Model for Balancing the Technological Advancement, Human Intervention and Business Sustainability in Industry 5.0 Adoption

Rahul Sindhwani[1], Sachin Kumar Mangla[2,3], Yigit Kazancoglu[4](✉), and Ayca Maden[5]

[1] Fortune Institute of International Business, New Delhi, India
[2] OP Jindal Global University, Sonipat, India
sachin.kumar@plymouth.ac.uk
[3] Plymouth Business School Plymouth, University of Plymouth, Plymouth, UK
[4] Department of Logistics Management, Yasar University, İzmir, Turkey
yigit.kazancoglu@yasar.edu.tr
[5] Industrial Engineering Department, Istanbul Beykent University, Istanbul, Turkey
aycamaden@beykent.edu.tr

Abstract. In Industry 5.0, humans and machines work together, using advanced technologies like Artificial Intelligence (AI), the Internet of Things (IoT), and automation to improve efficiency, productivity, and quality while also supporting sustainable practices and human values. There is a growing interest in learning about the challenges of Industry 5.0 and exploring these technologies to promote sustainability and responsible business practices. We need a hybrid decision model to strike a balance between technical progress, human values, and sustainable practices as we move toward Industry 5.0, which presents enormous challenges in the areas of technology, the environment, society and ethics, and business and economics. Through a literature analysis guided by the PRISMA technique and the Delphi method, the study highlighted challenges in the areas of technology, the environment, society and ethics, and business and economics, as well as solution measures to address them. The weightage of the challenges was determined using the Best Worst Method, and the ranking of the potential solutions was prioritized using the Elimination and Choice Expressing Reality method.

Keywords: Industry 5.0 · Artificial Intelligence · Technological Advancement · Human Values · Sustainable Practices · Challenges · Solution measure

1 Introduction

Ivanov [1] says that Industry 5.0 technologies have the ability to boost productivity, cut costs, and improve efficiency. But Feng et al. [2] the use of AI technologies also brings new problems linked to sustainability and human values. As the world tries to deal with climate change and other environmental problems, it is becoming more important to find

Z. Şen et al. (Eds.): IMSS 2023, LNME, pp. 693–698, 2024.
https://doi.org/10.1007/978-981-99-6062-0_65

a balance between technological advancement and environmentally friendly methods [3]. At the same time, we need to make sure that growth does not negatively impact important human values like privacy, safety, and security.

Ahmed et al. [4] says that AI's unique technological advances have caused big changes in the business world. Even though these changes have made businesses more efficient, productive, and profitable, they have also made people worry about how they will affect human values and the world. It is important to find a balance between advancements in technology, human values, and sustainable practices so that the benefits of technological progress are shared by many people and do not hurt society or the world [5].

There are two objectives, through which we aim to contribute in this study. First objective is to identify the challenges and solution measures related to balancing technological advancement, human values, and sustainable practices while adopting Industry 5.0. To identify the challenges and solution measures for providing a balance between technological advancement, human values, and sustainable practices while adopting Industry 5.0, we will use the Preferred Reporting Items for Systematic Reviews and Meta-Analyses (PRISMA) approach for systematic literature identification. We will conduct a comprehensive literature review by searching relevant databases and using search terms related to sustainability, human values, and technological advancements in Industry 5.0. We will also consult with experts to identify additional relevant sources. Through this process, we will identify the challenges and solution measures related to sustainability and human values in Industry 5.0.

Second objective is to analyzed the challenges and rank the solution measure to provide the implications for managers if Industry 5.0 is adopted. For attaining this objective, we will develop the framework through a hybrid Best Worst Method (BWM) - Elimination and Choice Expressing Reality (ELECTRE) approach. The initial inputs taken for BWM comparisons will be used to compute the weight of challenges, while further comparison of challenges and solution measures will be obtained for ELECTRE approach to compute the final ranking of the solution measures to overcome the challenges. This hybrid approach will help the authors provide implications and propositions for addressing the challenges to achieve a balance between technological advancement, human values, and sustainable practices in adoption process of Industry 5.0.

This study is structured as follows: Sect. 2 provides the literature review, Sect. 3 provides the Research gap, Sect. 4 provides the research design, Sect. 5 provides the conclusion.

2 Literature Review

The adoption of Industry 5.0 comes with its fair share of challenges, most of which have to do with finding a balance between advancements in technology, human values, and sustainable practices [6]. Industry 5.0 uses new technologies like artificial intelligence, robotics, and the Internet of Things to make output more efficient and better. But if these technologies are used quickly, they can have bad effects on human values and sustainable practices, like polluting the environment, using a lot of energy, and putting people out of work [7]. When adopting Industry 5.0, it is important to find and solve the challenges of balancing advancements in technology, human values, and sustainable practices.

One of the biggest challenges with implementing Industry 5.0 is that it might negatively impact human values. Sindhwani et al. [3] pointed out that the widespread use of advanced technologies in Industry 5.0 can lead to humans being replaced by robots, which can lead to a loss of jobs and a difference in income. The problem is made worse by the fact that there are not enough training and skill-building chances, which makes it hard for the workforce to keep up with the fast-paced changes in technology [1]. Therefore, it is important to make sure that implementing Industry 5.0 does not undermine human values and to come up with plans to help the workforce adjust to new technologies.

In addition to human values, adopting Industry 5.0 also makes it hard to do things in a way that is good for the environment. According to Sharma et al. [8], the integration of advanced technologies in Industry 5.0 may lead to increased energy consumption, carbon emissions, and waste generation, thereby exacerbating environmental problems. Furthermore, the production processes involved in Industry 5.0 may require the extraction of non-renewable resources, such as rare earth metals, leading to environmental degradation [4].

Several methods can be used to find a balance between advances in technology, human values, and practices that are good for the environment while using Industry 5.0. Maddikunta et al. [9] say that it is important to think about sustainability at every stage of implementing Industry 5.0, from planning to operation. This can be done by making a sustainability roadmap that lists the most important areas to focus on, such as energy efficiency, waste reduction, and assessing the effect on the environment, and sets up metrics to track progress toward sustainability goals. Also, everyone who has an interest in the process needs to be involved, including workers, customers, suppliers, and the wider community. Xu et al. [10] say that this can be done by coming up with a way to adopt Industry 5.0 that encourages collaboration and participation from all stakeholders. This method can help to find and address any worries about how Industry 5.0 will affect human values and the environment, as well as to promote practices that are ethical and good for society. In general, it is possible to find a balance between advancements in technology, human values, and sustainable practices when adopting Industry 5.0. This can be done by taking sustainability into account, involving stakeholders in the implementation process, and creating a culture of sustainable innovation.

3 Research Gap and Theoretical Lens

The study of the literature done on the adoption of Industry 5.0 shows that there are several gaps that need to be filled. First of all, there is not enough study on the challenges that organizations face when trying to balance technological progress, human values, and environmentally friendly practices as they move toward Industry 5.0. Even though there is a lot of writing about the benefits of Industry 5.0, not much of it talks about the trade-offs and problems that may come up. So, there is a need to learn more about the problems organizations may face as they move toward Industry 5.0.

Secondly, the role of human values in shaping the adoption and implementation of Industry 5.0 needs further examination. While some studies have examined the impact of Industry 5.0 on job displacement and skill development, little is known about how values such as ethics, social responsibility, and trust can guide the adoption and use of

Industry 5.0. Therefore, more research is needed to understand how human values can be integrated into the adoption process of Industry 5.0, which can help ensure that the technology aligns with human values and promotes positive outcomes for society.

The identified gaps emphasize the crucial requirement of a framework that can effectively balance technological progress, human values, and sustainable practices in the adoption of Industry 5.0. Moreover, these gaps also highlight the need to use the Multi-Criteria Decision-Making (MCDM) approach for assessing the challenges' severity and prioritizing solution measures. Therefore, the defined research objectives are validated by these findings.

The theoretical lens for this research will be grounded in the Multi-Criteria Decision-Making approach, which provides a framework for assessing and prioritizing the challenges and solutions related to balancing technological advancement, human values, and sustainable practices in the adoption of Industry 5.0. The MCDM approach enables decision-makers to evaluate multiple criteria simultaneously, including economic, social, environmental, and ethical factors, and make informed decisions based on a set of predetermined criteria. By using this approach, the research will be able to address the identified research gaps and provide a comprehensive understanding of the challenges and solution measures related to the adoption of Industry 5.0. Additionally, the MCDM approach can help policymakers and industry leaders make informed decisions about the future of Industry 5.0, based on a balanced consideration of technological advancement, human values, and sustainable practices.

4 Research Design

The study follows a positivism research paradigm as this study consists of two research designs. To develop a framework for balancing technological advancement, human values, and sustainable practices while adopting Industry 5.0, an exhaustive literature review was carried out to identify the key challenges and solution measures required to overcome them. These challenges and solution measure were then tabulated and presented before the decision panel for finalization. Based on the expert feedback, a framework was developed and tested using a combination of different methodologies [11]. A hybrid BWM-ELECTRE approach was used to rank the solution measures identified through the literature review. The BWM approach was utilized for computing the weight of the challenges, while the ELECTRE approach was used to identify high-priority solutions.

The study further discusses the relationship between the challenges and their solution measures by describing how these top-priority solutions will assist in balancing technological advancement, human values, and sustainable practices while adopting Industry 5.0. The implications of the study's findings for researchers and practitioners are also presented. The overall flow of present research work are as follows in Fig. 1.

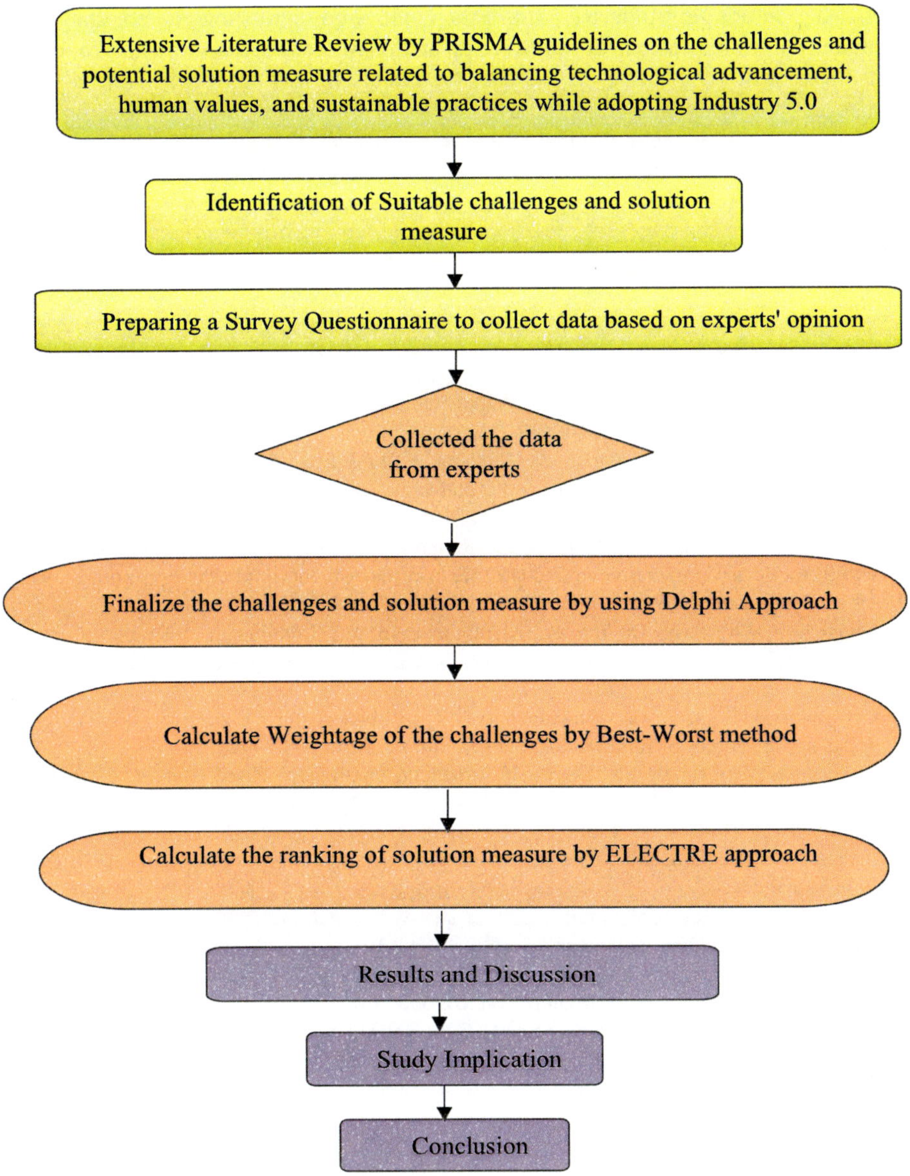

Fig. 1. The present research workflow

5 Conclusion

The results of this study will help us figure out how to balance advancements in technology, human values, and sustainable practices during the Industry 5.0 adoption process. The results of this study will be useful for policymakers, business leaders, and others who have a stake in bringing Industry 5.0 to action. The study will show the possible challenges and solution measures of Industry 5.0, and it will help decision-makers to decide about the Industry 5.0 future. In the end, the results of this study will help to make policies and plans that encourage Industry 5.0 to be used in a responsible way. By finding a balance between technological progress, human values, and sustainable practices, we can make sure that Industry 5.0 is good for society as a whole while minimizing the negative effects possible on individuals and the environment.

References

1. Ivanov, D.: The industry 5.0 framework: viability-based integration of the resilience, sustainability, and human-centricity perspectives. Int. J. Prod. Res. **61**(5), 1683–1695 (2023)
2. Feng, Y., Lai, K.H., Zhu, Q.: Green supply chain innovation: emergence, adoption, and challenges. Int. J. Prod. Econ. 108497 (2022)
3. Sindhwani, R., Afridi, S., Kumar, A., Banaitis, A., Luthra, S., Singh, P.L.: Can industry 5.0 revolutionize the wave of resilience and social value creation? A multi-criteria framework to analyze enablers. Technology in Society **68**, 101887 (2022)
4. Ahmed, T., Karmaker, C.L., Nasir, S.B., Moktadir, M.A., Paul, S.K.: Modeling the artificial intelligence-based imperatives of industry 5.0 towards resilient supply chains: a post-COVID-19 pandemic perspective. Comput. Industr. Eng. **177**, 109055 (2023)
5. Dwivedi, A., Agrawal, D., Jha, A., Mathiyazhagan, K.: Studying the interactions among industry 5.0 and circular supply chain: towards attaining sustainable development. Comput. Industr. Eng. **176**, 108927 (2023)
6. Carayannis, E.G., Morawska-Jancelewicz, J.: The futures of Europe: society 5.0 and industry 5.0 as driving forces of future universities. J. Knowl. Econ. 1–27 (2022)
7. Chander, B., Pal, S., De, D., Buyya, R.: Artificial intelligence-based internet of things for industry 5.0. Artif. Intell.-Based Internet Things Syst. 3–45 (2022)
8. Sharma, M., Sehrawat, R., Luthra, S., Daim, T., Bakry, D.: Moving towards industry 5.0 in the pharmaceutical manufacturing sector: challenges and solutions for Germany. IEEE Trans. Eng. Manage. (2022)
9. Maddikunta, P.K.R., et al.: Industry 5.0: a survey on enabling technologies and potential applications. J. Industr. Inf. Integr. **26**, 100257 (2022)
10. Xu, X., Lu, Y., Vogel-Heuser, B., Wang, L.: Industry 4.0 and industry 5.0—inception, conception and perception. J. Manuf. Syst. **61**, 530–535 (2021)
11. Yadav, G., Luthra, S., Jakhar, S.K., Mangla, S.K., Rai, D.P.: A framework to overcome sustainable supply chain challenges through solution measures of industry 4.0 and circular economy: an automotive case. J. Clean. Prod. **254**, 120112 (2020)

Prediction of Heart Disease Using Fuzzy Rough Set Based Instance Selection and Machine Learning Algorithms

Orhan Torkul, Safiye Turgay, Merve Şişci(✉), and Gül Babacan

Sakarya University, 54100 Serdivan, Sakarya, Turkey
mervesisci@sakarya.edu.tr

Abstract. In this study, instance selection was made using the fuzzy rough set based instance selection method, which is the main indicator of heart disease risk, with the finding of a certain narrowed main cardiovascular number and some other medical findings. Then, with the help of machine learning algorithms, a heart disease risk estimation model was developed over two different size and structure data sets. The heart disease dataset, formed by combining 5 different heart disease datasets, was taken from the IEEE dataport website [1]. In order to eliminate noisy instances, the 12-variable data of 1190 patients was reduced to 836 instances by fuzzy rough set based instance selection method. Qualitative variables used in the analysis are age, sex, chest pain type, resting bps, cholesterol, fasting blood sugar, resting ecg results, maximum heart rate, exercise induced angina, oldpeak, ST slope. Then, the data set was divided into two as 70% training and 30% test data sets, two-class averaged perceptron, two-class Bayes point machine, two-class logistic regression, two-class support vector machine, two-class neural network, two-class locally deep support vector machine and two-class boosted decision tree models were trained. As a result of the validity analysis carried out, the use of fuzzy rough set based instance selection method improved the prediction performance of all models. While the Two-Class Boosted Decision Tree method gave a higher accuracy than other methods, it gave an accuracy result between 89% and 93% in other methods.

Keywords: Heart Disease Risk · Fuzzy Rough Set · Instance Selection · Machine Learning

1 Introduction

Research on artificial intelligence has accelerated with the widespread use of information technologies in every sector in recent years. Artificial intelligence techniques frequently used in many areas such as engineering applications, education, defense industry and health. One of the most important areas of use of artificial intelligence is the field of health. Cardiovescular diseases are currently a major concern in healthcare, one of the chronic and deadly diseases that cause the most deaths. According to the data from the World Health Organization (WHO), approximately 20.5 million people die yearly due to cardiovascular diseases. It estimated that annual deaths due to such diseases increase [2].

Z. Şen et al. (Eds.): IMSS 2023, LNME, pp. 699–709, 2024.
https://doi.org/10.1007/978-981-99-6062-0_66

Heart disease and cardiovascular disease are terms that used interchangeably. Cardiovascular diseases usually refer to narrowed or blocked blood vessels that can lead to a heart attack, chest pain or stroke. Other diseases that affect the heart muscle, valves or rhythm can also be considered types of heart disease [3, 4]. More people die each year from heart disease than from any other cause. Although studies in the literature have determined the causes of heart disease risk, researchers state that more research needed to use this information [5]. Hypertension, diabetes, eating habits, smoking and alcohol use, high blood cholesterol, physical inactivity, gender, age and obesity characteristics of the patients are considered while conducting heart disease research [6]. In addition, the physician's experience and intuition taken into account when making decisions about clinical results. As a result, undesirable effects such as errors, cost-effectiveness, and negative patient care quality may occur. With data analysis methods such as artificial intelligence and data mining, it is possible to create support systems and eliminate undesirable factors [7]. The minimum number of medical measurements, etc., that should be considered in determining the risk of heart disease. Determining the parameter is essential for ease of implementation. Artificial intelligence classification algorithms used to solve this problem. In the literature, various approaches and high-performance studies are developed for detecting heart disease.

In this study, in order to develop a model under more realistic constraints, contrary to the widespread use of the Cleveland sub-dataset in the UCI Heart Disease database, a relatively well-prepared IEEE dataport heart disease dataset including the other four sub-datasets was included in the study as input [1]. Fuzzy Rough Set Based Instance Selection, one of the instance selection algorithms, used to eliminate noisy and inconsistent instances. The performance of the algorithms and the effect of the instance selection method evaluated by training seven different machine-learning algorithms on the datasets with and without the instance selection algorithm.

The rest of the paper is organized as follows. In Sect. 2, studies in the literature are included. Section 3 gives some information about fuzzy RST based instance selection and machine learning algorithms used in this study. Section 4 presents the experimental framework. Section 5 shows the results obtained. Section 6 presents the conclusions reached.

2 Literature Survey

This study aimed to predict heart disease effectively with machine learning methods. In order to improve the performance of the algorithms, the instance selection algorithm used in the methodology. To give examples of studies conducted in this context, Bircha et al. used machine-learning techniques for breast cancer data. Raja et al. applied classification methods based on artificial neural networks together with data mining algorithms [8].

Asencios and others have applied the XDBoost method to analyze credit and lending decisions [9]. Nanda et al. analyzed the CO_2 emission situation together with the multilayer perceptron neural network using the time series from 1980 to 2019 [10]. Jackson applied the support vector machine to multiclass classification problems with Bayesian classifiers [11]. Kumar and Vijaya applied the Naive Bayes machine learning approach to the image classification problem [12]. Using the real fire debris data, Bogdala and others tried to estimate the unknown 15% using Naive Bayes data using 85% of the data [13]. Santana et al., in their study of spatial system development for pattern recognition and intervention systems, examined the Naive Bayes approach together with the fuzzy logic method [14]. Maloney et al. used the two-class Bayes approach in the process of estimating the debt of low loan borrowers [15]. Hashemizadeh et al., adaptive Boosting with Bayesian ridge regression (BRR), K-nearest neighbors (KNN), support vector machine (SVM), decision tree (DT), and Decision Tree to prevent fracture pressure and pore pressure in oil and gas well drilling [16]. They used the regressor methods. Wang et al. aimed to contribute to incremental learning using the gradient-boosting decision tree method [17]. Liu et al. compared extreme gradient boosting (XGBoost) with machine learning-based individual classifiers in predicting financial distress [18]. Again, Liu et al. performed a two-class evaluation process with the tree-enhanced gradient boosting decision trees approach in their credit rating study [19]. Qian et al. developed gradient assisted decision trees with the feature selection approach [20]. Albano et al., in the Label Ranking task, aimed to construct preference models that learn to rank a finite set of labels based on a set of predictive features [21]. They implemented the AdaBoost boost algorithm. Louk et al. demonstrated the dual ensemble model by combining two existing ensemble techniques, such as bagging and gradient boosting decision tree (GBDT) [22]. Lee et al. applied the gradient boost decision tree to Parkinson's disease data [23]. Cardenas et al. applied the gradient boosting algorithm on decision trees using machine-learning algorithms in wireless networks on the intelligent transportation system [24]. Farzana et al. applied the gradient boosting decision tree algorithm for neuroimaging analysis such as small sample size, high dimensionality, and class imbalance [25]. Brenon and others discussed misclassification cases along with machine and deep learning algorithms [26]. Menagadevi et al. used Support Vector Machine (SVM), Extreme Learning Machine (ELM), and K-nearest neighbor algorithm (KNN) for octagonal histogram equalization analysis with Image preprocessing, modified optimal curve thresholding, and black-and-white Stretching for Alzheimer's patients [27]. Lahmiri et al. used deep learning convolutional neural networks (CNN) for automated feature extraction and adapted CNN-based features to a support vector machine (SVM) with Bayesian optimization to perform the classification task [28]. Abhishek and others used machine-learning techniques for binary classification [29]. Hasan et al. used support vector machines (SVM), extreme gradient boosting, artificial neural network (ANN) models, ensemble learning methods, and logistic regression to estimate mortality rates for COVID-19 patients [30]. Hong et al. used two-class kernel logistic regression and support vector machine [31]. El-Attaa et al. performed the analysis using two-class support vector machine and kernel functions [32]. Duran-Rosal et al. performed regression and classification with a feedforward neural network approach [33]. Godoy et al. used logistic regression and machine learning methods to diagnose stable ischemic heart disease [34].

3 Methods

In this study, as seen in Fig. 1, fuzzy rough set based instance selection approach [35] was preferred in instance selection and it aimed to select the data group that could give the most accurate estimate by choosing the active instance group [36]. Machine learning approaches in which instance selection is tested.

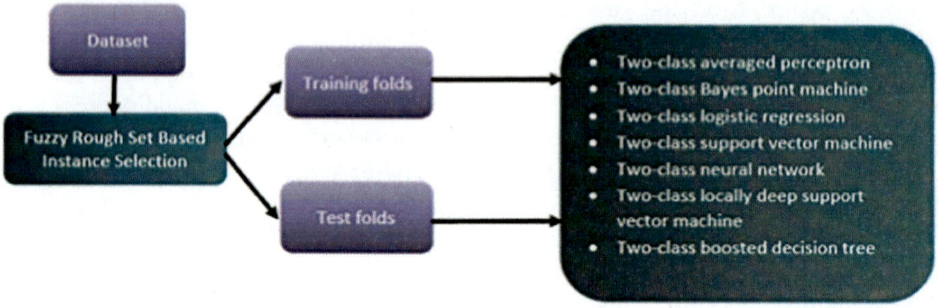

Fig. 1. Experimentation overview.

4 Implementation

The methodology proposed in the study consists of 4 main stages: data acquisition, data preprocessing, classification and evaluation of the performance of the models.

4.1 Dataset

The open source heart disease dataset (comprehensive) [1] on the IEEE dataport website was used to predict heart disease in the study. The dataset was created by combining 5 different heart disease datasets, namely Cleveland Dataset, Hungarian Dataset, Switzerland Dataset, Long Beach VA Dataset and Statlog Heart Data, which contain 303, 294, 123, 200 and 270 instances, respectively. The numbers of instances of these datasets can be seen in Fig. 2.

The qualifications in the dataset and detailed information about the qualifications given in Table 1. The dataset, which includes the clinical data of 1190 patients in total, consists of 12 attributes, including 5 numeric variables and 7 nominal variables. Of these patients, 629 diagnosed with heart disease and 561 were patients without heart disease. While the target attribute to be estimated was used as the output variable, the other 11 attributes were used as input variables.

4.2 Data Preprocessing

When the Comrehensive Heart Disease Dataset examined, it is understood that it does not contain missing data. The data preprocessing stage carried out with two different approaches as shown in Fig. 3. In the first approach, only normalization processes

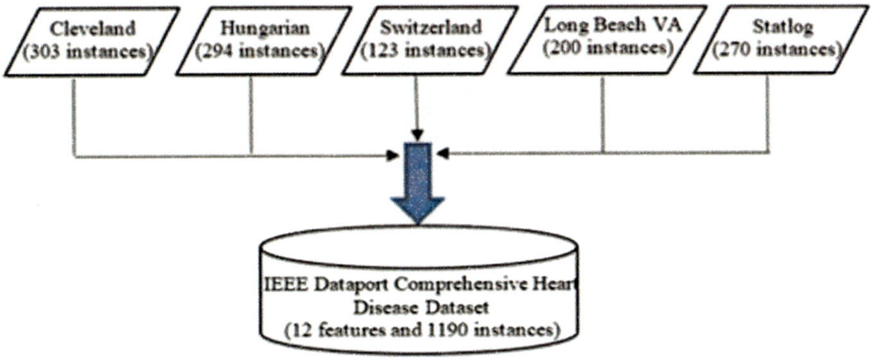

Fig. 2. Content of Comrehensive Heart Disease Dataset.

Table 1. Details of dataset features.

Feature	Data type	Values
Age	numeric	[28–77] (in years)
Sex	categorical (binary)	1 = male; 2 = female
Chest pain type	categorical	1 = typical angina; 2 = atypical angina; 3 = non-anginal pain; 4 = asymptomatic
Resting bps	numeric	[0, 200] (in mm Hg)
Cholesterol	numeric	[0, 603] (in mg/dl)
Fasting blood sugar	categorical (binary)	1 = true; 0 = false (>120 mg/dl)
Resting ecg results	categorical	1 = normal; 2 = having ST-T wave abnormality; 3 = showing probable or definite left ventricular hypertrophy by criteria of Estes; 4 = asymptomatic
Maximum heart rate	numeric	[60, 202]
Exercise induced angina	categorical (binary)	1 = yes; 0 = no
Oldpeak	numeric	[−2.6, 6.2] (in mg/dl)
ST slope	categorical	1 = upsloping; 2 = flat; 3 = downsloping
Target	categorical (binary)	1 = presence of heart disease, 0 = absence of heart disease

was applied to the Comrehensive Heart Disease Dataset and Dataset-1 was obtained. In the second approach, Dataset-2 were obtained by applying instance selection and normalization processes to the dataset.

The fuzzy rough instance selection was applied to the dataset containing 1190 cases with 12 features, in order to eliminate inconsistent instances. Instance selection application was carried out in RStudio program with the help of R programming language. Rıza et al. The 'RoughSets' package developed by [37] is used. The threshold value, which

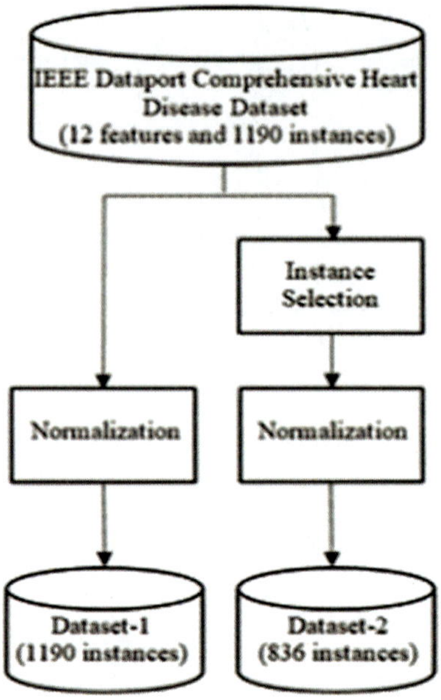

Fig. 3. Data preprocessing flowchart.

determines whether an object can be removed or not, was selected as 0.4, and the alpha parameter, which determines the level of detail of the fuzzy similarity measure, was selected as 0.8. For type of aggregation, the "lukasiewicz" t-norm was used. As a result of applying the fuzzy rough instance selection algorithm, a dataset with 836 instances was obtained.

As seen in Table 1, numerical variables named 'age', 'resting bps', cholesterol', 'maximum heart rate', 'oldpeak' in Dataset-1 and Dataset-2 are distributed in different scales. In order to prevent features with a wider value range than others from dominating the model [38] and to avoid generating biases in the training of machine learning algorithms [39], the min-max normalization process applied to these features. The min-max normalization given in Eq. 1 rescales the values found in the properties to be in the range of [0–1]. Here x_{mm}, min(x) and max(x) are the normalized, minimum and maximum values of the attribute, respectively. In this way, values in all numeric attributes converted to be in the same unit range [40].

$$x_{mm} = \frac{x - \min(x)}{\max(x) - \min(x)} \tag{1}$$

4.3 Classification

The training and testing phases of the models carried out on the Microsoft Azure Machine Learning Studio platform. Dataset-1 and Dataset-2 randomly divided into two as 70% training and 30% test dataset. 7 different two-class machine learning algorithms were used for training the models: Averaged Perceptron, Bayes Point Machine, Logistic Regression, Support Vector Machine, Neural Network, Locally Deep Support Vector Machine, Boosted Decision Tree. The 'Tune Model Hyperparameters' module used to determine the hyperparameters during the training of the models. Random parameter sweeping mode was selected for the use of the module and the maximum number of runs on random sweep was determined as 50. The hyperparameter values determined by the module as result of training the models on Dataset-1 and Dataset-2 given in Table 2.

Table 2. Hyperparameters and values of classifier models.

Model	Hyperparameter values for Dataset-1	Hyperparameter values for Dataset-2
Averaged Perceptron	learning rate = 0.173336, maximum number of iterations = 9	learning rate = 0.238288, maximum number of iterations = 1
Bayes Point Machine	number of training iterations = 30	number of training iterations = 30
Logistic Regression	optimization tolerance = 0.000007, L1Weight = 0.851971, L2Weight = 0.482102, memory size = 47	optimization tolerance = 0.000006, L1Weight = 0.182167, L2Weight = 0.428548, memory size = 15
Support Vector Machine	number of iterations = 34, lambda = 0.004656	number of iterations = 49, lambda = 0.000633
Neural Network	learning rate = 0.031553, loss function = cross entropy, number of iterations = 144	learning rate = 0.034342, loss function = cross entropy, number of iterations = 118
Locally Deep Support Vector Machine	LD-DVM tree depth = 4, lambda W = 0.081274, lambda theta = 0.029685, lambda theta prime = 0.027993, sigma = 0.825813, num iterations = 12438	LD-DVM tree depth = 2, lambda W = 0.027329, lambda theta = 0.04192, lambda theta prime = 0.001664, sigma = 0.096759, num iterations = 11413
Boosted Decision Tree	number of leaves = 91, minimum leaf instances = 15, learning rate = 0.169304, number of trees = 52	number of leaves = 14, minimum leaf instances = 11, learning rate = 0.291167, number of trees = 314

4.4 Evaluation of Performances

The next step after training the classifier machine learning models is to evaluate and compare the performances of the trained models on the test datasets. For this purpose, accuracy, precision, recall and F1-Score statistical performance criteria, which are widely used in classification studies, used in the study. Let TP be the true positive predictions,

FP the false positive predictions, TN the true negative predictions, and FN the false negative predictions. In this case, accuracy, precision, recall and F1-Score formulations are as in Eq. 2–5, respectively [41].

$$Accuracy = \frac{TP + TN}{TP + TN + FP + FN} \tag{2}$$

$$Precision = \frac{TP}{TP + FP} \tag{3}$$

$$Recall = \frac{TP}{TP + FN} \tag{4}$$

$$F1{-}secore = \frac{2 * Precision * Recall}{Precision + Recall} \tag{5}$$

5 Results

The performance criterion values of 7 models trained and tested on Dataset-1, which was developed without applying the instance selection method for heart disease risk estimation, are given in Table 3. The Boosted Decision Tree model, with values of 0.929, 0.921, 0.944 and 0.932, showed the best performance among all models in terms of four performance criteria: accuracy, precision, recall and F1-Score. The lowest performance, with values of 0.836, 0.810, 0.851 for accuracy, precision and F1-Score, belongs to the Bayes Point Machine model, while the Averaged Perceptron model for recall performance criteria. While the second highest accuracy value is obtained with 0.849 in Neural Network and Locally Deep Support Vector Machine models, Neural Network model's precision value is higher with 0.833 among these models, recall and F1-Score values of Locally Deep Support Vector Machine model are higher with 0.911 and 0.863 values.

Table 3. Performance criterion values of the models trained on Dataset-1.

Model	Accuracy	Precision	Recall	F1-Score
Averaged Perceptron	0,840	0,821	0,887	0,853
Bayes Point Machine	0,836	0,810	0,895	0,851
Logistic Regression	0,840	0,816	0,895	0,854
Support Vector Machine	0,845	0,822	0,895	0,857
Neural Network	0,849	0,833	0,887	0,859
Locally Deep Support Vector Machine	0,849	0,819	0,911	0,863
Boosted Decision Tree	0.929	0,921	0,944	0,932

The performance criterion values of the models trained and tested on the Dataset-2 obtained by applying the instance selection algorithm given in Table 4. When the table analyzed, it seen that the Boosted Decision Tree model provides the highest estimation performance among all models with 0.940, 0.928 and 0.947 values in accuracy, precision and F1-Score criteria, respectively. The highest value in terms of the Recall criterion obtained with the Locally Deep Support Vector Machine model with a value of 0.978. Averaged Perceptron and Support Vector Machine models had the lowest accuracy with 0.892, while the lowest precision and F1-Score values seen in the Averaged Perceptron model with 0.887 and 0.905 values, respectively. The lowest recall value obtained in the Logistic Regression model with a value of 0.914. When the results examined, it seen that the performance of all models for considered performance criteria have improved with the application of the Fuzzy Rough Instance Selection Algorithm. This improvement occurred with point increases between 0.011 and 0.085 in accuracy values, 0.007 and 0.091 in precision values, 0.024 and 0.067 in recall values, and 0.015 and 0.08 in F1-Score values.

Table 4. Performance criterion values of the models trained on Dataset-2.

Model	Accuracy	Precision	Recall	F1-Score
Averaged Perceptron	0,892	0,887	0,925	0,905
Bayes Point Machine	0,898	0,888	0,935	0,911
Logistic Regression	0,904	0,905	0,925	0,915
Support Vector Machine	0,892	0,879	0,935	0,906
Neural Network	0,904	0,914	0,914	0,914
Locally Deep Support Vector Machine	0,934	0,910	0,978	0,943
Boosted Decision Tree	0,940	0,928	0,968	0,947

6 Conclusions

In estimating the risk of heart disease, the Decision Trees Algorithm, with relatively few decision variables, provided simplicity in implementation and yielded acceptable (83%) accuracy. The scales of the data sets, which are considered as a future study, can be increased or other algorithms can be compared to alternative Decision Trees for this data set selection approach, and higher performance can be sought in heart disease risk estimation. This study used the Fuzzy Rough Instance Selection Algorithm for data reduction. Two class machine learning algorithms were used for heart disease prediction. As a result of the evaluation of the variables obtained, the validity of my diagnosis of the disease was tested with this study. Regarding future work, we can add to the two class algorithms and compare the performance results to improve other machine learning algorithms.

References

1. Siddhartha, M.: Heart disease dataset (comprehensive). IEEE Dataport (2020)

2. Reddy, K.V.V., Elamvazuthi, I., Aziz, A.A., Paramasivam, S., Chua, H.N., Pranavanand, S.: Heart disease risk prediction using machine learning classifiers with attribute evaluators. Appl. Sci **11**(18), 8352 (2021)
3. Dwivedi, A.K.: Performance evaluation of different machine learning techniques for prediction of heart disease. Neural Comput. Appl. **29**(10), 685–693 (2016). https://doi.org/10.1007/s00521-016-2604-1
4. Anbarasi, N.C.S.N.I.M., Anupriya, E.: Enhanced prediction of heart disease with feature subset selection using genetic algorithm enhanced prediction of heart disease with feature subset selection using genetic algorithm. Int. J. Eng. Sci. Technol. **2**(10), 5370–5376 (2010)
5. Kavitha, M., Gnaneswar, G., Dinesh, R., Sai, Y.R., Suraj, R.S.: Heart disease prediction using hybrid machine learning model. In: Proceedings of the 6th International Conference on Inventive Computation Technologies, ICICT 2021, pp. 1329–1333. IEEE, India (2021)
6. Dangare, C., Apte, S.: A data mining approach for prediction of heart disease using neural networks. Int. J. Comput. Eng. Technol. (IJCET) **3**(3) (2012)
7. Birchha, V., Nigam, B.: Performance analysis of averaged perceptron machine learning classifier for breast cancer detection. Procedia Comput. Sci. **218**, 2181–2190 (2023)
8. Aswathi, R.R., Jency, J., Ramakrishnan, B., Thanammal, K.K.: Classification based neural network perceptron modelling with continuous and sequential data. Microprocess. Microsyst. 104601 (2022)
9. Asencios, R., Asencios, C., Ramos, E.: Profit scoring for credit unions using the multilayer perceptron, XGBoost and TabNet algorithms: evidence from Peru. Expert Syst. Appl. **213**, 119201 (2023)
10. Nanda, A.K., Gupta, S., Saleth, A.L.M., Kiran, S.: Multi-layer perceptron's neural network with optimization algorithm for greenhouse gas forecasting systems. Environ. Challenges **11**, 100708 (2023)
11. Jackson, P.L.: Support vector machines as Bayes' classifiers. Oper. Res. Lett. **50**(5), 423–429 (2022)
12. Kumar, P.R., Vijaya, A.: Naïve Bayes machine learning model for image classification to assess the level of deformation of thin components. Mater. Today: Proc. **68**, 2265–2274 (2022)
13. Bogdal, C., Schellenberg, R., Höpli, O., Bovens, M., Lory, M.: Recognition of gasoline in fire debris using machine learning: part I, application of random forest, gradient boosting, support vector machine, and naïve bayes. Forensic Sci. Int. **331**, 111146 (2022)
14. Santana, É.R., Lopes, L., de Moraes, R. M.: Recognition of the effect of vocal exercises by fuzzy triangular naive bayes, a machine learning classifier: a preliminary analysis. J. Voice (2022)
15. Maloney, D., Hong, S.C., Nag, B.N.: Two class Bayes point machines in repayment prediction of low credit borrowers. Heliyon **8**(11), e11479 (2022)
16. Hashemizadeh, A., Maaref, A., Shateri, M., Larestani, A., Hemmati-Sarapardeh, A.: Experimental measurement and modeling of water-based drilling mud density using adaptive boosting decision tree, support vector machine, and K-nearest neighbors: a case study from the South Pars gas field. J. Petrol. Sci. Eng. **207**, 109132 (2021)
17. Wang, K., Lu, J., Liu, A., Song, Y., Xiong, L., Zhang, G.: Elastic gradient boosting decision tree with adaptive iterations for concept drift adaptation. Neurocomputing **491**, 288–304 (2022)
18. Liu, W., Fan, H., Xia, M., Pang, C.: Predicting and interpreting financial distress using a weighted boosted tree-based tree. Eng. Appl. Artif. Intell. **116**, 105466 (2022)
19. Liu, W., Fan, H., Xia, M.: Credit scoring based on tree-enhanced gradient boosting decision trees. Expert Syst. Appl. **189**, 116034 (2022)

20. Qian, H., Wang, B., Yuan, M., Gao, S., Song, Y.: Financial distress prediction using a corrected feature selection measure and gradient boosted decision tree. Expert Syst. Appl. **190**, 116202 (2022)
21. Albano, A., Sciandra, M., Plaia, A.: A weighted distance-based approach with boosted decision trees for label ranking. Expert Syst. Appl. **213**, 119000 (2023)
22. Louk, M.H.L., Tama, B.A.: Dual-IDS: a bagging-based gradient boosting decision tree model for network anomaly intrusion detection system. Expert Syst. Appl. **213**, 119030 (2023)
23. Lee, S.B., et al.: Predicting Parkinson's disease using gradient boosting decision tree models with electroencephalography signals. Parkinsonism Relat. Disord. **95**, 77–85 (2022)
24. Cárdenas, L.L., León, J.P.A., Mezher, A.M.: GraTree: a gradient boosting decision tree based multimetric routing protocol for vehicular ad hoc networks. Ad Hoc Netw. **137**, 102995 (2022)
25. Ali, F.Z., Wengler, K., He, X., Nguyen, M.H., Parsey, R.V., DeLorenzo, C.: Gradient boosting decision-tree-based algorithm with neuroimaging for personalized treatment in depression. Neurosci. Informatics, 100110 (2022)
26. Brenon, A., Moncla, L., McDonough, K.: Classifying encyclopedia articles: comparing machine and deep learning methods and exploring their predictions. Data Knowl. Eng. **142**, 102098 (2022)
27. Menagadevi, M., Mangai, S., Madian, N., Thiyagarajan, D.: Automated prediction system for Alzheimer detection based on deep residual autoencoder and support vector machine. Optik **272**, 170212 (2023)
28. Lahmiri, S.: Hybrid deep learning convolutional neural networks and optimal nonlinear support vector machine to detect presence of hemorrhage in retina. Biomed. Signal Process. Control **60**, 101978 (2020)
29. Abhishek, A., Jha, R.K., Sinha, R., Jha, K.: Automated classification of acute leukemia on a heterogeneous dataset using machine learning and deep learning techniques. Biomed. Signal Process. Control **72**, 103341 (2022)
30. Hasan, M., et al.: Pre-hospital prediction of adverse outcomes in patients with suspected COVID-19: Development, application and comparison of machine learning and deep learning methods. Comput. Biol. Med. **151** (2022)
31. Hong, H., Pradhan, B., Xu, C., Bui, D.T.: Spatial prediction of landslide hazard at the Yihuang area (China) using two-class kernel logistic regression, alternating decision tree and support vector machines. CATENA **133**, 266–281 (2015)
32. El-Atta, A.H.A., Hassanien, A.E.: Two-class support vector machine with new kernel function based on paths of features for predicting chemical activity. Inf. Sci. **403**, 42–54 (2017)
33. Durán-Rosal, A.M., Durán-Fernández, A., Fernández-Navarro, F., Carbonero-Ruz, M.: A multi-class classification model with parametrized target outputs for randomized-based feedforward neural networks. Appl. Soft Comput. **133**, 109914 (2023)
34. Godoy, C., et al.: Predicting left main stenosis in stable ischemic heart disease using logistic regression and boosted trees Lucas. Am. Heart J. **256**, 117–127 (2023)
35. Jensen, R., Cornelis, C.: Fuzzy-rough instance selection. In International Conference on Fuzzy Systems, pp. 1–7. IEEE, Spain (2010)
36. Derrac, J., Cornelis, C., García, S., Herrera, F.: Enhancing evolutionary instance selection algorithms by means of fuzzy rough set based feature selection. Inf. Sci. **186**(1), 73–92 (2012)
37. Riza, L.S., et al.: Package 'roughsets' (2015)
38. Lantz, B.: Machine Learning with R, 2nd edn. Packt Publishing, Birmingham (2015)
39. Subasi, A.: Practical Machine Learning for Data Analysis Using Python. Academic Press, London (2020)
40. Kubat, M.: An Introduction to Machine Learning, 3rd edn. Springer, Cham (2021)
41. Ghatak, A.: Machine Learning with R. Springer, Singapore (2017)

Optimization of Methylene Blue Adsorption on Olive Seed Activated Carbon Using Response Surface Methodology (RSM) Modeling-Artificial Neural Network

Tijen Over Ozcelik[1](✉), Mehmet Cetinkaya[1], Birsen Sarici[2], Dilay Bozdag[1], and Esra Altintig[3]

[1] Department of Industrial Engineering, Sakarya University, Serdivan 54050, Sakarya, Turkey
{tover,g200102356,y225006008}@sakarya.edu.tr
[2] Uğur Educational Institutions, Serdivan, Sakarya, Turkey
[3] Pamukova Vocational School, Sakarya University of Applied Sciences, Serdivan 54900, Sakarya, Turkey
altintig@subu.edu.tr

Abstract. In this study, olive seed activated carbon (OSAC), which is primarily agricultural waste, was used as an adsorbent. Response surface methodology (RSM) approach and Artificial Neural Network (ANN) was used to optimize/model the cationic dyestuff removal by adsorption. RSM was first applied to evaluate the process using four controllable operating parameters, namely the amount of OSAC, initial pH (pH initial), mixing time and dyestuff concentration, and optimal conditions for decolorization were determined. In the optimization method, initial dye concentration (50–150 mg/L), adsorbent dosage (0.1–0.5 g), pH (3–9) and contact time (10–90 min) were used as independent variables, percent removal efficiency was chosen as the dependent variable. Also, the value of R^2 ($R^2 = 0.9714$) shows that regression can predict the response for the adsorption process in the studied range. It shows that it is possible to optimize/estimate the dyestuff removal process using the RSM approach and shows that adsorption to OSAC can be used to remove colour from dye effluent.

Keywords: Response surface methodology · artificial neural network · dyes · optimization · adsorption · water treatment

1 Introduction

Along with the rapid development of human society, it brings environmental problems and different pollution problems. The increase in agricultural (70%), industrial (22%) and domestic (8%) water needs, among other needs, causes the formation of significant amounts of wastewater. Among the important pollutants found in wastewater are synthetic and complex industrial dyestuffs with low biodegradability [1]. Today, there are almost one million tons and more than one hundred thousand commercially produced dyestuffs annually [2, 3]. The wastewater of industries that consume large amounts of

Z. Şen et al. (Eds.): IMSS 2023, LNME, pp. 710–721, 2024.
https://doi.org/10.1007/978-981-99-6062-0_67

paint, such as textiles, paint, paper, food, cosmetics, plastics, solvents, furniture, leather, contain large amounts of dyestuffs [3]. Dyes are an important source of water pollution [2, 4]. There are different removal methods for removing organic dyes from the aquatic environment. These methods are a wide variety of physico-chemical and biological techniques such as photocatalysis, biodegradation, coagulation-flocculation, filtration, adsorption, and reverse osmosis [5, 6]. Due to its features such as ease of use, high efficiency, technology requiring low energy, availability of different adsorbents, efficiency in regeneration and reuse of adsorbent, adsorption is the most effective method used to remove dyes from aqueous solution compared to other techniques [7–10]. Modelling and optimization of treatment processes used in the removal of pollutants is of great importance in terms of operating these processes economically and technically [11]. When the literature is examined, traditional optimization studies cannot fully express the interaction between the variables affecting the process, but also require many experiments to determine the optimum operating conditions. Also, in traditional modelling/optimization methods, only one parameter is changed at a time [12]. This makes experiments time consuming and costly. Experimental results do not consider the interactions between the parameters affecting the process, making it difficult to obtain real optimum conditions [13, 14]. In the last twenty years, Among the multivariate experimental design-based methods, RSM is of great interest especially in the design, modelling, and optimization of environmental and chemical experiments. Response surface methodology (RSM) has been used by many researchers to study and optimize water and wastewater treatment processes [11, 15–17]. Surface response method (RSM) is a statistical modelling method that is used successfully especially in chemical processes and in which experimental parameters are optimized. By calculating the interaction between independent parameters, he designs an experimental set to determine the optimum conditions [18]. In addition to determining more accurate conditions for treatment with YYY, the number of experiments that need to be done is reduced compared to the traditional method, thus reducing the working time and cost. In particular, it is of great interest in the design, modelling and optimization of environmental and chemical experiments. Response surface methodology (RSM) has been used by many researchers to study and optimize water and wastewater treatment processes [11, 15–17]. Surface response method (RSM) is a statistical modelling method that is used successfully especially in chemical processes and in which experimental parameters are optimized. By calculating the interaction between independent parameters, he designs an experimental set to determine the optimum conditions [18]. In addition to determining more accurate conditions for treatment with RSM, the number of experiments that need to be done is reduced compared to the traditional method, thus reducing the working time and cost. The primary goal of RSM is to find out what the best operating variables of the process are and to evaluate the relative importance of the process variables in the presence of dynamic interactions. In this study, OSAC was used as an alternative adsorbent for the removal of MB dyestuff. The central compound design RSM was used to predict the effectiveness of the dye removal and the correlation between the dye removal and the four input parameters OSAC dosage, initial pH, contact time and initial dyestuff concentration, and develop a mathematical model for the dye removal performance.

2 Material and Method

2.1 Materials and Synthesis

Olive pit shells obtained from Sakarya province Geyve region were washed with distilled water (Minipure, Destup) and dried in an oven (JSR, JSOF-100) at 105 °C. After the dried materials were ground by a hammer mill, materials of different sizes were obtained by sifting with the help of sieves. These obtained materials were stored in a desiccator to be used in direct adsorption experiments. Methylene Blue (MB), a basic dye, was used in the study. The maximum wavelength of the dyestuff is 665 nm, its molecular weight is 319.85 g/mol, its chemical formula is $C_{16}H_{18}ClN_3S$ and its chemical structure is given in Fig. 1. A 1000 mg/L stock solution of the dyestuff was prepared and used from this solution by dilution in the experiments.

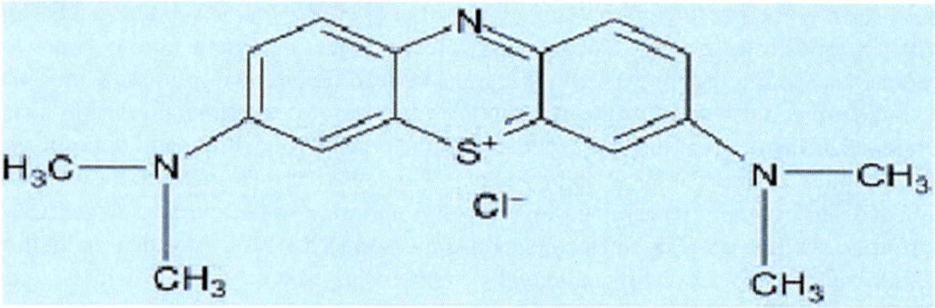

Fig. 1. Molecular structure of MB

Adsorption experiments were carried out in batch system under the conditions determined in the OSAC experimental design, and NaOH and HCl solutions were used for pH adjustments in the experiments. Equilibrium time was determined as 180 min in preliminary trials. At the end of the equilibrium period, the extracted samples were centrifuged at 6000 rpm for 5 min, the concentrations remaining in the solution were read in the spectrophotometer, and their adsorption capacities were calculated with the formulas shown in Eq. 1 and percent removal efficiencies in Eq. 2. Then, these values were used in RSM optimization.

$$q_e = \frac{(C_0 - C_e)V}{m} \tag{1}$$

$$removal(\%) = \frac{(C_0 - C_e)}{C_0} \times 100 \tag{2}$$

where qe (mg/g) is the adsorption capacity of MB, C_0 (mg/L) is the initial MB concentration, Ce (mg/L) is the equilibrium concentration of adsorbate, V is the volume of solution (L) and m is the weight of adsorbent (g).

2.2 Response Surface Method

Response surface method (RSM) can be summarized as the sum of statistical tools and techniques to establish a functional relationship between the response variable and independent variables. It is difficult or impossible to create complex mathematical equations that define many independent input variables to the result parameter with real values in RSM, which is widely used in many different fields. In this case, it may be possible to derive an expression that describes performance measures based on the response variable obtained from some combination of its independent variables. If the response of the independent variables in the system is independent, the response function is given in the first order. If there is non-linearity in the system, higher-order polynomials are used. A quadratic model is shown in Eq. 3.

$$Y = \beta_0 + \sum\nolimits_{i=1}^{k} \beta_i X_i + \sum\nolimits_{i=1}^{k} \beta_i X_i^2 + \sum\nolimits_{i=1}^{k-1} \sum\nolimits_{j=i+1}^{k} \beta_i X_i X_j + \varepsilon \tag{3}$$

Y; response variable, X_i ve X_i^2 $(i = 1, \ldots k)$; arguments $\beta_0, \beta_i, \beta_{ii}(i = 1, \ldots, k), \beta_{ij}(i = 1, \ldots, k-1)$; regression coefficients and ε; is the random error term.

Experimental design experiments were carried out in two stages. In the first stage, Placket-Burman (PB) scanning experiments were used, and in the second stage, the Face Centered Central Composite Design, in which the effective parameters and their effects were revealed. The independent variables and their corresponding actual values are given in Table 1.

Table 1. Value ranges and levels determined for independent variables

Variables	Low (−)	Middle (0)	High (+1)
pH	3	6	9
Adsorbent dose (g/100 mL)	0.1	0.3	0.5
Contact time (min)	10	50	90
MB concentration (mg/L)	50	100	150

The design matrix consisting of 31 experiments and the results obtained are given in Table 2.

2.3 Artificial Neural Network

Artificial neural networks are designed as structures with properties similar to the biological functions of the brain, such as the ability to learn, think, remember, understand reasons, and solve problems. Neural network models consist of neurons with simple processing elements, and weighted connections interconnect these neurons. Thanks to weighted connections, neurons interact and process. The neurons are arranged in different layers, and the input layer receives signals from external sources and transfers them to the intermediate layers. The intermediate layers process the incoming signals with the

Table 2. Experimental design of OSAC and obtained results

Experiment No	Factor 1 x_1: pH	Factor 2 x_2: : Amount (g/100 mL)	Factor 3 x_3: : Contact time (min)	Factor 4 x_4: : Concentration (mg/L)	% Y (MB Yield)
1	3	0.1	10	50	46.37
2	9	0.1	10	50	57.28
3	3	0.5	10	50	68.92
4	9	0.5	10	50	74.27
5	3	0.1	90	50	72.09
6	9	0.1	90	50	76.12
7	3	0.5	90	50	81.78
8	9	0.5	90	50	83.56
9	3	0.1	10	150	51.67
10	9	0.1	10	150	71.23
11	3	0.5	10	150	67.45
12	9	0.5	10	150	72.9
13	3	0.1	90	150	92.34
14	9	0.1	90	150	94.72
15	3	0.5	90	150	93.27
16	9	0.5	90	150	97.12
17	3	0.3	50	100	87.49
18	9	0.3	50	100	95.92
19	6	0.1	50	100	98.23
20	6	0.5	50	100	99.15
21	6	0.3	10	100	71.23
22	6	0.3	90	100	99.85
23	6	0.3	50	50	98.47
24	6	0.3	50	150	98.98
25	6	0.3	50	100	98.67
26	6	0.3	50	100	93.24
27	6	0.3	50	100	98.75
28	6	0.3	50	100	99.17
29	6	0.3	50	100	99.12
30	6	0.3	50	100	98.86
31	6	0.3	50	100	95.7

transfer function and send them to the output layer. The error is calculated by comparing the predicted values obtained from the output layer of the network with the actual values. This error is transferred to all network layers using the backward propagation algorithm. In this process, the weights are updated, and the training takes place. As a result, the predicted values become closer to the valid values when the same input data is given to the network [20].

3 Results and Discussion

3.1 Model Building and Statistical Analysis

RSM is an essential tool for optimizing the adsorption process when various operational individual parameters and their interactions affect the removal efficiency. An optimization technique called RSM with four variable interactions performed under various factors was used to optimize all responses and runs. Initial dye concentration (50–150 mg/L), adsorbent dose (0.1–0.5 g), pH (3–9) and time (10–90 min) to maximize dye removal efficiency chosen. 31 experiments designed with the Face Centered Central Composite Design method were carried out and the dyestuff removal efficiencies were measured. Modeling was done with Minitab statistical analysis program. Parameters that had little effect on the experimental setup were removed from the model-by-model reduction. The corrected quadratic regression model is given in Eq. 3. Y represents decolorization efficiency, x_1, x_2, x_3 and x_4 represent pH, amount (Absorbent dose), contact time and dyestuff concentration, respectively.

$$Y = -3.22 + 13.14 * x_1 + 58.1 * x_2 + 1.239 * x_3 + 0.0817 * x_4 - 0.936* x_1^2 - 0.00912 * x_3^2 - 0.01565 * x_1 * x_3 - 0.292 * x_2 * x_3 - 0.2193 * x_2* x_4 + 0.001459 * x_3 * x_4 \quad (4)$$

The results of the analysis of variance of the model are given in Table 3.

According to Table 3, it is seen that the regression model is significant ($p = 0.000 < 0.05$) and the lack of fit is insignificant ($p = 0.137 > 0.05$).

The graphs showing the interaction of the parameters with each other and the removal efficiency during the interaction are given in Figs. 2, 3, 4 and 5.

Each curve seen in Figs. 2, 3, 4 and 5 represents an infinite number of combinations between two variables when other variables are held at a constant level. Figures 2 and 5 confirm what we have already seen in Table 3, namely that time has a positive effect on response, usually an increase in efficiency is observed as contact time increases. In the literature, there are similar studies and results with the use of RSM in dye removal [15–19].

When the mathematical model obtained as a result of the experimental analyzes is used to obtain 100% removal, the graph shown in Fig. 6 emerges.

3.2 ANN Model

The artificial neural network model was created with Matlab (R2020a) software. The model used pH, Absorbent dose, MB concentration, and Contact time variables as inputs

Table 3. Analysis of Variance

Source	DF	Seq SS	Adj SS	Adj MS	F-Value	P-Value
Model	10	7216.70	7216.70	721.67	68.01	0.000
Linear	4	3354.03	3354.03	838.51	79.02	0.000
x_1	1	206.25	206.25	206.25	19.44	0.000
x_2	1	334.20	334.20	334.20	31.49	0.000
x_3	1	2457.94	2457.94	2457.94	231.63	0.000
x_4	1	355.64	355.64	355.64	33.51	0.000
Square	2	3506.03	3506.03	1753.01	165.20	0.000
x_1^2	1	2767.46	246.25	246.25	23.21	0.000
x_3^2	1	738.57	738.57	738.57	69.60	0.000
2-Way Interaction	4	356.64	356.64	89.16	8.40	0.000
$x_1 * x_3$	1	56.40	56.40	56.40	5.31	0.032
$x_2 * x_3$	1	87.14	87.14	87.14	8.21	0.010
$x_2 * x_4$	1	76.91	76.91	76.91	7.25	0.014
$x_3 * x_4$	1	136.19	136.19	136.19	12.83	0.002
Error	20	212.23	212.23	10.61		
Lack-of-Fit	14	180.79	180.79	12.91	2.46	0.137
Pure Error	6	31.44	31.44	5.24		
Total	30	7428.93				

R^2 : 0.9714; Adjusted R^2 : 0.9571; Predicted R^2 : 0.9223; MSE:6,8462.

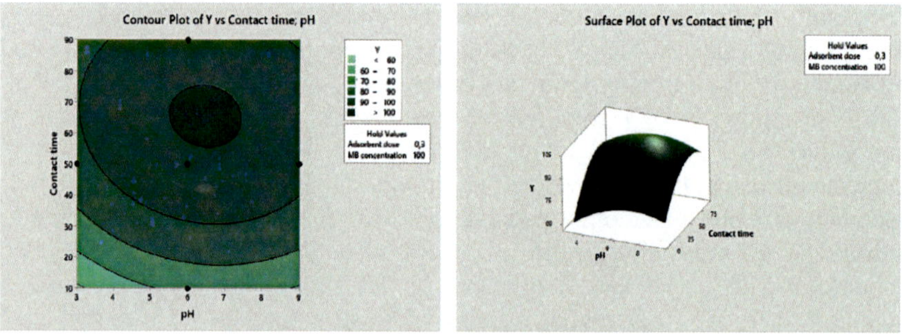

Fig. 2. pH – Contact time relationship.

and % removal variables as outputs. As a result of the experiments, it was decided to use an intermediate layer consisting of 8 neurons and the log-sigmoid transfer function. The artificial neural network model created is given in Fig. 7.

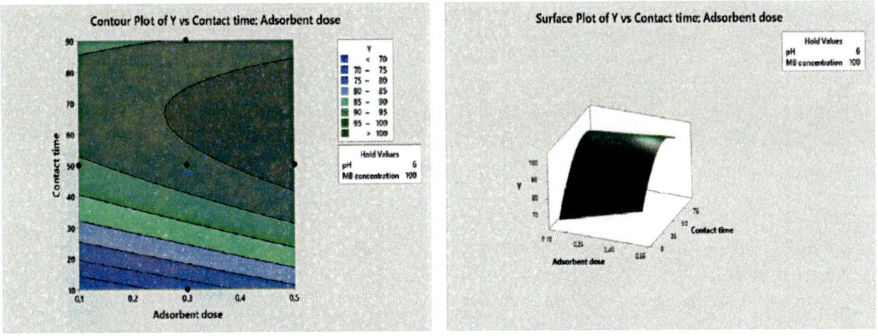

Fig. 3. Absorbent dose – Contact time relationship

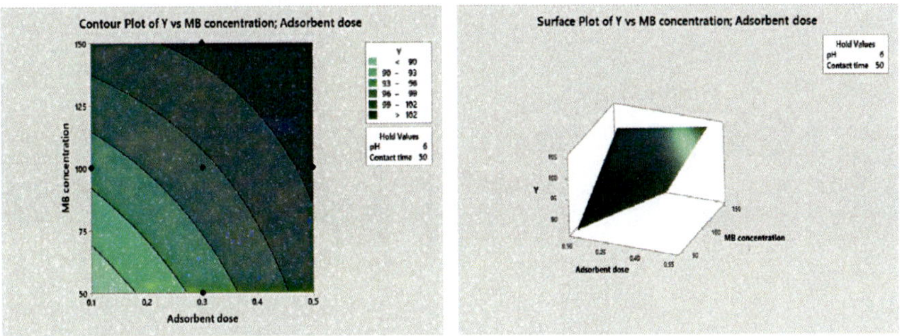

Fig. 4. Absorbent dose – MB concentration relationship

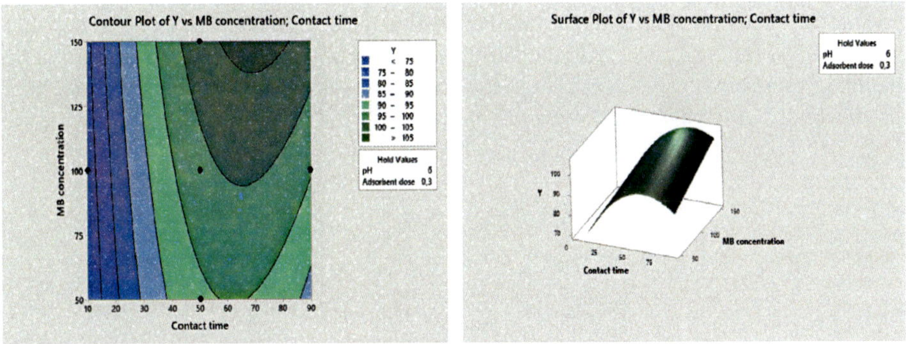

Fig. 5. Contact time – MB concentration relationship

All data are normalized in the range [−1,1]. The loss function was determined as mean squared error, and the Levenberg–Marquardt algorithm, suitable for training small-scale problems, was used to optimize the loss function [21]. Of the 31 data obtained by face-centered experimental design, 70% was used for training, 15% for validation, and 15% for testing. As a result of the training, high correlation coefficients were observed for

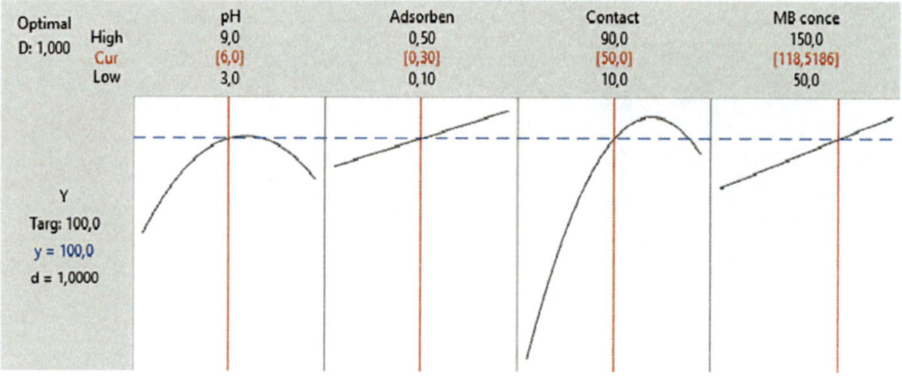

Fig. 6. Optimum parameters

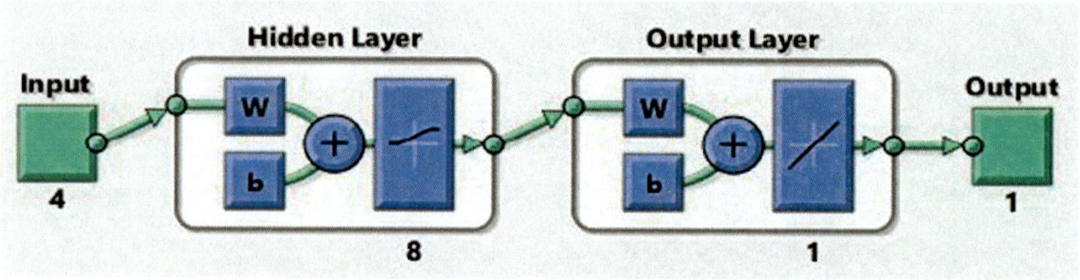

Fig. 7. Neural network model

the training, validation, and test datasets. These values are given in Fig. 8. The model's R^2 value was calculated as 0.9893, and the MSE value as 2.5669. The input values that provide 100% removal were calculated with the help of the Excel solver add-on. The calculated values were given as input to the artificial neural network, and their accuracy was checked. These values are given in Table 4.

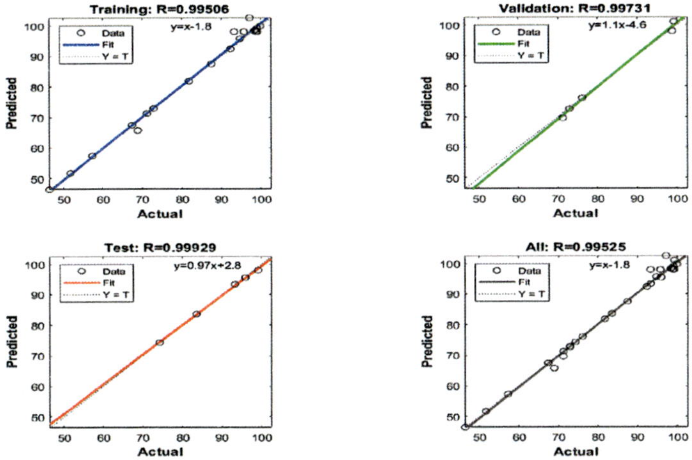

Fig. 8. The correlation coefficients of the data sets of the artificial neural network model

Table 4. Optimum values for ANN

Variables	pH	Adsorbent dose (g/100 mL)	Contact time (min)	MB concentration (mg/L)	%Removal
Values	6.6235	0.3246	54.2317	98.1320	100

4 Conclusion

This study used Response Surface Methodology (RSM) and Artificial Neural Network (ANN) to optimize MB adsorption on OSAC from an aqueous solution. For this purpose, the effect of four different parameters, such as pH, contact time, amount of adsorbent, and dyestuff concentration, on adsorption was investigated, and the percent removal efficiency was optimized. According to the results, the optimized process variables obtained from RSM and ANN for MB adsorption were in close agreement with the experimental data. However, the R2 value of the artificial neural network model was calculated higher than the R^2 value of the RSM model. According to the ANN model findings, the pH value was estimated as 6.6235; the adsorbent amount was 0.3246 g, the contact time was 54.2317 min, and the MB concentration was 98.1320 mg/L for optimum 100% removal. The results show that the ANN model is applicable to accurately describe and model the effect of the four experimental parameters on MB adsorption efficiency.

References

1. Pato, A.M., et al.: Fabrication of TiO_2@ITO-grown nanocatalyst as efficient applicant for catalytic reduction of Eosin Y from aqueous media. Environ. Sci. Pollut. Res. **28**, 947–959 (2021)

2. Buledi, J.A., et al.: Heterogeneous kinetics of CuO nanoflakes in simultaneous decolorization of Eosin Y and Rhodamine B in aqueous media. Appl. Nanosci. **11**, 1241–1256 (2021). Author, F., Author, S., Author, T.: Book title. 2nd edn. Publisher, Location (1999).
3. Zhang, W., Song, H., Zhu, L., Wang, G., Zeng, Z., Li, X.: High flux and high selectivity thin-film composite membranes based on ultrathin polyethylene porous substrates for continuous removal of anionic dyes. J. Env. Chem. Eng. **10**, 107202 (2022)
4. Salleh, M.A.M., Mahmoud, D.K., Karim, W.A.W.A., Idris, A.: Cationic and anionic dye adsorption by agricultural solid wastes: a comprehensive review. Desalination **280**, 1–13 (2011)
5. Ozdesa, D., Duran, C., Senturk, H.B., Avan, H., Bicer, B.: Kinetics, thermodynamics, and equilibrium evaluation of adsorptive removal of methylene blue onto natural illitic clay mineral. Desalination Water Treat. **52**, 1–3 (2014)
6. Ali, I., Asim, M., Khan, T.A.: Low cost adsorbents for the removal of organic pollutants from wastewater. J. Env. Manag. **113**, 170–183 (2012)
7. El Messaoudi, N., El Khomri, M., Bentahar, S., Dbik, A., Lacherai, A., Bakiz, B.: Evaluation of performance of chemically treated date stones: application for the removal of cationic dyes from aqueous solutions. J. Taiwan Inst. Chem. Eng. **67**, 244–253 (2016)
8. Akpomie, K.G., Conradie, J.: Banana peel as a biosorbent for the decontamination of water pollutants. A review. Environtal Chem. Lett. **18**, 1085–1112 (2020)
9. Bentahar, S., Dbik, A., Khomri, M.E., El Messaoudi, N., Lacherai, A.: Adsorption of methylene blue, crystal violet and congo red from binary and ternary systems with natural clay: kinetic, isotherm, and thermodynamic. J. Env. Chem. Eng. **5**, 5921–5932 (2017)
10. Onu, C.E., Nwabanne, J.T., Ohale, P.E., Asadu, C.O.: Comparative analysis of RSM, ANN and ANFIS and the mechanistic modeling in eriochrome black-T dye adsorption using modified clay. S. Afr. J. Chem. Eng. **36**, 24–42 (2021)
11. Montgomery, D.C.: Design and Analysis of Experiments, 7th edn. John Wiley-Sons, New York (2009)
12. Nair, A.T., Ahammed, M.M.: Coagulant recovery from water treatment plant sludge and reuse in post-treatment of UASB reactor effluent treating municipal wastewater. Environ. Sci. Pollut. Res. **21**, 10407–10418 (2014). https://doi.org/10.1007/s11356-014-2900-1
13. Shojaeimehr, T., Rahimpour, F., Khadivi, M.A., Sadeghi, M.: A modeling study by response surface methodology (RSM) and artificial neural network (ANN) on Cu^{2+} adsorption optimization using light expended clay aggregate (LECA). J. Ind. Eng. Chem. **20**, 870–880 (2014)
14. Agarwal, S., et al.: Rapid adsorption of ternary dye pollutants onto copper (I) oxide nanoparticle loaded on activated carbon: experimental optimization via response surface methodology. J. Env. Chem. Eng. **4**, 1769–1779 (2016)
15. Bagheri, R., Ghaedi, M., Asfaram, A., Dil, E.A., Javadian, H.: RSM-CCD design of malachite green adsorption onto activated carbon with multimodal pore size distribution prepared from Amygdalus scoparia: kinetic and isotherm studies. Polyhedron **171**, 464–472 (2019)
16. Igwegbea, C.A., Mohmmadib, L., Ahmadic, S., Rahdard, A., Khadkhodaiyb, D., Dehghanie, R., Rahdarc, S.: Modeling of adsorption of methylene Blue dye on Ho-$CaWO_4$ nanoparticles using response surface methodology (RSM) and artificial neural network (ANN) techniques. MethodsX **6**, 1779–1797 (2019)
17. Liang, L., Zhu, Q., Wang, T., Wang, F., Ma, J., Jing, L., Sun, J.: The synthesis of core-shell Fe_3O_4@mesoporous carbon in acidic medium and its efficient removal of dye. Microporous Mesoporous Mater. **197**, 221–228 (2014)
18. Gokcek, O.B., Uzal, N.: Arsenic removal by the micellarenhanced ultrafiltration using response surface methodology. Water Supply **20**(2), 574–585 (2020)

19. Kumar, M.R., King, P., Wolde, Z., Mulu, M.: Application of optimization response surface for the biosorption of crystal violet dye from textile wastewater onto Clerodendrum fragrans leaves. Biomass Convers. Biorefnery online press (2022)
20. Shanmugaprakash, M., Sivakumar, V.: Development of experimental design approach and ANN-based models for determination of Cr (VI) ions uptake rate from aqueous solution onto the solid biodiesel waste residue. Bioresour. Technol. **148**, 550–559 (2013)
21. Witek-Krowiak, A., Chojnacka, K., Podstawczyk, D., Dawiec, A., Pokomeda, K.: Application of response surface methodology and artificial neural network methods in modelling and optimization of biosorption process. Bioresour. Technol. **160**, 150–160 (2014)

Organizational Performance Evaluation Using Artificial Intelligence Algorithm

Elif Yıldırım[1] (✉), Kenan Aydoğdu[2], Ayten Yilmaz Yalciner[1], Tijen Over Ozcelik[1], and Mehmet Cetinkaya[1]

[1] Department of Industrial Engineering, Sakarya University, 54050 Sakarya, Turkey
{elifyildirim,ayteny,tover,g200102356}@sakarya.edu.tr
[2] Kayalar Group, İstanbul, Turkey
kenan@kas.com.tr

Abstract. The Balanced Scorecard (BSC) performance system application has been generally accepted and is frequently applied in evaluating and measuring company performance. This performance evaluation system envisages a system in which the company's performance is assessed in four dimensions, the indicators determined according to these four dimensions are weighted, and the performance calculation is made consequently. It is a performance system in which the performance of each dimension is independently examined and evaluated.

In this study, the "red area performance evaluation system" applied to Kayalar Pres company, which produces installations and fasteners, is modeled with fuzzy logic. The results are compared with the actual application results.

The application of this study in a performance system with actual data enables the accuracy of the performance indicator weights determined by the management to be compared. The data obtained on the performance system were formed as fuzzy sets based on expert opinions, and a fuzzy set matrix with 81 rules was developed with these sets. The created fuzzy set membership rules were made for all four indicators, and the "company performance score" was calculated. The 81 rules and output results determined in this direction were defined in the MATLAB fuzzy logic function, and the system was modeled on MATLAB accordingly. The results of the fuzzy logic function were obtained and interpreted by comparing the actual measurement results of the company.

Keywords: Balanced Scorecard · Performance System · Company Performance Score · Fuzzy Logic

1 Introduction

Developing an effective Performance Evaluation System (PDS) is necessary to obtain a precise evaluation and an effective productivity increase. The primary goal of the PDS is to provide a measure to compare units, produce accurate assessment documents, establish a remediation plan, and achieve a high level of quality and quantity in the unit's desired outputs. The following steps can help managers create an effective PDS for financial assessment:

Z. Şen et al. (Eds.): IMSS 2023, LNME, pp. 722–732, 2024.
https://doi.org/10.1007/978-981-99-6062-0_68

1. Determining performance criteria,
2. Developing an appropriate evaluation model,
3. Applying the model,
4. Drawing a rough outline for weaknesses,
5. Developing an improvement plan,
6. Setting up an evaluation program.

In the first step, standard performance measures should be drawn with the help of literature and sources such as expert opinion. While identifying these indicators is one of the time-consuming steps in PDS creation, careful identification of such indicators is essential to obtain a definitive assessment. In the second step, the PDS needs to be developed impartially. Therefore, a suitable model should be selected according to the specific conditions of the case study. Then the model should be implemented. According to the model's results, the fourth step is the recognition of performance weaknesses. Then an improvement plan is designed based on the weak elements. Finally, an evaluation procedure should be planned [1]

2 The Balanced Scorecard (BSC) Performance System

Given its extreme and quick development, the globalization age has seen fierce competition between commercial organizations in many marketplaces at all levels. With the aid of cutting-edge working practices and methodologies and the creation of effective plans, these firms have grown determined to improve their accounting and administration systems. In the last ten years, a fresh approach to assessing the strategic performance of businesses has emerged. The Balanced Scorecard, developed to evaluate an organization's performance from all angles, served as a business representation [2]. A revolution against traditional reliance on managerial accounting is seen to have occurred with the introduction of the Balanced Scorecard (BSC) methodology.

A group of non-financial and financial metrics known as BSC identifies and describes an organization's success. BSC is now a crucial strategic tool that firms can use to improve performance [3]. [4] says that BSC is a key tool for firms to communicate their strategy to various departments and organizations through a set of organizational plans. BSC aids in achieving all of the organization's goals. In addition, [5] saw BSC as a method for articulating a company's mission and strategy in a comprehensive set of performance indicators. It does this by providing a framework for adopting and putting into practice the organization's strategy while emphasizing achieving financial and non-financial objectives.

To analyze performance thoroughly from five aspects instead of relying solely on one measure, the financial measurement, BSC is thus intended to be one of the contemporary approaches and techniques [6]. The integration of non-financial and financial performance indicators is shown in the balanced scorecard. The operational metrics, such as clients, internal operations, growth, and learning, direct the company's future financial performance. The BSC has made the link between the organizational strategic aims and its operations more coherent. The managerial BSC connects short-term plans, processes, and events with strategic objectives [7].

Many academics are interested in the balanced scorecard method because it is widely used in literature. Scholars have considered the BSC technique for various industries. For instance, this method was applied by [8–10] to evaluate the efficacy of governments. Likewise, [11, 12] concentrated on enhancing local governments' performance using the BSC technique. A BSC study was carried out by [13] for charitable institutions. On the other hand, when assessing the performance of banks, [14–16] used the BSC approach's dimensions. Additionally, while examining energy firms, [17–20] developed the criteria following the BSC approach. A balanced scorecard approach was employed in the analysis process by [21] study to assess the performance of European airline firms. A related analysis of the Polish market by [22]. [23] also established a model for choosing sustainable technology using the balanced scorecard method. In addition, [24] applied the balanced scorecard concept to the logistics industry. To characterize the link between the indicators in the Esfahan Steel Complex Company, a mathematical model was used to determine the equilibrium among the four views of the balanced scorecard as four participants in a cooperative game [25].

3 Fuzzy Logic

Fuzzy logic is a logical structure that was first formed from an article published by Lütfü Aliasker Zade in 1965. Also, in his Fuzzy article Sets, he explained that fuzzy sets are different from known traditional sets. It is a branch based on thinking like a human and makes operations by converting them into mathematical functions.

Fuzzy logic is a mathematical order for expressing uncertainties and working with delays. Fuzzy logic is the logic of reasoning and allows to obtain approximate results by evaluating in an environment of uncertainty.

Classical (binary) logic is a logic system with two truth values (1 or 0, yes or no, true, or false), and a third condition is assumed to be impossible. Also, in binary logic, exact data is mentioned. Fuzzy logic has a broader application area, including fuzzy events that binary logic cannot handle. Fuzzy logic is also suitable for numerically describing ambiguous verbal expressions used daily.

In classical logic, does an element belong to a set or not. The membership value is 1 or 0, according to the relation.

Mehmet is thin1.
Mehmet is not thin0.
There is also scaling (grading) in fuzzy logic.
Mehmet is very weak 0.98.
Mehmet is a little weak 0.20 (Fig. 1).

Sharp lines separate classical logic. If 1.7 m is the limit for the tall cluster, Mehmet is tall because his height is 1.71 m. How true is it that someone who is 1.69 m does not enter the tall group and is in the regular set?

Fuzzy logic, as seen in the graphic above, it can be understood that someone who is 1.80 m tall is 0.70 m tall. It can be expressed with a more suitable logic for human nature, such as very tall, less tall, etc.

Fuzzy System: It processes multiple inputs with the rule base and inference unit and turns them into output.

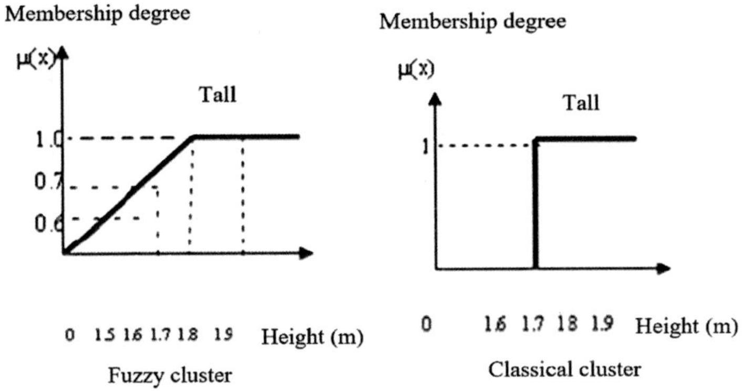

Fig. 1. Fuzzy versus classical clusters

- Knowledge Base: The part where the rule table is located, and data is stored.
- Blur Unit: Converts exact values to fuzzy values with the help of the membership function.
- Inference Unit: Infers conclusions from inputs and rules.
- Defuzzification Unit: Converts fuzzy results to numerical (precise) values.

4 Method

This study aims to compare actual data by modelling deterministic data with fuzzy logic. For this purpose, the data obtained with fuzzy logic will be compared with actual performance data. For this reason, it will be discussed how the preferred fuzzy logic method approaches the truth or can produce more efficient results.

There are various problems in which the fuzzy logic technique is applied. One of these problems is the decision-making problem. Here, it is seen that fuzzy logic studies can be done in various decision-making problems with the fuzzy logic technique to be applied here.

4.1 Problem Definition

Kayalar Group is a group of companies with six independent production and sales businesses. It was founded in Istanbul in 1970 and operates in the product groups of building elements. The business's main structure is based on working as independent companies and measuring and analyzing the performance of each company independently. For this reason, company performances are measured and evaluated by the management with the "Red Area Performance Table" applications as indicators specific to each company.

While making evaluations in line with the determined performance indicators, each indicator is multiplied by the weight determined by the management. As a result, a score is made from 100, and companies are expected to get a minimum of 85 points in their red zone performance programs every month. Company managers who cannot achieve this performance score have to submit a report to the management explaining the reason for this setback.

How realistic or successful the performance evaluation system, which has been established in this way, can measure according to the weights is a problem that needs to be examined and analyzed. For this reason, it is necessary to investigate how much the data reflects reality. How accurate is it to measure performance according to determined weights? This is the main result we want to achieve in this study. For this reason, we aim to make a comparative evaluation using various examination and analysis methods.

Based on the "Red Area Performance Table" of Kayalar Pres Döküm Ltd., which operates within the body of Kayalar Group and manufactures valves and fasteners, four indicators affecting the company's performance were studied. These four indicators, which we have chosen from among the ten indicators in the performance table that takes the company's red zone performance table, have been determined as the indicators that best express and show the company's performance. These indicators are as follows:

Table 1. Company performance indicators table

Indicator Name	Indicator Description	Calculation Method	Indicator Weight
Earnings	It is the net income of the company after deducting the general expenses	Operating Income – Overhead	0,256
Production per Unit Hour	It is calculated by the total working time in the company monthly and by dividing the Total Sales tonnage by this data. In other words, the amount of Tonnage the company converts into products in 1 h	Total Sales Tonnage (kg) / Total Working Time (hour)	0,256
Stock Turnover	It is a critical indicator that shows how quickly the company can circulate its stocks	Cost of Goods Sold / Average Stock	0,256
Scrap Ratio	It is an efficiency indicator that shows the rate of defective products after operations	Number of Defective Parts / Total Number of Production	0,232

As can be seen in Table 1, the performance weights of these four indicators are distributed in a total of 100, as seen in the rightmost column. "How accurate is it to evaluate according to these weights, or how suitable is the data obtained for the evaluation to be made this way?" We will seek an answer to your question.

As a method of finding answers to these questions, we have evaluated using fuzzy logic in this study. A fuzzy performance evaluation study based on expert opinion will be put forward in line with fuzzy sets established with fuzzy logic. A comparison will be made with actual data in this direction.

4.2 Application

The Kayalar Pres company uses a 10-indicator red area performance chart to determine its performance. In this study, this study was conducted on the four performance indicators that we think have the most impact on performance based on expert opinion. In line with the fuzzy logic theoretical knowledge shortly explained above, brief information will be given about how this technology will be applied in this problem (four criteria performance evaluation problem).

First, a fuzzy matching set was created based on expert opinion on these four criteria, and each criterion was evaluated by experts in line with a 3-level evaluation. A fuzzy matching set based on 34 = 81 rules was formed. In line with this matching, a matrix will be created, and these data will be entered into the MATLAB program, and calculation results related to fuzzy sets will be obtained by the MATLAB program. These fuzzy set rules will be as in Table 2 below.

Table 2. Company performance indicators table

Indicator Name	Fuzzy Membership			
Earnings	lower bound	Upper bound		
	−₺200.000	₺200.000	Low	L
	₺190.000	₺600.000	Medium	M
	₺590.000	₺1.500.000	High	H
Production per Unit Hour	lower bound	Upper bound		
	9,00 Ad/As	10,50 Ad/As	Low	L
	10,40 Ad/As	11,50 Ad/As	Medium	M
	11,00 Ad/As	13,00 Ad/As	High	H
Stock Turnover	lower bound	Upper bound		
	4,50	5,50	Low	L
	5,40	6,50	Medium	M
	6,00	7,00	High	H
Scrap Ratio	lower bound	Upper bound		

(continued)

Table 2. (*continued*)

Indicator Name	Fuzzy Membership				
	0,25%	0,30%	Low		L
	0,29%	0,41%	Medium		M
	0,40%	0,70%	High		H
Output: Performance Score	Very bad	Bad	Medium	Good	Very Good
	0–40	45–60	55–70	70–85	70–85
	VB	B	M	G	G

INPUT memberships of these performance criteria are as in Table 2 above. The result (OUTPUT) in line with these memberships is.

It will be evaluated in 5 stages VERY BAD – BAD – MEDIUM – GOOD – VERY GOOD. The Performance Result will be presented according to this calculation, in line with the 81 rules entered in the MATLAB matrix. The stages of the OUTPUT performance result are as seen in Table 3:

Table 3. Real Performance Calculation Chart (an Example)

No	Monthly Actual Data	Weight	Goal	January	Performance	Weighted performance
G.1	Earnings	0.256	₺282,400	₺828,525	100.00	25.60
G.2	Production per Unit Hour	0.256	11.85 unit/AS	10.92 unit/AS	92.11	23.58
G.3	Stock Turnover	0.256	6.00	6.05	100.00	25.60
G.4	Scrap Ratio	0.232	0.47%	0.63%	65.96	15.30
		1.00				90.08

Three of the four indicators determined are directly proportional, but the final indicator, G.4: Scrap Product Ratio, is inversely proportional. In other words, for a company to be successful, the first three indicators must be high, but the 4th indicator, the scrap product ratio indicator, must be low. In the fuzzy set matrix with 81 rules, which was prepared in line with expert opinions, the rules were determined by experts, taking this principle into account.

The resulting performance score will be calculated by entering the 81-ruled fuzzy sets matrix data determined in this direction in the MATLAB program, and these data were analyzed by comparing the actual data and revealing the difference between the two methods.

The results obtained compared the degree of accuracy by performing the Hypothesis Test.

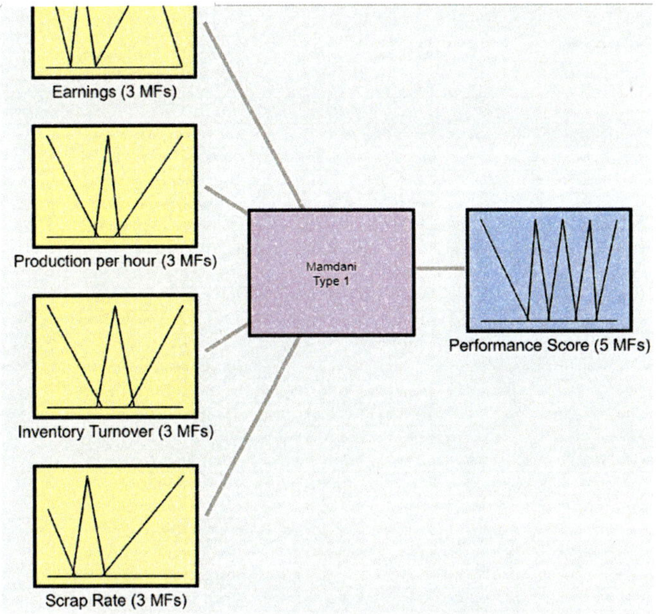

Fig. 2. General Structure of Fuzzy Logic Model

The Fig. 2 above shows the MATLAB General model structure of the established fuzzy logic model. A model with four inputs and one Output (Performance Score) was modeled on MATLAB, and then actual data and MATLAB results were compared.

Each Input has been defined as "Low – Medium – High," as seen in the Fuzzy Set Memberships table, and thus fuzzy set memberships have been created. Output performance based on expert opinion is also defined in the MATLAB program and associated with Input to MATLAB, and in this way, our performance model is defined on MATLAB. In short, this model was taught according to expert opinion.

5 Conclusion

The results of the actual performance indicators (i.e., management-weighted performances) obtained by the studies on the operation described above are calculated as seen in the table below. For each month, calculations were made as in the table below, thus, the following performance comparison results table was obtained.

If we briefly interpret the above table, the inversely proportional indicator G.7: Scrap Product Ratio should draw our attention to the performance calculation in this table. The performance calculation of this indicator is also calculated as inversely proportional. Weighted performance scores are calculated by multiplying the computed performance ratios by the Performance Weight Scores determined by the management. The "Main Performance Score" of the institution emerges by adding these calculated scores. The annual table of Main Performance Scores, which occurred after these calculations were made with actual monthly data, appeared in the table below.

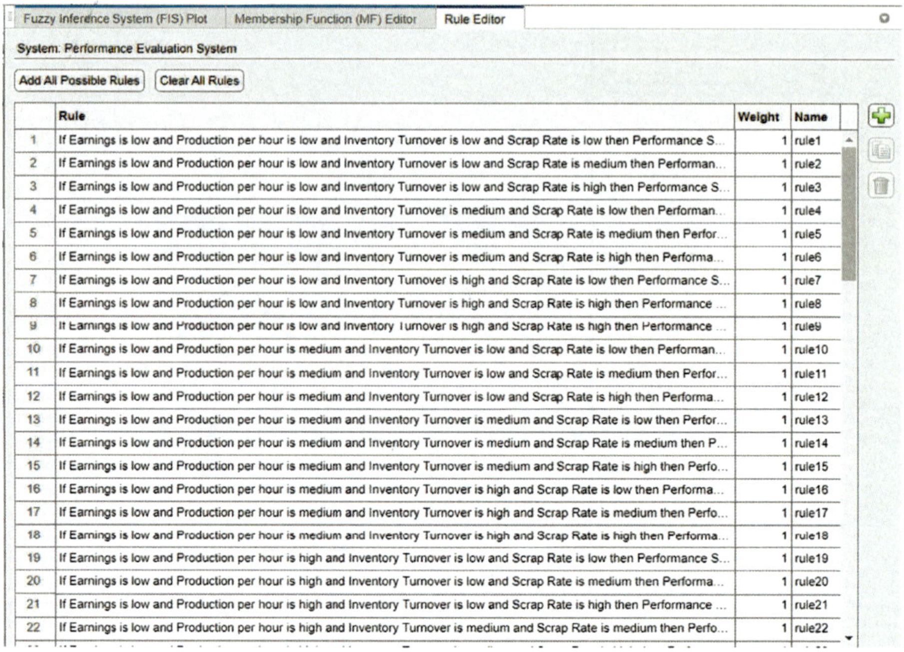

Fig. 3. MATLAB- Fuzzy Logic- January Performance Results Table

As seen in the Fig. 3, the fuzzy logic calculation score calculated with MATLAB for January-2021 is seen. By entering the actual realized indicator values for each month into the program, Fuzzy Logic calculations were made for each month. The comparison of the actual data formed in this direction with the Fuzzy Logic MATLAB data is shown in Table 4 below:

Table 4. Real Result – MATLAB Result Comparison Chart

Period	REEL PERF Score	FUZZY-MATLAB	Deviation
Jan	90,08 P	87,00 P	−3,42%
Feb	98,22 P	97,00 P	−1,24%
Mar	95,72 P	93,00 P	−2,84%
Apr	95,00 P	96,00 P	1,05%
May	87,84 P	88,00 P	0,18%
Jun	96,39 P	97,00 P	0,64%
Jul	69,40 P	71,00 P	2,31%

(continued)

Table 4. (*continued*)

Period	REEL PERF Score	FUZZY-MATLAB	Deviation
Aug	89,37 P	88,00 P	−1,53%
Sep	100,00 P	98,00 P	−2,00%
Avg	91,34 P	90,56 P	−0,85%

When we want to briefly evaluate the results seen in Table 5 above, it is seen that the Fuzzy Logic – MATLAB results are consistent. According to the comparison table, according to the performance evaluation calculation created with Fuzzy Logic, there is a 0.85% deviation between Real and Fuzzy data. Its distance from actual data is less than 1%, therefore, it can be said that it is a successful study.

In the following stages, the development and consistency of the model will be ensured by developing the rules and reflecting more expert opinions on the matrix. According to the rules, which will be reflected in line with the expert opinion, the resulting deviation value can be further reduced, and results that are precisely the same as the reality can be obtained.

References

1. Dalfard, V.M., Sohrabian, A., Najafabadi, A.M., Alvani, J.: Performance evaluation and prioritization of leasing companies using the super efficiency Data Envelopment Analysis model. Acta Polytechnica Hungarica **9**(9), 183–194 (2012)
2. Mio, C., Costantini, A., Panfilo, S.: Performance measurement tools for sustainable business: a systematic literature review on the sustainability balanced scorecard use. Corp. Soc. Responsib. Environ. Mgmt. **29**, 367–384 (2021)
3. Niven, R.: BSC Step by Step: Maximizing Performance and Maintaining Results. John Wiley & Sons Inc., Hoboken, NJ (2002)
4. Alsaadi, A., Ebrahim, M.S., Jaafar, A.: Corporate social responsibility, Shariah-compliance, and earnings quality. J. Financ. Serv. Res. **51**(2), 169–194 (2017)
5. Hemmington, S.M.: Soft Science. Univserity of Saskatchewan Press, Saskatoon (1997)
6. Horngren, C.T.: Accounting, 9th edn. Pearson Prentice Hall (2012)
7. Teichgräber, U., Sibbel, R., Heinrich, A., Güttler, F.: Development of a balanced scorecard as a strategic performance measurement system for clinical radiology as a cost center. Insights Imaging **12**(1), 1–8 (2021)
8. Sodan, S.: The impact of fair value accounting on earnings quality in Eastern European countries. Procedia Econ. Financ. **32**, 1769–1786 (2015)
9. Dimitropoulos, P., Kosmas, I., Douvis, I.: Implementing the balanced scorecard in a local government sport organization: evidence from Greece. Int. J. Prod. Perform. Manag. **66**(3), 362–379 (2017)
10. Yao, J., Liu, J.: E-government project evaluation: a balanced scorecard analysis. J. Electron. Commer. Organ. (JECO) **14**, 11–23 (2016)
11. Ndevu, Z.J., Muller, K.: Operationalising performance management in local government: the use of the balanced scorecard. S. Afr. J. Hum. Resour. Manag. **16**, 1–11 (2018)
12. Ge, L.: Exploration of balanced scorecard-based government performance management in China. In: Proceedings of the 2016 International Conference on Public Management. Atlantis Press, pp. 227–233 (2016)

13. Wang, Y., Tian, L., Chen, Z.: A reputation bootstrapping model for e-commerce based on fuzzy DEMATEL method and neural network. IEEE Access **7**, 52266–52276 (2019)
14. Ćwiklicki, M.: Applying balanced scorecard in non-government organizations. In: Bojkovic, V. (ed.) Decision Management: Concepts, Methodologies, Tools, and Applications. IGI Global, pp. 902–923 (2017)
15. Elkhouty, S.M., Ibrahim, M.M., ElFrargy, M.M., Kotb, A.S.: Measuring the effectiveness of banking risk balanced scorecard in enhancing bank value. Int. J. Econ. Financ. **7**(1), 139–152 (2015)
16. Sardjono, W., Pujadi, T.: Performance evaluation of systems Managed File Transfer in banking industry using IT Balanced Scorecard. In: Proceedings of the 2016 International Conference Information Managemnt Technology (ICIMTech), pp. 12–16. Bandung, Indonesia, 16–18 Nov 2016
17. Dinçer, H., Yüksel, S.: Comparative evaluation of BSC-based new service development competencies in Turkish banking sector with the integrated fuzzy hybrid MCDM using content analysis. Int. J. Fuzzy Syst. **20**(8), 2497–2516 (2018)
18. Dong, J., Xue, G., Li, X.: Value evaluation of integrated energy services based on balanced scorecard. In: Proceedings of the International Conference on Humanities and Social Science, pp. 233–235. Atlantis Press (2016)
19. Dincer, H., Yuksel, S.: Balanced scorecard-based analysis of investment decisions for the renewable energy alternatives: a comparative analysis based on the hybrid fuzzy decision-making approach. Energy **175**, 1259–1270 (2019)
20. Kahla, F.: Implementation of a balanced scorecard for hybrid business models–an application for citizen renewable energy companies in Germany. Int. J. Energy Sect. Manag. **11**, 426–443 (2017)
21. Dinçer, H., Hacıoğlu, Ü., Yüksel, S.: Balanced scorecard-based performance measurement of European airlines using a hybrid multicriteria decision making approach under the fuzzy environment. J. Air Transp. Manag. **63**, 17–33 (2017)
22. Tubis, A., Werbińska-Wojciechowska, S.: Balanced scorecard use in passenger transport companies performing at Polish market. Procedia Eng. **187**, 538–547 (2017)
23. Xia, D., Yu, Q., Gao, Q., Cheng, G.: Sustainable technology selection decision-making model for enterprise in supply chain: based on a modified strategic balanced scorecard. J. Clean. Prod. **141**, 1337–1348 (2017)
24. Agrawal, S., Singh, R.K., Murtaza, Q.: Outsourcing decisions in reverse logistics: sustainable balanced scorecard and graph theoretic approach. Res. Conserv. Recycl. **108**, 41–53 (2016)
25. Abedian, M., et al.: Determining the best combination of perspective indicators of balanced scorecard by using game theory. J. Optim. Ind. Eng. **14**(1), 185–194 (2021)

A Fuzzy Logic Approach for Corporate Performance Evaluation

Buşra Taşkan[1(✉)], Buket Karatop[2], and Cemalettin Kubat[3]

[1] Mus Alparslan University, Mus 49250, Turkey
b.taskan@alparslan.edu.tr
[2] Istanbul University Cerrahpasa, Istanbul 34500, Turkey
[3] Istanbul Gelisim University, Istanbul 34310, Turkey

Abstract. Today's brutal competition environment has made it a necessity for businesses to evaluate their performance in order to maintain their existence and gain sustainable competitive advantage. The current environment does not allow businesses that do not perform performance evaluations a chance to survive. Therefore, corporate performance evaluation is a multi-criteria, complex and uncertain real-life problem that is vital for businesses. Corporate performance should be measured multi-dimensionally, but performance indicators of these dimensions may not always be expressed with a numerical value or may have uncertainty. In this case, the closest results are obtained by using fuzzy logic. Based on the current importance of the subject, a fuzzy model has been developed for the evaluation of corporate performance in this study. The outputs of the model are total enterprise performance, corporate reputation and financial output. Although the number of inputs used for each output varies, the inputs used in the model are the environment, customers, society, government, competitors, suppliers, business processes and sustainability. The application of the model is demonstrated with a case study using the fuzzy logic toolbox of the MATLAB program. Performance values for "Total Enterprise Performance", "Corporate Reputation" and "Financial Output" were obtained as 16.9, 16.9 and 20.8, respectively.

Keywords: Performance · Performance Evaluation · Performance Measurement · Corporate Performance · Corporate Performance Evaluation · Corporate Performance Measurement · Artificial Intelligence · Fuzzy Logic

1 Introduction

Today, due to the globalization of the world, the commercial borders between countries have disappeared. This situation causes developments such as becoming the only market of the whole world, accelerating the flow of information and taking the place of current technological developments with new technological developments. Depending on these developments, the competition between businesses is increasing and becoming more fierce. In this ruthless competitive environment, the continuous evaluation of the performance of the enterprises is of great importance in terms of maintaining their existence and gaining sustainable competitive advantage. The

Z. Şen et al. (Eds.): IMSS 2023, LNME, pp. 733–743, 2024.
https://doi.org/10.1007/978-981-99-6062-0_69

motto "If you can't measure something, you can't understand it. If you can't understand it, you can't control it. If you can't control it, you can't improve it" clearly expresses the importance of the subject. Corporate performance measurement is not a simple process with its multi-dimensional, multi-criteria, complex and uncertain structure, and there is still no universally accepted method in the literature to solve this problem [1]. When the relevant literature is examined, the corporate performance evaluation methods developed in recent years are as follows;

- System Dynamics-Based Balanced Scorecard [2]
- Proactive Balanced Scorecard [3]
- Sustainability Performance Measurement System [4]
- A Test and Evaluation-Derived Approach to Organizational Performance Measurement [5]
- The Performance Wheel [6]
- The Small Business Performance Pyramid [6]
- The Comprehensive Reverse Logistics Enterprise Scorecard [7]
- An Integrated Fuzzy C-Means-TOPSIS Approach [8]
- A Model for Monitoring Organizational Performance [9]
- Comprehensive Reverse Logistics Enterprise Performance Measurement and

  Decision Making Model [10].

- An Adapted BSC Approach [11]
- A new integrated model for performance measurement in healthcare organ zations [12]
- Three-dimensional performance measurement system-SMD3D [13]
- An Integrated Framework for Organizational Performance Measurement [14]
- The sustainability balanced scorecard model for hybrid organizations [15]
- An Integrated BSC-Fuzzy AHP Approach [16]
- A Performance Measurement Scale for Higher Institutions [17]
- Sustainable Enterprise Excellence: Attribute-Based Assessment Protocol [18]
- Performance Management Assessment Model for Sustainable Development [19]
- Organizational Performance Assessment Instrument using Constructivist Multicriteria Decision Aid (MCDA-C) [20]
- The combined AHP and GRA method [21]
- A new SME self-assessment model [22]
- An Adapted BSC approach for social enterprises in emerging market [23]
- A DEA Malmquist model for productivity analysis of top global automobile manufacturers [24]

Looking at the above methods, it is seen that all of them were developed in the Industry 4.0 period. As it is known, the dominant feature of this period is the use of artificial intelligence. However, when **the mentioned methods were examined, artificial intelligence techniques were used** only in Proactive Balanced Scorecard, An Integrated Fuzzy C-Means-TOPSIS Approach and An Integrated BSC-Fuzzy AHP approach methods, was seen. For this reason, it can be said that the corporate performance field has not developed in parallel with Industry 4.0 [25]. Addressing all aspects of corporate performance and including fuzzy logic used in complex and uncertain situations in the calculation ensures accurate results.

Based on the current importance of the subject, a fuzzy logic approach is proposed for corporate performance evaluation in this study. By evaluating the corporate performance in terms of the determined performance dimensions, the results related to the total enterprise performance, corporate reputation and financial output were obtained.

2 Proposed Model

In the developed model, total enterprise performance, corporate reputation and financial output were evaluated through external stakeholders around the enterprise and processes within the enterprise. Since it is one of the popular concepts of recent times and is a very important concept for businesses to continue their existence, the sustainability dimension has also been taken into consideration. A business process is a group of coordinated activities performed by people or by means of tools to achieve a specific business result [26]. The business environment consists of elements such as customers ([27–29]), suppliers ([27–29]), government/communities [28], competitors ([27, 28]), environment [29], society ([28–30]) which are defined as the external stakeholders of the business. In this direction, the dimensions determined for the evaluation of corporate performance are business processes, environment, customer, society, government, competitors, suppliers and sustainability. The model has 3 outputs: total enterprise performance, corporate reputation and financial output. The number of performance dimensions used to evaluate each output is not the same. While all dimensions are used to evaluate the total enterprise performance output, the environment, customer, society, government, competitors, and sustainability dimensions are used to evaluate the corporate reputation output. Customer, competitors, business processes, suppliers, sustainability dimensions are used to evaluate the output named financial output.

In the study, the triangle membership function, which is the most used in practical applications, is preferred. The fuzzy ranges defined for the input and output variables are as follows;

- Low: Values between %0 and %45 (for $0 \leq x \leq 0{,}45$ μ (x))
- Medium: Values between %40 and %80 (for $0{,}40 \leq x \leq 0{,}80$ μ (x))
- High: %75 and higher values (for $x \geq 0{,}75$ μ (x))

A 3-point linguistic scale (Low, Medium, High) was used for input and output variables. In this case, $3^8 = 6561$ rules in total for total corporate performance, $3^6 = 729$ rules in total for corporate reputation, and $3^5 = 243$ rules in total for financial output are required. However, since this number of rules is very difficult to both create and interpret, rule reduction has been made. If we consider the total enterprise performance output, since all 8 performance dimensions are used, these dimensions are divided into 4 groups as in doubles (Environment + Sustainability; Customer + Society; Government + Competitors; Suppliers + Business Processes). It is necessary to create $3^2 = 9$ rules for each of the in doubles groups. At the current stage, since each of the 4 groups of in doubles is considered as a single indicator, $3^4 = 81$ rules should be created. The same processes were applied to corporate reputation and financial output. In this study, Mamdani was used as fuzzy inference method and centroid method was used for defuzzification process.

3 Application

The application of the proposed model was made in a manufacturing company. In the application, the performance values determined by the company in the form of percentages regarding the performance dimensions were used. Since the number of inputs used for each output of the model is not the same, separate data entries were made for each output. Since the total enterprise performance output uses all of the inputs, 8 inputs were first divided into 4 groups in doubles ("Environment + Sustainability"; "Customer + Society"; "Government + Competitors"; "Suppliers + Business Processes"). First of all, data entry was made for these doubled groups and then for 4 groups formed by doubles as seen in Fig. 1;

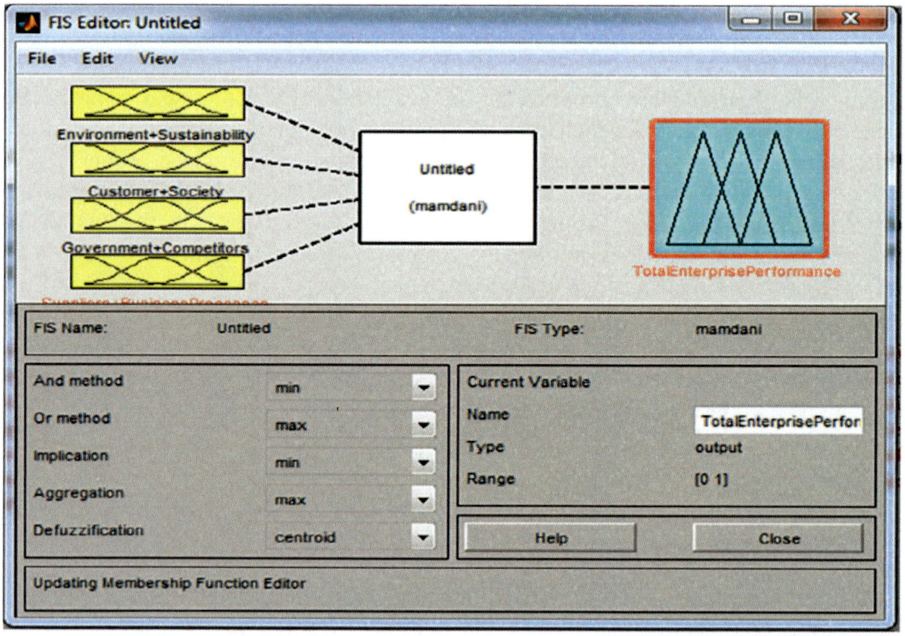

Fig. 1. Fuzzy system design related to the "total enterprise performance".

As a result of the data entries made into the model, the performance values are obtained for the "Environment + Sustainability" group as 18.5, the performance value for the "Customer + Society" as 20.3, the performance value for the "Government + Competitors" as 20.8, and the performance value for the "Suppliers + Business Processes" as 18.6. As seen in Fig. 2, when these values were entered into the model, the performance value for the "Total Performance Value" output was obtained as 16.9.

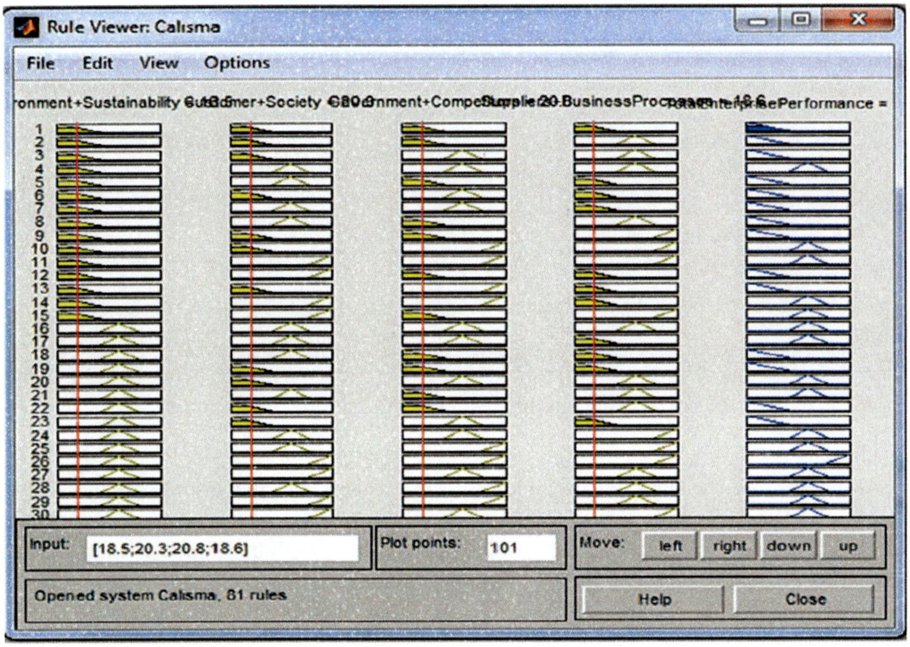

Fig. 2. Fuzzy evaluation interface for the "total enterprise performance".

Since the "Corporate Reputation" output uses 6 inputs, these inputs are first divided into 3 groups in doubles ("Environment + Sustainability"; "Customer + Society"; "Government + Competitors"). First of all, data entry was made for these doubled groups and then for 3 groups formed by these doubles as seen in Fig. 3.

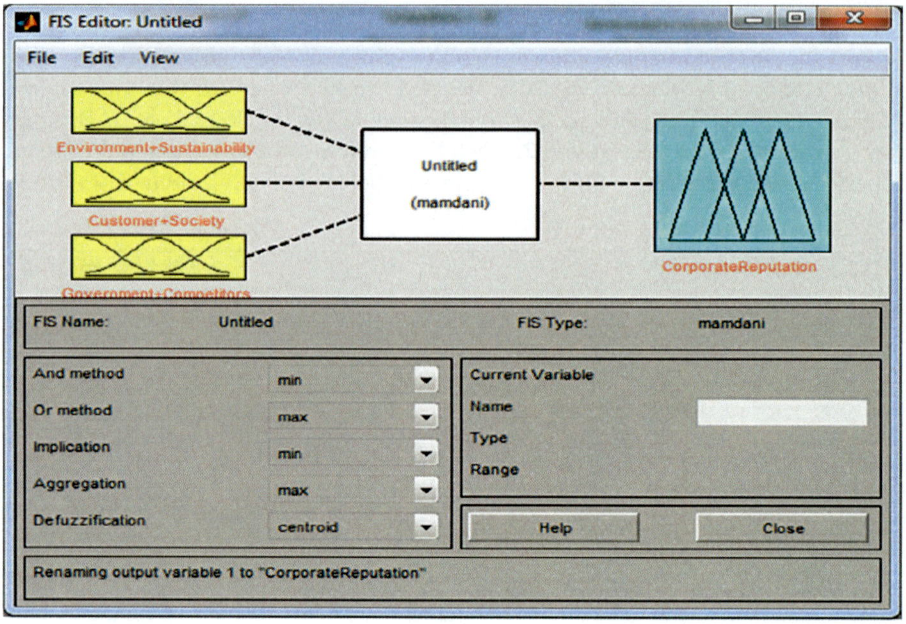

Fig. 3. Fuzzy system design related to the "corporate reputation".

As a result of the data entries made into the model, the performance value were obtained for the "Environment + Sustainability" group as 18.5, the performance value for the "Customer + Society" as 20.3, and the performance value for the "Government + Competitors" as 20.8. As seen in Fig. 4, when these values were entered into the model, the performance value for the "Corporate Reputation" output was obtained as 16.9.

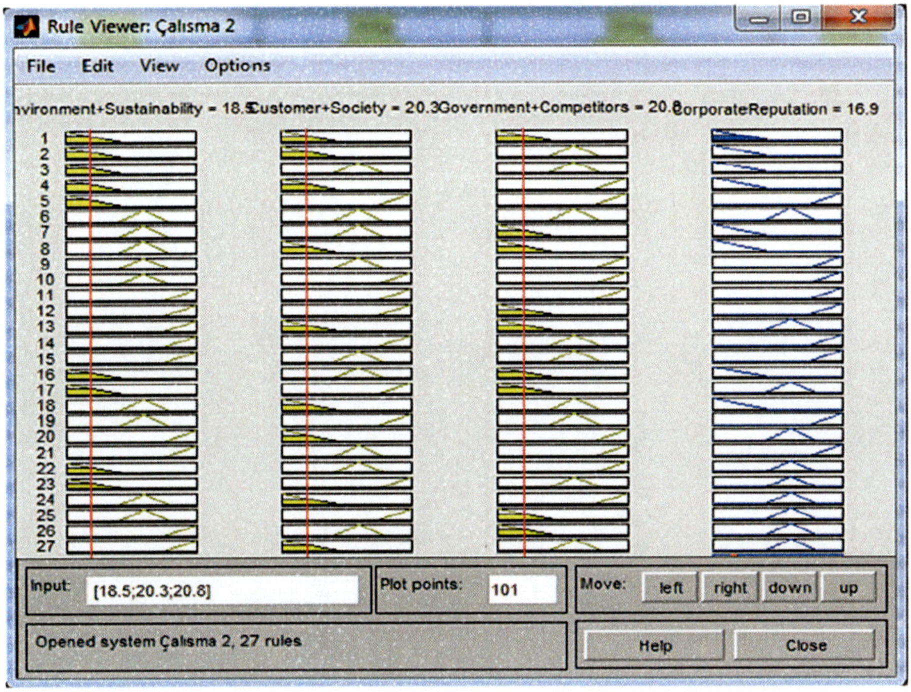

Fig. 4. Fuzzy evaluation interface for the "corporate reputation".

Since "Financial Output" uses 5 inputs, these inputs are first divided into 3 groups in doubles ("Customer + Sustainability"; "Business Processes + Suppliers"; "Competitors"). First of all, data entry was made for these doubled groups and then for 3 groups formed by these doubles as seen in Fig. 5.

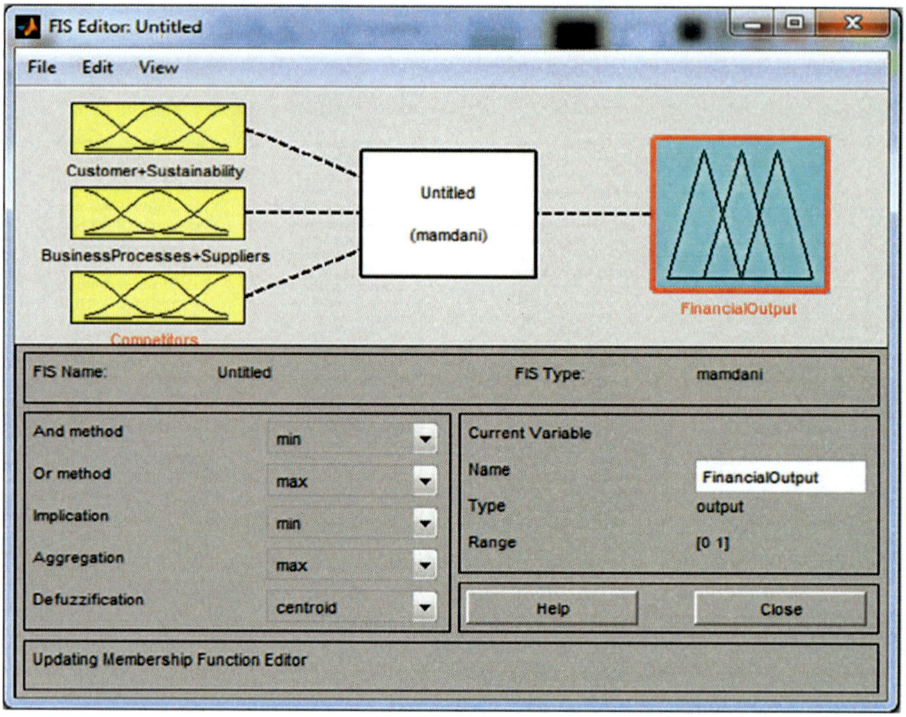

Fig. 5. Fuzzy system design related to the "financial output".

As a result of the data entries made into the model, the performance values were obtained for the "Customer + Sustainability" group as 20.3, the performance value for the "Business Processes + Suppliers" as 18.6, and the performance value for the "Competitors" as 39. As seen in Fig. 6, when these values were entered into the model, the performance value for "Financial Output" was obtained as 20.8.

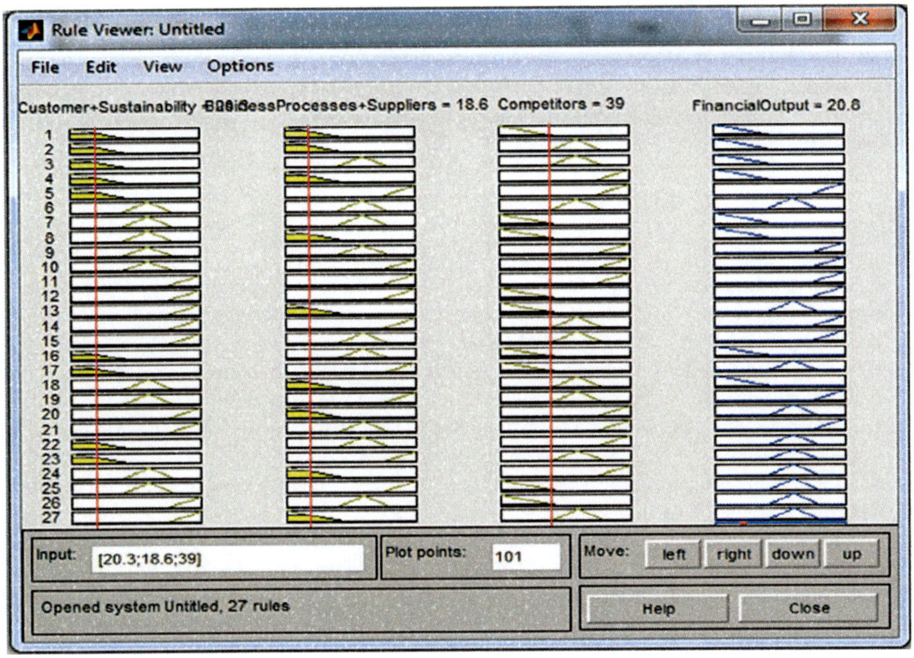

Fig. 6. Fuzzy evaluation interface for the "financial output".

Performance values for "Total Enterprise Performance", "Corporate Reputation" and "Financial Output" were obtained as 16.9, 16.9 and 20.8, respectively. While the performance of the enterprise for the "Total Enterprise Performance" and "Corporate Reputation" outputs is the same, it has been observed that the performance for the "Financial Output" is at the highest level.

4 Conclusion

Today's brutal competition environment has made it a necessity for businesses to evaluate their performance in order to maintain their existence and gain sustainable competitive advantage. Corporate performance evaluation is a multi-criteria, complex and uncertain real-life problem that is vital for businesses. Corporate performance should be measured multi-dimensionally, but performance indicators of these dimensions may not always be expressed with a numerical value or may have uncertainty. In this case, the closest results are obtained by using fuzzy logic. Based on the current importance of the subject, a fuzzy model has been developed for the evaluation of corporate performance in this study. The outputs of the model are total enterprise performance, corporate reputation and financial output. Although the number of inputs used for each output varies, the inputs used in the model are the environment, customers, society, government, competitors, suppliers, business processes and sustainability. The application of the model is demonstrated with a case study using the fuzzy logic toolbox of the MATLAB program. Performance values for "Total Enterprise Performance", "Corporate Reputation" and "Financial Output" were obtained as 16.9, 16.9 and 20.8, respectively. While the performance of the enterprise for the "Total Enterprise Performance" and "Corporate Reputation" outputs is the same, it has been observed that the performance for the "Financial Output" is at the highest level.

References

1. Trumpp, C., Endrikat, J., Zopf, C., Guenther, E.: Definition, conceptualization, and measurement of corporate environmental performance: a critical examination of a multidimensional construct. J. Bus. Ethics **126**, 185–204 (2015)
2. Barnabe, F.: A "system dynamics-based balanced scorecard" to support strategic decision making: insights from a case study. Int. J. Product. Perform. Manag. **60**(5), 446–473 (2011)
3. Chytas, P., Glykas, M., Valiris, G.: A proactive balanced scorecard. Int. J. Inf. Manage. **31**, 460–468 (2011)
4. Searcy, C.: Updating corporate sustainability performance measurement systems. Meas. Bus. Excell. **15**(2), 44–56 (2011)
5. Hester, P.T., Meyers, T.J.: Multi-Criteria performance measurement for public and private sector enterprises. In applications of management science, pp. 183–206. Emerald Group Publishing Limited (2012)
6. Watts, T., McNair-Connolly, C.J.: New performance measurement and management control systems. J. Appl. Acc. Res. **13**(3), 226–241 (2012)
7. Shaik, M., Abdul-Kader, W.: Performance measurement of reverse logistics enterprise: a comprehensive and integrated approach. Meas. Bus. Excell. **16**(2), 23–34 (2012)
8. Bai, C., Dhavale, D., Sarkis, J.: Integrating fuzzy c-means and TOPSIS for performance evaluation: an application and comparative analysis. Expert Syst. Appl. **41**(9), 4186–4196 (2014)
9. Draghici, A., Popescu, A.D., Gogan, L.M.: A proposed model for monitoring organizational performance. Procedia Soc. Behav. Sci. **124**, 544–551 (2014)
10. Shaik, M.N., Abdul-Kader, W.: Comprehensive performance measurement and causal-effect decision making model for reverse logistics enterprise. Comput. Ind. Eng. **68**, 87–103 (2014)
11. Elbanna, S., Eid, R., Kamel, H.: Measuring hotel performance using the balanced scorecard: a theoretical construct development and its empirical validation. Int. J. Hosp. Manag. **51**, 105–114 (2015)
12. Pirozzi, M.G., Ferulano, G.P.: Intellectual capital and performance measurement in healthcare organizations: an integrated new model. J. Intellect. Cap. **17**(2), 320–350 (2016)
13. Valenzuela, L., Maturana, S.: Designing a three-dimensional performance measurement system (SMD3D) for the wine industry: a Chilean example. Agric. Syst. **142**, 112–121 (2016)
14. Yaghoobi, T., Haddadi, F.: Organizational performance measurement by a framework integrating BSC and AHP. Int. J. Product. Perform. Manag. **65**(7), 959–976 (2016)
15. Ponte, D., Pesci, C., Camussone, P.F.: Between mission and revenue: measuring performance in a hybrid organization. Manag. Audit. J. **32**(2), 196–214 (2017)
16. Modak, M., Pathak, K., Ghosh, K.K.: Performance evaluation of outsourcing decision using a BSC and Fuzzy AHP approach: a case of the Indian coal mining organization. Resour. Policy **52**, 181–191 (2017)
17. Abubakar, A., Hilman, H., Kaliappen, N.: New tools for measuring global academic performance. SAGE Open **8**(3), 1–10 (2018)
18. Hussain, T., Edgeman, R., Eskildsen, J., Shoukry, A.M., Gani, S.: Sustainable enterprise excellence: attribute-based assessment protocol. Sustainability **10**(11), 4097 (2018)

19. Fechete, F., Nedelcu, A.: Performance management assessment model for sustainable development. Sustainability **11**(10), 2779 (2019)
20. Longaray, A.A., Ensslin, L., Dutra, A., Ensslin, S., Brasil, R., Munhoz, P.: Using MCDA-C to assess the organizational performance of industries operating at Brazilian maritime port terminals. Operat. Res. Perspect. **6**, 100109 (2019)
21. Liu, J.-W.: Developing GAHP concepts for measurement of travel agency organizational performance. Soft. Comput. **24**(11), 8051–8059 (2019)
22. Machado, M.C., Mendes, E.F., Telles, R., Sampaio, P.: Towards a new model for SME self-assessment: a Brazilian empirical study. Total Qual. Manag. Bus. Excell. **31**(9–10), 1041–1059 (2020)
23. Mamabolo, A., Myres, K.: Performance measurement in emerging market social enterprises using a balanced scorecard. J. Soc. Entrepreneurship **11**(1), 65–87 (2020)
24. Wang, C.N., Tibo, H., Nguyen, H.A.: Malmquist productivity analysis of top global automobile manufacturers. Mathematics **8**(4), 580 (2020)
25. Taşkan, B., Karatop, B.: Development of the field of organizational performance during the industry 4.0 period. Int. J. Res. Ind. Eng. **11**(2), 134–154 (2022)
26. Juric, M.B., et al.: Design principles for process-driven architectures using oracle bpm and soa suite 12c. Packt Publishing (2015)
27. Simpson, J., Taylor, J.: Corporate governance ethics and csr. Kogan Page (2013)
28. Page-Tickell, R.: Learning & development. Kogan Page (2014)
29. Whipple, B.: Trust in transition: navigating organizational change. American Society for Training & Development (2014)
30. Kulkarni, S.P.: Environmental ethics and information asymmetry among organizational stakeholders. J. Bus. Ethics **27**, 215–228 (2000)

Reverse Engineering in Electroless Coatings: An Application on Bath Parameter Optimization for User-Defined Ni-B-P Coating Properties

Abdullah Hulusi Kökçam[1], Mehmet Fatih Taşkın[1](✉), Özer Uygun[1], Harun Gül[2], and Ahmet Alp[3]

[1] Industrial Engineering Department, Sakarya University, Sakarya, Turkey
{akokcam,mftsakin,ouygun}@sakarya.edu.tr
[2] Metallurgical and Materials Engineering Department, Sakarya University of Applied Sciences, Sakarya, Turkey
harungul@subu.edu.tr
[3] Metallurgical and Materials Engineering Department, Sakarya University, Sakarya, Turkey
alp@sakarya.edu.tr

Abstract. This paper presents a study on the application of reverse engineering in electroless coatings, specifically in the optimization of bath parameters for the production of user-defined Ni–B-P coatings. Electroless coatings are widely used for their unique properties such as hardness, thickness, and corrosion resistance. However, the optimization of bath parameters for producing coatings with specific properties can be a complex and time-consuming process. In this study, reverse engineering is used to analyze the structure and properties of existing coatings and to determine the optimal bath parameters for the production of user-defined Ni–B-P coatings. The results show that the use of reverse engineering in the optimization of bath parameters can significantly reduce the time and cost associated with the development of coatings with specific properties. This study provides valuable insights into the application of reverse engineering in the optimization of electroless coatings and demonstrates its potential for enhancing the performance and functionality of coatings.

Keyword: Electroless Coatings · Artificial Neural Networks · Genetic Algorithms · Coating Properties

1 Introduction

The coating industry has long been focused on developing innovative solutions to protect and enhance the performance of materials and components. With the advent of AI, a new era of intelligent coating systems is emerging, offering unprecedented opportunities for optimizing coating formulations, application techniques, and maintenance strategies. The rapid advancements in artificial intelligence (AI) have led to significant breakthroughs in various fields, including materials science and surface engineering.

Z. Şen et al. (Eds.): IMSS 2023, LNME, pp. 744–752, 2024.
https://doi.org/10.1007/978-981-99-6062-0_70

Electroless coatings, specifically electroless nickel-phosphorus (Ni–P) coatings, are a type of metal plating process that does not require an external power source for deposition [1]. Instead, it relies on a chemical reduction process to deposit a uniform layer of nickel-phosphorus alloy onto a substrate. This method offers several advantages over traditional electroplating, such as improved corrosion resistance, wear resistance, and uniform thickness distribution, even on complex geometries [2].

The electroless Ni–P coating process involves immersing the substrate in a solution containing nickel ions, reducing agents, and a source of phosphorus. The chemical reaction between the reducing agent and the nickel ions leads to the deposition of the nickel-phosphorus alloy onto the substrate. The phosphorus content in the coating can vary, which influences the properties of the final coating, such as hardness and corrosion resistance [3].

Electroless Ni–P coatings have a wide range of applications, including automotive, aerospace, electronics, and oil and gas industries. They are used to enhance the performance and extend the service life of various components by providing increased wear resistance, corrosion protection, and improved surface properties [3, 4].

In addition to the benefits, electroless Ni–P coatings offer several other advantages. One of the key features is their ability to provide a consistent and uniform thickness across the entire surface of the substrate, regardless of its shape or complexity [5]. This uniformity ensures that the coating provides consistent protection and performance enhancement across the entire component.

Another advantage of electroless Ni–P coatings is their ability to be tailored to specific applications by adjusting the phosphorus content [6]. Low-phosphorus coatings (2–5% P) offer higher hardness and wear resistance, while high-phosphorus coatings (10–12% P) provide better corrosion resistance and are more ductile. Medium-phosphorus coatings (6–9% P) strike a balance between these properties, making them suitable for a wide range of applications.

Electroless Ni–P coatings can also be combined with other materials, such as nanoparticles or polymers, to create composite coatings with enhanced properties. These composite coatings can offer improved wear resistance, thermal conductivity, or electrical conductivity, depending on the added materials [7].

Despite their many advantages, electroless Ni–P coatings do have some limitations. The process can be sensitive to contamination, which may lead to issues such as poor adhesion or uneven deposition. Additionally, the chemicals used in the process can be hazardous, requiring proper handling and disposal procedures [8].

Electroless Ni–P coatings are a versatile and effective method for enhancing the performance and service life of various components across multiple industries. Their uniform thickness, adjustable properties, and potential for composite coatings make them a valuable tool for engineers and designers seeking to improve the performance of their products [9]. However, careful attention must be paid to process control and safety measures to ensure the successful application of these coatings.

Genetic algorithms (GAs) are a class of optimization and search techniques inspired by the process of natural selection and evolution. They are used to find approximate solutions to complex optimization problems, particularly those with large search spaces, multiple objectives, or non-linear constraints. GAs are well-suited for problems where traditional optimization methods may struggle or fail to find a suitable solution.

The core components of a genetic algorithm include a population of candidate solutions, a fitness function, and genetic operators such as selection, crossover, and mutation. The algorithm begins with an initial population of randomly generated solutions. The fitness function evaluates each solution's quality or suitability for the problem at hand [10].

During the selection process, solutions with higher fitness values are more likely to be chosen for reproduction. Crossover, also known as recombination, combines the genetic material of two parent solutions to create one or more offspring solutions. Mutation introduces small random changes to the offspring's genetic material, promoting diversity and exploration of the search space.

The algorithm iterates through multiple generations, applying selection, crossover, and mutation to create new populations. Over time, the population's overall fitness improves, converging towards an optimal or near-optimal solution [11].

In this paper, we study on reverse engineering with optimizing the properties of the coatings by using the mixed ANN-GA methods. In this way, both time and cost loss will be eliminated by determining the bath parameters and environmental conditions in advance and in a short time. In addition, strong optimization features of artificial intelligence techniques will be demonstrated in the field of coating.

2 Materials and Method

In the study, the Taguchi experimental design method was used to determine the average number of experiments for each bath type in order to carry out a minimum number of experiments in the laboratory. Ni–B, Ni–P and Ni–B-P coatings baths were prepared, and experiments were carried out in the laboratory environment. The coating process was carried out in a glass beaker using a magnetic stirrer. The substrate was immersed in the coating solution and then heated to the desired temperature using a heating mantle. The pH and the concentration of the coating solution were kept constant during the coating process. The coating process was carried out for 60 min and the coated substrate was then washed with deionized water and dried in an oven at 80 °C for 1 h. For example, in the Ni–B-P coating experiment, a Taguchi L9 orthogonal array is employed for an experiment with three factors and three levels each to optimize the process parameters for improving the corrosion resistance, hardness, and thickness of electroless Ni–B-P coatings. The specific process parameters that we varied were Sodium Hypophosphite (NaPO2H2) as reducing agent quantity, Dimethylamine Borane (DMAB – (CH3)2NHBH3) as reducing agent quantity, and temperature. These factors and their levels are given in Table 1.

There are three factors with three levels which can be expressed as 33. In terms of this experimental combination, there is 27 different combinations exists. Taguchi experimental design method is utilized using an L9 orthogonal array. According to this design, experiments are conducted and the obtained results for the response variables of hardness, corrosion, and thickness are given in Table 2.

After the Taguchi method was used to find out the best combination of the factors (process parameters) and to improve the quality of the outputs, namely coating hardness

Table 1 Factors and Levels

Factor	Unit	Levels		
		1	2	3
NaPO2H2	g/L	18.	20	22
(CH3)2NHBH3	g/L	1	2	3
Temperature	°C	70	80	90

Table 2. Values of design factors and response variables for L9 experiments

	Design Factors			Response Variables		
Exp	Sodium Hypophosphite (g/L)	Dimethylamine Borane (g/L)	Temperature (°C)	Hardness (GPa)	Thickness (μm)	Corrosion (mpy × 10^{-5})
1	18	1	70	719	14.44	51
2	18	2	80	731	15.00	56
3	18	3	90	742	16.00	64
4	20	1	80	703	17.01	36
5	20	2	90	719	20.17	41
6	20	3	70	734	23.01	69
7	22	1	90	598	17.98	41
8	22	2	70	656	23.24	49
9	22	3	80	676	26.81	50

(Hv), corrosion (mpy), and thickness (mm); we upload the database of experiments to Artificial Neural Networks (ANN) which is used for machine learning in MATLAB program.

3 Application

While creating ANN models, a model with a multi-layered feed-forward network is being studied. In the first stage, two-layer feed-forward ANN models, one hidden layer and one output layer, were created. Then, different models will be tried to be constructed by making changes on the parameters such as the number of input layer, hidden layer and output layer, transfer and activation functions of these models. These models are derived from different training algorithms which are Levenberg–Marquardt, BFGS Quasi-Newton, Resilient Backpropagation, Scaled Conjugate Gradient, Conjugate Gradient with Powell/Beale Restarts, Fletcher-Powell Conjugate Gradient, Polak-Ribiére Conjugate Gradient, One Step Secant, Variable Learning Backpropagation. etc. have been developed.

ANN model consisting of input and output layers with three input and three output variables and 1 hidden layer was used by making use of the experiments and the results of the Experimental Design. Since our ANN model is based on regression analysis according to the information learned from previous data, it has an output layer with linear output function and a hidden layer feed forward structure with sigmoid activation (see Fig. 1).

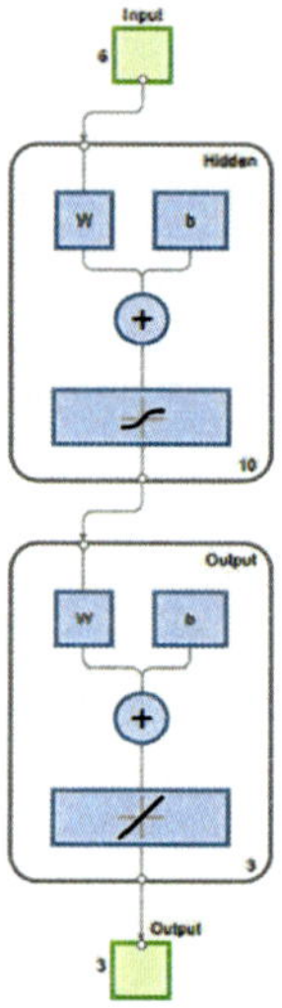

Fig. 1. Topology of ANN

After running the ANN models and determining the best learning model, it was observed that the prediction performance of the model was quite high when new data was entered to the program to predict. The screenshot with the relevant model outputs and performance indicators is given in Fig. 2.

Then we have been developed an interface over the MATLAB program to determine the electroless coating bath properties. Through this interface, the data is loaded with the screen given in Fig. 3 and the input and output values are selected by the user.

After the training is completed, metrics related to the training performance of the ANN model are presented on the "Results" tab on the same screen (see Fig. 4). Here also, the actual coating properties with the coating properties calculated by the model for the test data are shown on the graph.

After this step, to train the advanced model, the training screen of the advanced model is opened via the "Advanced Model" tab under the "Model" menu. Here, after the training parameters are determined for the advanced model, the model training is performed by clicking the "Train" button (see Fig. 5).

After the training is completed, the values of the best population obtained by the advanced model are listed on the "Results" tab on the same screen (see Fig. 6).

The results show us that the input bath parameters obtained against the output data at the determined value. So user defined parameters are the most important input in this

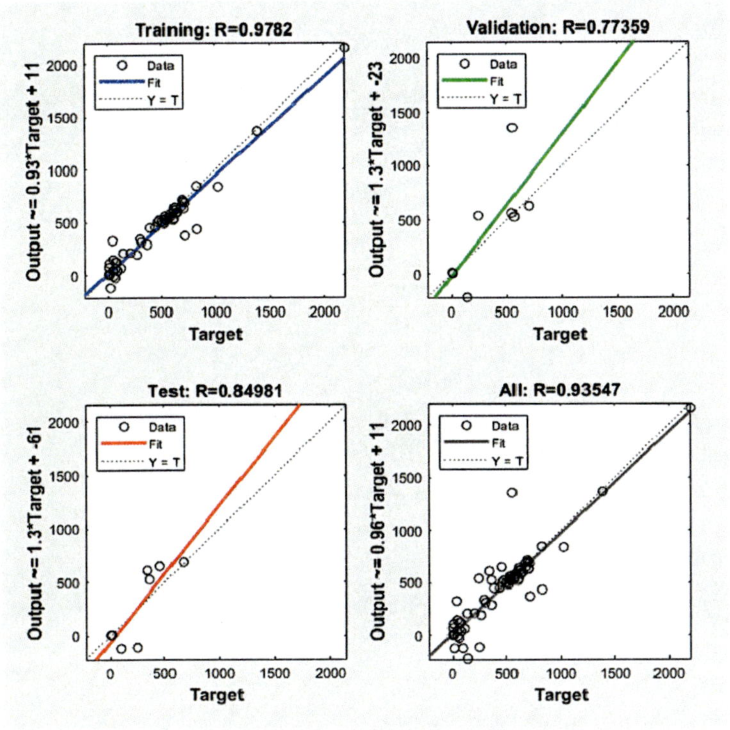

Fig. 2. Performance Outcomes of a Well-Learning ANN Model

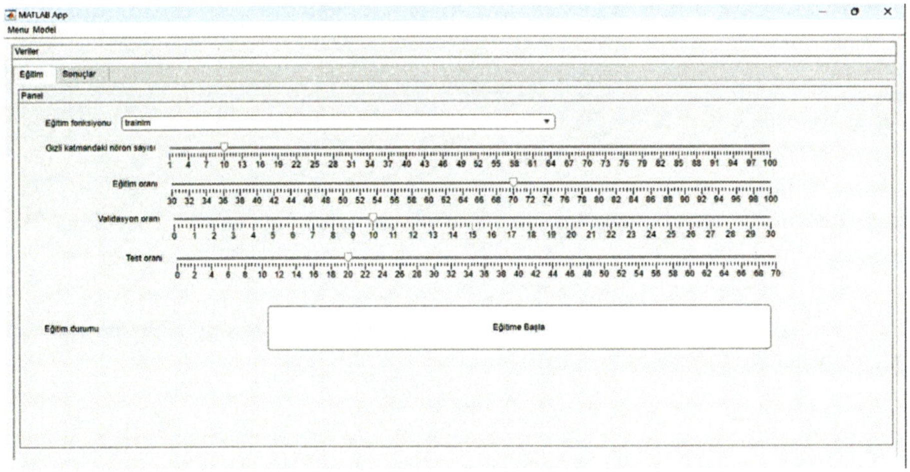

Fig. 3. Determining The Training Parameters of The ANN And Conducting The Training

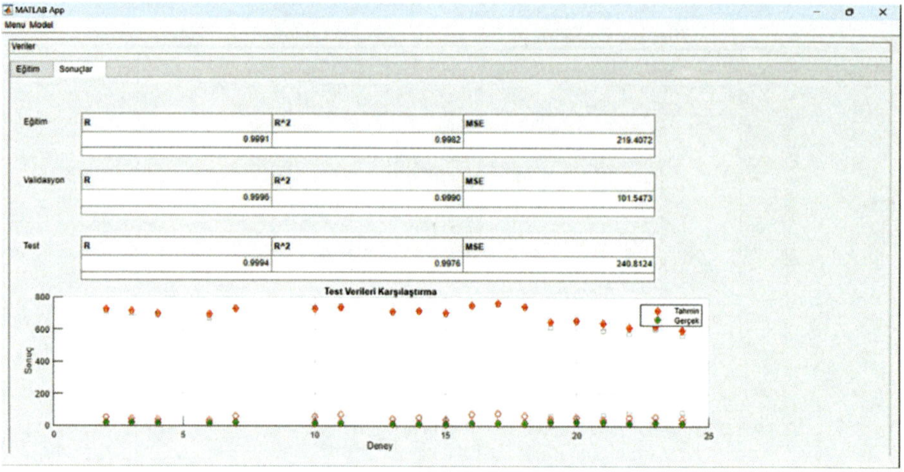

Fig. 4. Performance Metrics of The ANN Model

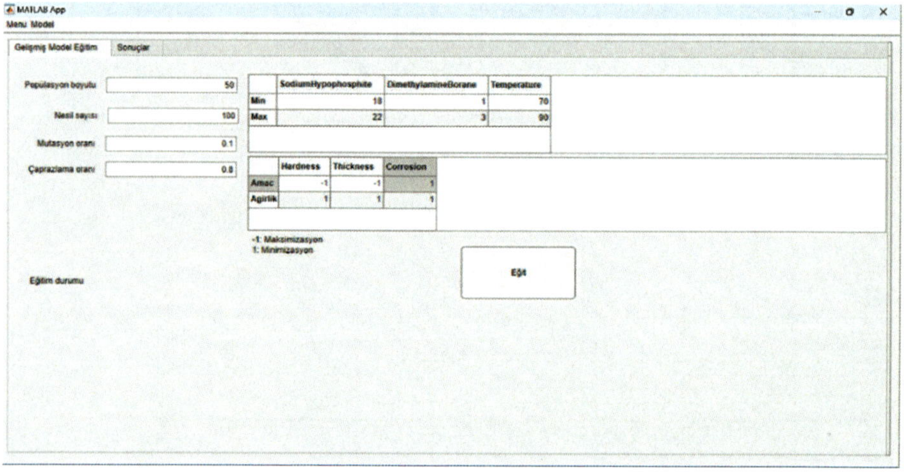

Fig. 5. Determining the Training Parameters of the Advanced Model and Conducting The Training

study. Normalized weight rankings are used to make the results even more specific and understandable to the user.

4 Results

It has been observed that the effect of the number of hidden nodes on the ANN performances has little effect on the prediction ability of the ANN.

Multiobjective GAs give better results in studies conducted by looking at the number of inputs and their values.

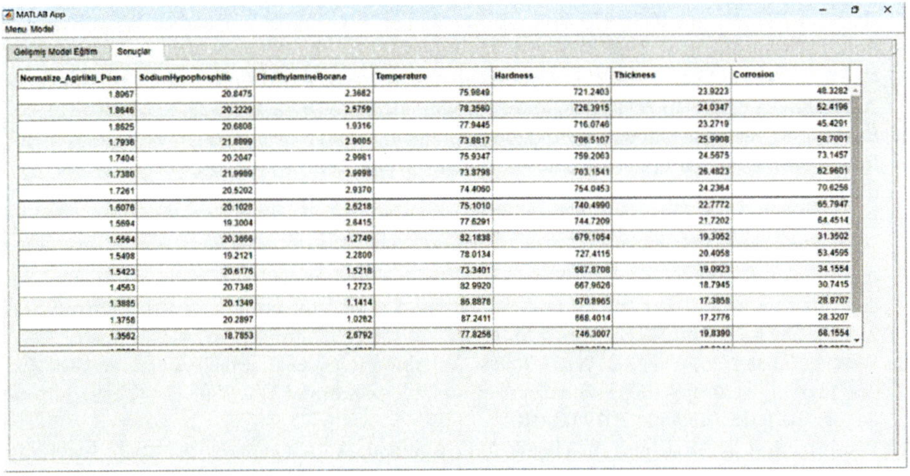

Normalize_Agirlikli_Puan	SodiumHypophosphite	DimethylamineBorane	Temperature	Hardness	Thickness	Corrosion
1.8967	20.8475	2.3682	75.9849	721.2403	23.9223	48.3282
1.8646	20.2229	2.5759	78.3560	726.3915	24.0347	52.4196
1.8625	20.6808	1.9316	77.9445	716.0746	23.2719	45.4291
1.7936	21.8899	2.9005	73.8817	708.5107	25.9908	58.7001
1.7404	20.2047	2.9981	75.9347	759.2063	24.5675	73.1457
1.7380	21.9989	2.9998	73.8798	703.1541	26.4823	62.9601
1.7261	20.5202	2.9370	74.4060	754.0453	24.2364	70.6256
1.6076	20.1028	2.6218	75.1010	740.4990	22.7772	65.7947
1.5694	19.3004	2.6415	77.6291	744.7209	21.7202	64.4514
1.5564	20.3666	1.2749	82.1838	679.1054	19.3052	31.3502
1.5498	19.2121	2.2800	78.0134	727.4115	20.4058	53.4595
1.5423	20.6176	1.5218	73.3401	687.8708	19.0923	34.1554
1.4563	20.7348	1.2723	82.9920	667.9626	18.7945	30.7415
1.3885	20.1349	1.1414	86.8878	670.8965	17.3858	28.9707
1.3758	20.2897	1.0282	87.2411	668.4014	17.2776	28.3207
1.3562	18.7853	2.6792	77.8256	746.3007	19.8390	68.1554

Fig. 6. The Advanced Model's Recommended Coating Parameters For The Best Coating Properties.

The GA's ability to make appropriate evaluations is achieved by choosing the better offspring from the past generation, with the maximization or minimization function (optimization of the input parameters). Here, the decision mechanism belongs to the previous experience and practices of the user.

Expected and actual results nearly the same because of establishing GA methods especially for coating problem parameters.

5 Future Works

It is aimed to reduce costs by adding detailed input and output range definition screens. These costs are important considering the usage amount of the bathroom components, which are among the input parameters. Also other bath types and parameters will be added in future studies. And finally the graphical user interfaces will be developed and more effective usage will be ensured in the future works.

References

1. Macit, Ş.: Investigating and Development of Electroless Plating Catalyst and Activation Solution in Electroplating Industry. Ege Üniversitesi (2017)
2. Chitty, J.A., Pertuz, A., Hintermann, H., Puchi, E.S.: Influence of electroless nickel-phosphorus deposits on the corrosion-fatigue life of notched and unnotched samples of an AISI 1045 steel. J. Mater. Eng. Perform. **8**(1), 83–86 (1999)
3. Yan, M., Ying, H.G., Ma, T.Y.: Improved microhardness and wear resistance of the as-deposited electroless Ni-P coating. Surf. Coatings Technol. **202**(24), 5909–5913 (2008)
4. Anik, M., Körpe, E.: Effect of alloy microstructure on electroless NiP deposition behavior on Alloy AZ91. Surf. Coatings Technol. **201**(8), 4702–4710 (2007)

5. Balaraju, J.N., Rajam, K.S.: Electroless deposition of Ni-Cu-P, Ni-W-P and Ni-WCu-P alloys. Surf. Coatings Technol. **195**(2–3), 154–161 (2005)
6. Chen, X.-M., Li, G.Y., Lian, J.S.: Deposition of electroless Ni-P/Ni-W-P duplex coatings on AZ91D magnesium alloy. Trans. Nonferrous Metals Soc. China **18**, s323–s328 (2008)
7. Sukkasi, S., Sahapatsombut, U., Sukjamsri, C., Saenapitak, S., Boonyongmaneerat, Y.: Electroless Ni-based coatings for biodiesel containers. J. Coatings Technol. Res. **8**(1), 141–147 (2011)
8. Sahoo, P., Das, S.K.: Tribology of electroless nickel coatings – A review. Mater. Des. **32**(4), 1760–1775 (2011)
9. Afroukhteh, S., Dehghanian, C., Emamy, M.: Preparation of the Ni-P composite coating co-deposited by nano TiC particles and evaluation of it's corrosion property. Appl. Surf. Sci. **258**(7), 2597–2601 (2012)
10. Wen, C., Zhang, Y., Wang, C., Xue, D., Bai, Y., Antonov, S., et al.: Machine learning assisted design of high entropy alloys with desired property. Acta Mater **170**, 109–117 (2019). https://doi.org/10.1016/j.actamat.2019.03.010
11. Ward, L., Liu, R., Krishna, A., Hegde, V.I., Agrawal, A., Choudhary, A., et al.: Including crystal structure attributes in machine learning models of formation energies via Voronoi tessellations. Phys Rev B **96**, 24104 (2017)

Multiple Time Series Analysis with LSTM

Hasan Şen(✉) and Ömer Faruk Efe

Bursa Technical University/Industrial Engineering, Bursa, Turkey
{hasan.sen,omer.efe}@btu.edu.tr

Abstract. Inflation is caused by the growing gap between the amount of money actively involved and the sum of products and services available for purchase. It is an economic and monetary process that manifests itself as a constant rise in prices, a fall in the current value of money. Inflation is a subject that keeps itself constantly updated in our country and around the world. The main purpose of the central banks, which are dependent on countries in the world and continue their activities, on the economy is to ensure price stability permanently. In recent years, artificial intelligence techniques have been used more and more in order to consistently predict the value of inflation in the future and to make future studies with the forecasts obtained. The aim of this study is to estimate inflation in the Turkish economy with time series analysis by using LSTM (Long Short Term Memory) model, which is one of the artificial neural networks types, on a python computer program. With this study, the estimation made by the LSTM model showed result when compared in terms of MAPE and MSE statistical analyses. It has been observed that the irregular increase in the inflation value within the country in the recent periods directly affects the success level of the models.

Keywords: Artificial Intelligence · Deep Learning · Inflation · LSTM

1 Introduction

Inflation is the situation in which the general situation of prices is constantly increasing. It is a word that means swelling in Latin. In economic terms, inflation occurs when the total demand at the current price level is more than the total supply. Until the 2000s, Turkey was constantly exposed to high inflation in terms of economy. With the policies implemented in the following years, the Central Bank followed the monetary policies towards price stability. Inflation is targeted as the main tool in monetary policy [1]. In recent years, many determinations have been made about how inflation affects economic growth. Economic growth is defined as the increase or decrease in the volume of production in an economy at the level of periods at specified frequencies [2]. While a decrease in inflation may affect the country's economy and economic growth positively, an increase in inflation negatively affects the country's economy and economic growth. High inflation causes a decrease in the purchasing power of money, increases the cost of living, and creates price instability. Inflation is a term that is affected by many external factors, and it emerges by reaching a certain value under the influence of these factors. It is not just a term

Z. Şen et al. (Eds.): IMSS 2023, LNME, pp. 753–760, 2024.
https://doi.org/10.1007/978-981-99-6062-0_72

affected by numerical concepts. Movements in domestic and foreign policies, decisions taken within the country, decisions that may affect the world in other countries directly affect inflation and play an important role in its decrease and increase. Particularly, the fluctuations in the exchange rate, in euro and dollar terms, the uncontrollability of the exchange rates and their disproportionate increase have emerged as factors affecting inflation. While this situation caused the Turkish lira to depreciate, consumers kept their money in foreign exchange accounts instead of Turkish lira in banks.

In recent years, it has become very popular to make forecasts according to the increase and decrease of inflation. Not only inflation, but also in many areas of economy and finance, future projections are made by using historical data. Making accurate and consistent forecasts for the future is very important for businesses and countries. Understanding past data and obtaining new future data in the light of this data creates a source of insight for businesses and countries. Thanks to these insights, countries and businesses can have the opportunity to define their plans, goals and strategies more clearly [3].

The aim of this study is to make an inflation forecast in the light of the decided parameters after examining the effects of gold, dollar, oil, exports, housing price index, unemployment parameters on inflation. With this study, the effect of the parameters used on inflation will be observed more clearly. It can be observed that the current rate of increase in inflation directly affects the success level of the model.

2 Method and Material

The data used in the study are the data prepared for time series analysis. Inflation data and data on gold, dollar rate, oil, exports, housing price index, unemployment data are used. In the problem in which the data between January 2007 and July 2022 were used, a total of 187 data were used and the value of the first 9 months data set of the data used is as in Table 1 in the table.

2.1 Augmented Dickey-Fuller (ADF) Test

In forecast models that occur in non-stationary time series, the results and forecast values do not reflect the true relationship. For this reason, stationarity tests are used to make analyzes performed in non-stationary time series more meaningful, and Augmented Dickey-Fuller (ADF) test is one of the tests used [4].

As a result of the ADF test, the resulting statistical values and the null hypothesis are tested against the alternative hypothesis. The null hypothesis is that the series is not stationary, while the alternative hypothesis is that the series is stationary.

2.2 Normalization of Data

Normalization of data is a process that can be performed with many different processes. When the data enters the normalization process, they are classified between a value of zero and a value and become normalized. In the conducted study, Min-Max normalization process was used and the data were normalized. The formula for the Min-Max

Table 1. Data used on a monthly basis

DATE	INFLATION	GOLD	DOLLAR	OİL	EXPORT	UNEMP	HPI
01.01.13	7,31	95	1,77	198	20,34M	8,6	127,7
01.02.13	7,03	95,3	1,77	197	21,99M	8,6	128
01.03.13	7,29	92,9	1,81	198	23,75M	8,6	129,3
01.04.13	6,13	87,02	1,79	195	22,43M	8,8	129,7
01.05.13	6,51	83,96	1,83	197	24,24M	8,8	130,4
01.06.13	7,31	95	1,77	206	23,53M	8,6	127,7
01.07.13	7,03	95,3	1,77	210	25,25M	8,6	128
01.08.13	7,29	92,9	1,81	213	21,77M	8,6	129,3
01.09.13	6,13	87,02	1,79	220	26,39M	8,8	129,7
01.10.13	6,51	83,96	1,83	216	24,03M	8,8	130,4
01.11.13	7,32	83,49	2,02	220	23,13M	8,8	132,8

normalization operation is shown in Eq. (1).

$$X' = \frac{X - X_{min}}{X_{max} - X_{min}} \tag{1}$$

2.3 Artificial Neural Networks

Basically, artificial neural networks (ANNs) are structures that can learn solutions by imitating the working principles of neurons in the human brain, developing methods for complex problems that are difficult to solve, and offering different suggestions for further problems with the help of these learnings [5]. In the ANN structure, there are basically 3 layers. From these, the data comes first to the input layer and proceeds through the hidden layer by performing the weighting process. The data coming on the hidden layer is processed with the determined hyperparameters and proceeds to the output layer in order to obtain output, and we obtain our results in the output layer. There are many types of artificial neural networks. The simplest single-layer artificial neural network is called a perceptron. Multi-layer neural networks are also more complex than single-layer neural networks. RNN structure, which is one of the machine learning algorithms, creates deep learning artificial neural networks together with deep learning artificial neural networks.

2.4 LSTM

Long short term memory (LSTM – Long Short Term Memory) is an iterative neural network (RNN) architecture used in deep learning and ANN. An exemplary LSTM architecture is shown in Fig. 1.

The LSTM architecture, unlike other standard neural network architectures, has feedback neural network structures. LSTM-based models are an extension of RNN structures

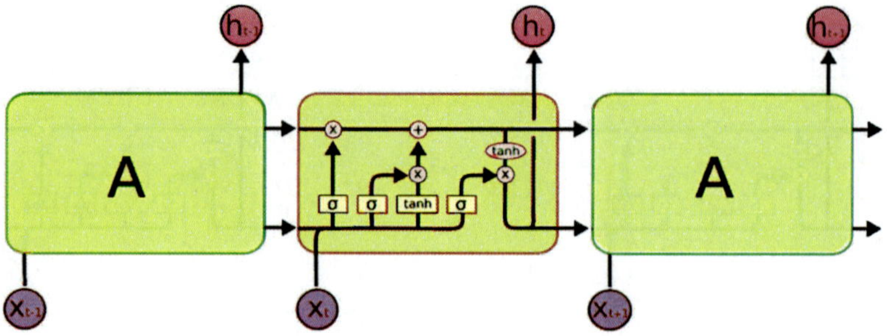

Fig. 1. Structure of the LSTM model

that can very cleanly solve the vanishing gradient problem. LSTM models expand the memory of RNN structures to enable them to retain long-term dependencies of information on input data and to learn more strongly from these dependencies. The LSTM model has the ability to remember the information sent into it for a longer period of time, allowing it to read, write and delete information from memories [6].

3 Research Findings

3.1 Preparation of Data

In this study on inflation forecasting, gold, dollar, oil, exports, unemployment and housing price index data were chosen to be used as independent variables in the model in order to make an inflation forecast. Inflation data was chosen as the dependent variable when establishing the model. After the data were prepared, the correlation between the data was examined first, and after the correlation was examined, the ADF stationarity test was performed to check whether the data were stationary. The correlation between the data is shown in Table 2.

Table 2. Correlation of data

	INFLATION	GOLD	DOLLAR	OİL	EXPORT	UNEMP	HPI
INFLATION	1	0,88	0,89	0,97	0,92	0,08	0,79
GOLD	0,88	1	0,99	0,93	0,98	0,29	0,88
DOLLAR	0,89	0,99	1	0,94	0,98	0,33	0,92
OİL	0,97	0,93	0,94	1	0,96	0,12	0,87
EXPORT	0,92	0,98	0,98	0,96	1	0,23	0,88
UNEMP	0,08	0,29	0,33	0,12	0,23	1	0,30
HPI	0,79	0,88	0,92	0,87	0,88	0,30	1

Values related to the interpretation range of correlation values are shown in Table 3.

Table 3. Correlation's table

Correlation Range	Relationship Level
(-0.25) – (0) and (0) – (0.25)	Very Weak
(-0.49) – (-0.26) and (0.26) – (0.49)	Weak
(-0.69) – (-0.50) and (0.50) – (0.69)	Medium
(-0.89) – (-0.70) and (0.70) – (0.89)	High
(-1) – (-0.90) and (0.90) – (1)	Very High

When the applied data set is examined according to the interpretation levels of the correlation coefficients, there is a positive correlation with a high correlation between the inflation value and the gold value. In the correlation between the inflation value and the dollar value, there is a positive correlation with a high correlation. The correlation between the inflation value and the oil value is a very high and positive correlation. As in the oil variable, there is a positive relationship with a very high correlation between the inflation value and the export value. There is a very weak positive correlation between the inflation value and the unemployment value. There is a highly positive correlation relationship between the inflation value and the housing price index.

3.2 Application of ADF (Augmented Dickey Fuller Test) Test

After examining the correlation between the dependent variable and the independent variables, the ADF test was performed until the data became stationary to test whether the variables were suitable for time series analysis. The results of the ADF test performed for each variable used are as in Table 4.

Table 4. ADF results for variables

	INFLATION	GOLD	DOLLAR	OIL	EXPORT	UNEMP	HPI
ADF T.S	-0,887	7,441	4,147	1,816	3,535	-2,269	1,543
(95%) p value	0,792	1	1	0,998	1	0,182	0,998

When the p value is greater than 0.05, the data set to be used in the time series is considered to be non-stationary. The series is considered stationary when the p value is less than 0.05. According to this inference, not all variables are stationary and they need to be made stationary. Differentiation will be performed to make the data stationary. The result of the ADF test performed for the all of variable data as a result of the 1st difference operation is as in Table 5.

When the 1st difference transactions in Table 5 are examined, the p values of the inflation data and unemployment data are made stationary because they are less than

Table 5. ADF results on data as a result of 1. differencing

Variable	ADF Test Statistic	(%95) p Value
Inflation	-3,560	0,006
Gold	0,150	0,969
Dollar	-1,129	0,703
Oil	-1,576	0,495
Export	-0,073	0,952
Unemployment	-5,679	8,551*e-07
HPI	-1,997	0,287

0.05. For gold, dollar, oil, export and HPI data that cannot be made stationary, the second difference is taken to make them stationary. The values of the variables as a result of the second difference process are shown in Table 6.

Table 6. ADF results on data as a result of 2. differencing

Variable	ADF Test Statistic	(%95) p Value
Gold	-6,096	1,01*e-07
Dollar	-6,096	1,72*e-07
Oil	-3,116	0,025
Export	-6,180	6,472*e-08
HPI	-8,130	1,092*e-12

As a result of the second difference process, all variables were made stationary and the data became available for time series analysis.

While creating the LSTM model, the 187 data used were divided into 80% training dataset and 20% test dataset.

3.3 Model Creation with LSTM

While creating the LSTM model, GridSearchCV hyperparameter analysis was performed to determine the hyperparameters. The hyperparameters obtained as a result of the hyperparameter analysis are shown in Table 7.

As a result of the created lSTM model, in the light of the results obtained, the comparison of the actual values with the test data of the LSTM model is shown in Fig. 2.

MAPE and MSE error values were calculated for the LSTM model, together with the model created in the light of the determined hyperparameters. As a result of the MAPE value, the success percentage of the model was also calculated. Obtained MAPE and MSE values are shown in Table 8.

Table 7. ADF results on data as a result of 2. differencing

Hyperparameter Variables	Optimum Hyperparameter
Batch-Size	40
Epochs	100
Neurons	100
Dropout Rate	0.2
Activation	Tanh
Optimizer	Adam

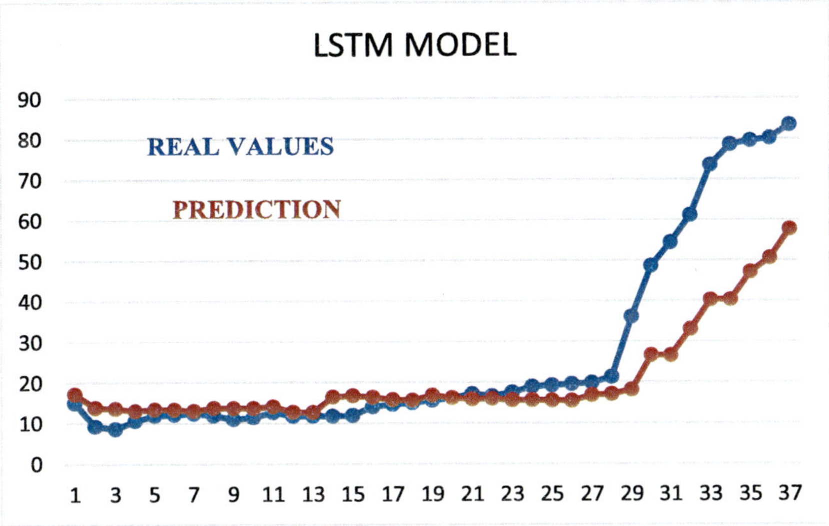

Fig. 2. Comparison of LSTM test data with actual values

Table 8. Statistical results for the LSTM model

Error Type	Value
MAPE	24,027
MSE	41,302

4 Conclusion

The concept of inflation is a very important concept in our country and for every country in the world. The most fundamental factor in the recovery of the economy of countries is inflation, which is at a constant level and at low levels.

In the study, inflation forecasting was carried out with multiple time series analysis using the LSTM model. In the established LSTM model, gold, dollar, oil, export, unemployment, housing price index data were used as independent variables. In total, 187 pieces of data were collected on a monthly basis and the difference between test data and training data was determined as 20% before the model was trained. The estimation results obtained in the model were compared according to the statistical evaluation criteria of MAPE and MSE. The LSTM model achieved a MAPE value of 24,027 and an MSE value of 41,302.

When the results of this study were compared with other studies in the literature, it contributed to the literature in terms of parameters and analyses used. In the future, the study can be further developed by using different time series analysis methods and different heuristic methods, the most accurate model selection can be made with different models to be used, and comparisons can be made between models.

References

1. Şengür, M.: Türkiye'de Enflasyonun Kaynağının Belirlenmesine Yönelik Ekonometrik Bir Analiz, Erciyes Üniversitesi İktisadi ve İdari Bilimler Fakültesi Dergisi, pp. 47–64. (2016)
2. Özel, H.A.: Ekonomik Büyümenin Teorik Temelleri. Çankırı Karatekin Üniversitesi İktisadi ve İdari Bilimler Fakültesi Dergisi **2**, 63–72 (2012)
3. Park, W.J., Park, J.-B.: History and application of artificial neural networks in dentistry. Eur. J. Dent **12**(04), 594–601 (2019). https://doi.org/10.4103/ejd.ejd_325_18
4. Peker, O., Göçer, İ: Yabancı Doğrudan Yatırımların Türkiye'deki İşsizliğe Etkisi: Sınır Testi Yaklaşımı. Ege Akademik Bakış Dergisi **10**, 1187–1194 (2010)
5. İnan, Ç.A.: Raınfall – runoff prediction based on artificial neural network, a case study in La Chartreux Sprıng (2017)
6. Namini, S.S., Tavakoli, N., Namini, A.S.: The performance of LSTM and BiLSTM in forecasting time series. In: 2019 IEEE International Conference on Big Data, pp. 3285–3292. (2019)

Measuring Product Dimensions with Computer Vision in Ceramic Sanitary Ware Sector

Murat Çöpoğlu[1,2](✉), Gürkan Öztürk[1], Emre Çimen[1], and Salih Can Akdemir[2]

[1] Department of Industrial Engineering, Eskisehir Technical University, Eskisehir, Turkey
muratcopoglu@ogr.eskisehir.edu.tr, {gurkan.o, ecimen}@eskisehir.edu.tr

[2] Eczacibasi Building Products, Bilecik, Turkey
salih.akdemir@vitra.com.tr

Abstract. Computer vision applications are a branch of science that aims to extract meaning from digital images. It is used in many fields such as automatic inspection, quality control, security, efficiency improvement, autonomous vehicle technology and health sector. Especially in quality control applications, the reliability of product measurement processes with traditional methods depends on the human factor. This study aimed to measure the product dimensions automatically and accurately with computer vision for the shower tray products of a company operating in the ceramic sanitary ware sector. In addition, low cost, high accuracy and high efficiency are targeted with this study. Object size measurement with computer vision consists of two main components: software and hardware. The choices related to these 2 components can directly affect the results of the application. The application stages can be listed as; calibration of the camera that is fully aligned with the product area, calculation of the pixel/centimeter ratio using the calibrated camera and ArUco marker, background segmentation on the collected image data, edge finding applications and measuring the product dimensions on the found edges. As a result of the application, for 79 shower tray images; For 43 data within 70% frame area, average 2.86 mm, for 36 shower tray images outside 70% frame area, average 4.44 mm and for all test data, measurement was made with an error of 3.67 mm on average. This error rate is below the company's 1 cm tolerance value in manual measurement and gives successful results.

Keywords: Computer Vision · OpenCV · Object Measurement · Camera Calibration

1 Introduction

Object size measurement with camera is an application of a branch of science called photogrammetry. Photogrammetry is a branch of science that aims to determine the geometric properties of objects from photographs. Although photo technology is a 200-year-old technology, the first studies in this field started with Leonardo Da Vinci's work on optical projection principles in 1492. The studies of scientists from different disciplines on photogrammetry gained a concrete meaning with the famous mathematician

Z. Şen et al. (Eds.): IMSS 2023, LNME, pp. 761–768, 2024.
https://doi.org/10.1007/978-981-99-6062-0_73

Henry Lambert using the word photogrammetry for the first time in his book written in 1759. Especially during and after World War II, it accelerated with military mapping and topography applications, and its development continued without interruption until today. Today, object size measurement applications with camera are used in many sectors. It can be seen in various applications in many fields such as autonomous vehicle technology, medical technology, food sector, heavy industry.

Quality control application of shower tray products in ceramic sanitary ware sector is the processes of determining surface defect detection and whether the dimensions of the products are within acceptable range at the end of production process. In traditional quality control applications, very heavy shower trays are transferred to another work environment for surface defect detection by eye after measurement mold for dimension measurement. In the current process that forms the basis of this study, while transferring the product between stations with vacuum levers; it can lead to negative consequences such as time loss, labor loss and measurement errors dependent on human, potential threat of very heavy and fragile products. The mentioned negative results constitute the reasons for this study. Measurement and error detection studies that can be done with computer vision technologies can enable more secure and fast applications if they can give results within the tolerance range that meet quality standards, and also enable quality control applications with lower cost.

In the first stage, it is aimed to measure the product dimensions with computer vision and digitalize this process, while similar studies in the literature generally focused on surface defect detection for tile products with the same raw material. In other sectors, there are many product size measurement studies. M. Yoldaş and C. Sungur performed the measurement of aluminum extrusion profiles with CMOS camera and resulted with 0.6% error [1]. Unlike our study, they worked with an average area of 4 cm and a higher pixel/centimeter ratio. Mehmet U. SALUR and others found 95–98% measurement performance in their study using Rasberry Camera for various objects [2]. Mogre and others proposed a system that detects objects first with YOLO algorithm and then measures dimensions with ArUco marker [3]. Patel and others tried to estimate the dimensions of objects at different distances using LIDAR sensor, camera and OpenCV library [4]. Bin Li successfully measured the dimensions of mill products with 0.1 mm error by examining a computer vision system entirely in his study [5]. Pu and his friends measured objects at different ranges with an average error rate of 1.66% without using auxiliary equipment such as stereo camera, infrared, using only a digital camera [6]. Asaad F. Said developed a method that measures distance and vehicle dimensions using camera external parameters to prevent collisions in autonomous vehicles [7]. The remarkable point in these studies is that the camera calibration was not performed while focusing on the product and the experimental environment. The cameras we collect image data have internal and external parameters. With the calibration of these parameters, when the effect of radial and tangential distortions that occur as they move away from the optical center is reduced, it is expected that the error rate in the measurement results will also decrease. Unlike similar studies in the literature, in this study, after camera calibration with checkboard, it is aimed to measure product dimensions with low error rate with various image processing techniques.

2 Aim and Scope

The aim of this study is to measure the product dimensions, which are a part of the quality control process, with computer vision methods within the specified tolerance values range. As a result of the study; it is aimed to reduce product transfer in the control process, reduce labor force used, reduce potential security problems caused by excessive transfer, shorten the time allocated for dimension measurement process, prevent bottleneck in quality control band, and create a dataset that can be used in various development and analysis studies later with measurement data generated in quality control processes related to dimension in the first stage.

Shower trays quality control process in ceramic sanitary ware sector consists of two stages. With this study, the first stage of checking the suitability of product dimensions will be performed automatically by the computer vision system.

3 Methods and Materials

The flow diagram of the methods used in the study is as shown in Fig. 1. The computer vision application consists of 2 main components: hardware and software. For hardware, Basler Ace – acA2040-55uc model 3-megapixel industrial camera, C23-0824-5M-P f8mm model Basler Lens, standard ambient lighting (Fluorescent), checkboard for camera calibration, ArUco marker to calculate pixel/centimeter ratio were used. For software, Visual Studio Code, Python programming language, OpenCV 4.6, NumPy 1.24, matplotlib 3.7.1 were used.

Fig. 1. Size Measurement Study Flow Diagram

3.1 Camera Calibration

Camera Calibration Camera calibration is an application that determines the internal and external parameters of the camera. Its purpose is to correct the distortions in the image and to establish the relationship between the real world coordinates and the camera coordinates. It is one of the most important factors to pay attention to in measurement studies with the camera. The length defined by a certain number of pixels in the image data should be the same or have the least possible error at every point of the frame area. The camera's internal parameters include parameters such as focal length (f_x,f_y) optical center (c_x, c_y) and lens distortion that define the optical and geometric characteristics of the camera. The camera matrix shown below (1), it shows the camera's internal parameters. If the camera matrix is optimized according to the current camera, lens distortions can be prevented. In addition, camera matrices are specific to a certain camera, they can be used repeatedly for the same camera if they are optimized once.

$$K = \begin{bmatrix} f_x & 0 & c_x \\ 0 & f_y & c_y \\ 0 & 0 & 1 \end{bmatrix} \quad (1)$$

However, the camera's external parameters can also significantly affect the measurement studies with the camera. The camera's external parameters should be optimized specifically for a certain camera and the camera's fixed position. With these parameters, radial and tangential distortions can be prevented. Radial distortion can cause bending in lines as the image moves away from the center. These distortions are usually in the form of barrel or pincushion. As seen in Fig. 2, in barrel distortion, the edges of the image are pulled outward, while in pincushion distortion, the edges are bent inward.

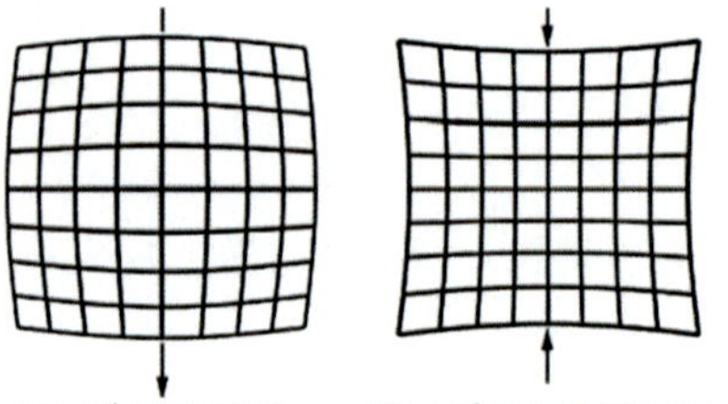

Fig. 2. The types of distortion

Tangential distortions are one of the distortions in the image caused by the camera lens. Axial distortion causes the points near the center of the image to shift. Tangential distortion is usually rectangular or trapezoidal. Tangential distortion is caused by the displacement of the camera's principal point and distorts the scaling of the image.

To overcome these problems, Open CV provides various methods to perform camera calibration. In this study, one of these methods, the checkboard method, was used. In this method, as seen in Fig. 3, images are taken from different points of the frame area with a reference object consisting of squares of certain sizes and the application is started.

Then the real world coordinates and image coordinates of the chessboard are kept in two separate lists. Corner points are found using the relevant functions and the camera matrix and distortion coefficients are calculated by considering the differences between the real world points and the image points. These parameters are then applied to all images taken with the camera used in the system to provide corrections.

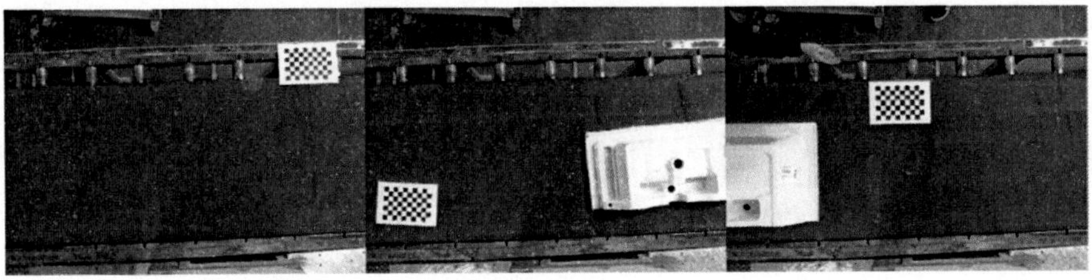

Fig. 3. Checkboard for camera calibration

3.2 Calculating the Pixel-Centimeter Ratio with ArUco

As seen in Fig. 4, the centimeter per pixel is found by using the ArUco marker on the plane to be measured with the calibrated camera. This stage should be done after the camera calibration. Because the test phase will be done on the calibrated images and the ArUco marker should also be used under the same conditions to find the correct pixel/centimeter ratio. In this stage, the pixel/centimeter ratio was found to be 11.6522. That is, for the established system, 1 mm in the real world corresponds to 1.16522 pixels in the camera environment.

Fig. 4. Using ArUco to calculate the pixel/centimeter ratio

3.3 Background Segmentation with Threshold Value

For the correct operation of the edge detection algorithms in the next stage, the object of interest should be separated from the background in the most accurate way. In similar image and measurement studies, clear distinctions have been created between the background and the object by providing maximum contrast. In the present study, the fact that the products are white and the background is black makes this stage quite easy.

3.4 Edge Detection

The edges of the images extracted from the background need to be detected for measuring their dimensions. In this stage of the study, the find Contours function of the OpenCV library was used. The most important factor in this stage will be to choose the appropriate method and approach from the 4 methods that can detect the edges. At this point, the RETR_TREE (Retrieval Tree) method and the CHAIN_APPROX_SIMPLE approach were applied. With RETR_TREE, the length and width detection was facilitated by keeping the relationship between the contours, and with CHAIN_APPROX_SIMPLE, the memory usage was reduced by keeping only the start and end points of the contours.

3.5 Size Measurement on Test Images

At this stage, as seen in Fig. 5, the test data collected were passed through all the above processes in order and it was determined how many millimeters in length the pixel count corresponding to the contours was in reality.

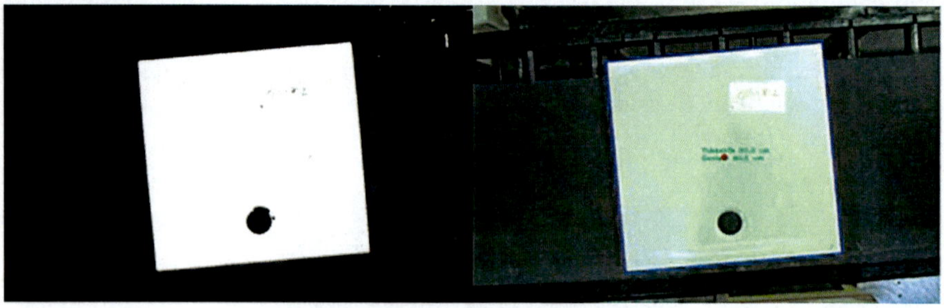

Fig. 5. Size measurement application

In the study, as shown in Fig. 6, a 70% frame area was created to see the measurement accuracy according to the position of the products. The data were classified as those inside and outside this area. In this way, it was aimed to test whether the accuracy changed as the products to be measured approached the optical center, and if it changed, how much it changed.

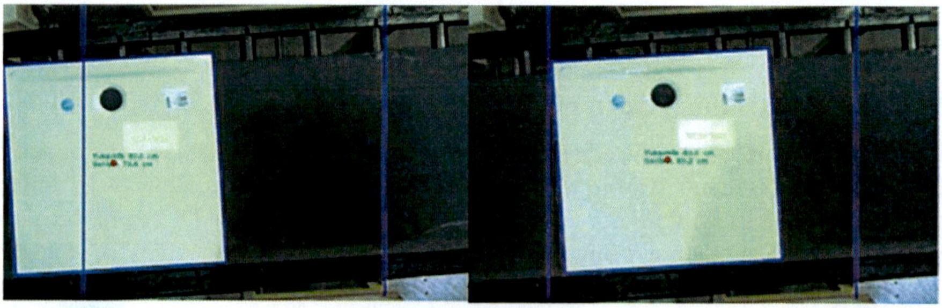

Fig. 6. Frame area (70%)

3.6 Results

The prepared application was applied on 79 test data with an average width of 855.38 mm and an average height of 819.5 mm. As a result of the study, 79 images were measured with an average error rate of 0.35% and an average error of 3.67 mm. As can be seen from Table 1, 89.9% of the errors remained 5 mm and below. Despite the camera calibration, it was seen that the error rate was lower in the measurements in the 70% frame area in the middle of the frame area. In this region, the average error was 2.86 mm in 43 images and 4.44 mm in 36 images taken outside this area, as the tangential distortion rate was lower.

Table 1. Mismeasurement Rate and Number of Error

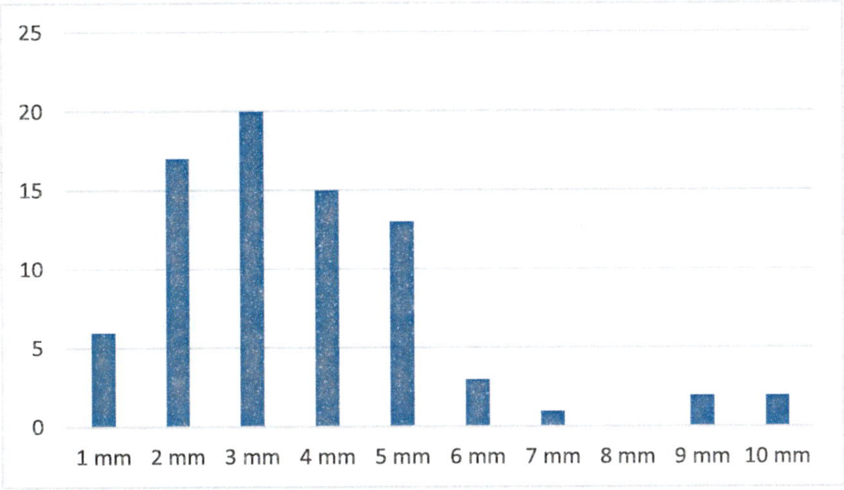

As seen in Table 2, all of the image data within the 70% frame area are within the dataset with measurement errors below 5 mm. This result proves that the distortion increases as it moves away from the optical center. In addition, as can be seen from Table 2 and Table 3, measurement errors can increase outside the 70% frame area.

Looking at the results of the application, very satisfactory values were obtained with an average error of 3.67 mm and an average error rate of 0.35%. These values, which are within the tolerance range of the quality control application, can enable the distribution of certain human labor in the quality control operation to other units, saving time by reducing the number of product transfers and eliminating potential occupational safety problems.

Table 2. Distribution of Errors Within 70% Frame Area

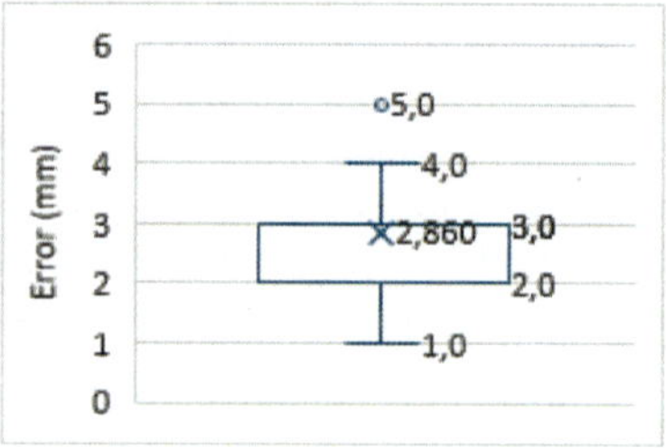

Table 3. Distribution of Errors Outside 70% Frame Area

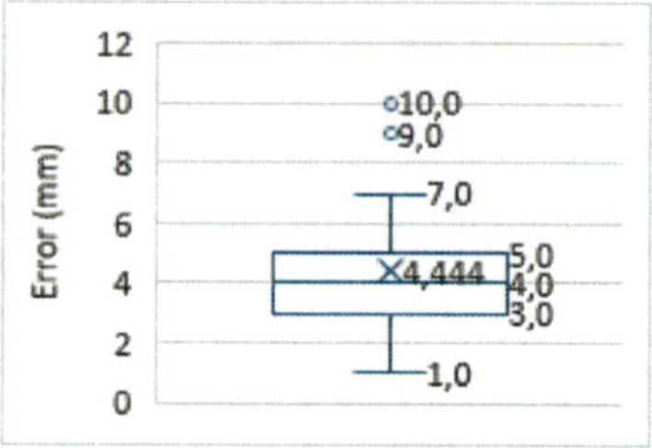

4 Conclusion

This study focuses entirely on the application of computer vision for size measurement in the ceramic sanitary ware sector. The most important output of the study is to create a low-cost control application through cameras. However, it will also provide an important information output and data set for similar studies in the production sector. In addition, using the outputs of this study, it is also planned to create a size estimation system within a certain tolerance range independent of the object's distance from the camera.

In the later stages of the application, studies will be conducted to find error detections with computer vision on bright surfaces, and it is planned to create a complete quality control station with low cost and high accuracy rate with a single camera.

References

1. Yoldaş, M., Sungur, C.: Alüminyum Ekstrüzyon Profillerinin Hassas Kesit Ölçümlerinin Görüntü işleme Teknolojisi ile Gerçekleştirilmesi, pp. 190–195 (2020)
2. Othman, N.A., Salur, M.U., Karaköse, M., Aydın, İ.: An Embedded Real-Time Object Detection and Measurement of its Size. In: International Conference on Artificial Intelligence and Data Processing (2018)
3. Mogre, N., Bhagat, S., Bhoyar, K., Hadke, H., Ingole, P.: Real time object detection and height measurement. Int. Res. J. Modern. Eng. Technol. Sci. **132**, 3825–3829 (2022)
4. Patel, B., Goswami, S.A., KaPatel, P.S., Dhakad, Y.M.: Realtime object's size measurement from distance using OpenCV and LiDAR. Turkish J. Comp. Math. Educ. **12**(4), 1044–1047 (2021)
5. Li, B.: Research on geometric dimension measurement system of shaft parts based on machine vision. EURASIP J. Image Video Process 1–9 (2018)
6. Pu, L., Tian, R., Wu, H.-C., Yan, K.: Novel Object-Size Measurement Using the Digital Camera. In: IEEE Advanced Information Management, Communicates, Electronic and Automation Control Conference (IMCEC) (2016)
7. Said, A.F.: Robust and Accurate Objects Measurement in Real-World Based on Camera System. In: IEEE Applied Imagery Pattern Recognition Workshop (AIPR) (2017)

Theory and Research Concerning the Circular Economy Model and Future Trend

Gülseli İşler[1] (✉), Derya Eren Akyol[1], and Harun Reşit Yazgan[2]

[1] Dokuz Eylül University, Izmir 35390, Turkey
gulseli.isler@hotmail.com, derya.eren@deu.edu.tr
[2] Sakarya University, Sakarya 54050, Turkey
yazgan@sakarya.edu.tr

Abstract. The aim of this study is to draw attention to Circular Economy (CE) approach, to address a future trend and to create a compilation of some important CE studies. In this context, we provide information to guide businesses in the transition to CE and to provide ideas for future studies. This study examines some studies about the concept, structure, principles, business models, benefits, and barriers of CE implementation, moreover differences between Linear Economy (LE) and CE. The failure to solve problems such as climate change, loss of biodiversity, overpopulation, and resource scarcity in the current LE Model has shown that sustainable economy models can be a successful solution. Artificial intelligence (AI), on the other hand, is an emerging technology with enormous potential to impact CE. The use of AI in integrating the Circular Economy into product lifecycle processes was investigated. This study collects many studies examining CE from different perspectives in a single study and also shows the usability areas of AI in applying CE in production. Another contribution of this study to the literature is, proposing the use of AI regarding CE's the 9R framework. This study has shown that, through artificial intelligence, the stated barriers to the transition to the circular economy can be removed or reduced. By using the business models outlined in the study, using Artificial Intelligence and implementing CE strategies, the stated benefits of the circular economy can begin to be achieved. The study guides the implementation of the CE concept.

Keywords: Circular Economy Model · Circular Economy Structure and Business Models · Circular Economy Benefits and Barriers · Product Life Cycle · Artificial Intelligence

1 Introduction

Our nature consists of processes that involve an endless cycle, such as a waste-free metabolism. Since an input of one process is the output of another process in nature, no waste is generated. The CE model, which is seen as an innovative alternative to the frequently used Linear Economy (LE) model, is a new and advanced system that optimizes natural resources used and keeps them in product life cycles as long as possible and transforms products that have completed their lifecycles into sources of other products and services cycles [40].

Z. Şen et al. (Eds.): IMSS 2023, LNME, pp. 769–782, 2024.
https://doi.org/10.1007/978-981-99-6062-0_74

There is no clear evidence that the CE concept is a source or creator from which it was first introduced. Most inspirations in the literature are based on the Industrial Ecology Model based on the closed-loop concept originating from Japan and Germany [27, 44]. However, the following names can be seen among those who contributed to the emergence of this concept: John Lyle [20], William McDonough [26], Chris Wise [43], Michael Braungart (2013), and Walter Stahel [37]. While Jawahir and Brandley (2016) and Homrich et al. (2018) claim that the term CE was first used by British economists Pearce and Turner in 1990, Pearce and Turner claim that this term was first used by Stahel and Reday-Mulvey in the Western literature in the 1980s to denote a system in which environment and economy have closed interactions [12]. Some sources state that the CE concept emerged with a report from Switzerland presented to the European Commission by Geneviève Reday Mulvey and Walter R. Stahel (1977), explaining the idea of using human power instead of energy.

This study presented a literature review on the origin, logic, structure, and principles of CE and gave detailed explanations. Differences between CE and LE are stated, and some studies on sustainability and CE worldwide are included. In addition, principles, processes, potential benefits, and potential barriers of the CE are listed. A literature review on the relationship between CE and the product life cycle is included. In literature, business model strategies of CE are explained as a preliminary preparation for adapting the CE to the product life cycle.

Artificial intelligence is an emerging technology with enormous potential to impact CE [41]. AI has the potential to enable and enhance the Circular Economy by providing new ways to optimize material and energy flows, reduce waste, and increase resource efficiency. The relationship between Circular Economy and Artificial Intelligence is an exciting and rapidly evolving field with the potential to unlock significant environmental and economic benefits by developing more efficient and sustainable systems. In this context, strategies and principles for transitioning to CE in product life cycle processes in the manufacturing sector were examined (9R). The usage areas of artificial intelligence in applying these principles were proposed [40]. Furthermore, this study proposes the use of AI in terms of CE's 9R (Refuse, Rethink, Reduce, Reuse, Repair, Refurbish, Remanufacture, Repurpose, Recycle and Recover) framework. This study has shown that, through artificial intelligence, the stated barriers to the transition to the circular economy can be removed or reduced. By using the business models outlined in the study and implementing CE strategies, the stated benefits of the circular economy can begin to be achieved.

2 Material and Method

2.1 Circular Economy Model Concept and Comparison with Linear Economy

LE and the "take-make-dispose" model that has shaped the global economy since the Industrial Revolution is a one-way system in which natural resources turn into waste through service or product production. Since this system is based on the principle that natural resources in our world are easily accessible, sufficient, and available, and the last step is cheap and trouble-free waste, we have faced a serious increase in the risk of world resource scarcity in many sectors (Steffen et al. 2015). Other factors that make this model unsustainable are the high demand for resources and the increase in population, inequality, demographic change, and consumption. With the effect of globalization, supply chains and production processes have become more complex, and price fluctuations and competition have increased in national and local markets. In addition, consequences such as the Covid-19 pandemic, which is under the influence the whole world, the rapid depletion of natural resources, and climate change push us to change the ecological boundaries of our world. The LE model being used worldwide cannot prevent these crises and cannot meet the needs of humanity, which has an increasing level of welfare and rapid population growth.

With the unsustainable production and consumption system, worldwide consumption has increased eightfold in the last few decades. As a result of this increase, it is predicted that global resource use will increase by more than three times by 2050. We cannot adequately manage waste generated due to the resources we use: Globally, 1/3 of plastic waste is not collected or managed, and greenhouse gas emissions are increasing over the years [39]. All the facts mentioned are a barrier to the Sustainable Development Goals (proclaimed by the United Nations in 2020 [38]). There is a need for a more inclusive economic model that increases inter-sectoral cooperation, ensures resource efficiency and reduces waste, and adopts the cycle's continuation principle by making end-of-consumption or production-based outputs as inputs of the same or another sector. CE, which ensures long and efficient use of resources, saves energy, and minimizes waste by keeping resources and products within product life cycles, is seen as a good alternative to LE in the European Union market and is increasing its reputation over time.

CE logic is different from LE. Because this concept transforms linear production into a competency model: it follows a path of maximum reuse, recycling of non-reusable goods, repair of damaged, etc., and reproduction of parts and products that are beyond repair [21].

The CE model is a model and tool with a holistic structure and a production system that enables the reuse of raw materials and semi-finished and finished products. At the same time, the CE model is a model in which wastes are recycled into the system, resources and energy are used very efficiently, produces as barely waste as possible, and minimizes environmental harm. The CE approach aims to maximize the added value of services and products in the value chain, minimize waste, and ensure that resources are in the cycle and the economy for much longer [27].

The European Union wants to reduce raw material production requirements, to obtain economic and environmental benefits, thanks to the transition to a new economic concept that adopts renewable energy and efficient use of resources. For this purpose, the European Commission adopted the CE Package on 2 December 2015, which sets out legislative proposals on waste and a comprehensive action plan for a circular and sustainable economy. In line with this plan, some countries that are members of the European Union have started to prepare and accept regional and national roadmaps for the transition from the LE model to the CE model [34].

Furthermore, the CE, adopted by the "European Green Deal" (adopted on 11 December 2019), is one of the priority policy areas in the new strategy. With this document, 2050 is targeted for the EU to offer new and better jobs, restore climate neutrality, improve people's well-being and increase growth. At the same time, they started the work needed to use resources more efficiently, stop biodiversity loss and climate change, and reduce environmental pollution by transitioning to a circular, clean economy. In the last decade, the United States, China, and South Korea have introduced research programs that promote circular economies in product and service markets by increasing reuse.

The 2030 Agenda for Sustainable Development was adopted by the United Nations General Assembly on 25 September 2015, thus taking a step towards implementing the CE model. In line with these studies, the EU's objectives for production are to develop some incentives that support cyclical product designs and to create efficient and innovative production-product life cycle processes. Strategies for recycling include promoting recycled raw materials, nutrients, and water, increasing knowledge about material flows in the cycle, and ensuring safe chemical management.

CE renews and repairs products and services through design. The overall goal of the CE is to maximize the value of a product at every point in its lifecycle. In this economy, it aims to optimize returns of resources, develop and protect natural capital, minimize risks in a system with flow chain management, and keep resources and products at the highest value with the highest benefit [36]. It delivers effective results in businesses of all sizes [40].

2.2 Circular Economy Structure

The CE is achieved by designing all industrial processes so that resources, services, and raw materials do not leave the product life cycle and are used continuously. In this direction, while wastes are minimized, residues that cannot be removed from the cycle are also recovered [40]. In CE, raw materials, components, and products are used in biological or technical cycles. In contrast, environmental impacts of services and products are minimized within the product lifecycle/value chain. The diagram in the Fig. 1. Described by the Ellen MacArthur Foundation [10] includes two cycles of the CE:

1. The biological cycle where residues and wastes are recycled to nature after use.
2. The technical cycle in which raw materials, components, and product designs are made to minimize waste.

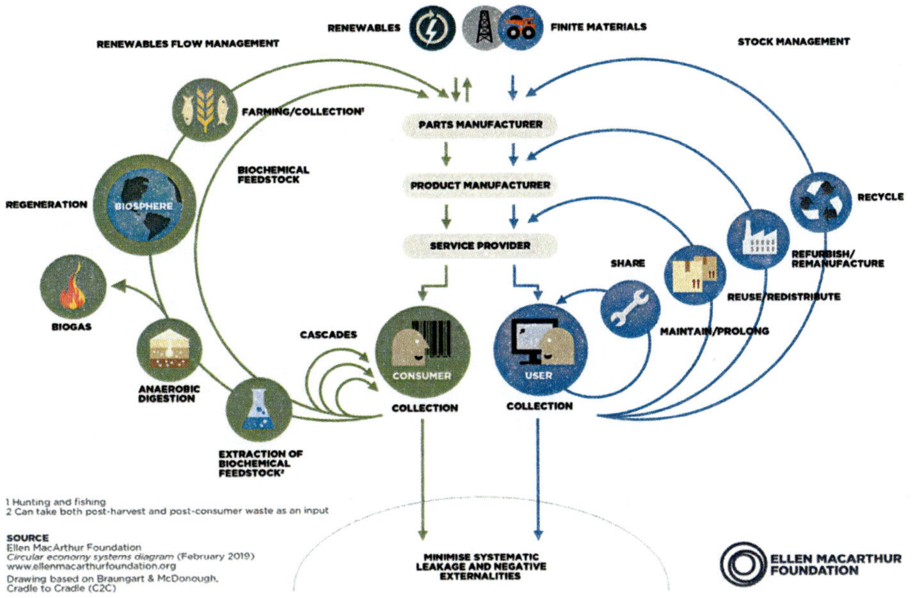

Fig. 1. Circular Economy diagram [10].

2.3 Benefits and Barriers of Applying Circular Economy to Product Life Cycle

Potential Benefits of CE. The CE is an economic concept that provides new strategies and business models to reuse resources with maximum efficiency and keep them in cycles within the confines of our planet. In this context, implementing the CE brings many benefits to businesses, countries, and our planet (in terms of sustainable use of its resources).

In line with the studies given in Table 1, it can be said that there are many observations and comments in the literature about the benefits of CE. In addition, it can be seen that the CE, a new concept, provides many economic, social, and environmental benefits. J. Korhonen et al. [23] group the dimensions in which sustainable development and implementation of a successful CE benefit under three headings: economic, social, and environmental. Making a radical and fundamental change and renewal in business processes, models, products, and factory designs is a huge step for a company. However, it is stated that this change will lead the company to the result of commercial success, and it will be an incentive factor for other companies [11]. Some benefits of applying CE:

- Globally $4.5 trillion growth opportunity [24]
- Creating new values for materials that have already been paid for [42]
- Minimization of costs [40]
- Efficiency opportunities in supply chains and industry in general [42]
- Saving by reducing dependence on energy and primary raw materials
- Ensuring the sustainability of the export of products and services [40]

Table 1. Benefits of Circular Economy.

Reference(s)	Objective of the study	Inferences about benefits
[8]	To describe opportunities to generate rapid and lasting economic benefits and enlist broad support for putting it into full-scale practice	Chance of innovation and R&D to create new opportunities
[9]	To provide a vision of how the Circular Economy could look and to highlight wide-ranging implications for government and business leaders	Greatly reducing greenhouse gas emissions and primary resource consumption in the short, medium, and long term
[11]	Examine how the circular model forces companies and organizations to create disruptive technology and business models based on longevity, reuse, repair, upgrade, renewability, refurbishment, etc	Creating much more job opportunities than a LE
[42]	To explore the potential for resource efficiency and to assess what the main benefits for society would be	Creating new productivity and labor opportunities in supply chains and industry
[5]	To give an overview of how the producers comply with the growing demands and to gain efficiency and increase profitability from Circular Economy	Providing economic value by reducing the ecological footprint and raw material costs and minimizing price variability
[18]	To explore the recovery and reuse of structural products from end-of-life buildings	New technologies could improve resource recovery systems and sharing systems of assets, also the benefits of reverse logistics application

- Increase in upgradability of products and services
- Growth despite using fewer resources [40]
- Increased security of supply
- Strengthening of reverse logistics activities
- Minimizing price fluctuations in raw materials [40]
- New employment and job opportunities
- Businesses have the goal of transitioning to the CE strengthens design and innovation [4]
- Increasing the competitiveness, flexibility, and endurance of businesses [40]

Potential Barriers of CE. The CE concept ensures that the life of products is extended by developing improvement, recycling, and repair processes instead of making waste after use. While the transition to the CE requires massive and radical changes in the production and consumption processes, it also has barriers at micro and macro scales. Although these barriers are written as separate items in the literature, it is stated that all these are related to each other. Some evaluations of removing a barrier will be a factor (catalyst-like) in removing barriers under other headings [14]. Table 2 shows some studies that classify the barriers to Circular Economy implementation.

Table 2. Barriers to Circular Economy implementation literature review.

Reference(s)	The objective of the study	Inferences about barriers
[29]	To explore the potential of the Circular Economy to help decouple increased wealth from growth in resource consumption	Lack of standardization of certification practices in the international market
[19]	To capitalize on eco-innovation to contribute to the design of policy guidelines and organizational strategies	Soft barriers (institutional and social) and hard barriers (technical and economic) classification
[22]	To present the first major N-study on Circular Economy barriers in the EU	Legal, technological, market, and cultural classification
[17]	To identify the barriers to and enablers for the Circular Economy within the built environment for facilitating the pathway towards circularity	Sectoral, legal, financial, and cultural classification
[14]	To examine how technological, market, institutional, and cultural barriers may hinder the implementation of a Circular Economy from a theoretical economic perspective	Technological, economic and market-based, legal and institutional, cultural and social classification

List of barriers to the use of the CE:

- Barriers to the implementation of a CE are similar to the barriers to the integration of sustainability strategies [30]
- The necessity of breaking down existing production and consumption patterns [16]
- For the reuse or recycling of products, as opposed to distribution, the flow of recovery from the market should also be improved
- Businesses need to update or completely renew their business models in line with the Circular Economy structure [1]
- Difficulty in disassembling parts of waste products that are in good condition
- Risk of damaging dismantled parts of the product that is in good condition when disassembled from the product
- The perception that the quality of remanufactured or recycled materials to be used as inputs to new systems will be low [3]
- The CE model requires changes in cultural structures and breaking resistance in the structure
- The model brings the necessity of advanced technological developments for cost minimization [22]
- CE logic, based on the permanence, does not fit the fashion concept

- Existing and potential customers are open to rapid changes
- Lack of sufficient awareness among current and potential consumers
- Lack of standardization of certification practices in the international market yet

2.4 Circular Economy Business Models

Changeover to the CE model requires many innovative and fundamental changes and improvements. In this part of the study, business model concepts are presented by Bocken et al. [4] that fit the approaches to slowing down resource and closing cycles. As it can be seen in Table 3, CE business models are broadly divided into two groups.

Table 3. Barriers to Circular Economy implementation literature review.

Business Model Strategy	Scope
Circular Economy Business Model Strategies Aimed at *Slowing* the Cycles	Business models that extend product or service life/encourage reuse; support the Circular Economy through repair, remanufacturing, upgrades, and retrofits
Circular Economy Business Model Strategies Aimed at *Closing* the Cycles	Business models that enable recycling of products and services; transform old/refused products and parts into new resources

Circular Economy Business Model Strategies to Slow Cycles. CE business model strategies to slow cycles are business models that extend product or service life and encourage reuse. These business models support the ce through repairs, remanufacturing, upgrades, and improvements. CE business models for slowing cycles are listed below [4].

Access/Performance (Sharing) Business Model. It is a business model that adopts a method of meeting the usage needs of consumers without owning physical products.

Business Model Aiming to Increase Product Value. Uncovering and using the value of waste materials is a business model that adopts a method of product residues generally in the form of production-consumer production.

Long Life (Classic) Business Model. It is a business model that adopts a method of designing products to allow repair and provide durability support, thereby increasing the product's lifetime.

The Business Model Aimed at Supporting Competence. It is a business model that adopts a method of developing and implementing solutions (such as service, durability, warranty and repair, and upgradeability) that will reduce the consumption of end consumers.

Circular Economy Business Model Strategies to Close Cycles. CE business model strategies to closing the cycles are business models that provide recycling of products and services. In business models that aim to close product and service life cycles, closing the cycle between supply/production and end-of-use is the basic element [33]. These business models enable the CE by transforming old/refused products and parts into new resources. CE business models for closing cycles are listed below [4].

The Business Model Aimed at Extending Resource Value. It is a business model that includes recycling products, collecting resources and materials that are seen as waste and thought to go to waste, etc., adding value to them by putting them in a new form.

Industrial Symbiosis Business Model. It is a business model for geographically close processes that involve using residues of one as a resource for the other and introducing them into the product lifecycle.

2.5 Relationship Between Circular Economy and Artificial Intelligence in Product Lifecycle Processes

The relationship between CE and Artificial Intelligence is an important and emerging area of research and development. The circular economy is an economic model that aims to minimize waste and maximize resource efficiency while creating a closed-loop system where materials are reused, repurposed, and recycled. Artificial Intelligence (AI), on the other hand, deals with developing intelligent machines that can perform tasks that typically require human intelligence. Artificial intelligence is an emerging technology with enormous potential to impact CE [43]. The combination of AI and the CE model presents exciting opportunities for sustainable innovation and growth. According to our limited knowledge, there is no any study regarding the CE processes where AI can contribute to the circular economy model and shape future trends. This section explores the use of Artificial Intelligence in integrating the Circular Economy into product lifecycle processes.

For this research, the following strategies and principles for the transition to CE in product lifecycle processes in the manufacturing sector have been considered: (9R) [35, 40].

Refuse. Converting the existing LE concept product into new and innovative products and services or pullback the existing product from circulation. AI can help identify products that no longer fit for purpose or are harmful to the environment and need to be withdrawn from the market [6]. By analyzing data on product safety, environmental impact, and customer feedback, AI can identify products that need to be recalled or withdrawn, enabling more sustainable and responsible business practices.

Rethink. Working towards frequently using the new product developed in line with the Innovative and CE, launching it, developing reuse, and sharing models. AI can personalize products and services, making them more attractive and relevant to customers. By analyzing data on individual customer behavior, using AI, companies can design products that meet specific customer needs and preferences and develop customized reuse and sharing models that encourage greater product use and engagement. By analyzing data on consumer behavior using AI, companies can identify the most effective marketing channels and messaging that resonate with target audiences, thereby increasing awareness and adoption of circular products and services [25].

Reduce. Increasing efficiency by reducing the consumption of resources and energy used during production or use of the product. With artificial intelligence, by analyzing data in the production process, including machine performance, energy use, and material waste, improvement areas can be identified, and production processes can be optimized to minimize waste and reduce energy consumption [2]. AI can analyze data on energy usage, including peak demand times and patterns; thus, companies can optimize energy use and reduce costs while minimizing environmental impact. It can be used to optimize the supply chain, reduce resource consumption, and predict when equipment and machinery will need maintenance, reducing downtime and optimizing resource usage.

Reuse. Reuse of a product that has not yet become a waste and is designed to be reused. Developing new business models for reuse with AI may be easier than it is now. By analyzing data on customer behavior and market trends using AI, companies can identify new opportunities for reuse, such as rental, leasing, or subscription-based models, which can help extend the life of products and reduce waste. Also, AI can help optimize the design of products for reuse and track the location and condition of assets designed for reuse by using sensors and tracking technology [31]. And also it can be used to optimize the logistics of reuse. Furthermore, integration of AI with CE can promote sustainability and waste reduction by supporting circular business models such as renting or sharing products instead of traditional selling in practice.

Repair. To ensure that the product that is broken or unable to function is maintained and repaired. By analyzing product performance and usage patterns, AI can predict when a product will likely break down or require maintenance (predictive maintenance) [28]. By analyzing data from sensors and other sources using AI, product problems can be easily diagnosed remotely, and the correct parts needed to repair a product can be identified. It can be used to optimize repair scheduling by analyzing data on repair times and availability.

Refurbish. Renewing/updating the product to a determined high-quality level. AI can be used to analyze products to determine the potential for refurbishment by analyzing product specifications, usage history, and performance data. AI can optimize the refurbishment process by analyzing refurbishment times, costs, and quality data. AI can help to ensure the quality of refurbished products and to personalize refurbished products for customers [13].

Remanufacture. Maintaining usable parts of waste and scrap products and reusing them in a new and original product. Using AI, waste and scrap products can be analyzed to identify usable parts and components that can be disassembled and reused using machine learning algorithms to identify these parts. Once the usable parts have been disassembled, by analyzing data on product specifications and performance, AI can test and sort them based on their quality and suitability for reuse [7]. It can match the usable parts with new product designs that require similar or identical components. By analyzing product specifications and performance data, AI can verify that the remanufactured products meet the required quality standards, reducing the risk of defects and ensuring customer satisfaction.

Repurpose. Reuse waste and scrap products or parts in some new products. AI can help support the Repurpose model in a Circular Economy by analyzing waste and scrap products, identifying usable materials, sorting and preparing materials for reuse, designing new products that use repurposed materials, and optimizing the manufacturing process for the repurposed materials. By repurposing waste and scrapped products or parts in new products, companies can reduce waste, conserve resources, and minimize the environmental impact of their operations [15].

Recycle. Ensuring that waste, scrapped products, and parts are converted into new items, parts, and products in new products or through reprocessing. AI can identify the materials in the waste, scrapped products, and parts by using machine learning algorithms to identify these materials [37]. After identifying the materials, AI can sort and separate them based on their properties and characteristics. In addition, with artificial intelligence, data on production parameters and performance can be analyzed, which can be used to optimize the recycling process. AI can create new and original products that use recycled materials efficiently and effectively.

In conclusion, the circular economy model and AI represent a promising approach to creating a more sustainable and equitable future. The circular economy model and AI have the potential to create a more sustainable and prosperous future. As more businesses and governments adopt circular economy principles, and AI technologies continue to advance, we can look forward to a more sustainable and equitable future.

3 Conclusion and Future Work

Today, customers do not evaluate the value of a company or a product based on its production process efficiency and low price. At the same time, customers evaluate a product's value according to the importance given to sustainability by the company that produces the product. CE, a sustainable development method, is an economic model that aims to maximize resource use efficiency by ensuring that materials and products circulate effectively and efficiently in biological and technical cycles.

This study explains the beginning and the concept of the CE in the literature and the differences of CE between the LE are stated. Studies conducted in the EU in line with Sustainability and CE are overviewed. The structure and principles of the CE are stated, and some studies on the benefits and barriers to implementing this concept are included. In addition, the benefits and barriers of the CE are listed. Classification and explanations of CE business models (models aimed at slowing and closing cycles) are given. Includes a review of the relationship between CE and product lifecycle. Finally, this study explores the use of Artificial Intelligence in integrating the Circular Economy into product lifecycle processes.

Applying the CE concept to systems can solve variability in raw material costs in line with the LE concept. This system, which aims to reach the optimum level of resource use, covers the efficient use of resources at the optimum level and the conversion of wastes into new resources through recycling or repair. In this sense, it not only affects a product's life cycle but also provides interactive resource use between products.

The study showed that AI could play a crucial role in promoting the frequent use of new products developed in line with the Innovative and Circular Economy. AI can help companies optimize production processes, manage energy consumption, optimize supply chains, and predict maintenance needs. In the transition to CE, product design and logistics can be optimized with AI. Artificial intelligence contributes to the Circular Economy application by providing predictive maintenance, remote diagnostics, parts identification, and repair planning. Overall, this study has shown that, through artificial intelligence, the stated barriers to the transition to the circular economy can be removed or reduced. By using the business models outlined in the study and implementing CE strategies, the stated benefits of the circular economy can begin to be achieved.

The impact of the CE on product life cycle processes is inevitable. To realize this concept, the structure of the processes in the product life cycle needs to be changed/improved. Future studies may contribute to the literature by making changes/improvements in processes to adopt the CE to a product life cycle. Research can be expanded with studies on the recognition and effectiveness of CE by companies and individuals. Strategies for the transition to the CE and the steps to be taken to implement the CE in the product life cycle can be determined, and the concept of the CE can be examined from the perspective of potential customers and employees. A real-life production model of the Circular Economy can be developed. Using CE and AI in the production sector can help companies reduce waste, save resources, increase efficiency, and create new business opportunities. Thus, a real-life model can be developed for the Circular Economy and to adapt Artificial Intelligence to the product life cycle.

References

1. Açıkalın, N.: Sürdürülebilir Pazarlama Bakış Açısı İle Döngüsel Ekonomi İncelemesi. Sakarya İktisat J. **9**(3), 238–257 (2020)
2. Nañez Alonso, S.L., Reier Forradellas, R.F., Pi Morell, O., Jorge Vazquez, J.: Digitalization, circular economy and environmental sustainability: the application of artificial intelligence in the efficient self-management of waste. Sustainability **13**, 2092 (2021)
3. Andrews, D.: The circular economy, design thinking and education for sustainability. Local Econ. **30**(3), 305–315 (2015)
4. Bocken, N.M.P., de Pauw, I., Bakker, C., van der Grinten, B.: Product design and business model strategies for a circular economy. J. Ind. Prod. Eng. **33**(5), 308–320 (2016)
5. Buruzs, A., Torma, A.: A review on the outlook of the circular economy in the automotive industry. Int. J. Environ. Ecol. Eng. **11**(6), 576–580 (2017)
6. Cowls, J., Tsamados, A., Taddeo, M., Floridi, L.: The AI gambit: leveraging artificial intelligence to combat climate change—opportunities, challenges, and recommendations. AI & Soc **38**, 283–307 (2023)
7. Chauhan, C., Parida, V., Dhir, A.: Linking circular economy and digitalisation technologies: a systematic literature review of past achievements and future promises. Technol. Forecast. Soc. Chang. **177**(3), 121508 (2022)
8. Towards the Circular Economy Vol.1: An Economic and Business Rationale for an Accelerated Transition, https://ellenmacarthurfoundation.org/towards-the-circular-economy-vol-1-an-economic-and-business-rationale-for-an. Last accessed 9 Jun 2023
9. Growth within: a circular economy vision for a competitive Europe. https://ellenmacarthurfoundation.org/growth-within-a-circular-economy-vision-for-a-competitive-europe. Last accessed 9 Jun 2023
10. The butterfly diagram: Visualising the circular economy. https://ellenmacarthurfoundation.org/circular-economy-diagram. Last accessed 9 Jun 2023

11. Esposito, M., Tse, T., Soufani, K.: Is the circular economy a new fast expanding market? Thunderbird Int. Business Rev. **59**(1), 9–14 (2015)
12. Gedik, Y.: Understanding the circular economy: a theoretical framework. Turkish Bus. J. **1**(2), 13–40 (2020)
13. Ghoreishi, M., Happonen, A.: New promises AI brings into circular economy accelerated product design: review on supporting literature. In: 7th International Conference on Environment Pollution and Prevention, pp. 1–10. Melbourne, Australia (2019)
14. Grafström, J., Aasma, S.: Breaking circular economy barriers. J. Clean. Prod. **292**, 1–14 (2021)
15. Gupta, P.K., Shree, V., Hiremath, L., Rajendran, S.: The use of modern technology in smart waste management and recycling: artificial intelligence and machine learning. Recent Adv. Comput. Intell. **823**, 173–188 (2019)
16. Haas, W., Krausmann, F., Wiedenhofer, D., Heinz, M.: How circular is the global economy?: An assessment of material flows, waste production, and recycling in the European Union and the World in 2005. J. Ind. Ecol. **19**(5), 765–777 (2015)
17. Hart, J., Adams, K., Giesekam, J., Tingley, D.D., Pomponi, F.: Barriers and drivers in a circular economy: the case of the built environment. Proc. CIRP **80**, 619–624 (2019)
18. Hopkinson, P., Chen, H.M., Zhou, K., Wong, Y., Lam, D.: Recovery and re-use of structural products from end of life buildings. Proceedings of the Institution of Civil Engineers - Engineering Sustainability **172**(3), 1–36 (2018)
19. Jesus, A., Mendonça, S.: Lost in transition? Drivers and barriers in the eco-innovation road to the circular economy. Ecol. Econ. **145**, 75–89 (2018)
20. Lyle, J.T.: Regenerative design for sustainable development. Rethinking the Mind in Nature. John Wiley, New York (1994)
21. Kirchherr, J., Reike, D., Hekkert, M.: Conceptualizing the circular economy: an analysis of 114 definitions. Resour. Conserv. Recycl. **127**, 221–232 (2017)
22. Kirchherr, J., et al.: Barriers to the circular economy: Evidence from the European Union (EU). Ecol. Econ. **150**, 264–272 (2018)
23. Korhonen, J., Honkasalo, A., Seppälä, J.: Circular economy: the concept and its limitations. Ecol. Econ. **143**, 37–46 (2018)
24. Lacy, P., Long, J., Spindler, W.: The Circular Economy Handbook. Palgrave Macmillan, London (2020)
25. John, L.A., Oloruntoba, A.S.: Artificial intelligence in the transition to circular economy. Am. J. Eng. Res. **9**(6), 185–190 (2019)
26. McDonough, W., Braungart, M., Clinton, B.: The Upcycle: Beyond Sustainability-Designing for Abundance. North Point Press, Albany (2013)
27. Murray, A., Skene, K., Haynes, K.: The Circular Economy: An Interdisciplinary Exploration of the Concept and Its Application in a Global Context. J. Bus. Ethics **140**(3), 369–380 (2017)
28. Rojek, I., Jasiulewicz-Kaczmarek, M., Piechowski, M., Mikołajewski, D.: An artificial intelligence approach for improving maintenance to supervise machine failures and support their repair. Appl. Sci. **13**(8), 4971 (2023)
29. Preston, F.: A global redesign? Shaping the circular economy briefing paper. Chatham House EERG BP 2012/02, London (2012)
30. Ritzén, S., Sandström, G.Ö.: Barriers to the circular economy – integration of perspectives and domains. Proc. CIRP **64**, 7–12 (2017)
31. Roberts, H., et al.: Artificial intelligence in support of the circular economy: ethical considerations and a path forward. AI & Soc. **1**, 1–14 (2022)
32. Sapmaz Veral, E.: Döngüsel Ekonomiye Geçiş Doğrultusunda Yeni Tedbirler ve AB Üye Ülkelerinin Stratejileri. Ankara Avrupa Çalışmaları J. **17**(2), 463–488 (2018)
33. Sapmaz Veral, E.: Döngüsel Ekonomi: Engeller, Stratejiler ve İş Modelleri. Ankara Üniversitesi Çevrebilimleri J **8**(1), 7–18 (2021)

34. Sapmaz Veral, E., Yiğitbaşıoğlu, H.: Avrupa Birliği atık politikasında atık yönetiminden kaynak yönetimi yaklaşımına geçiş yönelimleri ve döngüsel ekonomi modeli. Ankara Üniversitesi Çevrebilimleri J. **6**(1), 1–19 (2018)
35. Schempp, C., Hirsch, P.: Categorisation System for the Circular Economy. Publications Office of the European Union, Luxembourg (2020)
36. Sehnem, S., Vazquez-Brust, D., Pereira, S.C.F., Campos, L.M.S.: Circular economy: benefits, impacts and overlapping. Supply Chain Manag. Int. J. **24**(6), 784–804 (2019)
37. Stahel, W.R.: The circular economy. Nature **531**(7595), 435–438 (2016)
38. United Nations Sustainable Development Goals. https://sdgs.un.org/goals. Last accessed 24 Apr 2022
39. United Nations Environment Programme. https://wedocs.unep.org/bitstream/handle/20.500.11822/36830/RBMLCE.pdf. Last accessed 25 Apr 2022
40. Circular economy guide for businesses. https://business4goals.org/PDF/Dongusel_Ekonomi_Rehberi.pdf. Last accessed 25 Apr 2022
41. Wilson, M., Paschen, J., Pitt, L.: A circular economy meets artificial intelligence (Ai): understanding the opportunities of Ai for reverse logistics. Manag. Environ. Qual. Int. J. **33**(1), 9–25 (2022)
42. The Circular Economy and Benefits for Society. https://www.lagazettedescommunes.com/telechargements/etude-club-rome-eng.pdf. Last accessed 24 Apr 2022
43. Wise, C., Pawlyn, M., Braungart, M.: Eco-engineering: living in a materials world. Nature **494**, 172–175 (2013)
44. Yuan, Z., Bi, J., Moriguichi, Y.: The circular economy: a new development strategy in China. J. Ind. Ecol. **10**(1–2), 4–8 (2006)

Forecasting Electricity Prices for the Feasibility of Renewable Energy Plants

Bucan Türkmen[1], Sena Kır[2(✉)], and Nermin Ceren Türkmen[3]

[1] Turkiye İş Bank Inc., Adapazari, Sakarya, Turkey
[2] Sakarya University Engineering Faculty/Department of Industrial Engineering, Serdivan, Sakarya, Turkey
senas@sakarya.edu.tr
[3] Sakarya University of Applied Sciences Faculty of Applied Sciences, Department of International Trade and Business, Serdivan, Sakarya, Turkey

Abstract. Feasibility studies in assessing the viability of renewable energy investments involve conducting economic analyses to evaluate crucial performance criteria such as return on investment and payback period. In these analyses, it is essential to calculate the profitability derived from the electricity generated and sold, considering the investment and operational costs. At this point, electricity prices emerge as a determining parameter. However, accurately predicting electricity prices is challenging due to the need to account for production costs, as well as regulations and policies within the framework of the free market mechanism. Moreover, compared to other traded commodities, electricity prices are further complicated by electricity's inability to store, the necessity of instantaneous balancing of production and consumption, and the high seasonality of demand for domestic, industrial and commercial electricity. On the other hand, precise forecasting of future electricity prices is a critical factor that enhances the accuracy of economic analyses. This study employs the Prophet Algorithm, which utilizes time series analysis and considers historical electricity prices to make periodic predictions of future electricity prices. The Prophet algorithm is specifically designed to capture the evidence of deterministic trend and seasonality, and at the same time the effects of an econometric shock, providing reliable results in forecasting electricity prices. Our study has examined and compared the forecasting performance of Python's Prophet Algorithm and Excel Estimator. Although the Excel Estimator did not achieve the same level of precision, it produced results within an acceptable range.

Keywords: Electricity Prices · Predictive Forecasting · Prophet Algorithm · Time Series Analysis · Renewable Energy Investments

1 Introduction

The advent of Watt's steam engine, Ford's production line, Fayol's business knowledge, and post-war surge in demand, along with developments in logistics and globalization, have spurred industrial consumption. This, combined with the growth of fossil fuel-based energy and inattention to global warming, has led to irreversible environmental

Z. Şen et al. (Eds.): IMSS 2023, LNME, pp. 783–793, 2024.
https://doi.org/10.1007/978-981-99-6062-0_75

damage. Many organizations, including the United Nations, European Union, G8, G20, the World Bank, and Turkish Sustainable Energy Financing Facility, have promoted renewable energy sources as alternatives to fossil fuels. These initiatives have led to taxes or punitive financing of companies contributing to environmental degradation, pushing them to consider renewable energy sources for electricity generation.

Turkiye obtained approximately 40% of its 326.2 TWh electricity production in 2022 from renewable energy sources. As of the end of April 2023, 30.2% of the 104496 MW installed power produced from renewable resources is obtained from hydraulic energy, 11% from wind, 9.5% from solar and 1.6% from geothermal power plants [1] and new steps are being taken to increase renewable energy production. However, the installation of renewable energy power plants requires substantial investment. For example; the investment costs of a solar power plant (SPP) can be roughly classified as design stage cost, construction cost, interest during the work period. Among the components that most affect the installation cost, the construction cost includes photovoltaic module, connection parts, inverter, floating platform and anchoring system cost, electromechanical and power distribution/transmission work and monitoring system costs [2]. Likewise, similar cost items like design and project cost, turbines and foundations costs, electrical infrastructure cost are encountered for the installation of offshore wind farms similar to solar power plants [3].

These costs may deter potential investors, but electricity generation and sustainability advantages to be provided in the long term can cover these costs. For example; Taktak and Ilı's SPP project in Uşak's central district is expected to break even in 10 years and generate profit for 15 years [4]. Similarly, Bayrakçı and Gezer's analysis of an SPP project in Aydın's Çine district predicted a payback period of about 7 years [5]. A wind power plant installation project at Süleyman Demirel University estimated a payback time of approximately 8.5 years under ideal conditions [6].

In the installation of renewable power plants, accurately forecasting future electricity sales prices is pivotal for assessing net cash inflow, return on investment, and profitability. The dynamics of Turkey's strictly regulated energy market, influenced by supply-demand balance, necessitate reliable price predictions, factoring in market conditions, energy policies, and consumer demands. Such forecasts, vital for financial planning and investment decisions, guide power producers in strategic sales planning and pricing, boosting investor confidence and sectoral growth.

Advancements in technology and increased demand have led to a reduction in investment costs for renewable energy sources. Notably, solar cell prices fell from $4/Watt in 2008 to $0.22/Watt in 2019. The feasibility for investment decisions is typically handled by companies that supply necessary materials and oversee implementation. Once an investment is made, the projected cash flow from the generated energy is balanced against project expenses to establish a payback period. Current practices often rely on present electricity prices or rates inflated annually by a fixed multiplier for future cash flow calculations. This approach may rely on personal estimations and not necessarily reflect accurate econometric data. To address this issue, this study uses periodic estimations of future electricity prices, thereby facilitating a more realistic understanding of generation costs within the cash flow statement and a more accurate determination of payback periods.

In particular, as all major review publications have noted, comparisons between electricity price forecasting (EPF) methods are very difficult as studies use different datasets, different software implementations, and different measures of error; the lack of statistical rigor further complicates these analyses [7].

In this study, future electricity prices in Turkiye for renewable energy investments are forecasted using time series analysis based on historical data. The study's second part reviews international literature on electricity price forecasting techniques. The third part outlines our chosen method, time series analysis. The fourth part discusses implementation details, and the conclusion shares the study's findings.

2 Literature Review

Weron [8] extensively reviewed the literature on EPF. He classified EPF methods into five classes: multi-agent models, fundamental models, reduced-form models, statistical models, and computational intelligence models. Nowotarski and Weron [9], on the other hand, explored stochastic EPF. They suggested that smart grids and renewable integration requirements have amplified the uncertainty of future supply, demand, and prices, rendering stochastic electricity price forecasting increasingly critical for energy system planning and operations.

Building upon the extensive literature reviews, we succinctly examine a select number of studies addressing electricity price forecasting as per the classification system presented in [8]. An example of a study utilizing solely statistical models for price prediction is Kostrzewski and Kostrzewska's work [10]. They proposed a Bayesian Approach for stochastic forecasting of the next day's electricity prices using the Pennsylvania-New Jersey-Maryland (PJM) Interconnection dataset. In the study by Česnavičius [11], Autoregressive Integrated Moving Average (ARIMA) forecasting models were employed to predict electricity prices in the Lithuanian electricity market. The accuracy of the generated models was compared, and the most accurate model was determined. Jan et al. [12] utilized a functional autoregressive model for short-term price predictions, with the model size and lag structure automatically selected. Through an application in the Italian electricity market, it was found that the proposed method outperformed autoregressive and simpler models. Wang et al. [13] predicted electricity prices in the DE-LU bidding zone using time series analysis. They demonstrated the effectiveness of the Seasonal Autoregressive Integrated Moving-Average with Exogenous Regressors (SARIMAX) model supported by the Generalized Auto Regressive Conditional Heteroskedasticity (GARCH) model, which responded better to underlying trend deviations.

One of the studies that utilize both statistical models and computational intelligence techniques is presented by Karabiber and Xydis [14]. In this study, they perform one-day-ahead electricity forecasting for the Denmark-West region using ARIMA, Trend and Seasonal Components (TBATS), and ANN. While ARIMA provides the best results in terms of mean error, Artificial Neural Networks (ANN) exhibits the lowest minimum error and standard deviation. They find that TBATS outperforms ANN in terms of mean error. Bitirgen and Filik [15] forecast future electricity prices using Extreme Gradient Boosting (XGBoost) and ARIMA models. In terms of performance, they demonstrate that the XGBoost model is more effective compared to ARIMA in terms of computation speed and lower error.

Kuo and Huang [16] presented a study that employs computational intelligence methods for EPF. In this study, they proposed a system consisting of deep neural network models, such as Convolutional Neural Network (CNN) and Long Short-Term Memory (LSTM), which outperformed traditional machine learning methods for predicting electricity prices. Mohamed et al. [17] focused on the prediction of locational marginal prices (LMP) in several electricity markets in North America. They evaluated the accuracy of short-term LMP predictions using LSTM and Gated Recurrent Unit networks. The results indicated that LSTM achieved lower error rates and higher accuracy in 24-h ahead LMP forecasting compared to the Prophet Prediction model.

In addition to these classifications, there are studies that employ hybrid techniques for forecasting. Cheng et al. [18] proposed a hybrid prediction method using Empirical Wavelet Transform, Support Vector Regression, Bi-directional LSTM, and Bayesian optimization. The proposed methodology was applied to European Power Exchange Spot data. Zhang et al. [19] developed a new adaptive hybrid model based on Variational Mode Decomposition (VMD), Self-Adaptive Particle Swarm Optimization, Seasonal Autoregressive Integrated Moving Average (SARIMA), and Deep Belief Network (DBN) for short-term EPF. They demonstrated the effectiveness of the model using data obtained from the Australian, PJM, and Spanish electricity markets. In a recent study by Xiong and Qing [20], a new hybrid forecasting framework is proposed to improve the day-ahead prediction of electricity prices. The proposed model incorporates an adaptive copula-based feature selection algorithm, a new method of signal decomposition technique, and a Bayesian Optimization and Hyperband optimized LSTM model. Our study can also be included in the statistical models class within this classification.

3 Method: Prophet Algorithm

The Prophet algorithm is based on the assumption that time series data can be described as a combination of various components, including trends, seasonality, and events, with the objective of generating accurate predictions and estimating uncertainties [21].

Facebook highlights the importance of resource allocation and goal setting in capacity planning and performance assessment [22]. To meet these needs, they developed Prophet, an open-source software for forecasting time series of internal processes. Unlike traditional models like Holt Winters and ARIMA, Prophet employs a generalized additive model for more efficient smoothing and forecasting functions. Its fast-fitting process allows interactive model exploration. It can handle daily periodicity data with significant outliers and trend shifts, and it can model multiple periods of seasonality, thus enhancing forecasting capabilities [23].

The Prophet Algorithm also allows users to inspect the dataset and change the adjustable parameters (hyperparameters) at any time. The parameters that can be fine-tuned by users to make better predictions are changepoint_prior_scale, seasonality_prior_scale, holidays_prior_scale, seasonality_mode [24]. The Prophet algorithm follows the steps outlined below [25]:

- The algorithm expects input data in the form of a dataframe with two columns: 'ds' (datestamp) and 'y' (the target variable).
- An instance of the Prophet class is created, and the dataframe is fitted to the model using the fit() function.
- A new dataframe is created to specify the dates for which predictions are desired. This can be done using the make_future_dataframe() function, which determines the number of days to forecast based on the 'periods' parameter.
- The predict() function is called to generate predictions based on the input dataframe and the future dataframe. The predicted values, along with the lower and upper bounds of the uncertainty interval, are stored in the output dataframe. This allows for a review of the predictions and the associated level of uncertainty. The Prophet algorithm incorporates three components in its modelling Eq. (1):

  - The function g(t) models the underlying trend, capturing the overall pattern in the data. Prophet combines two trend models: a saturating growth model and a piece-wise linear model, depending on the nature of the forecasting problem.
  - The function s(t) models seasonality using Fourier series, capturing how the data is affected by periodic factors.
  - The function h(t) models the effects of events or significant occurrences that impact the time series. The term εt represents an irreducible error term.

$$y(t) = g(t) + s(t) + h(t) + \varepsilon_t \tag{1}$$

4 Application: Forecasting The Electricity Price

In existing feasibility studies, it is routine to utilize current electricity prices or arbitrarily inflated prices, rather than forecasted ones. This stems from practitioners installing these systems, who may not possess the needed economic and statistical proficiency or familiarity with academically accepted forecasting software. To mitigate this, we propose an Excel Estimator, which is simpler and requires minimal skills, without the need for coding. However, the validity of this approach must be confirmed. Hence, we compare the accuracy and explanatory power of forecasts from Python Prophet and the Excel Estimator.

4.1 Data Set

In this study, the Industrial MV TL-based price tariff released by the Energy Market Regulatory Authority is used for price forecasts. Monthly data for the period 2011.12-2022.10 were obtained in full from the data warehouse. Since the data in question does not contain seasonality, the deterministic component is not decomposed. Descriptive statistics and the graph of the data, calculated by Eviews, are given in the Table 1 and Fig. 1.

Table 1 indicates that the data series spans a minimum value of 200.12 and a maximum of 4094.67, with a mean and median of 556.8101 and 260.17 respectively. The

Table 1. Descriptive Statistics

	Mean	Median	Max	Mini	Std. Dev	Skewness	Kurtosis	Prob	Obs
Electricity Prices	556.81	260.17	4094.67	200.12	728.4	3.46	15.23	0.00	132

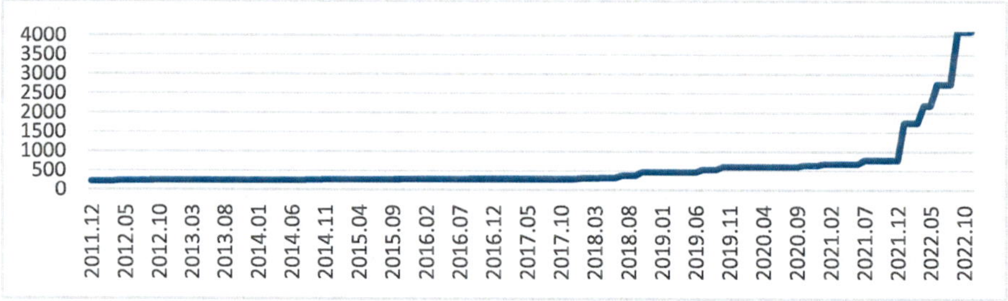

Fig. 1. Graph of Data Series (Electricity Prices)

standard deviation is calculated as 728.4037, providing an indication of the dataset's dispersion. The skewness value of 3.463428 suggests a rightward skew, deviating from a normal distribution. The kurtosis value, measured at 15.2314, suggests a leptokurtic distribution, indicating a sharper peak compared to a normal distribution. Since the Jarque-Bera probability value $= 0.00 < 0.05$, the null hypothesis H_0, which states that the errors have a normal distribution, is rejected.

4.2 Analysis

In order to ensure that the confidence interval of the forecast is high, 132 months of data are used in the data set, and the data used are included in the base electricity prices table. These data were then first forecasted using Excel Estimator. Here, the 33-month forecast is given as indicative. The results are given in the Excel price forecast table. The individual realizations of these forecasts are given in the error values table.

The necessary libraries were first identified and downloaded through the Python program. These are "fbprophet, matplotlib, plotly, pandas, scikit-learn". Afterward, the raw data set was transferred to the program, and the necessary transformations were made. Then, the number and type of the data set were checked. The data type transformation required by the program was made. Column names were changed to 'ds' and 'y'. Object conversion was done. The model was defined and created and the data was transferred to this model. Then the intervals we want to forecast were determined and the list formation was done by providing the necessary format to include the forecast values. Since monthly data forecasting was requested, necessary definitions were made. Forecasts were transferred. Forecasting was performed. Forecast results 'yhat' average monthly forecasts, 'yhat_lower' average monthly lowest forecasts, and 'yhat_upper' average monthly highest forecasts were included and displayed.

The pseudo code of the utilized Prophet Algorithm:

```
# Prerequisite library downloading Import necessary libraries
# Data obtaining
Read data from 'Book1.xlsx' into a DataFrame (df)
Plot the 'price' column of df
Display the first few rows of df
Display the shape of df
Create a boxplot of df
Rename columns of df to 'ds' and 'y'
Convert 'ds' column to datetime format
# Define the model
Create a Prophet model object (model)
# Fit the model
Fit the model using df
# Forecast period determination
Create an empty list (future)
For each month from 1 to 12
   Create a date string in the format '2023-MM'
   Append the date to the future list
Convert the future list to a DataFrame with column name 'ds'
Convert 'ds' column to datetime format
# Forecast summary
Use the model to predict future values based on the future DataFrame
Display the forecasted values for 'ds', 'yhat', 'yhat_lower', and 'yhat_upper'
Plot the forecasted values using the model
```

In addition, using the Python forecast as real data, excel forecasts were calculated to show the explanatory power on a 33-month basis, and the results were included in the Comparison error values table.

4.3 Results

We forecast the data by Excel Estimator and Facebook Prophet Algorithm. We use 132 data for electricity price and predict 33 future average prices. Moreover, with 5 realized values; we test this forecasting values separately. From Excel, the forecasted values are presented below at Table 2 and the performance criteria for the forecast is presented at Table 3.

Given that the study is focused on financial time series data, the Mean Absolute Percentage Error (MAPE) stands as the most attainable and preferred performance criterion. The use of MAPE for assessing forecasting performance is prevalent due to its interpretability and comparability across methods [26, 27]. Previous studies have shown Python-based forecasting to be effective in terms of MAPE [28]. This study investigates if Excel-based forecasts could serve as an effective alternative or supplement to Python-based ones.

The Python-based forecasting requires a certain level of mastery, making it potentially inaccessible to individuals lacking the requisite technical skills. In contrast, Excel is more user-friendly and widely accessible, making it an attractive option for users

Table 2. Excel and Python Prophet's Forecast Results

#	Excel Avg	Python Prophet's Avg	#	Excel Avg	Python Prophet's Avg
1	4109.345	2335.34	**18**	4354.794	3257.5
2	4123.783	2380.03	**19**	4369.232	3325.23
3	4138.221	2407.28	**20**	4383.67	3372.97
4	4152.659	2506.66	**21**	4398.108	3511.32
5	4167.098	2536.73	**22**	4412.546	3564.16
6	4181.536	2626.38	**23**	4426.984	3596.18
7	4195.974	2671.29	**24**	4441.423	3492.92
8	4210.412	2697.29	**25**	4455.861	3626.32
9	4224.85	2881.55	**26**	4470.299	3653.9
10	4239.288	2940.54	**27**	4484.737	3694.81
11	4253.727	2982.13	**28**	4499.175	3774.05
12	4268.165	2833.43	**29**	4513.613	3812.37
13	4282.603	2968.34	**30**	4528.052	3899.56
14	4297.041	3021.59	**31**	4542.49	3959.71
15	4311.479	3060.33	**32**	4556.928	3999.18
16	4325.917	3129.47	**33**	4571.366	4153.39
17	4340.355	3171.7	**-**	-	-

Table 3. Excel and Python Prophet's Forecast Performances

	Excel Forecast Performance	Python Prophet's Forecast Performance
	Average	Average
MAD	917.75	801.70
MSE	915,426.45	748,862.45
RMSE	956.78	865.37
MAPE	0.29	0.24

with limited coding skills [29]. To explore the potential efficacy of Excel in this context, Python-based forecasts were used as realized values against which Excel forecasts' performance could be evaluated. The evaluation of forecast accuracy was conducted using established Error Performance Criteria. The results of this analysis and graphical representation are presented in the following Table 4 and Fig. 2.

Based on the MAPE, the performance of Excel-based forecasts is not outstanding, yet it falls within acceptable limits.

Table 4. Python Prophet vs. Excel

	Average	Upper	Lower
MAE	1,687.5751	1687.5751	1687.5751
MSE	2,859,972.707	2,859,972.707	2,859,972.707
MedAE	1692.8861	1692.8861	1692.8861
RMSE	1691.1454	1691.1454	1691.1454
MAPE	31.4849	31.4849	31.4849

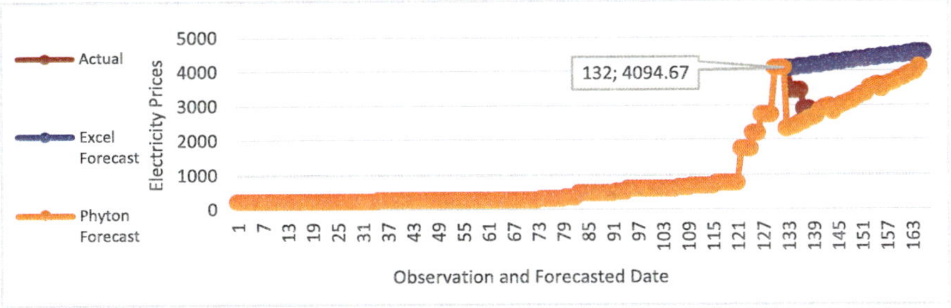

Fig. 2. Forecast and Actual Data

5 Conclusion

Our analysis has elucidated the comparative forecasting performance of Python's Prophet Algorithm and Excel. The Python Prophet Algorithm is recognized in academic circles for its high-quality forecasting capabilities, making it a benchmark for this study. Despite not achieving the same level of precision, the Excel forecasting method produced results within an acceptable range. The primary objective of this research was not to directly evaluate the quality or performance criteria of Excel's forecasts, but rather to examine their proximity to the forecasts produced with Python's Prophet algorithm. In essence, we aimed to demonstrate to individuals lacking extensive knowledge of economics, statistics, and software like Python whether forecasts generated using Excel can approximate the performance of those generated with Python. While Excel's forecasting performance seems reasonably acceptable on its own, according to the performance criteria assessed, it is not as potent as Python. However, considering the MAPE value, it suggests that Excel, while not preferred over Python, remains a viable alternative.

In conclusion, despite the superior forecasting quality of Python's Prophet Algorithm, there is merit in considering Excel as an estimation tool in feasibility studies, especially due to its ease of use. The utilization of complex econometric models like ARIMA and software such as Eviews, R, Stata, among others, can pose a considerable barrier to those without an academic background in economics and statistics or the ability to use these programs. Excel's advantages, such as its user-friendly interface, ease of learning, and no requirement for advanced technical skills, make it an alternative to more complex forecasting tools. This accessibility and convenience encourage a broader adoption of forecasting practices, thereby enhancing financial planning and decision-making in renewable energy investments. Excel can efficiently generate future electricity prices for reporting purposes. Furthermore, the Excel Estimator's relative simplicity and forecasting power provide a practical tool for cash inflow calculations, replacing intuitive coefficients or current prices that are commonly used. Despite the superior forecasting capabilities of Python's Prophet Algorithm, as demonstrated by this study, Excel's adoption can foster a more accessible and inclusive approach to economic analyses in renewable energy investments.

References

1. TR Ministry of Energy and Natural Sources. https://enerji.gov.tr/bilgi-merkezi-enerji-elektrik (2023). Last accessed 10 May 2023
2. Chiang, C.H., Young, C.H.: An engineering project for a flood detention pond surface-type floating photovoltaic power generation system with an installed capacity of 32,600.88 kWp. Energy Rep. **8**, 2219–2232 (2022)
3. Gonzalez-Rodriguez, A.G.: Review of offshore wind farm cost components. Energy Sustain. Dev. **37**, 10–19 (2017)
4. Taktak, F., Ilı, M.: Güneş enerji santrali (GES) geliştirme: Uşak örneği. Geomatik **3**, 1–21 (2018). (In Turkish)
5. Bayrakçı, H.C., Gezer, T.: Bir güneş enerjisi santralinin maliyet analizi: Aydın ili örneği. Teknik Bilimler Dergisi **9**, 46–54 (2019). (In Turkish)
6. Elibüyük, U., Yakut, A.K., Üçgül, İ.: Süleyman Demirel Üniversitesi rüzgâr enerjisi santrali projesi. Yekarum 3 (2016)
7. Lago, J., Marcjasz, G., De Schutter, B., Weron, R.: Forecasting day-ahead electricity prices: a review of state-of-the-art algorithms, best practices and an open-access benchmark. Appl. Energy **293**, 116983 (2021)
8. Weron, R.: Electricity price forecasting: a review of the state-of-the-art with a look into the future. Int. J. Forecast. **30**, 1030–1081 (2014)
9. Nowotarski, J., Weron, R.: Recent advances in electricity price forecasting: a review of probabilistic forecasting. Renew. Sustain. Energy Rev. **81**, 1548–1568 (2018)
10. Kostrzewski, M., Kostrzewska, J.: Probabilistic electricity price forecasting with Bayesian stochastic volatility models. Energy Econ. **80**, 610–620 (2019)
11. Česnavičius, M.: Lithuanian electricity market price forecasting model based on univariate time series analysis. Energetika **66** (2020)
12. Jan, F., Shah, Ismail, Ali, Sajid: Short-term electricity prices forecasting using functional time series analysis. Energies **15**(9), 3423 (2022)
13. Wang, D., Gryshova, I., Kyzym, M., Salashenko, T., Khaustova, V., Shcherbata, M.: Electricity price instability over time: time series analysis and forecasting. Sustainability **14**(15), 9081 (2022)
14. Karabiber, O.A., Xydis, G.: Electricity price forecasting in the Danish day-ahead market using the TBATS, ANN and ARIMA methods. Energies **12**, 928 (2019)
15. Bitirgen, K., Filik, Ü.B.: Electricity price forecasting based on XGBooST and ARIMA Algorithms". BSEU J. Eng. Res. Technol. **1**, 7–13 (2020)
16. Kuo, Ping-Huan., Huang, Chiou-Jye.: An electricity price forecasting model by hybrid structured deep neural networks. Sustainability **10**(4), 1280 (2018)

17. Mohamed, A.T., Aly, H.H., Little, T.A.: Locational marginal price forecasting based on deep neural networks and prophet techniques. In: IEEE Electrical Power and Energy Conference (EPEC), pp. 1–6 (2021)
18. Cheng, H., Ding, X., Zhou, W., Ding, R.: A hybrid electricity price forecasting model with Bayesian optimization for German energy exchange. Int. J. Electr. Power Energy Syst. **110**, 653–666 (2019)
19. Zhang, J., Tan, Z., Wei, Y.: An adaptive hybrid model for short term electricity price forecasting. Appl. Energy **258**, 114087 (2020)
20. Xiong, X., Qing, G.: A hybrid day-ahead electricity price forecasting framework based on time series. Energy **264**, 126099 (2023)
21. Shohan, M.J.A., Faruque, M.O., Foo, S.Y.: Forecasting of electric load using a hybrid LSTM-neural prophet model. Energies **15**(6), 2158 (2022)
22. Duarte, D., Faerman, J.: Comparison of time series prediction of healthcare emergency department indicators with ARIMA and Prophet. In: Computer Science & Information Technology (CS & IT) Computer Science Conference, pp. 123, 33 (2019)
23. Zhao, N., Liu, Y., Vanos, J.K., Cao, Guofeng: Day-of-week and seasonal patterns of PM2.5 concentrations over the United States: time-series analyses using the prophet procedure. Atmos. Environ. **192**, 116–127 (2018)
24. Regis Anne, W., Carolin Jeeva, S.: Machine learning modeling techniques and statistical projections to predict the outbreak of COVID-19 with implication to India. In: Lessons From COVID-19, pp. 289–311. Elsevier (2022)
25. Chadalavada, R.J., Raghavendra, S., Rekha, V.: Electricity requirement prediction using time series and facebook's prophet. Indian J. Sci. Technol. **13**, 4631–4645 (2020)
26. Chen, H., Li, B., Wang, C., Liu, L.: A new multi-step forecasting model for energy price based on improved PSR-BP neural network. IEEE Access **6**, 52789–52799 (2018)
27. Kumar, S., Jain, V., Kumar, D.: A hybrid machine learning approach for daily electricity price forecasting. IEEE Trans. Power Syst. **35**(1), 43–55 (2020)
28. Pereira, T., Silva, T., Mourelle, D.P.B., Guimarães, A., Sá, A.: A machine learning approach for wind power forecast. In: IEEE Milan PowerTech, pp. 1–6. Milan, Italy (2019)
29. Khosrow-Pour, M.: Excel as a professional tool: exploring the power of Excel. In: Encyclopedia of Information Science and Technology, pp. 4312-4321. 4th edn. IGI Global (2017)

Check for updates

A Clustering Approach for the Metaheuristic Solution of Vehicle Routing Problem with Time Window

Tuğba Gül Yantur, Özer Uygun(✉), and Enes Furkan Erkan

Sakarya University, Department of Industrial Engineering, Sakarya, Türkiye
tugba.yantur@ogr.sakarya.edu.tr, {ouygun, eneserkan}@sakarya.edu.tr

Abstract. Vehicle routing problems are one of the real-life problems studied extensively in the literature, especially in the logistics and transportation sectors, and consist of various constraints and parameters. Vehicle routing problems, the primary purpose of which is cost minimization, are solved with heuristic or metaheuristic methods within the scope of their content.

In this study, the problem is to plan the routes for delivering white goods from a main warehouse to homes or dealers in Ankara and surrounding cities, considering the delivery time window constraint. Deliveries can be made before or after the time window, but if there is a delay, it will incur penalty costs. Therefore, the problem examined is in the class of "Vehicle routing with flexible time windows" problems. The main focus in solving the problem is to minimize cost and deliver within the time window. A two-stage method based on "cluster-first route-second" approach has been proposed. Products to be delivered are divided into two groups regarding the size and product group-based placement constraint added to the DBSCAN clustering method. If there is capacity in the vehicle, the DBSCAN algorithm was revised to include the next point in the cluster. In the second stage, clusters with a high occupancy rate and a minimized number of vehicles are routed to deliver under time window constraints with the Ant Colony Algorithm approach. The results of the study are compared with the previous planning results, financially and operationally. The proposed approach achieved a 30% improvement in the number of vehicles. The vehicle occupancy rates have been increased to an average of 94.89%.

Keywords: Vehicle Routing Problem with Time Window · Ant Colony Algorithm · DBSCAN Algorithm

1 Introduction

In the last few years, one of the most important factors in the increasing competition in the global and local markets has been the delivery of the service or product to the end customer as soon as possible. In order to take part in this competition, companies focus on logistics processes in line with the expectations of their customers.

Z. Şen et al. (Eds.): IMSS 2023, LNME, pp. 794–809, 2024.
https://doi.org/10.1007/978-981-99-6062-0_76

In addition to delivering the product or service to the end customer as soon as possible, companies aim to keep their costs constant or reduce them with various strategies. Completing internal and external processes quickly and without sacrificing quality causes high costs. Fleet management and planning constitute one of the most financially significant processes within logistics operations. It is necessary to achieve optimum results by performing an accurate analysis of multi-criteria decision-making processes.

In the literature, various methods have been developed to solve vehicle routing problems depending on each criterion. Constraints such as capacity, distance, time, pick-and-drop, and demand can be one of these constraints. Vehicle routing problems are in the class of NP-hard problems and can be solved with heuristic and meta-heuristic solution methods.

During the shipment process, which vehicles will leave the warehouse at which capacity, which route they will follow, and in which order they will reach the delivery points can be planned as optimum and close to the optimum with the help of various techniques.

There are numerous studies in the literature that address Vehicle Routing Problems, initially proposed by Dantzig and Ramser in 1959 [1], considering various constraints and proposing various solution methods. Due to the wide range of vehicle routing problem types and constraints, as well as the diverse approaches proposed for their solutions, we will focus on recent studies that address the "cluster-first route-second" approach. The literature review is summarized as in Table 1.

2 Theoretical Framework

2.1 The Vehicle Routing Problem

Firms have attempted to solve vehicle routing problems using effective methods in order to reduce logistics costs. These problems involve constraints such as demand, capacity, time windows, and duration. Exact solution algorithms work well for small-scale problems, while heuristic/metaheuristic methods are used for larger-scale problems [13]. Each vehicle starts and ends its route at the same depot, with equal capacities and costs. The capacities and customer demands are known in advance [18].

The Vehicle Routing Problem with Time Window. In time window-constrained vehicle routing problems, target locations require deliveries in specified time windows. In vehicle routing problems with flexible time window, service can be provided to a target location before or after the time window, but in this case, a penalty cost is applied. In vehicle routing problem with strict time window, the vehicle must arrive at the target location within the time window; otherwise, service cannot be provided. These constraints can be incorporated into the mathematical models by adding or removing relevant expressions.

Table 1. Literature Review for Cluster First, Then Rotate

Özdağ et al. (2012) [2]	A weighted graph model has been developed to create the academic schedule at a university. They utilized the Ant Colony algorithm to achieve an optimal distribution
Zhang (2017) [3]	A routing approach using Density-based Clustering (DBSCAN) method and Ant Colony Algorithm was proposed for the Capacitated Vehicle Routing Problem
Ünsal and Yiğit (2018) [4]	The school bus routing problem is clustered using the K-Means algorithm, and routed with the Genetic Algorithm (GA) method
Cömert et al. (2019) [5]	For a Simultaneous Pickup and Delivery Vehicle Routing Problem, in the first stage, clustering methods such as K-Means and K-Medoids were used to cluster the points. In the second stage, the clusters were routed using integer linear programming
Bujel et al. (2019) [6]	Compared the performance of the Recursive-DBSCAN method using Google OR-Tools, and the classical DBSCAN method to improve the performance of the Time Windowed Capacitated Vehicle Routing Problem (TWCVRP)
Bozdemir et al. (2019) [7]	Three different clustering algorithms, namely K-Means, K-Medoids, and K-Modes, were applied to the nearest point service route problem
Göçken et al. (2019) [8]	Capacitated Vehicle Routing Problem with time window is clustered using K-Means, Center-Based Heuristic, Density-Based, and Shared Nearest Neighbor (SNN) clustering methods. Then routed with Genetic Algorithm and tested the solution using the Solomon benchmark dataset
Cömert et al. (2020) [9]	A Flexible Time Windowed Vehicle Routing Problem (FTWVRP) has been clustered using the K-Means and K-Medoids, considering penalties according to time windows. Mixed integer linear programming was utilized for the routing phase
Villalba and Rotta (2020) [10]	Representative customer selection has been investigated using density-based clustering (DBSCAN) and K-Means clustering algorithms. The clustering quality, computation time, number of clusters, and other factors were evaluated to compare the strengths and weaknesses of both methods
Akdaş et al. (2021) [11]	Clustering and Ant Colony algorithms have been applied to the vehicle routing problem of the solid waste collection process

(continued)

Table 1. (*continued*)

Villalba and Rotta (2022) [12]	A Vehicle Routing Problem with time window has been clustered using the K-Means and OPTICS clustering, routed using the Nearest Neighbor method
Choudhari et al. (2022) [13]	The Capacitated Vehicle Routing Problem has been examined using the DBSCAN algorithm for density-based clustering and employed integer linear programming for the single-depot Capacitated Vehicle Routing Problem with capacity constraints
Sanchez et al. (2022) [14]	For the vehicle routing problems with time window, a mixed integer linear programming model has been proposed for the battery unit placement problem, considering time windows and criteria such as travel time and battery levels. The points were clustered using the K-Means algorithm, and the optimal battery unit locations were determined
Le et al. (2022) [15]	During the COVID-19 pandemic when certain time windows were designated for the opening of business centers and markets. By applying the Capacitated K-Means Clustering Algorithm and the Branch-and-Bound Algorithm
Wang et al. (2023) [16]	For the multi-depot time-constrained vehicle routing problems, customers clustered into subgroups and then planned the routes using the Genetic Algorithm

2.2 Clustering Algorithms

Cluster algorithms are an analysis method that creates datasets by grouping the data with the highest similarity rates. In the most general form, we can group them under two headings as hierarchical and non-hierarchical clustering methods.

Hierarchical clustering is a clustering method that groups its elements in a hierarchical structure. Non-hierarchical clustering algorithms cluster the elements to be included in the cluster according to similarity or distance matrices, without hierarchy. The number of "*k*" clusters should be specified, which will be less than the number of "*n*" elements to be included in the cluster.

Density-Based Spatial Clustering of Applications with Noise (DBSCAN). The DBSCAN algorithm is a density-based clustering algorithm that groups data points based on their density. It is particularly suitable for datasets that contain regions of varying density and noise points. Unlike other clustering methods, DBSCAN is able to identify and handle noise points as it relies on the concept of density rather than distance or similarity measures.

One advantage of DBSCAN is that it does not require the initial specification of the number of clusters in the dataset. Instead, it defines two important parameters: epsilon (ε) and *MinPts*. ε determines the radius of the neighborhood around a data point, indicating the maximum distance for considering other points as neighbors. A core point in DBSCAN is a data point that has at least *MinPts* neighboring points within its epsilon distance. Core points play a crucial role in forming clusters by including other points within their neighborhood. On the other hand, if a data point does not have a sufficient number of neighboring points within its ε distance (at least *MinPts*), it is considered a noise point. Noise points are typically located in sparse regions of the dataset or outliers that do not belong to any cluster. The neighborhoods of a p ϵ D point between the $N_e(p)$ *epsilon* values are shown as in Fig. 1 [17].

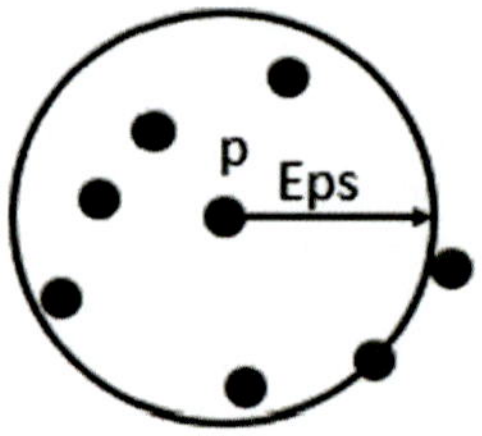

Fig. 1. The neighboring points

2.3 Ant Colony Optimization (ACO) Algorithm

The ant colony algorithm is an algorithm inspired by the ability of ant colonies to find the shortest path to a food source by providing pheromone trails. Pheromones are chemical substances used by ants to communicate with each other when they need to find directions. The pheromone reinforcement rate (alpha) determines the importance of the amount of pheromone between nodes. The heuristic reinforcement rate (beta) is a parameter that determines the importance of the distance between nodes. In the ant colony algorithm, which is used in the optimization of problems, the following parameters are used:

Number of Ants: It represents the number of ants used in the ant colony algorithm. The number of ants is chosen based on the complexity of the problem. As the number of ants increases, the algorithm will have a wider search capability.
Pheromone Update Coefficient (tau): It indicates the rate at which pheromones evaporate on each route. Increasing this value leads to faster convergence in the algorithm results.
Pheromone Evaporation Coefficient (rho): It represents the rate at which pheromones evaporate in each new iteration. A low parameter value will result in the repetition of ant tours and a decrease in search capabilities. The pheromone evaporation coefficient takes values in the range of [0,1] [2].

3 Method

3.1 Problem Description

The study is to plan the routing for delivering white goods from a logistics company's distribution warehouse to homes or dealers in Ankara and its surrounding provinces, taking into account the constraint of time windows. The provinces within the scope of the surrounding provinces are Kırıkkale, Çankırı, and Kırşehir. A dataset based on daily order and delivery addresses has been used. The delivery addresses and the location of the main warehouse are provided in Fig. 2.

Fig. 2. Customers and the main warehouse

There is a real-time problem where shipments are planned based on the experience and opinions of the planning personnel, without a systematic routing process.

In our problem, there are 75 delivery orders for home or dealer addresses in Ankara and its surrounding provinces. The main warehouse, where the delivery orders are located, is in Kahramankazan, Ankara, Turkiye. The most important constraint provided by the customer is the time windows, and services can be provided outside of these time windows, but in such cases, a penalty cost is applied. The penalty cost for unit lateness is set to ten times the duration of being late. Due to allowing late or early visits, the problem type is categorized in the literature as "Flexible Time Window Vehicle Routing Problem". The service time that the dealers will offer to the delivery points is considered 20 min. Time Windows are fixed and an appointment is made for one of the following time windows from sales dealers to customers for each delivery. The time windows are shown in Table 2.

Table 2. The time windows

Start of the time window	End of the time window
10:00	12:00
10:00	14:00
12:00	14:00
12:00	16:00
14:00	16:00
14:00	18:00
16:00	18:00

Delivery load and vehicle capacity are calculated based on area. The sum of the floor areas of the deliveries will not exceed the vehicle floor area. The ground and top area of the vehicle are constant and 30 m^2. In cases where the total area of the orders exceeds 30 m^2, the remaining orders must be loaded onto another vehicle. The upper floor area of the vehicle is an imaginary area designed for product groups that are placed on the floor but not too high and can be placed on top of each other. The ground and top floor design for the vehicle is given in Fig. 3.

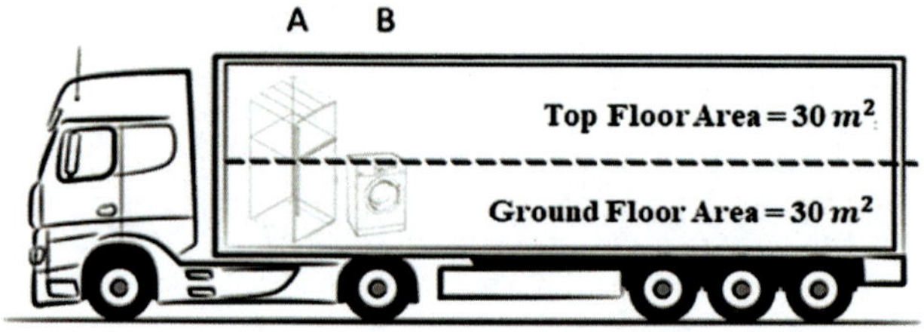

Fig. 3. The vehicle top and ground floor area

The products in the orders are evaluated in two groups. Group-A includes refrigerators that cover both the ground floor area and the top floor area of the vehicle. Group-B includes washing machines, dryers, and dishwashers which can be placed on top of each other in a maximum of 2 pieces. The area of Group-A products is taken as 1.3 m^2, the area of Group-B products is taken as 0.36 m^2. Although the sizes vary according to the model types, the values are calculated as the average of the product sizes in each group. The product groups and related areas are shown in Table 3.

Table 3. The product groups and areas of the products

	Group-A	Group-B
Product	Refrigerator	Dishwasher machine Washing machine Dryer machine
Ground area (m^2)	1.3 m^2	0.36 m^2
Top area (m^2)	1.3 m^2	0.36 m^2

3.2 First Phase: Clustering Delivery Points

It is observed that there is a density around the Ankara region in terms of the number of orders, number of dealers, and population. For this reason, it has been decided that the most suitable algorithm for the problem is the density-based clustering algorithm (DBSCAN). Each cluster we will obtain at the end of clustering is designed as a vehicle that will start from the main warehouse, deliver to each location in the cluster, and complete its route in the main warehouse. Therefore, while the variable k expresses the number of clusters, it also expresses the number of vehicles.

The fact that each cluster to be formed also represents a vehicle, requires the inclusion of a capacity constraint in the clustering process as well as the clustering algorithm to be applied. The algorithm steps are given in Table 4.

Table 4. DBSCAN Algorithm

```
START
Read distance matrix from Excel
ground_area=0; top_area=0; max_weight_per_cluster=30; top_cap=0;
Read the order list from Excel
Define ε and MinPts parameters
Assign the labels as "-1" for all points
            for i=1 to the number of points
                find points that are less than or equal to ε
                if there are as many points as MinPts around it
                   Assign the labels as" "0"
                end if
            end for
k = 0, create a cluster
for i=1 to the number of points
   if the label of point i is 0
        increase the number of clusters by 1
        ExpandCluster function
   end if
end for
for i=1 to the number of points
    if the label of point i is -1
       assign label of nearest cluster
    end if
end for
function ExpandCluster
END
```

First, the distance matrix of the main warehouse and delivery points is defined. The ground and top floor area of the vehicle is initially assumed to be 0, and the maximum capacity is defined as $max\_area\_per\_cluster = 30$. If there is only one Group-B product in the order, there will be a capacity enough for a Group-B product to be placed on top of that product placed on the vehicle. If the next order is a Group-B product, it will be placed on the previously placed product instead of on the ground. For this reason, the upper capacity is defined to be 0 at the beginning.

If there are at least *Minpts* points at ε distance, the label of that point is updated as "0", and defined as the core point. In the next step, the label of each point from *i* variable 1 to *n* delivery points is checked, if there is a core point with a label of "0", a cluster is created by increasing the number of *k* clusters, and the core point is included. Then, with the *ExpandCluster* function, neighboring points around the core point are clustered until they reach the point where they exceed the capacity constraint. In this function, the placement constraints of the vehicle are also provided. The algorithm flow chart of the *ExpandCluster* function, considering the vehicle capacity, is given in Fig. 4.

Defining Optimum Parameters. Response Surface Method (RSM) were used to optimize the *Minpts* and ε values in the clustering phase. While the factors were chosen as *Minpts* and ε values in the experimental design, the number of clusters, occupancy rate, and noise numbers were taken as the outputs. When determining the factor levels of the ε value, the data of the distance matrix between the delivery points and the warehouse were used. The mean of the overall distance in the whole matrix was taken as the average level of the factor, the mean of the points with distances below the average was taken as the low level and the average of the points with distances above the average was taken

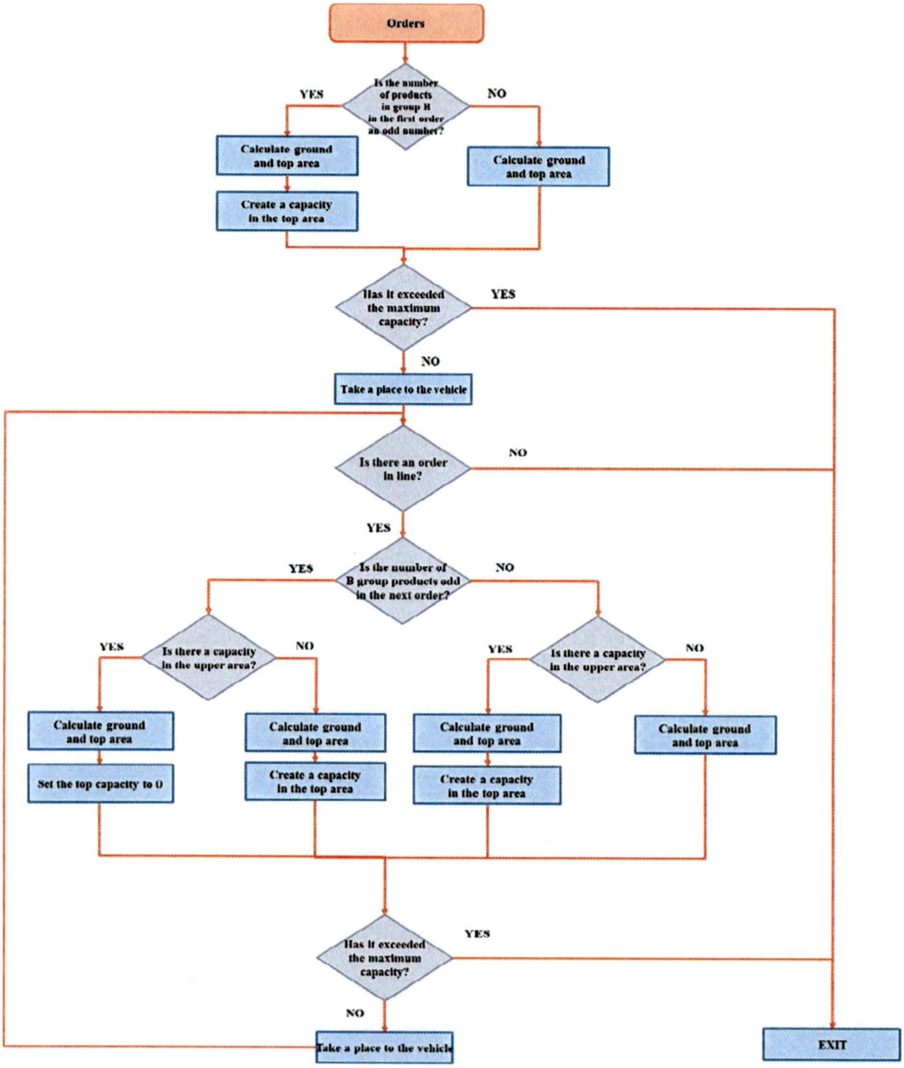

Fig. 4. *ExpandCluster* function algorithm

as the high level. In this case, according to the distance matrix, the average value is 64.6 km, the low-level value is 28.5 km, and the high-level value is 124.9 km. For the *Minpts* factor, the number of points on the map was observed and factor levels between 1 and 6 were defined. The central Composite Design option of the Response Surface Method was preferred. ε value is a continuous value as it indicates distance, are defined as a "continuous" variable The *Minpts* factor is defined as a "continuous" variable so that a higher *Minpts* value can be preferred when the density in a dataset is low, and a lower *Minpts* value can be used when the density is high. 14 experiments were carried out with different level values. Analysis results for Response A (the number of clusters), Response B (the occupancy rate), and Response C (the number of noise points) are given in Table 5. For each response, the R^2 value, which expresses the validity and significance of the model, was observed to be around 90% or higher.

Since the confidence level was 95% in the model, it shows that the P-Values below 0.05 have an effect on the response, therefore both parameters have an effect on the output parameters. The coefficients are statistically significant and have a value other than zero. If the number of clusters is minimized, the occupancy rates are maximized and the noise points number is minimized, the Response Optimizer detects the optimum value as 95.68 km for the ε parameter and as 3 points for the *Minpts* parameter.

As a result of clustering with optimum parameter values, 7 different clusters were created ($k = 7$). The results for each cluster are shown in Table 6. The average occupancy rates of the first 6 clusters are 94.89%, and since the last cluster includes the locations in the surrounding provinces, it is accepted to have a usually low occupancy rate.

Table 5. Responses for Clustering Parameters

Term	Coef	SE Coef	T-Value	P-Value	VIF
Response A					
Epsilon	−0,624	0,129	−4,84	0,001	1,02
MinPts	−0,764	0,141	−5,41	0,001	1,04
Response B					
Epsilon	0,0959	0,00980	9,79	0,000	1,02
MinPts	0,0323	0,0108	3,00	0,017	1,04
Response C					
Epsilon	−1,646	0,405	−4,07	0,004	1,02
MinPts	2,256	0,444	5,08	0,001	1,04

	S	R^2	R^2 (adj)	R^2 (pred)
Response A	0,381650	93,09%	88,77%	77,21%
Response B	0,0290513	95,69%	92,99%	82,56%
Response C	1,20012	89,97%	83,7%	57,47%

Table 6. The Clustering Results

Clusters	Points																					Total Cluster Orders	Occupancy Rate
1	1	2	3	4	5	6	9	29	30	31	32	33	34	35	36	37	38	39	51	53	55	59.32	98.87%
2	7	8	10	46	47	54	60	61	62	67	68	69	70	72	75							59.32	98.87%
3	11	12	13	14	15	16	17	18	19	41	43	44	45	73								59.24	98.73%
4	20	21	22	23	40	42	48	50	71	74												58.68	97.80%
5	24	25	26	27	28	49	52															55.24	92.07%
6	56	57	58	59																		49.8	83.00%
7	63	64	65	66																		11.48	19.13%

3.3 Second Phase: Vehicle Routing with Time Window

Due to the dynamic nature of the problem, Ant Colony Algorithm, which is one of the metaheuristic methods, has been proposed for the routing of 7 clusters obtained during the clustering phase. The number of delivery addresses varies, as the dimensions of the problem depend on customer orders. The Ant Colony Algorithm can deal with optimization problems involving combinations such as vehicle routing problems. However, unlike the traditional Ant Colony algorithm, the time windows of the problem are included in the problem at this stage, and the clusters that previously met the capacity constraint are evaluated under the priority and time window constraints in this step.

Defining Optimum Parameters. Response Surface Method (RSM) has been used to optimize the parameters. Maximum iteration number (*MaxIT*), number of ants (*Nant*), *alpha*, and *beta* are the input factors, which are the global parameters of the Ant Colony Algorithm. The outputs are the penalty cost and the distance traveled (*km*).

The levels of *alpha* and *beta* parameters are zero to 5. On the other hand, five levels for *Nant* and *MaxIT* parameters are taken (100, 125, 150, 175, and 200) as a result of the data obtained from the literature studies and trial-and-error methods.

The *rho* and *tau* parameters are the evaporation and pheromone regeneration rates that affect the ants' routes. Although not included in the experimental design, these values were chosen as the average of 0.5. The reason for this is that the pheromone renewal or evaporation rates are kept at average values, and the pheromone is not renewed at all, or the pheromone never evaporates, by taking extreme values.

30 experiments were carried out with different level values. Analysis results for Response A (*Penalty cost*) and Response B (*distance traveled*) are given in Table 7.

For each response, the R^2 value, which expresses the validity and significance of the model, was observed to be around 96%. As the R^2 value increases, we conclude that it expresses the experimental set well.

Since the confidence level was chosen as 95% in the model, it shows that the factors with P values below 0.05 have an effect on the response. Therefore the *beta* parameter has an effect on the output parameters. Response Optimizer has been run in order to minimize penalty cost and distance traveled (km) outputs.

When the results are examined, Factor A is 100 for *MaxIT* parameter; Factor B is 162 for parameter *Nant*; Factor C was determined as 5 for *alpha* parameter and 0 for *beta* parameter as optimum values.

Table 7. Responses for Routing Parameters

Term	Coef	SE Coef	T-Value	P-Value	VIF
Response A					
MaxIT	1571	2287	0,69	0,503	9,45
Nant	873	4804	0,18	0,858	35,64
alpha	2843	3376	0,84	0,413	14,35
Beta	14829	5121	2,90	0,011	36,35
Response B					
MaxIT	1593	2283	0,70	0,496	9,45
Nant	926	4797	0,19	0,849	35,64
alpha	2836	3371	0,84	0,413	14,35
Beta	14648	5113	2,86	0,012	36,35

	S	R^2	R^2 (adj)	R^2 (pred)
Response A	3064,47	%96,94	%94,09	%81,15
Response B	3059,83	%96,89	%93,99	%80,75

Implementing the Ant Colony Algorithm. In this study, unlike the classical Ant Colony algorithm, delivery constraints within time windows are included in the algorithm. Each vehicle will be routed taking time window constraints into account. The Ant Colony algorithm applied within the time window constraint is given in Table 8. The routes obtained as a result of the application and their costs are given in Table 9.

In the cost calculation, assuming that the vehicle burns 20 L of fuel per 100 km, the cost is calculated by adding the penalty cost.

All routes are given in Fig. 5.

The routing obtained as a result of the planning made by the company personnel based on their knowledge and experience is given in Table 10. The 10 vehicles clustered by the company as in the table, take into account the time windows and deliver them.

In the study, the number of 10 vehicles obtained from the routing based on the experience and observations of the planning personnel was reduced to 7 vehicles with an average vehicle occupancy rate of 94.89%.

Table 8. The Ant Colony Algorithm Steps

```
START
Read the distance and time matrix from Excel
Define MaxIT, Nant, alpha, beta, rho and tau parameters
Define the cost of best ant as infinitive
for i to MaxIT
    for k=1 to Nant
Select the first point as warehouse
Find the second point
Add the point to the tour
Add the travel time to the point to the total time of the ant
            If the ant come to the point before start time of the window
            The ant wait and a assign service time to ant time by adding it to start
time
            Elseif the ant come within the time window
            Add the service time to the ant
            Elseif the ant come after the time window
            Add penalty cost as 10 units of late time
            Add ervice time as 20 min
            end if
    end for
Add the penalty cost to the ant cost.
        If the cost of the ant is less than the cost of the best ant
        the cost of the ant is equal to the cost of the best ant
        end if
end for
pheromone update function
END
```

Table 9. The Routing Results

ROUTES																								Total km	Penalty cost (TL)	Total cost (TL)
1	76	36	30	31	6	53	1	38	39	55	9	29	5	2	3	35	37	33	34	32	51	4	76	35,436.4	34,780	176,525.40
2	76	8	10	60	75	68	47	62	67	46	54	70	69	61	7	72	76							4,367.8	4,100	21,571.20
3	76	14	19	73	43	15	16	18	13	11	12	44	45	41	17	76								9,982.2	9,510	49,438.80
4	76	71	21	20	42	50	23	40	22	48	74	76												159.7	0	638.80
5	76	28	26	49	24	27	25	52	76															124.8	0	499.20
6	76	59	56	57	58	76																		321.4	0	1,285.60
7	76	65	66	63	64	76																		473.7	0	1,894.80
																									TOTAL	**251,853.80**

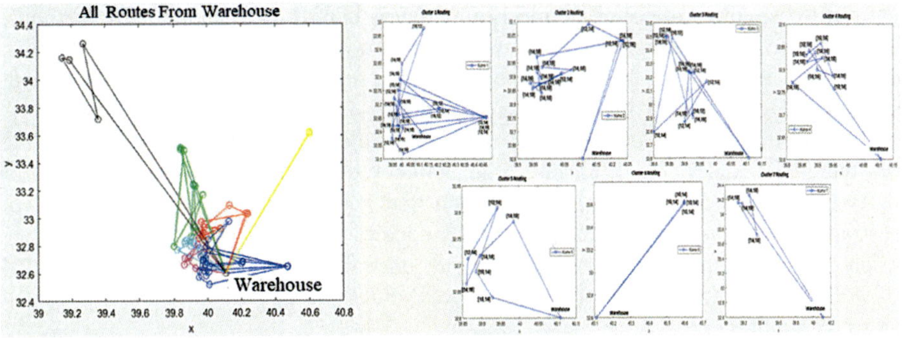

Fig. 5. Routes Graph

Table 10. Routes Before Situation

VEHICLES														
1	14	15	16	18										
2	4	47	59	60	61	65	69	70						
3	11	12	13	17	63	64	66	73						
4	7	8	9	10	72									
5	20	21	27	40	42	43	44	45	48					
6	56	57	58	62	67	68	75							
7	1	2	3	4	5	6	22	23	24	25	26	41	54	55
8	50	51	52	53	71									
9	28	29	30	31	32	33								
10	19	34	35	36	37	38	39	49	74					

4 Conclusion

In the study, the real-time problem of delivery from the main distribution warehouse in Ankara to 75 dealers in Ankara and surrounding provinces in certain time periods has been examined.

First of all, it was analyzed that the delivery addresses are concentrated around Ankara and the product groups are classified according to their placement constraints. The delivery points are clustered with the density-based DBSCAN algorithm. During clustering, in-vehicle placement was integrated so as not to exceed the capacity of the vehicle. The vehicle occupancy rate was increased by placing the product groups that meet the stackability criterion on top of each other.

Each cluster represents a vehicle. After the vehicles are clustered on the basis of their capacity, location constraints, and their proximity to each other, the problem arose in which order the delivery will be made according to the time windows.

In this case, the Ant colony algorithm, which is one of the metaheuristic methods, was chosen for the routing step due to the change in order size and address dynamics. At this stage, a routing constraint has been added to the algorithm, leaving the classical ant colony algorithm to provide time window constraints.

The constraint, which can wait if it arrives at the delivery point earlier than the time start, but adds penalty cost if it arrives late, is integrated into the algorithm.

As a result of this routing, an approach that provides both capacity, placement constraints, and time window constraints is proposed.

The proposed approach has integrated the constraints and algorithms and provided a 30% improvement in the number of vehicles. Vehicle occupancy rates were increased to an average of 94.89%.

In future studies, studies on the conditions assumed or accepted in this study are planned.

Variable-capacity vehicles instead of a single type of vehicle, and the stable floor area according to product groups are the default situations in this study.

References

1. Dantzig, G.B., Ramser, J.H.: The truck dispatching problem. Manage. Sci. **6**, 80–91 (1959)
2. Özdağ, H., Aygör, N., Parlak, A.: Karınca Kolonisi Algoritmasının Zaman Çizelgelemesi Üzerine: Bir Modellemesi ve Uygulaması. Akademik Bilişim **12**, 01–03 (2012)
3. Zhang, J.: An efficient density-based clustering algorithm for the capacitated vehicle routing problem. In: International Conference on Computer Network, Electronic and Automation, ICCNEA (2017)
4. Ünsal, Ö., Yiğit, T.: Yapay Zeka ve Kümeleme Tekniklerini Kullanarak Geliştirilen Yöntem ile Okul Servisi Rotalama Probleminin Optimizasyonu. Mühendislik Bilimleri ve Tasarım dergisi **6**(1), 7–20 (2018)
5. Cömert, S.E., Yazgan, H.R., Görgülü, N.: Eş Zamanlı Topla Dağıt Araç Rotalama Problemi için İki Aşamalı Bir Çözüm Yöntemi Önerisi. Int. J. Adv. Eng. Pure Sci. **2**, 107–117 (2019)
6. Bozdemir, M.K., Bozdemir, M., Özcan, B.: Route first-cluster second method for personal service routing problem. J. Eng. Stud. Res. **25**(2) (2019)
7. Bujel, K., Lai, F., Szczecinski, M., So, W., Fernandez, M.: Solving High Volume Capacitated Vehicle Routing Problem with Time Windows using Recursive-DBSCAN clustering algorithm, Dept. of Analytics, GoGo Tech Limited, Hong Kong (2019)
8. Gocken, T., Yaktubay, M.: Comparison of different clustering algorithms via genetic algorithm for VRPTW. Int. J. Simul. Model. **18**(4), 574–585 (2019)
9. Cömert, S.E., Yazgan, H.R., Çakır, B., Sarı, N.: Esnek Zaman Pencereli Araç Rotalama Probleminin Çözümü için Önce Kümele Sonra Rotala Temelli Bir Yöntem Önerisi, Bir Süpermarket Örneği. Konya J. Eng. Sci. **8**(1), 18–25 (2020)
10. Villalba, A.F.L., Rotta, E.C.G.L.: Comparison of Dbscan and K-means clustering methods in the selection of representative clients for a vehicle routing model. In: Congreso Internacional de Innovación y Tendencias en Ingeniería (2020)
11. Villalba, A.F.L., Rotta, E.C.G.L.: Clustering and heuristics algorithm for the vehicle routing problem with time windows. Int. J. Ind. Eng. Comput. **13**, 165–184 (2022)
12. Akdaş, H.Ş.: Vehicle route optimization for solid waste management: a case study of Maltepe, Istanbul. In: 13th International Conference on Electronics, Computers and Artificial Intelligence (ECAI). IEEE (2021)

13. Cordeau, J.F., Gendreau, M., Hertz, A., Laporte, G., Sormany, J.S.: New heuristics for the vehicle routing problem. In: Logistics Systems: Design and Optimization, pp. 279–297. Springer, Boston, MA (2005)
14. Sánchez, D.G., Tabares, A., Faria, L.T., Rivera, J.C., Franco, J Fredy: A clustering approach for the optimal siting of recharging stations in the electric vehicle routing problem with time windows. Energies **15**(7), 2372 (2022)
15. Wang, Y.: A clustering-based extended genetic algorithm for the multidepot vehicle routing problem with time windows and three-dimensional loading constraints. Appl. Soft Comput. **133**, 109922 (2023)
16. Le, T.D.C., Nguyen, D.D., Oláh, J., Pakurár, M.: Clustering algorithm for a vehicle routing problem with time windows. Transport **37**(1), 17–27 (2022)
17. Aydın, Ş.: Ülkelerin Yolsuzluk Göstergelerine Göre DBSCAN ile kümelenmesi. Marmara University, Institute of Social Sciences, Department of Econometrics, Department of Statistics, Master Thesis, (2022)
18. Kumar, S.N., Panneerselvam, R.: A survey on the vehicle routing problem and its variants. Intell. Inf. Manage. **04**(03), 66–74 (2012)

Author Index

Z. Şen et al. (Eds.): IMSS 2023, LNME, pp. 811–813, 2024.
https://doi.org/10.1007/978-981-99-6062-0